Gmelin Handbook of Inorganic and Organometallic Chemistry

8th Edition

Gmelin Handbook of Inorganic and Organometallic Chemistry

8th Edition

Gmelin Handbuch der Anorganischen Chemie

Achte, völlig neu bearbeitete Auflage

PREPARED AND ISSUED BY

Gmelin-Institut für Anorganische Chemie
der Max-Planck-Gesellschaft
zur Förderung der Wissenschaften

Director: Ekkehard Fluck

FOUNDED BY — Leopold Gmelin

8TH EDITION — 8th Edition begun under the auspices of the Deutsche Chemische Gesellschaft by R. J. Meyer

CONTINUED BY — E. H. E. Pietsch and A. Kotowski, and by Margot Becke-Goehring

Springer-Verlag
Berlin · Heidelberg · New York · London · Paris · Tokyo · Hong Kong · Barcelona · Budapest · Milan 1996

GMELIN HANDBOOK

Dr. J. von Jouanne

Dr. L. Berg, Dr. H. Bergmann, Dr. J. Faust, Dr. H. Katscher, Dr. A. Kubny, Dr. P. Merlet, Prof. Dr. W. Petz, Dr. H. Schäfer

Dr. R. Albrecht, D. Barthel, Dr. N. Baumann, Dr. K. Behrends, Dr. W. Behrendt, D. Benzaid, Dr. R. Bohrer, K. D. Bonn, Dr. U. Busch, A.-K. Castro, Dipl.-Ing. V. A. Chavizon, A. Dittmar, Dipl.-Geol. R. Ditz, R. Dowideit, Dr. H.-J. Fachmann, B. Fischer, Dr. D. Fischer, Dr. K. Greiner, Dipl.-Bibl. W. Grieser, Dr. R. Haubold, Dipl.-Min. H. Hein, H.-P. Hente, Dr. G. Hönes, Dr. W. Huisl, Dr. M. Irmler, B. Jaeger, Dr. R. Jotter, Dipl.-Chem. P. Kämpf, Dr. B. Kalbskopf, Dipl.-Phys. H. Keller-Rudek, Dipl.-Chem. C. Koeppel, R. Kolb, Dr. M. Kotowski, E. Krawczyk, Dr. W. Kurtz, Dr. B. Ledüc, H. Mathis, M. Meßer, C. Metz, K. Meyer, Dipl.-Chem. B. Mohsin, Dr. U. Neu-Becker, K. Nöring, Dipl.-Min. U. Nohl, Dr. U. Ohms-Bredemann, Dr. H. Pscheidl, Dipl.-Phys. H.-J. Richter-Ditten, Dr. B. Sarbas, Dr. R. Schemm, Dr. D. Schiöberg, V. Schlicht, A. Schwärzel, Dr. B. Schwager, Dr. F. Stein, Dr. C. Strametz, Dr. D. Tille, Dipl.-Phys. J. Wagner, R. Wagner, M. Walter, Dr. E. Warkentin, Dr. C. Weber, Dr. A. Wietelmann, Dr. M. Winter, Dr. B. Wöbke, K. Wolff

GMELIN ONLINE

Dr. G. Olbrich

Dr. P. Kuhn

Dr. R. Baier, Dr. B. Becker, Dipl.-Phys. R. Bost, Dr. A. Brandl, Dr. R. Braun, R. Hanz, Dipl.-Phys. C. Heinrich-Sterzel, Dr. M. Körfer, Dipl.-Chem. H. Köttelwesch, Dr. V. Kruppa, Dr. M. Kunz, Dipl.-Chem. R. Maass, Dr. A. Nebel, Dipl.-Chem. R. Nohl, Dr. B. Rempfer

Organometallic Compounds in the Gmelin Handbook

The following listing indicates in which volumes these compounds are discussed or are referred to:

Ag	Silber B5 (1975)
Au	Organogold Compounds (1980)
Be	Organoberyllium Compounds 1 (1987)
Bi	Bismut-Organische Verbindungen (1977)
Co	Kobalt-Organische Verbindungen 1, 2 (1973), Kobalt Erg.-Bd. A (1961), B 1 (1963), B 2 (1964)
Cr	Chrom-Organische Verbindungen (1971)
Cu	Organocopper Compounds 1 (1985), 2 (1983), 3 (1986), 4 (1987), Index (1987)
Fe	Eisen-Organische Verbindungen A 1 (1974), A 2 (1977), A 3 (1978), A 4 (1980), A 5 (1981), A 6 (1977), A 7 (1980), Organoiron Compounds A 8 (1986), A 9 (1989), A 10 (1991), A 11 (1995), Eisen-Organische Verbindungen B 1 (partly in English; 1976), Organoiron Compounds B 2 (1978), Eisen-Organische Verbindungen B 3 (partly in English; 1979), B 4, B 5 (1978), Organoiron Compounds B 6, B 7 (1981), B 8, B 9 (1985), B 10 (1986), B 11 (1983), B 12 (1984), B 13 (1988), B 14, B 15 (1989), B 16a, B 16b, B 17 (1990), B 18 (1991), B 19 (1992), Eisen-Organische Verbindungen C 1, C 2 (1979), Organoiron Compounds C 3 (1980), C 4, C 5 (1981), C 6a (1991), C 6b (1992), C 7 (1985), and Eisen B (1929–1932)
Ga	Organogallium Compounds 1 (1986)
Ge	Organogermanium Compounds 1 (1988), 2 (1989), 3 (1990), 4 (1994), 5 (1993), 6 (1996)
Hf	Organohafnium Compounds (1973)
In	Organoindium Compounds 1 (1991)
Mo	Organomolybdenum Compounds 5 (1992), 6 (1990), 7 (1991), 8 (1992), 9 (1993), 10 (1995), 11 (1996) **present volume**, 12 (1994), 13 (1996)
Nb	Niob B 4 (1973)
Ni	Nickel-Organische Verbindungen 1 (1975), 2 (1974), Register (1975), Nickel B 3 (1966), and C 1 (1968), C 2 (1969), Organonickel Compounds Suppl. Vol. 1 (1993), 2 (1994), 3 (1996)
Np, Pu	Transurane C (partly in English; 1972)
Os	Organoosmium Compounds A 1 (1992), A 2 (1993), B 3 (1994), B 4a (1995), B 5 (1994), B 6 (1993), B 8 (1995), B 9 (1995)
Pb	Organolead Compounds 1 (1987), 2 (1990), 3 (1992), 4 (1995)
Po	Polonium Main Volume (1941, reprint 1969), Suppl. Vol. 1 (1990)
Pt	Platin C (1939) and D (1957)
Re	Organorhenium 1, 2 (1989), 3 (1992), 4 (1996), 5 (1994), 7 (1996)
Ru	Ruthenium Erg.-Bd. (1970)
Sb	Organoantimony Compounds 1, 2 (1981), 3 (1982), 4 (1986), 5 (1990)
Sc, Y, La to Lu	Rare Earth Elements D 6 (1983)
Sn	Zinn-Organische Verbindungen 1, 2 (1975), 3, 4 (1976), 5 (1978), 6 (1979), Organotin Compounds 7 (1980), 8 (1981), 9 (1982), 10 (1983), 11 (1984), 12 (1985), 13 (1986), 14 (1987), 15, 16 (1988), 17 (1989), 18 (1990), 19 (1991), 20 (1993), 21 (1994), 22 (1995), 23 (1995), 24 (1996)
Ta	Tantal B 2 (1971)
Ti	Titan-Organische Verbindungen 1 (1977), 2 (1980), Organotitanium Compounds 3 (1984), 4 and Register (1984), 5 (1990)
U	Uranium Suppl. Vol. E 2 (1980)
V	Vanadium-Organische Verbindungen (1971), Vanadium B (1967)
Zr	Organozirconium Compounds (1973)

Gmelin Handbook of Inorganic and Organometallic Chemistry

8th Edition

Mo
Organomolybdenum Compounds

Part 11

With 37 illustrations

AUTHOR Lutz Dahlenburg (Erlangen-Nürnberg)

FORMULA INDEX Rainer Bohrer, Bernd Kalbskopf, Uwe Nohl, Hans-Jürgen Richter-Ditten

EDITORS Kerstin Behrends, Manfred Winter

CHIEF EDITOR Manfred Winter

Springer-Verlag
Berlin · Heidelberg · New York · London · Paris · Tokyo · Hong Kong · Barcelona · Budapest · Milan 1996

LITERATURE CLOSING DATE: END OF 1995

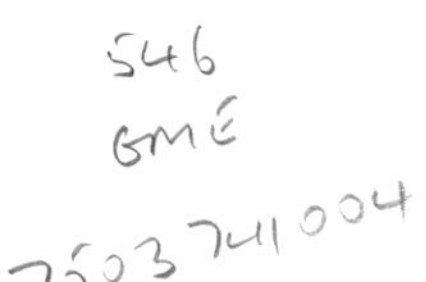

Library of Congress Catalog Card Number: Agr 25-1383

ISBN 3-540-93743-9 Springer-Verlag, Berlin · Heidelberg · New York · London · Paris · Tokyo · Hong Kong · Barcelona · Budapest · Milan
ISBN 0-387-93743-9 Springer-Verlag, New York · Heidelberg · Berlin · London · Paris · Tokyo · Hong Kong · Barcelona · Budapest · Milan

Printed in Germany

Typesetting, printing, and bookbinding: Wiesbadener Graphische Betriebe GmbH, Wiesbaden

Preface

This volume 11 continues the description in a series dealing with organomolybdenum compounds. An Empirical Formula Index provides ready access to the compounds covered.

Volume 5 describes mononuclear organomolybdenum compounds with isocyanide, carbene, alkynyl, alkyne, alkene, ^{3}L, and ^{4}L ligands with and without additional CO groups. Volume 6 starts the description of mononuclear organomolybdenum compounds with one ^{5}L ligand. The compounds contain either zero or one CO group bonded to the molybdenum atom. Volume 7 contains the description of ^{5}L-molybdenum compounds containing two CO groups, but no additional ^{n}L ligands. Volume 8 describes ^{5}L-molybdenum compounds with two CO groups and additional ^{1}L to ^{4}L ligands. Volume 9 describes first heteronuclear compounds containing one $^5LMo(CO)_2$ fragment and one or more other transition metal fragments. The second part starts the description of compounds containing the $C_5H_5Mo(CO)_3$ fragment including the $[C_5H_5Mo(CO)_3]^+$ cation and its sources of the types $C_5H_5Mo(CO)_3X$ (X is a weakly coordinated anion) and solvated forms of the $[C_5H_5Mo(CO)_3]^+$ cation. The radical $C_5H_5Mo(CO)_3^{\bullet}$, the anion $[C_5H_5Mo(CO)_3]^-$, and the hydride $C_5H_5Mo(CO)_3H$ are also presented. Volume 10 continues the description of compounds with one $C_5H_5Mo(CO)_3$ fragment including compounds of the types $C_5H_5Mo(CO)_3X$, $[C_5H_5Mo(CO)_3{}^2D]^+$, $[C_5H_5Mo(CO)_3{}^2D]^{\bullet}$, $[C_5H_5Mo(CO)_3EX_n]^-$, and heteronuclear compounds $C_5H_5Mo(CO)_3M(X,R)_n$ (M = transition metal).

This volume finishes the description of compounds with one $C_5H_5Mo(CO)_3$ fragment. These compounds contain one additional ^{1}L ligand including compounds of the types $C_5H_5Mo(CO)_3{}^1L$ and $[C_5H_5Mo(CO)_3{=}CR_2]^+$. The second part of this volume deals with compounds containing one $^5LMo(CO)_3$ ($^5L \neq C_5H_5$) fragment and additional X, ^{n}D, ^{1}L, or $M(D, X, R)_n$ (M = transition metal) ligands. The ^{5}L ligands in these sections are cyclopentadienyl derivatives, $C_5H_{5-n}R_n$ (n = 1 to 5) and ligands different from cyclopentadienyl, e.g. other cycloalkadienyl or open-chain pentadienyl derivatives.

Following the nomenclature used in the series of organomolybdenum compounds, ^{n}L is an organic ligand bonded by n C atoms to molybdenum, and ^{m}D is an electron donor with m donor electrons. Thus ^{2}D denotes a ligand such as PR_3.

Many of the data, particular those in tables, are given in an abbreviated form without units; for explanations see p. X. Additional information is given, if necessary, before the individual table.

Frankfurt am Main
October 1996

Manfred Winter

Remarks on Abbreviations and Dimensions

Many compounds in this volume are presented in tables in which numerous abbreviations are used, the dimensions are omitted for the sake of conciseness. This necessitates the following clarifications.

Abbreviations used with **temperatures** are m.p. for melting point, dec. for decomposition, and b.p. for boiling point.

NMR represents **nuclear magnetic resonance**. Chemical shifts are given as δ values in ppm and to low field from the following reference substances: $Si(CH_3)_4$ for ^1H and ^{13}C, $CFCl_3$ for ^{19}F, and H_3PO_4 for ^{31}P. ^{17}O and ^{95}Mo NMR chemical shifts are given relative to H_2O and $Na_2[MoO_4]$ standards, respectively (downfield positive). Otherwise the reference substances and signs are given. Multiplicities of the signals are abbreviated as s, d, t, q, qui, sext, sept, and oct (singlet to octet), m (multiplet), and br (broad); terms like dd are also used. Assignments referring to a labeled structural formula are given in the form C-4, H-3,5. Coupling constants J in Hz usually appear in parentheses after the δ value, along with the multiplicity and the assignment, and refer to the respective nucleus. If a more precise designation is necessary, it is given as, e.g., $^nJ(P, H)$ or J(H-1,3) referring to labeled formulas.

ESR represents **electron spin resonance**; the hyperfine interactions are given as a (^{95}Mo) or a(P).

Optical spectra are labeled as IR (infrared) and UV (electronic spectrum including the visible region). IR and Raman bands are given in cm^{-1}. The assigned bands are usually labeled with the symbols ν for stretching, δ for deformation, ρ for rocking, with the indexes sym and asym for symmetrical and asymmetrical. The UV absorption maxima, λ_{max}, are given in nm followed by the extinction coefficient ε (in $L \cdot cm^{-1} \cdot mol^{-1}$) or log ε in parentheses. The abbreviation sh is used for shoulder.

Solvents and **physical state** of the sample and the temperature (none is given if room temperature applies) are given in parentheses immediately after the spectral symbol, e.g., IR (solid), ^1H NMR (acetone-d_6, −30°C), or at the end of the data with their formula (C_6H_6 = benzene) or name (acetone-d_6), except THF, which represents tetrahydrofuran.

The abbreviation used with **electrochemical behavior** is SCE for saturated calomel electrode.

The bond distances in **Figures** are given in Å.

Further abbreviations:

D_{calc} for calculated density and D_{meas} for measured density.

Table of Contents

Organomolybdenum Compounds 11

1.5.1.4.1 Compounds Containing the Fragment $C_5H_5Mo(CO)_3$ (continued)

1.5.1.4.1.12 Compounds of the Type $C_5H_5Mo(CO)_3{}^1L$

1.5.1.4.1.12.1 Organization and General Remarks

The compounds collected under this heading contain a great variety of σ-bonded organo ligands ^1L. Hence the following subdivision based on the type of the pertinent residue ^1L connected to the molybdenum atom has been made: complexes with the metal bonded to an aliphatic sp^3-hybridized carbon atom such as in derivatives of the alkyl, allyl, or benzyl type are described in Section 1.5.1.4.1.12.2; compounds possessing Mo-sp^2-C linkages, e.g., $C_5H_5Mo(CO)_3$-vinyl, -acyl, -aryl, and the like, are summarized in Section 1.5.1.4.1.12.3 and subsections thereof; and derivatives with ^1L = -C≡CR, in the end, will be treated in the final Section 1.5.1.4.1.12.4. As to the further arrangement of the complexes in the subsections, explanatory remarks will be given at the beginning of the individual chapters, if necessary.

1.5.1.4.1.12.2 $C_5H_5Mo(CO)_3{}^1L$ Compounds Containing Mo-sp^3-C Bonds

All compounds are arranged in Table 1 with the derivatives containing linear and branched-chain as well as cyclic alkyl and heteroalkyl ligands as the first entries. The table also summarizes complexes where the molybdenum atom is bonded to open-chain and cyclic alk-2-en-1-yl or to 2,4-dien-1-yl residues as well as products resulting from the latter by cycloaddition or complexation reactions. $C_5H_5Mo(CO)_3{}^1L$ derivatives where ^1L represents an alk-2-yn-1-yl ligand are subsequently included, followed by a great variety of homo- and heteroarylmethyl derivatives, such as benzyl, naphthylmethyl, thienylmethyl, and the like.

General Reactivity. In organic solvents, the majority of alkyl derivatives $C_5H_5Mo(CO)_3R$ possessing various nonfluorinated organo ligands R are slowly decomposed by air, some more readily than others [2, 12, 39, 116], but their solutions can be stored in an inert atmosphere [2]. In contrast, complexes containing polyfluoro-substituted residues attached to the metal atom, e.g., $C_5H_5Mo(CO)_3C_2F_4H$ [11], show quite exceptional stability toward air [9].

Each of the reactions of $C_5H_5Mo(CO)_3R$ (R = H, CH_3, C_2H_5, $CH_2C_6H_5$, C_6H_5) with $[C_5H_5Fe(CO)_2]^-$, employed as its $[N(P(C_6H_5)_3)_2]^+$ salt in THF, resulted in alkyl transfer to the iron with formation of $[C_5H_5Mo(CO)_3]^-$. The reaction is first-order in $[C_5H_5Fe(CO)_2]^-$ and first-order in the alkyl complex used. The dependence of the rate constants (given parenthetically in $L \cdot mol^{-1} \cdot s^{-1}$) on R, H (too rapid for examination) $\gg$ $CH_2C_6H_5$ (also very fast) $>$ CH_3 $((1.1 \pm 0.1) \times 10^3)$ $>$ C_2H_5 $((4.0 \pm 0.1) \times 10^2)$ $\gg$ C_6H_5 (slow), is consistent with a nucleophilic attack mechanism. For methyl transfer, the rate correlates with the difference in nucleophilicity between the reactant and product anions. For the reaction of $C_5H_5Mo(CO)_3CH_3$ with $Na[C_5H_5Fe(CO)_2]$, the rate constant of 16 ± 2 $L \cdot mol^{-1} \cdot s^{-1}$, is substantially lower than for the $[N(P(C_6H_5)_3)_2]^+$ salt. This decreased reactivity was also observed for the reaction of $Na[C_5H_5Fe(CO)_2]$ with $C_5H_5Mo(CO)_3CH_2C_6H_5$ [389]. The self-exchange reaction "$C_5H_5Mo(CO)_3CH_3$ + $[CH_3C_5H_4Mo(CO)_3]^-$ ⇄ $CH_3C_5H_4Mo(CO)_3CH_3$ + $[C_5H_5Mo(CO)_3]^-$"

References on pp. 106/18

(THF, $[N(P(C_6H_5)_3)_2]^+$ salts) is first-order in $C_5H_5Mo(CO)_3CH_3$ and independent of $[CH_3C_5H_4Mo(CO)_3]^-$. The rate constant of $5.3 \times 10^{-6}\ s^{-1}$ is about one order of magnitude lower than that of methyl migration to CO under similar conditions [389].

For many of the complexes, this chapter deals with decarbonylation accompanied by a σ to π rearrangement of the ligand, which can be induced either by thermolysis, UV irradiation, or by treatment with $(CH_3)_3NO$. The latter amine oxide-catalyzed decarbonylation reaction is believed to occur by nucleophilic attack at the carbonyl carbon by the oxygen end of the amine oxide. The resulting complexes contain ligands mainly bonded as shown in Formulas Ia and Ib.

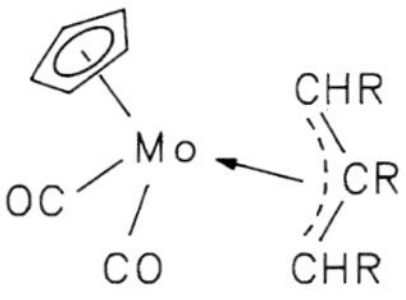

Ia

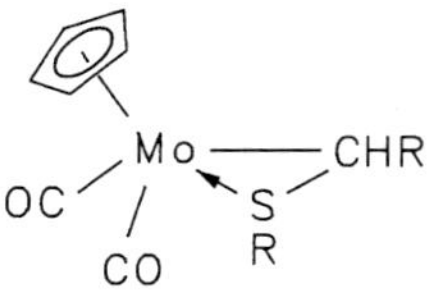

Ib

For more details and references, compare the individual compounds summarized in Table 1. By way of analogy, photolysis of the reaction mixture, obtained by treatment of $Na[C_5H_5Mo(CO)_3]$ with 1-(bromomethyl)cyclopentene, resulted in the smooth decarbonylation of the presumed σ-intermediate **$C_5H_5Mo(CO)_3CH_2(C{=}CH(CH_2)_3$-cyclo)** to produce the π-allyl derivative as the ultimately stable product [103].

Treatment of $C_5H_5Mo(CO)_3{}^1L$ (^1L = CH_3, C_2H_5, $CH_2C_6H_5$) in CH_2Cl_2 between −20 and 25 °C with the organometallic Lewis acids $C_5H_5Mo(CO)_3FBF_3$ and $C_5H_5Mo(CO)_3FPF_5$ efficiently promotes CO migratory insertions in the molybdenum alkyls to give acyl-bridged organometallic products. Therefore a principle product in all cases was the symmetrically π acyl-bridged cation $[(C_5H_5Mo(CO)_2)_2(\mu\text{-}RCO)]^+$ (see Formula II), isolated as the BF_4 or PF_6 salt [255, 263, 264, 290, 298]. If $C_5H_5Mo(CO)_3FPF_5$ was used as the Lewis acid in the case of R = CH_3 (No. 1 of Table 1), additionally $[(C_5H_5Mo(CO)_3)_2(\mu\text{-}RCO)]^+$ (Formula III) was formed [255]. By way of similarity, treatment of $C_5H_5Mo(CO)_3CH_3$ with the closely related Lewis acid $C_5H_5Mo(CO)_2(\eta^2\text{-}CH_2{=}CHCH_3)FBF_3$ in CH_2Cl_2 also produced the acetyl-bridged salt $[(C_5H_5Mo(CO)_2)_2(\mu\text{-}CH_3CO)]BF_4$ [298]. Additionally a reaction with $[(C_6H_5)_3C]^+$ has been mentioned that abstracts CH_3^- from $C_5H_5Mo(CO)_3CH_3$ to generate "$[C_5H_5Mo(CO)_3]^+$" as a reactive intermediate which rapidly adds to another molecule of the methyl complex forming again the π acyl-bridged cation [263]. Although the reactions with $[(C_6H_5)_2CH]^+$ and $[(C_6H_5)_3C]^+$ were described as proceeding in rather a sluggish fashion in [285].

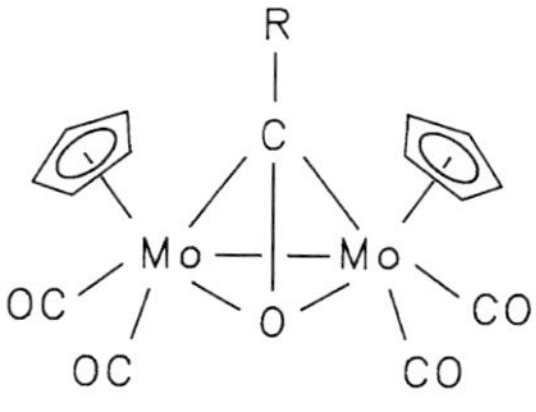

II

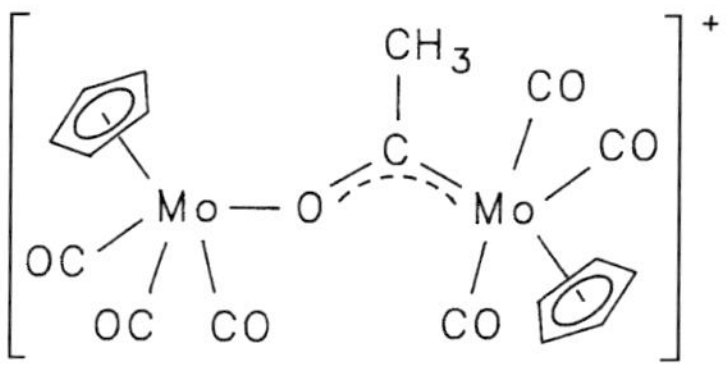

III

For $C_5H_5Mo(CO)_3CH_2OCH_3$ (No. 118), a similar reaction is described, but a methoxymethylidene salt was isolated as the main product along with $[(C_5H_5Mo(CO)_2)_2(\mu_2\text{-}CH_3\text{-}$

CO)]BF_4. Therefore this case is described more detailed in "Further information" at the end of Table 1 [290].

Many of the alkyl compounds $C_5H_5Mo(CO)_3R$ (R = CH_3, C_2H_5, $CH_2(CH(CH_2)_2$-cyclo), $CH_2CH{=}CH_2$, $CH_2CH{=}CHCH_3$, $CH_2C_6H_{5-n}X_n$, $CH_2CH(R')SO_2OCH_3$ (R' = H, CH_3), $(CH_2)_nFe(CO)_2C_5H_5$ (n = 3, 4, 6)) reacted with phosphanes or phosphites in a variety of solvents at ambient or elevated temperatures by migratory CO insertion to give substituted acyl derivatives as outlined in the following reaction:

$$C_5H_5Mo(CO)_3R + PR'_3 \rightleftarrows C_5H_5Mo(CO)_2(PR'_3)C(O)R.$$

In the case of $C_5H_5Mo(CO)_3CH_3$ at 60 °C in CH_3CN, an equilibrium between the acyl complex and the alkyl compound was established starting from $C_5H_5Mo(CO)_2(PR'_3)C(O)CH_3$ within 15 to 20 min without decomposition. Equilibrium constants were measured and data confirmed the steric demands of the phosphane to be the determining factor for formation of the acyl compounds and the position of the equilibrium, e.g., $K_{eq} = 0.1 \times 10^{-4}$ L/mol for $P(CH_3)_2C_6H_5$ and 90×10^{-4} L/mol for $P(C_6H_5)_2C_3H_7$-i, respectively. For details and more values, compare [104, 134]. In the contrary, the reaction of $C_5H_5Mo(CO)_3CH_3$ with $P(C_6H_5)_3$ in acetonitrile [68], THF or toluene [42] did not go to completion but approached equilibrium. The equilibrium constant, obtained from IR-spectrophotometric observation of the position of equilibrium in THF at 50.7°C, is K = 28.2 ± 6.0 L/mol, in reasonable agreement with the value calculated from the ratio of measured rate constants, $K = k_1 \cdot k_2/k_{-1} \cdot k_{-2} = 19.2 \pm 3.8$ L/mol [42]. Pressure only mildly accelerates the rates of migratory CO insertions in $C_5H_5Mo(CO)_3R$ and related transition metal alkyls, suggesting that $\Delta V^{\ddagger}$ is probably only slightly negative [260].

Kinetic measurements were done on the two competing reaction pathways (see below) leading to $C_5H_5Mo(CO)_2(PR'_3)C(O)R$ by treatment of $C_5H_5Mo(CO)_3R$ with various monotertiary phosphanes and phosphites and were discussed in detail. Generally it was established that the relative reactivity of the Mo complexes toward PR'_3 decreased in the order of R = $C_2H_5 > CH_3 > CH_2C_6H_5 > CH_2CH{=}CH_2$. The following table summarizes the cases investigated additionally with measured (k_{obs}) and calculated (k_1, k_3) rate constants [42, 52, 63, 241, 246, 252, 253, 316, 333, 342, 356]; see also [38, 271, 300]. The two competing pathways are outlined in Scheme 1. In some cases only the measured k_{obs} was given, which expressed the competition between the different reaction pathways. Calculations of k_1 and k_3 were made by the assumption that a steady-state concentration of the solvated intermediate (see below) existed and that the reaction went to completion. As demonstrated in

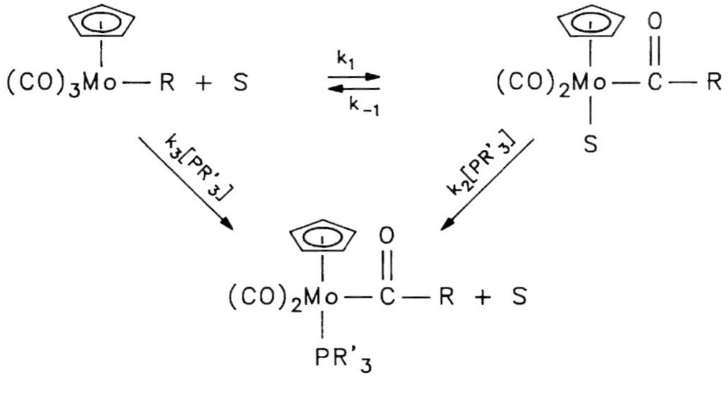

Scheme I

References on pp. 106/18

Scheme 1, one of the two reation pathways is solvent-assisted, and in the k_1 step, established as rate-determining, a molecule of a polar solvent induced the formation of an acyl intermediate from which the coordinated solvent is liberated. The second pathway (k_3, direct route) is preferred in nonpolar solvents, and in this case the tertiary phosphorus compounds performed the role as nucleophile in inducing the intramolecular migration of the alkyl group. At very high concentrations of the phosphorus compound and/or in less polar solvents a mixed pathway may be followed [42, 241, 246, 271, 300], see also [274].

The rate data obtained for the insertion reactions of the methyl derivatives $C_5H_5M(CO)_3CH_3$ (M = Cr, Mo, W) with different phosphane ligands reflect a generally observed decrease in metal-carbon bond strength, W $\gg$ Mo $\geq$ Cr [133].

Data Available for the Reaction of $C_5H_5Mo(CO)_3{}^1L$ with Phosphanes or Phosphites PR'_3.

^1L	PR'_3, reaction conditions	rate constants
CH_3	$P(CH_3)_3$, in CH_3CN [232] or THF [316, 342]	not studied
CH_3	$P(C_2H_5)_3$, in THF [27, 271]	in THF: $k_1 = 7.75\times10^{-4}\ s^{-1}$, $k_3 = 0.80\ L\cdot mol^{-1}\cdot s^{-1}$ at 60 °C [271]
CH_3	$P(C_4H_9\text{-}n)_3$, in CH_3CN [38, 52, 68], THF, toluene [42], or hexane [65]	in THF: $k_1 = 3.65\times10^{-4}\ s^{-1}$, $k_3 = 3.49\times10^{-3}\ L\cdot mol^{-1}\cdot s^{-1}$; in toluene: $k_1 = 5.80\times10^{-6}\ s^{-1}$, $k_3 = 6.44\times10^{-5}\ L\cdot mol^{-1}\cdot s^{-1}$, measured at 50.7 °C [42], see [65]; in CH_3CN: $k_{obs} = 6.93\times10^{-4}\ s^{-1}$ at 30.5 °C [52]; for hexane see [65]
CH_3	$P(CH_3)_2C_6H_5$, in CH_3CN [107] or THF [316, 342]	in toluene: $k_3 = 19.3\times10^{-6}\ L\cdot mol^{-1}\cdot s^{-1}$ at 30 °C [356]
CH_3	$PH(C_6H_5)_2$, in CH_3CN [383]	not studied
CH_3	$P(C_6H_5)_2C_2H_5$, in CH_3CN [383]	not studied
CH_3	$P(C_6H_5)_2CH_3$, in THF or $OC_4H_{8-n}(CH_3)_n$ (n = 1, 2) at 30 to 60 °C [246, 316, 342, 356]	in THF: $k_1 = 7.78\times10^{-4}\ s^{-1}$, $k_3 = 1.73\times10^{-4}\ L\cdot mol^{-1}\cdot s^{-1}$ at 60 °C [246], $k_1 = 1.53\times10^{-4}\ s^{-1}$, $k_3 = 0.3\times10^{-4}\ L\cdot mol^{-1}\cdot s^{-1}$ at 40 °C [356]
CH_3	$P(C_6H_5)_3$, in CH_3CN [38, 52], in THF or toluene [42] also for CD_3 [380]	in THF: $k_1 = 3.67\times10^{-4}\ s^{-1}$; in toluene: $k_3 = 3.22\times10^{-5}\ L\cdot mol^{-1}\cdot s^{-1}$; both at 50.7 °C [42]; in CH_3CN: $k_{obs} = 4.32\times10^{-4}\ s^{-1}$ at 30.5 °C [52]
CH_3	$P(N(CH_3)_2)_3$, no details [27]	not studied
CH_3	$P(OC(CH_3)_2CH_2)_2N$-bicyclo, in refluxing hexane [299]	not studied

^1L	PR'_3, reaction conditions	rate constants
CH_3	$P(OCH_3)_3$, in CH_3CN [38, 52] or THF [65, 316, 342]	in CH_3CN: $k_{obs} = 5.25 \times 10^{-4}\ s^{-1}$ at 30.5 °C [52], similar in THF [65]
CH_3	$P(OC_2H_5)_3$, in CH_3CN [38, 52]	not studied
CH_3	$P(OC_4H_9\text{-}n)_3$, in THF, toluene, CH_3NO_2, and $(CH_3)_2NCHO$ [42]	in THF: $k_1 = 3.94 \times 10^{-4}\ s^{-1}$; similar in CH_3NO_2; in toluene: $k_1 = 5.80 \times 10^{-6}\ s^{-1}$, $k_3 = 5.00 \times 10^{-5}\ L \cdot mol^{-1} \cdot s^{-1}$; in $(CH_3)_2NCHO$: $k_1 \cong 8.0 \times 10^{-2}\ s^{-1}$, all at 50.7 °C [42]; see also [271]; activation parameters for the THF-assisted methyl migration at 50.7 °C are $\Delta H^{\ddagger} = 16.1 \pm 0.6$ kcal/mol and $\Delta S^{\ddagger} = -25 \pm 2\ cal \cdot mol^{-1} \cdot K^{-1}$ [42]
CH_3	$P(OCH_2)_3CCH_3$-bicyclo, in CH_3CN [38, 52]	in CH_3CN: $k_{obs} = 5.20 \times 10^{-4}\ s^{-1}$ at 30.5 °C [52]
CH_3	$P(OC_6H_5)_3$, in THF or toluene [38, 42]	in THF: $k_1 = 3.10 \times 10^{-4}\ s^{-1}$; in toluene: $k_3 = 8.33 \times 10^{-6}\ L \cdot mol^{-1} \cdot s^{-1}$; both at 50.7 °C [42]; in CH_3CN: $k_{obs} = 6.45 \times 10^{-4}\ s^{-1}$ at 30.5 °C [52]
CH_3	$PF_2N(CH_3)_2$, in CH_3CN [111, 112]	not studied
C_2H_5	$P(CH_3)_3$, in refluxing heptane [235] or in THF [316, 342]	not studied
C_2H_5	$P(C_4H_9\text{-}n)_3$, in CH_3CN [38, 52] or in THF [316, 342]	not studied
C_2H_5	$PH(C_6H_5)_2$, in CH_3CN [383]	not studied
C_2H_5	$P(CH_3)_2C_6H_5$, in CH_3CN [107] or THF [316, 342]	not studied
C_2H_5	$P(C_6H_5)_2CH_3$, in THF [316, 342]	not studied
C_2H_5	$P(C_6H_5)_2C_2H_5$ [383]	not studied
C_2H_5	$P(C_6H_5)_3$, in CH_3CN [52], $CHCl_3$, or THF [316, 342]	in $CHCl_3$: $k_1 = 1.80 \times 10^{-4}\ s^{-1}$, $k_3 = 1.70 \times 10^{-3}\ L \cdot mol^{-1} \cdot s^{-1}$ at 30.5 °C [52]
C_2H_5	$P(OCH_3)_3$, in CH_3CN [52], $CHCl_3$, or THF [342]	in $CHCl_3$: $k_1 = 1.80 \times 10^{-4}\ s^{-1}$, $k_3 = 1.30 \times 10^{-3}\ L \cdot mol^{-1} \cdot s^{-1}$ at 30.5 °C [52]
C_2H_5	$P(OC_2H_5)_3$, in CH_3CN [38, 52]	not studied

References on pp. 106/18

^1L	PR'_3, reaction conditions	rate constants
C_2H_5	$P(OCH_2)_3CCH_3$-bicyclo, in CH_3CN or $CHCl_3$ [38, 52]	in $CHCl_3$: $k_1 = 1.75 \times 10^{-4}\ s^{-1}$, $k_3 = 2.50 \times 10^{-3}\ L \cdot mol^{-1} \cdot s^{-1}$ at 30.5 °C [52]
C_2H_5	$P(OC_6H_5)_3$, in CH_3CN [38, 52]	not studied
$CH_2(C_3H_5$-cyclo)	$P(C_6H_5)_3$, in THF [333]	in THF: $k_{obs} = 2.21 \times 10^{-1}\ s^{-1}$ at 50 °C; the rate constant is 6 times faster than for the CH_3 compound for steric reasons; activation parameters for the THF-assisted alkyl migration are $\Delta H^{\ddagger} = 11.2$ kcal/mol and $\Delta S^{\ddagger} = -28\ cal \cdot mol^{-1} \cdot K^{-1}$ [333]
$CH_2CH{=}CH_2$	$P(C_6H_5)_3$, in CH_3CN [38, 63]	in CH_3CN: $k_{obs} = 1.10 \times 10^{-4}\ s^{-1}$ at 45 °C [63]
$CH_2CH{=}CH_2$	$P(OC_6H_5)_3$, in CH_3CN [38, 63]	in CH_3CN: $k_{obs} = 1.70 \times 10^{-4}\ s^{-1}$ at 45 °C [63]
$CH_2CH{=}CH_2$	$P(OCH_2)_3CC_2H_5$-bicyclo, in CH_3CN at 70 °C [38, 63]	not studied
$CH_2CH{=}CHCH_3$	$P(C_6H_5)_3$, in CH_3CN [38, 63] or THF [394]	not studied
$CH_2C_6H_5$	$P(CH_3)_3$, in CH_3CN [342]	not studied
$CH_2C_6H_5$	$P(C_2H_5)_3$, in toluene [356]	in toluene: $k_3 = 1.92 \times 10^{-5}\ L \cdot mol^{-1} \cdot s^{-1}$ at 30 °C [356]
$CH_2C_6H_5$	$P(C_4H_9\text{-}n)_3$, in CH_3CN [38, 63]	not studied
$CH_2C_6H_5$	$P(C_4H_9\text{-}i)_3$, in toluene [356]	in toluene: $k_3 = 3.4 \times 10^{-6}\ L \cdot mol^{-1} \cdot s^{-1}$ at 30 °C [356]
$CH_2C_6H_5$	$P(CH_3)_2C_6H_5$, in CH_3CN [253, 342] or toluene [356]	in CH_3CN: $k_1 = 3.4 \times 10^{-4}\ s^{-1}$ [253]; in toluene: $k_3 = 1.93 \times 10^{-5}\ L \cdot mol^{-1} \cdot s^{-1}$, both at 30 °C [356]
$CH_2C_6H_5$	$P(C_6H_5)_2CH_3$, in CH_3CN [253, 342], THF, $OC_4H_{8-n}(CH_3)_n$ (n = 1, 2), or toluene [356]	in CH_3CN: $k_1 = 3.2 \times 10^{-4}\ s^{-1}$ at 30 °C [253]; in THF: $k_1 = 0.6 \times 10^{-4}\ s^{-1}$, $k_3 = 2.5 \times 10^{-5}\ L \cdot mol^{-1} \cdot s^{-1}$ at 40 °C; in toluene: $k_3 = 9.9 \times 10^{-6}\ L \cdot mol^{-1} \cdot s^{-1}$ at 30 °C [356]
$CH_2C_6H_5$	$P(C_6H_5)_2C_2H_5$, in CH_3CN [253]	in CH_3CN: $k_1 = 3.0 \times 10^{-4}\ s^{-1}$ at 30 °C [253]
$CH_2C_6H_5$	$P(C_6H_5)_2C_3H_7\text{-i}$, in CH_3CN [253]	in CH_3CN: $k_1 = 3.0 \times 10^{-4}\ s^{-1}$ at 30 °C [253]

References on pp. 106/18

^1L	PR'_3, reaction conditions	rate constants
$CH_2C_6H_5$	$P(C_6H_5)_2C_4H_9$-t, in CH_3CN [253] or toluene [356]	in CH_3CN: $k_1 = 3.0 \times 10^{-4}\ s^{-1}$ at 30 °C [253]; in toluene: $k_3 = 1.9 \times 10^{-6}\ L \cdot mol^{-1} \cdot s^{-1}$ at 30 °C [356]
$CH_2C_6H_5$	$P(C_6H_5)_3$, in CH_3CN [63, 241, 252, 253, 342] or toluene [356]	in CH_3CN: $k_{obs} = 2.56 \times 10^{-4}\ s^{-1}$ at 29 °C, $k_1 = 2.7 \times 10^{-4}\ s^{-1}$ [241]; see also [63, 252, 253]; in toluene: $k_3 = 4.3 \times 10^{-6}\ L \cdot mol^{-1} \cdot s^{-1}$ at 30 °C [356]
$CH_2C_6H_5$	$P(C_6H_2(OCH_3)_3$-2,4,6$)_3$ [356]	$k_3 = 3 \times 10^{-4}\ L \cdot mol^{-1} \cdot s^{-1}$; no details given [356]
$CH_2C_6H_5$	$P(OCH_3)_3$, in CH_3CN [342]	not studied
$CH_2C_6H_5$	$P(OCH_2)_3CC_2H_5$-bicyclo, in CH_3CN [38, 63]	not studied
$CH_2C_6H_5$	$P(OC_6H_5)_3$, in CH_3CN [38, 63]	in CH_3CN: $k_{obs} = 1.95 \times 10^{-4}\ s^{-1}$ at 30 °C [63]
$CH_2C_6H_4CH_3$-2	$P(C_6H_5)_2CH_3$, in THF or $OC_4H_{8-n}(CH_3)_n$ (n = 1, 2) [356]	in THF: $k_1 = 1.45 \times 10^{-4}\ s^{-1}$, $k_3 = 3.2 \times 10^{-5}\ L \cdot mol^{-1} \cdot s^{-1}$, at 40 °C [356]
$CH_2C_6H_4CH_3$-2	$P(C_6H_5)_3$, in CH_3CN [252]	in CH_3CN: $k_{obs} = 7.3 \times 10^{-4}\ s^{-1}$ at 30 °C [252]
$CH_2C_6H_4CH_3$-3	$P(C_6H_5)_3$, in CH_3CN [252]	in CH_3CN: $k_{obs} = 3.4 \times 10^{-4}\ s^{-1}$ at 30 °C [252]
$CH_2C_6H_4CH_3$-4	$P(C_6H_5)_3$, in CH_3CN [252]	in CH_3CN: $k_{obs} = 3.9 \times 10^{-4}\ s^{-1}$ [252], $k_1 = 3.6 \times 10^{-4}\ s^{-1}$ [241], both at about 30 °C
$CH_2C_6H_4C_3H_7$-i,2	$P(C_6H_5)_3$, in CH_3CN [252]	in CH_3CN: $k_{obs} = 150 \times 10^{-4}\ s^{-1}$ at 30 °C [252]
$CH_2C_6H_4C_3H_7$-i,4	$P(C_6H_5)_3$, in CH_3CN [252]	in CH_3CN: $k_{obs} = 3.7 \times 10^{-4}\ s^{-1}$ at 30 °C [252]
$CH_2C_6H_4OCH_3$-2	$P(C_6H_5)_3$, in CH_3CN [252]	in CH_3CN: $k_{obs} = 2.8 \times 10^{-4}\ s^{-1}$ at 30 °C [252]
$CH_2C_6H_4OCH_3$-3	$P(C_6H_5)_3$, in CH_3CN [252]	in CH_3CN: $k_{obs} = 2.5 \times 10^{-4}\ s^{-1}$ [252], $k_1 = 2.3 \times 10^{-4}\ s^{-1}$ [241], at ca. 30 °C
$CH_2C_6H_4OCH_3$-4	$P(C_6H_5)_3$, in CH_3CN [252]	in CH_3CN: $k_{obs} = 4.8 \times 10^{-4}\ s^{-1}$ [252], $k_1 = 4.5 \times 10^{-4}\ s^{-1}$ [241], at ca. 30 °C
$CH_2C_6H_4CF_3$-2	$P(C_6H_5)_3$, in CH_3CN [252]	in CH_3CN: $k_{obs} = 1.0 \times 10^{-4}\ s^{-1}$ at 30 °C [252]

References on pp. 106/18

^1L	PR'_3, reaction conditions	rate constants
$CH_2C_6H_4CF_3$-3	$P(C_6H_5)_3$, in CH_3CN [252]	in CH_3CN: $k_{obs} = 1.1 \times 10^{-4}\ s^{-1}$ [252], $k_1 = 1.1 \times 10^{-4}\ s^{-1}$ [241], at ca. 30 °C
$CH_2C_6H_4CF_3$-4	$P(C_6H_5)_3$, in CH_3CN [252]	in CH_3CN: $k_{obs} = 0.8 \times 10^{-4}\ s^{-1}$ [252], $k_1 = 0.7 \times 10^{-4}\ s^{-1}$ [241], at ca. 30 °C
$CH_2C_6H_4F$-2	$P(C_6H_5)_3$, in CH_3CN [252]	in CH_3CN: $k_{obs} = 0.9 \times 10^{-4}\ s^{-1}$ at 30 °C [252]
$CH_2C_6H_4F$-3	$P(C_6H_5)_3$, in CH_3CN [252]	in CH_3CN: $k_{obs} = 1.5 \times 10^{-4}\ s^{-1}$ [252], $k_1 = 1.4 \times 10^{-4}\ s^{-1}$ [241], at ca. 30 °C
$CH_2C_6H_4F$-4	$P(C_6H_5)_3$, in CH_3CN [252]	in CH_3CN: $k_{obs} = 3.1 \times 10^{-4}\ s^{-1}$ [252], $k_1 = 2.9 \times 10^{-4}\ s^{-1}$ [241], both at about 30 °C
$CH_2C_6H_4Cl$-2	$P(C_6H_5)_3$, in CH_3CN [252]	in CH_3CN: $k_{obs} = 1.1 \times 10^{-4}\ s^{-1}$ at 30 °C [252]
$CH_2C_6H_4Cl$-3	$P(C_6H_5)_3$, in CH_3CN [252]	in CH_3CN: $k_{obs} = 1.3 \times 10^{-4}\ s^{-1}$ at 30 °C [252]
$CH_2C_6H_4Cl$-4	$P(C_6H_5)_3$, in CH_3CN [252]	in CH_3CN: $k_{obs} = 1.7 \times 10^{-4}\ s^{-1}$ at 30 °C [252]
$CH_2C_6H_3Cl_2$-2,6	$P(C_6H_5)_3$, in CH_3CN [252]	in CH_3CN: $k_{obs} = 0.09 \times 10^{-4}\ s^{-1}$ at 30 °C [252]
$CH_2C_6H_3(OCH_3)_2$-2,6	$P(C_6H_5)_3$, in CH_3CN [252]	in CH_3CN: $k_{obs} = 0.09 \times 10^{-4}\ s^{-1}$ at 30 °C [252]
$CH_2C_6H_2(CH_3)_3$-2,4,6	$P(C_6H_5)_2CH_3$, in THF or $OC_4H_{8-n}(CH_3)_n$ (n = 1, 2) [356]	in THF: $k_1 = 1.31 \times 10^{-4}\ s^{-1}$, $k_3 = 5.8 \times 10^{-5}\ L \cdot s^{-1} \cdot mol^{-1}$ at 40 °C [356]
$CH_2CH_2SO_2OCH_3$	$P(C_6H_5)_3$, in THF [387]	not studied
$CH_2CH(CH_3)SO_2OCH_3$	$P(C_6H_5)_3$, in THF [387]	not studied
$(CH_2)_nFe(CO)_2C_5H_5$	$P(C_6H_5)_3$, for n = 3, 4, 6 [392]	not studied
$(CH_2)_nFe(CO)_2C_5H_5$	$P(CH_3)_2C_6H_5$, for n = 3, 4 [392]	not studied

The insertion products, $C_5H_5Mo(CO)_2(P(C_6H_5)_3)C(O)CH_2CH{=}CHR$ (R = H, CH_3), resulting from treatment of the σ-allyls $C_5H_5Mo(CO)_3CH_2CH{=}CHR$ with $P(C_6H_5)_3$ in THF under reflux conditions, are thermally unstable with respect to phosphane loss and decarbonylation and form $C_5H_5Mo(CO)_2(\eta^3\text{-}CH_2CHCHR)$ [394].

Investigations of the kinetics in polar solvents (CH_3CN, THF) indicated k_1 is first-order in the disappearance of $C_5H_5Mo(CO)_3R$ and independent of phosphane concentration. Contrarily in nonpolar solvents (e.g. toluene), the predominating direct pathway showed a linear dependence in phosphane concentration for k_3 according to second-order kinetics [42, 52, 246, 356]. Evidence for the direct attack of the polar solvent at the Mo center, as discussed in [42] in contrast to a suggested generalized stabilization of the migration insertion transition state by solvation [52], was established by studies of the reaction of $C_5H_5Mo(CO)_3CH_3$ with $P(C_6H_5)_2CH_3$ in tetrahydrofuran and methylsubstituted tetrahydrofurans, which resulted in a decrease of k_1 with increasing steric demand of the solvent molecules while k_3 remained nearly constant. Furthermore k_1 showed a linear dependence on solvent concentration itself; actually k_1 is a product of a second-order rate constant k_1' and solvent concentration [246]. Therefore the relative contribution of the nucleophilicity of the solvent to the k_1 step, and that of the ligand to the k_3 step, can be estimated if the rate constant k_1 (being effectively first-order due to the large excess of the solvent) is converted to a second-order constant, k_1', by dividing it by the solvent concentration [356]. Similar trends for values of k_1 were established for $C_5H_5Mo(CO)_3R$ (R = $CH_2C_6H_5$, $CH_2C_6H_4CH_3$-2, $CH_2C_6H_2(CH_3)_3$-2,4,6), but differentiation is reduced as the size of the alkyl ligand increases; for details and data, see [356]. As in the previous studies [246, 271], there is no significant variation in k_3 between the tetrahydrofurans, except for $C_5H_5(CO)_3MoCH_2C_6H_4CH_3$-2 in 2,5-dimethyltetrahydrofuran. A clear trend existed, however, with regard to the alkyl ligand: R = $CH_2C_6H_2(CH_3)_3$-2,4,6 > $CH_2C_6H_4CH_3$-2 > $CH_2C_6H_5$, and R = CH_3 > $CH_2C_6H_5$ [356].

The rates of reaction were also sensitive to changes in the nature of the entering nucleophile, as demonstrated for $C_5H_5Mo(CO)_3CH_3$ in the reaction with PR_3' in THF. For R' = n-C_4H_9 a linear dependence on phosphane concentration was proved, whereas for R = C_6H_5, OC_4H_9-n, OC_6H_5 a virtual independence of nucleophile and nucleophile concentration was established [42]; for contradictory results where R = OC_6H_5, see [271].

While the rate data obtained for 3- and 4-substituted benzyl ligands demonstrated a moderate increase in reactivity with increasing electron donation to the benzyl system, the values of k_{obs} obtained for the reaction of ortho-substituted benzyl derivatives $C_5H_5Mo(CO)_3$-$CH_2C_6H_4R$-2 with $P(C_6H_5)_3$ in CH_3CN at 30 °C provide striking evidence for increased reactivity with the size of the respective ortho substituents [252]. For $C_5H_5Mo(CO)_3CH_2$-$C_6H_2(CH_3)_3$-2,4,6, there is an additional small increase in the rate constant, which seems, however, more compatible with the electron-releasing effect of the three methyl groups rather than with the increased bulk. In striking contrast, disubstitution of the ortho positions by i-C_3H_7, OCH_3, and Cl residues causes a pronounced lowering of the reactivity for C_5H_5Mo-$(CO)_3CH_2C_6H_3(OCH_3)_2$-2,6 and $C_5H_5Mo(CO)_3CH_2C_6H_3Cl_2$-2,6, as well as a complete inhibition of migratory CO insertion for $C_5H_5Mo(CO)_3CH_2C_6H_2(C_3H_7\text{-i})_3$-2,4,6 [252, 300]. While the observed steric enhancement in the reactions of benzyl molybdenum complexes containing only one ortho substituent has been ascribed to the relatively greater release of steric strain in the transition state, as the large substituent is partially removed from the crowded environment at the metal center, it is believed that the second ortho substituent plays a specific role in restricting the access of the migrating benzylic carbon atom to the carbonyl atom [252].

Rate constants associated with the phosphane-induced formation of a number of acyls $C_5H_5Mo(CO)_2(PR_3')C(O)CH_2C_6H_5$ in CH_3CN at 30 °C show no correlation with the electronic properties of the respective phosphane, but demonstrate the high sensitivity of the k_2 stage of the reaction (reflected by the variation in k_2/k_{-1}) to the steric demand of the tertiary phosphanes discussed in terms of the increasing cone angle of the entering phosphane. This is in the opposite sense to the steric effect of the benzyl group on the k_1 stage, where the transition state involves the partial removal of that latter ligand. The decrease in rate at higher cone angles is consistent with the concept of a pocket of limited size around the metal center into which the phosphanes must fit in order to react [253]. For k_3 of the direct pathway,

References on pp. 106/18

an analogous trend was described; but for the very bulky and highly basic phosphane $P(C_6H_2(OCH_3)_3\text{-}2,4,6)_3$, a surprisingly high k_3 value was measured, suggesting that in extreme cases, electronic effects may override steric influence [356].

For further discussions on the influence of solvent effects, of steric and electronic factors of the phosphanes and phosphites, and of the kind of the alkyl substituent of the molybdenum precursor to direct the reaction more to the solvent-assisted or to the direct pathway, compare [42, 52, 63, 65, 241, 252, 356]. For comparison the rate constants of $C_5H_5Mo(CO)_3CH_2C_6H_5$ with those of compounds containing substituted cyclopentadienyl rings, see [291].

Thermochemical data for the phosphane-promoted carbonyl insertion indicated a dependence of the enthalpy of reaction on both the alkyl group and the phosphane ligand. For the latter, the exothermicity of the reactions spans a range of 5 to 6 kcal/mol in the order $P(CH_3)_3 > P(C_4H_9\text{-n})_3 > P(OCH_3)_3 > P(CH_3)_2C_6H_5 > P(C_6H_5)_2CH_3 > P(C_6H_5)_3$. For variations of the alkyl ligands, the CO insertion reaction becomes more exothermic in the order $C_2H_5 > CH_2C_6H_5 > CH_3$, which does not parallel the order of kinetic reactivity, where the benzyl complex is less reactive than the methyl and the ethyl derivatives [316, 342].

Enthalpies of Reaction (kcal/mol) of Selected Alkyl Complexes $C_5H_5Mo(CO)_3{}^1L$ with Phosphanes and Phosphites PR'_3 [316, 342].

$PR'_3/{}^1L$	CH_3 [a)]	C_2H_5 [a)]	$CH_2C_6H_5$ [b)]
$P(CH_3)_3$	−19.7(0.4)	−22.5(0.4)	−21.9(0.5)
$P(C_4H_9\text{-n})_3$	−18.6(0.4)	−21.8(0.2)	
$P(CH_3)_2C_6H_5$	−17.1(0.3)	−19.7(0.2)	−19.1(0.4)
$P(C_6H_5)_2CH_3$	−14.2(0.3)	−18.2(0.1)	−15.8(0.5)
$P(C_6H_5)_3$	c)	−15.4(0.5)	−14.1(1.2)
$P(OCH_3)_3$	−17.2(0.2)	−19.8(0.4)	−17.5(0.5)

a) In THF. — b) In CH_3CN. — c) Reaction did not proceed to completion.

Since the acyl complexes obtained from $C_5H_5Mo(CO)_3R$ (R = CH_3, C_2H_5, $CH_2CH{=}CH_2$, $CH_2C_6H_5$) and PR'_3 in CH_3CN possess trans stereochemistry [52, 63, 68, 107, 241] without any evidence of an intermediate cis isomer, it was suggested that the trans-acyls were probably produced directly by kinetic control, rather than in a subsequent thermodynamically controlled process [52, 63]. Nevertheless, a two-step process via a cis-acyl isomer with a ratio of rate constants k_{trans}/k_{cis} appreciably greater than unity (e.g., > 10) seems more appropriate [60], because the reactions of $C_5H_5Mo(CO)_3R$ (R = CH_3, CH_2CH_3, $CH_2C_6H_5$) with $P(C_6H_5)_3$ in acetone did furnish the respective cis-acyls as the initial products at the early stages of interaction [241]. The same phenomenon was noted for the systems $C_5H_5Mo(CO)_3CH_2C_6H_4OCH_3\text{-}3/P(C_4H_9\text{-n})_3$ in dimethyl sulfoxide and $C_5H_5Mo(CO)_3R/P(C_6H_5)_3$ in dimethylformamide (R = C_2H_5, $CH_2C_6H_4OCH_3\text{-}4$), whereas for the reactions occurring between $C_5H_5(CO)_3MoCH_2C_6H_4OCH_3\text{-}3$ and $P(OCH_3)_3$ or $P(OCH_2)_3CC_2H_5$-bicyclo in dimethyl sulfoxide, only the trans-acyl was observed [241].

In the conversion of $C_5H_5Mo(CO)_3CH_3$ by $P(C_6H_5)_3$ in refluxing hexane or THF, the acetyl complex $C_5H_5Mo(CO)_2(P(C_6H_5)_3)C(O)CH_3$, was isolated together with $C_5H_5Mo(CO)_2(P(C_6H_5)_3)CH_3$ as an accompanying by-product, with the mixture of two compounds arising from the initial formation of the acetyl derivative with subsequent decarbonylation to yield the methyl complex [40, 44]. A study of the course of the reaction with the donor ligands

$P(C_4H_9\text{-}n)_3$, $P(C_6H_5)_3$, $P(OCH_3)_3$, and $P(OC_6H_5)_3$ in CD_3CN by ^1H NMR revealed that the decarbonylation reaction occurred at a much slower rate than the initial substitution/insertion step [68]. Small amounts of substituted alkyl derivatives, $C_5H_5Mo(CO)_2(PR'_3)CH_2C_6H_5$, were also observed in the reactions of $C_5H_5Mo(CO)_3CH_2C_6H_5$ with phosphanes PR'_3, such as $P(C_2H_5)_3$, $P(C_4H_9\text{-}i)_3$, $P(CH_3)_2C_6H_5$, $P(C_6H_5)_2CH_3$, $P(C_6H_5)_2C_4H_9\text{-}t$, and $P(C_6H_5)_3$, in toluene at 30 °C; the extent of decarbonylation was most pronounced with the largest phosphanes $P(C_4H_9\text{-}i)_3$ and $P(C_6H_5)_2C_4H_9\text{-}t$ [356].

Analogous reactions of $C_5H_5Mo(CO)_3R$ with ditertiary phosphanes (^4D) also occurred and resulted in bisphosphane-bridged acyl derivatives of the general composition $(C_5H_5Mo(CO)_2C(O)R)_2(\mu\text{-}^4D)$ as proved for two selected compounds with R = CH_3 and $CH_2C_6H_5$ [40, 44, 67]. Treatment of both complexes with $R'_2PC_2H_4PR'_2$ (R' = CH_3, C_6H_5), trans-$(C_6H_5)_2PCH{=}CHP(C_6H_5)_2$, or $(C_6H_5)_2PC{\equiv}CP(C_6H_5)_2$ in CH_3CN at room temperature for several hours yielded $(C_5H_5Mo(CO)_2C(O)R)_2(\mu\text{-}R'_2PC_2H_4PR'_2)$, $(C_5H_5Mo(CO)_2C(O)R)_2(\mu\text{-trans-}(C_6H_5)_2PCH{=}CHP(C_6H_5)_2)$, and $(C_5H_5Mo(CO)_2C(O)CH_3)_2(\mu\text{-}(C_6H_5)_2PC{\equiv}CP(C_6H_5)_2)$ [67]. In the case of $(C_6H_5)_2PC_2H_4P(C_6H_5)_2$, the reaction proceeded also in refluxing THF or in hexane under UV irradiation [40, 44], but the reaction with cis-$(C_6H_5)_2PCH{=}CHP(C_6H_5)_2$ failed to give the corresponding phosphane-bridged acyl complex; instead cis-(cis-$(C_6H_5)_2PCH{=}CHP(C_6H_5)_2)_2Mo(CO)_2$ was produced [67]. This compound is identical to material prepared by UV irradiation of cis-$(C_6H_5)_2PCH{=}CHP(C_6H_5)_2$ with $C_5H_5Mo(CO)_3CH_2C_6H_5$ in benzene [59]. Similarly, exposure of the methyl or benzyl complexes in the presence of trans-$(C_6H_5)_2PCH{=}CHP(C_6H_5)_2$ to UV irradiation in benzene resulted in $(C_5H_5)_2Mo_2(CO)_3(\mu\text{-trans-}(C_6H_5)_2PCH{=}CHP(C_6H_5)_2)$ as illustrated in Formula IV [59].

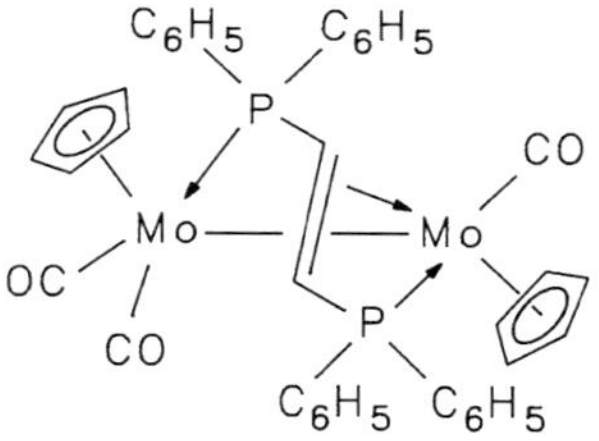

IV

Interaction of $C_5H_5Mo(CO)_3CH_3$ with $C_6H_5PC_2H_4(P(C_6H_5)_2)_2$, $P(C_2H_4P(C_6H_5)_2)_3$, $(C_6H_5)_2PC_2H_4P(C_6H_5)C_2H_4P(C_6H_5)C_2H_4P(C_6H_5)_2$, and $[(C_6H_5)_2P(C_2H_4)_2PCH_2]_2$ in CH_3CN at room temperature for several hours, accordingly resulted in the formation of oligometallic acetyls, $(C_5H_5Mo(CO)_2(D)C(O)CH_3)_n$ (n = 3, 4, 6) with D = $\mu\text{-}C_6H_5P(C_2H_4P(C_6H_5)_2)_2$ [97], $\mu\text{-}P(C_2H_4P(C_6H_5)_2)_3$, $\mu\text{-}(C_6H_5)_2PC_2H_4P(C_6H_5)C_2H_4P(C_6H_5)C_2H_4P(C_6H_5)_2$ [98], and $\mu\text{-}[((C_6H_5)_2PC_2H_4)_2PCH_2]_2$ [99], in which the phosphanes acted as triligate, tetraligate, and hexaligate ligands, respectively.

On the other hand, treatment of $C_5H_5Mo(CO)_3CH_3$ with the bidentate phosphine $(C_6H_5)_2PCH_2P(C_6H_5)_2$, either at ambient conditions or at 25 °C and 32 kbar, afforded only $C_5H_5Mo(CO)_2((C_6H_5)_2PCH_2P(C_6H_5)_2)C(O)CH_3$ containing a monoligate ditertiary phosphane ligand [260]. Analogous reactions of $C_5H_5Mo(CO)_3CH_3$ with the mixed ligands $(C_6H_5)_2PC_2H_4As(C_6H_5)_2$ and $C_6H_5P(C_2H_4As(C_6H_5)_2)_2$, even when the former was present in excess, similarly gave only the Mo-P-bonded monoligate monometallic acyls $C_5H_5Mo(CO)_2(\eta^1\text{-}(C_6H_5)_2AsC_2H_4P(C_6H_5)_2)C(O)CH_3$ and $C_5H_5Mo(CO)_2(\eta^1\text{-}P(C_6H_5)(C_2H_4As(C_6H_5)_2)_2)C(O)CH_3$, respectively, apparently because of the lower donor ability of arsenic relative to phosphorus [108].

References on pp. 106/18

The reaction between $C_5H_5Mo(CO)_3CH_3$ and N-piperidinodifluorophosphane in boiling CH_3CN is analogous to the conversion of the methylmolybdenum complex by cis-$(C_6H_5)_2$-$PCH{=}CHP(C_6H_5)_2$ (see above) in that the only organometallic derivative that could be isolated from the reaction mixture was cis-$(PF_2N(CH_2)_4CH_2$-cyclo$)_4Mo(CO)_2$, rather than the expected $C_5H_5Mo(CO)_2(PF_2N(CH_2)_4CH_2$-cyclo$)C(O)CH_3$ [111].

Compounds possessing strongly electron-attracting alkyl ligands failed to form acyl derivatives on treatment with phosphanes. Thus, no CO insertion product was observed when $C_5H_5Mo(CO)_3CF_3$ was reacted with $P(C_6H_5)_3$ or $(C_6H_5)_2PC_2H_4P(C_6H_5)_2$, respectively, under UV irradiation in hexane or benzene. Instead $C_5H_5Mo(CO)_2(P(C_6H_5)_3)CF_3$ or $C_5H_5Mo(CO)$-$((C_6H_5)_2PC_2H_4P(C_6H_5)_2)CF_3$ were isolated as a consequence of the strong Mo-CF_3 bond [60]. Furthermore, treatment of the trifluoromethyl complex with the highly nucleophilic $(CH_3)_2PC_2H_4P(CH_3)_2$ in CH_3CN solution also failed to give any evidence of forming an acyl derivative of the kind observed in the reaction of the bisphosphine with, e.g., C_5H_5Mo-$(CO)_3CH_3$ (see above) [67]. Similarly, the reaction of $C_5H_5Mo(CO)_3CH_2Cl$ with $P(C_6H_5)_3$ in CH_3CN at 25°C in the dark yielded only $C_5H_5Mo(CO)_2(P(C_6H_5)_3)Cl$, while in refluxing CH_3OH after 30 min, $C_5H_5Mo(CO)_2(P(C_6H_5)_3)C(O)CH_2OCH_3$ via $C_5H_5Mo(CO)_3CH_2OCH_3$ was obtained. Longer reaction times gave $C_5H_5Mo(CO)_2(P(C_6H_5)_3)Cl$ and $C_5H_5Mo(CO)$-$(P(C_6H_5)_3)_2Cl$ [238, 267, 293].

Analogous acyl compounds as discussed for the phosphanes and phosphites are also obtained in the reaction of $C_5H_5Mo(CO)_3CH_3$ with a 3-fold excess of $E(C_6H_5)_3$ (E = As, Sb) in CH_3CN at about 30°C [137, 154]. The compounds $C_5H_5Mo(CO)_2(E(C_6H_5)_3)C(O)CH_3$ were unequivocally characterized, although in an earlier work [38, 52] the lack of reaction between $C_5H_5Mo(CO)_3R$ and nitrogen, arsenic, and sulfur nucleophiles such as pyridines, $As(OCH_2)_3CCH_3$-bicyclo, or $S(C_4H_9\text{-t})_2$ was described. If a similar first-order reaction according to PR'_3 nucleophiles was taken as a basis, the failure of previous attempts at synthesis of, e.g., arsenic acyl products can be readily accounted for by the lower nucleophilicity of such compounds, which prevented reaction with the intermediate $C_5H_5Mo(CO)_2C(O)CH_3$ · solvate, and the ease of $C_5H_5Mo(CO)_2(E(C_6H_5)_3)C(O)CH_3$ to undergo thermal back-reaction, with ejection of the respective $E(C_6H_5)_3$ ligand, to recover the starting material already at temperatures as low as 40°C [154].

In analogy to phosphanes and phosphites, a variety of isocyanides R′NC also induced a carbonyl insertion reaction on $C_5H_5Mo(CO)_3R$ under mild conditions in CH_3CN or benzene to give the corresponding acyl derivatives $C_5H_5Mo(CO)_2(C(O)R)CNR'$ (R = CH_3, R′ = t-C_4H_9, cyclo-C_6H_{11}, $C_6H_3(CH_3)_2$-2,6; R = $CH_2C_6H_4X$ (X = H, CH_3, OCH_3, CF_3), R′ = t-C_4H_9). They existed as equilibrium mixtures of cis and trans isomers in solution and as trans isomers in the solid state. During preparation no change in the cis/trans ratio was observed indicating isomerization is rapid compared to CO insertion [74, 87, 141, 274].

The reaction of $C_5H_5Mo(CO)_3CH_2C_6H_5$ with t-C_4H_9NC was thoroughly studied in cyclohexane, THF, CH_3CN, and $(CH_3)_2SO$, and the mechanism and kinetics were revealed to be quite similar to that of phosphorus containing nucleophiles (see above). Again two competing pathways were proved in which one involves a solvent-stabilized intermediate in the first step, represented by k_1 (see the following table). In the second step (k_2), the solvent is replaced by R′NC. This mechanism is predominant in solvents of high donicity at low concentrations of a nucleophile of low strength. The second pathway, involving a direct attack of the isocyanide (k_3), is favored in nonpolar solvents and for high concentrations of strong nucleophiles. The reactions studied go to completion. For the solvent-assisted pathway, a steady state concentration of the intermediate was assumed. If the ligand concentration is constant, the reaction is first-order in the concentration of $C_5H_5Mo(CO)_3CH_2C_6H_5$. For $C_5H_5Mo(CO)_3CH_2C_6H_5$ interacting with t-C_4H_9NC at varying concentrations in different solvents at 28 °C, the following trends have been noted. In cyclohexane a linear relationship between the observed rate constant and the isocyanide concentration was found tending towards a limiting value in neat t-C_4H_9NC. This is consistent with second-order kinetics in noncoordinating solvents and an essentially zero k_1 value. The faster reaction in CH_3CN started with an initial increase of the rate constant corresponding to k_1; a subsequent linear increase reflected the competing influence of the second-order k_3, and a final decrease of the observed rate constant to the value for neat t-C_4H_9NC is due to the lowering of solvent concentration which reduced k_1, better represented as a second-order rate constant $k_1 = k_1'$ [solvent]. This decrease is exaggerated in solvents of high donicity for which k_1 is high, e.g., $(CH_3)_2SO$, but is insignificant in noncoordinating solvents, e.g., cyclohexane (see above), and also in solvents of comparatively low donor strength such as THF [274]. Analysis of the linear portions of the observed rate law provided values of k_1 and k_3 as given in the following table [274]:

Rate Constants for the Reaction of $C_5H_5(CO)_3MoCH_2C_6H_5$ with t-C_4H_9NC in Different Solvents at 28 °C [274].

solvent	$10^4 \cdot k_1$ (s^{-1})	$10^4 \cdot k_3$ ($L \cdot mol^{-1} \cdot s^{-1}$)
cyclo-C_6H_{12}	0	0.39
THF	ca. 0	0.4
CH_3CN	2.84	0.33
$(CH_3)_2SO$	40	–

As with the PR_3'-promoted CO insertion, the rate of formation of acyl complexes $C_5H_5Mo(CO)_2(C(O)C_6H_4R)CNC_4H_9$-t is mildly enhanced by electron-donating substituents R as shown in the following table. For more details, see [274].

Substituent Effect on Rate Constants for the Reaction of $C_5H_5Mo(CO)_3CH_2C_6H_4R$ with t-C_4H_9NC in CH_3CN at 29 °C [274].

R	$10^4 \cdot k_1$ (s^{-1})	$10^4 \cdot k_3$ ($L \cdot mol^{-1} \cdot s^{-1}$)
4-CF_3	0.58	0.06
3-CF_3	1.09	0.09
H	2.84	0.33
4-CH_3	3.94	0.45
4-OCH_3	4.55	1.65

Comparison of the reaction of $C_5H_5Mo(CO)_3CH_2C_6H_5$ with cyclo-$C_6H_{11}NC$ or t-C_4H_9NC resulted in significantly different k_3 values in CH_3CN at 28 °C, 0.22×10^{-4} vs. 0.33×10^{-4} $L \cdot mol^{-1} \cdot s^{-1}$, suggesting that similar to the PR_3'-induced CO insertions, steric factors are also important for the isocyanide-assisted reactions [274]. Moreover, if this reaction was carried out in benzene with cyclo-$C_6H_{11}NC$ as the nucleophile, a product mixture was obtained containing only small amounts of the expected CO insertion product, but mainly a chelated imino-acyl compound $C_5H_5Mo(CO)_2(\eta^2\text{-}C(CH_2C_6H_5)\text{=}NC_6H_{11}$-cyclo) [274] earlier formulated as $C_5H_5Mo(CO)_3(\eta^1\text{-}C(CH_2C_6H_5)\text{=}NC_6H_{11}$-cyclo) [74]. As demonstrated by ^1H NMR studies, the imino-acyl complex is formed via the CO insertion product, a trans-

References on pp. 106/18

formation, which is accelerated by light [274]. Similar compounds were prepared using p-chloro- and p-methoxybenzylmolybdenum derivatives. The respective yields indicated that the more polarizable benzyl groups favor the formation of the imino derivatives at the expense of the conventional CO insertion products [74]. In contrast, the reaction of $C_5H_5Mo(CO)_3CH_3$ with t-C_4H_9NC in boiling benzene did not lead to carbonyl or isocyanide insertion, but instead resulted in the losses of both CH_3 and C_5H_5 ligands to afford fac-$Mo(CO)_3(CNC_4H_9$-t$)_3$ [141]. Alkylmolybdenum derivatives $C_5H_5Mo(CO)_3R$ possessing highly electron-withdrawing organo ligands R, such as CH_2CN, CH_2Cl, or CF_3, are unreactive toward cyclo-$C_6H_{11}NC$ or t-C_9H_4NC except under conditions where complete decomposition is observed, presumably as a result of the enhanced strength of the Mo-C σ bonds [74, 141].

$C_5H_5Mo(CO)_3R$ (R = CH_3, C_2H_5, $CH_2(CH_2)_2CN$, $CH_2C_6H_5$) reacted with KCN in polar solvents like CH_3OH or CH_3OH/H_2O by stereospecific alkyl migration to give acylcyanometalates K[cis-$C_5H_5Mo(CO)_2(C(O)R)CN$] as the primary products which subsequently rearrange to their trans isomers [102, 115].

A CO insertion reaction also occurred in the reversible reaction of $C_5H_5Mo(CO)_3R$ (R = CH_3, C_2H_5) with an excess of CS_2 in refluxing benzene with the exclusion of light and cis-$C_5H_5Mo(CO)_2(\eta^2$-$CS_2)C(O)R$ was detected in solution. Use of different solvents, extended reaction time, or use of light did not have any effect on this reaction. Refluxing of the acyl product in benzene regenerated the starting material. Thus an interrelated equilibrium between the alkyl starting material and the CO insertion product was connected by CS_2 as the carrier molecule [201].

In contrast to their behavior towards CN^-, PR'_3, R′NC, and CS_2, the alkyls $C_5H_5Mo(CO)_3R$ (R = CH_3, C_2H_5, $CH_2CH{=}CH_2$, $CH_2C_6H_5$, CH_2SCH_3) reacted with ca. 300 atm CO at elevated temperatures in n-tetradecane by cleavage of the C_5H_5 ring and the CH_3 group to give $Mo(CO)_6$. For R = CH_3 and C_2H_5, no evidence of the intermediate formation of $(C_5H_5Mo(CO)_3)_2$ was observed between 55 to 153 °C, whereas for R = $CH_2C_6H_5$ at 100 °C, this dimer could be proved. Heating up to 175 °C also resulted in complete conversion to $Mo(CO)_6$. For R = $CH_2CH{=}CH_2$ and CH_2SCH_3, additionally π-bonded intermediates were found, e.g., $C_5H_5Mo(CO)_2(\eta^3$-$C_3H_5)$ and $C_5H_5Mo(CO)_2(\eta^2$-$CH_2SCH_3)$. This appeared to be a consequence of the greater ease of homolytic Mo-C cleavage for those alkyls where the R radicals liberated from the rapidly dimerizing $C_5H_5Mo(CO)_3^{\bullet}$ fragment are resonance-stabilized [189]. The sulphonic acid ester complexes $C_5H_5Mo(CO)_3CH_2CH(R)SO_2OCH_3$ (R = H, CH_3) are also degraded to the hexacarbonyl when reacted with CO at nonspecified conditions [387]. Similar results were obtained with $C_5H_5Mo(CO)_3{}^1L$ (^1L = CH_3, C_2H_5) in THF at 35 °C, where in a facile reduction $C_5H_5C(O)R$ was liberated along with $Mo(CO)_6$ [316]. Treatment of $C_5H_5Mo(CO)_3C_2H_5$ with 100 atm CO at room temperature has been reported to give the thermally unstable propionyl derivative $C_5H_5Mo(CO)_3C(O)C_2H_5$ as a red oil [14], although these results are doubtful.

$C_5H_5Mo(CO)_3{}^1L$ (^1L = CH_3, C_2H_5, $CH_2C_6H_5$) reacted readily with neat SO_2 under reflux to produce sulfinato-S derivatives, $C_5H_5Mo(CO)_3SO_2R$, but no reaction was observed for R = CF_3 [58]. The above transformation was studied by ^1H NMR at 37 °C, and observations revealed a two-step mechanism proceeding via O-sulfinated $C_5H_5Mo(CO)_3OS(O)R$ as intermediates. This was also confirmed from the facile replacement of RSO_2^- to give $C_5H_5Mo(CO)_3X$ (X = Br, I), if the reaction between $C_5H_5Mo(CO)_3R$ and SO_2 was done in the presence of X^-. Differences in the reactivity and the behavior of transition metal alkyls of the type $C_5H_5Mo(CO)_3R$, $C_5H_5Fe(CO)_2R$, and $(CO)_5MR$ (M = Mn, Re) toward liquid SO_2 were discussed [94, 120]. Furthermore, the rate of scission of the Mo-C σ bond by SO_2 was determined for R = CH_3, C_2H_5, $CH_2C_6H_5$ in neat sulfur dioxide. The methyl compound was studied in more detail, because for the other two compounds some decomposition occurred in less than two half-lives. From values of the pseudo-first-order rate constants k_{obs} for the disappearance of $C_5H_5Mo(CO)_3CH_3$ at temperatures ranging from −58 to −29 °C, the enthalpy of activation for the reaction with SO_2 was estimated as $\Delta H^{\ddagger} = 2.7 \pm 0.4$ kcal/mol, which was used to calculate the value for k_{obs} at −18 °C as $3.3 \times 10^{-4}\ s^{-1}$; analogous data for R = C_2H_5 and R = $CH_2C_6H_5$ were calculated as $1.4 \times 10^{-3}\ s^{-1}$ and $3.9 \times 10^{-4}\ s^{-1}$ using an assumed $\Delta H^{\ddagger} = 2.9$ kcal/mol. These data, as well as the large and negative value of $\Delta S^{\ddagger} = -63 \pm 2\ J \cdot mol^{-1} \cdot K^{-1}$, clearly confirmed the supposed reaction mechanism for the SO_2 insertion (see above). The kinetic data were also compared to the analogous SO_2 insertion reactions occurring with other transition metal alkyls [135]. A free-radical pathway could be neglected, because a simultaneous reaction of $C_5H_5Mo(CO)_3{}^1L$ (^1L = CH_3, $CH_2C_6H_4F$-4) and $C_5H_5Fe(CO)_3CH_2C_6H_5$ with SO_2 produced only the noncrossed sulfinato-S complexes. Otherwise almost statistical yields of crossed and uncrossed insertion products should be formed, because the Fe and Mo complexes reacted with SO_2 at comparable rates [124, 190].

$C_5H_5Mo(CO)_3CH_2CR{=}CR'R''$ also reacted readily with neat SO_2 at reflux to give, depending on the structure of the allyl group, $C_5H_5Mo(CO)_3SO_2CH_2CR{=}CR'R''$ and/or the 1,3-rearranged $C_5H_5Mo(CO)_3SO_2CR'R''CR{=}CH_2$ as principle products, except for R = R′ = R″ = H which yielded mainly the 1:1 cycloaddition compound $C_5H_5Mo(CO)_3CH(CH_2)_2SO_2$-cyclo. Furthermore for $C_5H_5Mo(CO)_3CH_2CH{=}CHCl$, the reaction gave mostly $C_5H_5Mo(CO)_3Cl$, and $C_5H_5Mo(CO)_3CH_2CCl{=}CHCl$ was only slowly transformed to the very unstable sulfinato-S product. Reactions were observed for R = R′ = H, R″ = CH_3 or C_6H_5; R = R″ = H, R′ = CH_3; and R = H, R′ = R″ = CH_3 [185, 186]. A reasonable mechanism for the formation of SO_2 insertion compounds indicated zwitterionic $C_5H_5Mo^+(CO)_3(\eta^2\text{-}CH_2{=}CRCR'R''SO_2^-)$ as a precursor. This is proved for $C_5H_5Mo^+(CO)_3(\eta^2\text{-}CH_2{=}CHCH_2SO_2^-)$ by IR and ^1H NMR spectroscopy at −60 °C, which in time was converted to $C_5H_5Mo(CO)_3SO_2CH_2CH{=}CH_2$ along with $C_5H_5Mo(CO)_3(CH(CH_2)_2SO_2$-cyclo). The latter complex was specifically prepared if $C_5H_5Mo(CO)_3CH_2CH{=}CH_2$ was dissolved in SO_2 at −10 °C followed by rapid removal of the solvent and was obtained in a mixture with its dimers and/or trimers [185]; see also [186]. For all other compounds, $C_5H_5Mo^+(CO)_3(\eta^2\text{-}CH_2{=}CRCR'R''SO_2^-)$ rearranges to the kinetically preferred ionpair $[C_5H_5Mo^+(CO)_3][CH_2{=}CRCR'R''SO_2^-]$, which is in equilibrium with the thermodynamically more stable and less sterically hindered $[C_5H_5Mo^+(CO)_3][CH_2(SO_2^-)\text{-}CR{=}CR'R'']$. Consequently the sterically less crowded crotyl complex showed 1,3-rearrangement, whereas the cinnamyl and 3,3-dimethylallyl precursor gave nonrearranged products [46, 77, 186]. The mechanism is depicted in the following Scheme 2.

Cycloaddition reactions of alk-2-en-1-yl complexes $C_5H_5Mo(CO)_3CH_2CR{=}CR'R''$, alk-2-yn-1-yl complexes $C_5H_5Mo(CO)_3CH_2C{\equiv}CR$, and alk-2,4-dien-1-yl complexes $C_5H_5Mo(CO)_3CH_2CH{=}CHCH{=}CRR'$ were investigated with various electrophiles as detailed in the following paragraph. A [3 + 2] cycloaddition occurred upon the reaction of $C_5H_5Mo(CO)_3CH_2C{\equiv}CR$ (R = H, CH_3, C_6H_5) with SO_2, either neat at low temperatures or in solution at ambient temperatures, to yield $C_5H_5Mo(CO)_3(C{=}C(R)S(O)OCH_2$-cyclo) (Section 1.5.1.4.1.12.3.1) [100]. In an earlier work, these vinyl sultines were erroneously assigned allenyl-sulfinato-O structures, $C_5H_5Mo(CO)_3OS(O)CR{=}C{=}CH_2$ [61]. An analogous 1:1 cycloadduct was isolated from $C_5H_5Mo(CO)_3CH_2C{\equiv}CR$ and SO_3 (neat or as an adduct with dioxane) in C_2Cl_4 resulting in $C_5H_5Mo(CO)_3(C{=}C(R)SO_2OCH_2$-cyclo) (Section 1.5.1.4.1.12.3.1) [92, 109, 119, 128]. These cyclizations appear to be unique to transition metalalk-2-yn-1-yl compounds, as the corresponding alk-2-en-1-yl (see above) and cyanomethyl complexes do not undergo addition reactions when allowed to interact with SO_2 [100]. A plausible mechanism involves initial attack by the electrophile SO_x on the electron-rich -C≡C- bond, presumably followed by a π-allenyl intermediate which then undergoes cyclization by internal attack of an oxygen lone-pair on the electrophilic α-carbon atom [92, 100, 109, 119, 128]. Treatment of $C_5H_5Mo(CO)_3CH_2C{\equiv}CR$ (R = CH_3, C_6H_5) with S_2O, liberated from its equivalent

References on pp. 106/18

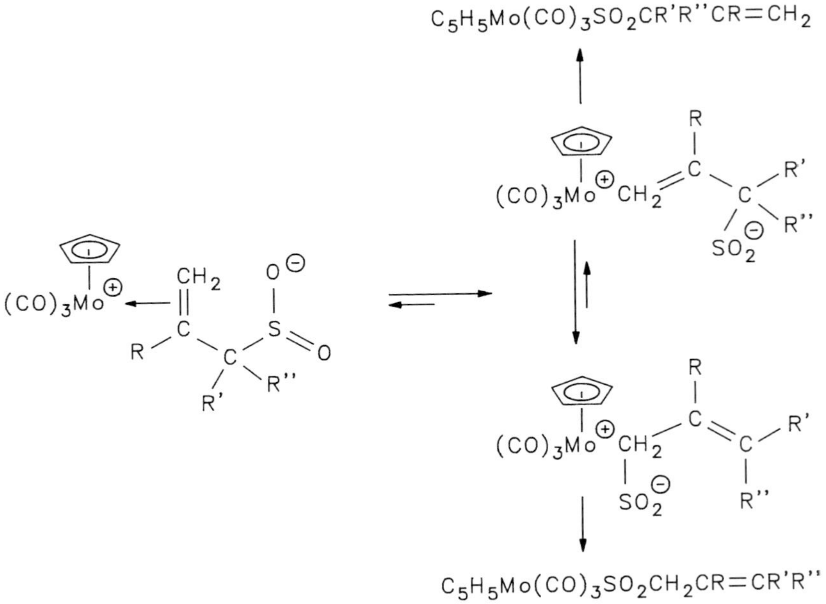

$C_5H_5Mo(CO)_3SO_2CR'R''CR{=}CH_2$

$C_5H_5Mo(CO)_3SO_2CH_2CR{=}CR'R''$

Scheme 2

4,5-diphenyl-3,6-dihydro-1,2-dithiin 1-oxide resulted also in a clean [3 + 2] cycloaddition in THF at 25°C and gave $C_5H_5Mo(CO)_3$(C=C(R)S(O)SCH$_2$-cyclo) (Section 1.5.1.4.1.12.3.1) [348, 382]. Cycloaddition of $ClSO_2NCO$ to the C≡C bond of $C_5H_5Mo(CO)_3CH_2C{\equiv}CC_6H_5$ as an example substrate in benzene at room temperature resulted in 1,2 Mo migration to give the metalated Δ^3-pyrrolinone, $C_5H_5Mo(CO)_3C{=}C(C_6H_5)C(O)N(SO_2Cl)CH_2$ (Section 1.5.1.4.1.12.3.1), but the use of $C_5H_5Mo(CO)_3CH_2CH{=}CHC_6H_5$ gave only high yields of $C_5H_5Mo(CO)_3Cl$ [125]. Similar [3 + 2] adducts were obtained from $C_5H_5Mo(CO)_3CH_2$-$C{\equiv}CC_6H_5$ and $C_5H_5Mo(CO)_3CH_2CH{=}CH_2$ by the reaction with CH_3SO_2NSO in CH_2Cl_2 at room temperature, resulting in $C_5H_5Mo(CO)_3$(C=C(C_6H_5)S(O)N(SO_2CH_3)CH$_2$-cyclo) (Section 1.5.1.4.1.12.3.1) and $C_5H_5Mo(CO)_3$(CHCH$_2$S(O)N(SO_2CH_3)CH$_2$-cyclo) (No. 14), respectively, whereas the analogous reaction with $C_5H_5Mo(CO)_3CH_2CCH_3{=}CH_2$ gave only a noncarbonyl decomposition product. Under these reaction conditions, use of $(CH_3SO_2N)_2S$ led only in the case of $C_5H_5Mo(CO)_3CH_2C{\equiv}CR$ (R = CH_3, C_6H_5) to a [3 + 2] cycloaddition affording $C_5H_5Mo(CO)_3$(C=C(R)S(=NSO$_2$CH$_3$)N(SO$_2$CH$_3$)CH$_2$-cyclo) (Section 1.5.1.4.1.12.3.1). For $C_5H_5Mo(CO)_3CH_2CH{=}CH_2$ or $C_5H_5Mo(CO)_3CH_2C(CH_3){=}CH_2$, novel [3 + 3] cycloadducts, $C_5H_5Mo(CO)_3$(C(R)(CH$_2$N(SO$_2$CH$_3$))$_2$S-cyclo) (Nos. 15, 16; see Formula V, R = H, CH_3) were obtained when treated with $S(NSO_2CH_3)_2$ at ambient conditions. A mechnism of these cycloadditions is discussed in detail [265, 318].

The [3 + 2] cycloaddition of $C_2(CN)_4$ to the σ-bonded allyl ligands of $C_5H_5Mo(CO)_3$-$CH_2C(R){=}CR'_2$ (R = H, R′ = CH_3; R = CH_3, R′ = H) or to $C_5H_5Mo(CO)_3CH_2C{\equiv}CC_6H_5$ in various solvents (CH_2Cl_2, CH_3CN, THF, C_6H_6, or C_5H_{12}) at 25°C gave the 1/1 adducts $C_5H_5Mo(CO)_3(\eta^1$-CRCR$'_2$(C(CN)$_2$)$_2$CH$_2$-cyclo) (Nos. 10 and 11) and $C_5H_5Mo(CO)_3$(C=C-(C_6H_5)(C(CN)$_2$)$_2$CH$_2$-cyclo) (Section 1.5.1.4.1.12.3.1), respectively. The rate dependence of these conversions on the polarity of the solvent (slow transformation of $C_5H_5(CO)_3MoCH_2$-

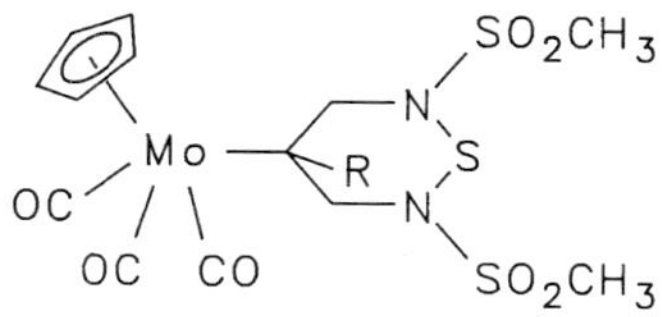

V

$CH{=}C(CH_3)_2$ by $C_2(CN)_4$ in pentane but very fast disappearance of the allyl compound in CH_2Cl_2) is consistent with a mechanism involving the intermediacy of a dipolar metalalkene complex [91, 101, 138]. $C_5H_5Mo(CO)_3CH_2CH{=}CHCH{=}CHR$ (R = H, CH_3) reacted with $C_2(CN)_4$ in stirred benzene at ambient temperature to give the [4 + 2] cycloadducts C_5H_5-$Mo(CO)_3(CH_2(CHCH{=}CHCHRC(CN)_2C(CN)_2$-cyclo) (Nos. 36 and 37), while treatment of $C_5H_5Mo(CO)_3CH_2CH{=}CHCH{=}C(CH_3)_2$ with $C_2(CN)_4$ in THF at 0°C produced $C_5H_5Mo(CO)_3$-$(CHCH(CH{=}C(CH_3)_2)(C(CN)_2)_2CH_2$-cyclo) (No. 35) as a [3 + 2] cycloadduct [343, 351]. Similar treatment of $C_5H_5Mo(CO)_3CH_2CH{=}CHCH{=}CH_2$ with maleic anhydride resulted in C_5H_5-$Mo(CO)_3CH_2(CHCH{=}CHCH_2CHC(O)OC(O)CH$-cyclo) (No. 38) [343, 345].

Similar [3 + 2] adducts were also isolated at ambient temperatures from the reaction of $C_5H_5Mo(CO)_3CH_2C{\equiv}CC_6H_5$ with $(t\text{-}C_4H_9)(NC)C{=}C{=}O$ in benzene yielding $C_5H_5Mo(CO)_3$-$(C{=}C(C_6H_5)C(O)C(CN)(t\text{-}C_4H_9)CH_2$-cyclo) (Section 1.5.1.4.1.12.3.1) [170]. Accordingly treatment of the same starting material with $(CF_3)_2CO$, neat or in CH_2Cl_2, resulted in C_5H_5Mo-$(CO)_3(C{=}C(C_6H_5)C(CF_3)_2OCH_2$-cyclo) [128, 147], and $C_5H_5Mo(CO)_3(C{=}C(C_6H_5)C(CCl_2F)$-$(CClF_2)OCH_2$-cyclo) was obtained using $(CCl_2F)(CClF_2)CO$ (both Section 1.5.1.4.1.12.3.1) [247]. Analogously $C_5H_5Mo(CO)_3(CHCH(C_6H_5)C(CF_3)_2OCH_2$-cyclo) (No. 12) was obtained using $C_5H_5Mo(CO)_3CH_2CH{=}CHC_6H_5$ and $(CF_3)_2CO$ in CH_2Cl_2. In neat hexafluoroacetone no reaction occurred [147].

The kinetics of the [3 + 2] cycloaddition of $C_5H_5Mo(CO)_3CH_2C{\equiv}CR$ (R = CH_3, C_6H_5) with $4\text{-}CH_3C_6H_4SO_2NCO$ was studied in CH_2Cl_2 at room temperature under pseudo-first-order conditions. Experiments with either the electrophilic agent or the Mo complex in a 10-fold excess revealed the reaction to be first-order in each reactant and second-order overall. The second-order rate constants were determined as 2.1×10^{-2} L · mol^{-1} · s^{-1} for R = CH_3 and as 7×10^{-4} L · mol^{-1} · s^{-1} for R = C_6H_5. Solvent influence on the rate of the cycloaddition occurring between $C_5H_5Mo(CO)_3CH_2C{\equiv}CCH_3$ and $4\text{-}CH_3C_6H_4SO_2NCO$ in CH_2Cl_2, toluene, and chlorobenzene is very small. For toluene as a reaction medium, the activation parameters for the latter reaction are E_a = 11.5 kcal/mol, $\Delta H^{\ddagger}$ = 10.9 kcal/mol, and $\Delta S^{\ddagger}$ = −31 J · mol^{-1} · K^{-1} [193, 247].

Reaction of $C_5H_5Mo(CO)_3R$ (R = CH_3, $CH_2C_6H_4Cl$-4) with CH_3SO_2NSO in CH_2Cl_2 or $CHCl_3$ at −25°C led to a mixture of N- or O-bonded insertion product $C_5H_5Mo(CO)_3N$-$(SO_2CH_3)S(O)R$ and $C_5H_5Mo(CO)_3OS(R){=}NSO_2CH_3$, which decomposed on warming up to 10°C to intractable materials. The room temperature reaction between $C_5H_5Mo(CO)_3CH_3$ and $(CH_3SO_2N)_2S$ gave a red product of low stability, identified as $C_5H_5Mo(CO)_3N$-$(SO_2CH_3)S(CH_3){=}NSO_2CH_3$ [317]. The alkyls $C_5H_5Mo(CO)_3CH_3$ and $C_5H_5Mo(CO)_3CH_2C_6H_5$ were found to react with $C_2(CN)_4$ in an extremely sluggish fashion and with considerable decomposition to produce red or purple oils of unknown composition together with trace quantities of $C_5H_5Mo(CO)_3CN$ [145].

Reactions of molybdenum alkyls $C_5H_5Mo(CO)_3R$ (R = CH_3, C_2H_5, $CH_2C_6H_5$) with $C_5H_4R'Mo(CO)_3H$ (R′ = H, CH_3) under mild thermal conditions smoothly afforded aldehydes RCHO (along with some toluene for R = $CH_2C_6H_5$) together with the dimers $C_5H_5(CO)_3Mo$-$Mo(CO)_3C_5H_4R'$ and $C_5H_5(CO)_2Mo{\equiv}Mo(CO)_2C_5H_4R'$. Clean second-order kinetics was re-

References on pp. 106/18

vealed with $k(CH_3) = 2.5 \times 10^{-4}$, $k(C_2H_5) = 4.0 \times 10^{-3}$, and $k(CH_2C_6H_5) = 2.5 \times 10^{-5}$ L · mol^{-1} · s^{-1} at 50 °C. Two different mechanisms must be operative, as shown by crossover experiments between $C_5H_5Mo(CO)_3CH_3$ and $CH_3C_5H_4Mo(CO)_3CD_3$ as well as between $C_5H_5Mo(CO)_3CH_2C_6H_5$ and $CH_3C_5H_4Mo(CO)_3CD_2C_6D_5$, respectively, in THF-$d_8$ at 50 °C. The formation of a crossover complex, $CH_3C_5H_4Mo(CO)_3CH_2C_6H_5$, only in the latter case, along with produced toluene in addition to aldehyde in the presence of $C_5H_5Mo(CO)_3H$, can be rationalized by assuming that the benzyl radical is stable enough so that molybdenum-carbon bond cleavage becomes competitive with alkyl/acyl isomerization ($C_5H_5Mo(CO)_3R \rightleftarrows C_5H_5Mo(CO)_2C(O)R$), the initial step of aldehyde formation in the reaction of $C_5H_5Mo(CO)_3R$ complexes with $C_5H_5Mo(CO)_3H$ [208, 227, 245]. For more details with respect to thermodynamics, see [342], and for mechanism, see [208, 211, 227 to 229, 245, 271].

Contrary to an earlier report [85], the reaction between $C_5H_5Mo(CO)_3H$ and C_2H_4 at 100 °C does not result in the insertion product $C_5H_5Mo(CO)_3C_2H_5$ as a stable compound, since under these conditions reaction of the ethyl complex with additional C_2H_4 is rapid and $(C_2H_5)_2CO$, accompanied by $(C_5H_5Mo(CO)_3)_2$ and $(C_5H_5Mo(CO)_2)_2$, are the final products. The alkyl compound was proved to be an intermediate in this transformation, because the reaction of independently prepared $C_5H_5Mo(CO)_3R$ (R = CH_3, C_2H_5) with C_2H_4 in THF-d_8 at 100 °C resulted in $CH_3C(O)C_2H_5$ and $C_2H_5C(O)C_2H_5$. The source of the new hydrogen atom in the final ketones is another molecule of C_2H_4 as was demonstrated by treatment of $C_5H_5Mo(CO)_3CH_3$ with C_2D_4 yielding completely ethyl-deuterated butane-2-one. The reaction pathway required one molecule of $CH_3C(O)CH{=}CH_2$ for each $CH_3C(O)C_2H_5$, but $CH_3C(O)CH{=}CH_2$ is easily polymerized and might not be expected to survive the reaction conditions [208, 211, 227, 245].

The complexes $C_5H_5Mo(CO)_3CH_3$ and $C_5H_5Mo(CO)_3C_2H_5$ are active catalysts for the polymerization and copolymerization of monosubstituted alkynes such as $C_6H_5C{\equiv}CH$ [305, 306]. $C_5H_5Mo(CO)_3CH_3$ has also been used for the bulk polymerization of $CH_2{=}C(CH_3)$-$C(O)OCH_3$ in the presence of CCl_4 as an activator [6]. Furthermore, for the $(C_5H_5Mo(CO)_3)_2$-catalyzed addition reaction of CCl_4 and CO to olefins, a mechanism has been proposed that involves a number of different $C_5H_5Mo(CO)_3R$ intermediates substituted by trichloromethyl, 1-alkyl-3,3,3-trichloropropyl, and 2-alkyl-4,4,4-trichlorobutyl groups, respectively [48, 86].

A variety of transition metal carbonyl compounds including $C_5H_5Mo(CO)_3C_nH_{2n+1}$ complexes with n = 1 to 6 have been proposed as especially useful photosensitive components for the preparation of silver-free, heat-developable photoimaging materials [176, 177]. Use of $C_5H_5Mo(CO)_3R$ compounds has also been made in photoimaging systems composed of crosslinked polymers, which can be applied to the formulation of photoresists and presensitized lithographic plates [248], as well as to the preparation of defect-free pattern films by selective irradiation with a laser or particle beam [284]. Additionally, $C_5H_5Mo(CO)_3CH_3$ is an active photosensitizing agent for the anhydride cure of epoxy resins [150].

Molybdenum-containing layers can be deposited by thermal decomposition of various alkyl complexes $C_5H_5Mo(CO)_3R$ in either liquid or vapor phase. Thermal decomposition in a reactive (carburizing, carbonitriding, oxidizing, or hydrolyzing) atmosphere results in the deposition of molybdenum carbides, nitrides, and oxides, respectively [391].

Some general methods for preparation are availabe:

Method I: Most of the complexes in this chapter can be prepared by the reaction of $[C_5H_5Mo(CO)_3]^-$ with an excess of electrophilic alkylhalogenides RX. In most cases, the sodium salt of the molybdenum precursor was treated at room temperature with the alkylchloride, but frequently the iodo or occasionally the bromo compound was used. The reaction was accomplished in THF as the solvent if not otherwise stated. For separation from $(C_5H_5Mo(CO)_3)_2$, a common by-product, and purification, either sublimation at high vacuum and temperatures below 100 °C or column chromatography on alumina, or frequently silica and Florisil, was used. For detailed reaction conditions and references, compare the individual compounds summarized in the following Table 1.

Method II: Starting from $C_5H_5Mo(CO)_3H$.

a. $C_5H_5Mo(CO)_3CH_2R$ (R = H, C_6H_5, $C_6H_4OCH_3$-4, $CO_2C_2H_5$) was prepared from $C_5H_5Mo(CO)_3H$ which was treated with N_2CHR in THF at ca. −80 °C for 30 to 60 min. Subsequently the reaction mixture was stirred for several hours at ambient temperature. Purification was accomplished by column chromatography on silica (R = H, C_6H_5, $C_6H_4OCH_3$-4) or alumina (R = $CO_2C_2H_5$) [26, 165].

b. $C_5H_5Mo(CO)_3CF_2CFXH$ (X = F, Cl) was prepared by treatment of $C_5H_5Mo(CO)_3H$ with $XCF{=}CF_2$ in n-pentane at ambient temperature in a stainless-steel bomb employing initial pressures in the range of 4 to 7 atm. After a most preferred reaction time of 20 hours to six days and subsequent evaporation of the solvent, the products can be isolated in modest yields by sublimation at 65 °C and 0.1 Torr [4, 5]. Similarly, $C_5H_5Mo(CO)_3(C_5H_2F_5$-cyclo) was prepared by condensing equimolar amounts of cyclo-C_5HF_5 to $C_5H_5Mo(CO)_3H$ in ether and subsequent stirring of the reaction mixture for 6 h at −30 °C. For purification medium-pressure column chromatography at −30 °C was used with pentane/CH_2Cl_2 as eluant [340].

Method III: Starting from $C_5H_5Mo(CO)_3CH_2CR{=}CR'R''$.

a. The reaction with $C_2(CN)_4$ induced ligand cyclization of $C_5H_5Mo(CO)_3CH_2CR{=}CR'R''$ at the β-position to give five-membered rings for R = H, R' = R'' = CH_3; and R = CH_3, R' = R'' = H and six-membered rings for R = R' = H, R'' = $CH{=}CH_2$, $CH{=}CHCH_3$ generating $C_5H_5Mo(CO)_3CRCR'R''(C(CN)_2)_2CH_2$-cyclo, $C_5H_5Mo(CO)_3CH_2(CHCH{=}CH_2(C(CN)_2)_2$-cyclo), or $C_5H_5Mo(CO)_3CH_2(CHCH{=}CHCH_3(C(CN)_2)_2$-cyclo). The reaction was accomplished in benzene at room temperature [91, 138, 343, 345]. In the case of $C_5H_5Mo(CO)_3CH_2CH{=}CHCH{=}C(CH_3)_2$ in THF at 0 °C, five-ring cyclization was observed after 2 h to give $C_5H_5Mo(CO)_3(CHCH(CH{=}C(CH_3)_2)(C(CN)_2)_2CH_2$-cyclo) [343, 351].

b. By treatment with an appropriate reactant, $C_5H_5Mo(CO)_3CH_2CR{=}CHR'$ (R = H, R' = C_6H_5; R = CH_3, R' = H) can be converted to $C_5H_5Mo(CO)_3(^1L$-cyclo), with ^1L = $CHCH(C_6H_5)C(CF_3)_2OCH_2$ [128, 147], $CH(CH_2)_2SO_2$ [185], $CH(CH_2)_2(SO)N(SO_2CH_3)$ [318], $CH(CH_2N(SO_2CH_3))_2S$ [318], or $CH_3C(CH_2N(SO_2CH_3))_2S$ [318]. CH_2Cl_2 was used as solvent unless otherwise stated. Similarly $C_5H_5Mo(CO)_3CH_2(CHCH{=}CHCH_2CHC(O)OC(O)CH$-cyclo) was prepared from $C_5H_5Mo(CO)_3CH_2CH{=}CHCH{=}CH_2$, which was stirred with an excess of maleic anhydride in benzene at room temperature for 4 h [345].

Method IV: Starting from $[C_5H_5Mo(CO)_3(CH_2{=}CHR)]^+$.

a. $C_5H_5Mo(CO)_3CH_2CH_2CH(CO_2C_2H_5)R$ (R = $CO_2C_2H_5$, $C(O)CH_3$) were prepared from $[C_5H_5Mo(CO)_3(CH_2{=}CH_2)]BF_4$ and $Na[CH(CO_2C_2H_5)R]$ (generated in situ by deprotonation of $CH_2(CO_2C_2H_5)R$ with NaH) in THF at −40 °C. After warming to 0 °C during 1 h, the reaction mixture was extracted with pentane for R = $CO_2C_2H_5$, but quenched with H_2O and then extracted with

References on pp. 106/18

ether for R = C(O)CH_3. By cooling the extracts to −60 °C, an oily liquid precipitated which could be further purified by chromatography on alumina [370].

b. $C_5H_5Mo(CO)_3CH_2CH_2CN$ and $[C_5H_5Mo(CO)_3CH_2CH_2{}^2D]PF_6$ (^2D = $N(CH_3)_3$, NC_5H_5, $P(C_6H_5)_3$) were prepared at room temperature by the treatment of $[C_5H_5Mo(CO)_3(CH_2{=}CH_2)]PF_6$ with $[(C_2H_5)_4N]CN$ or with ^2D. Except for NC_5H_5 which was used neat, CH_3CN was chosen as the solvent [148].

c. $C_5H_5Mo(CO)_3CH_2CH_2W(CO)_3C_5H_5$ and $C_5H_5Mo(CO)_3CH_2CH_2Re(CO)_5$ were prepared from $[C_5H_5Mo(CO)_3(CH_2{=}CH_2)]BF_4$ dissolved in CH_3CN which reacted with $K[C_5H_5W(CO)_3]$ or $Na[(CO)_5Re]$ in THF at −50 °C. The products precipitated from the reaction mixture [168, 231].

d. $C_5H_5Mo(CO)_3CH_2CHROCH_3$ (R = H, CH_3) was prepared by treating alkene-saturated CH_2Cl_2 solutions of the respective tetrafluoroborate, $[C_5H_5Mo(CO)_3(CH_2{=}CHR)]BF_4$, with methanol and a slight excess of sodium carbonate at −40 °C. The residue remaining after removal of the solvent at −10 °C in vacuum is extracted with petroleum ether and the product was isolated from the filtered extracts by evaporation to dryness [387].

Although a lot of complexes can be obtained as described by Method I, this way of preparation also failed occasionally, probably due to the instability of the desired σ-complex. This resulted in spontaneous decarbonylation and conversion of the σ-bonded ligand to a π-bonded substituent as demonstrated for the following examples.

The reaction between $[C_5H_5Mo(CO)_3]^-$ and $ClCH_2C(Si(CH_3)_3){=}CH_2$ led directly to the formation of a π-allyl product, $C_5H_5Mo(CO)_2(\eta^3\text{-}CH_2C(Si(CH_3)_3)CH_2)$; no intermediate σ-complex was observed [151]. Treatment of $Na[C_5H_5Mo(CO)_3]$ with $BrCH_2CH_2CH{=}CHR$ (R = H, CH_3) similarly resulted in spontaneous decarbonylation yielding allyl derivatives with the composition $C_5H_5Mo(CO)_2(\eta^3\text{-}RCHCHCHCH_3)$ [106].

Similar processes were confirmed by treatment of $[C_5H_5Mo(CO)_3]^-$ with $BrCH_2CH_2CR{=}C{=}CHR'$ (R = H, R' = H, CH_3; R = CH_3, R' = H; R = R' = CH_3), which resulted in the π-allyl products $C_5H_5Mo(CO)_2(\eta^3\text{-}R'CHCC(R)CH_2CH_2CO\text{-cyclo})$ [103, 199].

Although $\mathbf{C_5H_5Mo(CO)_3CH_2As(CH_2Si(CH_3)_3)_2}$, the primary product resulting from treatment of $Na[C_5H_5Mo(CO)_3]$ with $BrCH_2As(CH_2Si(CH_3)_3)_2$ in THF, has been identified in solution using IR spectroscopy (ν(CO): 1911 and 2006 cm^{-1}), it was spontaneously converted to form the three-membered metallaheterocycle $C_5H_5Mo(CO)_2CH_2As(CH_2Si(CH_3)_3)_2$ [262]. The reaction of $Na[C_5H_5Mo(CO)_3]$ with 2-$ClCH_2C_5H_4N$ similarly proceeds with spontaneous rearrangement of the initially formed σ-$CH_2C_5H_4N$ derivative to produce, via migratory CO insertion, the five-membered metallaheterocyclic acyl 2-$(C_5H_5Mo(CO)_2C(O)CH_2)C_5H_4N$ [33, 45]. In earlier work, similar rearrangement processes were claimed to occur on treatment of the $[C_5H_5Mo(CO)_3]^-$ anion with different 2-allenyl-substituted bromoalkanes, $BrCH_2CH_2C(R){=}C{=}CHR'$ (R = H, R' = H, CH_3; R = CH_3, R' = H; R = R' = CH_3), and produced heterocyclic acyls of composition $C_5H_5Mo(CO)_2(\eta^3\text{-}C(O)CH_2CH_2CR{=}C{=}CHR')$ [89]. A cyclic σ/π ester complex, $C_5H_5Mo(CO)_2(\eta^3\text{-}C(O)OCH_2CH{=}CH_2)$, was produced by reacting $Na[C_5H_5Mo(CO)_3]$ with epibromohydrin. In this transformation, the O atom of the three-membered ring of the initial intermediates, $\mathbf{C_5H_5Mo(CO)_3CH_2(CHOCH_2\text{-cyclo})}$ and $C_5H_5Mo(CO)_2C(O)CH_2(CHOCH_2\text{-cyclo})$, respectively, interacted with the central metal to generate an O-bonded allyloxy species, $C_5H_5Mo(CO)_3OCH_2CH{=}CH_2$, which on rapid migratory CO insertion formed the isolated product [334].

Table 1
$C_5H_5Mo(CO)_3{}^1L$ Compounds Containing Mo-sp^3-C Bonds.
An asterisk indicates further information at the end of the table.
For explanations, abbreviations, and units, see p. X.

No.	^1L	method of preparation (yield) properties and remarks
*1	-CH_3	I (58 to 85% for M = Na with the yield dependent on the preparation of the starting anion, between 25 to 40°C for 30 min and then some hours with CH_3I [1, 2, 214, 363]; 75 to 80% for M = K, in 1,2-dimethoxyethane at 25°C for 5 h (the starting material was prepared in situ from $Mo(CO)_6$ and cyclopentadienylpotassium [23, 55, 57]); 77% for M = Li, at −78°C for 2 h (the starting material was prepared in situ from $Li[(C_2H_5)_3BH]$ and $C_5H_5Mo(CO)_3Cl$ [169, 210]); deuterated complexes $C_5H_5Mo(CO)_3CH_nD_{3-n}$ (n = 0 to 3) were prepared analogously to Method I from $[C_5H_5Mo(CO)_3]^-$ and CH_2DBr [62] or from $K[C_5H_5Mo(CO)_3]$ and $CH_nD_{3-n}I$ (n = 0, 1) in 1,2-dimethoxyethane [295]; IIa (3%, for 12 h) [165], similarly in ether at 0°C for 24 h) [1, 2] yellow crystals with a camphoraceous odor [1, 2, 55], m.p. 119 to 121°C [214], 124°C (dec.) [1, 2, 23, 55, 178] ^1H NMR (toluene-d_8): 0.35 (s, CH_3), 4.53 (s, C_5H_5) [209, 222], similar in benzene-d_6 [239, 363]; see also [2] ^1H NMR (THF-d_8): 0.40 (s, CH_3), 5.45 (s, C_5H_5) [228]; quite similar in $CHCl_3$ [35], in $CDCl_3$ [216], and CCl_4 [13]; see also [7] for an early measurement not made with TMS as an internal reference see [3] the coupling constant J(H, D) in $C_5H_5Mo(CO)_3CH_2D$, determined in different solvents, has been used to calculate J(H, H) as ±8.8 (in C_6H_6), ±8.9 (in CH_3CN), and ±9.1 (in CS_2) (all values ±0.3 Hz); data were discussed according to a strong donor effect from the filled Mo d-orbitals to the methyl group [51, 62] ^{13}C NMR (C_6D_6): −22.1 (q, CH_3; J(C, H) = 138.0), 92.4 (dqui, C_5H_5; ^1J(C, H) = 177.5, ^{3}(C, H) = 6.9), 227.2 (CO cis MoC), 240.3 (CO trans MoC) [179]; see also [239], quite similar in CH_2Cl_2 [182] and $CDCl_3$ [216] ^{95}Mo NMR (C_6D_6): −1736; line width $\nu_{1/2}$ = 40 Hz [239] IR (cyclohexane): 1946 (A″, ν(CO)), 1950 (A′, ν(CO)), 2018 (A′, ν(CO)); k_1(CO diagonal to

References on pp. 106/18

Table 1 (continued)

No.	^1L	method of preparation (yield) properties and remarks
*1 (continued)		CH_3) = 15.49, k_2(CO diagonal to another) = 15.80, k_{id} = 0.51; for calculations the ratio of the diagonal interaction force constant (k_{id}) and the lateral interaction force constant (k_{is}) was assumed to be constant k_{id}/k_{is} = 2 [70], similar in n-$C_{16}H_{34}$ [27], in toluene [209, 222], and in i-octane [256] IR (vapor phase, 100 °C): 1150 (δ($MoCH_3$)), 1920, 1957, 2030 (ν(CO)), 2890, 2960 (ν(CH)) [2] IR (CCl_4): 1420 (ν(CC) of C_5H_5), 1930 (ν_{as}(CO)), 2018 (ν_s(CO)), 2814 ($2\delta_{as}$(CH_3)), 2904 (ν_s(CH_3)), 2984 (ν_{as}(CH_3)), 3115 (ν(CH) of C_5H_5), 3820, 3900, 4000 (CO overtones); additional bands at 1355, 1515 [2, 35, 240, 295]; similar in CS_2 with additional bands at 1643, 1730, 1828 [13]; see also [23, 178], and low-frequency bands at 810, 1005, 1010, 1060, 1110, 1167 [2]; similar ν(CO) values measured in THF [223]; see also [228, 363] IR (CO matrix at −261 °C): 1938.8, 1945.2 (matrix splitting), 2025.8 (ν(CO)), similar in an Ar, N_2, and CH_4 matrix; k_1 = 15.60, k_2 = 15.69, k_{12} = 0.44, k_{23} = 0.49, as defined by labeling the CO ligand trans to CH_3 as $CO^{(1)}$ and the CO groups cis to CH_3 as $CO^{(2)}$ and $CO^{(3)}$, respectively [240, 244]; for a paraffin matrix at −198 °C see [256]; (1,2-epoxyethylbenzene glass at −261 °C): 1912, 1926, 2012 (ν(CO)) [378]; (poly(vinyl chloride) matrix): 1918, 1928, 2015 (−261 °C); 1924, 2016 (25 °C) [279]; see also [240] UV (cyclohexane, ε): 315 (1960) [2], (CH_2Cl_2): 310 (2440) [35], similar in a poly(vinyl chloride) matrix [279] XPS (gas phase): 292.5 (C 1s, CO), 538.8 (O 1s) [172, 224]
*2	-C_2H_5	I (78%, with C_2H_5I for several hours at about 40 °C) [1, 2, 335] also regenerated from [$C_5H_5Mo(CO)_3(\eta^2$-C_2H_4)]BF_4 by hydride addition with an excess of Na[BH_4] in THF for 10 min at 25 °C, purification as in Method I (30%) [12] from $(C_5H_5Mo(CO)_3)_2$ in C_6D_6 by UV irradiation in the presence of $Zn(C_2H_5)_2$ (ca. 25%) or $In(C_2H_5)_3$ ($\geq$ 10%), along with $C_5H_5Mo(CO)_3M(C_2H_5)_n$ and $(C_5H_5Mo(CO)_3)_2M(C_2H_5)_n$ (M = Zn, n = 1, 0; M = In, n = 2, 1) [365]

References on pp. 106/18

Table 1 (continued)

No.	^{1}L	method of preparation (yield) properties and remarks
		also as by-products from the synthesis of a series of carbene derivatives $C_5H_5Mo(CO)_2(E(C_6H_5)_3)=C(OC_2H_5)CH_3$ (E = Ge, Sn), prepared by treatment of C_5H_5Mo-$(CO)_3E(C_6H_5)_3$ with CH_3Li and subsequent alkylation with $[(C_2H_5)_3O]^+$; compare also No. 1 [173, 379] yellow crystals with a camphoraceous odor, soluble in all common organic solvents, but insoluble in and unaffected by H_2O [1, 2, 12], m.p. 77.5 to 78.5°C (dec.) [1, 2] ^1H NMR (C_6D_6): 1.47, 1.56 (pseudo q, pseudo t, C_2H_5; J = 6), 4.40 (s, C_5H_5) [365]; see also [14], for toluene-d_8 see [335]; for an early measurement not made with TMS as an internal reference, see [2] ^1H NMR (CCl_4): 1.44 (CH_3), 1.69 (CH_2; J = 7.61), 5.26 (C_5H_5) [13], similar in THF-d_8 [228], and in CS_2 [20] ^{13}C NMR (C_6D_6): −3.9 (CH_2; ^1J(C, H) = 138.4, ^2J(C, H) = +7.6), 20.3 (CH_3; ^1J(C, H) = 125.2, ^3J(C, H) = −3.1), 92.7 (dqui, C_5H_5; ^1J(C, H) = 177.5, ^2J(C, H) = 6.7), 228.1 (CO cis MoC), 240.3 (CO trans MoC) [179]; see also [20] IR (CCl_4): 1370 ($\delta_s(CCH_3)$), 1420 (ν(CC) for C_5H_5), 1450 ($\delta_{as}(CCH_3)$), 1932, 2016 (ν(CO)), 2863, 2942, 3110 (ν(CH)), 3820, 3900, 4000 (CO overtones) [2], (CS_2): 1900 (sh), 1932, 2005 (sh), 2022 (ν(CO)) [13]; other bands at 810, 830, 910, 965, 1005, 1010, 1060, 1110, 1140, 1220 [2], 1643, 1727, and 1817 [13]; see also [335], similar ν(CO) values in i-octane [256] IR (CO matrix at −261°C): 1928.0, 1938.8 (overlapping A′ and A″, ν(CO)), 2019.3, 2018 (A′, ν(CO)) [277], quite similar in a CH_4 matrix [277, 366], in a paraffin wax at −196°C [256], or in a poly(vinyl chloride) matrix at 25°C [279] the IR spectrum of No. 2 in solution shows two ν(CO) frequencies, although three are expected according to the C_s symmetry of the $Mo(CO)_3C_2H_5$ fragment; in analogy to No. 1 the lower band was assumed to be the not resolved A″ and A′ bands arising from accidental coincidence, and the band at ca. 2020 was assigned to the symmetric stretch A′, these assumptions were confirmed from the IR spectra in CH_4 and CO matrices at −261°C, where the lower wavenumber band is split [277]

References on pp. 106/18

Table 1 (continued)

No.	^1L	method of preparation (yield) properties and remarks
*2 (continued)		UV (i-octane, ε): 312 (2190), 360 (sh) [256], similar in cyclohexane [2], (poly(vinyl chloride) matrix): 318, 365 (sh) [279] the strength of the Mo-C_2H_5 bond was estimated to be 37 kcal/mol from the assumption that a small difference of 10 ± 2 kcal/mol should exist between the bond strength of Mo-CH_3 (compare No. 1) and Mo-C_2H_5 [342]
3	-C_3H_7-n	I (10%, no details) [354] IR (THF): 1918, 2003 (νCO)) [354]
4	-C_3H_7-i	I (5%, with i-C_3H_7I for several hours at about 40 °C) [2] by slow decomposition of [$C_5H_5Mo(CO)_2$-$(P(OC_3H_7\text{-}i)_3)_2$] [$C_5H_5Mo(CO)_3$] in benzene by nucleophilic attack of the anion on the cation along with $C_5H_5Mo(CO)_2(P(OC_3H_7\text{-}i)_3)P(O)(OC_3H_7\text{-}i)_2$, the rate of formation increased in refluxing benzene [75, 78] yellow compound of camphoraceous odor, m.p. 29 to 30 °C [2] ^1H NMR (toluene): 4 to 4.5 (CH_3), 7.5 (C_5H_5), toluene as internal reference [2] IR (CS_2): fingerprint bands at 808, 1005, 1010, 1088, 1128, 1137, 1220; (CCl_4): 1420, 1430 (ν(CC) for C_5H_5), 1930, 2010 (ν(CO)), 2840, 2930, 3100 (ν(CH)); additional bands at 1360, 1445, 1455 [2] soluble in all common organic solvents, but solutions are only stable under vacuum or in an inert gas atmosphere, insoluble in and unaffected by H_2O, generally less stable than the ethyl compound No. 2 [2]
5	-C_5H_{11}-n	I (with n-$C_5H_{11}I$ overnight) [256] solid (from hexane at low temperatures) [256] IR (i-octane): 1937, 2020 (ν(CO)) [256] UV irradiation yielded $C_5H_5Mo(CO)_3H$ and pentene as principal products; studies at −196 °C revealed an analogous reaction pathway as for the corresponding ethyl complex No. 2 [256]
6	-$CH_2C_4H_9$-t	from $(C_5H_5Mo(CO)_3)_2$ and $(t\text{-}C_4H_9CH_2)_3In$ in benzene under narrow-band irradiation along with $C_5H_5Mo(CO)_3In(CH_2C_4H_9\text{-}t)_2$, identified by ^1H NMR analysis [365] unstable in solution, decomposed rapidly to unidentified products [365]

References on pp. 106/18

Table 1 (continued)

No.	^1L	method of preparation (yield) properties and remarks
		^1H NMR (C_6D_6): 1.16 (s, CH_3), 1.77 (s, CH_2), 4.41 (s, C_5H_5) [365]
7	$C(CH_3)_3$ H D D H	I (19%, by addition of threo-3,3-dimethylbutyl-1,2-d_2-triflate in ether at 0 °C followed by stirring at 25 °C for 1 h) [140] yellow solid, m.p. 80 to 82 °C [140] $^1H\{^2H\}$ NMR: 0.87 (s, CH_3), 1.44 (d, CH; J = 13.3), 1.56 (d, CH), 5.22 (s, C_5H_5) [140] IR (CCl_4): 1420 (ν(CC) of C_5H_5), 1930, 2015 (ν(CO)), additional bands at 910, 1230, 1360, 1460, 2860, 2950 [140]
8	$-CH_2(CH(CH_2)_2$-cyclo)	I (80%, with cyclopropylmethylbromide [333] or 67%, with cyclopropylcarbinyliodide [363] for more than 16 h; 30%, in $CH_3O(C_2H_4O)_2CH_3$ after ca. 8 h [106]) bright yellow solid [363]; isolated as an oil [106] ^1H NMR ($CDCl_3$): 0.25 to 1.3 (m, cyclo-$(CH_2)_2CH$), 1.65 (d, CH_2; J = 6), 5.32 (s, C_5H_5) [106, 366]; see also [333] ^{13}C NMR ($CDCl_3$): 9.03, 11.00, 18.81, 92.50 [363] IR (hexane): 1901, 1935, 2018 (ν(CO)) [333]; (THF): 1920, 2007 (ν(CO)) [363]; see also [106] mass spectrum: $[M]^+$, and fragments [333, 363] in refluxing hexane slow conversion to $C_5H_5Mo(CO)_2(\eta^3-CH_2CH{=}CHCH_3)$ occurred; under photochemical conditions reaction proceeded much faster; in the presence of $P(C_6H_5)_3$ also only this phosphane-free product was isolated under these conditions; for the reaction in THF, see p. 6 [333] reactions with $[(CO)_5CrH]^-$, $[(CO)_5WH]^-$, $[(CO)_4(P(OCH_3)_3)WH]^-$, and $[(CO)_4WH]^-$ in THF produced $[C_5H_5Mo(CO)_3]^-$ along with varying amounts of CH_4, $CH_3CH_2CH{=}CH_2$, and cyclo-$(CH_2)_2CHCH_3$; the product distribution was suggested to depend on the anionic metal hydride used and the different mechanistic pathways involved are discussed in detail [323, 363]
9	$-CH_2(CCH_3(CH^aH^b)_2$-cyclo)	I, using $CH_3C(CH_2I)_3$ and $K[C_5H_5Mo(CO)_3]$ at 65 °C for 1 h, no reaction occurred using the respective bromine compound even after prolonged heating; for this preparation a mechanism is discussed in detail [294] ^1H NMR (C_6D_6): 0.54 (br, H^a or H^b), 0.59 (br, H^a or H^b), 1.13 (s, CH_3), 1.70 (s, $MoCH_2$), 4.53 (C_5H_5), the two broad signals resulted from a second-

References on pp. 106/18

Table 1 (continued)

No.	^1L	method of preparation (yield) properties and remarks
9 (continued)		order effect due to the magnetic inequivalence of H^a and H^b [294] ^{13}C NMR (C_6D_6): 14.0 ($MoCH_2$), 20.3 (CH_2CH_2), 22.2 (C), 26.1 (CH_3), 92.7 (C_5H_5), 229.1 (CO cis MoC), 241.3 (CO trans MoC), the difference between H^a and H^b also appeared in the H-coupled ^{13}C NMR spectrum which exhibits a pattern close to a triplet for the respective cyclopropane ring carbons; the complexity results from the difference in coupling constants in the system $(CH^aH^b)_2$ [294] IR (hexane): 1936, 2018 (ν(CO)); despite the assumed C_s symmetry only two main bands were observed, obviously due to an accidental degeneracy of two of the three expected bands [294]
10	$-CHC(CH_3)_2(C(CN)_2)_2CH_2$-cyclo	IIIa (85 to 90% for 1 to 2 min, also in THF or CH_3CN; if the preparation was performed in pentane, No. 10 precipitated after 5 to 10 min from the reaction mixture; chromatography on alumina reduced the yield due to considerable decomposition) [91, 101, 138, 160] bright yellow solid, decomposition at 140°C without melting; stable in air at room temperature as solid, less in solution, moderately soluble in acetone and slightly in CH_2Cl_2 or $CHCl_3$ [91, 138] ^1H NMR (acetone-d_6): 1.36, 1.50 (both s, CH_3), 2.95 (m, $CHCH_2$), 5.45 (s, C_5H_5) [138] IR (CH_2Cl_2): 1946, 2034 (ν(CO)); (Nujol): 835 (δ(CH) of C_5H_5), 2245 (ν(CN)) [138]
11	$-C(CH_3)CH_2(C(CN)_2)_2CH_2$-cyclo	IIIa (45 to 50%, for detailed reaction conditions, compare No. 10) [91, 138, 160] lemon yellow solid [91, 138], decomposition at 142°C without melting; stable in air at room temperature as solid, less in solution, moderately soluble in acetone and slightly in CH_2Cl_2 or $CHCl_3$ [91, 101, 138] ^1H NMR (acetone-d_6): 1.84 (CH_3), 3.20, 3.68 (AB system, CH_2; J(H, H) = 15), 5.86 (C_5H_5) [91, 138] IR (Nujol): 835 (δ(CH) of C_5H_5), 1918, 1962, 2023 (ν(CO)), 2245 (ν(CN)) [138]
12	$-CHCH(C_6H_5)C(CF_3)_2OCH_2$-cyclo	IIIb (43%, with $(CF_3)_2CO$ for 3 h under a dry ice condenser, reaction did not succeed in neat $(CF_3)_2CO$) [128, 147]

References on pp. 106/18

Table 1 (continued)

No.	^{1}L	method of preparation (yield) properties and remarks
		yellow crystalline solid (by sublimation at 50 °C at 0.1 Torr), m.p. 135.5 to 136.5 °C; slightly soluble in pentane and very soluble in benzene, $CHCl_3$, and acetone [147] ^{1}H NMR ($CDCl_3$): 2.8 to 3.4 (m, MoCHCH), 3.75 to 4.75 (m, OCH_2), 4.99 (s, C_5H_5), 7.32 ("s", C_6H_5); except for the sharp singlet of the C_5H_5 fragment, the resonances are broad due to hydrogen-hydrogen-fluorine coupling [147] ^{19}F NMR: −74.4, −70.4 (both q; J = 9.4), the spectrum is characteristic of compounds containing nonequivalent CF_3 groups in a $C(CF_3)_2$ moiety; the chemical shift difference between the signals is due to the near asymmetric center [147] IR (pentane): 1944, 1951 (1960), 2028 (ν(CO)) [147]
13	$-CH(CH_2)_2SO_2$-cyclo	IIIb (16%) in neat SO_2, as a mixture with its dimer and/or trimer, instability precluded chemical studies on the constitution of the solid material, including the extent of aggregation [185] yellow-brown solid (from $CHCl_3$/pentane) [185], decomposition at ca. 105 °C [185] ^{1}H NMR (SO_2): 2.35 (m, CH), 3.44 (m, CH_2), 5.56, 5.62 (both s of variable intensity, C_5H_5) [185] IR (KBr): 1107, 1298 ($\nu(SO_2)$), 1918, 1959, 2021 (ν(CO)) [185]
14	NSO_2CH_3 SO	IIIb (67%, with CH_3SO_2NSO at 25 °C for 30 min) [318] yellow solid, m.p. 153 °C (dec.) [318] ^{1}H NMR ($CDCl_3$): 2.7 to 2.8 (m, CH), 3.10 (s, CH_3), 3.0 to 4.2 (m, CH_2), 5.42 (s, C_5H_5) [318] ^{13}C NMR ($CDCl_3$): 7.2 (CH), 41.8 (CH_3), 63.1 (NCH_2), 69.7 (SCH_2), 93.0 (C_5H_5), 227.9 (3CO) [318] IR (Nujol): 1080 (ν(SO)), 1160, 1320 ($\nu(SO_2)$); ($CHCl_3$): 1915, 1960, 2015 (ν(CO)) [318]
15	NSO_2CH_3 S NSO_2CH_3	IIIb (63%, with $S(NSO_2CH_3)_2$ at 25 °C; workup by column chromatography on alumina or Florisil) [318] yellow solid (from CH_2Cl_2/pentane), m.p. 140 °C (dec.) [318] ^{1}H NMR ($CDCl_3$): 3.14 (s, $2CH_3$), 3.75 to 4.48 (m, $2CH_2$), 5.47 (s, C_5H_5) [318] IR (Nujol): 1150, 1160, 1325, 1350 ($\nu(SO_2)$); ($CHCl_3$): 1940, 2025 (ν(CO)) [318]

References on pp. 106/18

Table 1 (continued)

No.	^{1}L	method of preparation (yield) properties and remarks
16	H_3C–C(CH$_2$NSO$_2$CH$_3$)$_2$S (ring: H_3C, –NSO$_2$CH$_3$, S, –NSO$_2$CH$_3$)	IIIb (15%), for details, compare No. 15 [318] yellow solid (from CH_2Cl_2/pentane), m.p. 70 °C (dec.) [318] ^{1}H NMR ($CDCl_3$): 1.58 (s, CCH_3), 2.93 (s, $2SCH_3$), 4.09 (br, $2CH_2$), 5.25 (s, C_5H_5) [318] IR (Nujol): 1155, 1160, 1320 ($\nu(SO_2)$), 1920, 1950, 2020 (ν(CO)) [318]
*17	-CH_2CH=CH_2	I (40%, with $ClCH_2CH$=CH_2 for 1.5 h; isolated along with small amounts of $C_5H_5Mo(CO)_2(\eta^3$-$CH_2CHCH_2)$ [12] by the phase-transfer-catalyzed reaction in CH_2Cl_2 at 24 °C between $C_5H_5(CO)_3MoCl$ and excess CH_2=$CHCH_2Br$ in the presence of $[(C_2H_5)_3NCH_2C_6H_5]Cl$ dissolved in 5 M NaOH (75%) [202, 237]; evidence has been presented for the intermediacy of $[C_5H_5Mo(CO)_3]^-$ [237] by slow decomposition of $[C_5H_5Mo(CO)_2(P(OCH_2CH$=$CH_2)_n(C_6H_5)_{3-n})_2]$ $[C_5H_5Mo(CO)_3]$ (n = 1 to 3) in benzene by nucleophilic attack of the anion on the cation along with $C_5H_5Mo(CO)_2(P(OCH_2CH$=$CH_2)_n$-$(C_6H_5)_{3-n})(P(O)(OCH_2CH$=$CH_2)_{n-1}(C_6H_5)_{3-r})_2$; the rate of formation increased in refluxing benzene [90] pale yellow oil, m.p. ca. −5 °C, decomposition at 60 °C [12] ^{1}H NMR (neat): 2.27 ("d", -CH_2-; J = 8.0), 3.5 (m, =CH_2), 5.11 (s, C_5H_5), 6.0 (m, -CH=) [12] IR (film): 1856 (probably $\nu(^{13}CO)$), 1911, 2021 (ν(CO)), 2860, 2930, 2965, 3077, 3105 (ν(CH)), 3904 (CO overtone) [12]; other bands at 735, 810, 879, 910, 983, 1012, 1060, 1108, 1195, 1260, 1298, 1355, 1375, 1427, 1550, 1612, 1640, 1760 including uncoordinated ν(CH) [12], similar ν(CO) in CH_2Cl_2 [90] and THF [354] readily soluble in common organic solvents [12] thermally unstable, in air rapid decomposition to $(C_5H_5Mo(CO)_3)_2$, a blue material containing Mo, and unidentified hydrocarbons [12, 394]
18	-CH_2CH=$CH_2(Fe^+(CO)_2C_5H_5)$	from $C_5H_5Mo(CO)_3(CH_2)_3Fe(CO)_2C_5H_5$ (No. 154) which was treated with $[(C_6H_5)_3C]PF_6$ in CH_2Cl_2 at room temperature for 77 h; the product precipitated from the reaction mixture as the PF_6^- salt (30%) [392] golden lustrous platelets (from acetone/hexane), m.p. >180 (dec.) [392]

References on pp. 106/18

Table 1 (continued)

No.	[1]L	method of preparation (yield) properties and remarks
		^{1}H NMR ($CDCl_3$, −30 °C): 1.91 (m, 1H, CH_2), 2.83 (m, 1H, CH_2), 3.28 (d, 1H, CH_2; J = 13.3), 3.30 (d, 1H, CH_2; J = 8.5), 6.41 (m, CH), 5.72 (s, C_5H_5Fe), 5.78 (s, C_5H_5Mo) [392] ^{13}C NMR ($CDCl_3$): 9.2 ($MoCH_2$), 40.1 (=CH_2), 88.9 (C_5H_5Fe), 95.0 (C_5H_5Mo), 120.0 (CH) [392] IR (CH_2Cl_2): 1909, 1944, 2021, 2067 (ν(CO)) [392] insoluble in ether, very sparingly soluble in CH_2Cl_2, and sparingly soluble in acetone [392] the structure of this complex is discussed to be a hybrid of two limiting forms: one with the Fe π- and the Mo σ-bonded to the allyl group, the other with both metals additionally weakly linked to the β-CH unit [392]
19	-CH_2CH=CHCl	the pattern of the ^{1}H NMR spectrum indicated a mixture of Z- and E-isomers [186] I (60 to 80%, with $ClCH_2CH$=CHCl for several hours) [186] yellow oil [186] ^{1}H NMR ($CDCl_3$): 2.18, 2.29 (both d, CH_2 of both isomers; J = 8 to 8.5), 5.21, 5.30 (both s, C_5H_5 of both isomers), 5.50 to 6.42 (m, CH=CH) [186] IR (Nujol): 814 (π(CH) of C_5H_5), 1660 (ν(C=C)), 1940, 2025 (ν(CO)) [186]
20	-CH_2CCl=CHCl	single isomer, distinction between Z- and E-isomer not possible I (60 to 80%, with $ClCH_2CCl$=CHCl for several hours) [186] yellow oil [186] ^{1}H NMR ($CDCl_3$): 2.47 (s, CH_2); 5.38 (s, C_5H_5), 5.71 (s, CH) [186] IR (neat): 8.22 (π(CH) of C_5H_5), 1570 (ν(C=C)), 1940, 2030 (ν(CO)) [186]
21	-CH_2CH=$CHCH_3$	the pattern of the ^{1}H NMR spectrum (see below) indicated a mixture of Z- and E-isomers [186]; pure E-isomer is believed to be obtained as described in [77]; pure Z-isomer is mentioned as being prepared using Z-$ClCH_2CH$=$CHCH_3$ as described in [63] I (60 to 80%, with XCH_2CH=$CHCH_3$, X = Cl, Br for several hours) [63, 77, 186], a mixture of Z- and E-isomers is isolated as a yellow oil that can be induced to crystallize [186, 323, 354] m.p. of pure E-isomer 64 °C [77]

References on pp. 106/18

Table 1 (continued)

No.	^1L	method of preparation (yield) properties and remarks
21 (continued)		^1H NMR ($CDCl_3$, mixture of Z-/E-isomers): 1.67 (d, CH_3; $^3J_{gem}$ = 6), 2.33 (d, CH_2; $^3J_{gem}$ = 7), 5.18, 5.25 (both s, C_5H_5 of both isomers), 4.97 to 6.20 (m, CH=CH) [186]; similar in toluene-d_8 [354] ^1H NMR ($CDCl_3$, pure E-isomer): 1.66 (CH_3; $^3J_{gem}$ = 6, $^4J_{vic}$ = 1), 2.38 (CH_2; $^3J_{gem}$ = 7), 5.20 (s, C_5H_5), 5.50 (CH=CH; trans-$^3J_{vic}$ = 15) [77] IR ($CHCl_3$, mixture of Z-/E-isomers): 811 (π(CH) of C_5H_5), 1640 (ν(C=C)), 1930, 2015 (ν(CO)) [186]; similar ν(CO) in THF [354] $(CH_3)_3NO$ catalyzed decarbonylation in THF resulting in a σ to π rearrangement and yielded $C_5H_5Mo(CO)_2(\eta^3\text{-}CH_2CHCHCH_3)$ in a one-pot reaction starting from $Na[C_5H_5Mo(CO)_3]$ and $ClCH_2CH{=}CHCH_3$ without isolation of No. 21 [288] protonation with simultaneous isomerization occurred by treatment with $[(C_6H_5)_3C]BF_4$ in $CHCl_3$ to give $[C_5H_5Mo(CO)_3CH_2{=}CHCH_2CH_3]^+$ between −10 and 50 °C; studies using $CDCl_3$ and D_2O, respectively, proved H_2O/D_2O as the proton source [195]
22	$-CH_2C(CH_3){=}CH_2$	I (35%, with $XCH_2C(CH_3){=}CH_2$; X = Cl, Br for 1 to 3 h below 25 °C) [138, 197] orange-yellow oil [138] ^1H NMR ($CDCl_3$): 1.79 (s, CH_3), 2.43 (s, $-CH_2-$), 4.58 (br, $=CH_2$) 5.23 (s, C_5H_5) [138] IR (pentane): 1940, 2023 (ν(CO)) [138] protonation by HBF_4 gave $[C_5H_5Mo(CO)_3(CH_2{=}C(CH_3)_2)]BF_4$ [197] $(CH_3)_3NO$ catalyzed decarbonylation resulted in $C_5H_5Mo(CO)_2(\eta^3\text{-}CH_2C(CH_3)CH_2)$ analogously to No. 21 [288]
23	-(E)-$CH_2CH{=}CHC_6H_5$	I (60 to 80%, with $ClCH_2CH{=}CHC_6H_5$ for several hours) [77, 186] orange solid contaminated with $(C_5H_5Mo(CO)_3)_2Hg$ which could not be completely removed [186], m.p. 115 °C [77] ^1H NMR ($CDCl_3$): 2.53 (CH_2; $^3J_{gem}$ = 7.5), 5.20 (s, C_5H_5), 6.40 (CH=CH; trans-$^3J_{vic}$ = 14), 7.26 (C_6H_5) [77]; see also [186] IR ($CHCl_3$): 814 (π(CH) of C_5H_5), 1620 (ν(C=C)), 1930, 2020 (ν(CO)) [186]
24	$-CH_2CH{=}C(CH_3)_2$	I (60 to 80%, with $ClCH_2CH{=}C(CH_3)_2$ for several hours) [186]

References on pp. 106/18

Table 1 (continued)

No.	^1L	method of preparation (yield) properties and remarks
		^1H NMR ($CDCl_3$): 1.65, 1.71 (both s, CH_3), 2.33 (d, CH_2; J = 9), 5.20 (s, C_5H_5), 5.40 (t, CH) [186] IR (CH_2Cl_2): 815 (π(CH) of C_5H_5), 1635 (ν(C=C)), 1930, 2015 (ν(CO)) [186]
25	-(E)-CH_2CH^1=$CH^2Si(CH_3)_3$	I (44%, with (E)-$ClCH_2CH^1$=$CH^2Si(CH_3)_3$ in 1,2-dimethoxyethane for 2 h) [151] solid (from hexane), m.p. 40 to 41 °C [151] ^1H NMR ($CDCl_3$): 0.02 (s, $Si(CH_3)_3$), 2.44 (dd, CH_2; $^3J_{gem}$ = 8, $^4J_{vic}$ = 1), 5.18 (s, C_5H_5), 5.32 (dt, H^2; trans-$^3J_{vic}$ = 18), 6.33 (dt, H^1) [151] IR: 1937, 1945, 2020 (ν(CO)) [151] UV irradiation in hexane gave $C_5H_5Mo(CO)_2(\eta^3$-$CH_2CHCHSi(CH_3)_3)$ [151]
26	-CH_2CH=$CHC(O)CH_3$	mixture of E- and Z-isomers in the ratio 81:19 as indicated by the ^1H NMR spectrum [368] I (58%, with $ClCH_2CH$=$CHC(O)CH_3$ for 6 h) [368] probably an air-stable oil ^1H NMR ($CDCl_3$): E-isomer: 2.18 (s, CH_3), 2.28 (d, CH_2; $^3J_{gem}$ = 9.4), 5.27 (s, C_5H_5), 5.87 (d, =CHC(O); trans-$^3J_{vic}$ = 15.2), 7.18 (dt, CCH=) [368] Z-isomer: 2.30 (d, CH_2; $^3J_{gem}$ = 10.0), 2.95 (s, CH_3), 5.32 (s, C_5H_5), 5.68 (d, =CHC(O); cis-$^3J_{vic}$ = 10.5), 6.42 (q, CCH=) [368] IR (CH_2Cl_2): 1653 (ν(C=O)), 1930, 2022 (ν(CO)) [368] mass spectrum: $[M-nCO]^+$ (n = 1 to 3) [368] smoothly lost one CO ligand when treated with excess $(CH_3)_3NO$ in THF [362] or CH_2Cl_2 [368] to give $C_5H_5Mo(CO)_2(\eta^3$-$CH_2CHCHC(O)CH_3)$ which exists as syn- and anti-diastereoisomers [362]; see also [355, 368]
27	-CH_2CH=$CHCO_2CH_3$	mixture of E- and Z-isomers in the ratio 48:52 as indicated by the ^1H NMR spectrum [368] I (56%, with $ClCH_2CH$=$CHCO_2CH_3$ for 6 h) [368] air-stable yellow oil [368] ^1H NMR ($CDCl_3$): E-isomer: 2.26 (d, CH_2; $^3J_{gem}$ = 9.7), 3.68 (s, OCH_3), 5.28 (s, C_5H_5), 5.55 (d, =CHC(O); trans-$^3J_{vic}$ = 14.9), 7.25 (dt, CCH=) [368] Z-isomer: 2.86 (d, CH_2; $^3J_{gem}$ = 10.5), 3.65 (s, OCH_3), 5.32 (s, C_5H_5), 5.34 (d, =CHC(O); cis-$^3J_{vic}$ = 11.1), 6.61 (q, CCH=) [368] IR (CH_2Cl_2): 1590 (ν(C=C)), 1696; 1934, 2022 (ν(CO)) [368]

References on pp. 106/18

Table 1 (continued)

No.	^1L	method of preparation (yield) properties and remarks
27 (continued)		mass spectrum: $[M-nCO]^+$ (n = 1, 2) [368] lost one CO ligand when treated with $(CH_3)_3NO$ to give $C_5H_5Mo(CO)_2(\eta^3\text{-}CH_2CHCHCO_2CH_3)$, compare also No. 22 [368]
28	$\text{-}CH_2C(C(O)CH_3)=CHCH_3$	I (65%, with $ClCH_2C(C(O)CH_3)=CHCH_3$ at −78 °C for 3 h) [385] yellow oil [385] ^1H NMR ($CDCl_3$): 1.23, 1.67 (each d, CH_2, $^2J_{gem}$ = 7.3), 2.26 (s, $OCCH_3$), 2.29 (d, $=CCH_3$; $^3J_{gem}$ = 6.9), 5.39 (s, C_5H_5), 6.35 (q, =CH) [385] ^{13}C NMR ($CDCl_3$): 25.4, 93.0, 133.5, 150.2, 200.0, 228.9, 240.7 [385] IR (Nujol): 1662 (ν(C=O)), 1924, 2012 (ν(CO)) [385] reaction with an excess of $(CH_3)_3NO$ in CH_2Cl_2 at 23 °C gave anti-$C_5H_5Mo(CO)_2(\eta^3\text{-}CH_2C(C(O)CH_3)CHCH_3)$ [388]; stirring alone in ether at 23 °C resulted in rearrangement to $C_5H_5Mo(CO)_2$-(η^3-(CH_3CH)$CCH_2C(O)OC(CH_3)$-cyclo) with a butyrolactone ring linked to the $C_5H_5Mo(CO)_2$ moiety [385]
29	$\text{-}CH_2C(CO_2CH_3)=CHCH_3$	I (65%, with $ClCH_2C(CO_2CH_3)=CHCH_3$ at −78 °C for 3 h) [381, 390] yellow oil [381, 390] ^1H NMR ($CDCl_3$): 1.62 (d, CH_3; $^3J_{gem}$ = 7.2), 2.37 (s, CH_2), 3.67 (s, OCH_3), 5.34 (s, C_5H_5), 6.48 (q, =CH) [390] ^{13}C NMR ($CDCl_3$): −8.6 (CH_2), 14.4 (CH_3), 51.2 (OCH_3), 93.0 (C_5H_5), 130.8, 140.5 (both C=C), 169.3 (OCO), 228.4, 240.6 (each MoCO) [390] IR (Nujol): 1706 (ν(C=O)), 1918, 2013 (ν(CO)) [390] reaction with an excess of $(CH_3)_3NO$ in CH_2Cl_2 at 23 °C gave anti-$C_5H_5Mo(CO)_2(\eta^3\text{-}CH_2C(CO_2CH_3)CHCH_3)$ [381, 390], but in contrast to No. 28, no intramolecular cyclization occurred even at 60 °C in THF [390]
30	$\text{-}CH_2C(C_5H_4FeC_5H_5)=C(CN)_2$	obtained in a one-pot reaction starting from $C_5H_5Mo(CO)_3CH_2C(O)C_5H_4FeC_5H_5$ (No. 107) which reacted with $[(C_2H_5)_3O]BF_4$ in CH_2Cl_2 for 2 h; the resulting $[C_5H_5Mo(CO)_3CH_2C^+(OC_2H_5)C_5H_4FeC_5H_5]BF_4$ (No. 108) was without isolation further treated with an excess of $CH_2(CN)_2$; after 5 min $(C_2H_5)_3N$ was added, and the crude product was

References on pp. 106/18

Table 1 (continued)

No.	^{1}L	method of preparation (yield) properties and remarks
		chromatographed on plates with neutral Al_2O_3 using ether/benzene as eluant to separate $C_5H_5FeC_5H_4C(CH_3)=C(CN)_2$ and No. 107 (9%) [212, 225] dark red solid (from ether/hexane), m.p. 117 to 119 °C (dec.), air-sensitive, highly soluble in polar organic solvents [212, 225] ^{1}H NMR ($CHCl_3$, with hexamethyldisiloxane as an internal standard): 2.73 (s, CH_2), 4.20 (s, C_5H_5Fe), 4.57, 4.90 (both t, C_5H_4), 5.48 (s, C_5H_5Mo) [225] IR (KBr): 1495 (ν(C=C)), 1930, 1955, 2030 (ν(CO)), 2210 (ν(CN)) [225]
31	F F H^2 F F F H^1	IIb (45%) [340] yellow crystals which are extremely sensitive to air and moisture, m.p. 35 °C (dec.) [340] ^{1}H NMR: 3.1 (H^1), 4.5 (H^2) [340] ^{19}F NMR: −130.9 (F^A; $J(F^A, F^D) = 19.5$), −168.2 (F^B), −179.2 (F^C), −197.8 (F^D; $J(F^D, H^1) = 47.1$, $J(F^D, F^E) = 21.3$), −208.9 (F^E; $J(F^E, H^2) = 53.2$) [340] the NMR spectra were assigned to an ABCDEXY system IR (n-pentane): 1944, 1963, 2005, 2053 (ν(CO)) [340] mass spectrum: $[M]^+$, $[M-nCO]^+$ (n = 1, 2), $[M-nCO-Mo]^+$ (n = 2, 3), $[M-3CO-Mo-HF]^+$, and other organic fragments [340]
*32	$-CH_2{}^1CH^2{=}CH^3CH^4{=}CH^5H^6$	I (72%, with (E)-$ClCH_2CH{=}CHCH{=}CH_2$ at −78 °C for 6 h) [319] yellow crystalline solid which decomposes slowly at room temperature [319] ^{1}H NMR (benzene-d_6): 2.30 (d, H^1; $J(H^1, H^2) = 7.2$), 4.39 (s, C_5H_5), 4.92 (dd, H^6; $J(H^4, H^6) = 9.6$, $J(H^5, H^6) = 1.2$), 5.05 (dd, H^5; $J(H^4, H^5) = 16.1$), 6.00 to 6.05 (m, H^2, H^3), 6.25 to 6.40 (m, H^4) [319] IR (pentane): 1620 (ν(C=C)), 1940, 1960, 2018 (ν(CO)) [319] mass spectrum: $[M-nCO]^+$ (n = 1, 2), $[M-nCO-C_5H_7]^+$ (n = 0 to 3) [319]
*33	$-CH_2{}^1CH^2{=}CH^3CH^4{=}CH^5CH^6_3$	I (57%, with (E,E)-$ClCH_2CH{=}CHCH{=}CHCH_3$ at −78 °C for 3 h) [345] yellow needlelike crystals [345] ^{1}H NMR (benzene-d_6): 1.64 (d, H^6; $J(H^5, H^6) = 6.4$), 2.38 (d, H^1; $J(H^1, H^2) = 8.4$), 5.50 (m, H^5;

References on pp. 106/18

Table 1 (continued)

No.	^1L	method of preparation (yield) properties and remarks
*33 (continued)		$J(H^4, H^5) = 16.4$), 5.90 to 6.10 (m, H^2, H^3, H^4) [345] IR (pentane): 1942, 1962, 2015 (ν(CO)) [345] mass spectrum: $[M-nCO]^+$ (n = 1 to 3), $[M-nCO-C_6H_9]^+$ (n = 0 to 3) [345]
34	$-CH_2CH{=}CHCH{=}C(CH_3)_2$	I (72%, with $ClCH_2CH{=}CHCH{=}C(CH_3)_2$ at 0°C for 3 h) [351] yellow solid [351] ^1H NMR ($CDCl_3$): 1.68, 1.69 (both s, CH_3), 2.46 (d, CH_2; $^3J_{gem} = 8.4$), 4.44 (s, C_5H_5), 5.91 (m, =CHCH=), 6.23 (dt, CH_2CH; trans-$^3J_{vic} = 16$) [351] IR ($CHCl_3$): 1615 (ν(C=C)), 1934, 1960, 2010 (ν(CO)) [351]
*35	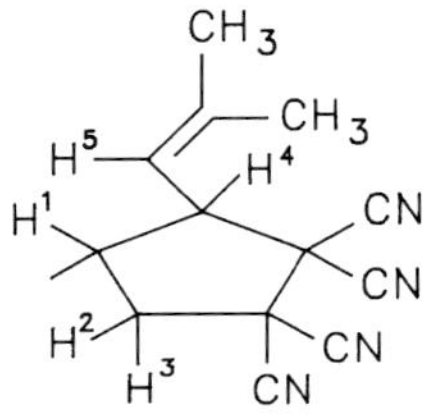	IIIa (39%) [343, 351] pale yellow crystals (from $CHCl_3$) [351] ^1H NMR ($CDCl_3$): 1.81, 1.86 (both s, CH_3), 2.71 (dd, H^2; $J(H^1, H^2) = 12.3$, $J(H^2, H^3) = 11.9$), 2.76 (m, H^1), 3.12 (dd, H^3; $J(H^1, H^3) = 7.8$), 3.63 (dd, H^4; $J(H^1, H^4) = 10.4$, $J(H^4, H^5) = 10.1$), 5.07 (d, H^5), 5.40 (s, C_5H_5) [351] ^{13}C NMR ($CDCl_3$): 15.8 (CH^1), 45.0 ($C(CN)_2$), 51.7 (CH^2H^3), 52.0 ($C(CN)_2$), 81.2 (CH^4), 111.0, 111.9, 112.6, 112.7 (all CN), 120.2 (CH^5), 143.4 ($C(CH_3)_2$) [351] IR ($CHCl_3$): 1670 (ν(C=C)), 1930, 1950, 2010 (ν(CO)), 2250 (ν(CN)) [351]
36	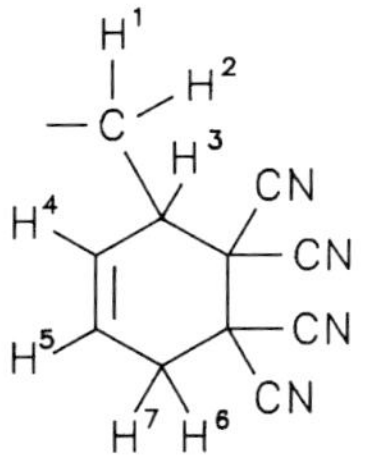	IIIa (66%, for 2 h) [343, 345] yellow crystals (from CH_2Cl_2/hexane) [343, 345] ^1H NMR ($CDCl_3$): 1.40 (dd, H^1; $J(H^1, H^2) = 10.4$, $J(H^1, H^3) = 12.0$), 2.00 (dd, H^2; $J(H^2, H^3) = 3.2$), 3.0 to 3.3 (m, H^3, H^6, H^7; $J(H^6, H^7) = 19.2$), 5.41 (s, C_5H_5), 5.75 (m, H^5; $J(H^4, H^5) = 10.4$), 5.79 (m, H^4) [345] IR (Nujol): 1656 (ν(C=C)), 1920, 1950, 2010 (ν(CO)), 2232 (ν(CN)) [345]
37	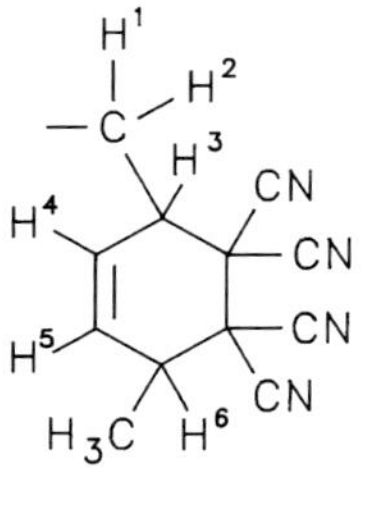	IIIa (8%, for 2 h) [343, 345] yellow crystals (from CH_2Cl_2/hexane) [343] ^1H NMR ($CDCl_3$): 1.52 (d, CH_3; $J(CH_3, H^6) = 6.8$), 1.71 (dd, H^1; $J(H^1, H^2) = 10.0$, $J(H^1, H^3) = 10.4$), 2.01 (dd, H^2; $J(H^2, H^3) = 4.4$), 3.12 (m, H^3, H^6), 5.42 (s, C_5H_5), 5.62 (dd, H^5; $J(H^4, H^5) = 9.8$, $J(H^5, H^6) = 3.3$), 5.85 (dd, H^4; $J(H^3, H^4) = 5.2$) [345] IR (KBr): 1654 (ν(C=C)), 1923, 1948, 2010 (ν(CO)), 2230 (ν(CN)) [345]

References on pp. 106/18

Table 1 (continued)

No.	^1L	method of preparation (yield) properties and remarks
38	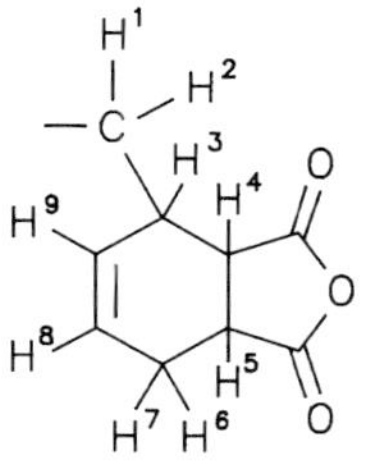	IIIb (79%) [343, 345] orange blocks (from CH_2Cl_2/hexane) [345] ^1H NMR ($CDCl_3$): 1.65 (dd, H^1; $J(H^1, H^2)$ = 10.1, $J(H^1, H^3)$ = 10.1), 1.92 (dd, H^2; $J(H^2, H^3)$ = 4.0), 2.25 (m, H^6), 2.63 (ddd, H^7; $J(H^5, H^7)$ = 2.9, $J(H^6, H^7)$ = 16.0, $J(H^7, H^8)$ = 5.8), 2.70 (m, H^3; $J(H^3, H^6)$ undetermined), 3.30 (dd, H^4; $J(H^3, H^4)$ = 5.8, $J(H^4, H^5)$ = 9.6); 3.38 (ddd, H^5; $J(H^5, H^6)$ = 8.8), 5.32 (s, C_5H_5), 5.85 (m, H^8; $J(H^3, H^8)$ = 1.5, $J(H^6, H^8)$ = 4.4), 5.95 (ddd, H^9; $J(H^3, H^9)$ = 3.2, $J(H^6, H^9)$ = 2.1, $J(H^8, H^9)$ = 9.8) [345] ^{13}C NMR ($CDCl_3$): 2.2 (CH^1H^2), 23.2 (CH^6H^7), 40.6, 40.7, 48.5 (CH^3, CH^4, CH^5), 125.9, 136.3 (CH^8, CH^9), 172.4, 174.8 (-C(O)OC(O)-), 228.8, 229.5, 239.2 (MoCO) [345] IR (Nujol): 1650 (ν(C=C)), 1775 (ν(C(O)OC(O)-)), 1910, 1935, 2005 (ν(CO)) [345]
*39	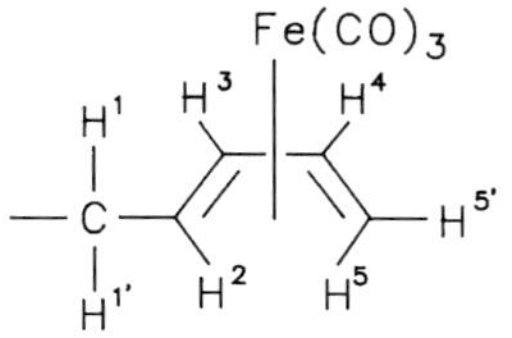	similar to I (47%, using $[N(P(C_6H_5)_3)_2]$-$[C_5H_5Mo(CO)_3]$ and $[(\eta^5\text{-}CH_2CHCHCHCH_2)$-$Fe(CO)_3]PF_6$ in CH_3CN, THF or CH_2Cl_2 as the reaction medium for 5 min) [360, 369] yellow needles (from CH_3CN at −20 °C after 24 h) [369] ^1H NMR (CD_3NO_2): 0.52 (ddd, H^5; $J(H^3, H^5)$ = 1.0, $J(H^4, H^5)$ = 9.4, $J(H^5, H^{5'})$ = 2.2), 1.74 (dd, H^1; $J(H^1, H^{1'})$ = 9.3, $J(H^1, H^2)$ = 12.3), 1.82 (ddd, $H^{5'}$; $J(H^3, H^{5'})$ = 1.0, $J(H^4, H^{5'})$ = 7.0), 1.99 (dddd, H^2; $J(H^{1'}, H^2)$ = 3.0, $J(H^2, H^3)$ = 9.2, $J(H^2, H^4)$ = 1.0), 2.44 (dd, $H^{1'}$), 5.26 (dddd, H^4; $J(H^3, H^4)$ = 5.0), 5.48 (s, C_5H_5), 5.54 to 5.55 (m, H^3) [369] IR (hexane): 1937, 1947, 1966, 1973, 2019, 2043 (ν(CO)) [369]
*40	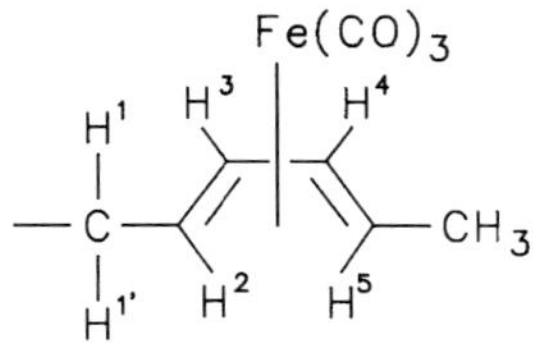	similar to I (46%, using $[N(P(C_6H_5)_3)_2][C_5H_5Mo(CO)_3]$ with $[(\eta^5\text{-}CH_2CHCHCHCHCH_3)Fe(CO)_3]PF_6$ at 0 °C for 1 h) [360, 369] yellow crystals (from CH_3CN at −20 °C) [369] ^1H NMR (CD_3NO_2): 1.33 to 1.43 (m, H^5 and CH_3; $J(H^4, H^5)$ = 9.4), 1.78 (dd, H^1; $J(H^1, H^{1'})$ = 9.3, $J(H^1, H^2)$ = 12.2), 1.95 (dddd, H^2; $J(H^{1'}, H^2)$ = 3.0, $J(H^2, H^3)$ = 9.3, $J(H^2, H^4)$ = 1.0), 2.42 (dd, $H^{1'}$), 5.07 (m, H^4), 5.30 (dd, H^3; $J(H^3, H^4)$ = 5.0), 5.49 (s, C_5H_5) [369] IR (hexane): 1937, 1946, 1968, 2018, 2037 (ν(CO)) [369]

References on pp. 106/18

Table 1 (continued)

No.	^{1}L	method of preparation (yield) properties and remarks
*41	$Fe(CO)_3$, H^3, H^4, H^2, CH_3, CH^1, H^5, $H^{1'}$	for preparation, see "Further information" at the end of the table not isolated, established only in solution ^1H NMR (acetone-d_6, −35 °C): 1.47 (d, CH_3; J(CH_3, H^5) = 6.3), 1.99 (dd, H^1; J(H^1, $H^{1'}$) = 9.3, J(H^1, H^2) = 13.5), 2.31 (dd, $H^{1'}$; J($H^{1'}$, H^2) = 4.0), 2.99 (dq, H^5; J(H^4, H^5) = 9.3), 3.57 (ddd, H^2; J(H^2, H^3) = 7.2), 5.24 (dd, H^3; J(H^3, H^4) = 5.1), 5.44 (dd, H^4), 5.59 (s, C_5H_5) [369] IR (CH_2Cl_2, −35 °C): 1918, 1960, 2008, 2033 (ν(CO)) [369] UV (CH_2Cl_2, ε): 340 (10000) [369] reacted rapidly in THF at −25 °C with 2 equivalents of $P(C_6H_5)_3$ to form [$Fe(CO)_3$(η^4-$(C_6H_5)_3$-PCH_2CH=CH-CH=$CHCH_3$)][$C_5H_5Mo(CO)_3$] [369]
*42	$Fe(CO)_3$, H^1, H^3, H^4, C, $H^{5'}$, H^2, H^5, CH_3	for preparation see "Further information" at the end of the table [369] not isolated, established only in solution ^1H NMR (acetone-d_6, −35 °C): 0.58 (br d, H^5; J(H^4, H^5) = 10), 1.57 (d, CH_3; J(CH_3, H^1) = 7), 1.74 (br d, $H^{5'}$; J(H^4, $H^{5'}$) = 7), 2.25 (m, H^2), 2.82 (m, H^1), 5.33 (m, H^4), 5.53 (m, H^3), 5.64 (s, C_5H_5) [369]
*43	—$Fe(CO)_3$	similar to I (high yield, using [N$(P(C_6H_5)_3)_2$][$C_5H_5Mo(CO)_3$] and [(η^5-C_6H_7-cyclo)$Fe(CO)_3$]PF_6 at −78 °C) [369] pale yellow solid, unstable above −18 °C [369] IR (hexane/ether, −25 °C): 1927, 1971, 2010, 2041 (ν(CO)) [369]
*44	-CH_2C≡CH	I (with $BrCH_2C$≡CH for several hours); because of the instability of No. 44 compared to its homologues No. 45 and 47, no attempt at its isolation was made [61, 100] IR (THF): 1930, 2020 (ν(CO)), 2080 (ν(C≡C)) [61]
*45	-CH_2C≡CCH_3	I (71%, with $BrCH_2C$≡CCH_3 for 3.5 h) [61, 100] yellow crystals (from pentane at −78 °C) [100], m.p. 93 °C (dec.) [100]; see also [61] ^1H NMR ($CDCl_3$): 1.88 (s, CH_2 and CH_3), 5.52 (s, C_5H_5) [100]; similar in CS_2 [61] IR (cyclohexane): 1936, 1944, 2022 (ν(CO)), 2206 (ν(C≡C)) [100]
*46	-CH_2C≡CC≡CCH_3	I (57%, with $ClCH_2C$≡CC≡CCH_3 at −10 °C for 15 h) [273]

Table 1 (continued)

No.	^1L	method of preparation (yield) properties and remarks
		amber oil [273] ^1H NMR ($CDCl_3$): 1.85 (CH_2), 1.92 (CH_3), 5.3 (C_5H_5) [273] IR ($CHCl_3$): 1930 to 1940, 2020 (ν(CO)) [273]
*47	$-CH_2C{\equiv}CC_6H_5$	I (with $BrCH_2C{\equiv}CC_6H_5$) [61, 100] m.p. 79 °C [61] ^1H NMR (CS_2): 2.0 (CH_2) [61] IR: 2190 (ν(C≡C)) [61] UV (CH_2Cl_2, ε) = 266 (sh), 315 (3500) [247]
48	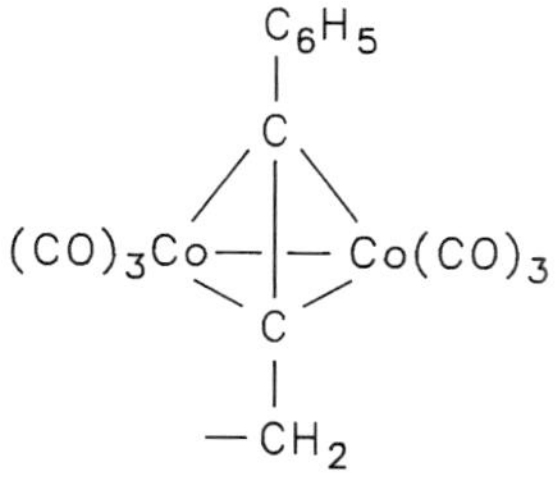 C_6H_5 C $(CO)_3Co$—|—$Co(CO)_3$ C $-CH_2$	from $C_5H_5Mo(CO)_3CH_2C{\equiv}CC_6H_5$ (No. 47) and $Co_2(CO)_8$ in equimolar amounts at room temperature in pentane; since the compound rapidly decomposed in solution on both alumina and Florisil columns and during attempts at crystallization, it was isolated by evaporation of the crude reaction mixture to dryness and was not further purified [344] ^1H NMR (CD_2Cl_2): 3.20 (s, CH_2), 5.33 (s, C_5H_5), 7.29 to 7.48 (m, C_6H_5) [344] IR (pentane): 1941, 1953, 2025, 2045, 2069, 2086 (ν(CO)) [344] treatment with CF_3CO_2H in pentane at 25 °C furnished $(C_5H_5Mo(CO)_2)(\mu,\eta^2\text{-}C_6H_5CCCH_3)$-$Co(CO)_3$ (structure similar to Formula XXV on p. 91) along with a paramagnetic trifluoroacetato derivative of Co(II); on alumina a similar reaction was proposed [344]
*49	$-CH_2C{\equiv}CCH_2OH$	I (with $ClCH_2C{\equiv}CCH_2OH$ for 12 h) [197, 286] amber oil (from CH_2Cl_2) [286] ^1H NMR ($CDCl_3$): 1.90 (CH_2), 4.30 (CH_2O), 5.30 (C_5H_5) [197] IR ($CHCl_3$): 1930, 1940, 2020 (ν(CO)) [197]
*50	$-CH_2C{\equiv}CCH(C_6H_5)OH$	I (with $ClCH_2C{\equiv}CCH(C_6H_5)OH$ for 12 h) [197] ^1H NMR ($CDCl_3$): 2.0 (CH_2), 5.26 (C_5H_5), 5.65 (CH), 7.35 (C_6H_5) [197] IR ($CHCl_3$): 1930, 1940, 2020 (ν(CO)) [197]
*51	$-CH_2C{\equiv}CC(CH_3)_2OH$	I (with $ClCH_2C{\equiv}CC(CH_3)_2OH$ for 12 h) [197] ^1H NMR ($CDCl_3$): 1.70 (CH_3), 1.95 (CH_2), 2.55 (OH), 5.65 (C_5H_5) [197] IR ($CHCl_3$): 1930, 1940, 2020 (ν(CO)) [197]
*52	$-CH_2C{\equiv}CC(CH_3)(C_6H_5)OH$	I (80 to 90%, no details) [218, 292] ^1H NMR ($CDCl_3$): 1.72 (CH_3), 1.87 (CH_2), 2.5 (OH), 5.22 (C_5H_5), 7.3 to 7.8 (C_6H_5) [218] IR ($CHCl_3$): 1940, 2010 (ν(CO)) [218]

References on pp. 106/18

Table 1 (continued)

No.	¹L	method of preparation (yield) properties and remarks
*53	$-CH_2C{\equiv}CC(C_6H_5)_2OH$	I (80 to 90%, no details) [218, 292] ^{1}H NMR ($CDCl_3$): 1.80 (CH_2), 2.7 (OH), 5.15 (C_5H_5), 7.3 to 7.8 [218] IR ($CHCl_3$): 1940, 2010 (ν(CO)) [218]
*54	$-CH_2C{\equiv}CC(OH){=}$(fluorenylidene)	I (80 to 90%, no details) [218, 292] ^{1}H NMR ($CDCl_3$): 1.70 (CH_2), 2.2 (OH), 5.0 (C_5H_5), 7.1 to 7.4 (fluorenylidene H) [218] IR ($CHCl_3$): 1940, 2010 (ν(CO)) [218]
*55	$-CH_2C{\equiv}CCH_2CH_2OH$	I (not isolated, with $ClCH_2C{\equiv}CCH_2CH_2OH$ for 12 h) [197] IR ($CHCl_3$): 1930, 1940, 2020 (ν(CO)) [197]
*56	$-CH_2C_6H_5$	I (50% to 68%, with $ClCH_2C_6H_5$ for 16 h) [22, 30, 45, 74, 136, 214, 241]; similarly from $ClCH_2C_6H_5$ and $[C_{10}H_{24}N_4][C_5H_5Mo(CO)_3]_2$ ($C_{10}H_{24}N_4$ = octamethyloxamidinium) suspended in THF for 66 h (ca. 13%, the low yield is probably a consequence of the insolubility of the oxamidinium salt) [25] IIa (84%, for 12 h) [165] also by UV irradiation of $C_5H_5Mo(CO)_3SO_2CH_2C_6H_5$ in benzene which resulted in partial SO_2 elimination (35%), but this formation is of little preparative value as the starting material has to be synthesized from No. 56 by SO_2 insertion [58] strongly refracting bright yellow needles (from ether at −78°C) [22, 30, 136, 165], m.p. 86 to 89°C [22, 30, 74, 136, 165, 214, 241] ^{1}H NMR (SO_2, -37 ± 7°C): 2.88 (s, CH_2), 5.43 (s, C_5H_5), 7.18 (m, C_6H_5) [94, 120]; similar in THF at room temperature [228, 389] ^{1}H NMR (CS_2): 2.88 (s, CH_2), 5.14 (s, C_5H_5), 7.13 (C_6H_5) [30]; see also [136]; similar in $CDCl_3$ [74, 241] ^{13}C NMR ($CDCl_3$): 4.8 (CH_2), 94 (C_5H_5), 123.6 (C-4), 127.4 (C-2,6), 127.9 (C-3,5), 151.1 (C-1), 228 (CO cis MoC), 239 (CO trans MoC); ^{13}C signals of the phenyl ring were shifted to higher field compared to free benzene or toluene which demonstrates the electron donor character of the $C_5H_5Mo(CO)_3CH_2$ moiety [242, 258]; see also [113, 126] ^{95}Mo NMR (CH_3CN): −1598.8 [217]; ($CHCl_3$): −1577 [258]; inversion-recovery measurements

References on pp. 106/18

Table 1 (continued)

No.	^{1}L	method of preparation (yield) properties and remarks
		for ^{95}Mo were used to determine the spin-lattice relaxation time T_1 = (4.76 ± 0.13) ms; spin-spin relaxation time was calculated from line widths as T_2 = 4.76 ms, and results indicated that quadrupole relaxation is the only significant mechanism involved [304] IR (cyclohexane): 1942 (A″, ν(CO)), 1951 (A′, ν(CO)), 2014 (A′, ν(CO)); k_1(CO diagonal to CH_3) = 15.51, k_2(CO diagonal to another) = 15.73, k_{id} = 0.50; for calculations the ratio of the diagonal interaction force constant (k_{id}) and the lateral interaction force constant (k_{is}) was assumed to be constant, k_{id}/k_{is} = 2 [70]; see also [30]; similar in CH_3CN [241] IR (KBr): 1910, 1933, 1999 (ν(CO)), 2920, 2980, 3030 (ν(CH)); other bands at 698, 724, 755, 818, 830, 841, 852, 901, 922, 961, 974, 995, 1000, 1010, 1026, 1041, 1057, 1060, 1107, 1150, 1199, 1345, 1411, 1420, 1445, 1483, 1586 [30, 74]; see also [136] UV (no solvent given, ε): 314 (10638) [329] polarographic reduction (Hg electrode; millimolar 1,2-dimethoxyethane solutions containing 10^{-1} mol/L $[N(C_4H_9\text{-}n)_4]ClO_4$) resulted in $[C_5H_5Mo(CO)_3]^-$ and $C_6H_5CH_2^{\bullet}$, the fate of the latter is unknown; half-wave potential $E_{p/2}$ is −2.1 ± 0.1 V vs. Ag/10^{-3} M $AgClO_4$ [29, 32] thermal behavior and mass spectrum are similar to that of the 1-methylnaphthyl complex $C_5H_5Mo(CO)_3CH_2C_{10}H_7$ (No. 89) [116]
*57	$\text{-}CH_2C_6H_4CH_3\text{-}2$	I (ca. 50%, with $ClCH_2C_6H_4CH_3$-2 for several hours) [241, 258] solid (from hexane) [241] ^{13}C NMR ($CDCl_3$): 0.2 (CH_2), 20.1 (CH_3), 94 (C_5H_5), 123.6 (C-4), 125.9 (C-5), 127.9 (C-6), 129.8 (C-3), 133.8 (C-2), 149.9 (C-1), 228 (CO cis MoC), 239 (CO trans MoC) [258] ^{95}Mo NMR ($CHCl_3$): −1572 [258]; spin-lattice relaxation time and spin-spin relaxation time were determined as T_1 = (4.86 ± 0.12) ms and T_2 = 4.67 ms; for details, see preceding No. 56 [304]
*58	$\text{-}CH_2C_6H_4CH_3\text{-}3$	I (ca. 50%, with $ClCH_2C_6H_4CH_3$-3 for several hours) [241, 258] ^{13}C NMR ($CDCl_3$): 4.8 (CH_2), 94 (C_5H_5), 124.6 (C-4), 124.6 (C-6), 127.8 (C-5), 128.3 (C-2), 137.2 (C-3),

References on pp. 106/18

Table 1 (continued)

No.	^1L	method of preparation (yield) properties and remarks
*58 (continued)		151.0 (C-1), 228 (CO cis MoC), 239 (CO trans MoC) [258] ^{95}Mo NMR ($CHCl_3$): −1579 [258]; spin-lattice relaxation time and spin-spin relaxation time were determined as T_1 = (5.74 ± 0.09) ms and T_2 = 5.45 ms; for details, see No. 56 [304]
*59	$-CH_2C_6H_4CH_3$-4	I (40%, with $ClCH_2C_6H_4CH_3$-4 for several hours) [66]; see also [241] bright yellow solid (from CH_2Cl_2/petroleum ether at −78 °C), m.p. 94 to 96 °C (with slight dec.) [66, 241] ^1H NMR ($CDCl_3$): 2.27 (s, CH_3), 2.92 (s, CH_2), 5.17 (s, C_5H_5), 7.04 (m, C_6H_4) [66]; see also [241] ^{13}C NMR ($CDCl_3$): 4.7 (CH_2), 20.9 (CH_3), 94 (C_5H_5), 127.4 (C-2,6), 128.6 (C-3,5), 133.1 (C-4), 147.6 (C-1), 228 (CO cis MoC), 239 (CO trans MoC) [258] ^{95}Mo NMR ($CHCl_3$): −1583 [258]; spin-lattice relaxation time and spin-spin relaxation time were determined as T_1 = (6.00 ± 0.10) ms and T_2 = 5.88 ms; for details see, No. 56 [304] IR (cyclohexane): 1935, 2015 (ν(CO)) [66]; (CH_3CN): ca. 1922, ca. 2008 (ν(CO)) [241] on heating at 105 °C/6.5 × 10^{-5} atm decarbonylation occurred to give $C_5H_5Mo(CO)_2(\eta^3\text{-}CH_2C_6H_4CH_3\text{-}4)$ (see Formula XXX on p. 95 as the analogous reaction product of the unsubstituted benzyl complex No. 56) [66]
*60	$-CH_2C_6H_2(CH_3)_3$-2,4,6	I (ca. 50%, with $ClCH_2C_6H_2(CH_3)_3$-2,4,6 for several hours) [241, 258] solid (from hexane) [241] ^{13}C NMR ($CDCl_3$): −2.7 (CH_2), 94 (C_5H_5), 128.7 (C-3,5), 132.5 (C-4), 134.5 (C-2,6), 145.3 (C-1), 228 (CO cis MoC), 239 (CO trans MoC) [258] ^{95}Mo NMR ($CHCl_3$): −1599 [258]; spin-lattice relaxation time and spin-spin relaxation time were determined as T_1 = (2.51 ± 0.04) ms and T_2 = 2.52 ms; for details, see No. 56 [304]
61	$-CH_2C_6H_4(C_3H_7\text{-}i)$-2	no details given [252]
62	$-CH_2C_6H_4(C_3H_7\text{-}i)$-4	no details given [252]
63	$-CH_2C_6H_4(CH_2N(O^\bullet)C_6H_2(C_4H_9\text{-}t)_3\text{-}2,4,6)$-4	was generated upon irradiation (λ > 420 nm) of $(C_5H_5Mo(CO)_3)_2(\mu\text{-}CH_2C_6H_4CH_2\text{-}4)$ in benzene in

Table 1 (continued)

No.	^1L	method of preparation (yield) properties and remarks
		the presence of 2,4,6-tris(t-butyl)nitrosobenzene as indicated by ESR spectroscopy [329] ESR (C_6H_6): g = 2.0059; a(N) = 1.371, a(CH_2) = 1.484, a($H^{3,5}$) = 0.08 mT [329] as this nitroxyl adduct is sensitive to visible light, prolonged irradiation resulted first in $C_5H_5Mo(CO)_3ONCH_2C_6H_2(C_4H_9\text{-t})_3\text{-2,4,6}^{\bullet}$ and 2,4,6-(t-$C_4H_9)_3C_6H_2NOC_4H_9\text{-t}^{\bullet}$; further photolysis induces gradual disappearance of the radicals from the photolyzed mixture [329]
64	$-CH_2C_6H_4CH{=}CH_2$	I (64%, with a mixture of meta and para $ClCH_2C_6H_4CH{=}CH_2$ at 0°C for 8 h) [364] bright yellow crystals (from benzene/ligroin as 1:1 solvates with benzene), m.p. 102°C [364] ^1H NMR (C_6D_6): 2.33 (s, CH_2), 4.10 (s, C_5H_5), 4.48 to 5.64 (m, $CH{=}CH_2$), 6.64 (br d, C_6H_4) [364] IR (CH_2Cl_2): 1920, 2010 (ν(CO)) [364] free radical-assisted addition polymerization with azo-isobutyronitrile using $CH_2{=}C(CH_3)CO_2CH_3$ and No. 64 in a 4:1 molar ratio in refluxing benzene occurred to give a copolymer $(C_5H_5(CO)_3\text{-}MoCH_2C_6H_4CH\text{-}[\text{-}CH_2C(CH_3)(CO_2CH_3)\text{-}]_n\text{-}CH_2\text{-})_x$ with an incorporation ratio, organometallic monomer:organic monomer, of ca. 1:7; on the other hand, little incorporation of the compound was observed on attempted copolymerization with $C_6H_5CH{=}CH_2$ [364]
65	$-CH_2C_6H_4CF_3\text{-2}$	probably I, but no details given spin-lattice relaxation time and spin-spin relaxation time were determined as T_1 = (5.15 ± 0.04) ms and T_2 = 5.04 ms; for details, see No. 56 [304]
*66	$-CH_2C_6H_4CF_3\text{-3}$	I (ca. 50%, with $ClCH_2C_6H_4CF_3$-3 for several hours) [241] solid (from hexane), m.p. 115 to 116°C [241] ^1H NMR ($CDCl_3$): ca. 2.9 (s, CH_2), ca. 5.2 (s, C_5H_5) [241] ^{13}C NMR ($CDCl_3$): 3.3 (CH_2), 94 (C_5H_5), 120.2 (C-4; J(F, C) = 3.9), 123.8 (C-2; J(F, C) = 3.9), 124.2 (CF_3; J(F, C) = 272.5), 128.2 (C-5), 130.1 (C-3; J(F, C) = 31.5), 130.5 (C-6), 152.3 (C-1), 228 (CO cis MoC), 239 (CO trans MoC) [258] ^{95}Mo NMR ($CHCl_3$): −1555 [258]; similar in CH_3CN [217]; spin-lattice relaxation time and spin-spin relaxation time were determined as

References on pp. 106/18

Table 1 (continued)

No.	^1L	method of preparation (yield) properties and remarks
*66 (continued)		T_1 = (5.35 ± 0.12) ms and T_2 = 5.24 ms; for details see No. 56 [304] IR (CH_3CN): ca. 1922, ca. 2008 (ν(CO)) [241]
*67	$-CH_2C_6H_4CF_3$-4	I (ca. 50%, with $ClCH_2C_6H_4CF_3$-4 for several hours) [241] solid (from hexane), m.p. 96 to 98°C [241] ^1H NMR ($CDCl_3$): ca. 2.9 (s, CH_2), ca. 5.2 (s, C_5H_5) [241] ^{13}C NMR ($CDCl_3$): 3.3 (CH_2), 94 (C_5H_5), 124.7 (CF_3; J(F, C) = 271.5), 124.8 (C-3,5; J(F, C) = 3.9), 125.4 (C-4; J(F, C) = 31.0), 127.3 (C-2,6), 156.0 (C-1), 228 (CO cis MoC), 239 (CO trans MoC) [258] ^{95}Mo NMR ($CHCl_3$): −1551 [258]; spin-lattice relaxation time and spin-spin relaxation time were determined as T_1 = (5.29 ± 0.09) ms and T_2 = 5.13 ms; for detail see No. 56 [304] IR (CH_3CN): ca. 1922, ca. 2008 (ν(CO)) [241]
68	$-CH_2C_6H_4N{\equiv}C$-2	I (48%, with $ICH_2C_6H_4N{\equiv}C$-2 in 1,2-dimethoxyethane at −78°C for 30 min, with subsequent warming to −20°C over 3 h) [283] yellow, crystalline, slightly foul-smelling solid, m.p. >90°C (with darkening); best stored under N_2 in the cold and in the dark [283] ^1H NMR (CD_2Cl_2): 2.84 (s, CH_2), 5.42 (s, C_5H_5) [283] ^{13}C NMR ($CDCl_3$, with $Cr(CH_3C(O)CH{=}C(O)CH_3)_3$ as relaxation agent): −3.2 (CH_2), 93.0 (C_5H_5), 123.3, 125.8, 127.7, 128.6, 148.9 (each C_6H_4), 164.5 (NC, broad signal due to ^{14}N coupling), 227.5 (CO cis MoC), 238.5 (CO trans MoC) [283] IR (hexane): 1939, 1951, 2026 (ν(CO)), 2117 (ν(NC)); similar in CH_2Cl_2 [283] mass spectrum: $[M]^+$, $[M-nCO]^+$ (n = 2, 3); no detailed examination of the fragmentation process [283] soluble in most organic solvents, solutions in CH_2Cl_2 or benzene undergo rapid darkening [283] stirring the complex in benzene at room temperature or at reflux did not give any evidence for either the formation of iminoacyl or acyl products or for species derived from carbonyl substitution; instead, a slightly soluble brown material was recovered from the reaction mixture [283] reaction with $M(P(C_6H_5)_3)_2Cl_2$ (M = Pd, Pt) in benzene at 25°C produced the heterodimetallic

Table 1 (continued)

No.	^1L	method of preparation (yield) properties and remarks
		complexes $C_5H_5Mo(CO)_3CH_2C_6H_4(N{\equiv}CMCl_2P(C_6H_5)_3\text{-cis})\text{-}2$ (Nos. 71, 72) [283]
69	$\text{-}CH_2C_6H_4N{\equiv}CCr(CO)_5\text{-}2$	I (79%, with $ICH_2C_6H_4N{\equiv}CCr(CO)_5\text{-}2$ in 1,2-dimethoxyethane at −78 °C for 1 h; subsequently the reaction was stirred overnight at 25 °C) [270, 280] yellow crystalline solid (from ether/hexane at −15 °C), m.p. 119 to 120 °C [280] ^1H NMR (CD_2Cl_2): 2.84 (s, CH_2), 5.37 (s, C_5H_5) [280] ^{13}C NMR ($CDCl_3$): −3.1 (CH_2), 93.0 (C_5H_5), 123.6, 126.1, 128.1, 128.5, 148.8 (all C_6H_4), 171.2 (NC), 214.3 ($Cr(CO)_5$; CO cis CrCN), 216.4 ($Cr(CO)_5$; CO trans CrCN), 227.6 ($Mo(CO)_3$; CO cis MoC), 238.0 ($Mo(CO)_3$; CO trans MoC) [280] IR (hexane): 1947, 1952, 1965, 1998, 2031, 2061 (ν(CO)), 2139 (ν(NC)); similar in CH_2Cl_2 [280] mass spectrum: $[M]^+$, $[M-nCO]^+$ (n = 1 to 8) [280]
70	$\text{-}CH_2C_6H_4N{\equiv}CW(CO)_5\text{-}2$	I (74% with $ICH_2C_6H_4N{\equiv}CW(CO)_5\text{-}2$ analogous to No. 69) [270, 280] yellow crystals (from ether/hexane at −15 °C), m.p. 121 to 122 °C [280] ^1H NMR (CD_2Cl_2): 2.85 (s, CH_2), 5.39 (s, C_5H_5) [280] ^{13}C NMR ($CDCl_3$): −3.0 (CH_2), 93.1 (C_5H_5), 123.5, 123.7, 126.3, 128.1, 128.8, 149.4 (each C_6H_4), 193.8 ($W(CO)_5$; CO cis WCN; J(W, C) = 126), 195.9 ($W(CO)_5$; CO trans WCN), 227.7 ($Mo(CO)_3$; CO cis MoC), 238.0 ($Mo(CO)_3$; CO trans MoC), NC signal not observed [280] IR (hexane): 1937, 1949, 1962, 1993, 2027, 2061 (ν(CO)), 2138 (ν(NC)); (CH_2Cl_2): 1929, 1949, 2023, 2062 (ν(CO)), 2145 (ν(NC)) [280] mass spectrum: $[M]^+$, $[M-nCO]^+$ (n = 1 to 8) [280]
71	$\text{-}CH_2C_6H_4(N{\equiv}CPdCl_2P(C_6H_5)_3\text{-cis})\text{-}2$	by adding an excess of $C_5H_5Mo(CO)_3CH_2C_6H_4N{\equiv}C\text{-}2$ (No. 68) in benzene to a suspension of $Pd(P(C_6H_5)_3)_2Cl_2$ in the same solvent and subsequent stirring of the yellow-orange solution at 25 °C for 1 h; the product precipitated from the concentrated reaction mixture on addition of ether (55 to 70%) [283] yellow solid (from CH_2Cl_2/ether), m.p. >200 °C (with darkening), the complex is stable as a solid

References on pp. 106/18

Table 1 (continued)

No.	^1L	method of preparation (yield) properties and remarks
71 (continued)		but decomposes in solution at ambient temperature [283] ^1H NMR (CD_2Cl_2): 2.84 (s, CH_2), 5.58 (s, C_5H_5) [283] ^{31}P NMR (CD_2Cl_2): 23.58 [283] IR (Nujol): 295, 338 (ν(PdCl)); (CH_2Cl_2): 1924, 2019 (ν(CO)), 2209 (ν(NC)) [283]
72	$-CH_2C_6H_4(N{\equiv}CPtCl_2P(C_6H_5)_3$-cis)-2	analogously to No. 71 using $Pt(P(C_6H_5)_3)_2Cl_2$ (55 to 70%) [283] yellow solid (from CH_2Cl_2/ether), m.p. >210°C (with darkening), the complex is stable as a solid but slowly decomposes in solution at ambient temperature [283] ^1H NMR (CD_2Cl_2): 2.79 (s, CH_2), 5.55 (s, C_5H_5) [283] ^{31}P NMR (CD_2Cl_2): 8.50 (J(Pt, P) = 3362) [283] IR (Nujol): 306, 340 (ν(PtCl)); (CH_2Cl_2): 1925, 2019 2205 (ν(NC)) [283]
*73	$-CH_2C_6H_4OCH_3$-2	I (with $ClCH_2C_6H_4OCH_3$-2 for several hours) [258] ^{13}C NMR ($CDCl_3$): −2.2 (CH_2), 54.8 (CH_3), 94 (C_5H_5), 109.8 (C-3), 120.4 (C-5), 124.5 (C-4), 129.0 (C-6), 139.8 (C-1), 155.3 (C-2), 228 (CO cis MoC), 239 (CO trans MoC) [258] ^{95}Mo NMR ($CHCl_3$): −1573 [258]; spin-lattice relaxation time and spin-spin relaxation time were determined as T_1 = (5.55 ± 0.18) ms and T_2 = 5.34 ms; for details, see No. 56 [304]
*74	$-CH_2C_6H_4OCH_3$-3	I (ca. 50%, with $ClCH_2C_6H_4OCH_3$-3 for several hours) [241] solid (from hexane), m.p. 102 to 104°C [241] ^1H NMR ($CDCl_3$): ca. 2.9 (s, CH_2), ca. 5.2 (s, C_5H_5) [241] ^{13}C NMR ($CDCl_3$): 4.8 (CH_2), 94 (C_5H_5), 109.3 (C-4), 112.9 (C-2), 120.2 (C-6), 128.7 (C-5), 152.9 (C-1), 159.2 (C-3), 228 (CO cis MoC), 239 (CO trans MoC) [258] ^{95}Mo NMR ($CHCl_3$): −1574 [258]; spin-lattice relaxation time and spin-spin relaxation time were determined as T_1 = (4.82 ± 0.09) ms and T_2 = 4.59 ms; for details, see No. 56 [304] IR (CH_3CN): ca. 1922, ca. 2008 [241]
*75	$-CH_2C_6H_4OCH_3$-4	I (63%, with $ClCH_2C_6H_4OCH_3$-4 for several hours) [74, 241]; IIa (56%, for 3 h) [165] solid (from hexane), m.p. 103 to 105°C (dec.) [74, 165, 241]

Table 1 (continued)

No.	^1L	method of preparation (yield) properties and remarks
		^1H NMR ($CDCl_3$): 2.94 (s, CH_2), 3.75 (s, OCH_3), 5.15 (s, C_5H_5), 6.91 (q, AB type, C_6H_4; J(H, H) = 9.0) [74]; see also [241] ^{13}C NMR ($CDCl_3$): 4.6 (CH_2), 55.2 (CH_3), 94 (C_5H_5), 113.5 (C-3,5), 128.4 (C-2,6), 142.8 (C-1), 156.3 (C-4), 228 (CO cis MoC), 239 (CO trans MoC) [258] ^{95}Mo NMR ($CHCl_3$): −1587 [258]; spin-lattice relaxation time and spin-spin relaxation time were determined as T_1 = (5.11 ± 0.12) ms and T_2 = 5.04 ms; for details, see No. 56 [304] IR (KBr): 1895, 1925, 2002 (ν(CO)) [74]; similar in CH_3CN [241]
*76	-$CH_2C_6H_4F$-2	I (with $ClCH_2C_6H_4F$-2 for several hours) [258] ^{13}C NMR ($CDCl_3$): −5.0 (CH_2; J(F, C) = 1), 94 (C_5H_5), 114.8 (C-3; J(F, C) = 22.5), 123.8 (C-5; J(F, C) = 3.9), 124.7 (C-4; J(F, C) = 8.8), 129.9 (C-6; J(F, C) = 4.9), 138.8 (C-1; J(F, C) = 15.6), 159.8 (C-2; J(F, C) = 242.2), 228 (CO cis MoC), 239 (CO trans MoC) [258] ^{95}Mo NMR ($CHCl_3$): −1560 [258]; spin-lattice relaxation time and spin-spin relaxation time were determined as T_1 = (7.00 ± 0.20) ms and T_2 = 6.72 ms; for details, see No. 56 [304]
*77	-$CH_2C_6H_4F$-3	I (56%, with $ClCH_2C_6H_4F$-3 at 0°C for 1 h, then 72 h at 25°C) [84], see also [241] bright yellow crystals (from hexane at −78°C) [84], m.p. 102 to 104°C (dec.) [84]; compare also [241] ^1H NMR ($CDCl_3$): ca. 2.9 (s, CH_2), ca. 5.2 (s, C_5H_5) [241] ^{13}C NMR ($CDCl_3$): 3.8 (CH_2; J(F, C) = 1.9), 94 (C_5H_5), 110.3 (C-4; J(F, C) = 21.5), 114.1 (C-2; J(F, C) = 21.5), 123.1 (C-6; J(F, C) = 2.9), 129.1 (C-5; J(F, C) = 8.8), 154.2 (C-1; J(F, C) = 7.8), 162.6 (C-3; J(F, C) = 244.1), 228 (CO cis MoC), 239 (CO trans MoC) [258] ^{19}F NMR ($CDCl_3$, rel. to C_6H_5F): −1.09 [258]; similar in CH_2Cl_2 [84] ^{95}Mo NMR ($CHCl_3$): −1562 [258] IR ($CHCl_3$): 1937, 2017 (ν(CO)) [84]; similar in CH_3CN [241]
*78	-$CH_2C_6H_4F$-4	I (ca. 50%, with $ClCH_2C_6H_4F$-4 for several hours) [123, 241] solid (from hexane), m.p. 88 to 89°C [241] ^1H NMR ($CDCl_3$): 2.88 (s, CH_2), ca. 5.2 (s, C_5H_5) [123, 241]; see also [220]

References on pp. 106/18

Table 1 (continued)

No.	^{1}L	method of preparation (yield) properties and remarks
*78 (continued)		^{13}C NMR ($CDCl_3$): 3.5 (CH_2), 94 (C_5H_5), 114.6 (C-3,5; J(F, C) = 20.3), 128.5 (C-2,6; J(F, C) = 7.8), 146.5 (C-1; J(F, C) = 2.9), 159.8 (C-4; J(F, C) = 242.2), 228 (CO cis MoC), 239 (CO trans MoC) [258] ^{19}F NMR ($(CH_3)_2NCHO$, rel. to C_6H_5F): −7.14; similar in CH_2Cl_2 [123] ^{95}Mo NMR ($CHCl_3$): −1577 [258]; spin-lattice relaxation time and spin-spin relaxation time were determined as T_1 = (5.99 ± 0.17) ms and T_2 = 6.09 ms; for details, see No. 56 [304] IR ($CHCl_3$): 1928, 1936, 2019 (ν(CO)) [123], similar in CH_3CN [241]
*79	$-CH_2C_6H_4Cl$-2	probably I, but no details given spin-lattice relaxation time and spin-spin relaxation time were determined as T_1 = (5.68 ± 0.15) ms and T_2 = 5.45 ms; for details, see No. 56 [304]
*80	$-CH_2C_6H_4Cl$-3	I (with $ClCH_2C_6H_4Cl$-3 for several hours) [258] ^{13}C NMR ($CDCl_3$): 3.4 (CH_2), 94 (C_5H_5), 123.6 (C-4), 125.5 (C-6), 127.1 (C-2), 129.1 (C-5), 133.5 (C-3), 153.6 (C-1), 228 (CO cis MoC), 239 (CO trans MoC) [258] ^{95}Mo NMR ($CHCl_3$): −1559 [258]; spin-lattice relaxation time and spin-spin relaxation time were determined as T_1 = (5.59 ± 0.11) ms and T_2 = 5.59 ms; for details, see No. 56 [304]
*81	$-CH_2C_6H_4Cl$-4	I (75%, with $ClCH_2C_6H_4Cl$-4 for several hours) [74] m.p. 97.5 to 99.5°C (dec.) [74] ^{1}H NMR ($CDCl_3$): 2.84 (s, CH_2), 5.19 (s, C_5H_5), 7.11 ("s", C_6H_4) [74] ^{13}C NMR ($CDCl_3$): 3.5 (CH_2), 94 (C_5H_5), 127.9 (C-3,5), 128.6 (C-2,6), 128.9 (C-4), 149.7 (C-1), 228 (CO cis MoC), 239 (CO trans MoC) [258] ^{95}Mo NMR ($CHCl_3$): −1566 [258]; spin-lattice relaxation time and spin-spin relaxation time were determined as T_1 = (5.64 ± 0.07) ms and T_2 = 5.49 ms; for details, see No. 56 [304] IR (KBr): 1899, 1930, 2002 (ν(CO)) [74]
82	$-CH_2C_6H_3(C_3H_7\text{-}i)_2$-3,5	I (64%, with $ClCH_2C_6H_3(C_3H_7\text{-}i)_2$-3,5) [66] bright yellow solid (from CH_2Cl_2/petroleum ether), m.p. 91 to 92°C (slight dec.) [66] ^{1}H NMR ($CDCl_3$): 1.21 (d, CH_3), 2.81 (sept, CH), 2.93 (s, CH_2), 5.15 (s, C_5H_5), 6.72 (m, H-4), 6.90 (d, H-2,6) [66] IR (cyclohexane): 1930 (2 components), 2010 (ν(CO)) [66]

Table 1 (continued)

No.	^1L	method of preparation (yield) properties and remarks
		pyrolysis under high vacuum at 110 °C resulted in smooth decarbonylation to yield $C_5H_5Mo(CO)_2$-(η^3-$CH_2C_6H_4(C_3H_7$-i$)_2$-3,5) (see Formula XXX on p. 95) as the analogous reaction product of the unsubstituted benzyl complex No. 56 [66]
83	-$CH_2C_6H_3(OCH_3)_2$-2,6	no details given [252]
84	-$CH_2C_6H_3Cl_2$-2,6	no details given [252]
85	-$CH_2C_6(CH_3)_5Fe^+(C_5H_5)$	from $C_5H_5Mo(CO)_3I$ in THF and two equivalents of the methylenecyclohexadienyl complex (CH_2=$C_6(CH_3)_5)FeC_5H_5$ dissolved in benzene at −20 °C; the precipitated iodide was purified by TLC (59%); preparation at 20 °C gave significant quantities of $(C_5H_5Mo(CO)_3)_2$ and a decreased yield of 30% [398] the iodide is soluble in H_2O, acetone, and CH_3CN and can be metathesized to the PF_6 salt with aqueous HPF_6 which was used for characterization [398] crystalline solid (from C_2H_5OH) [398] ^1H NMR (CD_3CN): 2.37, 2.39, 2.42 (all s, CH_3), 2.65 (s, CH_2), 4.39 (s, C_5H_5Fe), 5.53 (s, C_5H_5Mo) [398] ^{13}C NMR (CD_2Cl_2): −6.1 (CH_2), 17.8 (m, CH_3), 78.2 (C_5H_5Fe), 93.8 (C_5H_5Mo), 94.8, 96.7, 98.1 (C_6 ring), 229.6 (CO) [398] IR (Nujol): 1930, 1945, 2050 (ν(CO)) [398]
86	-$CH(CH_3)C_6H_5$	the kinetics of thermolysis was studied in benzene in the absence and presence of free radical traps; activation parameters for the homolytic bond dissociation into $C_5H_5Mo(CO)_3^{\bullet}$ and $C_6H_5C(CH_3)H^{\bullet}$ were determined as $\Delta H^{\ddagger} = 26.6$ kcal/mol and $\Delta S^{\ddagger} = 15$ cal · mol^{-1} · K^{-1}, and the corresponding bond dissociation energy was deduced to be 25 kcal/mol [386]; no further details given
87	-$CH(C_6H_5)_2$	I (with $ClCH(C_6H_5)_2$ for several hours) [116] no further details given [116]
88	-$C(C_6H_5)_3$	I (with $ClC(C_6H_5)_3$ for several hours) [116] no further details given [116]
89	—CH_2—(1-naphthyl)	I (52%, with $ClCH_2C_{10}H_7$ for 6 h) [116], (62%) [322] yellow needle-shaped crystals (from hexane at −78 °C) [116, 322], m.p. 92 to 94 °C (dec.) [116], 125 to 127 °C (dec.) [322] ^1H NMR (CS_2): 3.30 (s, CH_2), 5.11 (s, C_5H_5), 7.02 to 7.92 (m, $C_{10}H_7$) [116], ($CDCl_3$): 3.39 (s, CH_2), 5.15 (s, C_5H_5), 7.25 to 7.95 (m, $C_{10}H_7$) [322]

Table 1 (continued)

No.	^1L	method of preparation (yield) properties and remarks
89 (continued)		IR (CS_2): 1933, 2016 [116], (KBr): 1930, 2005 (ν(CO)) [322] mass spectrum (> 100 °C): $[C_5H_5Mo(CO)_2CH_2C_{10}H_7]^+$; this ion either loses its CO ligands to generate $[C_5H_5MoCH_2C_{10}H_7]^+$ or undergoes cleavage at the metal-carbon σ-bond to give $[C_5H_5Mo(CO)_2]^+$; further fragmentation resulted in $[C_5H_5Mo]^+$, $[C_3H_3Mo]^+$, and Mo^+ [116]; see also [322] air-stable but somewhat light-sensitive as a solid, in solution precipitation of a bluish decomposition product occurred [116, 322] the complex is diamagnetic and its molar magnetic susceptibility was determined at 22 °C as $\chi_{mol} = -204.6$ cm³/mol [116] heating in a sublimer at 110 to 120 °C or UV irradiation resulted in $C_5H_5Mo(CO)_2(\eta^3\text{-}CH_2C_{10}H_7)$ along with Mo-Mo bonded dimers [322] differential thermal analysis revealed endothermic melting and decarbonylation steps occurring between 90 and 100 °C as well as exothermic molecular rearrangement processes in the range of 105 to 115 °C [116]
*90	$\text{-}(CH_2C_6H_4[\text{-}CHCH_2\text{-}])_n$ (linear)	I (by stirring for 36 h with chloromethylated polystyrene with the exclusion of light, precipitated from the reaction mixture by hexane) [136] orange solid [136] ^1H NMR (CS_2): 1.0 to 2.1 (br m, chain protons), 4.94 (s, C_5H_5), 6.45 (H-2), 6.95 (br s, H-3,4) [136] IR (Nujol): 1940, 2020 (ν(CO)) [136]
*91	$\text{-}(CH_2C_6H_4[\text{-}CHCH_2\text{-}])_n$ (cross-linked with 2% divinylbenzene)	I (by refluxing for 48 h with chloromethylated styrene-divinylbenzene with the exclusion of light) [132] yellow solid [136] IR (KBr): 1930, 2020 (ν(CO)) [136]; see also [132]
*92	$\text{-}CH_2C_5H_4N\text{-}3$	I (70%, with 3-chloromethylpyridine) [81] yellow crystals (from CH_2Cl_2/hexane) [81] ^1H NMR ($CDCl_3$): 2.76 (CH_2), 5.30 (C_5H_5), 7.0 to 9.0 (C_5H_4N) [81] UV (alkaline CH_3OH, log ε): 303 (4.1) [81]
*93	$\text{-}CH_2C_5H_4N\text{-}4$	I (50%, with 4-chloromethylpyridine) [81] yellow crystals (from CH_2Cl_2/hexane) [81]

Table 1 (continued)

No.	^1L	method of preparation (yield) properties and remarks
		^1H NMR ($CDCl_3$): 2.61 (CH_2), 5.14 (C_5H_5), 6.96 to 8.22 (C_5H_4N) [81] UV (alkaline CH_3OH, log ε), 298 (4.1) [81]
*94	$-CH_2C_5H_4NH\text{-}3^+$	see "Further information" at the end of the table
*95	$-CH_2C_5H_4NH\text{-}4^+$	see "Further information" at the end of the table
96	$-CH_2C_5H_4NCH_3\text{-}4^+$	prepared from $C_5H_5Mo(CO)_3CH_2C_5H_4N$-4 (No. 93) by treatment with an excess of 4-$CH_3C_6H_4SO_2OCH_3$ in CH_2Cl_2 at 0°C, extraction of the mixture with H_2O and subsequent addition of $Na[B(C_6H_5)_4]$ gave the $[B(C_6H_5)_4]$ salt (75%) [81] yellow solid (from aqueous acetone) [81] UV (0.1 M aqueous H_2SO_4): 322 [81]
97	—CH_2–(2-thienyl, C_4H_3S)	I (85%, with 2-chloromethylthiophene for 25 h) [69] yellow crystals (from hexane at −15°C), m.p. 63 to 64°C [69] ^1H NMR ($CDCl_3$): 3.12 (s, CH_2), 5.23 (s, C_5H_5), 6.77, 6.83, 6.90 (C_4H_3S) [69] IR (cyclohexane): 1940, 1944, 2027 (ν(CO)); (KBr): 3040, 3105 (ν(CH)); other bands at 635, 676, 798, 815, 838, 1000, 1020, 1056, 1082, 1236, 1350, 1418 [69] mass spectrum: $[M]^+$, $[C_5H_5Mo(CO)_nCH_2C_4H_3S]^+$ (n = 0 to 2), $[C_5H_5Mo(CO)_n]^+$ (n = 0 to 3); for further fragmentation, see [69] UV irradiation in hexane resulted in $C_5H_5Mo(CO)_2(\eta^3\text{-}CH_2C_4H_3S\text{-}2)$ (structure analogous to Formula XXX on p. 95) [69]
98	—CH_2–(3-thienyl, C_4H_3S)	I (75%, with 3-chloromethylthiophene for 25 h) [69] yellow crystals (from hexane at −15°C), m.p. 85 to 86°C [69] ^1H NMR ($CDCl_3$): 2.88 (s, CH_2), 5.10 (s, C_5H_5), 6.78, 7.0 to 7.1 (C_4H_3S) [69] IR (cyclohexane): 1935, 1945, 2025 (ν(CO)), (KBr): 2990, 3114 (ν(CH)); other bands at 652, 680, 695, 790, 830, 837, 870, 877, 900, 945, 1017, 1030, 1076, 1098, 1172, 1256, 1395, 1440 [69] mass spectrum: $[M]^+$, $[C_5H_5Mo(CO)_nCH_2C_4H_3S]^+$ (n = 0 to 2), $[C_5H_5Mo(CO)_n]^+$ (n = 0 to 3); for further fragmentation, see [69] UV irradiation in hexane resulted in $C_5H_5Mo(CO)_2(\eta^3\text{-}CH_2C_4H_3S\text{-}3)$ (structure analogous to Formula XXX on p. 95) [69]

References on pp. 106/18

Table 1 (continued)

No.	^1L	method of preparation (yield) properties and remarks
99	$-CH_2C_5H_4Mn(CO)_3$	I (21%, by stirring with $ClCH_2C_5H_4Mn(CO)_3$ at 20 °C for 1 h, and subsequent refluxing for the same period, the mixture left overnight) [174] yellow-green crystals, m.p. 127 to 128 °C [174] ^1H NMR (CCl_4): 2.38 (s, CH_2), 4.65 ("s", C_5H_4), 5.40 (s, C_5H_5) [174] IR (hexane): 1940, 1948, 2017, 2024, 2029 (ν(CO)); from the local symmetry only four bands are expected in the carbonyl region, the additional band is believed to be due to the presence of geometric isomers according to the rotation about the Mo-C bond [174] in the mass spectrum three intense ions, $[C_5H_5Mo(CO)_3Mn]^+$, $[C_6H_6Mn(CO)_3]^+$, and $[C_5H_5MoC_6H_6]^+$ were formed which proved preferential loss of the first three CO groups from Mn and rupture of the Mo-CH_2 bond as the main fragmentation pathways; furthermore, complete decarbonylation and removal of Mn resulted in the latter of the three fragments; for more details, see [219] reaction with $HgCl_2$ in refluxing acetone yielded $C_5H_5Mo(CO)_3HgCl$ [174] in contrast to the benzyl analogue (No. 56), no migration of the σ-bonded ligand into the C_5H_5 ring of the $C_5H_5Mo(CO)_3$ moiety and formation of the corresponding binuclear derivatives was observed in refluxing xylene; moreover, no reaction occurred with gaseous HCl in heptane/benzene at 80 °C [174]
100	$-CH_2CN$	I (52%) [74], (70%, with $ClCH_2CN$ for several hours) [100] also formed as a side product in the synthesis of $C_5H_5Mo(CO)_3C_6H_4NO_2$-4 from p-$O_2NC_6H_4I$ in the presence of $[C_5H_5Mo(CO)_3]^-$ in CH_3CN by electrolysis [336] yellow solid [100], m.p. 109 °C (dec.) [100], 140 to 145 °C (dec.) [74] ^1H NMR ($CDCl_3$): 1.42 (s, CH_2), 5.50 (s, C_5H_5) [74]; see also [100] IR (KBr): 1924, 2010 (ν(CO)), 2190 (ν(CN)) [74]; (cyclohexane): 1952, 1963, 2046 (ν(CO)), 2211 (ν(CN)) [100] no reaction occurred with SO_2 [100] or with cyclo-$C_6H_{11}NC$ [74]

References on pp. 106/18

Table 1 (continued)

No.	^1L	method of preparation (yield) properties and remarks
101	$-CH_2CH{=}NC_3H_7$-i	prepared by condensation of $C_5H_5Mo(CO)_3CH_2CHO$ (No. 103) with an equimolar amount of i-$C_3H_7NH_2$ in the presence of 1 equivalent of $BF_3 \cdot O(C_2H_5)_2$ in THF at −78 °C during 6 h (75%); the crude product was chromatographed through a silica column using ether/hexane as eluant to remove $(C_5H_5Mo(CO)_3)_2$ [372] yellow solid [372] ^1H NMR ($CDCl_3$): 1.28 (d, CH_3; J = 6.6), 2.37 (d, CH_2; J = 8.6), 3.82 (m, =NCH), 5.56 (s, C_5H_5), 7.89 (t, CH=) [372] ^{13}C NMR ($CDCl_3$): 1.61 (CH_2), 21.79 (CH_3), 53.8 (=NCH), 94.0 (C_5H_5), 178 (C=N), 227.0, 236 (both CO) [372] IR (CH_2Cl_2): 1615 (ν(C=N)), 1950, 2048 (ν(CO)) [372] mass spectrum: $[M]^+$ [372] decarbonylation, analogously to No. 102, gave the π-azaallyl complex $C_5H_5Mo(CO)_2(\eta^3$-$CH_2CHNC_3H_7$-i) [372]
102	$-CH_2C(CH_3){=}NC_3H_7$-i	analogously to No. 101 with $C_5H_5Mo(CO)_3CH_2C(O)CH_3$ (No. 104) as starting material (70%) [372] yellow solid [372] ^1H NMR ($CDCl_3$): 1.28 (d, $NCCH_3$; J = 6.6), 2.08 (s, $C(CH_3)$=N), 2.09 (s, CH_2), 3.40 (m, =NCH), 5.35 (s, C_5H_5) [372] ^{13}C NMR ($CDCl_3$): 3.91 (CH_2), 21.5, 22.5 (both CH_3), 49.4 (=NCH), 94.5 (C_5H_5), 197.0 (C=N), 228.7, 229.7 (both CO) [372] IR (CH_2Cl_2): 1620 (ν(C=N)), 1952, 2044 (ν(CO)) [372] mass spectrum: $[M]^+$ [372] decarbonylation occurred on treatment with 1 equivalent of $(CH_3)_3NO$ in CH_2Cl_2 to give the π-azaallyl complex $C_5H_5Mo(CO)_2(\eta^3$-$CH_2C(CH_3)NC_3H_7$-i) [372]
*103	$-CH_2CHO$	I (71%, with $ClCH_2CHO$ at −80 °C; after reaching 25 °C, the reaction mixture was stirred for 3 h) [152] brown-yellow solid, m.p. 175 °C (dec.) [152] ^1H NMR (CS_2): 1.80 (d, CH_2), 5.20 (s, C_5H_5), 9.35 (t, C(O)H) [152] IR (neat): 1648 (ν(C=O)), 1900, 1945, 2020 (ν(CO)) [152]

References on pp. 106/18

Table 1 (continued)

No.	^{1}L	method of preparation (yield) properties and remarks
*104	$-CH_2C(O)CH_3$	I (with $ClCH_2C(O)CH_3$ at −40 °C, after reaching 25 °C the reaction mixture was stirred for 3 h; 67% [152], 29% [331]) brown wax [152]; reddish crystalline solid, m.p. 28 to 29 °C [331] ^{1}H NMR (benzene-d_6): 1.98 (s, CH_2), 2.03 (s, CH_3), 4.62 (s, C_5H_5) [331]; (CS_2): 0.70 (s, CH_2), 2.00 (s, CH_3), 5.36 (s, C_5H_5) [152] ^{13}C NMR (benzene-d_6): 8.54 (CH_2), 30.13 (CH_3), 94.02 (C_5H_5), 211.91 (acyl-CO), 228.10 (CO cis MoC), 239.74 (CO trans MoC) [331] IR (THF): 1658 (ν(C=O)), 1933, 1978, 2026 (ν(CO)) [331]; see also [151]
105	$-CH_2C(O)C_6H_5$	I (ca. 30%, with $ClCH_2C(O)C_6H_5$ at −30 °C for 20 min and 25 °C for 1 h) [50] orange-yellow air-sensitive crystals, m.p. 85 to 86 °C, soluble in common organic solvents [50] ^{1}H NMR ($CDCl_3$): 2.63 (s, CH_2), 5.35 (s, C_5H_5), 7.43 (m, C_6H_5, H-3,4,5), 7.82 (m, C_6H_5, H-2,6) [50] IR (KBr): 1615 (ν(C=O)), 1918, 1948, 2021 (ν(CO)), 2941, 2998, 3107 (ν(CH)); additional bands at 711, 798, 825, 848, 929, 940, 1005, 1015, 1029, 1079, 1119, 1175, 1268, 1295, 1310, 1418, 1428, 1444, 1571, 1591, and a shoulder at 1895 [50] reaction with $[C_5H_5Fe(CO)_2(OC_4H_8)]BF_4$ in CH_2Cl_2 yielded $[C_5H_5Mo(CO)_3CH_2C(OFe(CO)_2C_5H_5)C_6H_5]^+$ during 1 h; the latter apparently rearranged to $[C_5H_5(CO)_2FeCH_2C(OMo(CO)_3C_5H_5)C_6H_5]^+$ by migration of the $C_5H_5Mo(CO)_3$ moiety from the methylene carbon to the benzoyl oxygen atom; further treatment of these adducts with $P(C_6H_5)_3$ results in $C_5H_5Mo(CO)_3CH_2C(O)C_6H_5$ and $C_5H_5Fe(CO)_2CH_2C(O)C_6H_5$, the ratio of which changes from 6.4:1 after 1 h to 1:2.5 after 72 h [349]
106	$-CH_2C(O)C_5H_4Mn(CO)_3$	I (25%, with $ClCH_2C(O)C_5H_4Mn(CO)_3$ at −70 °C) [192] yellow crystalline solid, m.p. 130 to 131 °C (dec.), soluble in benzene, ether, $CHCl_3$, acetone, and THF [192] ^{1}H NMR ($CHCl_3$, with hexamethyldisiloxane as an internal standard): 2.15 (s, CH_2), 4.73, 5.30 (both t, C_5H_4), 5.40 (s, C_5H_5) [192]

References on pp. 106/18

Table 1 (continued)

No.	^1L	method of preparation (yield) properties and remarks
		IR (paraffin oil): 1628 (ν(C=O)); (CCl_4): 1950, 1955, 2035 (ν(CO)); the small number of bands may be due to superposition [192] mass spectrum: $[M-nCO]^+$ (n = 1, 2, 4 to 7), $[M-6CO-COCH_2]^+$, $[C_5H_5Mo(CO)_n]^+$ (n = 3 to 0), $[CH_3C(O)C_5H_4Mn(CO)_3]^+$; for further fragmentation, see [192]; see also [219] unstable in solution, but also decomposition occurred in the solid state under argon; less stable than the corresponding iron compound No. 107 [192]
107	$-CH_2C(O)C_5H_4FeC_5H_5$	I (10%, with $ClCH_2C(O)C_5H_4Fe(CO)_3$ in ether) [192, 212], for an early report see [161] regenerated as by-product in about 20% from the reaction of $[C_5H_5Mo(CO)_3CH_2C(OC_2H_5)C_5H_4$-$FeC_5H_5]BF_4$ (No. 108) with $CH_2(CN)_2$ in CH_2Cl_2 in the presence of $N(C_2H_5)_3$ [225] orange crystals, m.p. 76 to 78°C (dec.), soluble in benzene, ether, $CHCl_3$, acetone, and THF [192] ^1H NMR ($CHCl_3$, with hexamethyldisiloxane as an internal standard): 2.40 (s, CH_2), 4.12 (s, C_5H_5Fe), 4.40, 4.70 (both t, C_5H_4), 5.33 (s, C_5H_5Mo) [192] IR (paraffin oil): 1610 (ν(C=O)); (CCl_4): 1940, 2028 (ν(CO)); compare No. 106 [192] only ions corresponding to decomposition products were observerd in the mass spectrum [192] unstable in solution, but also as a solid under argon the complex decomposed by one third in a day [192] reaction with I_2/CH_2Cl_2 or $CH_3C(O)Cl$ led to cleavage of the Mo-C bond to give $C_5H_5Mo(CO)_3Cl$ along with $C_5H_5FeC_5H_4C(=CH_2)OC(O)CH_3$ [212]; treatment with $[O(C_2H_5)_3]BF_4$ in CH_2Cl_2 resulted in the formation of $[C_5H_5Mo(CO)_3CH_2C(OC_2H_5)$-$C_5H_4FeC_5H_5]BF_4$ (No. 108) [212, 225]
108	$-CH_2C^+(OC_2H_5)C_5H_4FeC_5H_5$	prepared from $C_5H_5Mo(CO)_3CH_2COC_5H_4FeC_5H_5$ (No. 107) and 2 equivalents of $[O(C_2H_5)_3]BF_4$ in CH_2Cl_2 for 2 h [212, 225] the complex, formed as the BF_4^- salt, is extremely unstable in air and was only characterized in solution [212, 225] ^1H NMR (acetone-d_6): 4.50 (s, C_5H_5Fe), 5.00, 5.14 (both m, C_5H_4), 6.08 (s, C_5H_5Mo);

References on pp. 106/18

Table 1 (continued)

No.	^1L	method of preparation (yield) properties and remarks
108 (continued)		in addition signals of decomposition products were observed [225] IR (film): 1505 (ν(C-O)), 1970, 2050 (ν(CO)) [225] treatment with $CH_2(CN)_2$ in CH_2Cl_2 in the presence of $N(C_2H_5)_3$ resulted mainly in $C_5H_5Mo(CO)_3CH_2C(C_5H_4FeC_5H_5)$=$C(CN)_2$ (No. 30), along with $C_5H_5Mo(CO)_3CH_2C(O)C_5H_4FeC_5H_5$ (No. 107) and $C_5H_5FeC_5H_4C(CH_3)$=$C(CN)_2$ [212, 225]
109	$-CH_2C(O)NH_2$	I (<40%, with $ClCH_2C(O)NH_2$ at −78 °C and subsequent warming to 25 °C for 1.5 h) [37, 64] yellow crystals, m.p. 129 °C [64] ^1H NMR (acetone): 1.78 (s, CH_2), 5.38 (s, C_5H_5), 5.89 (br, NH_2) [64] IR (mull): 1652 (ν(C=O)), 1909, 1938, 2026 (ν(CO)), 2950, 2988, 3110 (ν(CH)), 3142, 3252, 3306, 3482 (ν(NH)); additional bands at 832, 840, 902, 1002, 1007, 1059, 1147, 1350, 1355, 1416, 1426, 1428, 1606 [64] soluble in $CHCl_3$, acetone, or aqueous acids, moderately soluble in H_2O, and moderately to slightly soluble in ether [64] fairly stable to oxidation by O_2, and even its solutions in organic solvents decomposed only slowly [64] treatment with aqueous HCl at 80 °C yielded $C_5H_5Mo(CO)_3CH_2CO_2H$ (No. 110) [37, 64]
*110	$-CH_2CO_2H$	I (68%, with $ClCH_2CO_2Si(CH_3)_3$ at −78 °C for 1 h and 3 h at 25 °C, the resulting residue was dissolved in ether and treated with silica gel to hydrolyze the primary formed silyl ester) [341] also from hydrolysis of $C_5H_5Mo(CO)_3CH_2C(O)NH_2$ (No. 109) in a boiling 2:1 mixture of H_2O and concentrated HCl for 10 min; the complex precipitated from the reaction mixture on cooling (80%) [37, 64] yellow crystals (from $CHCl_3$), m.p. 164 °C [37, 64] ^1H NMR ($CDCl_3$): 1.75 (s, CH_2), 5.14 (s, C_5H_5); the absence of the OH resonance is due to a rapid proton exchange [64] IR (mull): 1283 (ν(C-O)), 1650 (ν(C=O)), 1885, 1955, 2026 (ν(CO)), 2539, 2641, 2693 (ν(OH)), 2794, 2959, 3122 (ν(CH)); additional bands at 826, 833, 846, 926, 936, 1004, 1011, 1045, 1060, 1106, 1400, 1418, 1429, the band of the carboxyl group was found at a lower frequency than for other aliphatic acids which may be due to

Table 1 (continued)

No.	^1L	method of preparation (yield) properties and remarks
		hydrogen bonding or other interactions of the CO_2H group; the position of $\nu(OH)$ is consistent with considerable hydrogen bonding; furthermore, $\nu(OH)$ is solvent sensitive [64] in contrast to an earlier publication [37], the complex is indicated to be an unusually weak acid by its pK_a value of 8.26 ± 0.03 in 50% aqueous dioxane; accordingly no CO_2 was liberated from $NaHCO_3$ in aqueous alcoholic solution [64] soluble in acetone, $CHCl_3$, slightly in ether, almost insoluble in H_2O, but does dissolve in dilute alkali in which it will decompose within minutes unless recovering was accomplished by rapid neutralization; the compound is only slowly oxidized by O_2 [64] treatment with excess CH_2N_2 in ether resulted in $C_5H_5Mo(CO)_3CH_2CO_2CH_3$ (No. 111) [37, 64]; the same methylester is obtained from the reaction with $(C(O)Cl_2)_2$ in aprotic solvents like CH_2Cl_2, THF, or ether which gave $C_5H_5Mo(CO)_3CH_2C(O)Cl$ as an intermediate; further reaction with CH_3OH in the presence of $N(C_2H_5)_3$ resulted in No. 111 [341] treatment with $(C_6H_6Mo(\eta^3\text{-}CH_2CHCH_2))_2(\mu\text{-}Cl)_2$ in aqueous KOH at elevated temperatures yielded $C_5H_5Mo(CO)_3CH_2CO_2Mo(C_6H_6)(\eta^3\text{-}CH_2CHCH_2)$ (see Formula VI on p. 74) [122]
111	$-CH_2CO_2CH_3$	I (37%, with $ClCH_2CO_2CH_3$ for several hours) [26] from $C_5H_5Mo(CO)_3CH_2CO_2H$ (No. 110) and excess CH_2N_2 in ether at 25 °C for 40 min; chromatography on an alumina column using ether as eluant gave an 80% yield [37, 64] from $C_5H_5Mo(CO)_3CH_2CO_2H$ (No. 110) which was reacted with excess $(C(O)Cl_2)_2$ in THF at 25 °C to give $C_5H_5Mo(CO)_3CH_2C(O)Cl$; further treatment with CH_3OH at −78 °C in the presence of $N(C_2H_5)_3$ gave No. 111 after stirring at 25 °C for 1 h in 75% yield [341] yellow crystalline solid (from light petroleum at low temperatures [26, 64] or ether/hexane [341]), m.p. 47 to 49 °C [26], 70 °C (dec.) [64], sublimes at 100 °C/0.1 Torr [26] ^1H NMR (CS_2): 1.70 (s, CH_2), 3.48 (s, CH_3), 5.35 (s, C_5H_5) [26] IR (mull): 1669 (ν(C=O)), 1948, 2046 (ν(CO)), 2980, 3028, 3120, 3131 (ν(CH)); other bands at 721, 826, 836, 841, 861, 946, 1009, 1015, 1027, 1055,

References on pp. 106/18

Table 1 (continued)

No.	^1L	method of preparation (yield) properties and remarks
111 (continued)		1065, 1102, 1249, 1411, 1420, 1430, 2029 [64]; (cyclohexane): 1955 (A″, ν(CO)), 1961 (A′, ν(CO)), 2027 (A′, ν(CO)); k_1(CO diagonal to CH_3) = 15.67, k_2(CO diagonal to another) = 15.94, k_{id} = 0.51; for calculations the ratio of the diagonal interaction force constant (k_{id}) and the lateral interaction force constant (k_{is}) was assumed to be constant k_{id}/k_{is} = 2 [70]; for KBr as medium, see [26] slightly soluble in ether and $CHCl_3$, insoluble in water [64] reaction with methanolic KOH gave $[C_5H_5Mo(CO)_3]^-$ which was isolated by addition of $[N(C_2H_5)_4]Br$ [26] reaction with $(CH_3)_2PC_2H_4P(CH_3)_2$ in CH_3CN gave a yellow precipitate, indicated by IR to be a $C_5H_5Mo(CO)_2(CH_2CO_2CH_3)((CH_3)_2PC_2H_4P(CH_3)_2)$ derivative [67] polarographic reduction (Hg electrode; millimolar 1,2-dimethoxyethane solutions containing 10^{-1} mol/L $[N(C_4H_9\text{-}n)_4]ClO_4$) resulted in $[C_5H_5Mo(CO)_3]^-$ and $CH_2CO_2CH_3^{\bullet}$, the fate of the latter being unknown; half-wave potential $E_{p/2}$ = -2.0 ± 0.1 V vs. Ag/10^{-3} M $AgClO_4$ [29, 32]
*112	$-CH_2CO_2C_2H_5$	I (73%, with $ClCH_2CO_2C_2H_5$ for 4 h) [331]; IIa (3%, for 16 h) [26] yellow-orange crystals (from pentane at −78°C) [26], m.p. 32.5 to 34°C [26, 331] ^1H NMR (benzene-d_6): 1.09 (t, CH_3; J = 7), 1.84 (s, $MoCH_2$), 4.02 (q, OCH_2), 4.68 (s, C_5H_5) [331]; (CS_2): 1.18 (t, CH_3; J = 7), 1.69 (s, $MoCH_2$), 3.93 (q, OCH_2), 5.37 (s, C_5H_5) [26] ^{13}C NMR (benzene-d_6): −3.94 ($MoCH_2$), 14.73 (CH_3), 59.32 (OCH_2), 93.51 (C_5H_5), 181.02 (acyl CO), 227.04 (CO cis MoC), 240.63 (CO trans MoC) [331] IR (KBr): 1708 (ν(C=O)), 2875, 2940, 3060 (ν(CH)); other bands at 755, 820, 847, 922, 949, 1003, 1010, 1035, 1077, 1088, 1235, 1293, 1357, 1380, 1416, 1422, 1435, 1460, 1475 [26] IR (cyclohexane): 1954 (A″, ν(CO)), 1960 (A′, ν(CO)), 2027 (A′, ν(CO)); k_1(CO diagonal to CH_3) = 15.66, k_2(CO diagonal to another) = 15.93, k_{id} = 0.52; for calculations the ratio of the diagonal interaction force constant (k_{id}) and the lateral interaction force constant (k_{is}) was

References on pp. 106/18

Table 1 (continued)

No.	^1L	method of preparation (yield) properties and remarks
		assumed to be constant k_{id}/k_{is} = 2 [70]; for KBr as medium, see also [26]; for THF, see [331]
113	$-CH_2Si(CH_3)_3$	I (3 to 5%, with $ICH_2Si(CH_3)_3$ for 5.5 h along with $C_5H_5Mo(CO)_3CH_3$ (No. 1) as the principle product) [76, 80] yellow liquid [76] proved as a precursor of $C_5H_5Mo(CO)_3CH_3$ (No. 1) which was formed from No. 113 and $[C_5H_5Mo(CO)_3]^-$ under remarkably mild conditions; this behavior is in accord with the strong electron-releasing quality of the $C_5H_5Mo(CO)_3CH_2$ group [80]
114	$-CH_2N(CH_3)_2$	I (53%, with $[(CH_3)_2N{=}CH_2]I$ overnight [143]); use of the respective $[(CH_3)_2N{=}CH_2]Cl$ reagent yielded only in the redox product $(C_5H_5Mo(CO)_3)_2$ [43, 143], while treatment of $Na[C_5H_5Mo(CO)_3]$ with $[(CH_3)_2N{=}CHCl]Cl$ produced the salt $[C_5H_5Mo(CO)_3({=}CH(N(CH_3)_2))][C_5H_5Mo(CO)_3]$ by NaCl elimination with concurrent C-Cl oxidative addition [142, 187] yellow crystals (from light petroleum at −70°C) [143] ^1H NMR ($CDCl_3$): 2.32 (CH_2), 2.36 ($N(CH_3)_2$), 5.64 (C_5H_5) [143] IR (cyclohexane): 1953, 1969, 2037 (ν(CO)) [143] refluxing in light petroleum produced the metallaheterocyclic derivative $C_5H_5Mo(CO)_2(\eta^2\text{-}CH_2N(CH_3)_2)$ [143]
115	$-CH_2N^+C_5H_5$	from $C_5H_5Mo(CO)_3CH_2Cl$ (No. 128) and pyridine in the presence of $TlBF_4$ in CH_2Cl_2 at 25°C for 2 to 4 d, isolation as BF_4 salt by addition of hexane to the filtrated concentrated reaction mixture (72%) [302] m.p. of the BF_4 salt: 137 to 138°C (dec.) [302] ^1H NMR (acetone-d_6): 5.26 (s, CH_2), 5.89 (s, C_5H_5), 8.05 (H-3,5), 8.42 (H-4; J = 8), 9.01 (d, H-2,6; J = 6) [302] IR (CH_2Cl_2): 1945, 2036 (ν(CO)) [302]
116	$-CH_2NCO$	I (18 to 42% with $ClCH_2NCO$ for 16 h; yield obviously depends on the preparation method of $Na[C_5H_5Mo(CO)_3]$) [34] yellow crystals (from pentane or pentane/ether at −78°C), soon darkening on standing, m.p. 54 to 56°C, sublimes at 100°C/0.2 Torr [34] ^1H NMR (CS_2): 3.69 (s, CH_2), 5.43 (s, C_5H_5) [34]

References on pp. 106/18

Table 1 (continued)

No.	^1L	method of preparation (yield) properties and remarks
116 (continued)		IR (cyclohexane): 1950, 1959, 2033 (ν(CO)); (KBr): 2270 (ν(NCO)), 2820, 2870, 3050 (ν(CH)), other bands at 812, 819, 836, 848, 1001, 1005, 1054, 1142, 1225, 1415, 1460 [34] mass spectrum: $[C_5H_5Mo(CO)_2CH_2NCO]^+$ undergoing three stepwise losses of CO followed by the loss of CH_2N to give $[C_5H_5Mo]^+$, $[C_3H_3Mo]^+$, and Mo^+; furthermore $[C_6H_6Mo]^+$ and $[C_5H_5Mo_2(CO)_n]^+$ (n = 0, 2 to 4) were observed; for more details, see [72] sublimation at 100°C or UV irradiation in hexane gave only a mixture of $(C_5H_5Mo(CO)_3)_2$ and unchanged starting material, but no dicarbonyl product; treatment with boiling CH_3OH resulted in $(C_5H_5Mo(CO)_3)_2$ as the only product [34]
117	—$CH_2N(CO)_2C_6H_4$ (phthalimidomethyl)	I (50 to 60%, with $ClCH_2N(CO)_2C_6H_4$ for 24 h) [167, 272] yellow solid [167, 272], m.p. 135 to 137°C (dec.) [167, 272] ^{1}H NMR ($CHCl_3$): 4.21 (s, CH_2), 5.66 (s, C_5H_5), 7.71 (C_6H_4) [167, 272] IR (THF): 1922, 2018 (ν(CO)) [167, 272]; (CH_3OH): 1706 (ν(C=O)) [167]
*118	-CH_2OCH_3	I (ca. 40%, with $ClCH_2OCH_3$ first at −60°C and subsequently at 25°C for 50 min) [39] pale yellow oil which can be distilled in vacuum without decomposition, but is oxidized in air, very soluble in common organic solvents [39] ^{1}H NMR (CS_2): 3.25 (s, CH_3), 4.59 (s, CH_2), 5.29 (s, C_5H_5) [39]; see also [301] IR (neat): 1054, 1158 (ν(COC)), 1933, 1950, 2029 (ν(CO)), 2808, 2878, 2931, 2975, 3114 (ν(CH)); other bands at 738, 812, 908, 1106, 1110, 1226, 1354, 1424, 1445, 1832 [39]; similar data for (ν(CO)) in $CDCl_3$ [301] and cyclohexane [267]
*119	-$CH_2O_2CCH_3$	I (91%, with $BrCH_2O_2CCH_3$ for 24 h; the resulting oil crystallized on cooling) [293] dark yellow crystals, air-sensitive, indefinitely stable at −20°C under an inert atmosphere [293] ^{1}H NMR (benzene-d_6): 1.73 (s, CH_3), 4.59 (s, C_5H_5), 5.40 (s, CH_2) [293] ^{13}C NMR (benzene-d_6): 20.93 (CH_3), 50.97 (CH_2), 92.61 (C_5H_5), 170.58 (OCO), 226.90 (CO cis MoC), 239.40 (CO trans MoC) [293]

Table 1 (continued)

No.	${}^1L$	method of preparation (yield) properties and remarks
		IR (Nujol): 1725 (ν(C=O)), 1945, 2010 (ν(CO)); additional bands at 630, 705, 810, 820, 830, 855, 935, 950, 995, 1045, 1095, 1165, 1250, 3120, 3940 [293] mass spectrum: $[M]^+$, $[M-nCO]^+$ (n = 0 to 3), $[C_5H_5MoCH_2O]^+$, $[C_5H_5MoCH_2]^+$, $[C_5H_5Mo]^+$, in addition, metastable peaks were observed [293]
*120	$-CH_2O_2CC_4H_9$-t	I (good yield, with $ClCH_2O_2CC_4H_9$-t overnight) [216] yellow plates (from hexane at −78 °C), m.p. 67 to 71 °C [216] ${}^1H$ NMR (no solvent given): 1.04 (s, CH_3), 5.24 (s, CH_2), 5.37 (s, C_5H_5) [216] ${}^{13}C$ NMR ($CDCl_3$): 27.5 (qsept, C($\mathbf{C}H_3)_3$; ${}^1J$(C, H) = 126.9, ${}^3J$(C, H) = 4.4), 39.2 (m, $\mathbf{C}(CH_3)_3$), 50.9 (t, CH_2; ${}^1J$(C, H) = 155.1), 92.7 (dqui, C_5H_5; ${}^1J$(C, H) = 177.8; ${}^2J$(C, H) = ${}^3J$(C, H) = 6.7) [216] IR (Nujol): 1709 (ν(C=O)), 1908, 1984, 2016 (ν(CO)) [216]
121	$-CH_2O_2CC_6H_5$	by treatment of $C_5H_5Mo(CO)_3CH_2O_2CCH_3$ (No. 119) with a twofold excess of $C_6H_5CO_2H$ in benzene-d_6 for several days at 25 °C (80%); only characterized in solution [293] ${}^1H$ NMR (benzene-d_6): 4.62 (s, C_5H_5), 5.39 (s, CH_2), 6.88 to 7.16 (m, 3H of C_6H_5), 8.105 (dd, 2H of C_6H_5) [293]
122	$-CH(CH_3)O_2CCH_3$	I (16%, with $BrCH(CH_3)O_2CCH_3$ for 24 h; for purification chromatography on alumina at 0 °C with ether/petroleum ether was used) [293] orange-red oil of high thermal lability even at ambient temperature [293] ${}^1H$ NMR (benzene-d_6): 1.69 (s, O_2CCH_3), 1.83 (d, $CHC\mathbf{H}_3$; J = 7), 4.68 (s, C_5H_5), 7.06 (q, CH); compared to No. 119, a downfield shift of the α-hydrogen of acetoxyalkyl group was observed [293] ${}^{13}C$ NMR (benzene-d_6): 21.28 (CH$\mathbf{C}H_3$), 30.23 ($O_2C\mathbf{C}H_3$), 64.34 (CH), 92.90 (C_5H_5), 169.35 (OCO), 227.05 (CO cis MoC), 240.03 (CO trans MoC) [293] IR (film): 1720 (ν(C=O)), 1850, 1885, 1940, 1995, 2020 (ν(CO)); additional bands at 775, 815, 930, 1010, 1075, 1100, 1190, 1240, 1370, 1425, 2870, 2920, 2950, 3120, 3940 [293] under dark, anaerobic conditions at 25 °C, decomposition occurred to give vinyl acetate and

References on pp. 106/18

Table 1 (continued)

No.	^1L	method of preparation (yield) properties and remarks
122 (continued)		$(C_5H_5Mo(CO)_3)_2$ involving a binuclear elimination of H_2 [293]
*123	$-CH_2-O\overset{\oplus}{\cdots}C(CH_3)\cdots Fe(CO)(P(C_6H_5)_3)(C_5H_5)$	by quick addition of 1 equivalent of $C_5H_5Fe(CO)(P(C_6H_5)_3)C(O)CH_3$ to a solution of $[C_5H_5Mo(CO)_3{=}CH_2]PF_6$ (prepared in situ from $C_5H_5Mo(CO)_3CH_2Cl$ (No. 128) and $AgPF_6$) in CH_2Cl_2 at −78°C; after warming to room temperature; the filtered and concentrated reaction mixture was added dropwise into a large volume of ether to precipitate the PF_6^- salt (68%) [301] yellow powder (from CH_2Cl_2/ether), soluble in CH_2Cl_2, insoluble in ether [301] ^1H NMR (acetone-d_6): 2.68 (s, CH_3), 4.82 (br s, CH_2), 5.01 (d, C_5H_5Fe; J(P, H) = 1.5), 5.56 (s, C_5H_5Mo), 7.51 (br, $P(C_6H_5)_3$) [301] ^{13}C NMR (CD_2Cl_2): 45.04 (CH_3), 72.48 (CH_2), 87.76 (C_5H_5Fe), 94.00 (C_5H_5Mo), 129 to 133 ($P(C_6H_5)_3$), 228.08 (CO cis MoC), 237.28 (CO trans MoC) [301] IR (CH_2Cl_2): 1955, 2036 (ν(MoCO)), 1979 (ν(FeCO)) [301]
*124	$-CH_2SCH_3$	I (40 to 50%, with $ClCH_2SCH_3$ for 12 to 24 h) [332], (ca 75%, impure) [24]; contrarily no complex was obtained using $ClCH_2CH_2SCH_3$ and $[C_5H_5Mo(CO)_3]^-$ as starting materials [24, 73] yellow-orange crystals (from pentane at −78°C) [18, 24, 332], m.p. 66 to 67°C [18, 24] ^1H NMR (CS_2): 2.18 (s, CH_3), 2.38 (s, CH_2), 5.38 (s, C_5H_5) [24]; similar in $CDCl_3$ [332] ^{13}C NMR (CH_2Cl_2): 92.9 (C_5H_5, d; J(C, H) ≅ 175) [20] IR (cyclohexane): 1948 (A″, ν(CO)), 1959 (A′, ν(CO)), 2023 (A′, ν(CO)); k_1(CO diagonal to CH_3) = 15.65, k_2(CO diagonal to another) = 15.85, k_{id} = 0.52; for calculations the ratio of the diagonal interaction force constant (k_{id}) and the lateral interaction force constant (k_{is}) was assumed to be constant k_{id}/k_{is} = 2 [70]; for hydrocarbon oil as medium, see [24] IR (KBr): 2850, 2875, 2925, 3100 (ν(CH)); other bands at 684, 693, 728, 825, 954, 1008, 1080, 1106, 1300, 1415 [24] UV (cyclohexane, ε): 221 (18500), 313 (4250) [24]
125	$-CH_2S^+(CH_3)_2$	as the BF_4^- salt from $C_5H_5Mo(CO)_3CH_2Cl$ (No. 128) and $(CH_3)_2S$ in the presence of $TlBF_4$ at room

References on pp. 106/18

Table 1 (continued)

No.	^1L	method of preparation (yield) properties and remarks
		temperature in CH_2Cl_2 for 2 to 4 d; the compound precipitated from the filtrated and concentrated solution by the addition of hexane (61%) [302] from No. 124 and $[O(CH_3)_3]BF_4$ in CH_2Cl_2 at 25 °C for ca. 4 h; attempts to purify the crude BF_4^- salt failed (see below) [332] m.p. of the light brown BF_4^- salt: 116 to 118 °C (dec.) [302] ^1H NMR (acetone-d_6): 2.88 (s, CH_2), 3.02 (s, CH_3), 5.86 (s, C_5H_5) [332]; see also [302] IR (CH_2Cl_2): 1955, 2040 (ν(CO)) [302] recrystallization from CH_3OH gave amber-colored crystals which decomposed rapidly during drying in vacuum [332] an attempt to use the salt for cyclopropanation of 1,1-diphenylethene failed; only 5% conversion of the alkene to 1,1-diphenylcyclopropene could be established; similar results occurred with cyclooctene [322]
126	$-CH_2S^+(CH_3)C_6H_5$	as the BF_4^- salt from $C_5H_5Mo(CO)_3CH_2Cl$ (No. 128) and $CH_3SC_6H_5$ analogously to No. 125 (61%) [302] m.p. 109 to 112 °C (dec.) [302] ^1H NMR (acetone-d_6): 3.26 (AB system, CH_2; J(A, B) = 11), 3.37 (s, CH_3), 5.80 (s, C_5H_5), 7.73 to 7.84 (m, H-3,4), 8.12 to 8.14 (m, H-2) [302] IR (CH_2Cl_2): 1968, 2042 (ν(CO)) [302]
127	2-tetrahydrothienyl (ring: C2(H)–C3(H,H)–C4(H,H)–C5(H^{endo},H^{exo})–S)	I (30%, with chlorotetrahydrothiophene/benzene for 1 h) [352] yellow oil [352] ^1H NMR (benzene-d_6): 1.35 (m, H^4-endo), 1.74 (m, H^3-endo), 1.90 (m, H^4-exo), 2.59 (m, H^3-exo), 2.68 (m, H^5-exo, H^5-endo), 3.47 (dd, H^2-exo; J = 4.8, 11.6), 4.76 (s, C_5H_5) [352] IR (hexane): 1935, 1951, 2033 (ν(CO)) [352] mass spectrum: $[M-nCO]^+$ (n = 1, 2), $[M-2CO-2H]^+$, [M−3CO−2H], $[C_4H_7S]^+$ [352]
*128	$-CH_2Cl$	I (60 to 70%, with ICH_2Cl for 5 h) [117]; see also [74] from $C_5H_5Mo(CO)_3CH_2OCH_3$ (No. 118) by treatment with gaseous HCl in light petroleum at −40 °C (ca. 80%) [39]; also by protolytic cleavage of the CH_2-O bond of $C_5H_5Mo(CO)_3CH_2O_2CCH_3$ (No. 119) with dry HCl (94%) [293] yellow crystalline material (from light petroleum at −40 °C) [39], orange crystals (from pentane at

References on pp. 106/18

Table 1 (continued)

No.	^1L	method of preparation (yield) properties and remarks
*128 (continued)		−78 °C) [117], m.p. 105 to 107 °C (dec.) [74], 136 to 137 °C [117], sublimes at 100 °/0.1 Torr [117] ^1H NMR (CS_2): 4.08 (s, CH_2), 5.42 (s, C_5H_5) [39]; (CCl_4): 4.22 (s, CH_2), 5.65 (s, C_5H_5) [117]; for $CDCl_3$, see also [74]; (acetone-d_6): 4.00 (s, CH_2), 4.56 (s, C_5H_5) [293] IR (mull): (1887 sh), 1924, 1949, (1966 vw), 2037 (ν(CO)), 2937, 3100, 3120 (ν(CH)), other bands at 722, 830 [39]; see also [293]; similar in KBr [74]; (cyclohexane): 1947, 1961, 2037 (ν(CO)) [117]
129	$-CH_2Br$	from $C_5H_5Mo(CO)_3CH_2OCH_3$ (No. 118) by treatment with HBr in light petroleum (ca. 40%) [39]; traces were also produced from $C_5H_5Mo(CO)_3C(O)CH_2Br$ on exposure of petroleum ether solutions to direct sunlight; the main product of this photochemical decomposition was $C_5H_5Mo(CO)_3Br$ [337] ^1H NMR (CS_2): 3.72 (s, CH_2), 5.15 (s, C_5H_5) [39] IR (mull): 1920, 1945, 2031, (2032 sh) (ν(CO)), 3110, 3129 (ν(CH)); other bands at 715, 814, 821, 830, 835, 846, 923, 1003, 1011, 1058, 1066, 1103, 1422, 1426 [39] less stable than its chloromethyl analogue No. 128; similar to the latter, it readily dissolves in ether and benzene [39] reacted with CH_3OH in the presence of NaOH to regenerate low yields of $C_5H_5Mo(CO)_3CH_2OCH_3$ (No. 118), but produced mainly $C_5H_5Mo(CO)_3Br$ [39]
130	$-CH_2I$	I (17%, with ICH_2I for 0.25 h) [117]; also from nucleophilic displacement of $C_5H_5Fe(CO)(P(C_6H_5)_3)C(O)CH_3$ from $[C_5H_5Mo(CO)_3CH_2OC(CH_3)Fe(CO)(P(C_6H_5)_3)$-$C_5H_5]PF_6$ (No. 123) using $[N(C_4H_9$-n$)_4]I$ in CH_2Cl_2 [301] yellow crystalline solid (from pentane at −78 °C) [117] ^1H NMR (CCl_4): 2.92 (s, CH_2), 5.64 (s, C_5H_5) [117] IR (pentane): 1938, 1957, 2040 (ν(CO)) [117] decomposed at 25 °C within minutes, especially in the presence of light [117]
131	$-CH_2CH_2CH{=}C(CH_3)_2$	I (27%, with $BrCH_2CH_2CH{=}C(CH_3)_2$ at 50 °C for 4 d) [106] m.p. 40 °C [106]

Table 1 (continued)

No.	^{1}L	method of preparation (yield) properties and remarks
		^{1}H NMR ($CDCl_3$): 1.58 (m, $MoCH_2$-), 1.61, 1.65 (both s, CH_3), 2.20 (m, CCH_2C), 5.10 (m, CH), 5.28 (s, C_5H_5) [106] IR (KBr): 1910, 2010 (ν(CO)); additional prominent band at 830 [106] refluxing in 1,2-dimethoxyethane resulted in decomposition to $(C_5H_5Mo(CO)_3)_2$ [106]
132	$-CH_2CH_2CH{=}CH_2(Fe^+(CO)_2C_5H_5)$	as PF_6^- salt from $C_5H_5Mo(CO)_3(CH_2)_4Fe(CO)_2C_5H_5$ (No. 156) and $[C(C_6H_5)_3]PF_6$ in CH_2Cl_2 at 25 °C for 48 h (26%); compare No. 18 [392] orange solid, unstable in solution and in the solid state [392] IR (CH_2Cl_2): 1922, 2017, 2038, 2073 (ν(CO)) [392]
133	$-CH_2CH_2CH(CO_2C_2H_5)_2$	IVa (68%) [370] yellow oil [370] ^{1}H NMR (benzene-d_6): 0.87 (m, CH_3), 1.51 (m, $MoCH_2$), 2.41 (m, CCH_2C), 3.45 (t, CH; J = 7), 4.00 (m, OCH_2), 4.61 (s, C_5H_5) [370] IR (THF): 1730 (ν(C=O)), 1900, 2000 (ν(CO)) [370]
134	$-CH_2CH_2CH(CO_2C_2H_5)C(O)CH_3$	IVa (57%) [370] yellow oil [370] ^{1}H NMR (benzene-d_6): 0.91 (t, CH_3 of ester group; J = 7), 1.40 (m, $MoCH_2$), 1.92 (s, $C(O)CH_3$), 2.20 (m, CCH_2C), 3.33 (t, CH; J = 7), 3.89 (q, OCH_2), 4.64 (s, C_5H_5) [370] IR (benzene): 1715, 1745 (ν(C=O)), 1924, 2040 (ν(CO)) [370]
135	$-CH_2CH_2CN$	IVb (33%, for 15 min) [148] yellow crystals (from benzene/hexane) [148] ^{1}H NMR (benzene-d_6): 1.3, 2.2 (both m, CH_2), 4.5 (C_5H_5) [148] IR (Nujol): 1900 to 2040 (ν(CO)), 2290 (ν(CN)) [148]
136	$-CH_2CH_2N^+(CH_3)_3$	IVb (52%, for 5 min, precipitated as PF_6^- salt from the concentrated reaction mixture by dilution with benzene); the isolated salt contains 0.2 equivalent of benzene [148] ^{1}H NMR (CD_3CN): 1.5 (m, $MoCH_2$), 2.92 (CH_3), 3.35 (m, CH_2N), 5.48 (C_5H_5) [148] IR (Nujol): 1940, 2040 (ν(CO)) [148] decomposition occurred above 80 °C and in aqueous acetonitrile; from the latter experiment $(C_5H_5Mo(CO)_3)_2$ was isolated [148]

References on pp. 106/18

Table 1 (continued)

No.	^1L	method of preparation (yield) properties and remarks
137	$-CH_2CH_2N^+C_5H_5$	IVb (50%, precipitated as the PF_6^- salt from the reaction mixture by dilution with H_2O) [148] yellow crystals [148] ^1H NMR (CD_3CN): 1.8 (m, $MoCH_2$), 4.65 (m, CH_2N), 5.5 (C_5H_5), 7.7 to 8.8 ($N^+C_5H_5$) [148] IR (Nujol): 1930, 2030 (ν(CO)) [148] no decomposition was observed in aqueous CH_3CN [148]
138	$-CH_2CH_2P^+(C_6H_5)_3$	IVb (80%, for 10 min, precipitated as the PF_6^- salt from the reaction mixture by dilution with ether) [148] yellow crystals [148] ^1H NMR (CD_3CN): 1.5 (m, $MoCH_2$), 3.3 (m, CH_2P), 5.6 (C_5H_5), 7.6 to 7.9 (C_6H_5) [148] IR (Nujol): 1940, 2030 (ν(CO)) [148]
139	$-CH_2CH_2OH$	formed on bubbling ethylene oxide for 2.5 h through a solution of $Na[C_5H_5Mo(CO)_3]$ in 1,2-dimethoxyethane, the intermediate $Na[C_5H_5Mo(CO)_2$-(η^2-CH_2CH_2OCO-cyclo)] · $CH_3OC_2H_4OCH_3$ was precipitated by dilution of the filtered reaction mixture with ether; protonation by stirring with H_2O led to precipitation of the β-hydroxyethyl derivative (14%) [148] IR (Nujol): 1940, 2020 (ν(CO)) [148] somewhat light-sensitive resulting in slow decomposition at 25 °C, but in the dark the complex is stable for weeks [148] converted to $[C_5H_5Mo(CO)_3(CH_2{=}CH_2)]BF_4$ by treatment with strong acids [148] treatment with NaH in 1,2-dimethoxyethane regenerated $[C_5H_5Mo(CO)_2(\eta^2$-CH_2CH_2OCO-cyclo)] · $CH_3OC_2H_4OCH_3$ [148]
140	$-CH_2CH_2OCH_3$	IVd (74%) [387] orange microcrystals, m.p. 37 °C [387] ^1H NMR (benzene-d_6): 1.69 (m, $MoCH_2$), 3.25 (s, OCH_3), 3.60 (m, CH_2O), 4.45 (s, C_5H_5) [387] ^{13}C NMR ($CDCl_3$): −0.7 ($MoCH_2$), 57.4 (OCH_3), 78.5 (CH_2O), 92.4 (C_5H_5), 228.4 (CO cis MoC), 240.2 (CO trans MoC) [387] IR (pentane): 1936, 2024 (ν(CO)) [387] condensing SO_2 onto a petroleum ether solution at −40 °C gave $C_5H_5Mo(CO)_3CH_2CH_2SO_2OCH_3$ (No. 144) [387]

Table 1 (continued)

No.	^1L	method of preparation (yield) properties and remarks
141	$-CH_2CH(CH_3)OCH_3$	IVd (66%) [387] orange oil [387] ^1H NMR ($CDCl_3$): 1.19 (d, CCH_3; ^3J = 6.0), 1.36 (dd, 1H, CH_2; $^2J_{gem}$ = 11.2, $^3J_{vic}$ = 5.1), 1.82 (dd, 1H, CH_2; $^3J_{vic}$ = 7.5), 3.27 (s, OCH_3), 3.40 (m, CH), 5.32 (s, C_5H_5) [387] ^{13}C NMR ($CDCl_3$): 6.5 (CH_2), 22.8 (**C**CH_3), 55.8 (OCH_3), 82.0 (CH), 92.6 (C_5H_5) [387] IR (pentane): 1936, 2020 (ν(CO)) [387] condensing SO_2 onto a petroleum ether solution at −40 °C gave $C_5H_5Mo(CO)_3CH_2CH(CH_3)SO_2OCH_3$ (No. 145) [387]
142	$-CH_2CH_2O(CH_2)_2W(CO)_3C_5H_5$	from $[C_5H_5W(CO)_3(CH_2{=}CH_2)]PF_6$ and a slight excess of $Na[C_5H_5Mo(CO)_2(\eta^2\text{-}CH_2CH_2OCO\text{-cyclo})] \cdot CH_3OC_2H_4OCH_3$ in acetone for 10 min at room temperature; precipitated from the filtered and concentrated reaction mixture along with $(C_5H_5Mo(CO)_3)_2$ as the lesser soluble fraction [148] cream-colored solid [148] ^1H NMR (CD_3CN): 1.96 (m, $MoCH_2$ and WCH_2), 3.85 (m, CH_2OCH_2), 5.32, 5.47 (both C_5H_5) [148] IR (Nujol): 1900, 2000 (ν(CO)) [148]
143	$-CH_2CH_2O(CH_2)_2Fe(CO)_2C_5H_5$	from $[C_5H_5Fe(CO)_2(CH_2{=}CH_2)]PF_6$ and equimolar amounts of $Na[C_5H_5Mo(CO)_2(\eta^2\text{-}CH_2CH_2OCO\text{-cyclo})] \cdot CH_3OC_2H_4OCH_3$ in CH_3CN at room temperature for 5 min; precipitated from the concentrated reaction mixture by dilution with hexamethyldisiloxane to furnish the title compound as analytically pure yellow crystals (53%, crude product; 22% after recrystallization) [148] yellow crystals (from CH_2Cl_2/hexamethyldisiloxane) [148] ^1H NMR (CD_3CN): 1.5, 1.9 (both m, $MoCH_2$ and $FeCH_2$), 3.5 (m, CH_2OCH_2), 4.88, 5.42 (both C_5H_5) [148] IR (Nujol): 1880, 1920, 2000 (ν(CO)) [148]
144	$-CH_2CH_2SO_2OCH_3$	obtained by condensing SO_2 onto a petroleum ether solution of $C_5H_5Mo(CO)_3CH_2CH_2OCH_3$ (No. 140) chilled to −40 °C; warming to 25 °C followed by removal of all volatiles in vacuum left the insertion product (90%) [387] ocher microcrystals, m.p. 91 (dec.) [387]

References on pp. 106/18

Table 1 (continued)

No.	^{1}L	method of preparation (yield) properties and remarks
144 (continued)		^1H NMR ($CDCl_3$): 1.55 (m, $MoCH_2$), 3.26 (s, OCH_3), 3.41 (m, CH_2S), 5.33 (s, C_5H_5) [387] ^{13}C NMR ($CDCl_3$): −12.4 ($MoCH_2$), 55.2 (OCH_3), 56.2 (CH_2S), 93.0 (C_5H_5), 227.6 (CO cis MoC), 237.6 (CO trans MoC) [387] IR (Nujol): 1164, 1344 (ν(SO)), 1928, 2012 (ν(CO)) [387]
145	$-CH_2CH(CH_3)SO_2OCH_3$	from SO_2 and $C_5H_5Mo(CO)_3CH_2CH(CH_3)OCH_3$ (No. 141) (83%); for details, compare No. 144 [387] orange oil [387] ^1H NMR ($CDCl_3$): 1.25 (m, 1H, CH_2), 1.38 (d, CCH_3; ^3J = 6.9), 2.23 (m, 1H, CH_2), 3.40 (m, CH), 3.85 (s, OCH_3), 5.36 (s, C_5H_5) [387] ^{13}C NMR ($CDCl_3$): −6.2 (CH_2), 17.0 (C**C**H_3), 55.0 (OCH_3), 62.7 (CH), 95.8 (C_5H_5), 224.2 (CO cis MoC), 233.5 (CO trans MoC) [387] IR (pentane): 1160, 1340 (ν(SO)), 1936, 2020 (ν(CO)) [382]
146	$-CH_2CH_2W(CO)_3C_5H_5$	IVc (83%) [168, 231]; also from $K[C_5H_5Mo(CO)_3]$ and $[C_5H_5W(CO)_3(CH_2{=}CH_2)]BF_4$ in a CH_3CN/THF mixture at −50 °C (82%) [168, 231] yellow crystals [231] IR (KBr): 1890, 1950, 1990 (ν(CO)), 2842, 2931, 2940 ($\nu(CH_2)$) [168] in the mass spectrum CO, $(C_5H_5M(CO)_3)_2$ (M = Mo, W) and fragments, $CH_2{=}CH_2$, as well as N_2 were proved [168] only slightly soluble which prevented characterization by ^1H NMR [168, 231] stable as solid, but in solution spontaneous decomposition to $CH_2{=}CH_2$, $(C_5H_5Mo(CO)_3)_2$, and $(C_5H_5W(CO)_3CH_2)_2$ was observed [168]
147	$-CH_2CH_2Re(CO)_5$	IVc [168, 231] yellow crystals decomposition at temperatures above −15 °C, only isolable at low temperature because of its instability [168, 231] reacted further with $[Re(CO)_5]^-$ to give $(CO)_5ReC_2H_4Re(CO)_5$ and probably $(C_5H_5Mo(CO)_3)_2$ [168, 231]
*148	$-CH_2(CH_2)_nC{\equiv}CCH_3$ (n = 3, 4)	I (from $[N(C_4H_9\text{-}n)_4][C_5H_5Mo(CO)_3]$ and $ICH_2(CH_2)_nC{\equiv}CCH_3$; the Mo precursor must be used in excess to increase the rate of the

Table 1 (continued)

No.	¹L	method of preparation (yield) properties and remarks
		binuclear alkylation relative to the rate of subsequent reactions; after 1 h at 20 °C the solution contained 45% of the desired product, 22% starting material, and 33% $[N(C_4H_9\text{-}n)_4][C_5H_5Mo(CO)_2\text{-}IC(O)(CH_2)_{n+1}C{\equiv}CCH_3]$ petroleum ether was added to precipitate the latter two complexes, and subsequent removal of the solvent gave $C_5H_5Mo(CO)_3CH_2(CH_2)_nC{\equiv}CCH_3)$ [206] for n = 3 the complex was isolated as an orange oil, while for n = 4 the complex was only generated in solution [206] ^{1}H NMR (benzene-d_6): n = 3: 1.74 (m, all alkyl protons), 4.65 (s, C_5H_5); n = 4: 1.75 (m, all alkyl protons), 4.66 (s, C_5H_5) [206] IR (benzene, n = 3, 4): 1920, 2010 (ν(CO)) [206]
149	$\text{-}CH_2(CH_2)_2CN$	I [115] no details given [115]
*150	$\text{-}CH_2(CH_2)_2Br$	I (26 to 33%, with $Br(CH_2)_3Br$ for 16 h) [36, 95], (30 to 80%, with $Br(CH_2)_3Br$ by stirring at 25 °C for 19 h followed by refluxing for 1 h) [268] light yellow crystals (from hexane at −78 °C), readily soluble in organic solvents [36, 95], m.p. 85 to 87 °C [36], 90 to 92 °C [95] ^{1}H NMR (CS_2): ca. 1.5 (m, CH_2), ca. 2.0 (m, CH_2), 3.17 (t, CH_2Br; J = 7), 5.13 (s, C_5H_5) [36] IR (cyclohexane): 1945, 2026 (ν(CO)) [36, 95]; (KBr): 2920, 3070 (ν(CH)); other bands at 820, 835, 845, 1001, 1014, 1115, 1205, 1295, 1415, 1423 [36]
151	$\text{-}CH_2(CH_2)_2I$	I (with $I(CH_2)_3I$, only proved as an intermediate which was converted further into $C_5H_5Mo(CO)_2(=CO(CH_2)_3\text{-cyclo})I$; see Formula XXXIVb on p. 103) [268] IR (THF): 1921, 2018 (ν(CO)) [268] mass spectrum: $[M]^+$, $[M-nCO]^+$ (n = 1 to 3), $[M-3CO-nCH_2]^+$ (n = 1 to 3), and $[C_5H_5MoI]^+$ which is assumed to be generated from $[M]^+$ via a five-membered cyclic transition state; $[C_5H_5MoCH_2I]^+$ presumably arises from an alternative fragmentation pathway involving elimination of C_2H_4 from $[M]^+$ [49]
152	$\text{-}CH_2(CH_2)_3Br$	I (17%, with $Br(CH_2)_4Br$ for 16 h) [36] light yellow crystals (from pentane at −78 °C), m.p. 48 to 50 °C [36]

References on pp. 106/18

Table 1 (continued)

No.	^1L	method of preparation (yield) properties and remarks
152 (continued)		^1H NMR (CS_2): 1.66 (br, $(CH_2)_2$), 1.89 (br, CH_2), 3.39 (t, CH_2Br; J = 6), 5.32 (s, C_5H_5) [36] IR (KBr): 2815, 2910, 3060 (ν(CH)); other bands at 816, 830, 1003, 1010, 1115, 1280, 1418, 1430 [36] IR (cyclohexane): 1938 (A″, ν(CO)), 1945 (A′, ν(CO)), 2013 (A′, ν(CO)); k_1(CO diagonal to CH_3) = 15.42, k_2(CO diagonal to another) = 15.69, k_{id} = 0.53; for calculations the ratio of the diagonal interaction force constant (k_{id}) and the lateral interaction force constant (k_{is}) was assumed to be constant k_{id}/k_{is} = 2 [70]; see also [36] mass spectrum: $[M]^+$, [M−nCO] (n = 1 to 3), $[M-nCO-4CH_2]^+$ (n = 0 to 3); for further fragments, see [53] thermolysis in vacuum gave $(C_5H_5Mo(CO)_3)_2$ and $C_5H_5Mo(CO)_3Br$ [36] interaction with $Na[C_5H_5Fe(CO)_2]$ in THF at 25 °C gave $(C_5H_5Fe(CO)_2)_2(\mu\text{-}(CH_2)_4)$ [36] attempts at the synthesis of a six-membered ring homologue of Formula XXXIII (see p. 103) by treating a THF solution of $C_5H_5Mo(CO)_3CH_2(CH_2)_3Br$ with $Li[(C_2H_5)_3BH]$ were unsuccessful but yielded $[C_5H_5Mo(CO)_2(\eta^2\text{-}OCH(CH_2)_4Br)]^-$ (compare Formula VIII on p. 80) [315]
153	$-CH_2(CH_2)_3I$	I [254] although no details concerning the preparation and characterization of this particular complex are available, the compound has briefly been reported to interact with LiI in refluxing THF to produce $C_5H_5Mo(CO)_3I$ rather than a 2-oxacyclohexylidene product, $C_5H_5Mo(CO)_2(=CO(CH_2)_4\text{-cyclo})I$ [254]
*154	$-CH_2(CH_2)_2Fe(CO)_2C_5H_5$	I (66%, with $ICH_2(CH_2)_2Fe(CO)_2C_5H_5$ for 73 h) [358]; only an 18% yield was obtained using $BrCH_2(CH_2)_2Fe(CO)_2C_5H_5$ as the alkylating agent in refluxing THF [251] yellow needles (from hexane at −78 °C), m.p. 97 to 99 °C [251, 358] ^1H NMR ($CDCl_3$): 1.44 (m, $FeCH_2$), 1.62 (m, $MoCH_2$), 1.71 ("s", CCH_2C), 4.74 (s, C_5H_5Fe), 5.24 (s, C_5H_5Mo) [358]; similar in benzene [251] ^{13}C NMR ($CDCl_3$): 6.4 ($MoCH_2$), 8.4 ($FeCH_2$), 45.5 (C**C**H_2C), 85.2 (C_5H_5Fe), 92.7 (C_5H_5Mo), 217.7 (FeCO), 227.7 (CO cis MoC), 240.0 (CO trans MoC) [358]

Table 1 (continued)

No.	^1L	method of preparation (yield) properties and remarks
		IR (hexane): 1925, 1937, 1951, 2005, 2016 (ν(CO)) [358], similar in cyclohexane [251]
*155	$-CH_2(CH_2)_3Fe(CO)_2C_5H_5$	I (80%, with $ICH_2(CH_2)_3Fe(CO)_2C_5H_5$ for 72 h) [358] yellow crystalline solid (from hexane at −78 °C), m.p. 110 to 115 °C [358] ^1H NMR ($CDCl_3$): 1.48 ("s", $Fe(CH_2)_2$), 1.68 ("s", $Mo(CH_2)_2$), 4.72 (s, C_5H_5Fe), 5.28 (s, C_5H_5Mo) [358] ^{13}C NMR ($CDCl_3$): 2.8 ($MoCH_2$), 3.3 ($FeCH_2$), 42.0, 44.4 ($FeCH_2\mathbf{C}H_2$, $MoCH_2\mathbf{C}H_2$), 85.1 (C_5H_5Fe), 92.6 (C_5H_5Mo), 217.3 (FeCO), 227.1 (CO cis MoC), 235.0 (CO trans MoC) [358] IR (hexane): 1933, 1954, 2008, 2018 (ν(CO)) [358]
*156	$-CH_2(CH_2)_4Fe(CO)_2C_5H_5$	I (41%, with $ICH_2(CH_2)_4Fe(CO)_2C_5H_5$ for 120 h) [358] yellow crystalline solid (from hexane at −78 °C), m.p. 87 to 88 °C [358] ^1H NMR ($CDCl_3$): 1.36 ("s", $CCCH_2CC$), 1.44 ("s", $Fe(CH_2)_2$), 1.60 ("s", $Mo(CH_2)_2$), 4.72 (C_5H_5Fe), 5.26 (s, C_5H_5Mo) [358] ^{13}C NMR ($CDCl_3$): 3.0 ($MoCH_2$), 3.8 ($FeCH_2$), 36.4 ($MoCH_2\mathbf{C}H_2$), 38.1 ($FeCH_2\mathbf{C}H_2$), 40.8 ($CC\mathbf{C}H_2CC$), 85.3 (C_5H_5Fe), 92.7 (C_5H_5Mo), 217.7 (FeCO), 227.7 (CO cis MoC), 239.9 (CO trans MoC) [358] IR (hexane): 1933, 1954, 2008, 2017 (ν(CO)) [358]
*157	$-CH_2(CH_2)_5Fe(CO)_2C_5H_5$	I (45%, with $ICH_2(CH_2)_5Fe(CO)_2C_5H_5$ for 21 h) [358] yellow crystalline solid (from hexane at −78 °C), m.p. 56 to 59 °C [358] ^1H NMR ($CDCl_3$): 1.34 ("s", $FeCH_2CH_2C\mathbf{H}_2$, $MoCH_2CH_2C\mathbf{H}_2$), 1.46 ("s", $Fe(CH_2)_2$), 1.60 ("s", $Mo(CH_2)_2$), 4.73 (C_5H_5Fe), 5.32 (s, C_5H_5Mo) [358] ^{13}C NMR ($CDCl_3$): 2.8 ($MoCH_2$), 3.7 ($FeCH_2$), 34.3 ($FeCH_2CH_2\mathbf{C}H_2$), 35.1 ($MoCH_2CH_2\mathbf{C}H_2$), 36.5 ($MoCH_2\mathbf{C}H_2$), 38.3 ($FeCH_2\mathbf{C}H_2$), 85.3 (C_5H_5Fe), 92.7 (C_5H_5Mo), 217.8 (FeCO), 227.7 (CO cis MoC), 239.9 (CO trans MoC) [358] IR (hexane): 1933, 1954, 2008, 2018 (ν(CO)) [358]
*158	$-CF_3$	from $C_5H_5Mo(CO)_3C(O)CF_3$ which was heated at atmospheric pressure to 120 °C for 2 h until no further CO evolution was observed; analytically pure $C_5H_5Mo(CO)_3CF_3$ was isolated by direct sublimation of the pentane-washed and dried crude residue at 60 to 80 °C and 1.3×10^{-4} atm (60%, if the starting material was already partially decarbonylated by sublimation, otherwise only a 24%

References on pp. 106/18

Table 1 (continued)

No.	^{1}L	method of preparation (yield) properties and remarks
*158 (continued)		yield was obtained); also from a one-pot reaction starting from $Na[C_5H_5Mo(CO)_3]$ and $(CF_3CO)_2O$ [16] from $C_5H_5Mo(CO)_3Cl$ and excess $(CF_3)_2Cd \cdot CH_3O$-$C_2H_4OCH_3$ in $(CH_3)_2SO$ at $-30\,^\circ C$ in the dark for several days; the crude product was purified by sublimation in high vacuum (54%) [346] bright yellow crystalline material; darkens on exposure to light [16, 346], sensitive to X-ray [28, 41] and laser beams [346], stable against oxidation by air and hydrolysis [346]; m.p. 153 °C [16, 346] ^{1}H NMR ($CHCl_3$): 5.55 (C_5H_5) [16, 20, 346] ^{13}C NMR (benzene): 91.6 (C_5H_5, d; J(C, H) ≅ 175) [20] ^{19}F NMR (($(CH_3)_2SO$): +13.34; similar in $(CH_3)_2NCHO$; (CH_3CN): +12.23 [346]; ($CDCl_3$): +11.89 (^1J(F, C) = 366.7) [346] (see also [16]); similar in CH_2Cl_2 [60] IR (cyclohexane): 1978 (A″, ν(CO)), 1984 (A′, ν(CO)), 2049 (A′, ν(CO)); k_1(CO diagonal to CH_3) = 16.04, k_2(CO diagonal to another) = 16.31, k_{id} = 0.51; for calculations the ratio of the diagonal interaction force constant (k_{id}) and the lateral interaction force constant (k_{is}) was assumed to be constant k_{id}/k_{is} = 2 [70]; similar in halocarbon oil [16] and hexane [281] IR (hexane): 1021, 1030, 1055 (ν(CF)) [281]; (KBr): 976, 1004, 1044 (ν(CF)), 3100 (ν(CH)), additional bands at 693, 825, 860, 907, 935, 1350, 1405, 1417, 1425; No. 158 exhibits much lower C-F stretching frequencies than, e.g., its parent acyl $C_5H_5Mo(CO)_3C(O)CF_3$; this fact and the enhanced thermal stability of the title complex has been attributed to metal d orbitals donating electron density to empty C-F σ* antibonding orbitals [17], which, in valence bond terms, is equivalent to significant contributions of hyperconjugative resonance structures such as $C_5H_5Mo(CO)_3{}^+{=}CF_2F^-$ (for carbene formation, see "Further information" at the end of the table [16]) IR (Ar matrix at 12 K): 1024.0, 1036.0, 1058.5 (ν(CF)), 1974.5 (A″, ν(CO)), 1980.0 (A′, ν(CO)), 2058.0 (A′, ν(CO)); similar in a N_2 matrix at 12 K [359] IR (CO matrix at 12 K): 1023.5, 1034.0, 1058.5 (ν(CF)), 1971.5 (A″, ν(CO)), 1979.0 (A′, ν(CO)),

Table 1 (continued)

No.	^1L	method of preparation (yield) properties and remarks
		2057.5 (A′, ν(CO)); similar in a CH_4 matrix at 12 K [359] IR (5% ^{13}CO-doped CH_4 matrix at 12 K): 1973.0 (A″, ν(CO)), 1980.0 (A′, ν(CO)), 2059.5 (A′, ν(CO)) plus bands at 1940.5 and 2041.0 originating from cis-$C_5H_5Mo(^{12}CO)_2(^{13}CO)CF_3$ [359] UV (cyclohexane, ε) = 220 (22200), 308 (2940) [16]; (Ar matrix at 12 K): 295 (vs), 510 (vw) [359] mass spectrum: $[M]^+$, $[M-nCO]^+$ (n = 1 to 3), $[M-3CO-CF_2]^+$, $[M-3CO-HF]^+$; further fragmentation resulted in $[C_5H_5Mo]^+$, $[C_3H_3MoF]^+$, and $[C_3H_3Mo]^+$ (the latter two presumably originating from elimination of acetylene from the C_5H_5 group), as well as $[MoF]^+$, and Mo^+ as the most abundant metal-containing ions; the most prominent metal-free ion is $[C_6H_5]^+$ which apparently arises by insertion of a carbon atom from a CF_3 or CO ligand into the C_5H_5 ring; for further information see [71] polarographic reduction (Hg electrode; millimolar 1,2-dimethoxyethane solutions containing 10^{-1} mol/L $[N(C_4H_9\text{-}n)_4]ClO_4$) resulted in $[C_5H_5Mo(CO)_3]^-$ and $CF_3^{\cdot}$, the fate of the latter being unknown; half-wave potential $E_{p/2}$ = -2.1 ± 0.1 V vs. Ag/10^{-3} M $AgClO_4$ [29, 32]
159	$-CCl_3$	prepared by condensing slightly more than 1 equivalent of BCl_3 onto a CH_2Cl_2 solution of $C_5H_5Mo(CO)_3CF_3$ (No. 158) at −196 °C; the reaction proceeded rapidly and quantitatively upon warming to room temperature; formed BF_3 and the solvent were then removed under vacuum (84%) [281] yellow crystalline material (from pentane at −78 °C) [281] ^1H NMR (benzene-d_6): 4.65 (C_5H_5) [281] IR (Nujol): 722 (ν(CCl)); (hexane): 1967, 1979, 2053 (ν(CO)) [281] mass spectrum: $[M]^+$, $[M-2CO]^+$ [281] thermally sensitive in solution and in the solid state; therefore the complex can be sublimed at 80 to 100 °C under dynamic high vacuum, although some decomposition occurs under these conditions; hydrocarbon solutions decomposed within 1 h at 60 °C evolving HCl [281]

References on pp. 106/18

Table 1 (continued)

No.	^1L	method of preparation (yield) properties and remarks
160	$-CBr_3$	analogously to No. 159, but no details were given [281] ^1H NMR (benzene): 4.65 (C_5H_5) [281] IR (hexane): 1965, 1980, 2052 (ν(CO)) [281] for chemical behavior, see preceding No. 159
161	$-CF_2CF_2H$	IIb (11%) [5, 11] yellow-orange air-stable crystals, m.p. 53 to 54 °C [4, 5, 11] ^{19}F NMR: −128.0 (CF_2H; J(F, H) = 58), −59.9 ($MoCF_2$; J(F, H) < 1) [11] IR (CS_2): 1945, 2000, 2045 (ν(CO)), 2920, 3080 (ν(CH)); additional bands at 780, 822, 915, 972, 1005, 1100, 1165, 1172, 1350 [5, 11] effective antiknock agent when added to petroleum hydrocarbons and may also be used as a supplement to lead antiknocks such as $Pb(CH_3)_4$ or $Pb(C_2H_5)_4$; further useful in metal plating applications and as an additive for lubricant oils and greases to increase their antiwear activity; may also be used to control the rate of combustion of pyrophoric materials; being biologically active, it has also been proposed as a fungicide, herbicide, pesticide, and the like [5]
162	$-CF_2CF(Cl)H$	IIb (4%) [5] IR (CS_2): 1969 (br), 2053 (ν(CO)); additional bands at 754, 776, 824, 957, 1065, 1091, 1147, 1346 [5] somewhat unstable at ambient temperature, decomposing with a rate of approximately 20% per day [5] for technical use compare preceding No. 161
163	$-C_2F_5$	by pyrolysis of $C_5H_5Mo(CO)_3C(O)C_2F_5$ (Section 1.5.1.4.1.12.3.2) as described for Nos. 158 and 165, but no details are given [16, 281] IR (CH_2Cl_2): 989, 1028, 1176, 1192, 1291 (ν(CF)); (hexane): 1970, 1979, 2055 (ν(CO)) [281] easily converted to $C_5H_5Mo(CO)_3CCl_2CF_3$ (No. 164) via α-regiospecific halogen exchange upon interaction with BCl_3 in CH_2Cl_2 at −78 °C; the reaction pathway involves a cationic carbene intermediate providing the basis for the α-regiochemistry of this transformation; see also the analogous reaction for No. 158 [281]; proton acid-promoted hydrolysis by aqueous HBF_4 in CH_3NO_2 formed $C_5H_5Mo(CO)_3C(O)CF_3$ [282]

Table 1 (continued)

No.	^1L	method of preparation (yield) properties and remarks
*164	$-CCl_2CF_3$	from $C_5H_5Mo(CO)_3C_2F_5$ (No. 163) with 1 equivalent of BCl_3 as detailed for the trihalomethyl derivative $C_5H_5Mo(CO)_3CCl_3$ (No. 159) [281] ^1H NMR ($CDCl_3$): 5.56 (C_5H_5) [281] IR (CH_2Cl_2): 1159, 1219 (ν(CF)); (hexane): 1965, 1983, 2055 (ν(CO)) [281]
*165	$-C_3F_7$	from $C_5H_5Mo(CO)_3C(O)C_3F_7$ (Section 1.5.1.4.1.12.3.2) which was heated at atmospheric pressure to 120°C for 2 h until no further CO evolution was observed; the crude residue was extracted with pentane from which the product precipitated at −78°C; final purification was accomplished by sublimation at 60°C and 1.3×10^{-4} atm (41%) [16] bright yellow air-stable crystalline material, m.p. 81 to 83°C [16, 41] ^1H NMR ($CHCl_3$): 5.58 (C_5H_5) [16] ^{19}F NMR ($CHCl_3$): −113.3 (CF_2), −79.0 (CF_3), −56.2 ($MoCF_2$) [16] IR (KBr): 982, 986, 1017, 1020, 1075, 1166, 1180, 1220, and 1321 (ν(CF)), 3100 (ν(CH)); additional bands at 713, 793, 832, 852, 863, 888, 935, 1156, 1355, and 1425 [16]; similar in hexane [281] IR (cyclohexane): 1975 (A″, ν(CO)), 1985 (A′, ν(CO)), 2049 (A′, ν(CO)); k_1(CO diagonal to CH_3) = 16.07, k_2(CO diagonal to another) = 16.28, k_{id} = 0.52; for calculations the ratio of the diagonal interaction force constant (k_{id}) and the lateral interaction force constant (k_{is}) was assumed to be constant k_{id}/k_{is} = 2 [70], similar in halocarbon oil [16] or hexane [281] UV (cyclohexane, ε): 218 (23800), 310 (2490) [16] mass spectrum: $[M]^+$, [M−nCO] (n = 1 to 3), $[M-C_3F_7]$, $[M-C_3F_7-nCO]$ (n = 1 to 3); the most abundant metal-containing ions are $[C_5H_5MoF]^+$ and $[C_5H_5MoF_2]^+$, the most prominent metal-free ions are $[C_8H_5F_4]^+$ (formed from $[C_5H_5MoC_3F_7]^+$ by elimination of MoF_3) and $[C_7H_5F_2]^+$, arising from the former ion by CF_2 loss; for other fragmentation pathways, see [71]
166	$-CCl_2C_2F_5$	from $C_5H_5Mo(CO)_3C_3F_7$ (No. 165) with 1 equivalent of BCl_3 as detailed for the trihalomethyl derivative $C_5H_5Mo(CO)_3CCl_3$ (No. 159) [281] ^1H NMR (acetone-d_6): 5.97 (C_5H_5) [281]

References on pp. 106/18

Table 1 (continued)

No.	^1L	method of preparation (yield) properties and remarks
166 (continued)		^{19}F NMR (acetone-d_6): −96.61 (CF_2), −71.79 (CF_3); both singlets are slightly broadened due to unresolved $^3J(F, F)$ coupling (<2) [281] IR (hexane): 1040, 1135, 1163, 1204, 1219 (ν(CF)), 1967, 1980, 2054 (ν(CO)) [281]

VI

* Further information:

$C_5H_5Mo(CO)_3CH_3$ (Table **1**, No. **1**) was also obtained from $[(C_5H_5Mo(CO)_2P(CH_3)_3)_2$-($\mu$-$As(CH_3)_2)][C_5H_5Mo(CO)_3]$ and CH_3I in benzene at 25 °C for 30 min in 85% yield [215]. [cyclo-$(S(CH_2)_2SC)N(CH_3)_2][C_5H_5Mo(CO)_3]$ [233] or $[N(CH_3)_4][(C_5H_5Mo(CO)_3)_2M]$ (M = Cu, Ag) [397] were also reacted with CH_3I. No. 1 was also isolated as the principle product from the reaction of $Na[C_5H_5Mo(CO)_3]$ with $ICH_2Si(CH_3)_3$ in THF at 20 °C with $C_5H_5Mo(CO)_3CH_2$-$Si(CH_3)_3$ (No. 113) as a proved intermediate [76, 80]. A similar reaction occurred with $ICH_2Sn(CH_3)_3$ producing $C_5H_5Mo(CO)_3CH_3$ along with mainly $C_5H_5Mo(CO)_3Sn(CH_3)_3$ [130].

In addition to Method I and II, compound No. 1 was obtained in many other reactions but in general accompanied by other complexes. Facile reduction of $[C_5H_5Mo(CO)_3{=}CH_2]^+$ (generated in situ from $C_5H_5Mo(CO)_3CH_2Cl$ (No. 128) and $AgPF_6$ at −78 °C) in CH_2Cl_2, using $C_5H_5Mo(CO)_3H$ or $C_5H_5Mo(CO)_3CH_2OCH_3$ (No. 118) as hydride donors gave $C_5H_5Mo(CO)_3$-CH_3 in 43% or 63% yield along with a mixture of $[(C_5H_5Mo(CO)_3)_2(\mu\text{-}CH_3CO)]PF_6$ (see Formula III on p. 2) and $[(C_5H_5Mo(CO)_2)_2(\mu,\eta^2\text{-}CH_3CO)]PF_6$ (see Formula II on p. 2) for the hydride or $[C_5H_5Mo(CO)_3{=}CH(OCH_3)]PF_6$ for the methoxymethyl complex [290]. Treatment of $[C_5H_5Mo(CO)_3{=}CH_2]^+$ in CH_2Cl_2 with $C_5H_5Fe(CO)(P(C_6H_5)_3)C(O)CH_3$ resulted in a 44% yield of No. 1 along with $[C_5H_5Mo(CO)_3CH_2OC(CH_3)Fe(CO)(C_5H_5)P(C_6H_5)_3]PF_6$ (No. 123). Nucleophilic hydride donors such as $Li[(C_2H_5)_3BH]$, $[CH_3P(C_6H_5)_3][H_3BCN]$, or $C_5H_5Fe(CO)$-$(P(C_6H_5)_3)H$ displaced $C_5H_5Fe(CO)(P(C_6H_5)_3)C(O)CH_3$ from No. 123 to generate C_5H_5Mo-$(CO)_3CH_3$ as an accompanying product. In CH_2Cl_2, No. 123 also degenerated to No. 1, $C_5H_5Fe(CO)(P(C_6H_5)_3)C(O)CH_3$, and $[C_5H_5Mo(CO)_3OC(CH_3)Fe(CO)(C_5H_5)P(C_6H_5)_3]PF_6$ [301].

$C_5H_5Mo(CO)_3CH_3$ was isolated from $C_5H_5Mo(CO)_3FPF_5$ (generated in situ from C_5H_5Mo-$(CO)_3H$ and $[(C_6H_5)_3C]^+$ in CH_2Cl_2 at −78 °C) and $C_5H_5Mo(CO)_3CH_2OCH_3$ (No. 118) in 5% yield along with mainly $[C_5H_5Mo(CO)_3{=}CHOCH_3]PF_6$ and $[(C_5H_5Mo(CO)_2)_2(\mu,\eta^2\text{-}CH_3$-$CO)]PF_6$, or from $[C_5H_5Mo(CO)_3]PF_6$ and $C_5H_5Fe(CO)(L)CH_2OR$ (L = CO, R = CH_3; L = $P(C_6H_5)_3$, R = CH_3, C_2H_5) in <5% yield for L = CO or in 18% yield for L = $P(C_6H_5)_3$, together with $[C_5H_5Fe(CO)(L){=}CH(OR)]PF_6$, $C_5H_5Fe(CO)(L)CH_3$, and $[C_5H_5Fe(CO)_2(L)]^+$. Formation of No. 1 in the latter case accounted for CH_3 transfer from the methyl iron complexes onto the the Lewis-acidic molybdenum atom. A mechanism is discussed in detail (compare also No. 118) [290]. A similar CH_3 migration was observed in the reaction of $C_5H_5Mo(CO)_3FSbF_5$ with $C_5H_5Fe(CO)_2CH_3$, or by treatment of $C_5H_5Mo(CO)_3X$ (X = FBF_3, FPF_5, $FAsF_5$, $FSbF_5$)

with $M(CO)_5CH_3$ (M = Mn, Re) in CH_2Cl_2 at about −35 °C to give No. 1 among other complexes. Analogously $C_5H_5Mo(CO)_3CH_3$ was obtained using $(CO)_5MCD_3$ [263, 264]. A 13 to 14% yield of $C_5H_5Mo(CO)_3CH_3$ was also seen in the reactions of $C_5H_5M(CO)_3FBF_3$ (M = Mo, W) with $[C_5H_5Mo(CO)_2(\eta^2\text{-}CH_3CHO)]^-$ along with $(C_5H_5Mo(CO)_2)_2(\mu,\eta^2\text{-}CH_3CHO)$, $(C_5H_5M(CO)_3)_2$ (M = Mo, W), and $(C_5H_5)_2MoW(CO)_6$. This reflects the tendency of the $[C_5H_5M(CO)_3]^+$ cations to abstract hydride from the $[C_5H_5Mo(CO)_2(\eta^2\text{-}CH_3CHO)]^-$ anion, the result of which is the formation of the acetyl intermediate $C_5H_5Mo(CO)_3C(O)CH_3$ known to rapidly undergo decarbonylation with formation of $C_5H_5Mo(CO)_3CH_3$ [367].

Formation of $C_5H_5Mo(CO)_3CH_3$ from $C_5H_5Mo(CO)_2(PR_2R')C(O)CH_3$ (R = R' = OC_4H_9, OC_6H_5, C_4H_9, C_6H_5; R = CH_3, R' = C_6H_5) has been noticed on removal of the phosphane ligands by CH_3I as $[R_3PCH_3]I$ [308]; No. 1 has been furthermore isolated in minor amounts by the decarbonylation of $C_5H_5Mo(CO)_2(P(C_6H_5)_3)C(O)CH_3$ in refluxing THF for 48 h, the main product being $C_5H_5Mo(CO)_2(P(C_6H_5)_3)CH_3$. An analogous reaction occurred under CO [40]. $[C_5H_5Mo(CO)_2(P(OCH_3)_3)_2][C_5H_5Mo(CO)_3]$ slowly decomposed in benzene solution by nucleophilic attack of the anion on the cation to give $C_5H_5Mo(CO)_3CH_3$ along with $C_5H_5Mo(CO)_2(P(OCH_3)_3)P(O)(OCH_3)_2$. The rate of this reaction can be increased in refluxing benzene [75, 78].

In contrast to the phosphane-substituted compounds, facile ejection of $E(C_6H_5)_3$ (E = As, Sb) from the substituted acetyls $C_5H_5Mo(CO)_2(E(C_6H_5)_3)C(O)CH_3$ in THF or CH_3CN has been found to be the preferred pathway of degradation at 40 °C, producing $C_5H_5Mo(CO)_3CH_3$ as the exclusive organometallic material. In the case of E = As, UV irradiation gave in addition $C_5H_5Mo(CO)_2(As(C_6H_5)_3)CH_3$ [154]. Accordingly $C_5H_5Mo(CO)_3CH_3$ has been isolated as by-product from the syntheses of a series of carbene derivatives $C_5H_5Mo(CO)_2(E(C_6H_5)_3)=C(OCH_3)CH_3$ (E = Ge, Sn) prepared by treatment of $C_5H_5Mo(CO)_3E(C_6H_5)_3$ with $LiCH_3$ and subsequent alkylation with $[O(CH_3)_3]^+$. This probably indicates some loss of the $E(C_6H_5)_3$ ligand by the reaction of the starting material with CH_3Li to give $[C_5H_5Mo(CO)_3]^-$, which then is alkylated by the trialkyloxonium ion [173, 379].

The photoreaction of $C_5H_5Mo(CO)_2(^2D)CH_3$ (^2D = $P(C_4H_9)_3$, $P(C_6H_5)_3$) in a 1,2-ethoxyethylbenzene glass at −261 °C gave $C_5H_5Mo(CO)_3CH_3$ [378]. $C_5H_5Mo(CO)_3Pb(CH_3)_3$ was converted to $C_5H_5Mo(CO)_3CH_3$ (No. 1) in high yield by UV irradiation in hexane. The transformation also succeeded thermally in refluxing hexane but with a prolonged reaction period of 5 d and a decreased yield of 30% [236]. Treatment of $C_5H_5Mo(CO)_3E(CH_3)_3$ (E = Sn, Pb) with $LiN(C_3H_7\text{-}i)_2$ in THF at 0 °C for 15 min and subsequent quenching with excess CH_3I resulted in a mixture of No. 1 and $(CH_3)_3EC_5H_4Mo(CO)_3CH_3$ with the anions $[C_5H_5Mo(CO)_3]^-$ and $[(CH_3)_3EC_5H_4Mo(CO)_3]^-$, respectively, as intermediates. Use of $LiC_4H_9\text{-}n$ instead of $LiN(C_3H_7\text{-}i)_2$ resulted in an increased yield of No. 1 [350]. Treatment of $C_5H_5Mo(CO)_3Tl(CH_3)_2$ with CH_3I for 24 h resulted in a 76% yield of the title compound after sublimation (80 °C, 2 Torr) [178]. $C_5H_5Mo(CO)_3CH_3$ was also obtained quantitatively from $C_5H_5Mo(CO)_3CH_2O_2CC_4H_9\text{-}t$ (No. 120) which was treated with $LiCH_3$ in ether at −78 °C. After warming to room temperature, the reaction mixture was quenched with excess CH_3I [216]. $(C_5H_5Mo(CO)_2CH_2P(C_6H_5)_3)_2$ reacted with excess CH_3I in CH_2Cl_2 for 4 h at room temperature forming $C_5H_5Mo(CO)_3CH_3$ in a 72% yield together with small quantities of $C_5H_5Mo(CO)_2(P(C_6H_5)_3)I$ [339]. $C_5H_5Mo(CO)_3CH_3$ was also isolated from a reaction mixture, produced from $[C_5H_5Mo(CO)_2(\eta^2\text{-}CH_3CHO)]^-$ and $Fe_2(SO_4)_3$ in THF at 25 °C after 3 h, containing $(C_5H_5Mo(CO)_2)_2(\mu,\eta^2\text{-}CH_3CHO)$ and $[C_5H_5Mo(CO)_3]^-$. The latter was converted to 19% of No. 1 and $C_5H_5Mo(CO)_3I$ by the addition of CH_3I [367]. The title compound was also isolated in an 85% yield from a solution of $[C_5H_5Mo(CO)_2(\eta^2\text{-}CH_3CHO)]^-$ in THF which was treated with CH_3I under CO at room temperature for 4 h [296].

References on pp. 106/18

Iodide ions cleaved the cation of $[(C_5H_5Mo(CO)_2)_2(\mu,\eta^2\text{-}CH_3CO)]PF_6$ into an 88% yield of $C_5H_5Mo(CO)_3CH_3$ and $(Mo(CO)_2C_5H_5)_2(\mu\text{-}I)_2$, as well as varying amounts of $C_5H_5Mo(CO)_3I$ and $[C_5H_5Mo(CO)_2I_2]^-$ [255]. $C_5H_5Mo(CO)_3CH_3$ is furthermore formed in 50% yield by electrophilic cleavage of the C-O bond of $C_5H_5Mo(CO)_3CH_2OCH_3$ (No. 118) using 40% aqueous HF [39].

The ^{13}C NMR relaxation mechanisms of $C_5H_5Mo(CO)_3CH_3$ were studied in detail. Spin anisotropy and scalar relaxation of the second kind as contributions to the relaxation mechanism can be neglected, and contribution to the spin-lattice relaxation time T_1 were only the C-H dipolar-dipolar mechanism and the internal spin-rotation mechanisms, represented in terms of T_1^{DD} and T_1^{SR}. The relative magnitudes of these contributions are determined by the rate of overall molecular tumbling and the barrier to methyl rotation. $T_1 = 5.13 \pm 5$ s and the nuclear Overhauser enhancement at 25.1 MHz, $\eta = 1.99 \pm 0.02$, were determined at 38 °C (similar values at 29 °C), from which $T_1^{DD} = 5.13 \pm 0.3$ s and $T_1^{SR} > 510$ s were calculated. Data revealed that the spin-rotation contribution can be neglected, which reflects the substantial steric congestion as well as the low symmetry of the environment of the CH_3 group due to the short $CH_3-C_5H_5$ distance and the resulting depth of the potential well in the most stable configuration. The methyl rotation is hindered, the carbon hydrogen reorientation correlation time τ_c is long, and therefore T_1^{DD} decreased. From a plot of $\ln(1/T_1^{DD})$ vs. 1/T, the reorientation activation energy was determined to be 2.90 kcal/mol. Thus there is a rather high barrier to methyl rotation in $C_5H_5Mo(CO)_3CH_3$ of more than 6 kcal/mol. At 70 °C relaxation due to spin rotation becomes more significant as shown by $T_1 = 7.22$ s, $\eta = 1.81$, $T_1^{DD} = 7.94$ s, and $T_1^{SR} = 80.0$ s [207, 249]. For an earlier report, see [180].

The IR spectrum of $C_5H_5Mo(CO)_3CH_3$ was investigated thoroughly. Although the C_s symmetry of the compound demands three stretching vibrations (2A′ + A″) in the carbonyl region, usually only two bands were observed. The lack of the third band may be due to accidental cancellation of dipole moments or, more probably, to the coincidence of the A″ and the lower A′ band, and careful measurements revealed that the lower more intensive $\nu(CO)$ band can be resolved into two closely separated bands [2, 70]. The carbonyl stretching frequencies of No. 1 were compared to the respective bands of a ^{13}CO enriched sample, prepared by photolyzing $C_5H_5Mo(CO)_3CH_3$ in hexane under 4 atm ^{13}CO for 2 h. Data were determined from samples in a CH_4 matrix at −261 °C. Frequencies due to all possible $C_5H_5Mo(^{12}CO)_{3-n}(^{13}CO)_nCH_3$ isotopomers were proved by the excellent correspondence between observed and calculated band positions [244]. The IR spectra in CCl_4 for $C_5H_5Mo(CO)_3CHD_2$ and $C_5H_5Mo(CO)_3CD_3$ were determined in the CH and CD stretching region to be 2145 ($\nu_s(CD_2)$), 2229 ($\nu_{as}(CD_2)$), 2946 ($\nu_s(CH^{is})$), 2964 ($\nu_{as}(CH^{is})$), and 3115 (ν(CH) of C_5H_5) for the former and to be 2061, 2117 ($\nu_s(CD_3) + 2d_{as}(CD_3)$), 2234, 2240 ($\nu_{as}(CD_3)$), and 3118 ($\nu$(CH) of C_5H_5) for the latter. In keeping with the appreciable barrier to internal methyl rotation found by NMR studies (see above), the appearance of two "isolated" $\nu(CH^{is})$ absorptions for the CHD_2 species, the splitting of $\nu_{as}(CD_3)$ for the CD_3 derivative, and the broad profile of $\nu_{as}(CH_3)$ for the parent CH_3 complex, indicate the presence of two distinct types of C-H bonds: CH_s lying in the symmetry plane and CH_a lying out of it as shown in

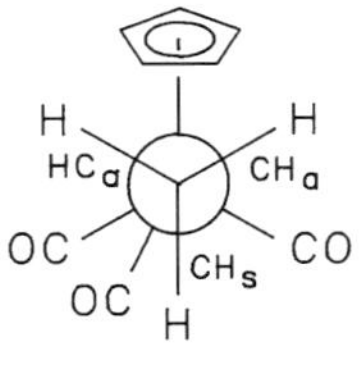

VII

Formula VII. The CH_s bond is proved to be stronger than the CH_a bond as indicated by the relative intensities of $\nu_s(CH^{is})$ bands and the sum rule calculations. From energy-factored normal coordinate treatments of the CH and CD stretching frequencies, the individual bond-stretching force constants were calculated to be $k_s = 4.801$, $k_a = 4.742$, and the interaction force constant $k_i = 0.003$ mdyn/Å. The low value for k_i may be due to the dissolved state of the molecule. Alternatively the predicted HCH angle may be too small. Bond lengths and dissociation energies were determined to be $r(CH_s) = 1.094(0)$, $r(CH_a) = 1.095(8)$ Å, $D^{\circ}_{298}(CH_s) = 425.7$, and $D^{\circ}_{298}(CH_a) = 419.0$ kJ/mol (see Formula VII) [295, 320].

A qualitative MO scheme for $C_5H_5Mo(CO)_3CH_3$ has been constructed and is shown in Scheme 3 [156].

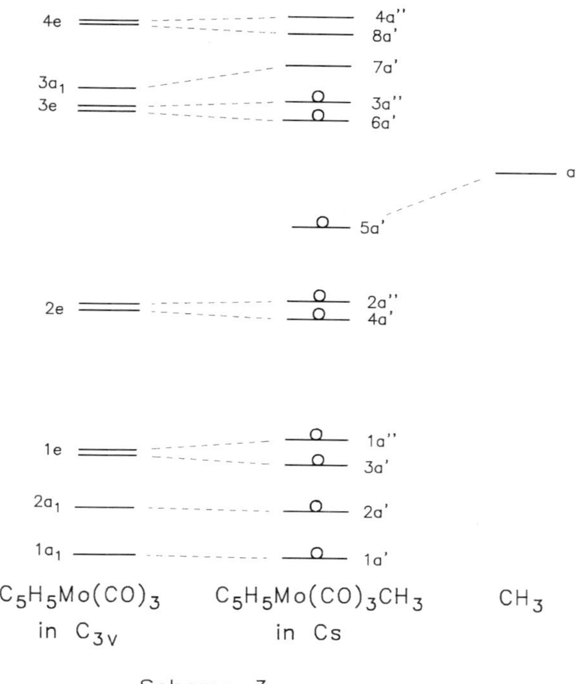

Scheme 3

The MO scheme is substantiated by the photoelectron spectra of $C_5H_5Mo(CO)_3CH_3$ depicted in [156]. The absorption at the lowest energy, occurring at 7.78 eV, is assigned to ionization from the close-spaced highest occupied metal d orbitals (3a″ and 6a′ in symmetry group C_s). Bands at 9.7 and 10.0 eV are similar in shape and ionization energy to the bands attributed to ionization from the C_5H_5 e orbitals in $C_5H_5Mo(CO)_3$ (point group C_{3v}) and so are assigned to the corresponding 2a″ and 4a′ ionizations of $C_5H_5Mo(CO)_3CH_3$. An additional band observed at 9.07 eV is ascribed to ionization from the 5a′ molecular orbital, corresponding to the metal-carbon s bond, while a shoulder appearing on the high-energy side of the 10.0 eV band at 12.2 eV may be assigned to ionization from the C-H bonding orbitals of the methyl group. The PES spectrum of $C_5H_5Mo(CO)_3CH_3$ was compared to the related alkyl complexes $C_5H_5W(CO)_3CH_3$ and $C_5H_5M(CO)_2CH_2R$ (M = Fe, Ru; R = H, CN) [156]. For the XPS carbon 1s and oxygen 1s binding energies of the CO ligands [172, 224], see Table 1.

References on pp. 106/18

Optimized angular parameters of the four-legged piano-stool structure of $C_5H_5Mo(CO)_3$-CH_3 and the orientation preference of the π-bonded C_5H_5 ligand were derived from extended Hückel calculations. The two occupied d block orbitals z^2 and xy of No. 1 are destabilized relative to $[C_5H_5Mo(CO)_4]^+$ by substitution of CO by CH_3, which agrees with the photoelectron spectrum [374, 377].

$C_5H_5Mo(CO)_3CH_3$ is readily soluble in petroleum ether and also in all common organic solvents but is insoluble in and unaffected by water. It was converted to $(C_5H_5Mo(CO)_3)_2$ in organic solvents in air after a few days at 25 °C along with some tarry material. This transformation can be accelerated by heating the complex in vacuum at 110 °C. Solutions in organic solvents are stable under vacuum or in an inert gas atmosphere [2]. The heat of solution in THF has been determined as $+2.6 \pm 0.1$ kcal/mol [316]. Upon increasing the temperature to 155 °C, decomposition occurred via an initial insertion of CO into the Mo-CH_3 bond, followed by subsequent homolytic cleavage and hydrogen abstraction. Observed products were CH_3CHO, together with CO and hydrogen [96].

UV irradiation of $C_5H_5Mo(CO)_3CH_3$ was studied in detail. In an early report [164], $(C_5H_5Mo(CO)_3)_2$ was detected as the principle product in inert solvents like pentane, benzene, or THF along with other photodegraded products which could not be identified. Additionally CH_4 was produced as the only gaseous product. No deuterated methane was found during photolysis in benzene-d_6 indicating the solvent was not the hydrogen source. Hydrogen abstraction from the C_5H_5 ring by the departing CH_3 was demonstrated to be a principle reaction pathway which is in accord with CHD_3, CH_2D_2, and CH_3D produced during the photolysis of $C_5H_5Mo(CO)_3CD_3$ in benzene-d_6 [164]. For investigating the reaction mechanism, photolysis was studied in an ESR cavity in toluene at 20 °C and the observed clearly resolved signal was first attributed to the paramagnetic anion $[C_5H_5Mo(CO)_2CH_3]^{\cdot-}$ [204], but a later in-depth ESR study in $(t\text{-}C_4H_9)_2O_2$, under conditions excluding the formation of radical anions proved the same paramagnetic species indicating it was a result of the oxidation of the starting complex by the t-butoxy radical [321]. In full agreement with this assumption, closely related paramagnetic species with extremely similar g-factors and coupling constants for the unpaired electron with the central metal were recorded by ESR in the photolysis of C_5H_5Mo-$(CO)_3CH_3$ in $(t\text{-}C_4H_9)_2O_2$ in the presence of $P(OC_4H_9\text{-}n)_3$ or $P(OCH_3)_3/H_2SO_4$ to give C_5H_5Mo-$(CO)(PR_3)_2CH_3$ [321]. Gas evolution during the photolysis in the presence of $P(C_6H_5)_3$ also indicated carbonyl substitution by phosphane [204]. Conclusive results for the involvement of $CH_3^{\cdot}$ radicals in the photo-degradation of $C_5H_5Mo(CO)_3CH_3$ have been obtained by spin-trapping this radical with $C_6H_5CH{=}N(O)C_4H_9$-t, since the ESR signal of the spin adduct appeared immediately on irradiation of the methyl complex in toluene and disappeared after termination of the photolysis [204]. Fragmentation processes may be temperature-dependent, because photolysis in CH_2Cl_2 at -30 °C in the presence of 2,3,5,6-$(CH_3)_4C_6HNO$ gave only $C_5H_5Mo(CO)_3(N(O)C_6H(CH_3)_4\text{-}2,3,5,6)^{\cdot}$, with no evidence for the expected $CH_3^{\cdot}$; however at $+20$ °C, conversely the spin adducts derived from the corresponding alkyl radicals were clearly detected but no molybdenum-containing radicals [243].

Photolysis of No. 1 in a poly(vinyl chloride) film was thoroughly studied by IR between -261 and 25 °C and a reaction pathway established. Final products at room temperature were $C_5H_5Mo(CO)_3H$, $C_5H_5Mo(CO)_3Cl$, and $(C_5H_5Mo(CO)_3)_2$ as well as some $Mo(CO)_6$. Evidence that the species $C_5H_5Mo(CO)_2CH_3$, $C_5H_5Mo(CO)_2(OC_4H_8)CH_3$ in the presence of THF, and $C_5H_5Mo(CO)_2^{\cdot}$, generated at -261 °C, are intermediates is provided by the fact that the room temperature photoproducts are seen to be formed if the matrix is warmed up to 25 °C [249, 279]. $C_5H_5Mo(CO)_3Cl$ was also formed nearly quantitatively upon irradiation in $CHCl_3$ [164]. In contrast, no evidence was obtained for the homolysis of the Mo-CH_3 bond upon low-temperature photolysis at -261 °C of No. 1 isolated at high dilution in Ar, N_2, CO, and

CH_4 matrices [244] or in 1,2-epoxyethylbenzene glass [378] and at −198 °C in paraffin wax [256]. These photoreactions afforded the coordinatively unsaturated 16e species $C_5H_5Mo(CO)_2CH_3$ as the principal photoproduct [244, 256, 378] which reacted reversely with the carbon monoxide, liberated in the primary photolysis step after irradiation with long-wavelength or annealing the matrix to ca. −243 °C and then recooling to −261 °C, to regenerate the starting complex [244]; see also [256, 378]. The corresponding reaction of the 16e photointermediate in the N_2 matrix produced $C_5H_5Mo(CO)_2(N_2)CH_3$ as demonstrated by ^{13}CO labeling and energy-factored CO force-field fitting [244]. The failure to observe radical species in gas matrices has been attributed to both the absence of secondary reaction processes, e.g., Cl abstraction, suited for radical stabilization and to the creation of tight cages made up from efficiently packed small gas matrix molecules, in which radicals produced on photolysis are unable to diffuse far enough apart to prevent thermal back-reactions [244].

$C_5H_5Mo(CO)_3CH_3$ was readily attacked by iodine, but in contrast to an early report [2], no formation of $C_5H_5Mo(CO)_3I$ and CH_3I was proved rather than $C_5H_5Mo(CO)_2I_3$ and CH_3COI after reinvestigation of this reaction [312]; see also [342]. In order to determine the Mo-CH_3 bond strength, the enthalpy of reaction for "$C_5H_5Mo(CO)_3CH_3 + I_2 \rightarrow C_5H_5Mo(CO)_3I + CH_3I$" was calculated to be $\Delta H = -24.6 \pm 1.6$ kcal/mol based on the difference between the reaction enthalpies of the rapid and quantitative reactions "$Na[C_5H_5Mo(CO)_3] + I_2 \rightarrow C_5H_5Mo(CO)_3I + NaI$ ($\Delta H = -32.3 \pm 1.3$ kcal/mol)" and "$Na[C_5H_5Mo(CO)_3] + CH_3I \rightarrow C_5H_5Mo(CO)_3CH_3 + NaI$ ($\Delta H = -7.7 \pm 0.3$ kcal/mol)". Using these data the Mo-CH_3 bond strength could be estimated to be 47.0 kcal/mol [312]; see also [342]. With HCl gas at room temperature, an essentially quantitative yield of $C_5H_5Mo(CO)_3Cl$ was obtained, in addition to a large quantity of a noncondensable gas, assumed to be CH_4 [96].

Treatment with LiC_4H_9-n- or LiC_4H_9-i at −78 °C in THF resulted in metalation of the cyclopentadienyl ring, readily and specifically. Neither deprotonation of the CH_3 group nor any attack at the CO ligands were observed. The compound, formulated as $LiC_5H_4Mo(CO)_3CH_3$ was evidenced from the subsequent incorporation of deuterium into the cyclopentadienyl ligand [162], or by preparation of $IC_5H_4Mo(CO)_3CH_3$ from the lithiated intermediate and 1,2-diiodoethane at −78 °C. An analogous reaction with iodine led only to a product mixture [353]. No appreciable reaction was observed between No. 1 and LiI in refluxing CH_3OH or THF [296].

Treatment of a solution of No. 1, first with HBF_4/ether and subsequently with LiSeH/THF at −30 °C, gave $C_5H_5Mo(CO)_3SeH$ with $[C_5H_5Mo(CO)_3O(C_2H_5)_2]^+$ as a supposed intermediate [330]. UV irradiation in THF at room temperature led to $C_5H_5Mo(CO)_2^{\cdot}$, and in the presence of $((NO)_2Co(\mu\text{-}Br))_2$, stabilization by coordination of the three-electron donor NO gave $C_5H_5Mo(CO)_2NO$ along with $Co(CO)_3NO$ [324]. Contrarily the methyl compound is completely unreactive toward $(CH_3)_3NO$ in solution at ambient conditions [205]. One carbonyl group was replaced from $C_5H_5Mo(CO)_3CH_3$ by PF_3 using an excess of $Ni(PF_3)_4$ in refluxing toluene to furnish $C_5H_5Mo(CO)_2(PF_3)CH_3$ without any migration of the methyl ligand [110]. Treatment of the methyl complex, produced in situ from $Na[C_5H_5Mo(CO)_3]$ and CH_3I, with excess PCl_5 or PBr_5 in refluxing CH_2Cl_2 gave $C_5H_5MoCl_4$ [376] and $C_5H_5MoBr_4$ [396], respectively.

Reactions of $C_5H_5Mo(CO)_3CH_3$ with $[(CO)_5CrH]^-$, $[(CO)_5WH]^-$, $[(CO)_4W(P(OCH_3)_3)H]^-$, and $[(CO)_4WH]^-$ used as $[N(P(C_6H_5)_3)_2]^+$ salts in THF yielded CH_4 as the major and CH_3CHO as the minor organic product, in addition to $[C_5H_5Mo(CO)_3]^-$. In the case of $[(CO)_5CrH]^-$, $(\mu\text{-}H)(Cr(CO)_5)_2$ was also identified [323, 363].

From the reaction of $C_5H_5Mo(CO)_3CH_3$ with liquid ammonia at 20 °C, $C_5H_5Mo(CO)_2(NH_3)H$ was obtained along with $CH_3C(O)NH_2$. At a reaction temperature of −33 °C, only a mixture

References on pp. 106/18

of products were produced, even after a prolonged reaction period. Using ND_3 analogously, $C_5H_5Mo(CO)_2(ND_3)D$ was isolated [163]. No appreciable interaction was observed when $C_5H_5Mo(CO)_3CH_3$ was combined with both dimethyl amine and trimethyl amine at room temperature [96]. Different products were isolated by the reaction of No. 1 with $P(CH_3)_3$ under UV irradiation in pentane at 25 °C, depending on the ratio of reactants and the reaction period. Equimolar amounts of No. 1 and the phosphane gave the monosubstitution product $C_5H_5Mo(CO)_2(P(CH_3)_3)CH_3$ in less than 20 min. If the reaction period was prolonged up to 2 h and/or an excess of $P(CH_3)_3$ was used, a second CO group was substituted yielding $C_5H_5Mo(CO)(P(CH_3)_3)_2CH_3$. In a side reaction, the asymmetrically disubstituted binuclear complex $C_5H_5Mo(CO)(P(CH_3)_3)_2$-$Mo(CO)_3C_5H_5$ was additionally formed. $C_5H_5Mo(CO)$-$(P(CH_3)_3)_2$-$Mo(CO)_3C_5H_5$ became the main product after 3 h [183]. Similar UV supported substitution reactions were described for PR_3 (R = OC_6H_5, C_6H_5) in alkane solution (e.g. hexane) to give $C_5H_5Mo(CO)_2(PR_3)CH_3$ at 25 °C [40, 256], or for PR_3 (R = C_4H_9-r, C_6H_5) in 1,2-epoxyethylbenzene at −173 °C yielding small quantities of cis-$C_5H_5Mo(CO)_2(PR_3)CH_3$ [378]. Formation of $C_5H_5Mo(CO)_2(P(OC_6H_5)_3)CH_3$ was also proved if No. 1 was irradiated at −198 °C in rigid media containing $P(OC_6H_5)_3$ to induce CO abstraction followed by subsequent warming [256]. Similar photo-induced substitutions were accomplished using the 4-electron donor ligands cis-$(C_6H_5)_2PCH{=}CHP(C_6H_5)_2$ or 1,2-$(C_6H_5)_2EC_6H_4P(C_6H_5)_2$ (E = P, As) in THF at room temperature yielding the chelate complexes cis-$C_5H_5Mo(CO)$-$(C_6H_5)_2PCH{=}CHP(C_6H_5)_2)CH_3$ or $C_5H_5Mo(CO)(1,2$-$(C_6H_5)_2EC_6H_4P(C_6H_5)_2)CH_3$ [230]. However, UV irradiation in the presence of 1,4-$C_6H_4(P(N(CH_3)_2)_2)_2$ at 0 to 10 °C in pentane/THF afforded the phosphine-bridged bis(dicarbonyl) complex $(C_5H_5Mo(CO)_2CH_3)_2(\mu$-1,4-$C_6H_4$-$(P(N(CH_3)_2)_2)_2)$ [325].

$C_5H_5Mo(CO)_3CH_3$ (No. 1) reacted with many hydride sources like LiH, $LiAlH_4$, $Na[R_3BH]$, and $M[(C_3H_7$-$i)_3BH]$ in THF to give $[C_5H_5Mo(CO)_3]^-$, but $Li[(C_2H_5)_3BH]$ behaves quite differently. Although no evidence was found for the formation of any formyl species on treatment of the methyl complex with $Li[(C_2H_5)_3BH]$ [234], reinvestigation using $Li[(C_2H_5)_3BH]$ at room temperature in THF proved the metallaoxirane $[C_5H_5Mo(CO)_2(\eta^2$-$CH_3CHO)]^-$ (see Formula VIII) as the product of an immediate reaction [275, 296].

VIII

For a detailed study the reaction was started at −78 °C. After addition of a slight excess of the hydride to the reaction mixture followed by subsequent warming to −66 °C, the anionic formyl complex $[C_5H_5Mo(CO)_2(CHO)CH_3]^-$ resulted, proved by ^1H and ^{13}C NMR spectra. Above −40 °C conversion to $[C_5H_5Mo(CO)_2(C(O)CH_3)H]^-$ occurred; further warming to room temperature gave finally $[C_5H_5Mo(CO)_2(\eta^2$-$CH_3CHO)]^-$, along with some $[C_5H_5Mo(CO)_3]^-$. The metallaoxirane can be isolated as the Li^+ or $[Li(12$-crown-$4)]^+$ salt. The analogous reactions using $Li[(C_2H_5)_3BD]$ and $C_5H_5Mo(CO)_3CD_3$, respectively, gave the corresponding deuterated products [275, 276, 296, 314, 328]; see also [367]. A stoichiometric cycle for the synthesis of acetaldehyde was additionally shown by the reaction of $[C_5H_5Mo(CO)_2$-$(\eta^2$-$CH_3CHO)]^-$ with CH_3I under CO, which regenerated $C_5H_5Mo(CO)_3CH_3$ by evolution of CH_3CHO with $C_5H_5Mo(CO)_2(CH_3CHO)CH_3$ as an undetected intermediate [296, 314].

Addition of the strong Lewis acid $AlBr_3$ to a solution of $C_5H_5Mo(CO)_3CH_3$ in toluene at 0 °C induced a rapid CO insertion into the Mo-CH_3 bond to give the thermally unstable cyclic acyl complex $C_5H_5Mo(CO)_2C(CH_3)OAlBrBr_2$ (see Formula IX). Further treatment with 0.5 atm CO at 0 °C resulted in a rapid conversion to $C_5H_5Mo(CO)_3C(CH_3)OAlBr_3$. The latter reaction is thermodynamically favored, because the Lewis acid stabilized the resultant acyl product in contrast to the parent molecule $C_5H_5Mo(CO)_3C(O)CH_3$, which is formed from compound IX by hydrolysis under mild conditions, but decomposed further to No. 1 along with other complexes [209, 222, 226]. $C_5H_5Mo(CO)_3CH_3$ reacts much less readily with $AlCl_3$ [285].

A similar facile methyl migration in $C_5H_5Mo(CO)_3CH_3$ was induced by $(C_2H_5)_2Al$-$N(C_3H_7$-i)-$P(C_6H_5)_2$, which contains both a Lewis acidic and a Lewis basic center, in toluene to give $C_5H_5Mo(CO)_2O(CH{=}CH_2)$-$Al(C_2H_5)_2$-$N(C_3H_7$-i)-$P(C_6H_5)_2$ (see Formula X). A reaction mechanism is discussed [257, 269]. Interaction with $(C_2H_5)_2Al$-$N(C_4H_9$-t)-$P(C_6H_5)_2$ gave only a product mixture which could not be separated [269].

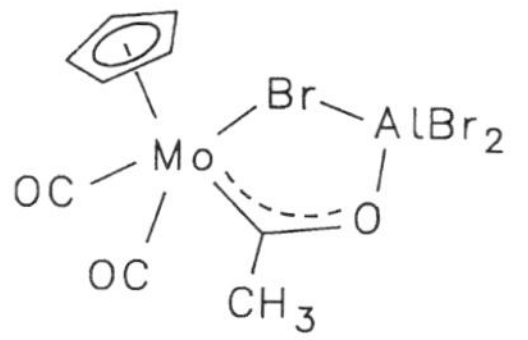

IX

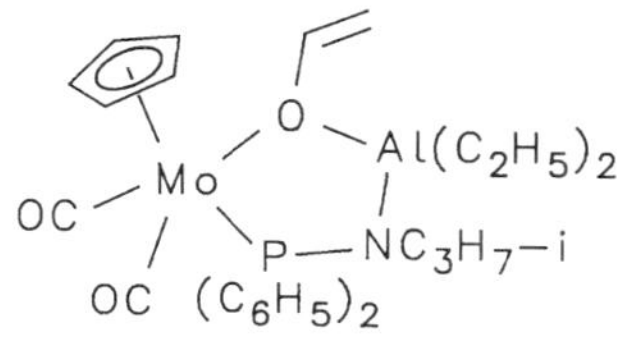

X

$C_5H_5Mo(CO)_3CH_3$ reacted with $[N(C_2H_5)_4]GeCl_3$ in acetone to give $[N(C_2H_5)_4]$[trans-$C_5H_5Mo(CO)_2(GeCl_3)C(O)CH_3$] that was sufficiently stable to be isolated, but would also react further with $[R_3O]^+$ to yield the carbene derivatives trans-$C_5H_5Mo(CO)_2(=C(OR)CH_3)$-$GeCl_3$ (R = CH_3, C_2H_5). Monitoring the $C_5H_5Mo(CO)_3CH_3/[GeCl_3]^-$ reaction by IR and NMR spectroscopy gave no evidence of any intermediate cis isomer, indicating that in this particular system formation of the thermodynamically stable trans compound occurred very rapidly [153].

Between $C_5H_5Mo(CO)_3CH_3$ and SnX_2 (X = Cl, Br) no insertion reaction was detected in refluxing CH_3OH, but with $SnBr_2$, $C_5H_5Mo(CO)_3Br$ and small quantities of $C_5H_5Mo(CO)_3$-$SnBr_3$ were isolated. Using $SnCl_2$ only low pressure UV irradiation afforded $C_5H_5Mo(CO)_3Cl$ along with small amounts of $C_5H_5Mo(CO)_3SnCl_3$. The amount of the latter compound increased with continued irradiation. Medium-pressure UV irradiation resulted in a substantial acceleration of the reaction and gave $(C_5H_5Mo(CO)_3)_2SnCl_2$ as a further product [129]. On the other hand the bulky monomeric tin(II) alkyl $Sn(CH(Si(CH_3)_3)_2)_2$ underwent oxidative insertion into the Mo-CH_3 bond in hexane at room temperature resulting in $C_5H_5Mo(CO)_3$-$Sn(CH(Si(CH_3)_3)_2)_2CH_3$ [139, 158].

$C_5H_5Mo(CO)_3CH_3$ was converted into $[C_5H_5Mo(CO)_3(CH_3CO_2H)]BF_4$ by treatment with ethereal HBF_4 and glacial acetic acid at 0 °C in CH_2Cl_2. Further reaction with K_2CO_3 gave a product mixture consisting of $C_5H_5Mo(CO)_3OC(O)CH_3)$ and its decarbonylation product $C_5H_5Mo(CO)_2O_2CCH_3$ along with $(C_5H_5Mo(CO)_3)_2$. An analogous reaction sequence using $C_6H_5CO_2H$ gave $C_5H_5Mo(CO)_3OC(O)C_6H_5$ and $C_5H_5Mo(CO)_2O_2CC_6H_5$; with t-$C_4H_9CO_2H$ only $C_5H_5Mo(CO)_2O_2CC_4H_9$-t was isolated [326]. In a one-pot reaction starting from No. 1 and $HBF_4/C_6H_5CO_2H$, the reaction mixture was subsequently treated with $CuO_2CC_6H_5$ and 4-$CH_3C_5H_4N$ at room temperature to give $C_5H_5Mo(CO)_2(\mu$-$O_2CC_6H_5)_2CuNC_5H_4CH_3$-4 (Formula XI). Analogously $C_5H_5Mo(CO)_2(\mu$-$O_2CCH_3)_2CuNC_5H_4CH_3$-4 (Formula XI) was obtained from $C_5H_5Mo(CO)_3CH_3$ and $(Cu(\mu$-$O_2CCH_3)_2NC_5H_4CH_3$-$4)_2$ under UV irradiation in THF

References on pp. 106/18

along with $C_5H_5Mo(CO)_3OC(O)CH_3$ and $C_5H_5Mo(CO)_2O_2CCH_3$. No reaction occurred under thermal conditions between 20 and 50 °C [326].

XI

Other heterodimetallic complexes were also formed starting from $C_5H_5Mo(CO)_3CH_3$. Reaction with $(C_5H_5)_2NbBH_4$ in refluxing toluene in the presence of $N(C_2H_5)_3$ gave $C_5H_5(CO)$-$Mo(\mu\text{-}CO)_2Nb(C_5H_5)_2$, shown in Formula XII [191, 198], and treatment with equimolar quantities of $C_5H_5W(CO)_2{\equiv}CC_6H_4R$-4 (R = CH_3, OCH_3) under UV irradiation in toluene at 0 °C produced $C_5H_5(CO)_2Mo(\mu,\eta^3\text{-}4\text{-}RC_6H_4CC(CH_3)O)W(CO)_2C_5H_5$, illustrated in Formula XIII [338]. $C_5H_5Fe(CO)_2P(C_6H_5)_2$ slowly reacted with $C_5H_5Mo(CO)_3CH_3$ in CH_3CN at 22 °C to yield an approximate 4:1 mixture of $C_5H_5(CO)_2Mo(\mu\text{-}OCCH_3)(\mu\text{-}P(C_6H_5)_2)Fe(CO)C_5H_5$ (see Formula XIV) and $C_5H_5(CO)_2Mo(\mu\text{-}H)(\mu\text{-}P(C_6H_5)_2)Fe(CO)C_5H_5$ (Formula XV). No deuterated products were isolated from CD_3CN after 4 d which ruled out the solvent as being the μ-H source. Use of $C_5H_5Mo(CO)_3CD_3$ gave $C_5H_5(CO)_2Mo(\mu\text{-}OCCD_3)(\mu\text{-}P(C_6H_5)_2)Fe(CO)$-$C_5H_5$, together with a 3:1 mixture of $C_5H_5(CO)_2Mo(\mu\text{-}H)(\mu\text{-}P(C_6H_5)_2)Fe(CO)C_5H_5$ and $C_5H_5(CO)_2Mo(\mu\text{-}D)(\mu\text{-}P(C_6H_5)_2)Fe(CO)C_5H_5$, implying that some, but not all, of the bridging hydrido ligand was derived from the $MoCH_3$ group [307].

XII

XIII

XIV

XV

Methyl transfer reactions "m-CH_3 + $[m']^-$ ⇄ $[m]^-$ + m'-CH_3" (m, m' = $C_5H_5Mo(CO)_3$, $C_5H_5W(CO)_3$, $Mn(CO)_5$, $Re(CO)_5$, $C_5H_5Fe(CO)_2$) were only successful if the carbonyl metalate $[m']^-$ is more nucleophilic than the anion $[m]^-$ from which the methyl compound is derived [127, 144].

References on pp. 106/18

Photo-induced CO dissociation from $C_5H_5Mo(CO)_3CH_3$ at $-78\,°C$ in C_2H_4-purged i-octane led to $C_5H_5Mo(CO)_2(C_2H_4)CH_3$ [256]. UV irradiation in n-pentane in the presence of cyclopenta-1,3-diene at $-60\,°C$ resulted in the conversion to the thermolabile $C_5H_5Mo(CO)(\eta^4\text{-}C_5H_6)C(O)CH_3$ and more stable $C_5H_5Mo(CO)(\eta^4\text{-}C_5H_6)CH_3$, along with the secondary product $(C_5H_5)_2Mo(C(O)CH_3)H$. Also isolated $C_5H_5Mo(CO)_2(\eta^3\text{-}C_5H_7)CH_3$ is probably formed during workup. If the reaction mixture was warmed up to 36 °C after photolysis, $C_5H_5Mo(CO)(\eta^4\text{-}C_5H_6)C(O)CH_3$ rearranged further to $C_5H_5Mo(\eta^5\text{-}C_5H_4C(O)CH_3)H_2$, $C_5H_5Mo(CO)(\eta^5\text{-}C_5H_4C(O)CH_3)$, $C_5H_5Mo(CO)_2(\eta^3\text{-exo-}C_5H_6C(O)CH_3)$, and $(C_5H_5Mo(CO)_3)_2$ [310, 311]. Treatment of $C_5H_5Mo(CO)_3CH_3$ with excess $(CH_3)_5C_5H$ in boiling n-decane led to $(C_5H_5Mo(CO)_3)_2$ as the only identifiable organometallic compound. No evidence of any (pentamethylcyclopentadienyl)molybdenum derivative was obtained from this reaction [200].

No simple diene complexes could be detected from the reaction of $C_5H_5Mo(CO)_3CH_3$ with buta-1,3-diene. The mixture contained enyl derivatives, $C_5H_5Mo(CO)_2(\eta^3\text{-R})$ (R = $CH_2CHCHCH_2CH_3$, $CH_2CHCHCH_2C(O)CH_3$, $CH_3C(O)CHCHCHCH_3$), formed by methyl and acyl group transfer, respectively [384].

$C_5H_5Mo(CO)_3CH_3$ reacted with an excess of the respective alkyne RC≡CR′ (R = R′ = H, $Si(CH_3)_3$, C_6H_5; R = H, R′ = CH_3, C_6H_5) in pentane or hexane under UV irradiation to give compounds of the type $C_5H_5Mo(CO)_2(\eta^2\text{-}CR'{=}CRC(O)CH_3)$ as illustrated in Formula XVI [166, 188]; see also [149, 155, 289]. Treatment with $CH_3C{\equiv}CCH_3$ in refluxing hexane resulted analogously in $C_5H_5Mo(CO)_2(\eta^2\text{-}C(CH_3){=}C(CH_3)C(O)CH_3)$ [188], but the thermal reaction with acetylene produced black polyacetylene together with unidentified organomolybdenum derivatives [8]. However treatment with $C_6H_5C{\equiv}CR$ (R = H, C_6H_5) under thermal conditions in refluxing p-xylene at 140 to 150 °C gave $(C_5H_5Mo(CO)_2)_2(\mu\text{-}C_6H_5C{\equiv}CR)$ [8, 15]. A comprehensive discussion of these cyclizations is given in [375].

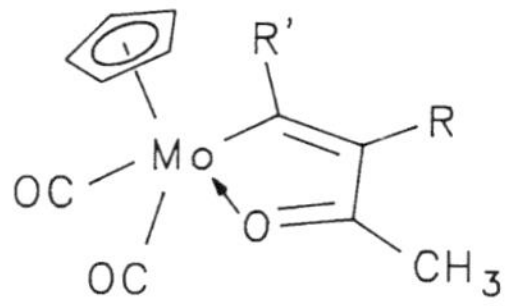

XVI

$C_5H_5Mo(CO)_3CH_3$ reacted with an excess of $CH_3C{\equiv}CCH_3$ in the presence of CF_3CO_2H to form $[C_5H_5Mo(CO)(CH_3C{\equiv}CCH_3)_2]^+$, isolated as the PF_6 salt by addition of methanolic NH_4PF_6. The reaction also proceeded in two steps starting first from No. 1 and CF_3CO_2H in benzene at 20 °C. The resulting $C_5H_5Mo(CO)_3OC(O)CF_3$ was then treated with an excess of $CH_3C{\equiv}CCH_3$ in CH_3CN at 65 °C for 8 h to give the cationic complex. Because of the forcing conditions of the second step, the acyl complex cannot be the intermediate in the direct route [221]; see also [206]. Contrarily prompt decomposition occurred in 98% H_2SO_4 or 100% CF_3CO_2H with CH_4 evolution [7, 245], while treatment with CH_3CO_2H did not led to any detectable reaction [227, 245]. Interaction of $C_5H_5Mo(CO)_3CH_3$ with the acids $HOSO_2F$, $HOSO_2CF_3$, $HOSO_2C_6F_{13}$, $HOC(O)C_3F_7$ [327], and $[HO(C_2H_5)_2]BF_4$ [313, 327] resulted in the formation of $C_5H_5Mo(CO)_3$ derivatives that are best regarded as adducts containing the organometallic Lewis acid $[C_5H_5(CO)_3Mo]^+$ coupled to extremely labile or weakly attached anions, FBF_3^-, OSO_2F^-, and the like.

$C_5H_5Mo(CO)_3CH_3$ reacted with azobenzene in refluxing light petroleum to give small amounts of $C_5H_5Mo(CO)_2(\eta^2\text{-}C_6H_4N{=}NC_6H_5)$ (Formula XVII) and $(C_5H_5Mo(CO)_3)_2$. The main product of the reaction was a not-further-characterized complex believed to contain ortho-semidine ligands [83, 88]. Treatment with benzo[h]quinoline gave small amounts of C_5H_5Mo-

$(CO)_2(\eta^2\text{-}C_{13}H_8N)$ (Formula XVIII) along with mostly $(C_5H_5Mo(CO)_3)_2$ [118]. The reaction with sodium 1,3-diphenyltriazenide resulted in $C_5H_5Mo(CO)_2(\eta^2\text{-}N_3(C_6H_5)_2)$ (Formula XIX) [146].

XVII XVIII XIX

The phototransformation of $C_5H_5Mo(CO)_3CH_3$ by carboxylic acid azides, $N_3C(O)R$, and azido carboxylic acid esters, N_3CO_2R, respectively, was shown to be a general route to nitrene-bridged complexes. Reaction with $N_3C(O)C_6H_5$ in CH_2Cl_2 at 20°C under UV irradiation gave $(C_5H_5MoO)_2(\mu\text{-}NC(O)C_6H_5)(\mu\text{-}O)$. Using identical conditions, treatment with $(N_3CO_2C_2H_4)_2O$ or $N_3CO_2C_2H_5$ gave $(C_5H_5MoO)_2(\mu\text{-}NCO_2CH_2CH_2OCH_2CH_2O_2R)(\mu\text{-}O)$ (R = N_3, NH_2) and $(C_5H_5MoO)_2(\mu\text{-}NCO_2C_2H_5)(\mu\text{-}O)$, respectively. The structure of the complexes are illustrated in Formula XX. In the latter case, additionally small amounts of $(C_5H_5MoO)_2$-$(\mu\text{-}NCO_2C_2H_5)_2$ were isolated. $C_5H_5Mo(CO)_3Cl$ was generally found as a by-product proving that the solvent takes part in these reactions. Secondary organic products, such as $H_2NCO_2C_2H_5$, $(H_2NCO_2C_2H_4)_2O$, and $H_2NC(O)C_6H_5$ isolated from the reaction mixture, arose from hydrogen abstraction from the solvent by free nitrenes. The bis(μ-$NCO_2C_2H_5$)-bridged complex $(C_5H_5MoO)_2(\mu\text{-}NCO_2C_2H_5)_2$ was also exclusively obtained from No. 1 and N_3CO_2-C_2H_5 in refluxing C_2Cl_4 [196, 213].

XX

$C_5H_5Mo(CO)_3C_2H_5$ (Table **1**, No. **2**) was also formed by slow decomposition of $[C_5H_5Mo(CO)_2(P(OC_2H_5)_3)_2][C_5H_5Mo(CO)_3]$ in benzene solution by nucleophilic attack of the anion on the cation, along with $C_5H_5Mo(CO)_2(P(OC_2H_5)_3)P(O)(OC_2H_5)_2$. The rate of this reaction can be increased in refluxing benzene [75, 78]. No. 2 was also proved as an intermediate in the reactions of $C_5H_5Mo(CO)_3H$ with C_2H_4 at 100°C in THF-d_8, but under these conditions the reaction with additional ethene is rapid and $C_2H_5C(O)C_2H_5$ is isolated as the final product [208, 211, 227, 245] rather than No. 2 as mentioned in [85].

$C_5H_5Mo(CO)_3C_2H_5$ has been fully characterized by X-ray crystallography, but cell parameters were not given; for the molecule with selected bond lengths and angles see **Fig. 1**. The average Mo-C(C_5H_5) bond distances were determined as 2.38 ± 0.02 and the mean C-C bond length in the cyclopentadienyl ring as 1.43 ± 0.041 Å. The angle between the planes of the cyclopentadienyl ring and that of the three carbonyl-carbon atoms is 9.2° [10]; compare also [21, 28].

Based on the geometrical parameters of the $C_5H_5Mo(CO)_3C_2H_5$ molecule, the splitting of the d orbitals for this sandwich-type compound possessing a cyclopentadienyl ring and four

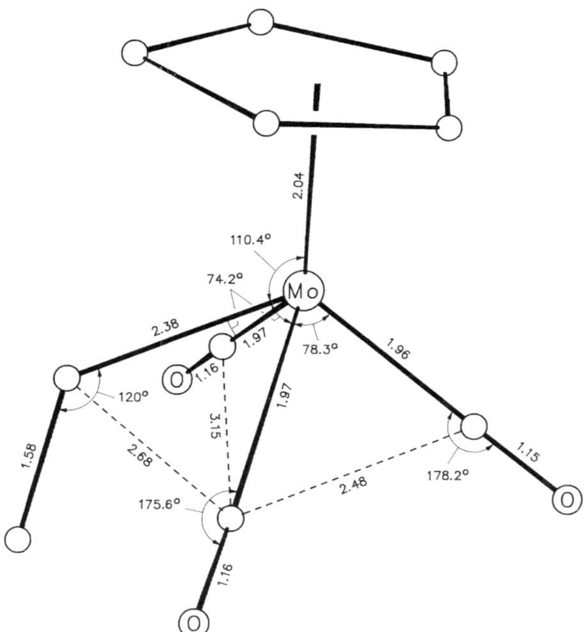

Fig. 1. The molecular structure of $C_5H_5Mo(CO)_3C_2H_5$ [10].

monodentate ligands at the corners of a square pyramid has been investigated. It was shown that although the molecule belongs to the symmetry point group C_s characterized by a single mirror plane, the effective symmetry was C_{4v} [19]. Using HMO theory and the assumption that the cyclopentadienyl ligand may be considered as an independent system not distorted by the bonding to the metal, a reasonable explanation of the bond lengths in the C_5H_5 ring has been given in terms of the flow of electrons from the occupied π molecular orbitals of the five-membered ring onto the central metal, as well as from the Mo atom into the unoccupied molecular orbitals of the ring [43].

Solutions in organic solvents are stable under vacuum or an inert gas atmosphere, but $C_5H_5Mo(CO)_3C_2H_5$ is less stable than the methyl compound No. 1. The heat of solution in THF has been determined to be $+4.6 \pm 0.1$ kcal/mol [316]. After heating $C_5H_5Mo(CO)_3C_2H_5$ above its melting point up to 100 °C in a sealed evacuated tube, ethyl group transfer occurred to give the dimers $(C_2H_5C_5H_4Mo(CO)_3)_2$ and $C_5H_5Mo(CO)_3$-$Mo(CO)_3C_5H_4C_2H_5$, together with nonsubstituted $(C_5H_5Mo(CO)_3)_2$. In the reaction gases ethene and/or ethane, n-butane, and two C_5 fractions as condensable, and methane along with traces of hydrogen as noncondensable hydrocarbons were detected. Accordingly, a radical mechanism was postulated for the reaction involving Mo-C_2H_5 homolysis allowing the ethyl radical either to attack the C_5H_5 ring of $C_5H_5Mo(CO)_3^{\bullet}$ or to dimerize forming n-C_4H_{10} [14]; see also [335].

UV irradiation of $C_5H_5Mo(CO)_3C_2H_5$ in pentane resulted in dealkylation by β-elimination with evolution of C_2H_4 and C_2H_6 along with H_2 and CO_2 to give $C_5H_5Mo(CO)_3H$ and $(C_5H_5Mo(CO)_3)_2$. The latter may be created either from the reaction of the hydride complex with No. 2 or from $C_5H_5Mo(CO)_3H$ alone by H_2 evolution [211, 227, 235, 261]. The mechanism of the photochemical conversion of $C_5H_5Mo(CO)_3C_2H_5$ into $C_5H_5Mo(CO)_3H$ was investi-

References on pp. 106/18

gated in detail by photolysis in paraffin wax at −196°C [256, 259] or in CH_4, N_2, or CO matrices at about −261°C [277, 366]. In addition, time resolved IR spectroscopy at room temperature in heptane was used to identify intermediates [366]. Two distinct 16e intermediates were established in CH_4 matrices at −261°C as well as in heptane solution at 25°C: the matrix-stabilized $C_5H_5Mo(CO)_2C_2H_5 \cdots CH_4$ as the primary product which lost the coordinated CH_4 thus forming $C_5H_5Mo(CO)_2C_2H_4(\mu\text{-}H)$, stabilized by a β-hydrogen of the ethyl group forming an agostic bond to Mo. This latter compound is the immediate precursor to β-hydrogen transfer to give trans-$C_5H_5Mo(CO)_2(C_2H_4)H$ which coexists in an equilibrium with its precursor $C_5H_5Mo(CO)_2C_2H_4(\mu\text{-}H)$. cis-$C_5H_5Mo(CO)_2(C_2H_4)H$ could not be detected. The consecutive loss of alkene and uptake of CO subsequently yielded $C_5H_5Mo(CO)_3H$ as one of the final products. The results imply that β-hydrogen transfer is a relative facile process which does not involve a substantial change in free energy or significant activation barriers [366]. For analogous findings see [250, 256, 259, 277]. Photolysis of $C_5H_5Mo(CO)_3C_2H_5$ in poly(vinyl chloride) at ambient temperature with near-UV or visible light led to $Mo(CO)_6$ as the principle product along with $C_5H_5Mo(CO)_3H$, subsequently converted to $(C_5H_5Mo(CO)_3)_2$ and $C_5H_5Mo(CO)_3Cl$. Studies at −261°C again revealed $C_5H_5Mo(CO)_2C_2H_5$ as the first photoproduct which reacted at −183°C either with ejected CO to regenerate the starting complex, or with available THF to give $C_5H_5Mo(CO)_2(OC_4H_8)C_2H_5$. Warming up to 25°C gave the final products with trans-$C_5H_5Mo(CO)_2(\eta^2\text{-}C_2H_4)H$ as the intermediate along with matrix stabilized $C_5H_5Mo(CO)_2H$ [279]. Fragmentation induced by UV irradiation in CH_2Cl_2 was shown to be temperature-dependent, because at −30°C in the presence of 2,3,5,6-$(CH_3)_4C_6HNO$, only $C_5H_5Mo(CO)_3(N(O)C_6H(CH_3)_4\text{-}2,3,5,6)^{\bullet}$ was produced, but at 20°C only $CH_3^{\bullet}$ was detected [243].

The reaction of $C_5H_5Mo(CO)_3C_2H_5$ with $Li[(C_2H_5)_3BH]$ in THF resulted in the anionic aldehyde complex $[C_5H_5Mo(CO)_2(\eta^2\text{-}C_2H_5CHO)]^-$; compare Formula VIII for the corresponding methyl complex studied in considerably more detail [314].

UV irradiation of $C_5H_5Mo(CO)_3C_2H_5$ in pentane solution in the presence of $P(CH_3)_3$ produces a cis/trans mixture of $C_5H_5Mo(CO)(P(CH_3)_3)_2Mo(CO)_3C_5H_5$, along with $C_5H_5Mo(CO)_2(P(CH_3)_3)H$ and $C_5H_5Mo(CO)_2(P(CH_3)_3)C_2H_5$ [235]. A similar reaction is described in the presence of PR_3 (R = OC_6H_5, C_6H_5) at high concentrations to give $C_5H_5Mo(CO)_2(PR_3)C_2H_5$, but careful investigation of this substitution using nearly equimolar amounts of the reactants at −78°C revealed the simultaneous formation of $C_5H_5Mo(CO)_2(C_2H_4)H$ from the primary irradiation product $C_5H_5Mo(CO)_2C_2H_5$. This may result from a light saturation of the 16-valence-electron species $C_5H_5Mo(CO)_2C_2H_5$ via an interaction of the β-hydrogen with the metal center. Irradiation in solutions in the presence of $P(OC_6H_5)_3$ leads to clean, quantum-efficient substitution of CO by the phosphite, giving $C_5H_5Mo(CO)_2(P(OC_6H_5)_3)C_2H_5$ [256].

Reversible abstraction of a hydride ion occurred if $C_5H_5Mo(CO)_3C_2H_5$ was treated with stoichiometric amounts of $[C(C_6H_5)_3]BF_4$ in $CHCl_3$ at ambient conditions to give salts of $[C_5H_5Mo(CO)_3(C_2H_4)]^+$ [12, 148]. $[C(C_6H_5)_3]PF_6$ may be also a suitable reagent, but with $[C(C_6H_5)_3]ClO_4$ in THF, only rapid decomposition was observed [12].

Reaction with ethanolic $HgCl_2$ produced an unidentified orange, rather insoluble material, but compare $C_5H_5Mo(CO)_3CH_2SCH_3$ (No. 124) [24]. Treatment with $C_6H_5C{\equiv}CC_6H_5$ affords, among other products, 5-methyl-1,2,3,4-tetraphenylcyclopentadiene accompanied by $(1,2,3,4\text{-}(C_6H_5)_4C_6H_2)Mo(C_5H_5)$ [8, 15].

$C_5H_5Mo(CO)_3CH_2CH{=}CH_2$ (Table **1**, No. **17**). Heating pure $C_5H_5Mo(CO)_3CH_2CH{=}CH_2$ up to 60°C in vacuum or in dilute refluxing xylene solution resulted mainly in the formation of $(C_5H_5Mo(CO)_3)_2$ along with small amounts of $C_5H_5Mo(CO)_2(\eta^3\text{-}CH_2CHCH_2)$. However, the

References on pp. 106/18

latter is obtained as the main product by irradiating the neat parent σ-allyl compound, No. 17, under a high vacuum with intense UV light [12], or by $(CH_3)_3NO$-catalyzed decarbonylation in a one-pot reaction started from $Na[C_5H_5Mo(CO)_3]$ and $BrCH_2CH{=}CH_2$ in THF without isolation of No. 17 [288].

In contrast to $C_5H_5Mo(CO)_3CH_3$ (No. 1), treatment of $C_5H_5Mo(CO)_3CH_2CH{=}CH_2$ with I_2, probably in THF, resulted in direct iodination to give $C_5H_5Mo(CO)_3I$ and $ICH_2CH{=}CH_2$. The enthalpy of this reaction was determined to be $\Delta H = -24.1 \pm 0.7$ kcal/mol, and the $Mo{-}CH_2C{=}CH_2$ bond strength can be estimated to be 35 kcal/mol [342].

Light petroleum solutions of $C_5H_5Mo(CO)_3CH_2CH{=}CH_2$ reacted with equimolar amounts of gaseous HCl to form $[C_5H_5Mo(CO)_3(CH_3CH{=}CH_2)]Cl$ as an instantaneous yellow precipitate which can be isolated from an aqueous solution as the PF_6^- or $PtCl_6^{2-}$ salt. Using DCl the respective deuterated $[C_5H_5Mo(CO)_3(CH_2DCH{=}CH_2)]^+$ was obtained. With an excess of HCl, the reaction proceeded further to give neutral $C_5H_5Mo(CO)_3Cl$ [12].

UV irradiation of a solution of $C_5H_5Mo(CO)_3CH_2CH{=}CH_2$ and excess $CF_3C{\equiv}CCF_3$ in hexane afforded the bis-insertion product $C_5H_5Mo(CO)(\eta^4\text{-}(CF_3C{=}CCF_3)_2CH_2CH{=}CH_2)$ (see Formula XXI), along with $(C_5H_5Mo(CO)_2)_2(\mu,\eta^2\text{-}F_3CC{\equiv}CCF_3)$ and minor amounts of $(C_5H_5\text{-}Mo(CO)_3)_2$ [157].

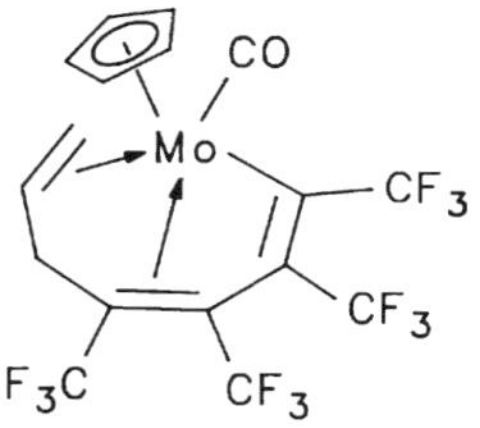

XXI

$C_5H_5Mo(CO)_3CH_2CH{=}CHCH{=}CHR$ (R = H, CH_3; Table **1**, Nos. **32** and **33**) undergo σ- to π-conversion of their η^1-bonded ligands when irradiated in ether at $-20\,°C$. Under these conditions the pentadienyl complex No. 32 produced a mixture consisting mainly of $C_5H_5\text{-}Mo(CO)_2(syn\text{-}\eta^3\text{-}CH_2CHCHCH{=}CH_2)$ (exo and endo isomer) and small amounts of $C_5H_5Mo(CO)(\eta^5\text{-}CH_2CHCHCHCH_2)$ along with $(C_5H_5Mo(CO)_3)$ [319, 345]. Compound No. 33, on the other hand, gave only small quantities of an isomeric mixture consisting of $C_5H_5Mo(CO)_2(\eta^3\text{-}syn\text{-}CH_2CHCHCH{=}CHCH_3)$ (exo and endo isomer) and $C_5H_5Mo(CO)_2\text{-}(\eta^3\text{-}CH_3CHCHCHCH{=}CH_2)$, where the latter was produced by photolytic isomerization of the former. The main products of this photolytic isomerization were $(C_5H_5Mo(CO)_3)_2$ and dodecatetraene [345]. A more convenient method to the $\eta^3\text{-}CH_3CHCHCHCH{=}CH_2$ and $\eta^3\text{-}CH_2\text{-}CHCHCH{=}CHCH_3$ complexes was shown to be decarbonylation by $(CH_3)_3NO$ in CH_2Cl_2 at 0 °C [345].

$C_5H_5Mo(CO)_3CH_2CH{=}CHCH{=}CHR$ (R = H, CH_3) readily underwent carbonyl substitution with PR'_2R'' ($R' = R'' = C_2H_5$; $R' = CH_3$, $R'' = C_6H_5$) in diethyl ether at ambient conditions to give high yields of $C_5H_5Mo(CO)_2(PR'_2R'')CH_2CH{=}CHCH{=}CHR$, but attempts to obtain the $P(CH_3)_3$ analogue in a similar reaction were not successful [345].

$C_5H_5Mo(CO)_3CHCH(CH{=}C(CH_3)_2)(C(CN)_2)_2CH_2$-cyclo (Table **1**, No. **35**) crystallizes in the monoclinic space group $P2_1/n{-}C_{2h}^5$ (No. 14) with a = 10.760(8), b = 15.898(6), c = 12.57(4) Å, $\beta = 96.97(14)°$; Z = 4 molecules per unit cell, $D_{calc} = 1.457$ g/cm³. The molecu-

References on pp. 106/18

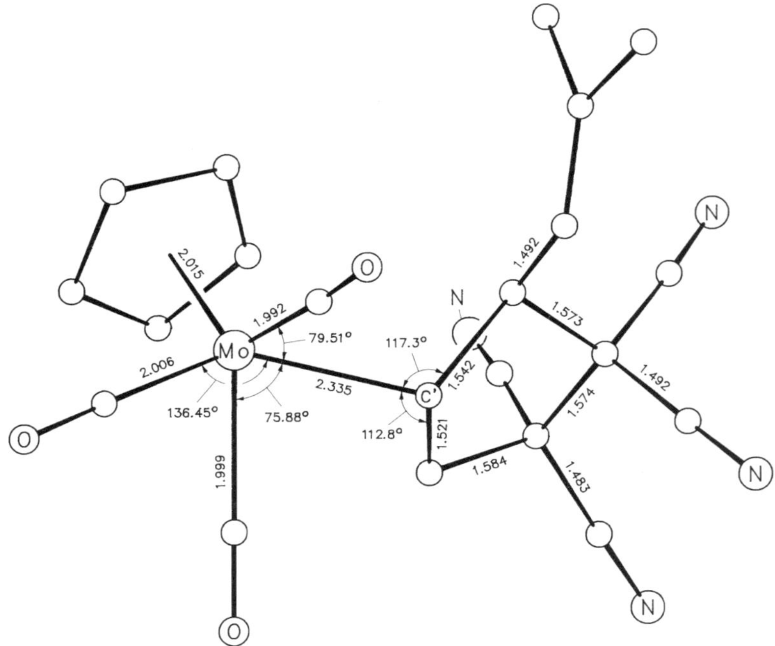

Fig. 2. The molecular structure of $C_5H_5Mo(CO)_3CHCH(CH{=}C(CH_3)_2)(C(CN)_2)_2CH_2$-cyclo [343, 351].

lar structure, shown in **Fig. 2** confirms the [3 + 2] cycloaddition between No. 34 and $C_2(CN)_4$ with the $C_5H_5Mo(CO)_3$ moiety migrating to the β-carbon atom of the dienyl residue to give No. 35. The vinyl double group lies trans to the Mo-C′ bond, suggesting that the reaction proceeds via an ionic mechanism with the zwitterionic $C_5H_5Mo(CO)_3^+[CH_2{=}CHCH$-$(CH{=}C(CH_3)_2)C(CN)_2C^-(CN)_2]$ as a supposed intermediate. The coordination geometry about the central molybdenum atom is approximately a distorted square pyramid [343, 351]; see also [131].

$C_5H_5Mo(CO)_3((\eta^4$-CHRCH=CHCH=CHR′)Fe(CO)$_3)$ (R = H, R′ = H, CH_3; R = CH_3, R′ = H; Table **1**, Nos. **39** to **42**) and **$C_5H_5Mo(CO)_3((\eta^4$-C_6H_7-cyclo)Fe(CO)$_3)$** (Table **1**, No. **43**). The preparation of E,E-1-$C_5H_5Mo(CO)_3(\eta^4$-$CH_2CH{=}CHCH{=}CHCH_3)Fe(CO)_3)$ (No. 40) was studied in detail. Precooled solutions of $[N(P(C_6H_5)_3)_2][C_5H_5Mo(CO)_3]$ and $[(\eta^5$-$CH_2CHCHCHCH$-$CH_3)Fe(CO)_3]PF_6$ in acetone-d_6 (similar in CH_2Cl_2) were mixted at −78 °C in equimolar amounts resulting in the spontaneous formation of Z,E-1-$C_5H_5Mo(CO)_3((\eta^4$-$CH_2CH{=}CH$-$CH{=}CHCH_3)Fe(CO)_3)$ (No. 41). A ^1H COSY spectrum proved this complex to be the Z-isomer of No. 40, established through a typical downfield shift for H^1 and H^5. Moreover, the vicinal coupling constant $J(H^2, H^3)$ is characteristically reduced for No. 41. After warming up to −15 °C, three new compounds were produced in an approximate molar ratio of 2:1:1 and 3:1:3 at 50% and 90% conversion of No. 41, respectively. They were identified by ^1H COSY measurements as $C_5H_5Mo(CO)_3Fe(CO)_3(\eta^3$-$CH_2CH{=}CHCH{=}CHCH_3)$ with a terminal η^3-allyl iron group, $C_5H_5Mo(CO)_3((\eta^4$-$CH(CH_3)CH{=}CHCH{=}CH_2)Fe(CO)_3)$ (No. 42, the less stable form of No. 40), and No. 40. When the solution was allowed to equilibrate at room temperature, only No. 40 persisted. The transformation of No. 41 to $C_5H_5Mo(CO)_3Fe$-

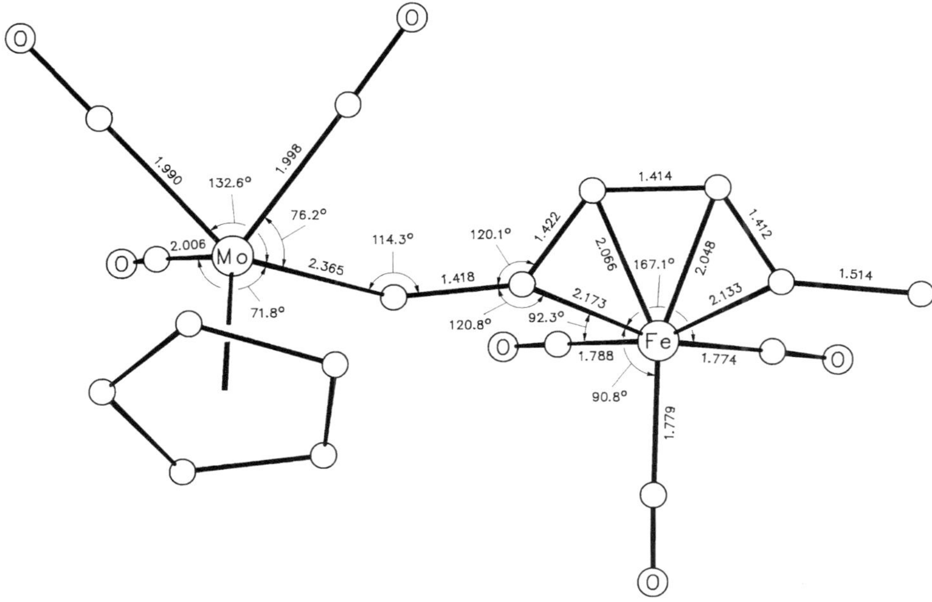

Fig. 3. The molecular structure of E,E-1-$C_5H_5Mo(CO)_3((\eta^4$-CH_2CH=$CHCH$=$CHCH_3)Fe(CO)_3)$ [369].

$(CO)_3(\eta^3$-$CH_2CHCHCH$=$CHCH_3)$ was shown to be similar in solvents which differ widely in polarity indicating it to be a nonionic pathway [369].

E,E-1-$C_5H_5Mo(CO)_3((\eta^4$-CH_2CH=$CHCH$=$CHCH_3)Fe(CO)_3)$ (No. 40) crystallizes in the monoclinic space group $P2_1/n-C^5_{2h}$ (No. 14) with a = 10.642(2), b = 7.658(1), c = 22.440(4) Å, β = 93.44°; Z = 4 molecules per unit cell; D_{calc} = 1.70 g/cm^3. As shown in **Fig. 3**, the addition of the $C_5H_5Mo(CO)_3$ nucleophile has occurred at the methylene terminus of the hexadienyl moiety. The hydrogen atoms at the diene were all located in the difference Fourier map, but were not given in the molecular structure shown [369].

Solutions of both E-1-$C_5H_5Mo(CO)_3((\eta^4$-CH_2CH=$CHCH$=$CHR)Fe(CO)_3)$ (R = H, CH_3, Nos. 39, 40) are thermally unstable and decomposed slowly even at room temperature in the absence of air to give $(C_5H_5Mo(CO)_3)_2$ and an equimolar amount of the iron dimer $((CO)_3$-$Fe(\eta^4$-CH_2CH=$CHCH$=$CHR))_2$ [360, 369]. This decomposition was shown to follow clean first-order kinetics, and values for the first-order rate constants were determined at 50°C in THF to be 2.9×10^{-6} s^{-1} for No. 39 and 2.97×10^{-7} s^{-1} for No. 40. From the temperature dependence of the rate constants between 50 to 60°C for No. 39 and 23 to 50°C for No. 40, activation parameters were determined to be $\Delta H^\ddagger$ = 31.3 ± 1.8 kcal/mol and $\Delta S^\ddagger$ = 20 ± 6 cal · mol^{-1} · K^{-1}, and to be $\Delta H^\ddagger$ = 27.7 ± 1.6 kcal/mol and $\Delta S^\ddagger$ = 14 ± 5 cal · mol^{-1} · K^{-1}, respectively. The averaged first-order rate constants, obtained at ca. 30°C from both the decay of the hexadienyl adduct No. 40 and the growth of its decomposition products, exhibit only limited variation with solvent polarity, but solvent viscosity reveals a significantly more pronounced effect, as indicated by the decrease in the rate constant on going from hexane to dodecane to Nujol [369].

References on pp. 106/18

As shown by ESR spectra (compare Nos. 39, 40), decomposition of $C_5H_5Mo(CO)_3$-$((\eta^4$-$CH_2CH{=}CHCH{=}CHR)Fe(CO)_3)$ occurred via the radical intermediates $C_5H_5Mo(CO)_3^{\bullet}$ and $Fe(CO)_3(\eta^5$-$CH_2CHCHCHCHR)^{\bullet}$. For R = CH_3 this transformation was studied thoroughly in THF containing additives such as $P(C_6H_5)_3$, CBr_4, CCl_4, $CHBr_3$, $CHCl_3$, and 2-nitroso-2-methylpropane. An excess of $P(C_6H_5)_3$ resulted in $((CO)_2P(C_6H_5)_3Fe(\eta^4$-$CH_2CH{=}CH$-$CH{=}CHCH_3))_2$ and $(C_5H_5)_2Mo_2(CO)_5P(C_6H_5)_3$. Use of CBr_4 gave $BrFe(CO)_3(\eta^4$-$CH_2CH{=}CH$-$CH{=}CHCH_3)$ and $C_5H_5Mo(CO)_3Br$, and similar results were observed with CCl_4. In contrast, $CHBr_3$ was less active in trapping formed radicals, and in the presence of $CHCl_3$, only $(C_5H_5Mo(CO)_3)_2$ and $((CO)_3Fe(\eta^4$-$CH_2CH{=}CHCH{=}CHCH_3))_2$ were obtained. The spin trap 2-nitroso-2-methylpropane intercepts all molybdenum radicals. Note that decomposition also followed first-order kinetics in all cases also at high concentrations of the radical traps [369].

$C_5H_5Mo(CO)_3((\eta^4$-C_6H_7-cyclo$)Fe(CO)_3)$ (No. 43) was thermally unstable and decomposed quickly in solution at temperatures above −20 °C via the same radical mechanism as for Nos. 39 and 40 to give $((CO)_3FeC_6H_7$-cyclo$)_2$ and $(C_5H_5Mo(CO)_3)_2$ [369].

$C_5H_5Mo(CO)_3CH_2C{\equiv}CR'$ (R′ = H, CH_3, $C{\equiv}CCH_3$, C_6H_5; Table **1**, Nos. **44** to **47**). Treatment of $C_5H_5Mo(CO)_3CH_2C{\equiv}CH$ (No. 44) with alcohols ROH (R = CH_3, t-C_4H_9, $CH_2C_6H_5$, C_6H_5) led to chelated carbenoid derivatives $C_5H_5Mo(CO)_2{=}C(CH_3)CH{=}C(OR)O$-cyclo (see Formula XXII), but if the same reaction was done in the presence of $BrCH_2C{\equiv}CH$ in THF at −20 °C, π-bonded allyl complexes $C_5H_5Mo(CO)_2(\eta^3$-$CH_2C(CO_2R)CH_2)$ (see Formula XXIII, R = CH_3, t-C_4H_9, $CH_2C_6H_5$, $CH_2C{\equiv}CH$, C_6H_5) were obtained. The analogous RSCO-substituted π-bonded allyl complexes $C_5H_5Mo(CO)_2(\eta^3$-$CH_2C(C(O)SR)CH_2)$ (Formula XXIII, R = CH_3, t-C_4H_9, $CH_2C_6H_5$, C_6H_5) were obtained under similar reaction conditions but irrespective of whether $BrCH_2C{\equiv}CH$ is added or not [105, 184].

XXII

XXIII

Similar C-C bond formation which resulted in $C_5H_5Mo(CO)_2(\eta^3$-$CH_2C(COCl)CH_2)$ occurred when $C_5H_5Mo(CO)_3CH_2C{\equiv}CH$ was reacted with equimolar $AlCl_3$/fumaric acid in 1:1 stoichiometry in THF between 0 °C and room temperature [388].

The reaction of $C_5H_5Mo(CO)_3CH_2C{\equiv}CR'$ (R′ = CH_3, $C{\equiv}CCH_3$, C_6H_5; Nos. 45 to 47) with alcohols ROH (R = CH_3, C_2H_5) in THF at −20 °C or with water in CH_3CN also gave compounds of the π-allyl type $C_5H_5Mo(CO)_2(\eta^3$-$CH_2C(CO_2R)CHR')$ (Formula XXIII; R′ = CH_3, R = CH_3, C_2H_5; R′ = $C{\equiv}CCH_3$, C_6H_5, R = CH_3) [105, 184, 273]. syn-$C_5H_5Mo(CO)_2$-

XXIV

(η^3-$CH_2C(CO_2CH_3)CHC{\equiv}CCH_3$) reacted further with an excess of CH_3OH undergoing 1,5 addition in the presence of $BrCH_2C{\equiv}CH$ at 50°C to give $C_5H_5Mo(CO)_2(\eta^3$-CH_3OCH_2-$C(CO_2CH_3)CHC{=}CHCH_3)$ (Formula XXIV). If $C_5H_5Mo(CO)_3CH_2C{\equiv}CC{\equiv}CCH_3$ was protonated with HBF_4 in ether at −40°C, $[C_5H_5Mo(CO)_3(\eta^2$-$CH_2{=}C{=}CHC{\equiv}CCH_3)]^+$ resulted which was converted into a mixture of syn-$C_5H_5Mo(CO)_2(\eta^3$-$CH_2C(CO_2CH_3)CHC{\equiv}CCH_3)$ and C_5H_5Mo-$(CO)_2(\eta^3$-$CH_3OCH_2C(CO_2CH_3)CHC{=}CHCH_3)$ after the addition of $NaOCH_3$ in methanolic solution [273].

Protonation of $C_5H_5Mo(CO)_3CH_2C{\equiv}CC_6H_5$ (No. 47) by $HClO_4$ in benzene accordingly formed the ionic complex $[C_5H_5Mo(CO)_3(\eta^2$-$CH_2{=}C{=}CHC_6H_5)]ClO_4$ [93], and alkylation of $C_5H_5Mo(CO)_3CH_2C{\equiv}CH$ (No. 44) by $[C(C_6H_5)_3]PF_6$ resulted in $[C_5H_5Mo(CO)_3(\eta^2$-$CH_2{=}C{=}CHC(C_6H_5)_3)]PF_6$ [395].

The reaction of $C_5H_5Mo(CO)_3CH_2C{\equiv}CC_6H_5$ (No. 47) with $(C_5H_5Mo(CO)_2)_2$ in toluene between 0 and 5°C led to a cluster compound suspected to be $(C_5H_5Mo(CO)_2)_2$-$(\mu,\eta^2$-$C_6H_5CCCH_2Mo(CO)_3C_5H_5)$ (Formula XXV) which rapidly decomposed in solution above 10°C, or as solid at room temperature, to $(C_5H_5Mo(CO)_3)_2$; no attempt was made to unequivocally characterize it [357]. Treatment with $Co_2(CO)_8$ in pentane at room temperature gave $((CO)_3Co)_2(\mu_2,\eta^2$-$C_6H_5CCCH_2Mo(CO)_3C_5H_5)$ (No. 48; see Formula XXV) [344].

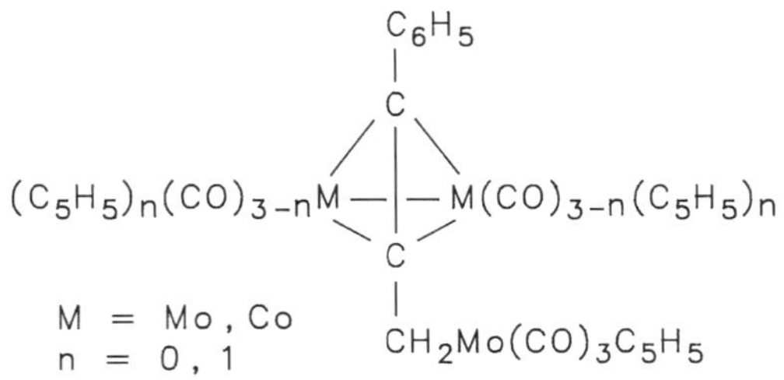

C_6H_5; C; $(C_5H_5)_n(CO)_{3-n}M$ — $M(CO)_{3-n}(C_5H_5)_n$; C; $CH_2Mo(CO)_3C_5H_5$

M = Mo, Co
n = 0, 1

XXV

Stirring equimolar quantities of $Fe_2(CO)_9$ and $C_5H_5Mo(CO)_3CH_2C{\equiv}CR$ (R = CH_3, C_6H_5; Nos. 45, 47) as pentane suspensions at room temperature provided $C_5H_5Mo(CO)_2$-$(\mu,\eta^3$-$RC{=}C{=}CH_2)Fe(CO)_3$ (Formula XXVI) as the major product along with small quantities of $C_5H_5Mo(CO)_2Fe_2(CO)_6(\mu_3,\eta^3$-$RC{=}C{=}CH_2)$ (Formula XXVII). With a 2:1 molar ratio of $C_5H_5Mo(CO)_3CH_2C{\equiv}CR$ relative to $Fe_2(CO)_9$, only compound XXVI was isolated; using a 2:3 molar ratio caused the yield of compound XXVII to increase. The same reaction in refluxing benzene resulted in compound XXVI and $(C_5H_5Mo(CO)_3)_2$ for R = C_6H_5, but only in decomposition to the dimeric molybdenum complex for R = CH_3, due to the thermal lability of No. 45 under these conditions. Similar results gave the reaction of $C_5H_5Mo(CO)_3CH_2C{\equiv}CR$ with $Fe_3(CO)_{12}$ in refluxing benzene leading mainly to $(C_5H_5Mo(CO)_3)_2$ along with compound XXVI in the case of R = C_6H_5. A possible reaction mechanism is discussed [347, 373].

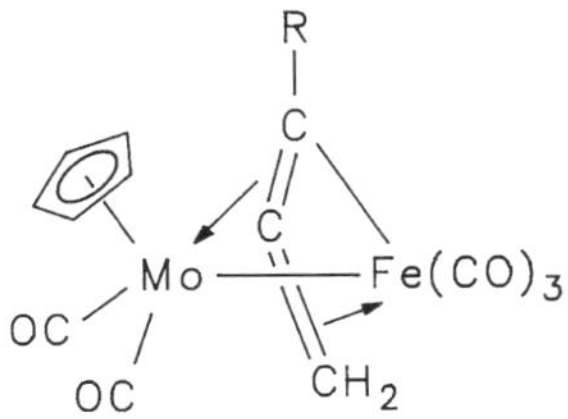

R; C; C; Mo — $Fe(CO)_3$; OC; OC; CH_2

XXVI

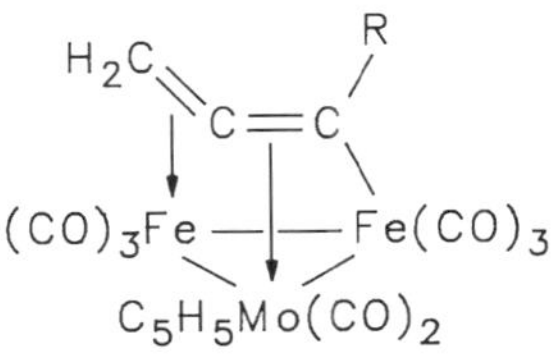

H_2C; R; C=C; $(CO)_3Fe$ — $Fe(CO)_3$; $C_5H_5Mo(CO)_2$

XXVII

References on pp. 106/18

$C_5H_5Mo(CO)_3CH_2C{\equiv}CCRR'OH$ (R = R' = H, CH_3, C_6H_5; Table **1**, Nos. **49**, **51**, and **53**; R = H, CH_3, R' = C_6H_5; Table **1**, Nos. **50** and **52**; R, R' = fluorenylidene; Table **1**, No. **54**) and **$C_5H_5Mo(CO)_3CH_2C{\equiv}CCH_2CH_2OH$** (Table **1**, No. **55**). None of these complexes could be isolated analytically pure since all of them tend to isomerize when warmed in solution or worked up by column chromatography on alumina, but the type of isomerization product depends on R and R'. Compounds which contain $-CH_2C{\equiv}CCH_2OH$ (No. 49), $-CH_2C{\equiv}CCH(C_6H_5)OH$ (No. 50), $-CH_2C{\equiv}CC(CH_3)_2OH$ (No. 51), and $-CH_2C{\equiv}CCH_2CH_2OH$ (No. 55) as σ-bonded ligands lactonized to give compounds of the type $C_5H_5Mo(CO)_2(\eta^3\text{-}CH_2C(CO_2\text{-}CRR'(CH_2)_n)CH\text{-cyclo})$ as shown in Formula XXVIII [197, 286], while the complexes with more bulky ligands, $-CH_2C{\equiv}CC(CH_3)(C_6H_5)OH$ (No. 52), $-CH_2C{\equiv}CC(C_6H_5)_2OH$ (No. 53), and $-CH_2C{\equiv}CC(OH){=}C(C_{12}H_8)$ (No. 54), afforded complexes of the type $C_5H_5Mo(CO)_3\text{-}(\eta^1\text{-}C{=}CHCRR'OCH_2\text{-cyclo})$ (Section 1.5.1.4.1.12.3.1) depicted in Formula XXIX [218].

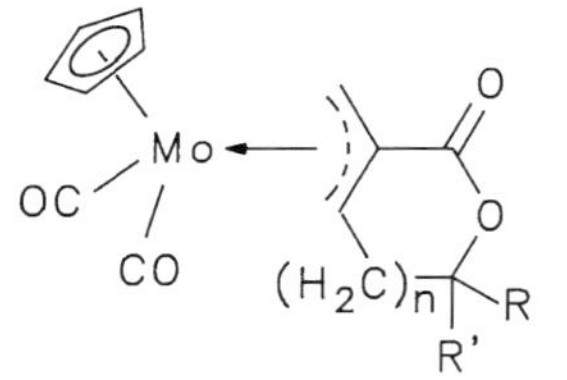

XXVIII

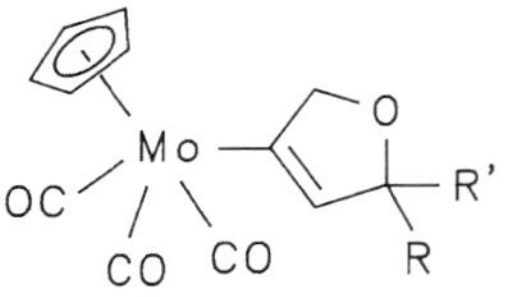

XXIX

A reaction mechanism is discussed. Zwitterionic species are considered as common intermediates which are capable of intermolecular nucleophilic attack by their alcoholate function at either the electrophilic CO carbon, producing XXVIII, or the electrophilic CH_2 carbon, affording XXIX [197, 218]. Protonation of $C_5H_5Mo(CO)_3CH_2C{\equiv}CCRR'OH$ (R = CH_3, R' = C_6H_5; R = R' = C_6H_5; R, R' = fluorenylidene, Nos. 52 to 54) at the alcohol function followed by dehydration, using ethereal HBF_4 at temperatures below −20 °C, gave cationic π-bonded butatriene complexes of the type $[C_5H_5Mo(CO)_3(CH_2{=}C{=}C{=}CRR')]BF_4$ [292].

$C_5H_5Mo(CO)_3CH_2C_6H_4R$ (R = H, CH_3-2, -3, -4; Table **1**, Nos. **56 to 59**), **$C_5H_5Mo(CO)_3\text{-}CH_2C_6H_2(CH_3)_3\text{-}2,4,6$** (No. **60**), and **$C_5H_5Mo(CO)_3CH_2C_6H_4R$** (R = CF_3-3, -4, Nos. **66**, **67**; R = OCH_3-2, -3, -4, F-2, -3, -4, Cl-2, -3, -4, Nos. **73** to **81**). Complexes No. 56, 57, and 60 were investigated by X-ray analysis and results were compared. Cell parameters for the three compexes are given in the following table [309].

No.	compound	space group, cell parameters
56	$C_5H_5Mo(CO)_3CH_2C_6H_5$	$Pbca-D_{2h}^{15}$ (No. 61), a = 8.863(3), b = 18.080(5), c = 16.806(4) Å; Z = 8, D_{meas} = 1.61, D_{calc} = 1.65 g/cm³
57	$C_5H_5Mo(CO)_3CH_2C_6H_4CH_3$-2	$P\bar{1}-C_i^1$ (No. 2), a = 14.929(5), b = 14.418(5), c = 6.918(2) Å, α = 100.29(2)°, β = 90.55(2)°, γ = 94.68(3)°; Z = 4, D_{calc} = D_{meas} = 1.58 g/cm³
60	$C_5H_5Mo(CO)_3CH_2C_6H_2(CH_3)_3$-2,4,6	$P2_12_12_1-D_2^2$ (No. 19), a = 20.867(3), b = 11.088(2), c = 7.132(1) Å; Z = 4, D_{calc} = 1.52 g/cm³

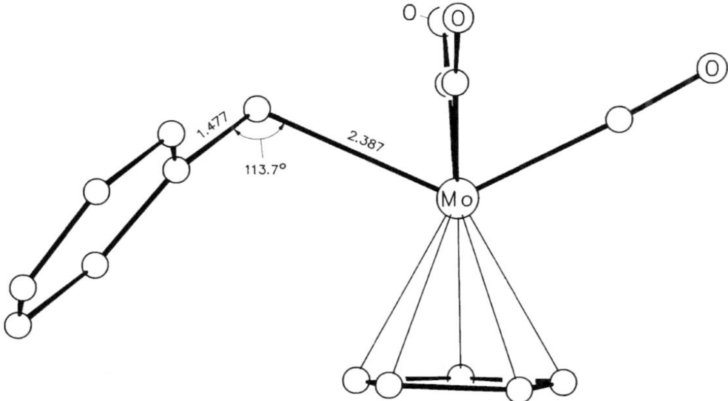

Fig. 4. The molecular structure of $C_5H_5Mo(CO)_3CH_2C_6H_5$ [309].

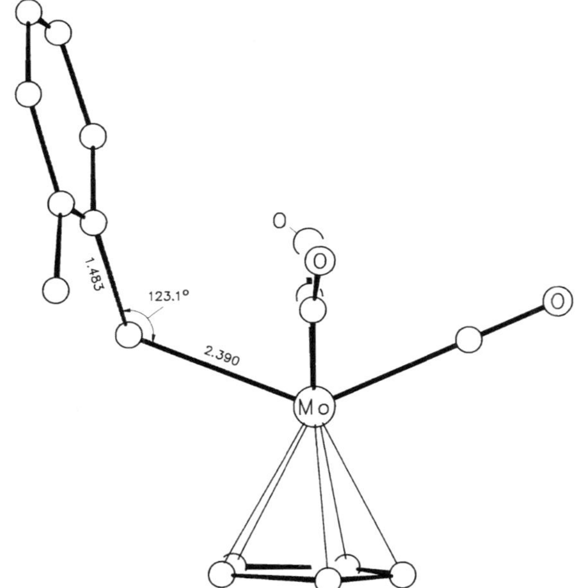

Fig. 5. The molecular structure of $C_5H_5Mo(CO)_3CH_2C_6H_4CH_3$-2 [309].

As shown in **Fig. 4 to Fig. 6**, the complexes exhibit a distorted pseudo-square pyramidal geometry around Mo with the cyclopentadienyl ring in the apical position and CO as well as benzyl in the plane. The CH_2 group lies above the plane of the three CO carbon atoms. In No. 57 the Mo-CH_2 bond is staggered with respect to the cyclopentadienyl ring carbons, but in Nos. 56 and 60 it is nearly eclipsed. The tilt of the cyclopentadienyl ring with respect to the plane of the carbonyl carbon atoms varies regularly from 8.93° to 10.74° to 13.73° across the series as the amount of steric congestion in the molecule increases. For Nos. 57 and 60

References on pp. 106/18

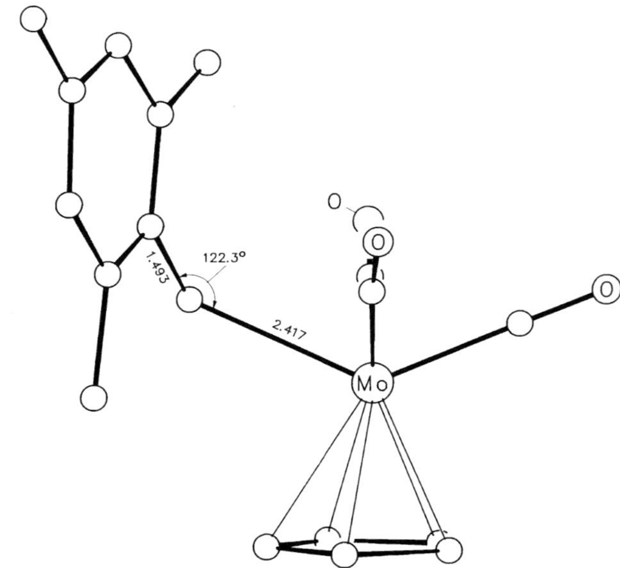

Fig. 6. The molecular structure of $C_5H_5Mo(CO)_3CH_2C_6H_2(CH_3)_3$-2,4,6 [309].

some disorder in the cyclopentadienyl ring was indicated by high thermal parameters, probably resulting indirectly from intramolecular interactions through the o-substitution. In the case of No. 57, two distinct, but similar molecules were found in the asymmetric unit, and bond lengths as well angles given in Fig. 5 are average value of both molecules.

The general structural features for the three benzyl derivatives are in line with the increase of the rate constants k_1 for the first stage of $P(C_6H_5)_3$-assisted migratory CO insertion in CH_3CN at 30 °C. This is an expected correlation as a progressive increase in steric congestion in the ground state, relative to the transition state of the k_1 step, will serve to continuously lower the activation energy [309].

The phenyl hydrogens of $C_5H_5Mo(CO)_3CH_2C_6H_5$ (No. 56) easily enter into isotopic exchange under acidic conditions in CF_3CO_2D or in mixtures of CF_3CO_2D and CH_3CO_2D at 25 °C to give $C_5H_5Mo(CO)_3CH_2C_6H_4D$. It was shown that the $C_5H_5Mo(CO)_3CH_2$ moiety increased the acidic hydrogen exchange rate compared to benzene or toluene and directed mostly to the 4- and 2-position. The rate constant was determined to be 0.6×10^{-6} s^{-1}. Data proved the $C_5H_5Mo(CO)_3CH_2$ moiety to be a strong electron donor on the benzene ring, and it was shown that the hydrogen exchange followed the electrophilic aromatic substitution mechanism involving as an intermediate the formation of a carbenium ion stabilized by σ and π conjugation. The hydrogen exchange is accompanied by the cleavage of the Mo-C σ bond to give toluene and the corresponding trifluoroacetate, and the absence of hydrogen exchange in the cyclopentadienyl ring was proved by the fact that the resultant $C_5H_5Mo(CO)_3OC(O)CF_3$ did not contain deuterium [126, 220, 242].

Substitution effects on an aromatic ring can be described in terms of resonance interaction with a π system (the σ_R^0 constant) and a polar or inductive effect (the σ_I constant); values for the $C_5H_5Mo(CO)_3CH_2$ moiety were derived to be $\sigma_R^0 = -0.21$ and $s_I = -0.07$, using a combination of ^{19}F and ^{13}C NMR spectra. It was shown that nearly equal data were

References on pp. 106/18

calculated for a wide range of 2-, 3-, and 4-substituted benzyl derivatives $C_5H_5Mo(CO)_3$-$CH_2C_6H_4R$. The values for σ_R^0 and s_I indicated $C_5H_5Mo(CO)_3CH_2$ to be a weak inductive donor similar to a CH_3 group, but a strong π-electron donor comparable in magnitude to the Cl substituent. This is reasonable in terms of the electropositive nature of the Mo group and the donation of the electrons from C_5H_5 to stabilize the positive charge of Mo [84, 258]. For the series of benzyl derivatives No. 56 to 60, Nos. 66, 67, and Nos. 73 to 81, systematic variations for the chemical shift values are seen for the aromatic and CH_2 carbon atoms and for the molybdenum atom. $\delta(^{95}Mo)$ is shown to be highly sensitive to substituents, and, as expected, electron donors resulted in a high field shift proving the electron density on Mo to be of primary importance. Since the ^{95}Mo shifts show an excellent correlation with Hammett substituent constants in this series, they appear to be controlled by the paramagnetic term for this spin 5/2 nucleus [258]. These electronic effects, however, do not result in a variation in the electric field gradient large enough to significantly affect the relaxation times of the monosubstituted compounds, having T_1 mostly in the 5 to 6 ms range [304]. Investigation of the ground state substituent effects of the $C_5H_5(CO)_3MoCH_2$ residue and related carbonyl(transition metallo)-methyl groups by examination of the para-C_6H_4F ^{19}F NMR chemical shifts indicates that these substituents are of comparable electron-donating ability to the most effective main group metallomethyl substituents, although the substituent effects in the ground state are markedly smaller than those detected in transition states and in studies of equilibria [123].

On the other hand, the molybdenum-benzyl bonding is dominated by inductive effects as indicated by only small variations in σ_R^0 values for m- and p-fluorophenyl transition metal complexes, while the inductive parameter σ_I varied greatly. π-Delocalization from Mo to the σ-bonded benzyl ring does occur but is of secondary importance. The group electronegativity for the $C_5H_5Mo(CO)_3$ residue is determined to be $\chi = 2.28$ [84].

UV irradiation of a concentrated solution of $C_5H_5Mo(CO)_3CH_2C_6H_5$ in hexane for 5 d gave small amounts of $C_5H_5Mo(CO)_2(\eta^3$-$CH_2C_6H_5)$ (Formula XXX) along with mainly $(C_5H_5Mo(CO)_3)_2$. The same product distribution but less pure was obtained by heating No. 54 in a sublimer at 100 °C at 1.3×10^{-4} atm [22, 30]. The same π-allyl complex was obtained from a $(CH_3)_3NO$-catalyzed decarbonylation in a one-pot reaction starting from $Na[C_5H_5$-$Mo(CO)_3]$ and $C_6H_5CH_2Cl$ in THF without isolation of No. 54 [288].

$(CO)_2Mo$ CH_2

XXX

Under more dilute conditions in pentane, photolysis is described as being more effective and led only to dimeric $(C_5H_5Mo(CO)_3)_2$, because any η^3-derivative formed would undergo further photochemical degradation to give toluene and unidentified decomposition products [261, 287]. Thermal decomposition at 140 °C, neat or solutions in benzene, decane and cumene, respectively, resulted in the formation of the dimers $(C_5H_5Mo(CO)_3)_2$, $(C_6H_5CH_2$-$C_5H_4Mo(CO)_3)_2$, and $C_5H_5Mo(CO)_3Mo(CO)_3C_5H_4CH_2C_6H_5$ along with bibenzyl on thermolysis in benzene and decane. The latter two complexes originated from direct benzyl migration from the metal to the endo side of the cyclopentadienyl ring, which was described as concerted homolysis of the Mo-$CH_2C_6H_5$ bond. The $^{\bullet}CH_2C_6H_5$ radical simultaneously attacks the

References on pp. 106/18

C_5H_5 ring, but never becomes free of the rest of the molecule [136]. Thermolysis of the deuterated analogue of No. 56, $C_5H_5Mo(CO)_3CD_2C_6D_5$, in THF-d_8, produces some deuterated toluene, $C_6D_5CD_2H$ [227].

Fragmentation induced by UV irradiation in CH_2Cl_2 was shown to be temperature-dependent, because at −30 °C in the presence of 2,3,5,6-$(CH_3)_4C_6HNO$, only $C_5H_5Mo(CO)_3$-$N(O)C_6H(CH_3)_4$-2,3,5,6 was produced, but at 20 °C only $C_6H_5CH_2^{\bullet}$ was detected [243].

In contrast to $C_5H_5Mo(CO)_3CH_3$ (No. 1), treatment of $C_5H_5Mo(CO)_3CH_2C_6H_5$ (No. 56) with I_2, probably in THF, resulted in direct iodination to give $C_5H_5Mo(CO)_3I$ and $C_6H_5CH_2I$. The enthalpy of this reaction was determined to be $\Delta H = -28.8 \pm 1.0$ kcal/mol and the Mo-$CH_2C_6H_5$ bond strength can be estimated to be 32 kcal/mol [342].

$C_5H_5Mo(CO)_3CH_2C_6H_5$ (No. 56) reacted like $C_5H_5Mo(CO)_3CH_3$ (No. 1) and $C_5H_5Mo(CO)_3$-C_2H_5 (No. 2) with $Li[(C_2H_5)_3BH]$ to form an anionic π-aldehyde derivative $[C_5H_5Mo(CO)_2$-$(\eta^2\text{-}C_6H_5CH_2CHO)]^-$ (compare Formula VIII on p. 80) [314]. A reaction at 60 °C with $CH_3C{\equiv}CCH_3$ in hexane gave a σ-bonded vinylketone derivative $C_5H_5Mo(CO)_2C(CH_3)$=C-$(CH_3)COCH_2C_6H_5$ in which the keto group is coordinated to the molybdenum atom; compare Formula XVI on p. 83. A mechanism is briefly discussed proposing a side-bonded acyl species which react with the alkyne to form a η^2-bonded acetylene complex. Subsequent migration of the formed acyl group onto the coordinated $CH_3C{\equiv}CCH_3$ gives the product [155, 188].

A solution of $C_5H_5Mo(CO)_3CH_2C_6H_4F$-4 (No. 78) and LiCl in CH_3OH was treated with a fivefold excess of Ce^{4+} ion at room temperature to afford 4-$FC_6H_4CH_2CO_2CH_3$ rapidly and quantitatively. A mechanism is discussed employing oxidized $[C_5H_5Mo(CO)_3CH_2C_6H_4F\text{-}4]^{+\bullet}$ and the subsequent CO insertion product $[C_5H_5Mo(CO)_2C(O)CH_2C_6H_4F\text{-}4]^{+\bullet}$ as proposed intermediates in which the latter easily reacts with CH_3OH [121].

$C_5H_5Mo(CO)_3(CH_2C_6H_4[\text{-}CHCH_2\text{-}])_n$ (Table **1**, Nos. **90**, **91**). The polymer-supported compounds, derived from both linear chloromethylated polystyrene and chloromethylated styrene/divinylbenzene cross-linked resins, pertinaciously retain impurities such as undisplaced chloride and hydrated molybdenum oxides produced from traces of moisture [132, 136]. Both linear and cross-linked polymers No. 90 and 91 decompose at 140 °C very slowly neat in the solid phase, faster in solution. In both modes the dimer $(C_5H_5Mo(CO)_3)_2$ is formed in addition to polymers resulting from benzyl migration, viz., $([\text{-}CH_2CH\text{-}]C_6H_4CH_2)_nC_5H_4Mo$-$(CO)_3Mo(CO)_3C_5H_5$ and $([\text{-}CH_2CH\text{-}]C_6H_4)_nCH_2C_5H_4Mo(CO)_3Mo(CO)_3C_5H_4(CH_2C_6H_4[\text{-}CH$-$CH_2\text{-}])_n$ [136]; cf. the thermolysis of the benzyl monomer, $C_5H_5(CO)_3MoCH_2C_6H_5$ (No. 56).

$C_5H_5Mo(CO)_3CH_2C_5H_4N$-3 and **$C_5H_5Mo(CO)_3CH_2C_5H_4N$-4** (Table **1**, Nos. **92**, **93**). Both compounds are only slightly soluble in water and undergo base-catalyzed decomposition in alkaline solution. Alkylation of $C_5H_5Mo(CO)_3CH_2C_5H_4N$-4 (No. 93) at the pyridine nitrogen using 4-$CH_3C_6H_4SO_2OCH_3$ gave $[C_5H_5Mo(CO)_3CH_2C_5H_4NCH_3\text{-}4]^+$ (No. 96) which can be precipitated with $Na[B(C_6H_5)_4]$ as the corresponding salt [81]. Rapid oxidation of $C_5H_5Mo(CO)_3$-$CH_2C_5H_4N$-3 (No. 92) in the presence of LiCl by Ce^{4+} in CH_3OH resulted in 3-NC_5H_4-$CH_2CO_2CH_3$. A mechanism is discussed involving $[C_5H_5Mo(CO)_3CH_2C_5H_4N\text{-}3]^{+\bullet}$ which was converted to $[C_5H_5Mo(CO)_2C(O)CH_2C_5H_4N\text{-}3]^{+\bullet}$ which reacted further with CH_3OH [121]. Protonation of both complexes dissolved in CH_3OH with dilute aqueous mineral acids, e.g., 0.1 M H_2SO_4, yielded $[C_5H_5Mo(CO)_3CH_2C_5H_4NH\text{-}3]^+$ (No. 94) and $[C_5H_5Mo(CO)_3CH_2$-$C_5H_4NH\text{-}4]^+$ (No. 95) [81].

$[C_5H_5Mo(CO)_3CH_2C_5H_4NH\text{-}3]^+$ and **$C_5H_5Mo(CO)_3CH_2C_5H_4NH\text{-}4]^+$** (Table **1**, Nos. **94**, **95**) were prepared from $C_5H_5Mo(CO)_3CH_2C_5H_4N$-3 and $C_5H_5Mo(CO)_3CH_2C_5H_4N$-4 (Nos. 92, 93) by protonation with dilute aqueous mineral acids, e.g., 0.1 M H_2SO_4 [81]. The two cations are

References on pp. 106/18

reported to be relatively insensitive to oxygen, but light-sensitive [82]. They are stable in dilute mineral acids for several days at room temperature in the dark [56].

UV spectra were measured for both cations in 0.1 M aqueous H_2SO_4, and data are given as λ_{max} (log ε) = 293(4.2) for No. 94 and λ_{max} (log ε) = 318(4.1) for No. 95 [81].

Spectrophotometric pK_a measurements in 0.1 M phosphate, carbonate, or borate buffer solutions, accurate within 0.1 pH unit, gave values of pK_a = 6.5 for $[C_5H_5Mo(CO)_3CH_2C_5H_4NH\text{-}3]^+$ and pK_a = 7.9 for $[C_5H_5Mo(CO)_3CH_2C_5H_4NH\text{-}4]^+$, from which the associated σ Hammett parameters of acidity have been derived as −0.21 and −0.45, respectively. Data revealed the pyridinium ions are very weak acids, the 4-substituted one more so than the 3-substituted compound, and proved the $C_5H_5Mo(CO)_3CH_2$ moiety to be a very strong electron donor; for similar conclusion compare No. 56 and related benzyl compounds. From the comparison with other substituted pyridinium ions, a large conjugative electron interaction with the pyridyl moiety was ascribed to the alkylmetal carbonyl-containing substituents, best interpreted as a hyperconjugative effect in view of the CH_2 group. Accordingly the large electron donation is almost equally divided between the inductive and the hyperconjugative effect [81].

Decomposition with the exclusion of light occurred at about 65 °C, and the mechanism was investigated by kinetic studies and is discussed with respect to the reaction products. For both complexes a clean first-order kinetics in aqueous $HClO_4$, HCl, and HBr acids unaffected by oxygen was observed. For rate constants at several temperatures and acid concentrations, see [82]. For $[C_5H_5Mo(CO)_3CH_2C_5H_4NH\text{-}4]^+$ a rate-determining unimolecular heterolysis of the Mo-C bond (S_E1 at carbon, S_N1 at molybdenum) was indicated by the absence of any major dependence of the rate of decomposition of the complex on acid or halide ion concentration and the formation of $NC_5H_4CH_3$-4 as the sole organic product. $C_5H_5Mo(CO)_3Cl$ was isolated as the inorganic compound. A general salt effect was not observed. On the basis of reduced decomposition rates and the formation of NC_5H_4CHO-3 along with $Mo(CO)_6$ as additional products, homolysis of the Mo-C bond was indicated as the rate-determining step for $[C_5H_5Mo(CO)_3CH_2C_5H_4NH\text{-}3]^+$. The supposed radicals formed, $C_5H_5Mo(CO)_3^{\cdot}$ and $[CH_2C_5H_4NH\text{-}3]^{+\cdot}$, can then react with an excess of oxygen to give the above oxidation products. In the absence of a sufficient amount of oxygen, the radicals reacted with themselves or the medium. These characteristics of decomposition are not surprising, because a much higher energy of the transition state for the heterolysis is expected for the 3-substituted complex [56, 82]. Activation energy and entropy for the decomposition of $[C_5H_5Mo(CO)_3CH_2C_5H_4NH\text{-}3]^+$ were determined as $\Delta H^{\ddagger}$ = 31.4 kcal/mol and $\Delta S^{\ddagger}$ = 10 cal · mol^{-1} · K^{-1}, and the data are consistent with a rate-determining homolysis. Homolysis of the 3- and 4-pyridiomethylmolybdenum cations was also observed during several days in $CDCl_3$ to afford $NC_5H_4CH_2D$ rather than $NC_5H_4(CH_2)_2C_5H_4N$ [82].

The reactions of both the meta- and para-substituted cationic complexes No. 94 and 95 with $Tl(ClO_4)_3$ in aqueous perchloric acid involves a rapid (probably two-electron) oxidation step followed by a series of reactions resulting in the formation of the corresponding N-protonated pyridinioacetic acids, 3- and 4-$HN^+C_5H_4CH_2CO_2H$, as the main organic products. The significant increase in the rate of oxidation with decreasing acid concentration has been ascribed to the reactivity order $[Tl(OH)_2]^+ > [Tl(OH)]^{2+} \gg Tl^{3+}$. This overall redox behavior closely resembles the corresponding oxidations of both No. 94 and 95 by $[IrCl_6]^{2-}$, although the latter are believed to involve two sequential electron transfers from the hexachloroiridate(IV) ion to the two organomolybdenum complexes. The reactions of dilute solutions of Nos. 94 and 95 with $Hg(ClO_4)_2$ in aqueous acidic solution appeared to comprise initial rapid but reversible coordination of the organomolybdenum substrates to an electrophilic mercury(II) species, followed by a breakdown of the intermediates thus generated to the corre-

References on pp. 106/18

sponding meta- and para-substituted pyridiniomethyl mercury(II) ions, $[HN^+C_5H_4CH_2Hg]^{2+}$, as the final organometallic products. The rates of these conversions are indicative of normal bimolecular electrophilic substitution reactions [159]. For further details related to the kinetics of the degradation of Nos. 94 and 95 by Hg(II), Tl(III), and Ir(IV), including the influence of added chloride ion on the above reactions, cf. the original reference [159].

$C_5H_5Mo(CO)_3CH_2C(O)R$ (R = H, CH_3; Table **1**, Nos. **103, 104**) reacted with i-$C_3H_7NH_2$ in the presence of $BF_3 \cdot O(C_2H_5)_2$ in THF at −78°C to produce $C_5H_5Mo(CO)_3CH_2C(R)$=NC_3H_7-i (R = H, CH_3, Nos. 101, 102) along with $(C_5H_5Mo(CO)_3)_2$.

Different products were isolated from the reaction of $C_5H_5Mo(CO)_3CH_2CHO$ (No. 103) and $C_5H_5Mo(CO)_3CH_2C(O)CH_3$ (No. 104), respectively, with hydrazine in the presence of $BF_3 \cdot O(C_2H_5)_2$ at −78°C in CH_2Cl_2. No. 103 was converted to a dimetalloazine, C_5H_5Mo-$(CO)_3CH_2CHNNCHCH_2Mo(CO)_3C_5H_5$, while reaction of No. 104 led to the formation of $C_5H_5Mo(CO)_2(\eta^2$-C(O)NHN=C$(CH_3)_2)$ as the final product with **$C_5H_5Mo(CO)_3CH_2C(CH_3)$=$NNH_2$** (IR: 1605 ($\nu$(C=N)), 1936, 2012 ($\nu$(CO))) as the proved intermediate according to Scheme 4. Attempts to isolate the η^1-imine failed because of its kinetic instability in solution, and its conversion to the cyclic acyl derivative is preferred since further condensation of the labile primary metallohydrazone with the corresponding metalloketone No. 104 to give a dimetalloazine is less likely than with metalloaldehyde No. 103. Treatment of No. 103 with methyl hydrazine likewise produced also a cyclic acyl derivative, $C_5H_5Mo(CO)_2(\eta^2$-C(O)N(CH_3)-N=$CHCH_3$) [372].

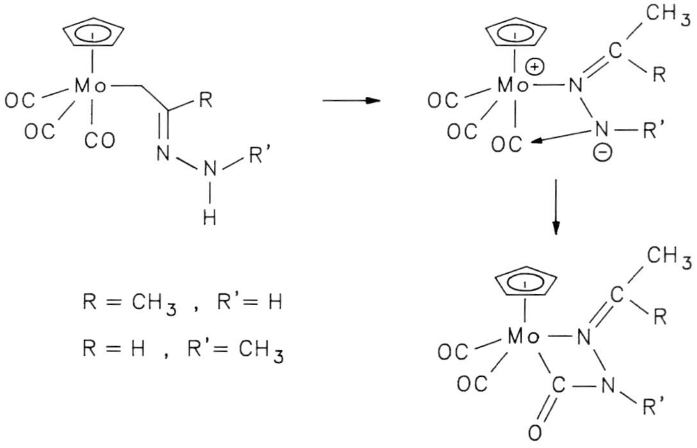

R = CH_3 , R' = H
R = H , R' = CH_3

Scheme 4

$C_5H_5Mo(CO)_3CH_2C(O)CH_3$ (No. 104) was converted into $C_5H_5Mo(CO)_2(\eta^3$-$CH_2C(CH_3)O)$ by UV irradiation at $\lambda > 360$ nm [297, 303, 331].

Attempts to protonate $C_5H_5Mo(CO)_3CH_2C(O)CH_3$ (No. 104) by HPF_6 in ether or dilute HCl in petroleum ether remained unsuccessful, resulting in decomposition and no reaction, respectively. No vinyl alcohol compound was detected [152].

$C_5H_5Mo(CO)_3CH_2CO_2H$ (Table **1**, No. **110**) crystallizes in the monoclinic space group $P2_1/n-C_{2h}^5$ (No. 14) with a = 6.82(2), b = 13.17(3), c = 12.17(3) Å, β = 92.7(2)°; Z = 4 molecules per unit cell, D_{meas} = 1.853, D_{calc} = 1.835 g/cm^3. The crystal contains hydrogen-bonded dimers of $C_5H_5Mo(CO)_3CH_2CO_2H$ as shown in **Fig. 7**, and each dimer is symmetric

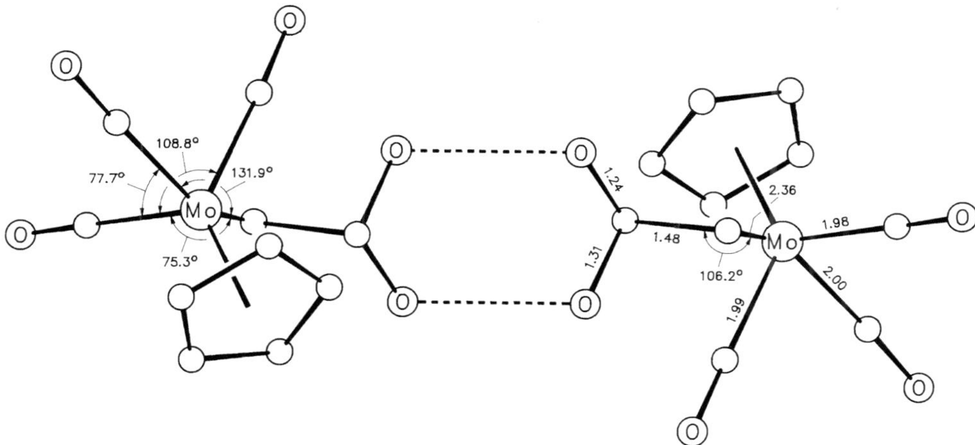

Fig. 7. The molecular structure of the hydrogen-bonded dimer of $C_5H_5Mo(CO)_3CH_2CO_2H$ [37, 64].

about an inversion center. The average molybdenum-to-ring-carbon contact distance (2.35 Å) and the distance from the Mo atom to the ring centroid (2.01 Å) agree well with similar distances in analogous compounds [37, 64].

$C_5H_5Mo(CO)_3CH_2CO_2C_2H_5$ (Table **1**, No. **112**). The mass spectrum of No. 112 was investigated thoroughly and three fragmentation patterns were distinguished starting from the highest peak $[C_5H_5Mo(CO)_2CH_2CO_2C_2H_5]^+$. One fragmentation pathway involves further decarbonylation resulting in $[C_5H_5Mo(CO)CH_2CO_2C_2H_5]^+$, $[C_5H_5MoCH_2CO_2C_2H_5]^+$, $[C_5H_5MoCH_2CO_2H]^+$, $[C_5H_5MoOH]^+$, and $[C_5H_5MoO]^+$. In a second competing pathway the highest fragment loses ketene to give $[C_5H_5Mo(CO)_2OC_2H_5]^+$, $[C_5H_5Mo(CO)OC_2H_5]^+$, $[C_5H_5MoOCH{=}CH_2]^+$, and $[C_5H_5MoOH]^+$. A third mode of decomposition of the highest dicarbonyl ion generated $[C_5H_5Mo(CO)_2(HC{\equiv}CO_2C_2H_5)]^+$, $[C_5H_5Mo(CO)_2]^+$, $[C_5H_5Mo]^+$, $[C_3H_3Mo]^+$, and $[Mo]^+$. Appreciable concentrations of dipositive ions, $[C_5H_5MoOC_2H_3]^{2+}$ and $[C_5H_5Mo]^{2+}$, were observed. Additionally, weak ions corresponding to the then still unknown $(C_5H_5Mo(CO)_2)_2$ were found; for more detail see [53].

Early studies on thermal or photochemical treatment resulted in only $(C_5H_5Mo(CO)_3)_2$ just as in the reaction with CH_3MgBr or $LiAlH_4$ [26], but later investigations showed $C_5H_5Mo(CO)_3CH_2CO_2C_2H_5$ was converted into $C_5H_5Mo(CO)_2(\eta^3\text{-}CH_2C(OC_2H_5)O)$ by UV irradiation at $\lambda > 360$ nm [297, 303, 331, 399].

$C_5H_5Mo(CO)_3CH_2OCH_3$ (Table **1**, No. **118**) was also obtained in trace amounts from the reaction of $C_5H_5Mo(CO)_3FPF_5$ and $C_5H_5Fe(CO)_2CH_2OCH_3$ in CH_2Cl_2 at $-20\,°C$ along with mainly $(C_5H_5Mo(CO)_3)_2$, $[C_5H_5Fe(CO)_2{=}CHOCH_3]^+$, and some other by-products [290]. No. 118 is additionally regenerated in poor yield from $C_5H_5Mo(CO)_3CH_2X$ (X = Cl, Br; Nos. 128, 129) which reacted with CH_3OH alone for 2 h [238, 267], or in the presence of NaOH [39], but gave $C_5H_5Mo(CO)_3X$ as main products. The chloro- and bromomethyl derivatives themselves, however, are daughter products of the methoxymethyl parent compound No. 118, which reacted with HCl or HBr in light petroleum to give CH_3OH and $C_5H_5Mo(CO)_3CH_2X$ through cleavage of the carbon-oxygen bond [39].

The reaction of $C_5H_5Mo(CO)_3CH_2OCH_3$ (No. 118) with $C_5H_5Mo(CO)_3FPF_5$ in CH_2Cl_2 at $-20\,°C$ and subsequent warming to room temperature resulted in two major products,

$[C_5H_5Mo(CO)_3{=}CHOCH_3]PF_6$ and $[(C_5H_5Mo(CO)_2)_2(\mu_2,\eta^2\text{-}CH_3CO)]PF_6$ (compare Formula II on p. 2, along with variable amounts of $[(C_5H_5Mo(CO)_2)_2(\mu\text{-}H)]PF_6$, $[C_5H_5Mo(CO)_4]PF_6$, $C_5H_5Mo(CO)_3CH_3$, and $(C_5H_5Mo(CO)_3)_2$. An excess of $C_5H_5Mo(CO)_3FPF_5$ favored the formation of the μ-H compound, whereas an excess of No. 118 favored the methoxymethylidene salt. The amount of the μ_2,η^2-acyl compound remained invariant. $[C_5H_5Mo(CO)_3{=}CH\text{-}OCH_3]PF_6$ was also obtained from $C_5H_5Mo(CO)_3CH_2OCH_3$ and $[C(C_6H_5)_3]PF_6$. The involvement of $[C_5H_5Mo(CO)_3{=}CH_2]^+$ as the intermediate was proved from an independent reaction of this ionic carbene (generated in situ from $C_5H_5Mo(CO)_3CH_2Cl$ (No. 128) and $AgPF_6$ at −80 °C in CH_2Cl_2) with No. 118 to give equal quantities of $[C_5H_5Mo(CO)_3{=}CHOCH_3]^+$ and $C_5H_5Mo(CO)_3CH_3$. For a detailed mechanism, see [290]. The same intermediate was suggested based on the reaction of No. 118 with HBF_4 in the presence of cyclohexene affording good yields of norcarane by CH_2 transfer to the carbon-carbon double bond [31]. Treatment of $C_5H_5Mo(CO)_3CH_2OCH_3$ with 40% aqueous hydrogen fluoride produces $C_5H_5Mo(CO)_3CH_3$ [39].

$C_5H_5Mo(CO)_3CH_2O_2CCH_3$ (Table **1**, No. **119**). Thermolysis led to a complex array of organic products including small amounts of CH_3CO_2H, but $CH_3CO_2CH_3$ was not observed [293].

Reactions with protic acids were investigated and showed that the C-O bond is more susceptible to protolytic cleavage than the Mo-C bond, resulting in CH_3CO_2H and the highly reactive carbene species $[C_5H_5(CO)_3Mo{=}CH_2]^+$ as an intermediate. With a compatible nucleophilic counterion, a covalent bond to the electrophilic carbene carbon was formed, such as with HCl in petroleum ether, resulting in $C_5H_5Mo(CO)_3CH_2Cl$ (No. 128). Reaction with a twofold excess of $C_6H_5CO_2H$ gave $C_5H_5Mo(CO)_3CH_2O_2CC_6H_5$ (No. 121) at ambient temperature. Acetate exchange using CD_3CO_2D is slow at room temperature, but at 60 °C, CH_3CO_2D is liberated. This transcarboxylation is also complete in less than 24 h at ambient temperature if catalytic amounts of HBF_4 are added which facilitate the formation of the migrating $[C_5H_5Mo(CO)_3{=}CH_2]^+$. A reaction with HBF_4 alone gave $C_5H_5Mo(CO)_3FBF_3$, $(C_5H_5Mo(CO)_3)_2$, and CH_3CO_2H. Treatment with excess $4\text{-}CH_3C_6H_4SO_3H$ may result in the respective tosylate ester, but attempts to isolate it led only to decomposition [293].

The complex is completely resistant to hydrolysis, transesterification with methanol (either in the presence or absence of catalytic amounts of acid), and carboxylate displacement by nucleophiles such as HO^-, CH_3O^-, H_2N^-, or $(C_2H_5)_2N^-$. Treatment with 120 atm CO or reaction with $P(CH_3)_3$ or $P(C_6H_5)_3$ failed to induce migratory insertion of CO at temperatures up to 90 °C [293].

$C_5H_5Mo(CO)_3CH_2O_2CC_4H_9$-t (Table **1**, No. **120**) was treated with excess $LiCH_3$ in ether at −78 °C. After warming to room temperature the mixture was quenched with ethereal HCl to give mainly $C_5H_5Mo(CO)_3H$ and $t\text{-}C_4H_9C(CH_3)_2OH$. In this reaction, ester cleavage is suggested to be the first step resulting in $t\text{-}C_4H_9C(O)CH_3$ which reacted further with $LiCH_3$. The remaining $[C_5H_5Mo(CO)_3CH_2O]^-$ loses formaldehyde to give $[C_5H_5Mo(CO)_3]^-$ which is subsequently protonated to furnish the stable hydrido complex, rather than the hoped for $C_5H_5Mo(CO)_3CH_2OH$. Quenching with CH_3I accordingly produces $C_5H_5Mo(CO)_3CH_3$ and not $C_5H_5Mo(CO)_3CH_2OCH_3$ [216].

$[C_5H_5Mo(CO)_3CH_2OC(CH_3)Fe(CO)(P(C_6H_5)_3)C_5H_5]PF_6$ (Table **1**, No. **123**) is stable as a solid but unstable in solution under all conditions. In polar solvents degradation occurred very easily, but is more complicated in nonpolar solvents such as CH_2Cl_2. Accordingly in acetone only several diamagnetic materials of unknown composition were observed; in CH_3NO_2 $[C_5H_5Mo(CO)_3OC(CH_3)Fe(CO)(P(C_6H_5)_3)C_5H_5]PF_6$ (see Formula XXXI) was also identified. Decomposition in CH_2Cl_2 occurred cleanly via unknown pathways to a 1 : 1 : 1 mixture of $C_5H_5Fe(CO)(P(C_6H_5)_3)C(O)CH_3$, $C_5H_5Mo(CO)_3CH_3$ (No. 1), and $[C_5H_5Mo(CO)_3OC(CH_3)$-

$Fe(CO)(P(C_6H_5)_3)C_5H_5]PF_6$ (see Formula XXXI). $[C_5H_5Mo(CO)_3CH_2OC(CH_3)Fe(CO)(P$-$(C_6H_5)_3)C_5H_5]PF_6$ reacted readily with a slight excess of $[C_6H_5CH_2N(C_2H_5)_3]Cl$ in CH_2Cl_2 at room temperature to give $C_5H_5Fe(CO)(P(C_6H_5)_3)C(O)CH_3$ and $C_5H_5Mo(CO)_3CH_2Cl$ (No. 128). An analogous reaction occurred with the iodide. A variety of nucleophilic hydride donors such as $[(C_6H_5)_3PCH_3][BH_3CN]$ and $C_5H_5Fe(CO)(P(C_6H_5)_3)H$ in CH_2Cl_2 at room temperature or $Li[(C_2H_5)_3BH]$ in THF at −78 °C similarly attacked the methylene group to give $C_5H_5Fe(CO)(P(C_6H_5)_3)C(O)CH_3$ and $C_5H_5Mo(CO)_3CH_3$ (No. 1) in all cases, but a reaction with $BH_3 \cdot OC_4H_8$ in THF failed [301].

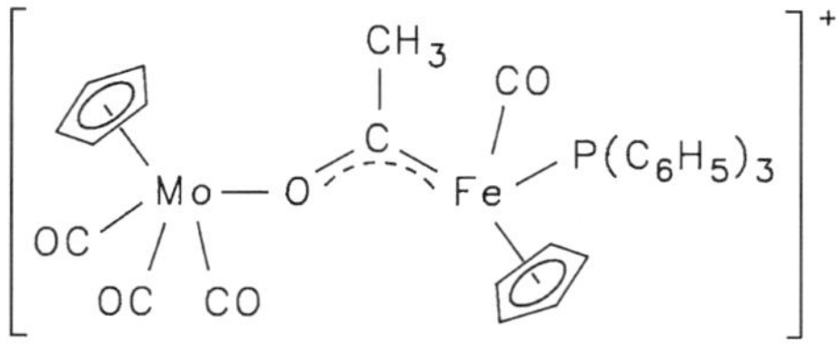

XXXI

$C_5H_5Mo(CO)_3CH_2SCH_3$ (Table **1**, No. **124**) gradually darkens on storage under nitrogen at room temperature, and lost one CO group during sublimation (see below) [24].

The mass spectrum of the tricarbonyl complex is as expected for the dicarbonyl derivatives, particularly on the basis of the highest fragment ion corresponding to $[C_5H_5Mo(CO)_2$-$CH_2SCH_3]^+$, which is in accordance with the facile CO loss observed on heating [54].

The compound underwent smooth decarbonylation on heating at 70 to 80 °C and 6.5×10^{-4} atm, or less preferably on prolonged UV irradiation in benzene, to form C_5H_5Mo-$(CO)_2(\eta^2$-$CH_2SCH_3)$ with the organic ligand bonded via the CS moiety [18, 24]. Treatment with ethanolic $HgCl_2$ at room temperature formed $C_5H_5Mo(CO)_3HgCl$ rather than the expected adduct $C_5H_5Mo(CO)_3CH_2SCH_3 \cdot HgCl_2$ [24]. Methyl iodide reacted only very slowly to form a yellow-brown not-further-identified ether-insoluble product containing metal carbonyl groups [24]. Alkylation by 1 equivalent of $[O(CH_3)_3]BF_4$ in CH_2Cl_2 yielded $[C_5H_5Mo(CO)_3$-$CH_2S(CH_3)_2]BF_4$ (No. 125) [332].

$C_5H_5Mo(CO)_3CH_2Cl$ (Table **1**, No. **128**) was also obtained by displacement of $C_5H_5Fe(CO)$-$(P(C_6H_5)_3)C(O)CH_3$ from $[C_5H_5Mo(CO)_3CH_2OC(CH_3)Fe(CO)(P(C_6H_5)_3)C_5H_5]PF_6$ (No. 123) by treatment with $[C_6H_5CH_2N(C_2H_5)_3]Cl$ in CH_2Cl_2 to give the complex in 97% yield [301]. Traces were found in the photochemical decomposition of $C_5H_5Mo(CO)_3C(O)CH_2Cl$, which resulted in $C_5H_5Mo(CO)_3Cl$ as the major product [194, 203].

$C_5H_5Mo(CO)_3CH_2Cl$ is soluble in ether and benzene. It is photosensitive and solutions decomposed in sunlight after few hours, but even when kept in the dark and under nitrogen, the complex is relatively unstable [39].

Reaction with methanolic NaOH regenerated $C_5H_5Mo(CO)_3CH_2OCH_3$ in poor yield, but gave mainly $C_5H_5Mo(CO)_3Cl$ [39]. When the complex was allowed to stand in CH_3OH alone, $C_5H_5Mo(CO)_3CH_2OCH_3$ was also formed [238, 267]. The compound undergoes a complex reaction with NaI in acetone at room temperature to slowly form $C_5H_5Mo(CO)_3I$, $Na[C_5H_5$-$Mo(CO)_3]$, and $ClCH_2I$, but is recovered unchanged after treatment with CH_3Li or $HN(CH_3)_2$ at room temperature, suggesting that the Mo-C bond to the chloromethyl group is more readily attacked by nucleophilic reagents than is the carbon-chlorine bond [117].

Slow reaction with an excess of nucleophiles ^2D like $S(CH_3)_2$, $CH_3SC_6H_5$, and NC_5H_5 in the presence of $TlBF_4$ at room temperature in CH_2Cl_2 gave $[C_5H_5Mo(CO)_3CH_2{}^2D]^+$ (Nos. 115,

References on pp. 106/18

125, and 126); at higher temperatures other compounds were formed, e.g., CO substitution products if $N(C_2H_5)_3$ or $P(CH_3)_2C_6H_5$ was present in the reaction mixture. No reaction occurred with $TlBF_4$ in the absence of an added nucleophile [302]. Treatment with $AgPF_6$ in CH_2Cl_2 at −80 °C afforded the thermally highly unstable methylidene salt $[C_5H_5(CO)_3$-$Mo{=}CH_2]PF_6$ [290, 301].

Products obtained from the reaction of $C_5H_5Mo(CO)_3CH_2Cl$ with $P(C_6H_5)_3$ were dependent on the reaction conditions. In CH_3CN at ambient temperature in the dark resulted $C_5H_5Mo(CO)_2(P(C_6H_5)_3)Cl$ after 28 days. Refluxing in CH_3OH for 30 min afforded C_5H_5Mo-$(CO)_2(P(C_6H_5)_3)C(O)CH_2OCH_3$, but after a prolonged reaction period up to 4 h, C_5H_5Mo-$(CO)_2(P(C_6H_5)_3)Cl$ along with some $C_5H_5Mo(CO)(P(C_6H_5)_3)_2Cl$ and $C_5H_5Mo(CO)_2(P(C_6H_5)_3)$-$C(O)CH_2OCH_3$ was isolated [238, 267].

$C_5H_5(CO)_3MoCH_2(CH_2)_nC{\equiv}CCH_3$ (n = 3, 4; Table **1**, No. **148**). The complex could be isolated pure for n = 3, but the orange oil obtained decomposed within 1 to 2 h at 20 °C, possibly by the intermolecular reaction of the acetylenic moieties with Mo centers to give polymeric products. Diluted solutions in polar solvents are more stable, although decomposition occurred after 12 h. Attempted isolation of $C_5H_5Mo(CO)_3CH_2(CH_2)_4C{\equiv}CCH_3$ by removal of the solvent under vacuum at room temperature gave only a brown polymeric solid which could not be re-dissolved in THF, CH_2Cl_2, or CH_3CN. $C_5H_5Mo(CO)_3CH_2(CH_2)_nC{\equiv}CCH_3$ generated in situ by alkylation of $[C_5H_5Mo(CO)_3]^-$ gave $C_5H_5Mo(CO)_2(\eta^2$-$C(CH_3){=}C(CH_2)_{n+1}CO$-cyclo) (Formula XXXII) through intramolecular cyclization of the acetylenic bond across the Mo-acyl bond. This cyclization was also accomplished for n = 2, but was shown to be too rapid to prove $C_5H_5Mo(CO)_3CH_2(CH_2)_2C{\equiv}CCH_3$ as an intermediate. $C_5H_5Mo(CO)_3CH_2(CH_2)_3C{\equiv}CCH_3$ was observed to build up in solution during cyclization, while $C_5H_5Mo(CO)_3CH_2(CH_2)_4$-$C{\equiv}CCH_3$ was completely formed before cyclization started [206].

XXXII

$C_5H_5Mo(CO)_3CH_2(CH_2)_2Br$ (Table **1**, No. **150**) is volatile in high vacuum at ca. 80 °C, but decomposition occurred under these conditions to give $(C_5H_5Mo(CO)_3)_2$ and $C_5H_5Mo(CO)_3Br$ [36].

Treatment of $C_5H_5Mo(CO)_3CH_2(CH_2)_2Br$ with $Na[Mn(CO)_5]$ in THF at room temperature resulted in transmetallation to afford "$Mn_2(CO)_{10}(CH_2)_3$" [36], later established as the cyclic carbene complex $Mn(CO)_5Mn(CO)_4{=}CO(CH_2)_3$-cyclo rather than the trimethylene-bridged $(Mn(CO)_5)_2(\mu$-$(CH_2)_3)$ [79]. The analogous substitution employing $Na[C_5H_5Fe(CO)_2]$ as a nucleophile did, however, produce genuine $(C_5H_5Fe(CO)_2)_2(\mu$-$(CH_2)_3)$ [36].

Addition of the hydride donor $Li[(C_2H_5)_3BH]$ to a THF solution of the complex led to $C_5H_5Mo(CO)_2(\eta^2$-$OCH(CH_2)_3$-cyclo) (Formula XXXIII) along with small quantities of the allyl derivative $C_5H_5Mo(CO)_2(\eta^3$-$CH_2CHCHCH_3)$ [266, 314, 315].

Furthermore, $C_5H_5Mo(CO)_3CH_2(CH_2)_2Br$ reacted with a number of anionic or neutral nucleophiles ^2D to smoothly produce derivatives containing the 2-oxacyclopentylidene ligand $=COCH_2CH_2CH_2$-cyclo. The key initial step of this transformation appears to involve migration of the alkyl chain to an adjacent carbonyl group and coordination of the nucleophile

References on pp. 106/18

XXXIII

at the metal atom. The acyl intermediates $[C_5H_5Mo(CO)_2(C(O)(CH_2)_3Br)^2D]^{n-}$ (n = 0, 1) so formed then undergo rapid ring-forming elimination of Br^- by virtue of the nucleophilicity of the acyl oxygen atom to give the observed cyclized products. For details of the reaction mechanism, see [268, 315].

A reaction with LiI · 3 H_2O in refluxing THF gave $C_5H_5Mo(CO)_2(=CO(CH_2)_3\text{-cyclo})I$; in CH_3OH, treatment with C_6H_5SH/KOH at room temperature, or with equimolar amounts of KCN in the refluxing solvent, analogously gave $C_5H_5Mo(CO)_2(=CO(CH_2)_3\text{-cyclo})SC_6H_5$ and $C_5H_5Mo(CO)_2(=CO(CH_2)_3\text{-cyclo})CN$. As shown in Formula XXXIVb, these compounds were all isolated as the thermodynamically preferred trans isomer [114, 254, 268]. In the case of KCN, $K[cis\text{-}C_5H_5Mo(CO)_2((CH_2)_3CN)CN]$ was obtained if the cyanide was used in 100% excess [114].

If $P(C_6H_5)_3$ was used as the nucleophile in a concentrated solution of CH_3CN at room temperature, $[cis\text{-}C_5H_5Mo(CO)_2(P(C_6H_5)_3)(=CO(CH_2)_3\text{-cyclo})]^+$ (Formula XXXIVa) precipitated within minutes. Prolonged reaction time in a more dilute solution again produced the respective trans isomer (Formula XXIVb) [95, 268]. A similar preparation of $[C_5H_5Mo(CO)_2(PR_{3-n}H_n)(=CO(CH_2)_3\text{-cyclo})]^+$ started from No. 150 and $PR_{3-n}H_n$ (n = 1, 2) in CH_3OH in the presence of $Na[B(C_6H_5)_4]$, and the cis isomer precipitated from the reaction mixture for $P(C_6H_5)_2H$ and $P(C_6H_{11}\text{-cyclo})H_2$. In the case of $P(C_6H_{11}\text{-cyclo})_2H$, again only the trans isomer was obtained. Generally the cis isomer can be transformed into the trans isomer, and although in some case not observed, all these reactions are assumed to first produce the kinetically preferred cis configuration which is further isomerized into the thermodynamically more stable trans form [361].

Treatment with $[N(C_2H_5)_4]GeCl_3$ in acetone yielded the cyclic carbene complex $C_5H_5Mo(CO)_2(=CO(CH_2)_3\text{-cyclo})GeCl_3$, probably via $[C_5H_5Mo(CO)_2(C(O)CH_2(CH_2)_2Br)GeCl_3]^-$ as an intermediate [153].

$C_5H_5Mo(CO)_3CH_2(CH_2)_2Br$ reacted with $[C_5H_5W(CO)_3]^-$ in THF to give $C_5H_5Mo(CO)_3W(CO)_2(=CO(CH_2)_3\text{-cyclo})C_5H_5$. Analogously $C_5H_5Mo(CO)_3Mo(CO)_2(=CO(CH_2)_3\text{-cyclo})C_5H_5$ was obtained in a one-pot reaction starting from $Br(CH_2)_3Br$ and two equivalents of $[C_5H_5Mo(CO)_3]^-$ in THF at ambient temperature for 3 d with No. 150 as an intermediate [278].

XXXIVa XXXIVb n = 0, 1

$C_5H_5Mo(CO)_3CH_2(CH_2)_nFe(CO)_2C_5H_5$ (n = 2 to 5; Table **1**, Nos. **154** to **157**). These heterobimetallic alkanediyl complexes are stable in air for extended periods but are mildly light-

References on pp. 106/18

sensitive. The compounds are generally stable in solution under nitrogen, but decompose when the solutions are exposed to air. Their hexane solubility increases with an increasing length of the connecting alkyl chain [358].

Mass spectra were investigated in detail. For a compilation of the major ions observed in the electron impact mass spectra of these derivatives and possible fragmentation sequences giving rise to these ions, as well as for a comparison of these $C_5H_5Mo(CO)_3CH_2(CH_2)_nFe(CO)_2C_5H_5$ heterobimetallics with related alkanediyl derivatives of the $C_5H_5Fe(CO)_2$ fragment containing $C_5H_5W(CO)_3$ and $Re(CO)_5$ moieties instead of $C_5H_5Mo(CO)_3$, see [358].

$C_5H_5Mo(CO)_3CH_2(CH_2)_nFe(CO)_2C_5H_5$ (n = 2, 3; Nos. 154, 155) reacted with $[C(C_6H_5)_3]PF_6$ in CH_2Cl_2 at room temperature for several hours to produce $[C_5H_5Mo(CO)_3CH_2CH{=}CH_2(Fe(CO)_2C_5H_5)]PF_6$ (No. 18) and $[C_5H_5Mo(CO)_3CH_2CH_2CH{=}CH_2(Fe(CO)_2C_5H_5)]PF_6$ (No. 132), where the molybdenum atom is believed to remain σ-bonded to the hydrocarbon chain, whereas the iron atom tends to become π-bonded to the terminal -$CHCH_2$ unit [392].

Attempts to decarbonylate complex No. 154 by heating in toluene or by treating it with $(CH_3)_3NO$ remained unsuccessful [392].

$C_5H_5Mo(CO)_3CF_3$ (Table **1**, No. **158**) was produced in situ from $C_5H_5Mo(CO)_3C(O)CF_3$ in frozen gas matrices (Ar, CH_4, CO, N_2) at ca. −261 °C under UV irradiation. Details of the whole photodegeneration process are described for the compound $C_5H_5Mo(CO)_3C(O)CF_3$ (see Section 1.5.1.4.1.12.3.2) [359].

$C_5H_5Mo(CO)_3CF_3$ did not act as a polar trifluoromethylation reagent in attempted reactions with $AgNO_3$, $Cd(CH_3)_2$, and $C_6H_5C(O)Cl$ [346]. The great stability of the Mo-C bond prevented reaction with ethanolic $HgCl_2$, and unlike $C_5H_5Mo(CO)_3CH_2SCH_3$ (No. 124), no $C_5H_5Mo(CO)_3HgCl$ was obtained [24]. The exothermic reaction with $Na[C_5H_5Fe(CO)_2]$ afforded $(C_5H_5Fe(CO)_2)_2$, together with unchanged $C_5H_5Mo(CO)_3CF_3$ instead of a desired fluorocarbon transition metal derivative containing two different metal atoms [16]. $C_5H_5Mo(CO)_3CF_3$ was proved as an intermediate in the reaction of $C_6H_5C{\equiv}CC_6H_5$ with $C_5H_5Mo(CO)_3C(O)CF_3$ in refluxing hexane resulting in $C_5H_5Mo(CO)(\eta^2\text{-}C_6H_5C{\equiv}CC_6H_5)CF_3$. The corresponding reaction with $CH_3C{\equiv}CCH_3$ proceeded under UV irradiation to give $C_5H_5Mo(CO)(CH_3C{\equiv}CCH_3)CF_3$ [155, 171].

$C_5H_5Mo(CO)_3CF_3$ could be converted to the difluorocarbene complex $[C_5H_5Mo(CO)_3{=}CF_2]^+$, as demonstrated by the reaction with a 2-fold excess of $(CH_3)_3SiOSO_2CF_3$ in benzene at room temperature in nonpolar solvents. Use of solvents like CH_2Cl_2 resulted in rapid decomposition of $[C_5H_5Mo(CO)_3{=}CF_2]O_3SCF_3$ and regeneration of $C_5H_5Mo(CO)_3CF_3$ as the major organometallic compound [371]. Use of SbF_6 in liquid SO_2 at −78 °C generated the corresponding $[C_5H_5Mo(CO)_3{=}CF_2]SbF_6$, but this salt could not be isolated. Any attempt resulted in further reaction giving $[C_5H_5Mo(CO)_4]SbF_6$. Accordingly, the reaction with gaseous BF_3 in CH_2Cl_2 at −78 °C leads to $[C_5H_5Mo(CO)_4]BF_4$ immediately, but transformation is probably initiated by removal of fluoride ion from the neutral CF_3 compound to generate $[C_5H_5(CO)_4Mo{=}CF_2]BF_4$ which then disproportionates [175, 181].

The ability of electrophiles to promote carbene formation in $C_5H_5Mo(CO)_3CF_3$ is also evident in the reaction of No. 158 with aqueous HBF_4 in CH_3NO_2, which results in quantitative hydrolytic conversion to $[C_5H_5Mo(CO)_4]BF_4$ via $[C_5H_5Mo(CO)_3{=}CF_2]^+$ [282]; the tetracarbonyl cation is also formed on treatment of $C_5H_5Mo(CO)_3CF_3$ with excess $BF_3 \cdot O(C_2H_5)_2$, $[(CC_6H_5)_3]BF_4$, or $[C(C_6H_5)_3]PF_6$ [371].

Treatment of $C_5H_5Mo(CO)_3CF_3$ with slightly more than 1 equivalent of BX_3 (X = Cl, Br) in noncoordinating CH_2Cl_2 at −78 °C resulted in rapid and quantitative halogen exchange to give $C_5H_5Mo(CO)_3CX_3$ (Nos. 159, 160). The halogen exchange is thermodynamically favored

References on pp. 106/18

because of the great strength of the B-F bond in developing BF_3. Furthermore, the ability of the CF_3 group to form the difluorocarbene ligand by electrophilic fluoride abstraction suggests that cationic dihalocarbene complexes are also reasonable intermediates in these halogen exchange reactions. The following reaction sequence, in which the last step should be irreversible for thermodynamical reasons, was suggested. Complete replacement of F by X resulted from further repetition of this sequence [281]:

$$C_5H_5Mo(CO)_3CF_3 + BX_3 \rightarrow [C_5H_5Mo(CO)_3{=}CF_2]BX_3F \rightarrow C_5H_5Mo(CO)_3CF_2X + BX_2F$$

$C_5H_5Mo(CO)_3CCl_2CF_3$ (Table **1**, No. **164**) exhibits a temperature-dependent ^{19}F NMR spectrum which is attributed to a hindered rotation about the C-C single bond; see Scheme 5. In acetone at 20 °C, a single resonance is observed at −71.24 ppm which broadens upon cooling. Coalescence is observed at −69 °C. At −94 °C two signals have appeared, and the approximate chemical shifts are measured as −79 ppm for the unique fluorine anti to Mo and as −68 ppm for the two gauche fluorine atoms, but the slow-exchange limiting spectrum could not be attained. At the coalescence temperature, the approximate free energy of activation for this process was determined to be 8.7 kcal/mol at −69 °C. Coupling was not resolved at this low temperature, and J was assumed to be 100 Hz. For a series of formally related compounds, $XCCl_2CF_3$, the facility of C-C bond rotation increases in the order X = $Mn(CO)_5 < Re(CO)_5 < C_5H_5Mo(CO)_3 < I < CCl_3$ [281].

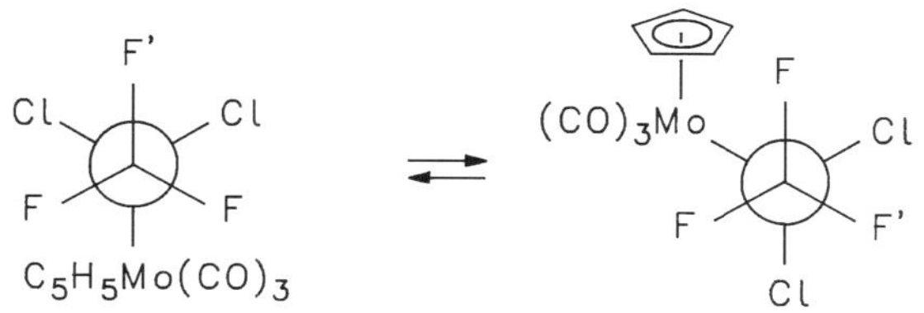

F'
Cl Cl
F F
$C_5H_5Mo(CO)_3$
⇄
F
$(CO)_3Mo$ Cl
F F'
Cl

Scheme 5

$C_5H_5Mo(CO)_3C_3F_7$ (Table **1**, No. **165**) crystallizes in the monoclinic space group $P2_1/c-C^5_{2h}$ (No. 14) with a = 8.301(7), b = 15.240(14), c = 11.249(12) Å, β = 106.25°(5); Z = 4 molecules per unit cell, D_{calc} = 2.01 g/cm³. The molecular structure, which is similar to that of $C_5H_5Mo(CO)_3C_2H_5$ (No. 2), is shown in **Fig. 8**. An approximate mirror plane passes through C_α, C_4, Mo, and the center of the C_1-C_2 bond. The Mo atom lies 1.992 ± 0.005 Å below the cyclopentadienyl ligand, with the respective individual Mo-C distances varying systematically around the ring in such a way as to conserve the pseudomirror symmetry of the coordination sphere. The Mo-C_3F_7 linkage of 2.288 Å after correction for thermal motion showed a contraction of 0.109 Å from the final Mo-C value for the Mo-C bond length of $C_5H_5Mo(CO)_3C_2H_5$ (No. 2), and thus provides a direct estimation of the shortening of the metal-carbon bond length that occurs on going from a transition metal alkyl to a transition metal perfluoroalkyl. This contradiction can be divided in two parts, the strengthened σ bonding due to the high electronegativity of C_3F_7, and the π bonding due to back donation from the Mo d_π orbital to σ* of the C-F bonds. The average α-carbon-fluorine bond length is 1.40 Å, which is 0.02 Å longer than expected and thus is also consistent with π interaction between filled metal d orbitals and empty carbon orbitals. The C_3F_7 group is in the customary staggered conformation with an angle Mo-CF_2-CF_2 of 123.3° and an average F-C-F angle of 104.8° [28, 41, 47].

In full agreement with the enhanced strength of the Mo-C bond and the apparent weakness of the α-carbon-fluorine bonds of $C_5H_5Mo(CO)_3C_3F_7$ and similar perfluoroalkyls, a fluoride ion is readily abstracted from the C_3F_7 group on the addition of SbF_5 to solutions of the

References on pp. 106/18

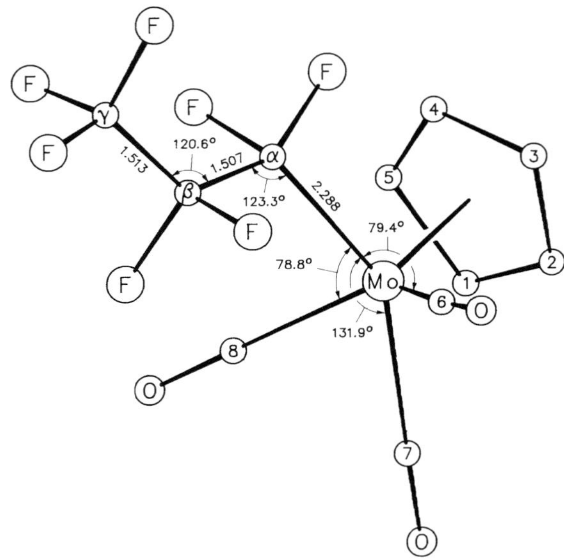

Fig. 8. The molecular structure of $C_5H_5Mo(CO)_3C_3F_7$ [28, 41].

title compound in liquid SO_2 at −78°C to give isolable red $[C_5H_5Mo(CO)_3{=}CFC_2F_5]SbF_6$. Use of BF_3 in CH_2Cl_2 at −78°C also generated a red precipitate which is thought to be $[C_5H_5Mo(CO)_3{=}CFC_2F_5]BF_4$, but warming up to −20°C resulted in gas evolution and regeneration of the starting material $C_5H_5Mo(CO)_3C_3F_7$ [175, 181]. Treatment with BCl_3 in an inert solvent at −78°C resulted in regiospecific halogen exchange at the α-carbon atom to produce $C_5H_5Mo(CO)_3CCl_2C_2F_5$ (No. 166); for analogous reaction compare Nos. 158 and 163 [281].

References:

[1] Piper, T. S.; Wilkinson, G. (Naturwissenschaften **42** [1955] 625).
[2] Piper, T. S.; Wilkinson, G. (J. Inorg. Nucl. Chem. **3** [1956] 104/24).
[3] Strohmeier, W.; Lemmon, R. M. (Z. Naturforsch. **14a** [1959] 109/12).
[4] King, R. B.; Treichel, P. M.; Stone, F. G. A. (Proc. Chem. Soc. **1961** 49/70).
[5] Stone, F. G. A.; Treichel, P. M.; Ethyl Corporation (U.S. 3290343 [1961/66]; C.A. **66** [1967] No. 105054).
[6] Bamford, C. H.; Finch, C. A.; Courtaulds Ltd. (Br. 1033161 [1961/66]; C.A. **76** [1972] No. 25832).
[7] Davison, A.; McFarlane, W.; Pratt, L.; Wilkinson, G. (J. Chem. Soc. **1962** 3653/66).
[8] Nakamura, A. (Mem. Inst. Sci. Ind. Res. Osaka Univ. **19** [1962] 81/95; C.A. **59** [1963] 8786).
[9] Treichel, P. M. (Diss. Harvard Univ., Ann Arbor, Mich., 1962; Diss. Abstr. Int. B **23** [1963] 2318).
[10] Bennett, M. J.; Mason, R. (Proc. Chem. Soc. **1963** 273/4).

[11] Treichel, P. M.; Morris, J. H.; Stone, F. G. A. (J. Chem. Soc. **1963** 720/3).
[12] Cousins, M.; Green, M. L. H. (J. Chem. Soc. **1963** 889/94).
[13] Davison, A.; McCleverty, J. A.; Wilkinson, G. (J. Chem. Soc. **1963** 1133/8).

[14] McCleverty, J. A.; Wilkinson, G. (J. Chem. Soc. **1963** 4096/9).
[15] Nakamura, A.; Hagihara, N. (Nippon Kagaku Zasshi **84** [1963] 344/8; C.A. **59** [1963] 14021).
[16] King, R. B.; Bisnette, M. B. (J. Organomet. Chem. **2** [1964] 15/37).
[17] Cotton, F. A.; McCleverty, J. A. (J. Organomet. Chem. **4** [1964] 490).
[18] King, R. B.; Bisnette, M. B. (J. Am. Chem. Soc. **86** [1964] 1267/8).
[19] Randić, M. (Theor. Chim. Acta **2** [1964] 468/72).
[20] Lauterbur, P. C.; King, R. B. (J. Am. Chem. Soc. **87** [1965] 3266/7).

[21] Bennett, M. J. (Ph.D. Thesis Sheffield 1965; cited in [28]).
[22] King, R. B.; Fronzaglia, A. (Proc. 2nd Int. Symp. Organomet. Chem., Madison, Wisc., 1965, p. 77).
[23] King, R. B. (Organomet. Synth. **1** [1965] 145/7).
[24] King, R. B.; Bisnette, M. B. (Inorg. Chem. **4** [1965] 486/93).
[25] King, R. B. (Inorg. Chem. **4** [1965] 1518/20).
[26] King, R. B.; Bisnette, M. B.; Fronzaglia, A. (J. Organomet. Chem. **5** [1966] 341/56).
[27] Capron-Cotigny, G.; Poilblanc, R. (C.R. Seances Acad. Sci. C **263** [1966] 885/7).
[28] Churchill, M. R.; Fennessey, J. P. (Chem. Commun. **1966** 695/6).
[29] Dessy, R. E.; Stary, F. E.; King, R. B.; Waldrop, M. (J. Am. Chem. Soc. **88** [1966] 471/6).
[30] King, R. B.; Fronzaglia, A. (J. Am. Chem. Soc. **88** [1966] 709/12).

[31] Jolly, P. W.; Pettit, R. (J. Am. Chem. Soc. **88** [1966] 5044/5).
[32] Dessy, R. E.; Weissman, P. M.; Pohl, R. L. (J. Am. Chem. Soc. **88** [1966] 5117/21).
[33] King, R. B.; Bisnette, M. B. (Inorg. Chem. **5** [1966] 293/300).
[34] King, R. B.; Bisnette, M. B. (Inorg. Chem. **5** [1966] 306/8).
[35] Patil, H. R. H.; Graham, W. A. G. (Inorg. Chem. **5** [1966] 1401/5).
[36] King, R. B.; Bisnette, M. B. (J. Organomet. Chem. **7** [1967] 311/9).
[37] Green, M. L. H.; Ariyaratne, J. K. P.; Bjerrum, A. M.; Ishaq, M.; Prout, C. K. (Chem. Commun. **1967** 430/2).
[38] Craig, P. J.; Green, M. (J. Chem. Soc. Chem. Commun. **1967** 1246/7).
[39] Green, M. L. H.; Ishaq, M.; Whiteley, R. N. (J. Chem. Soc. A **1967** 1508/15).
[40] Barnett, K. W.; Treichel, P. M. (Inorg. Chem. **6** [1967] 294/9).

[41] Churchill, M. R.; Fennessey, J. P. (Inorg. Chem. **6** [1967] 1213/20).
[42] Butler, I. S.; Basolo, F.; Pearson, R. G. (Inorg. Chem. **6** [1967] 2074/9).
[43] Randić, M.; Trinajstić, N. (J. Chem. Phys. **46** [1967] 1469/74).
[44] Barnett, K. W. (Diss. Univ. Wisconsin, Madison, Wisc., 1967; Diss. Abstr. Int. B **28** [1968] 3203).
[45] King, R. B. (U.S. Clearinghouse Fed. Sci. Tech. Inform. AD-665429 [1967] 36 pp.; U.S. Gov. Res. Dev. Rep. **68** [1968] 48; C.A. **69** [1968] No. 56601).
[46] Downs, R. L. (Diss. Ohio State Univ., Columbus, Ohio, 1968; Diss. Abstr. Int. B **29** [1968] 1593/4).
[47] Fenessey, J. P. (Diss. Harvard Univ., Cambridge, Mass., 1968; Diss. Abstr. Int. B **29** [1968] 1594).
[48] Susuki, T.; Tsuji, J. (Tetrahedron Lett. **1968** 913/5).
[49] Hawthorne, J. D.; Mays, M. J.; Simpson, R. N. F. (J. Organomet. Chem. **12** [1968] 407/10).
[50] Ishaq, M. (J. Organomet. Chem. **12** [1968] 414/6).

[51] Duncan, J. D.; Green, J. C.; Green, M. L. H.; McLauchlan, K. A. (Chem. Commun. **1968** 721/2).
[52] Craig, P. J.; Green, M. (J. Chem. Soc. A **1968** 1978/81).

[53] King, R. B. (J. Am. Chem. Soc. **90** [1968] 1417/29).
[54] King, R. B. (J. Am. Chem. Soc. **90** [1968] 1429/37).
[55] Jolly, W. L. (Inorg. Synth. **11** [1968] 116/9).
[56] Johnson, M. D.; Winterton, N.; Shortland, A. C. (New Aspects Chem. Met. Carbonyls Deriv. 1st Int. Symp. Proc., Venice 1968, C4, 5 pp.; C.A. **72** [1970] No. 11758).
[57] Jolly, W. L.; Birchall, T.; Rustad, D. S.; United States Atomic Energy Commission (U.S. 3535355 [1968/70]; C.A. **74** [1971] No. 42493).
[58] Graziani, M.; Bibler, J. P.; Montesano, R.; Wojcicki, A. (J. Organomet. Chem. **16** [1969] 507/11).
[59] King, R. B.; Kapoor, P. N. (J. Organomet. Chem. **18** [1969] 357/60).
[60] King, R. B.; Kapoor, R. N.; Pannell, K. H. (J. Organomet. Chem. **20** [1969] 187/93).

[61] Roustan, J. L.; Charrier, C. (C.R. Hebd. Seances Acad. Sci. C **268** [1969] 2113/6).
[62] Duncan, J. D.; Green, J. C.; Green, M. L. H.; McLauchlan, K. A. (Discuss. Faraday Soc. **47** [1969] 178/82).
[63] Craig, P. J.; Green, M. (J. Chem. Soc. A **1969** 157/60).
[64] Ariyaratne, J. K. P.; Bjerrum, A. M.; Green, M. L. H.; Ishaq, M.; Prout, C. K.; Swanwick, M. G. (J. Chem. Soc. A **1969** 1309/21).
[65] Hart-Davis, A. J.; Mawby, R. J. (J. Chem. Soc. A **1969** 2403/7).
[66] Cotton, F. A.; Marks, T. J. (J. Am. Chem. Soc. **91** [1969] 1339/46).
[67] King, R. B.; Houk, L. W.; Kapoor, P. N. (Inorg. Chem. **8** [1969] 1792/4).
[68] Barnett, K. W. (Inorg. Chem. **8** [1969] 2009/11).
[69] King, R. B.; Kapoor, R. N. (Inorg. Chem. **8** [1969] 2535/9).
[70] King, R. B.; Houk, L. W. (Can. J. Chem. **47** [1969] 2959/64).

[71] King, R. B. (Appl. Spectrosc. **23** [1969] 137/47).
[72] King, R. B. (Org. Mass Spectrom. **2** [1969] 387/99).
[73] King, R. B. (Acc. Chem. Res. **3** [1970] 417/27).
[74] Yamamoto, Y.; Yamazaki, H. (J. Organomet. Chem. **24** [1970] 717/24).
[75] Haines, R. J.; Nolte, C. R. (J. Organomet. Chem. **24** [1970] 725/36).
[76] Collier, M. R.; Lappert, M. F.; Truelock, M. M. (J. Organomet. Chem. **25** [1970] C36/C38).
[77] Mérour, J. Y. (C.R. Hebd. Seances Acad. Sci. C **271** [1970] 1397/9).
[78] Haines, R. J.; Marais, I. L.; Nolte, C. R. (J. Chem. Soc. D **1970** 547/8).
[79] Casey, C. P. (J. Chem. Soc. D **1970** 1220/1).
[80] Collier, M. R.; Kingston, B. M.; Lappert, M. F. (J. Chem. Soc. D **1970** 1498).

[81] Johnson, M. D.; Winterton, N. (J. Chem. Soc. A **1970** 507/11).
[82] Johnson, M. D.; Winterton, N. (J. Chem. Soc. A **1970** 511/6).
[83] Bruce, M. I.; Iqbal, M. Z.; Stone, F. G. A. (J. Chem. Soc. A **1970** 3204/9).
[84] Stewart, R. P.; Treichel, P. M. (J. Am. Chem. Soc. **92** [1970] 2710/8).
[85] Schunn, R. A. (Inorg. Chem. **9** [1970] 2567/72).
[86] Susuki, T.; Tsuji, J. (J. Org. Chem. **35** [1970] 2982/6).
[87] Yamamoto, Y.; Yamazaki, H. (Bull. Chem. Soc. Jpn. **43** [1970] 143/7).
[88] Stone, F. G. A. (U.S. Clearinghouse Fed. Sci. Tech. Inform. AD-714623 [1970] 21 pp.; U.S. Gov. Res. Dev. Rep. **71** [1971] 72; C.A. **75** [1971] No. 20455).
[89] Benaïm, J.; Mérour, J. Y.; Roustan, J. L. (Tetrahedron Lett. **1971** 983/6).
[90] Haines, R. J.; Du Preez, A. L.; Marais, I. L. (J. Organomet. Chem. **28** [1971] 97/104).

[91] Su, S. R.; Wojcicki, A. (J. Organomet. Chem. **31** [1971] C34/C36).
[92] Lichtenberg, D. W.; Wojcicki, A. (J. Organomet. Chem. **33** [1971] C77/C79).
[93] Benaïm, J.; Mérour, J. Y.; Roustan, J. L. (C.R. Hebd. Seances Acad. Sci. C **272** [1971] 789/91).

[94] Jacobson, S. E.; Reich-Rohrwig, P.; Wojcicki, A. (J. Chem. Soc. D **1971** 1526/7).
[95] Cotton, F. A.; Lukehart, C. M. (J. Am. Chem. Soc. **93** [1971] 2672/6).
[96] Hagen, A. P.; Higgins, C. R.; Russo, P. J. (Inorg. Chem. **10** [1971] 1657/60).
[97] King, R. B.; Kapoor, P. N.; Kapoor, R. N. (Inorg. Chem. **10** [1971] 1841/50).
[98] King, R. B.; Kapoor, R. N.; Saran, M. S.; Kapoor, P. N. (Inorg. Chem. **10** [1971] 1851/60).
[99] King, R. B.; Saran, M. S. (Inorg. Chem. **10** [1971] 1861/7).
[100] Thomasson, J. E.; Robinson, P. W.; Ross, D. A.; Wojcicki, A. (Inorg. Chem. **10** [1971] 2130/7).

[101] Su, S. R. (Diss. Ohio State Univ., Columbus, Ohio, 1971; Diss. Abstr. Int. B **32** [1972] 6238).
[102] Kruck, T.; Höfler, M.; Liebig, L. (Chem. Ber. **105** [1972] 1174/83).
[103] Roustan, J. L.; Benaïm, J.; Charrier, C.; Mérour, J. Y. (Tetrahedron Lett. **1972** 1953/6).
[104] Barnett, K. W.; Pollman, T. G.; Solomon, T. W. (J. Organomet. Chem. **36** [1972] C23/C26).
[105] Roustan, J. L.; Charrier, C.; Mérour, J. Y.; Benaïm, J.; Gianotti, C. (J. Organomet. Chem. **38** [1972] C37/C40).
[106] Mérour, J. Y.; Charrier, C.; Benaïm, J.; Roustan, J. L.; Commereuc, D. (J. Organomet. Chem. **39** [1972] 321/8).
[107] Craig, P. J.; Edwards, J. (J. Organomet. Chem. **46** [1972] 335/7).
[108] King, R. B.; Kapoor, P. N. (Inorg. Chim. Acta **6** [1972] 391/4).
[109] Roustan, J. L.; Mérour, J. Y.; Benaïm, J.; Charrier, C. (C.R. Hebd. Seances Acad. Sci. C **274** [1972] 537/40).
[110] King, R. B.; Efraty, A. (J. Am. Chem. Soc. **94** [1972] 3768/73).

[111] King, R. B.; Zipperer, W. C.; Ishaq, M. (Inorg. Chem. **11** [1972] 1361/70).
[112] Zipperer, W. C. (Diss. Univ. Georgia, Athens, Ga., 1972; Diss. Abstr. Int. B **33** [1972] 2930).
[113] Nesmeyanov, A. N.; Fedin, É. I.; Fedorov, L. A.; Petrovskii, P. V. (Zh. Strukt. Khim. **13** [1972] 1033/42; J. Struct. Chem. [Engl. Transl.] **13** [1972] 964/72).
[114] Kruck, T.; Liebig, L. (Chem. Ber. **106** [1973] 1055/61).
[115] Kruck, T.; Liebig, L. (Chem. Ber. **106** [1973] 3588/94).
[116] Seidel, W.; Reichardt, W. (Z. Chem. **13** [1973] 106/7).
[117] King, R. B.; Braitsch, D. M. (J. Organomet. Chem. **54** [1973] 9/14).
[118] Bruce, M. I.; Goodall, B. L.; Stone, F. G. A. (J. Organomet. Chem. **60** [1973] 343/9).
[119] Lichtenberg, D. W.; Wojcicki, A. (Inorg. Chim. Acta **7** [1973] 311/4).
[120] Jacobson, S. E.; Reich-Rohrwig, P.; Wojcicki, A. (Inorg. Chem. **12** [1973] 717/23).

[121] Anderson, S. N.; Fong, C. W.; Johnson, M. (J. Chem. Soc. Chem. Commun. **1973** 163).
[122] Green, M. L. H.; Mitchard, L. C.; Silverthorn, W. E. (J. Chem. Soc. Dalton Trans. **1973** 1403/8).
[123] Fong, C. W.; Johnson, M. D. (J. Chem. Soc. Perkin Trans. II **1973** 986/90).
[124] Jacobson, S. E.; Wojcicki, A. (J. Am. Chem. Soc. **95** [1973] 6962/70).
[125] Yamamoto, Y.; Wojcicki, A. (Inorg. Chem. **12** [1973] 1779/88).
[126] Orlova, T. Yu.; Churanov, S. S.; Fedorov, L. A.; Setkina, V. N.; Kursanov, D. N. (Izv. Akad. Nauk SSSR Ser. Khim. **1973** 1652/4; Bull. Acad. Sci. USSR Div. Chem. Sci. [Engl. Transl.] **1973** 1602/4).
[127] Cyr, C. R. (Diss. Univ. Wisconsin, Madison, Wis., 1973; Diss. Abstr. Int. B **34** [1973] 1937/8).

[128] Lichtenberg, D. W. (Diss. Ohio State Univ., Columbus, Ohio, 1973; Diss. Abstr. Int. B **34** [1974] 5372).
[129] Cole, B. J.; Cotton, J. D.; McWilliam, D. (J. Organomet. Chem. **64** [1974] 223/7).
[130] King, R. B.; Hodges, K. C. (J. Organomet. Chem. **65** [1974] 77/80).

[131] Rosenblum, M. (Acc. Chem. Res. **7** [1974] 122).
[132] Beck, W.; Höfler, R.; Erbe, J.; Menzel, H.; Nagel, U.; Platzen, G. (Z. Naturforsch. **29b** [1974] 567/8).
[133] Barnett, K. W.; Beach, D. L.; Gaydos, S. P.; Pollman, T. G. (J. Organomet. Chem. **69** [1974] 121/30).
[134] Barnett, K. W.; Pollman, T. G. (J. Organomet. Chem. **69** [1974] 413/21).
[135] Jacobson, S. E.; Wojcicki, A. (J. Organomet. Chem. **72** [1974] 113/20).
[136] Pittman, C. U., Jr.; Felis, R. F. (J. Organomet. Chem. **72** [1974] 399/413).
[137] Harris, A.; Rest, A. J. (J. Organomet. Chem. **78** [1974] C29/C30).
[138] Su, S. R.; Wojcicki, A. (Inorg. Chim. Acta **8** [1974] 55/60).
[139] Cotton, J. D.; Davison, P. J.; Goldberg, D. E.; Lappert, M. F.; Thomas, K. M. (J. Chem. Soc. Chem. Commun. **1974** 893/5).
[140] Bock, P. L.; Whitesides, G. M. (J. Am. Chem. Soc. **96** [1974] 2826/9).

[141] King, R. B.; Saran, M. S. (Inorg. Chem. **13** [1974] 364/7).
[142] Hartshorn, A. J.; Lappert, M. F.; Turner, K. (J. Chem. Soc. Chem. Commun. **1975** 929/30).
[143] Fong, C. W.; Wilkinson, G. (J. Chem. Soc. Dalton Trans. **1975** 1100/4).
[144] Casey, C. P.; Cyr, C. R.; Anderson, R. L.; Marten, D. F. (J. Am. Chem. Soc. **97** [1975] 3053/9).
[145] Su, S. R.; Wojcicki, A. (Inorg. Chem. **14** [1975] 89/98).
[146] King, R. B.; Nainan, K. C. (Inorg. Chem. **14** [1975] 271/4).
[147] Lichtenberg, D. W.; Wojcicki, A. (Inorg. Chem. **14** [1975] 1295/301).
[148] Knoth, W. H. (Inorg. Chem. **14** [1975] 1566/72).
[149] Alt, H. G. (Angew. Chem. **88** [1976] 800/1; Angew. Chem. Int. Ed. Engl. **15** [1976] 759).
[150] Brown, D. L. S.; Connor, J. A.; Dobinson, B.; Stark, B. P. (Angew. Makromol. Chem. **50** [1976] 9/14).

[151] Pannell, K. H.; Lappert, M. F.; Stanley, K. (J. Organomet. Chem. **112** [1976] 37/48).
[152] Hillis, J.; Ishaq, M.; Gorewit, B.; Tsutsui, M. (J. Organomet. Chem. **116** [1976] 91/7).
[153] Dean, W. K.; Graham, W. A. G. (J. Organomet. Chem. **120** [1976] 73/86).
[154] Gingell, A. C.; Harris, A.; Rest, A. J.; Turner, R. N. (J. Organomet. Chem. **121** [1976] 205/10).
[155] Davidson, J. L.; Green, M.; Nyathi, J. Z.; Scott, C.; Stone, F. G. A.; Welch, A. J.; Woodward, P. (J. Chem. Soc. Chem. Commun. **1976** 714/5).
[156] Green, J. C.; Jackson, S. E. (J. Chem. Soc. Dalton Trans. **1976** 1698/702).
[157] Davidson, J. L.; Green, M.; Stone, F. G. A.; Welch, A. J. (J. Chem. Soc. Dalton Trans. **1976** 2044/53).
[158] Cotton, J. D.; Davidson, P. J.; Lappert, M. F. (J. Chem. Soc. Dalton Trans. **1976** 2275/86).
[159] Chrzastowski, J. Z.; Johnson, M. D. (J. Chem. Soc. Dalton Trans. **1976** 2456/63).
[160] Cutler, A.; Ehntholt, D.; Giering, W. P.; Lennon, P.; Raghu, S.; Rosan, A.; Rosenbaum, M.; Tancrede, J.; Wells, D. (J. Am. Chem. Soc. **98** [1976] 3495/507).

[161] Pannell, K. H.; Cassias, J. B.; Crawford, G. M.; Flores, A. (Inorg. Chem. **15** [1976] 2671/5).

[162] Beach, D. L. (Diss. Univ. Missouri, St. Louis, Miss., 1976; Diss. Abstr. Int. B **38** [1978] 5358).
[163] Behrens, H.; Pfister, A.; Moll, M.; Sepp, E. (Z. Anorg. Allg. Chem. **428** [1977] 61/7).
[164] Rausch, M. D.; Gismondi, T. E.; Alt, H. G.; Schwärzle, J. A. (Z. Naturforsch. **32b** [1977] 998/1000).
[165] Herrmann, W. A.; Biersack, H. (Chem. Ber. **110** [1977] 896/915).
[166] Alt, H. G. (Chem. Ber. **110** [1977] 2862/6).
[167] Beck, W.; Petri, W. (J. Organomet. Chem. **127** [1977] C40/C44).
[168] Beck, W.; Olgemöller, B. (J. Organomet. Chem. **127** [1977] C45/C47).
[169] Gladysz, J. A.; Williams, G. M.; Tam, W.; Johnson, D. L. (J. Organomet. Chem. **140** [1977] C1/C6).
[170] Chen, L. S.; Lichtenberg, D. W.; Robinson, P. W.; Yamamoto, Y.; Wojcicki, A. (Inorg. Chim. Acta **25** [1977] 165/72).
[171] Davidson, J. L.; Green, M.; Nyathi, J. Z.; Stone, F. G. A.; Welch, A. J. (J. Chem. Soc. Dalton Trans. **1977** 2246/55).
[172] Jolly, W. L.; Avanzino, S. C.; Rietz, R. R. (Inorg. Chem. **16** [1977] 964/6).
[173] Dean, W. K.; Graham, W. A. G. (Inorg. Chem. **16** [1977] 1061/7).
[174] Nesmeyanov, A. N.; Kolobova, N. E.; Valueva, Z. P.; Solodova, M. Ya.; Anisimov, K. N. (Izv. Akad. Nauk SSSR Ser. Khim. **1977** 1138/42; Bull. Acad. Sci. USSR Div. Chem. Sci. [Engl. Transl.] **1977** 1044/8).
[175] Dukes, M. D. (Diss. Univ. South Carolina, Columbia, S.C., 1977; Diss. Abstr. Int. B **38** [1978] 5925/6).
[176] Gardner, S. A.; Lelental, M.; Eastman Kodak Co. (U.S. 4097281 [1977/78]; C.A. **90** [1979] No. 79133).
[177] Gardner, S. A.; Lelental, M. (Res. Discl. **166** [1978] 8/10).
[178] Walther, B.; Albert, H.; Kolbe A. (J. Organomet. Chem. **145** [1978] 285/302).
[179] Braun, S.; Dahler, P.; Eilbracht, P. (J. Organomet. Chem. **146** [1978] 135/41).
[180] Jordan, R. F.; Tsang, E.; Norton, J. R. (J. Organomet. Chem. **149** [1978] C53/C56).
[181] Reger, D. L.; Dukes, M. D. (J. Organomet. Chem. **153** [1978] 67/72).
[182] Todd, L. J.; Wilkinson, J. R.; Hickey, J. P.; Beach, D. L.; Barnett, K. W. (J. Organomet. Chem. **154** [1978] 151/7).
[183] Alt, H. G.; Schwärzle, J. A. (J. Organomet. Chem. **162** [1978] 45/56).
[184] Charrier, C.; Collin, J.; Mérour, J. Y.; Roustan, J. L. (J. Organomet. Chem. **162** [1978] 57/66).
[185] Chen, L. S.; Su, S. R.; Wojcicki, A. (Inorg. Chim. Acta **27** [1978] 79/89).
[186] Downs, R. L.; Wojcicki, A. (Inorg. Chim. Acta **27** [1978] 91/103).
[187] Hartshorn, A. J.; Lappert, M. F.; Turner, K. (J. Chem. Soc. Dalton Trans. **1978** 348/56).
[188] Green, M.; Nyathi, J. Z.; Scott, C.; Stone, F. G. A.; Welch, A. J.; Woodward, P. (J. Chem. Soc. Dalton Trans. **1978** 1067/80).
[189] King, R. B.; King, A. D., Jr.; Iqbal, M. Z.; Frazier, C. C. (J. Am. Chem. Soc. **100** [1978] 1687/94).
[190] Crease, A. E.; Johnson, M. D. (J. Am. Chem. Soc. **100** [1978] 8013/4).
[191] Pasynskii, A. A.; Skripkin, Yu. V.; Suvorova, K. M.; Kuz'micheva, O. N.; Kalinnikov, V. T. (Izv. Akad. Nauk SSSR Ser. Khim. **1978** 1226; Bull. Acad. Sci. USSR Div. Chem. Sci. [Engl. Transl.] **1978** 1069).
[192] Nesmeyanov, A. N.; Perevalova, E. G.; Leont'eva, L. I.; Eremin, S. A. (Dokl. Akad. Nauk SSSR **243** [1978] 1208/11; Dokl. Chem. [Engl. Transl.] **238/243** [1978] 609/12).
[193] Bell, P. B. (Diss. Ohio State Univ., Columbus, Ohio, 1978; Diss. Abstr. Int. B **39** [1979] 4878).

[194] Dilgassa, M. (Diss. Univ. Michigan, Ann Arbor, MI, 1978; Diss. Abstr. Int. B **39** [1979] 4879/80).

[195] Fadel, S.; Weidenhammer, K.; Ziegler, M. L. (Z. Anorg. Allg. Chem. **453** [1979] 98/106).

[196] Korswagen, R.; Weidenhammer, K.; Ziegler, M. L. (Z. Naturforsch. **34b** [1979] 1507/11).

[197] Benaïm, J.; Giuliéri, F. (J. Organomet. Chem. **165** [1979] C28/C32).

[198] Pasynskii, A. A.; Skripkin, Yu. V.; Eremenko, I. L.; Kalinnikov, V. T.; Aleksandrov, G. G.; Andrianov, V. G.; Struchkov, Yu. T. (J. Organomet. Chem. **165** [1979] 49/56).

[199] Roustan, J. L.; Mérour, J. Y.; Charrier, C.; Benaïm, J.; Cadiot, F. (J. Organomet. Chem. **169** [1979] 39/52).

[200] King, R. B.; Iqbal, M. Z.; King, A. D., Jr. (J. Organomet. Chem. **171** [1979] 53/63).

[201] Coville, N. J.; Albers, M. O. (J. Organomet. Chem. **172** [1979] C1/C3).

[202] Gibson, D. H., Hsu, W.-L.; Lin, D.-S. (J. Organomet. Chem. **172** [1979] C7/C12).

[203] Dilgassa, M.; Curtis, M. D. (J. Organomet. Chem. **172** [1979] 177/84).

[204] Samuel, E.; Rausch, M. D.; Gismondi, T. E.; Mintz, E. A.; Giannotti, C. (J. Organomet. Chem. **172** [1979] 309/15).

[205] Blumer, D. J.; Barnett, K. W.; Brown, T. L. (J. Organomet. Chem. **173** [1979] 71/6).

[206] Watson, P. L.; Bergman, R. G. (J. Am. Chem. Soc. **101** [1979] 2055/62).

[207] Jordan, R. F.; Norton, J. R. (J. Am. Chem. Soc. **101** [1979] 4853/8).

[208] Jones, W. D.; Bergman, R. G. (J. Am. Chem. Soc. **101** [1979] 5447/9).

[209] Butts, S. B.; Holt, E. M.; Strauss, S. H.; Alcock, N. W.; Stimson, R. E.; Shriver, D. F. (J. Am. Chem. Soc. **101** [1979] 5864/6).

[210] Gladysz, J. A.; Williams, G. M.; Tam, W.; Johnson, D. L.; Parker, D. W.; Selover, J. C. (Inorg. Chem. **18** [1979] 553/8).

[211] Jones, W. D. (Diss. California Inst. Technol., Pasadena, CA, 1979; Diss. Abstr. Int. B **40** [1979] 1708).

[212] Eremin, S. A. (Deposited Doc. VINITI-3782 [1979] **1979** 174/6; C.A. **94** [1981] No. 192440).

[213] Korswagen, R.; Ziegler, M. L. (Z. Naturforsch. **35b** [1980] 1196/200).

[214] Gompper, R.; Bachmann, E. (Liebigs Ann. Chem. **1980** 229/40).

[215] Janta, R.; Albert, W.; Rößner, H.; Malisch, W.; Langenbach, H.-J.; Röttinger, E.; Vahrenkamp, H. (Chem. Ber. **113** [1980] 2729/38).

[216] Labinger, J. A. (J. Organomet. Chem. **187** [1980] 287/96).

[217] Masters, A. F.; Brownlee, R. T. C.; O'Connor, M. J.; Wedd, A. G.; Cotton, J. D. (J. Organomet. Chem. **195** [1980] C17/C20).

[218] Benaïm, J.; Giuliéri, F. (J. Organomet. Chem. **202** [1980] C9/C14).

[219] Sizoi, V. F.; Nekrasov, Yu. S.; Sukharev, Yu. N.; Leontyeeva, L. I.; Eremin, S. A.; Kolobova, N. E.; Solodova, M. Ya.; Valueva, Z. P. (J. Organomet. Chem. **202** [1980] 83/90).

[220] Anderson, S. N.; Cooksey, C. J.; Holton, S. G.; Johnson, M. D. (J. Am. Chem. Soc. **102** [1980] 2312/8).

[221] Watson, P.; Bergman, R. G. (J. Am. Chem. Soc. **102** [1980] 2699/703).

[222] Butts, S. B.; Strauss, S. H.; Holt, E. M.; Stimson, R. E.; Alcock, N. W.; Shriver, D. F. (J. Am. Chem. Soc. **102** [1980] 5093/100).

[223] Frommer, J. E.; Bergman, R. G. (J. Am. Chem. Soc. **102** [1980] 5227/34).

[224] Avanzino, S. C.; Bakke, A. A.; Chen, H.-W.; Donahue, C. J.; Jolly, W. L.; Lee, T. H.; Ricco, A. J. (Inorg. Chem. **19** [1980] 1931/6).

[225] Nesmeyanov, A. N.; Perevalova, E. G.; Leont'eva, L. I.; Eremin, S. A. (Dokl. Akad. Nauk SSSR **252** [1980] 119/23; Dokl. Chem. [Engl. Transl.] **250/255** [1980] 215/8).

[226] Butts, S. B. (Diss. Northwestern Univ., Evanston, IL, 1980; Diss. Abstr. Int. B **41** [1980] 2170/1).

[227] Jones, W. D. (Report LBL-10214 [1979] 1/178; Energy Res. Abstr. **5** [1980] Abstr. No. 32658; C.A. **95** [1981] No. 6002).

[228] Huggins, J. M. (Report LBL-11697 [1980] 1/206; Energy Res. Abstr. **6** [1981] Abstr. No. 5132; C.A. **95** [1981] No. 41785).

[229] Huggins, J. M. (Diss. California Inst. Technol., Pasadena, CA, 1981; Diss. Abstr. Int. B **41** [1980] 2172).

[230] Talay, R.; Rehder, D. (Z. Naturforsch. **36b** [1981] 451/62).

[231] Olgemöller, B.; Beck, W. (Chem. Ber. **114** [1981] 867/76).

[232] Malisch, W.; Blau, H.; Haaf, F. J. (Chem. Ber. **114** [1981] 2956/70).

[233] Komiya, S.; Sano, K.; Yamamoto, T. (Transition Met. Chem. [London] **6** [1981] 336/8).

[234] Selover, J. C.; Marsi, M.; Parker, D. W.; Gladysz, J. A. (J. Organomet. Chem. **206** [1981] 317/29).

[235] Alt, H. G.; Eichner, M. E. (J. Organomet. Chem. **212** [1981] 397/403).

[236] Pannell, K. H.; Kapoor, R. N. (J. Organomet. Chem. **214** [1981] 47/52).

[237] Gibson, D. H.; Hsu, W.-L.; Ahmed, F. U. (J. Organomet. Chem. **215** [1981] 379/401).

[238] Botha, C.; Moss, J. R.; Pelling, S. (J. Organomet. Chem. **220** [1981] C21/C24).

[239] Le Gall, J.-Y.; Kubicki, M. M.; Pétillon, F. Y. (J. Organomet. Chem. **221** [1981] 287/90).

[240] Hitam, R. B.; Hooker, R. H.; Mahmoud, K. A.; Narayanaswamy, R.; Rest, A. J. (J. Organomet. Chem. **222** [1981] C9/C13).

[241] Cotton, J. D.; Crisp, G. T.; Daly, V. A. (Inorg. Chim. Acta **47** [1981] 165/9).

[242] Orlova, T. Yu.; Setkina, V. N.; Kursanov, D. N. (Inorg. Chim. Acta **51** [1981] 131/4).

[243] Hudson, A.; Lappert, M. F.; Lednor, P. W.; MacQuitty, J. J.; Nicholson, B. K. (J. Chem. Soc. Dalton Trans. **1981** 2159/63).

[244] Mahmoud, K. A.; Narayanaswamy, R.; Rest, A. J. (J. Chem. Soc. Dalton Trans. **1981** 2199/204).

[245] Jones, W. D.; Huggins, J. M.; Bergman, R. G. (J. Am. Chem. Soc. **103** [1981] 4415/23).

[246] Wax, M. J.; Bergman, R. G. (J. Am. Chem. Soc. **103** [1981] 7028/30).

[247] Bell, P. B.; Wojcicki, A. (Inorg. Chem. **20** [1981] 1585/92).

[248] Wagner, H. M.; Purbrick, M. D. (J. Photogr. Sci. **29** [1981] 230/5).

[249] Jordan, R. F. (Diss. Princeton Univ., Princeton, N.J., 1982; Diss. Abstr. Int. B **42** [1982] 3679/80).

[250] Alt, H. G.; Eichner, M. E. (Angew. Chem. **94** [1982] 77; Angew. Chem. Int. Ed. Engl. **21** [1982] 78; Angew. Chem. Suppl. **1982** 121).

[251] Moss, J. R. (J. Organomet. Chem. **231** [1982] 229/35).

[252] Cotton, J. D.; Kimlin, H. A.; Markwell, R. D. (J. Organomet. Chem. **232** [1982] C75/C77).

[253] Cotton, J. D.; Markwell, R. D. (Inorg. Chim. Acta **63** [1982] 13/6).

[254] Bailey, N. A.; Chell, P. L.; Mukhopadhyay, A.; Tabbron, H. E.; Winter, M. J. (J. Chem. Soc. Chem. Commun. **1982** 215/7).

[255] LaCroce, S. J.; Cutler, A. R. (J. Am. Chem. Soc. **104** [1982] 2312/4).

[256] Kazlauskas, R. J.; Wrighton, M. S. (J. Am. Chem. Soc. **104** [1982] 6005/15).

[257] Labinger, J. A.; Miller, J. S. (J. Am. Chem. Soc. **104** [1982] 6856/8).

[258] Brownlee, R. T. C.; Masters, A. F.; O'Connor, M. J.; Wedd, A. G.; Kimlin, H. A.; Cotton, J. D. (Org. Magn. Reson. **20** [1982] 73/7).

[259] Wrighton, M. S.; Graff, J. L.; Kazlauskas, R. J.; Mitchener, J. C.; Reichel, C. L. (Pure Appl. Chem. **54** [1982] 161/76).

[260] Gladysz, J. A. (Report AFOSR-TR-82-0522, Order No. AD-A 117079 [1982] 1/9; Gov. Rep. Announce Index [U.S.] **82** [1982] 4842; C.A. **98** [1983] No. 89593).

[261] Gismondi, T. E. (Diss. Univ. Massachusetts, Amherst, 1982; Diss. Abstr. Int. B **43** [1983] 2555).
[262] Meyer, A.; Hartl, A.; Malisch, W. (Chem. Ber. **116** [1983] 348/59).
[263] Sünkel, K.; Schloter, K.; Beck, W.; Ackermann, K.; Schubert, U. (J. Organomet. Chem. **241** [1983] 333/42).
[264] Sünkel, K.; Nagel, U.; Beck, W. (J. Organomet. Chem. **251** [1983] 227/43).
[265] Leung, T. W.; Christoph, G. G.; Wojcicki, A. (Inorg. Chim. Acta **76** [1983] L281/L282).
[266] Adams, H.; Bailey, N. A.; Cahill, P.; Rogers, D.; Winter, M. J. (J. Chem. Soc. Chem. Commun. **1983** 831/3).
[267] Pelling, S.; Botha, C.; Moss, J. R. (J. Chem. Soc. Dalton Trans. **1983** 1495/501).
[268] Bailey, N. A.; Chell, P. L.; Manuel, C. P.; Mukhopadhyay, A.; Rogers, D.; Tabbron, H. E.; Winter, M. J. (J. Chem. Soc. Dalton Trans. **1983** 2397/403).
[269] Labinger, J. A.; Bonfiglio, J. N.; Grimmett, D. L.; Masuo, S. T.; Shearin, E.; Miller, J. S. (Organometallics **2** [1983] 733/40).
[270] Michelin, R. A.; Facchin, G.; Uguagliati, P. (Atti 16th Congr. Naz. Chim. Inorg., Ferrara, Italy, 1983, pp. 18/21; C.A. **100** [1984] No. 85894).

[271] Wax, M. J. (Report LBL-16586 [1983] 1/71; Energy Res. Abstr. **8** [1983] Abstr. No. 57462; C.A. **100** [1984] No. 209995).
[272] Petri, W.; Beck, W. (Chem. Ber. **117** [1984] 3265/9).
[273] Giuliéri, F.; Benaïm, J. (J. Organomet. Chem. **276** [1984] 367/76).
[274] Cotton, J. D.; Dunstan, P. R. (Inorg. Chim. Acta **88** [1984] 223/7).
[275] Gauntlett, J. T.; Taylor, B. F.; Winter, M. J. (J. Chem. Soc. Chem. Commun. **1984** 420/1).
[276] Adams, H.; Bailey, N. A.; Gauntlett, J. T.; Winter, M. J. (J. Chem. Soc. Chem. Commun. **1984** 1360/1).
[277] Mahmoud, K. A.; Rest, A. J.; Alt, A. G.; Eichner, M. E.; Jansen, B. M. (J. Chem. Soc. Dalton Trans. **1984** 175/86).
[278] Adams, H.; Bailey, N. A.; Winter, M. J. (J. Chem. Soc. Dalton Trans. **1984** 273/8).
[279] Hooker, R. H.; Rest, A. J. (J. Chem. Soc. Dalton Trans. **1984** 761/70).
[280] Michelin, R. A.; Facchin, G.; Uguagliati, P. (Inorg. Chem. **23** [1984] 961/9).

[281] Richmond, T. G.; Shriver, D. F. (Organometallics **3** [1984] 305/14).
[282] Richmond, T. G.; Crespi, A. M.; Shriver, D. F. (Organometallics **3** [1984] 314/9).
[283] Facchin, G.; Uguagliati, P.; Michelin, R. A. (Organometallics **3** [1984] 1818/22).
[284] Yamamoto, M.; Sato, M.; Nakagawa, Y.; Seiko Instruments and Electronics Ltd. (Jpn. Kokai Tokkyo Koho 60245227 [85224227] [1984/85]; C.A. **105** [1986] No. 33945).
[285] Martin, D. T. (Diss. Univ. British Columbia, Vancouver, BC, 1984; Diss. Abstr. Int. B **46** [1985] 835).
[286] Benamou, C.; Benaïm, J. (J. Organomet. Chem. **280** [1985] 377/87).
[287] Gismondi, T. E.; Rausch, M. D. (J. Organomet. Chem. **284** [1985] 59/71).
[288] Luh, T.-Y.; Wong, C. S. (J. Organomet. Chem. **287** [1985] 231/3).
[289] Alt, H. G.; Engelhardt, H. E.; Thewalt, U.; Riede, J. (J. Organomet. Chem. **288** [1985] 165/77).
[290] Markham, J.; Tolman, W.; Menard, K.; Cutler, A. (J. Organomet. Chem. **294** [1985] 45/58).

[291] Cotton, J. D.; Kimlin, H. A. (J. Organomet. Chem. **294** [1985] 213/7).
[292] Giuliéri, F.; Benaïm, J. (Nouv. J. Chim. **9** [1985] 335/40).

[293] Himmel, S. E.; Young, G. B.; Fung, D. C. M.; Hollinshead, C. (Polyhedron **4** [1985] 349/56).
[294] Poli, R.; Wilkinson, G.; Motevalli, M.; Hursthouse, M. B. (J. Chem. Soc. Dalton Trans. **1985** 931/9).
[295] McKean, D. C.; McQuillan, G. P.; Morrisson, A. R.; Torto, I. (J. Chem. Soc. Dalton Trans. **1985** 1207/12).
[296] Gauntlett, J. T.; Taylor, B. F.; Winter, M. J. (J. Chem. Soc. Dalton Trans. **1985** 1815/20).
[297] Doney, J. J.; Bergman, R. G.; Heathcock, C. H. (J. Am. Chem. Soc. **107** [1985] 3724/6).
[298] Markham, J.; Menard, K.; Cutler, A. (Inorg. Chem. **24** [1985] 1581/7).
[299] Febvay, J.; Casabianca, F.; Riess, J. G. (Inorg. Chem. **24** [1985] 3235/9).
[300] Cotton, J. D.; Markwell, R. D. (Organometallics **4** [1985] 937/9).

[301] Bodnar, T. W.; Cutler, A. R. (Organometallics **4** [1985] 1558/65).
[302] Barefield, E. K.; McCarten, P.; Hillhouse, M. C. (Organometallics **4** [1985] 1682/4).
[303] Heathcock, C. H.; Doney, J. J.; Bergman, R. G. (Pure Appl. Chem. **57** [1985] 1789/98).
[304] Brownlee, R. T. C.; O'Connor, M. J.; Shehan, B. P.; Webb, A. G. (J. Magn. Reson. **61** [1985] 516/25).
[305] Liu, D.; Ye, D.; Liu, D. (Gaodeng Xuexiao Huaxue Xuebao **6** [1985] 851/3; C.A. **103** [1985] No. 178711).
[306] Liu, D.; Ye, D.; Zhou, W.; Liu, D. (Wuhan Daxue Xuebao Ziran Kexueban **1985** 71/6; C.A. **104** [1986] No. 19863).
[307] Targos, T. S.; Geoffroy, G. L.; Rheingold, A. L. (J. Organomet. Chem. **299** [1986] 223/31).
[308] Akita, M.; Kondoh, A. (J. Organomet. Chem. **299** [1986] 369/76).
[309] Cotton, J. D.; Kennard, C. H. L.; Markwell, R. D.; Smith, G.; White, A. H. (J. Organomet. Chem. **309** [1986] 117/23).
[310] Kreiter, C. G.; Kögler, J.; Nist, K. (J. Organomet. Chem. **310** [1986] 35/46).

[311] Kreiter, C. G.; Kögler, J.; Sheldrick, W. S.; Nist, K. (J. Organomet. Chem. **311** [1986] 125/43).
[312] Nolan, S. P.; López de la Vega, R.; Hoff, C. D. (J. Organomet. Chem. **315** [1986] 187/99).
[313] Green, M. (Polyhedron **5** [1986] 427/33).
[314] Gauntlett, J. T.; Winter, M. J. (Polyhedron **5** [1986] 451/9).
[315] Adams, H.; Bailey, N. A.; Cahill, P.; Rogers, D.; Winter, M. J. (J. Chem. Soc. Dalton Trans. **1986** 2119/26).
[316] Nolan, S. P.; López de la Vega, R.; Mukerjee, S. L.; Hoff, C. D. (Inorg. Chem. **25** [1986] 1160/5).
[317] Leung, T. W.; Christoph, G. G.; Galucci, J.; Wojcicki, A. (Organometallics **5** [1986] 366/74).
[318] Leung, T. W.; Christoph, G. G.; Galucci, J.; Wojcicki, A. (Organometallics **5** [1986] 846/53).
[319] Lee, G.-H.; Peng, S.-M.; Lee, T.-W.; Liu, R.-S. (Organometallics **5** [1986] 2378/80).
[320] McKean, D. C.; McQuillan, G. P.; Torto, I.; Morrisson, A. R. (J. Mol. Struct. **141** [1986] 457/64).

[321] Solodovnikov, S. P.; Tumanskii, B. L.; Bubnov, N. N.; Kabachnik, M. I. (Izv. Akad. Nauk SSSR Ser. Khim. **1986** 2147/50; Bull. Acad. Sci. USSR Div. Chem. Sci. [Engl. Transl.] **1986** 1960).

[322] Lin, S.; Lin, C. Y.; Wang, H. Y. (J. Chin. Chem. Soc. **33** [1986] 261/3).
[323] Park, Y. K. (Diss. Texas A & M Univ., College Station, TX, 1986; Diss. Abstr. Int. B **47** [1986] 1545/6).
[324] Brauel, A.; Rehder, D. (Z. Naturforsch. **42b** [1987] 605/9).
[325] Schmitt, G.; Pritzkow, H.; Latscha, H. P. (Z. Naturforsch. **42b** [1987] 1115/20).
[326] Werner, H.; Roll, J.; Zolk, R.; Thometzek, P.; Linse, K.; Ziegler, M. L. (Chem. Ber. **120** [1987] 1553/64).
[327] Appel, M.; Schloter, K.; Heidrich, J.; Beck, W. (J. Organomet. Chem. **322** [1987] 77/88).
[328] Mann, B. E.; Shaw, S. D. (J. Organomet. Chem. **326** [1987] C13/C16).
[329] Isobe, K.; Tero-Kubota, S.; Tanaka, H.; Nakamura, Y.; Okeya, S.; Saito, K. (J. Organomet. Chem. **329** [1987] C21/C24).
[330] Fischer, R. A.; Kneuper, H.-J.; Herrmann, W. A. (J. Organomet. Chem. **330** [1987] 365/76).

[331] Burkhardt, E. R.; Doney, J. J.; Bergman, R. G.; Heathcock, C. H. (J. Am. Chem. Soc. **109** [1987] 2022/39).
[332] O'Connor, E. J.; Brandt, S.; Helquist, P. (J. Am. Chem. Soc. **109** [1987] 3739/47).
[333] Pannell, K. H.; Kapoor, R. N.; Wells, M.; Giasolli, T.; Parkanyi, L. (Organometallics **6** [1987] 663/7).
[334] Pannell, K. H.; Cea-Olivares, R.; Toscano, R. A.; Kapoor, R. N. (Organometallics **6** [1987] 1821/2).
[335] Lucas, C. R.; Walsh, K. A. (J. Chem. Educ. **64** [1987] 265/6).
[336] Denisovich, L. I.; Ustynyuk, N. A.; Peterleitner, M. G.; Vinogradova, V. N.; Kravtsov, D. N. (Izv. Akad. Nauk SSSR Ser. Khim. **1987** 2635/6; Bull. Acad. Sci. USSR Div. Chem. Sci. [Engl. Transl.] **1987** 2450).
[337] Tafesse, F.; Dilgassa, M. (Bull. Chem. Soc. Ethiop. **1** [1987] 83/6).
[338] Hart, I. J.; Jeffery, J. C.; Lowry, R. M.; Stone, F. G. A. (Angew. Chem. **100** [1988] 1769/75; Angew. Chem. Int. Ed. Engl. **27** [1988] 1703).
[339] Endrich, K.; Alburquerque, P.; Korswagen, R. P.; Ziegler, M. L. (Z. Naturforsch. **43b** [1988] 1293/306).
[340] Paprott, G.; Lehmann, S.; Seppelt, K. (Chem. Ber. **121** [1988] 727/33).

[341] Akita, M.; Kakinuma, M.; Moro-oka, Y. (J. Organomet. Chem. **348** [1988] 91/4).
[342] Nolan, S. P.; López de la Vega, R.; Mukerjee, S. L.; Gonzales, A. A.; Zhang, K.; Hoff, C. D. (Polyhedron **7** [1988] 1491/8).
[343] Lee, G.-H.; Peng, S.-M.; Lush, S.-L.; Liu, R.-S. (J. Chem. Soc. Chem. Commun. **1988** 1513/4).
[344] Wido, T. M.; Young, G. H.; Wojcicki, A.; Calligaris, M.; Nardin, G. (Organometallics **7** [1988] 452/8).
[345] Lee, T.-W.; Liu, R.-S. (Organometallics **7** [1988] 878/83).
[346] Naumann, D.; Varbelow, H.-G. (J. Fluorine Chem. **41** [1988] 415/9).
[347] Young, G. H.; Wojcicki, A.; Calligaris, M.; Nardin, G.; Bresciani-Pahor, N. (J. Am. Chem. Soc. **111** [1989] 6890/1).
[348] Raseta, M. E.; Cawood, S. A.; Welker, M. E.; Rheingold, A. L. (J. Am. Chem. Soc. **111** [1989] 8268/70).
[349] Akita, M.; Kawahara, T.; Terada, M.; Kakinuma, N.; Moro-oka, Y. (Organometallics **8** [1989] 687/93).
[350] Cervantes, J.; Vincenti, S. P.; Kapoor, R. N.; Pannell, K. H. (Organometallics **8** [1989] 744/8).

[351] Lee, G.-H.; Peng, S.-M.; Yang, G.-M.; Lush, F.-S.; Liu, R.-S. (Organometallics **8** [1989] 1106/11).

[352] Glavee, G. N.; Daniels, L. M.; Angelici, R. J. (Organometallics **8** [1989] 1856/65).
[353] Lo Sterzo, C.; Miller, M. M.; Stille, J. K. (Organometallics **8** [1989] 2331/7).
[354] Park, Y. K.; Han, I. S.; Huh, T. S.; Darensbourg, M. Y. (Bull. Korean Chem. Soc. **10** [1989] 134/7).
[355] Lin, H.-S.; Vong, W.-J.; Cheng, C.-Y.; Wang, S.-L.; Liu, R.-S. (Tetrahedron Lett. **31** [1990] 7645/8).
[356] Cotton, J. D.; Kroes, M. M.; Markwell, R. D.; Miles, E. A. (J. Organomet. Chem. **388** [1990] 133/42).
[357] Young, G. H.; Wojcicki, A. (J. Organomet. Chem. **390** [1990] 351/60).
[358] Friedrich, H. B.; Moss, J. R.; Williamson, B. K. (J. Organomet. Chem. **394** [1990] 313/27).
[359] Campen, A. K.; Mahmoud, K. A.; Rest, A. J.; Willis, P. A. (J. Chem. Soc. Dalton Trans. **1990** 2817/23).
[360] Lehmann, R. E.; Bockman, T. M.; Kochi, J. K. (J. Am. Chem. Soc. **112** [1990] 458/9).

[361] Powell, J.; Fuchs, E.; Sawyer, J. F. (Organometallics **9** [1990] 1722/9).
[362] Vong, W.-J.; Peng, S.-M.; Liu, R.-S. (Organometallics **9** [1990] 2187/9).
[363] Park, Y. K.; Kim, S. J.; Ash, C. (Bull. Korean Chem. Soc. **11** [1990] 109/14).
[364] Kher, S. S.; Nile, T. A. (Transition Met. Chem. [London] **16** [1991] 28/30).
[365] Thorn, D. L. (J. Organomet. Chem. **405** [1991] 161/71).
[366] Johnson, F. P. A.; Gordon, C. M.; Hodges, P. M.; Poliakoff, M.; Turner, J. J. (J. Chem. Soc. Dalton Trans. **1991** 833/9).
[367] Adams, H.; Bailey, N. A.; Gauntlett, J. T.; Winter, M. J.; Woodward, S. (J. Chem. Soc. Dalton Trans. **1991** 2217/21).
[368] Vong, W.-J.; Peng, S.-M.; Lin, S.-H.; Lin, W.-J.; Liu, R.-S. (J. Am. Chem. Soc. **113** [1991] 573/82).
[369] Lehmann, R. E.; Kochi, J. K. (Organometallics **10** [1991] 190/202).
[370] Kegley, S. E.; Bergstrom, D. T.; Crocker, L. S.; Weiss, E. P.; Berndt, W. G.; Rheingold, A. L. (Organometallics **10** [1991] 567/73).

[371] Koola, J. D.; Roddick, D. M. (Organometallics **10** [1991] 591/7).
[372] Yang, G.-M.; Lee, G.-H.; Peng, S.-M.; Liu, R.-S. (Organometallics **10** [1991] 1305/10).
[373] Young, G. H.; Raphael, M. V.; Wojcicki, A.; Calligaris, M.; Nardin, G.; Bresciani-Pahor, N. (Organometallics **10** [1991] 1934/45).
[374] Kubacek, P.; Hoffmann, R.; Havlas, Z. (Organometallics **1** [1982] 180/8).
[375] Alt, H. G. (J. Organomet. Chem. **383** [1990] 125/42).
[376] Krueger, S. T.; Owens, B. E.; Poli, R. (Inorg. Chem. **29** [1990] 2001/6).
[377] Poli, R. (Organometallics **9** [1990] 1892/900).
[378] Hill, R. H.; Debad, J. D. (Polyhedron **10** [1991] 1705/12).
[379] Adams, H.; Bailey, N. A.; Bentley, G. W.; Hough, G.; Winter, M. J.; Woodward, S. (J. Chem. Soc. Dalton Trans. **1991** 749/58).
[380] Adams, H.; Bailey, N. A.; Gauntlett, J. T.; Harkin, I. M.; Winter, M. J.; Woodward, S. (J. Chem. Soc. Dalton Trans. **1991** 1117/28).

[381] Yang, G.-M.; Lee, G.-H.; Peng, S.-M.; Liu, R.-S. (Organometallics **10** [1991] 2531/2).
[382] Raseta, M. E.; Mishra, R. K.; Cawood, S. A.; Welker, M. E.; Rheingold, A. L. (Organometallics **10** [1991] 2936/45).
[383] Blenkiron, P.; Lavender, M. H.; Morris, M. J. (J. Organomet. Chem. **426** [1992] C28/C32).
[384] Kreiter, C. G. (Adv. Organomet. Chem. **26** [1986] 277/375).
[385] Yang, G.-M.; Lee, G.-H.; Peng, S.-M.; Liu, R.-S. (J. Chem. Soc. Chem. Commun. **1991** 478/9).

[386] Manusco, C.; Halpern, J. (J. Organomet. Chem. **428** [1992] C8/C11).
[387] Schenk, W. A.; Pfeffermann, J. (J. Organomet. Chem. **440** [1992] 341/51).
[388] Su, G.-M.; Lee, G.-H.; Peng, S.-M.; Liu, R.-S. (J. Chem. Soc. Chem. Commun. **1992** 215/7).
[389] Wang, P.; Atwood, J. D. (J. Am. Chem. Soc. **114** [1992] 6424/7).
[390] Yang, G.-M.; Su, G.-M.; Liu, R.-S. (Organometallics **11** [1992] 3444/51).

[391] Döllein, G.; Solvay Deutschland (Ger. Offen. 4136321 [1991/93]; C.A. **119** [1993] No. 54209).
[392] Friedrich, H. B.; Moss, J. R. (J. Chem. Soc. Dalton Trans. **1993** 2863/9).
[393] Wu, I.-Y.; Tseng, T.-W.; Lin, Y.-C.; Cheng, M.-C.; Wang, Y. (Organometallics **12** [1993] 478/85).
[394] Michelini-Rodriquez, I.; Romero, A. L.; Kapoor, R. N.; Cervantes-Lee, F.; Pannell, K. H. (Organometallics **12** [1993] 1221/4).
[395] Wu, I.-Y.; Tsai, J.-H.; Huang, B.-C.; Chen, S.-C.; Lin, Y.-C. (Organometallics **12** [1993] 3971/8).
[396] Gordon, J. C.; Lee, V. T.; Poli, R. (Inorg. Chem. **32** [1993] 4460/3).
[397] Hackett, P.; Manning, A. R. (J. Chem. Soc. Dalton Trans. **1975** 1606/9).
[398] Astruc, D.; Hamon, J.-R.; Román, E.; Michaud, P. (J. Am. Chem. Soc. **103** [1981] 7502/14).
[399] Burkhardt, E. R.; Doney, J. J.; Slough, G. A.; Stack, J. M.; Heathcock, C. H.; Bergman, R. G. (Pure Appl. Chem. **60** [1988] 1/6).

1.5.1.4.1.12.3 $C_5H_5Mo(CO)_3{}^1L$ Compounds Containing Mo-sp^2-C Bonds

1.5.1.4.1.12.3.1 Derivatives with Vinyl Type Ligands

All complexes this chapter deals with contain a $C_5H_5Mo(CO)_3C{=}C(R)$- moiety either as open-chain en-1-yl or as cycloalken-1-yl. All of the compounds are summarized in Table 2, and most of them were prepared by the following methods:

Method I: Starting from $[C_5H_5Mo(CO)_3]^-$.

a. A limited number of vinyl-substituted $C_5H_5Mo(CO)_3R$ complexes can be prepared by electrophilic substitution of $M[C_5H_5Mo(CO)_3]$ (M = Li, Na, K) with vinylic halides in THF as solvent. Addition of $(NC)_2C{=}CXCl$ (X = H, Cl, CN) to the starting material at −78 °C and subsequent stirring at room temperature for 1.5 h yielded compounds of the type $C_5H_5Mo(CO)_3CX{=}C(CN)_2$ which can be purified by column chromatography or sublimation in vacuum followed by recrystallization from CH_2Cl_2/hexane [6, 9]. Similarly $C_5H_5Mo(CO)_3(C{=}C(Cl)C(O)C(O)$-cyclo) was prepared using dichlorocyclobutenedione at −55 °C [38], and with perfluoro-1-methylcyclopent-1-ene as a precursor at −30 °C, after 30 min $C_5H_5Mo(CO)_3(C{=}C(CF_3)(CF_2)_3$-cyclo) was obtained. Purification of the crude product was accomplished by column chromatography on silica. An analogous reaction using perfluoro-2-methylpent-2-ene resulted in a mixture of isomeric substitution products and was not further investigated [39].

b. $C_5H_5Mo(CO)_3(C{=}CH(CH_2)_nC(O)O$-cyclo) (n = 1, 2) was prepared from two equivalents of $[C_5H_5Mo(CO)_3]^-$ (obtained in situ from $Mo(CO)_6$ and Na-$[C_5H_5]$) which were reacted with $ClC(O)CH_2(CH_2)_nC(O)Cl$ in THF at room

References on pp. 130/1

temperature for ca. 30 min. Separation from $C_5H_5Mo(CO)_3H$ and $(C_5H_5$-$Mo(CO)_3)_2$ was accomplished by column chromatography on silica [41].

Method II: Compounds containing cyclic en-1-yl ligands were obtainable by the [3 + 2] cycloaddition of electrophiles like $C_2(CN)_4$, t-$C_4H_9(NC)C{=}C{=}O$, $(CF_3)_2CO$, $FCl_2CC(O)CF_2Cl$, SO_2, SO_3, S_2O, $ClSO_2NCO$, 4-$CH_3C_6H_4SO_2NCO$, CH_3SO_2-NSO, or $S(NSO_2CH_3)_2$ to $C_5H_5Mo(CO)_3CH_2C{\equiv}CR$ (R = H, CH_3, C_6H_5) [3, 4, 7, 11, 12, 19, 23, 33, 35, 42]. For detailed reaction conditions, compare the individual compounds of Table 2.

Method III: Compounds of the type $C_5H_5Mo(CO)_3(C{=}CHCR(R')OCH_2$-cyclo) (R = CH_3, R' = C_6H_5; R = R' = C_6H_5; R = R' = fluorenylidene) were obtained from $C_5H_5Mo(CO)_3CH_2C{\equiv}CCR(R')OH$ which were easily isomerized into the cyclized η^1-3,5-dihydrofuran-1-yl complexes by refluxing in $CHCl_3$ for several hours or by chromatography on alumina [30].

Not included in the following table are compounds like **$C_5H_5Mo(CO)_3C(NR'_2){=}CHR$** mentioned as intermediates in the insertion reaction of $RC{\equiv}CNR'_2$ (R = H: R' = CH_3, C_2H_5; R'_2 = $(CH_2CH_2)_2O$; R = CH_3: R' = C_2H_5; R = $CO_2C_2H_5$: R' = C_2H_5; R = $B(N(CH_3)CH_2)_2$: R' = C_2H_5) into $C_5H_5Mo(CO)_3H$ as these complexes are highly unstable toward migratory CO insertion and hence promptly isomerize to give π-acryloyl compounds of composition $C_5H_5Mo(CO)_2(\eta^3$-$C(O)C(NR'_2){=}CHR)$ [21, 34]. Furthermore, the product obtained by treatment of $[C_5H_5Mo(CO)_3]^-$ with tetrakis(trifluoromethyl)allene was misformulated as the σ-vinyl derivative $C_5H_5Mo(CO)_3C({=}C(CF_3)_2)C({=}CF_2)CF_3$ in an early report [20], but was later shown to be a π-allylidene complex, $C_5H_5Mo(CO)_2(\eta^3$-$F_2CC(CF_3)CC(CF_3)_2)$ [25]. The reaction of $Na[C_5H_5Mo(CO)_3]$ with α-chloroenamines, $(CH_3)_2C{=}C(NR_2)Cl$, also did not lead to vinyl derivatives, but instead resulted in $C_5H_5Mo(CO)_2(\eta^2$-$C(NR_2){=}C(CH_3)_2)$ [15 to 17].

Table 2
$C_5H_5Mo(CO)_3{}^1L$ Complexes Containing Vinyl Ligands.
An asterisk indicates further information at the end of the table.
For explanations, abbreviations, and units, see p. X.

No.	^1L	method of preparation (yield) properties and remarks
1	$-C(CH_3){=}C(CH_3)_2$	preparation probably analogous to Method Ia ^1H NMR ($CDCl_3$): 5.37 (C_5H_5) [32] ^{95}Mo NMR ($CDCl_3$): −607; line width $w_{1/2}$ = 110 Hz [32]
2	trans-$C(CN){=}CHCN$	from $C_5H_5Mo(CO)_3H$ with excess $C_2(CN)_2$ in THF at 25°C by stirring overnight; purification was accomplished by column chromatography (50%); the analogous reaction with $HC{\equiv}CCN$ resulted in no defined Mo compound [36] yellow microcrystals, m.p. 105°C [36] ^1H NMR ($CDCl_3$): 5.60 (s, C_5H_5), 6.55 (s, =CH) [36] ^{13}C NMR ($CDCl_3$): 95.5 (C_5H_5), 118.0, 123.1 (both CN), 127.6 (=CH), 149.4 (MoC), 224.8, 233.7 (both CO) [36]

References on pp. 130/1

Table 2 (continued)

No.	^{1}L	method of preparation (yield) properties and remarks
2 (continued)		IR (Nujol): 1520 (ν(C=C)), 1950, 2060 (ν(CO)), 2200, 2220 (ν(CN)) [36] irradiation in the presence of $P(C_6H_5)_3$ in THF gave $C_5H_5Mo(CO)_2CH(CN)C(CN)(P^+(C_6H_5)_3)$-cyclo (Formula I on p. 127) along with some $C_5H_5Mo(CO)_2C(CN){=}CH(CN)(P(C_6H_5)_3)$ [36]
*3	$-CH{=}C(CN)_2$	Ia (37%, along with a deep blue product) [6, 9] yellow solid (sublimed at 100 °C at 1.5×10^{-4} atm), dec. 127 °C [6, 9] ^{1}H NMR ($CDCl_3$): 5.64 (s, C_5H_5), 10.24 (s, CH) [9] ^{13}C NMR (CH_2Cl_2 or CS_2): 94.9 (C_5H_5), 99.2 (=C), 114.0, 115.5 (both CN), 223.4 (CO cis MoC), 232.7 (CO trans MoCH) [13] (converted from the CS_2 scale by taking $\delta(CS_2)$ to be 192.8 ppm) IR (KBr): 1474 (ν(C=C)), 3114, 3132 (ν(CH)); (CH_2Cl_2): 1971, 2055 (ν(CO)), 2226, 2233 (ν(CN)) [6, 9]
*4	$-C(CN){=}C(CN)_2$	Ia (37%) [6, 9] yellow solid (sublimed at 100 °C at 1.5×10^{-4} atm with decomposition), m.p. 133 °C (dec.) [6, 9] ^{1}H NMR ($CDCl_3$): 5.72 (s, C_5H_5) [9] IR (KBr): 1472 (ν(C=C)), 3100 (ν(CH)); (CH_2Cl_2): 1983, 2068 (ν(CO)), 2190, 2209, 2231 (ν(CN)) [6, 9]
*5	$-CCl{=}C(CN)_2$	Ia (89 to 95%; chromatography considerably decreased the yield by decomposition) [6, 9] yellow solid [6, 9], m.p. 134 to 136 °C [6]; see also [9] ^{1}H NMR ($CDCl_3$): 5.72 (s, C_5H_5) [9] IR (KBr): 1450 (ν(C=C)), 3110 (ν(CH)); (CH_2Cl_2): 1981, 2064 (ν(CO)), 2226 (ν(CN)) [9]; see also [6]
*6	-trans-$C(CF_3){=}CHCF_3$	for preparation see "Further information" at the end of the table [31] yellow solid [31] ^{1}H NMR ($CDCl_3$): 5.58 (s, C_5H_5), 6.75 (m, CH) [31] ^{19}F NMR ($CDCl_3$): −59.3 (m), −58.0 (dq; J(F, H) = 7.75) [31] IR (KBr): 1430 (C_5H_5), 1610 (ν(C=C)); (CCl_4): 1955, 2000, 2055 (ν(CO)) [31]
*7	-trans-$C(CF_3){=}C(SCH_3)CF_3$	for preparation see "Further information" at the end of the table [31] ^{1}H NMR ($CDCl_3$): 2.50 (s, CH_3), 5.48 (s, C_5H_5) [31]

Table 2 (continued)

No.	^1L	method of preparation (yield) properties and remarks
		^{19}F NMR ($CDCl_3$): −54.1, −51.4 (each q; J(F, F) = 13.4) [31] IR (KBr): 1425 (C_5H_5), 1490 (ν(C=C)); (CCl_4): 1930, 1995, 2060 (ν(CO)) [31]
8	$-C(N^+H(C_2H_5)_2)=CHCH_3$	isolated as the BF_4^- salt from the π-acryloyl complex $C_5H_5Mo(CO)_2(\eta^3$-$C(O)C(N(C_2H_5)_2)=CHCH_3)$ in THF by protonating with 54% ethereal HBF_4 [21, 34] yellow crystalline solid (from CH_2Cl_2/ether), m.p. 90 °C [34] ^1H NMR (acetone-d_6): 1.34 (t, CH_3 of NC_2H_5; J = 7), 1.80 (d, $=CCH_3$; J = 7), 3.5 (q, NCH_2), 6.08 (s, C_5H_5), 6.58 (q, =CH) [34] IR (CH_2Cl_2): 1564 (ν(C=C)), 1945, 1973, 2044 (ν(CO)), 3150 (ν(NH)) [34]
9	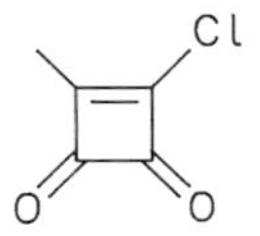 Cl, O, O	Ia (64%) [38] brown solid (from ether) [38] ^1H NMR (CD_2Cl_2): 5.98 (s, C_5H_5) [38] ^{13}C NMR (CD_2Cl_2): 92.84 (C_5H_5) [38] IR (KBr): 1419, 1735, 1772 (ν(C=CC=O)), 1955, 2045 (ν(CO)) [38]
10	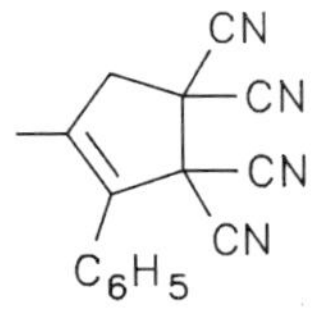 CN, CN, CN, CN, C_6H_5	II (70%, rapidly in CH_3CN, THF, or benzene as solvent at 25 °C; chromatography on alumina should be avoided due to considerable decomposition) [12] yellow solid (from CH_2Cl_2/pentane), decomposition at 140 °C without melting [12] ^1H NMR ($CDCl_3$): 3.65 (s, CH_2), 5.31 (s, C_5H_5), 7.45 (m, C_6H_5) [12] IR (Nujol): 835 (δ(CH) of C_5H_5), 1930, 1970, 2040 (ν(CO)); 2256 (ν(CN)) [12] stable to air at 25 °C; the stability in solution is much lower, therefore characterization in solution must be accomplished within 30 min; the complex is moderately soluble in acetone, and slightly soluble in $CHCl_3$ and CH_2Cl_2 [12]
11	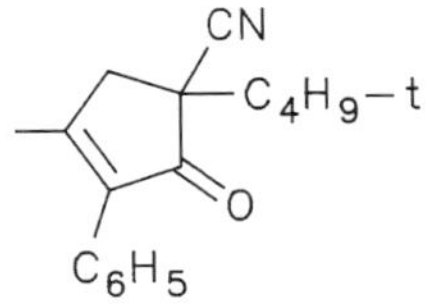 CN, C_4H_9−t, O, C_6H_5	II (37%, by stirring in benzene at 25 °C for 3 h; purification by column chromatography on alumina with pentane/CH_2Cl_2 and CH_2Cl_2 as eluant) [23] yellow air-stable solid, soluble in organic solvents; m.p. 143 to 145 °C (dec.) [23] ^1H NMR ($CDCl_3$): 1.10 (s, $C(CH_3)_3$), 3.27, 3.43 (AB system, CH_2; J(A, B) = 17), 5.41 (s, C_5H_5) [23] IR (CH_2Cl_2): 1681 (C=O), 1943, 1966, 2038 (ν(CO)); (Nujol): 2226 (ν(CN)); the ketonic absorption is

References on pp. 130/1

Table 2 (continued)

No.	^1L	method of preparation (yield) properties and remarks
11 (continued)		consistent with conjugation of the carbonyl group with the C=C linkage [23]
12	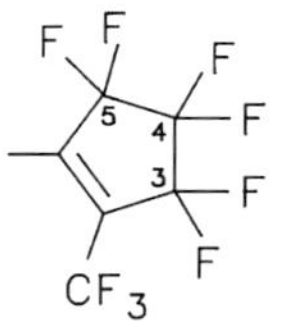 F, F, F, F, F, F, 5, 4, 3, CF_3	Ia (77%; at 25 °C $[C_5H_5Mo(CO)_3]^-$ behaves more as a reducing agent than as a nucleophile) [39] yellow crystals (from pentane/benzene), m.p. 88 to 89 °C [39] ^{19}F NMR (C_6D_6): −130.4 (m, F-4), −108.08 (m, F-3), −98.03 (m, F-5), −56.11 (m, CF_3) [39] IR (THF): 1905 (sh), 1948, 1973, 2043 (ν(CO)) [39]
13	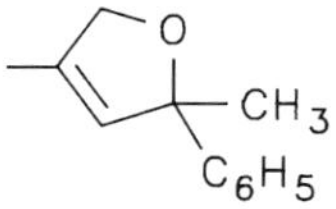 O, CH_3, C_6H_5	III (50 to 70%) [30] ^1H NMR ($CDCl_3$): 1.80 (CH_3), 4.65 (d, CH_2; J = 2), 5.30 (C_5H_5), 5.75 (t, =CH), 7.2 to 7.4 (C_6H_5) [30] ^{13}C NMR ($CDCl_3$): 27.67 (CH_3), 76.99 (**C**(CH_3)C_6H_5), 86.51 (CH_2), 92.34 (C_5H_5), 145.28 (=CH), 146.90 (MoC=), 225.68, 237.07 (both CO) [30] IR ($CHCl_3$): 1940, 2010 (ν(CO)) [30]
14	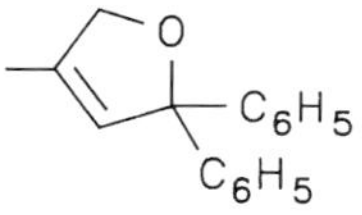 O, C_6H_5, C_6H_5	III (50 to 70%) [30] ^1H NMR ($CDCl_3$): 4.65 (d, CH_2; J = 2), 5.25 (C_5H_5), 5.95 (t, =CH), 7.2 to 7.4 (C_6H_5) [30] IR ($CHCl_3$): 1940, 2010 (ν(CO)) [30]
15	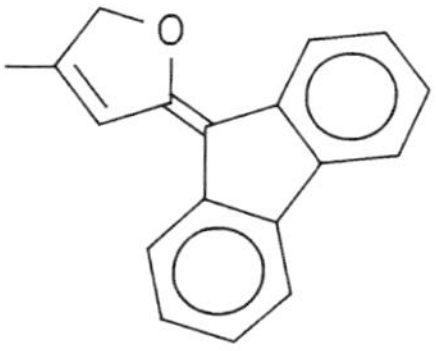 O	III (50 to 70%) [30] ^1H NMR ($CDCl_3$): 4.85 (d, CH_2; J = 2), 5.25 (C_5H_5), 7.1 to 7.4 (fluorenylidene) [30] IR ($CHCl_3$): 1940, 2010 (ν(CO)) [30]
16	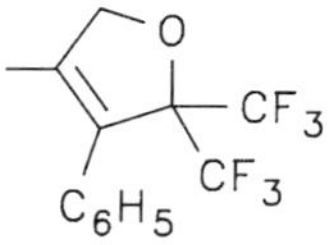 O, CF_3, CF_3, C_6H_5	II (57%; in CH_2Cl_2 or benzene for ca. 30 min under a dry ice condenser; 18%; in neat hexafluoroacetone; purification by column chromatography on Florisil or alumina) [19] yellow crystalline solid, m.p. 150 to 151 °C (dec.) [19] ^1H NMR ($CDCl_3$): 4.87 (s, CH_2), 5.18 (s, C_5H_5), 7.28 ("s", C_6H_5) [19] ^{13}C NMR ($CDCl_3$): 89.58 (CH_2), 99.28 (C_5H_5), 122.66 (CF_3; J(F, C) = 287), 127.80, 131.06 (C-2 to C-6 of C_6H_5), 135.67 (C_6H_5C=), 136.29 (C-1 of C_6H_5), 148.65 (MoC=), 218.56 (CO) [24] ^{19}F NMR ($C_6H_5CF_3$): −73.6 (s) [19] IR (pentane): 1952, 1972, 2038 (ν(CO)) [19] sublimed at 50 °C at ca. 1.3×10^{-3} atm; slightly soluble in pentane, but very soluble in benzene, $CHCl_3$, and acetone [19]

References on pp. 130/1

Table 2 (continued)

No.	^1L	method of preparation (yield) properties and remarks
17	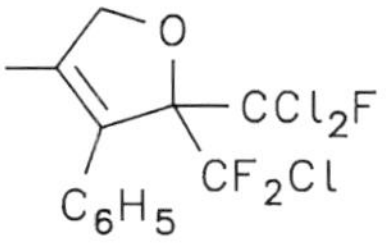 O, CCl_2F, CF_2Cl, C_6H_5	II (58%, in CH_2Cl_2 at 25 °C for 2 to 3 h; purification was accomplished by column chromatography on alumina with CH_2Cl_2/pentane as eluant) [33] orange solid, m.p. 183 °C [33] ^1H NMR ($CDCl_3$): 4.95 (s, CH_2), 5.23 (s, C_5H_5), 7.40 (m, C_6H_5) [33] IR (CH_2Cl_2): 1939, 1962, 2025 (ν(CO)) [33] mass spectrum: $[M]^+$, $[M-CO]^+$, $[C_5H_5Mo]^+$, $[C_9H_7]^+$ [33]
18	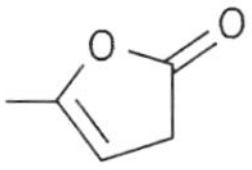 O, O	Ib (29%; during preparation $C_5H_5Mo(CO)_3C(O)CH_2CH_2C(O)Cl$ is supposed as intermediate; subsequent cyclization is induced by a second $[C_5H_5Mo(CO)_3]^-$) [41] yellow air-stable crystalline solid, highly air-sensitive in solution [41] ^1H NMR ($CDCl_3$): 3.10 (d, CH_2; J = 2.1), 5.53 (s, C_5H_5), 5.64 (t, CH; J = 2.1) [41] ^{13}C NMR ($CDCl_3$): 35.0 (CH_2), 93.0 (C_5H_5), 122.7 (=CH), 164.8 (MoC=), 180.1 (CO_2), 225.9, 236.6 (both MoCO) [41] IR (KBr): 1753, 1778 ($\nu(CO_2)$), 1920, 1954, 2030 (ν(CO)) [41]
19	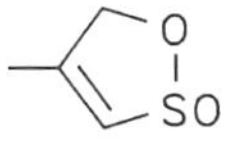 O, SO	II (42%, by bubbling SO_2 for 30 min into a THF solution of the starting material prepared in situ from $[C_5H_5Mo(CO)_3]^-$ and $BrCH_2C{\equiv}CH$; isolated along with $(C_5H_5Mo(CO)_3)_2$; separation by chromatography on alumina) [1, 3] yellow solid (from $CHCl_3$/pentane), which decomposed slowly to a green material even at 0 °C [3]; m.p. probably 135 °C (dec.) [1] ^1H NMR ($CDCl_3$): 5.10, 5.48, and 6.42 (ABX system, $CH_{2(A,B)}$ and $=CH_{(X)}$; J(A, B) = 15, J(A, X) = J(B, X) = 2.5), 5.58 (s, C_5H_5); the magnetic nonequivalence of the $-OCH_2$ protons is ascribed to the diamagnetic anisotropy of the S=O bond which deshields the proton cis to the sulfinyl oxygen [3]; see also [1] IR (KBr): 906, 1105 ($\nu(SO_2)$); ($CHCl_3$): 1950, 1970, 2042 (ν(CO)) [3]; see also [1]
20	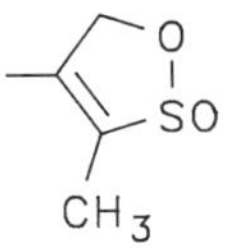 O, SO, CH_3	II (91%, using liquid SO_2 at −78 °C for 30 min with subsequent warming up to −10 °C to remove the solvent) [1, 3] yellow stable solid (from $CHCl_3$/pentane), m.p. 151 °C (dec.) [3]; see also [1]

References on pp. 130/1

Table 2 (continued)

No.	^{1}L	method of preparation (yield) properties and remarks
20 (continued)		^{1}H NMR ($CDCl_3$): 5.06, 5.43, and 2.05 (ABX_3 system, $CH_{2(A,B)}$ and $=CCH_{3(X)}$; J(A, B) = 14.5, J(A, X) = J(B, X) = 2), 5.58 (s, C_5H_5); compare also No. 19 [3]; see also [1] IR (KBr): 900, 1100 ($\nu(SO_2)$); ($CHCl_3$): 1949, 1973, 2040 (ν(CO)) [3]; see also [1]
21	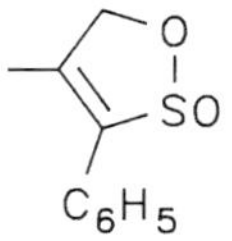 O, SO, C_6H_5	II (50%; for reaction conditions, see No. 20) [1, 3] yellow solid (from $CHCl_3$/pentane), m.p. 118 to 119°C (dec.) [3]; see also [1] ^{1}H NMR ($CDCl_3$): 5.24, 5.72 (AB system, CH_2; J(A, B) = 15), 5.29 (s, C_5H_5), 7.48 ("s", C_6H_5); compare also No. 19 [3]; see also [1] IR (KBr): 895, 1115 ($\nu(SO_2)$); ($CHCl_3$): 1944, 1964, 2031 (ν(CO)) [3]; see also [1]
22	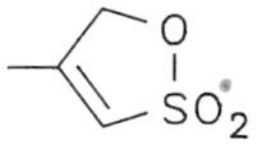 O, SO_2	II (traces only, using $SO_3 \cdot$ dioxane in CH_2Cl_2 at −70°C for 2 h; neutralization was accomplished with $BaCO_3$) [4] unstable, even in the solid state, dec. ca. 150°C [4] ^{1}H NMR ($CDCl_3$): 5.1 (d, CH_2; J = 2), 5.55 (s, C_5H_5), 6.35 (t, =CH) [4] IR (KBr): 970, 1150, 1310 (ν(SO)), 1955, 1960, 2040 (ν(CO)) [4]
23	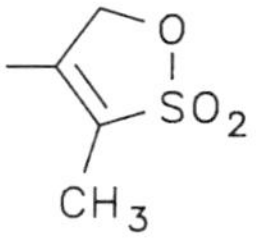 O, SO_2, CH_3	II (ca. 40%; for reaction conditions, see No. 22) [4] solid (from $CHCl_3$/pentane or CH_2Cl_2/hexane), dec. ca. 150°C [4] ^{1}H NMR ($CDCl_3$): 2.05 (t, CH_3; J = 2), 4.9 (q, CH_2), 5.55 (s, C_5H_5) [4] IR (KBr): 920, 1180, 1300 (ν(SO)), 1590 (ν(C=C)), 1950, 2040 (ν(CO)) [4]
24	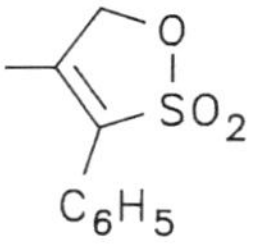 O, SO_2, C_6H_5	II (51 to 54%; for reaction conditions, see No. 22; at a temperature of 0°C the reaction period can be reduced to 10 min; also with uncomplexed SO_3 in C_2Cl_4 at 25°C; neutralization with $NaHCO_3$) [2, 4, 7], (ca. 40%) [4] yellow crystalline air-stable solid (from hot benzene [7] or from $CHCl_3$/pentane or CH_2Cl_2/hexane [4]), dec. ca. 145°C without melting [2, 7]; see also [4] ^{1}H NMR ($CDCl_3$): 5.07 (s, CH_2), 5.30 (s, C_5H_5), 7.47 ("s", C_6H_5) [2, 7]; for $(CD_3)_2SO$ as solvent, see [4] IR (KBr): 918, 1170, 1310 (ν(SO)) [4, 7], 1930, 1975, 2030 (ν(CO)) [4]; for additional bands, see [2]; for $CHCl_3$ as medium, see [7]

References on pp. 130/1

Table 2 (continued)

No.	^1L	method of preparation (yield) properties and remarks
		insoluble in pentane, but sparingly soluble in benzene and readily soluble in $CHCl_3$, acetone, and CH_3OH; remarkably stable to acids [2, 7]
25	(ring: S–SO, CH_3)	II (45%, with 4,5-diphenyl-3,6-dihydro-1,2-dithiin 1-oxide as S_2O source in THF at 25 °C for 24 h; separation from ca. 50% unreacted $C_5H_5Mo(CO)_3CH_2C{\equiv}CCH_3$ by column chromatography on silica) [40, 42] dark green-brown air-stable solid, m.p. 131 to 132 °C [40, 42] ^1H NMR ($CDCl_3$): 2.28 ("t", CH_3), 4.24, 4.78 (both dq, CH_2; $^2J_{gem}$ = 16, J(CH_2, CH_3) = 2), 5.60 (s, C_5H_5) [40, 42] IR ($CDCl_3$): 1055 (ν(SO)), 1970, 2055 (ν(CO)), 2890, 2941, 2985 (ν(CH)) [40, 42] treatment with $[NH_4]_2[Ce(NO_3)_6]$ in CH_2Cl_2/C_2H_5OH under 1 atm CO at −78 °C resulted in the five-membered-ring thiolsulfinate ester $C_2H_5O_2C(C{=}C(CH_3)S(O)SCH_2\text{-cyclo})$ [40, 42]
26	(ring: S–SO, C_6H_5)	II (43%; for reaction conditions, see No. 25) [40, 42] dark yellow-brown air-stable solid [42] ^1H NMR ($CDCl_3$): 4.51, 5.09 (both d, CH_2; $^2J_{gem}$ = 16), 5.35 (s, C_5H_5), 7.35 to 7.52 (m, C_6H_5) [40, 42] IR ($CDCl_3$): 1061 (ν(SO)), 1965, 2058 (ν(CO)), 3060 (ν(CH)); additional bands at 1439, 1618 [40, 42] treatment with $[NH_4]_2[Ce(NO_3)_6]/CO/C_2H_5OH$ analogously to No. 25 resulted in $C_2H_5O_2C(C{=}C(C_6H_5)S(O)SCH_2\text{-cyclo})$ [40, 42]
27	(ring: NSO_2Cl, C=O, C_6H_5)	II (85%, in benzene for 30 min at 25 °C; purification by chromatography on alumina) [11] yellow-orange solid, m.p. 148 to 152 °C (dec.) [11] ^1H NMR ($CDCl_3$): 5.31 (br s, CH_2), 5.50 (s, C_5H_5), 7.25 to 7.35 (m, C_6H_5) [11] IR (KBr): 1330 (ν(SO)?), 1670 (ν(C=O)), 1945, 1975, 2025 (ν(CO)); another prominent absorption was observed at 1220 [11]
28	(ring: $NSO_2C_6H_4CH_3$-4, C=O, CH_3)	II (93%, in CH_2Cl_2 at 25 °C; after 30 min the compound was precipitated from the reaction mixture using hexane) [33] orange air-stable solid, m.p. 167 °C (dec.) [33] ^1H NMR ($CDCl_3$): 1.74 (t, $=CCH_3$; J = 1.6), 2.42 (s, $C_6H_4\mathbf{CH_3}$), 4.36 (q, CH_2), 5.58 (s, C_5H_5), 7.30, 7.94 (both "d", each 1/2 C_6H_4; J = 14) [33] IR (CH_2Cl_2): 1699 (ν(C=O)), 1947, 1960, 2038 (ν(CO)) [33]

References on pp. 130/1

Table 2 (continued)

No.	^1L	method of preparation (yield) properties and remarks
29	$-C=C(C_6H_5)-C(=O)-N(SO_2C_6H_4CH_3-4)-CH_2-$ (ring)	II (90%, in CH_2Cl_2 at 25 °C; after 8 h the crude product was purified by chromatography on alumina with hexane and subsequently with CH_2Cl_2/hexane as eluant) [33] orange air-stable solid, m.p. 176 °C (dec.) [33] ^1H NMR ($CDCl_3$): 2.43 (s, CH_3), 4.59 (s, CH_2) 5.42 (s, C_5H_5), 7.3 (m, 1/2 C_6H_4), 8.06 ("d", 1/2 C_6H_4; J = 15) [33] IR (CH_2Cl_2): 1708 (ν(C=O)), 1953, 1970, 2048 (ν(CO)) [33] UV (CH_2Cl_2, ε): 275 (sh), 289 (10800), 322 (sh), 370 (1900) [33]
30	$-C=C(C_6H_5)-SO-N(SO_2CH_3)-CH_2-$ (ring)	II (50%, by stirring in CH_2Cl_2 for 30 min at 25 °C; purification by column chromatography on alumina or Florisil with acetone as eluant) [37] yellow solid (from CH_2Cl_2/pentane), m.p. 105 °C (dec.) [37] ^1H NMR ($CDCl_3$): 3.19 (s, CH_3), 4.70, 4.85 (AB system, CH_2; J(A, B) = 14), 5.30 (s, C_5H_5), 7.39 ("s", C_6H_5) [37] IR (Nujol): 1085 (ν(SO)), 1170, 1345 ($\nu(SO_2)$), 1960, 2015 (ν(CO)) [37]
31	$-C=C(CH_3)-S(=NSO_2CH_3)-N(SO_2CH_3)-CH_2-$ (ring)	II (57%; for reaction conditions, see No. 30) [37] yellow solid (from CH_2Cl_2/pentane), m.p. 150 °C (dec.) [37] ^1H NMR ($CDCl_3$): 2.12 (s, CCH_3), 2.90, 3.17 (both s, SCH_3), 4.91 (br, CH_2), 5.46 (s, C_5H_5) [37] ^{13}C NMR ($CDCl_3$): 16.6 (CCH_3), 41.9, 43.4 (both SCH_3), 69.7 (CH_2), 93.3 (C_5H_5) [37] IR (Nujol): 1000 (ν(SN)), 1130, 1165, 1280, 1345 ($\nu(SO_2)$); ($CHCl_3$): 1945, 1965, 2040 (ν(CO)) [37]
32	$-C=C(C_6H_5)-S(=NSO_2CH_3)-N(SO_2CH_3)-CH_2-$ (ring)	II (55%; for reaction conditions, see No. 30) [35, 37] yellow solid (from CH_2Cl_2/pentane), m.p. 150 °C (dec.) [37] ^1H NMR ($CDCl_3$): 2.90, 3.35 (both s, SCH_3), 4.83, 4.96 (AB system, CH_2; J(A, B) = 13), 5.46 (s, C_5H_5), 7.35 to 7.8 (C_6H_5) [37] IR (Nujol): 1010 (ν(SN)), 1130, 1165, 1280, 1340 ($\nu(SO_2)$); ($CHCl_3$): 1950, 1970, 2040 (ν(CO)) [37]
33	$-C=CH-CH_2-CH_2-C(=O)-O-$ (ring)	Ib (33%; during preparation $C_5H_5Mo(CO)_3C(O)CH_2(CH_2)_2C(O)Cl$ is supposed as intermediate; subsequent cyclization is induced by a second $[C_5H_5Mo(CO)_3]$

References on pp. 130/1

Table 2 (continued)

No.	^1L	method of preparation (yield) properties and remarks
		yellow air-stable crystalline solid; highly air-sensitive in solution [41] ^1H NMR ($CDCl_3$): 2.35 (m, $=CCH_2$), 2.53 (t, CH_2CO_2; J = 6.4), 5.49 (s, C_5H_5), 5.51 (t, CH; J = 4.6) [41] ^{13}C NMR ($CDCl_3$): 22.2, 29.2 (both CH_2), 93.4 (C_5H_5), 124.4 (CH), 160.4 (MoC), 170.4 (CO_2), 227.1, 238.1 (both MoCO) [41] IR (KBr): 1738 ($\nu(CO_2)$), 1930, 1951, 2025 (ν(CO)) [41]

*Further information:

$C_5H_5(CO)_3MoCX{=}C(CN)_2$ (X = H, CN, Cl; Table **2**, Nos. **3** to **5**). These relatively air-stable polycyanovinyl complexes can be sublimed under vacuum (ca. 1.5×10^{-4} atm) at about 100 °C only with some decomposition which prevented satisfactory mass spectra from being obtained [9].

The tendency of the molybdenum group to donate electrons from filled Mo d orbitals into suitable antibonding orbitals of the double bond in the 2,2-dicyanovinyl groups is proved by the relative large decrease in the ν(C=C) frequency upon bonding to Mo in contrast to free $(NC)_2C{=}CXCl$. Furthermore, electron density can be transmitted into antibonding orbitals of the C≡N bond resulting in a decrease of ν(CN) compared to the free polycyanovinyl compound. The strong deshielding of the 2,2-dicyanovinyl proton in No. 3 also confirmed this electron donating effect of the transition metal group [9].

$C_5H_5Mo(CO)_3C(CN){=}C(CN)_2$ (No. 4) reacted with $N(C_2H_5)_3$ in CH_2Cl_2 at room temperature to give cyclic $C_5H_5Mo(CO)_2(\eta^2\text{-}C(CN){=}C(CN)C(NH_2))$ (Formula II). No water needs to be added to the reaction mixture, because $N(C_2H_5)_3$ in reagent grade CH_2Cl_2 contains enough for this transformation. The analogous reaction using $C_5H_5Mo(CO)_3CCl{=}C(CN)_2$ (No. 5) as starting material yielded $C_5H_5Mo(CO)_2(\eta^2\text{-}CH{=}C(CN)C(OH){=}NH)$ as shown in Formula III. The same product was obtained when No. 5 was treated with alumina in stirred CH_2Cl_2. This method of reaction could not be transferred to No. 4 for which no tractable product resulted, although some compound II was proved in the product mixture [18].

The reaction of $C_5H_5Mo(CO)_3CX{=}C(CN)_2$ (X = CN, Cl) with alcohols R′OH (R′ = CH_3, C_2H_5) in the presence of NaOH or with the corresponding sodium alkoxides was investigated. For X = Cl, $C_5H_5Mo(CO)_2(\eta^2\text{-}C(CO_2R'){=}C(CN)C(OR'){=}NH)$ (see Formula III) was isolated, but yields and rates in these base-catalyzed alcoholyses decrease with the increasing length and size of the alkyl chain of the alcohol, and, e.g., only trace quantities of the corresponding

I

References on pp. 130/1

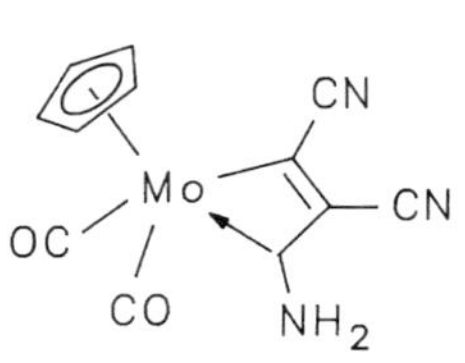

II

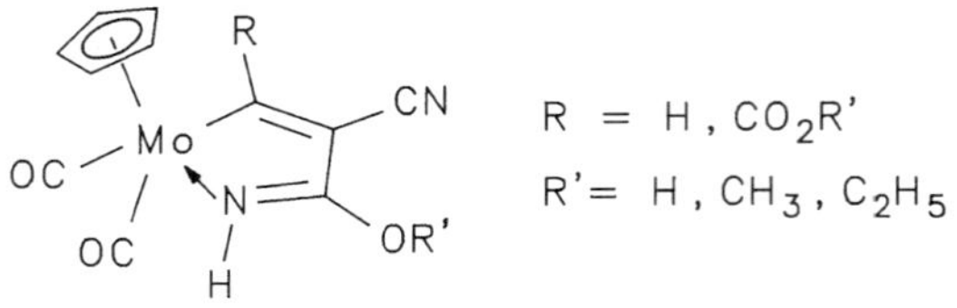

III

$R = H, CO_2R'$
$R' = H, CH_3, C_2H_5$

isopropyl derivative can be obtained. A general reaction scheme for the formation of these unusual derivatives from $C_5H_5(CO)_3MoCCl{=}C(CN)_2$ involving dicyanovinylidene intermediates of the type $C_5H_5Mo(CO)_2(C{=}C(NC)_2)H$ and $C_5H_5Mo(CO)_2(C{=}C(NC)_2)CO_2R'$ has been proposed and is discussed in the original literature. If, on the contrary, $C_5H_5Mo(CO)_3$-$C(CN){=}C(CN)_2$ was treated with $NaOH/CH_3OH$, a large amount of product insoluble in CH_2Cl_2 was isolated along with some $(C_5H_5Mo(CO)_3)_2$ and $C_5H_5Mo(CO)_3CN$ [18].

Compound $C_5H_5Mo(CO)_2(\eta^2\text{-}(CH_3)_2N{=}C{=}C(CN)_2)$ (Formula IV) resulted when C_5H_5Mo-$(CO)_3CCl{=}C(CN)_2$ was reacted with $HN(CH_3)_2$ in CH_2Cl_2 at 25°C or with $P(N(CH_3)_2)_3$, $C_6H_5P(N(CH_3)_2)_2$, and $(C_6H_5)_2PC_2H_4P(N(CH_3)_2)_2$, respectively, in refluxing THF. Similarly, $C_5H_5Mo(CO)_2(\eta^2\text{-cyclo-}(CH_2)_5N{=}C{=}C(CN)_2)$ (Formula IV) was obtained using cyclo-$(CH_2)_5NH$ in CH_2Cl_2 or $P(N(CH_2)_5\text{-cyclo})_3$ in refluxing THF as the reactant [18, 26, 29].

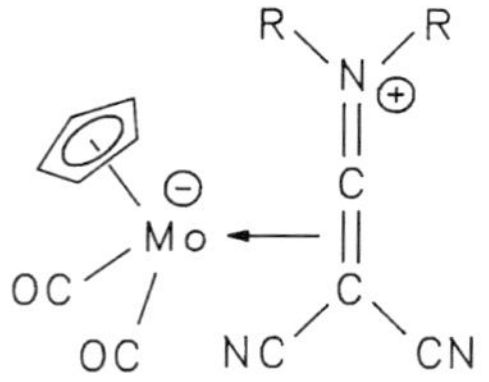

IV

Simple carbonyl substitution occurred in the reaction of $C_5H_5Mo(CO)_3CX{=}C(CN)_2$ with $P(C_6H_5)_3$ in refluxing benzene, and $C_5H_5Mo(CO)_2(CX{=}C(CN)_2)(P(C_6H_5)_3)$ was isolated for X = H [10] and X = CN [18]. Treatment of $C_5H_5Mo(CO)_3CCl{=}C(CN)_2$ (No. 5) with $P(OCH_3)_3$ similarly gave $C_5H_5Mo(CO)_2(CCl{=}C(CN)_2)(P(OCH_3)_3)$, when a short reaction time and small scale was considered (for the reaction pathway under more rigorous conditions, see below) [28], but however with $P(C_6H_5)_3$, No. 5 was converted to a mixture of the cis and trans isomer of $C_5H_5Mo(C{=}C(CN)_2)(P(C_6H_5)_3)_2Cl$ in refluxing heptane, octane, or other inert solvents. When the reaction was carried out in refluxing benzene or toluene, adducts of composition $C_5H_5Mo(C{=}C(CN)_2)(P(C_6H_5)_3)Cl \cdot$ arene were obtained. The substitution product C_5H_5Mo-$(CO)_2(CCl{=}C(CN)_2)(P(C_6H_5)_3)$ could only be detected during reaction when the transformation was carried out under mild conditions in CH_2Cl_2 at 37°C [5, 10]. The complete decarbonylation of No. 5 by $P(C_6H_5)_3$, which accompanies the conversion of the $MoCCl{=}C(CN)_2$ moiety into coordinated chloro and terminal dicyano-vinylidene ligands, has been explained in terms of the very strong electron-demanding properties of the $C{=}C(CN)_2$ group precluding the simultaneous bonding of this ligand system and one or more carbonyl groups to the same central metal [8]. This conversion of the 1-chloro-2,2-dicyanovinyl group to coordinated dicyanovinylidene and chloride has been shown to be general since the use of analogous ligands ^2D such as $E(C_6H_5)_3$ (E = As, Sb), $P(CH_3)_2C_6H_5$, and $P(OR)_3$ (R = CH_3, C_2H_5, C_6H_5) as the reactants in refluxing benzene resulted in compounds of the type $C_5H_5Mo(C{=}C(CN)_2)(^2D)_2Cl$, but only for $^2D = Sb(C_6H_5)_3$ were two isomers found. In the reaction of $C_5H_5Mo(CO)_3CCl{=}C(CN)_2$ with the very basic $P(CH_3)_2C_6H_5$, $C_5H_5Mo(C{=}C(CN)_2)(P(CH_3)_2C_6H_5)_2Cl$ was isolated along with cis-$C_5H_5Mo(CO)_2(P(CH_3)_2C_6H_5)Cl$. The latter complex became the main product when the reaction was carried out in THF [10].

Reactions of $C_5H_5Mo(CO)_3CCl{=}C(CN)_2$ (No. 5) with di(tertiary phosphanes) were studied. $(C_6H_5)_2PC_2H_4P(C_6H_5)_2$ was shown to act as a monoligate or biligate ligand depending on reaction conditions. $C_5H_5Mo(C{=}C(CN)_2)((C_6H_5)_2PC_2H_4P(C_6H_5)_2)_2Cl$ was obtained in refluxing THF using a twofold excess of the phosphane, but in refluxing benzene and with equimolar amounts of the reactants, $C_5H_5Mo(C{=}C(CN)_2)(\eta^2\text{-}(C_6H_5)_2PC_2H_4P(C_6H_5)_2)Cl$ was formed. Treatment of No. 5 with an excess of cis-$(C_6H_5)_2PCH{=}CHP(C_6H_5)_2$ in refluxing THF gave only biligated $C_5H_5Mo(C{=}C(CN)_2)$(cis-$(C_6H_5)_2PCH{=}CHP(C_6H_5)_2)Cl$, while the reaction with $C_6H_5P(C_2H_4P(C_6H_5)_2)_2$ in refluxing octane resulted in displacement not only of all the carbonyl groups, but also of the chloro ligand, to produce $[C_5H_5Mo(C{=}C(CN)_2)(\eta^3\text{-}C_6H_5P(C_2H_4P(C_6H_5)_2)_2)]^+$ which is best isolated as its hexafluorophosphate salt [6, 10].

Monosubstitution of carbonyl groups was observed in the reaction of $C_5H_5Mo(CO)_3C(CN){=}C(CN)_2$ (No. 4) with t-C_4H_9NC in refluxing benzene yielding $C_5H_5Mo(CO)_2(C(CN){=}C(CN)_2)CNC_4H_9$-t. No tendency of further decarbonylation and rearrangement to give tricyanoethene derivatives was observed. This contrasted with $C_5H_5Mo(CO)_3CCl{=}C(CN)_2$ (No. 5) which only gave $C_5H_5Mo(CO)_2(CCl{=}C(CN)_2)CNC_4H_9$-t under controlled reaction conditions involving a short reaction time and a small reaction scale. In refluxing benzene, further transformation by substitution of all carbonyl groups and apparent elimination of $(CH_3)_2C{=}CH_2$ afforded lateral and diagonal isomers of the tricyanoethene complex $C_5H_5Mo((NC)_2C{=}C(CN)H)(CNC_4H_9\text{-t})_2Cl$ (Formula V) as well as a third isomer containing a nitrogen-hydrogen bond, which has been formulated tentatively as a π-allyl derivative, $C_5H_5Mo^-(\eta^3\text{-}(NC)_2CC(CN)C{=}N^+HC_4H_9\text{-t})(CNC_4H_9\text{-t})Cl$. The supposed structure is shown in Formula VI [28].

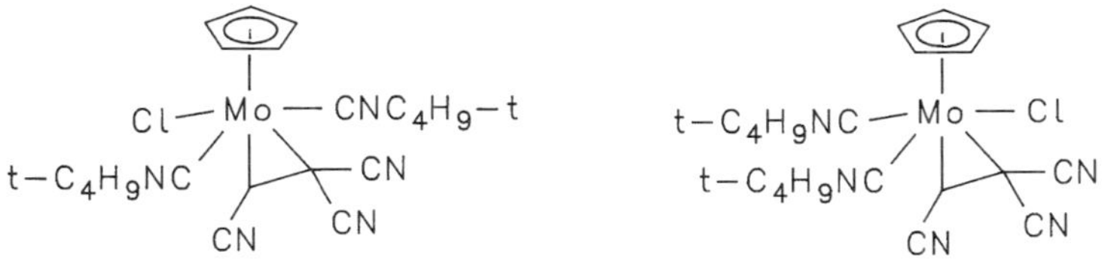

V

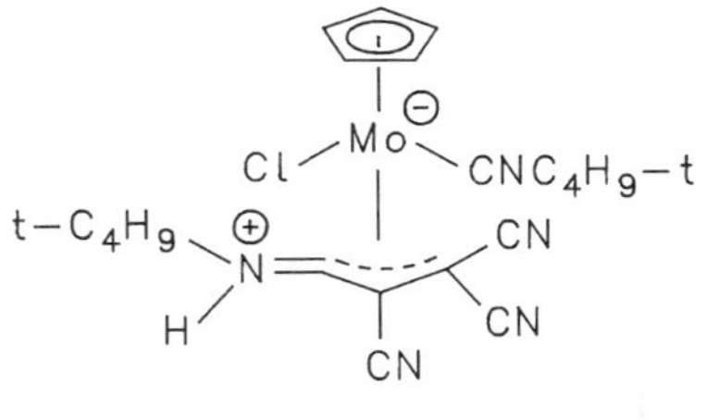

VI

The reaction of $C_5H_5Mo(CO)_3CCl{=}C(CN)_2$ (No. 5) with $C_6H_5C{\equiv}CC_6H_5$ in refluxing benzene or other aprotic solvents like THF or xylene at temperatures in the range 65 to 135 °C led to cyclization of the dicyanovinylidene fragment with two diphenylacetylene units to form 6,6-dicyano-1,2,3,4-tetraphenylfulvene. The yield of the cyclization product can be improved by the addition of ethanolic ceric ammonium nitrate to the reaction mixture suggesting the pres-

References on pp. 130/1

ence of a molybdenum complex of $(NC)_2C$=$(CC(C_6H_5)$=$C(C_6H_5)C(C_6H_5)$=$C(C_6H_5)$-cyclo) in the crude product [14].

$C_5H_5Mo(CO)_3C(CF_3)$=$CRCF_3$ (R = H, SCH_3; Table **2**, Nos. **6** and **7**) were obtained upon refluxing a THF solution of $C_5H_5Mo(CO)_3H$ and $CF_3C{\equiv}CCF_3$ in the presence of excess $S_2(CH_3)_2$, along with $C_5H_5Mo(CO)_2(\eta^2$-$C(CF_3)$=$C(CF_3)C(O)SCH_3)$ and an isomer of No. 6, presumed to be $C_5H_5Mo(CO)_2(C(O)C(CF_3)$=$CHCF_3)H$. Column chromatography of this mixture did not allow No. 6 to be separated from its isomer (total yield of the two complexes: 15%) or vinyl derivative No. 7 to be isolated in an analytically pure state (approximate yield: 10%) [31]. In solution between 293 and 333 K, the vinyl complex No. 6 co-exists in an equilibrium with its isomer, the latter being favored at lower temperatures [31].

$C_5H_5(CO)_3MoC(CF_3)$=$C(CF_3)SCH_3$ (No. 7) had previously been claimed to result from photolyzing a mixture of $(C_5H_5(CO)_3Mo)_2$, $CF_3C{\equiv}CCF_3$, and $S_2(CH_3)_2$ in pentane solution [22], but subsequent X-ray analysis indicated that the product so obtained was actually $C_5H_5Mo(CO)(\eta^4$-$C(O)C(CF_3)$=$C(CF_3)C(O)SCH_3$-cyclo), another isomer of this family of compounds [27].

References:

[1] Roustan, J. L.; Charrier, C. (C.R. Seances Acad. Sci. C **268** [1969] 2113/6).
[2] Lichtenberg, D. W.; Wojcicki, A. (J. Organomet. Chem. **33** [1971] C77/C79).
[3] Thomasson, J. E.; Robinson, P. W.; Ross, D. A.; Wojcicki, A. (Inorg. Chem. **10** [1971] 2130/7).
[4] Roustan, J. L.; Mérour, J. Y.; Benaïm, J.; Charrier, C. (C.R. Seances Acad. Sci. C **274** [1972] 537/40).
[5] King, R. B.; Saran, M. S. (J. Chem. Soc. Chem. Commun. **1972** 1053/4).
[6] King, R. B.; Saran, M. S. (J. Am. Chem. Soc. **94** [1972] 1784/5).
[7] Lichtenberg, D. W.; Wojcicki, A. (Inorg. Chim. Acta **7** [1973] 311/4).
[8] King, R. B. (Inorg. Nucl. Chem. Lett. **9** [1973] 457/60).
[9] King, R. B.; Saran, M. S. (J. Am. Chem. Soc. **95** [1973] 1811/7).
[10] King, R. B.; Saran, M. S. (J. Am. Chem. Soc. **95** [1973] 1817/24).

[11] Yamamoto, Y.; Wojcicki, A. (Inorg. Chem. **12** [1973] 1779/88).
[12] Su, S. R.; Wojcicki, A. (Inorg. Chim. Acta **8** [1974] 55/60).
[13] Gansow, O. A.; Burke, A. R.; King, R. B.; Saran, M. S. (Inorg. Nucl. Chem. Lett. **10** [1974] 291/5).
[14] King, R. B.; Saran, M. L. (J. Chem. Soc. Chem. Commun. **1974** 851/2).
[15] King, R. B.; Hodges, K. C. (J. Am. Chem. Soc. **96** [1974] 1263/4).
[16] Hodges, K. C. (Diss. Univ. Georgia, Athens, Ga., 1974; Diss. Abstr. Int. B **35** [1975] 4822).
[17] King, R. B.; Hodges, K. C. (J. Am. Chem. Soc. **97** [1975] 2702/12).
[18] King, R. B.; Saran, M. L. (Inorg. Chem. **14** [1975] 1018/25).
[19] Lichtenberg, D. W.; Wojcicki, A. (Inorg. Chem. **14** [1975] 1295/301).
[20] Nesmeyanov, A. N.; Kolobova, N. E.; Zlotina, I. B.; Solodova, M. Ya.; Anisimov, K. N. (Tezisy Dokl. 12th Vses. Chugaevskoe Soveshch. Khim. Kompleksn. Soedin., Novosibirsk, USSR, 1975, Vol. 3, pp. 474/5; C.A. **86** [1977] No. 5580).

[21] Beck, W.; Brix, H.; Köhler, F. H. (J. Organomet. Chem. **121** [1976] 211/23).
[22] Pétillon, F. Y.; Sharp, D. W. A. (J. Fluorine Chem. **8** [1976] 323/7).
[23] Chen, L. S.; Lichtenberg, D. W.; Robinson, P. W.; Yamamoto, Y.; Wojcicki, A. (Inorg. Chim. Acta **25** [1977] 165/72).
[24] Williams, J. P.; Wojcicki, A. (Inorg. Chem. **16** [1977] 2506/12).

[25] Kolobova, N. E.; Zlotina, I. B.; Solodova, M. Ya. (Izv. Akad. Nauk SSSR Ser. Khim. **1977** 2334/6; Bull. Acad. Sci. USSR Div. Chem. Sci. [Engl. Transl.] **1977** 2168/70).
[26] Diefenbach, S. P. (Diss. Univ. Georgia, Athens, GA, 1978; Diss. Abstr. Int. B **39** [1979] 3319).
[27] Guerchais, J. E.; Le Floch-Perennou, F.; Pétillon, F. Y.; Keith, A. N.; Manojlovic-Muir, Lj.; Muir, K. (J. Chem. Soc. Chem. Commun. **1979** 410/1).
[28] King, R. B.; Saran, M. S.; McDonald, D. P.; Diefenbach, S. P. (J. Am. Chem. Soc. **101** [1979] 1138/42).
[29] King, R. B.; Diefenbach, S. P. (Inorg. Chem. **18** [1979] 69/74).
[30] Benaïm, J.; Giuliéri, F. (J. Organomet. Chem. **202** [1980] C9/C14).

[31] Pétillon, F. Y.; Le Floch-Perennou, F.; Guerchais, J. E.; Sharp, D. W. A.; Manojlovic-Muir, Lj.; Muir, K. (J. Organomet. Chem. **202** [1980] 23/37).
[32] Le Gall, J.-Y.; Kubicki, M. M.; Pétillon, F. Y. (J. Organomet. Chem. **221** [1981] 287/90).
[33] Bell, P. B.; Wojcicki, A. (Inorg. Chem. **20** [1981] 1585/92).
[34] Brix, H.; Beck, W. (J. Organomet. Chem. **234** [1982] 151/74).
[35] Leung, T. W.; Christoph, G. G.; Wojcicki, A. (Inorg. Chim. Acta **76** [1983] L281/L282).
[36] Scordia, H.; Kergoat, R.; Kubicki, M. M.; Guerchais, J. E.; L'Haridon, P. (Organometallics **2** [1983] 1681/7).
[37] Leung, T. W.; Christoph, G. G.; Galucci, J.; Wojcicki, A. (Organometallics **5** [1986] 846/53).
[38] Beck, W.; Schweiger, M. J.; Müller, G. (Chem. Ber. **120** [1987] 889/93).
[39] Artamkina, G. A.; Mil'chenko, A. Yu.; Beletskaya, I. P.; Reutov, O. A. (Metalloorg. Khim. **1** [1988] 908/12; Organomet. Chem. USSR [Engl. Transl.] **1** [1988] 501/3).
[40] Raseta, M. E.; Cawood, S. A.; Welker, M. E.; Rheingold, A. L. (J. Am. Chem. Soc. **111** [1989] 8268/70).

[41] Wong, A.; Morgan, R. L. II; Golder, J. M.; Quimbita, G. E.; Pawlick, R. V. (Organometallics **8** [1989] 844/6).
[42] Raseta, M. E.; Mishra, R. K.; Cawood, S. A.; Welker, M. E.; Rheingold, A. L. (Organometallics **10** [1991] 2936/45).

1.5.1.4.1.12.3.2 Derivatives with Acyl Type Ligands

This chapter summarizes compounds where the $C_5H_5Mo(CO)_3$- moiety is bonded to a -C(O)R ligand. Compounds of the type $C_5H_5Mo(CO)_3C(CH_2C_6H_4R\text{-}4){=}NC_6H_{11}$-cyclo (R = H, Cl, OCH_3; Nos. 19 to 21) are also listed and described in Table 3, although their structure, derived in [11], was called into question, and they may contain more probably a chelating rather than a mono-hapto-coordinated amino acyl ligand as outlined in [37]. The reaction of $[C_5H_5Mo(CO)_3]^-$ with $C_6H_5CCl{=}NC_6H_5$ was believed to give initially $C_5H_5Mo(CO)_3C(C_6H_5){=}NC_6H_5$ which spontaneously eliminates CO to yield $C_5H_5Mo(CO)_2(\eta^2\text{-}C(C_6H_5){=}NC_6H_5)$ [19, 22]. Treatment of $Na[C_5H_5Mo(CO)_3]$ with $(CH_3)_2NC(S)Cl$ similarly resulted in $C_5H_5Mo(CO)_2(\eta^2\text{-}C(N(CH_3)_2)S)$, but no $C_5H_5Mo(CO)_3C(S)N(CH_3)_2$ was found [15, 16].

Compound $\mathbf{C_5H_5Mo(CO)_3C(O)CH{=}CHC_6H_5}$ is not included in Table 3, because this complex is only observed as an intermediate in the reaction of $K[C_5H_5Mo(CO)_3]$ with C_6H_5-$CH{=}CHC(O)Cl$ in THF at low temperatures, as indicated from an IR spectrum of the reaction mixture (IR (THF): 1935, 2025 ν(CO)). It immediately reacted further to C_6H_5-$CH{=}CHC(O)C_5H_4Mo(CO)_3Mo(CO)_3C_5H_5$ [26]. Complexes of the type $C_5H_5Mo(CO)_3C(O)CH_2$-$(CH_2)_nC(O)Cl$ (n = 1, 2) were supposed as intermediates in the transformation of $Na[C_5H_5$-

References on pp. 142/3

$Mo(CO)_3]$ with $ClC(O)CH_2(CH_2)_nC(O)Cl$ to metallolactones, $C_5H_5Mo(CO)_3(C{=}CH(CH_2)_n$-$C(O)O$-cyclo) (Section 1.5.1.4.1.12.3.1), but the mono-hapto acyl compounds could not be proved [43].

No evidence for the existence of $C_5H_5Mo(CO)_3C(O)C_5H_4FeC_5H_5$ was obtained in the reaction of $Na[C_5H_5Mo(CO)_3]$ with $C_5H_5FeC_5H_4C(O)Cl$ [18], in contrast to the analogous manganese complex, $C_5H_5Mo(CO)_3C(O)C_5H_4Mn(CO)_3$ (No. 10), obtained by a similar reaction.

Although carbonylation of $C_5H_5Mo(CO)_3{}^1L$ (^1L = alkyl; Section 1.5.1.4.1.12.2) seems to be a convenient method to prepare compounds of the type $C_5H_5Mo(CO)_3C(O)R$, only a very limited number of complexes were obtained by this way; compare Nos. 2, 3, and 11 of Table 3. Thus, treatment of the derivatives containing R = CH_3, C_2H_5, $CH_2CH{=}CH_2$, $CH_2C_6H_5$, and CH_2SCH_3 with CO at normal or elevated pressure appeared to result in the loss of both the cyclopentadienyl and the alkyl ligands to produce $Mo(CO)_6$ without any evidence of the intermediate formation of the corresponding acyls $C_5H_5Mo(CO)_3C(O)R$ [24, 40]. It has been demonstrated that in the absence of CO pressure, the acyl complexes are thermodynamically unstable with respect to decarbonylation, whereas in the presence of CO, they are unstable with respect to further cleavage, forming $C_5H_5C(O)R$ and $Mo(CO)_6$ [40]. A similar degradation under CO has been mentioned for $C_5H_5Mo(CO)_3CH_2CH(R)$-SO_2OCH_3 (R = H, CH_3) [45].

Compounds this chapter deals with are summarized in Table 3, and most of them were prepared by the following methods:

Method I: Compounds of the type $C_5H_5Mo(CO)_3C(O)R$ (R = CH_3, CH_2Cl, CH_2Br, CF_3, C_3F_7, $C{\equiv}CC_6H_5$, $C_5H_4Mn(CO)_3$, C_6H_5, $C(O)C_6H_5$, CO_2CH_3) were prepared from $M[C_5H_5Mo(CO)_3]$ (M = Li, Na, K) which was reacted with the appropriate acid chloride R(O)Cl for several hours, unless stated otherwise. The reactants were usually combined at −78 °C in THF as the solvent [1, 3, 6, 20, 27, 28, 30, 39, 46], except for R = $C{\equiv}CC_6H_5$ or C_6H_5, where the reactants were mixed at −50 °C [12, 26] and for R = CH_2Br, where reaction started at −10 °C [41]. For more detailed reaction conditions, see the individual complexes in Table 3.

Method II: Complexes $C_5H_5Mo(CO)_3C(O)R$ (R = C_6F_5, $NHCH_3$, $N(CH_3)_2$, $N(CH_2)_5$-cyclo) were prepared by treatment of $[C_5H_5Mo(CO)_4]PF_6$ with LiC_6F_5 in THF at −78 °C [5] or with CH_3NH_2, $(CH_3)_2NH$, or cyclo-$(CH_2)_5NH$ in ether [13].

Method III: Complexes $C_5H_5Mo(CO)_3C(CH_2C_6H_4R\text{-}4){=}NC_6H_{11}$-cyclo (R = H, Cl, OCH_3) were prepared from $C_5H_5Mo(CO)_3CH_2C_6H_4R$-4 (Section 1.5.1.4.1.12.2) which were reacted with cyclo-$C_6H_{11}NC$ in benzene at room temperature for 8 h. The complexes were obtained along with small amounts of $C_5H_5Mo(CO)_2(CNC_6H_{11}$-cyclo)$C(O)CH_2C_6H_4R$-4, and these by-products could be separated by chromatography on alumina with benzene as the eluant [11].

References on pp. 142/3

Table 3
$C_5H_5Mo(CO)_3{}^1L$ Complexes Containing Acyl Ligands and Related Compounds.
An asterisk indicates further information at the end of the table.
For explanations, abbreviations, and units, see p. X.
Force constants k and interaction constants k_i, in mdyn/Å, were calculated according to the method of Cotton and Kraihanzel.

No.	^1L	method of preparation (yield) properties and remarks
*1	$-C(O)CH_3$	I; only proved as a by-product along with $C_5H_5Mo(CO)_3Cl$ [6] also by hydrolyzing the stable cyclic insertion product $C_5H_5Mo(CO)_2(\eta^2\text{-}C(CH_3)OAl(Br_2)Br\text{-cyclo})$ under mild conditions [29, 33] resulted also along with $C_5H_5Mo(CO)_3I$ from cleavage of $[(C_5H_5Mo(CO)_3)_2(\mu,\eta^2\text{-}CH_3CO)]^+$ by the addition of one equivalent of $[N(n\text{-}C_4H_9)_4]I$ to a CH_2Cl_2 solution of the acyl complex [36] yellow oil [6], only characterized in solution because of its instability ^1H NMR (toluene-d_8): 2.44 (s, CH_3), 4.68 (s, C_5H_5) [6, 29, 33] IR (toluene): 1661 (ν(C=O)), 1923, 1940, 2017 (ν(CO)) [29, 33] thermally unstable, decomposing to $C_5H_5Mo(CO)_3CH_3$ along with other unidentified products [29, 33]
2	$-C(OAlBr_3)CH_3$	from the cyclic adduct $C_5H_5Mo(CO)_2(\eta^2\text{-}C(CH_3)O\text{-}Al(Br_2)Br\text{-cyclo})$ which reacted with 0.5 atm CO in toluene at 0°C to replace the bromo ligand of the Lewis acid from Mo; isolated in 95% yield with respect to the cyclic adduct; in order to get a clean product, preparation was accomplished as a one-pot reaction started from $C_5H_5Mo(CO)_3CH_3$ dissolved in toluene at 0°C to which $AlBr_3$ was added during 20 min whilst CO was bubbled through the solution; after warming to 25°C to induce precipitation, the reaction mixture was stirred further for 1.5 h at 0°C (51%) [29, 31, 33] mustard yellow solid, stable at room temperature [29, 31, 33] ^1H NMR (toluene-d_8): 2.81 (s, CH_3), 4.70 (s, C_5H_5) [29, 33] IR (toluene): 1457 (ν(C=O)), 1963, 1986, 2052 (ν(CO)) [29, 33] the rate of forming adduct No. 2 was discussed and compared to the analogous reactions of related transition metal acyls [33]
3	$-C(O)C_2H_5$	small amounts by CO insertion into $C_5H_5Mo(CO)_3C_2H_5$ which was treated with 100 atm CO for 3 h at room

References on pp. 142/3

Table 3 (continued)

No.	^1L	method of preparation (yield) properties and remarks
3 (continued)		temperature; the isolated red oil can be crystallized on cooling; it was shown to contain No. 3 along with mainly $(C_2H_5C_5H_4Mo(CO)_3)_2$ and probably $C_2H_5C_5H_4Mo(CO)_3Mo(CO)_3C_5H_5$ and $(C_5H_5Mo(CO)_3)_2$ [2]; only traces were formed from $Na[C_5H_5Mo(CO)_3]$ and $C_5H_5C(O)Cl$ [2] red crystals (from petroleum ether) [2] only isolated impure; chromatography or recrystallization in petroleum ether must be avoided because decomposition to $(C_5H_5Mo(CO)_3)_2$ and unidentified material occurrs [2] ^1H NMR (benzene): 1.02 (CH_3), 2.18 (CH_2), 4.59 (C_5H_5) [2] IR (CS_2): 1675, 1708 (ν(C=O)); 1930, 2016 (ν(CO)) [2] very sensitive to oxygen and thermally unstable [2]
*4	$-C(O)CH_2Cl$	I (69%, by stirring at 0 °C for 2 h) [27, 28] orange crystals (from CH_2Cl_2/hexane on cooling), m.p. 82 to 84 °C [28] ^1H NMR (benzene-d_6): 3.59 (CH_2), 4.01 (C_5H_5) [28] IR (KBr): 1648 (ν(C=O)), 1927, 1935, 1954, 2029 (ν(CO)) [28] mass spectrum: $[M]^+$, $[C_5H_5Mo(CO)_3Cl]^+$; absence of peaks originating from $[C_5H_5Mo(CO)_3CH_2Cl]^+$ indicated the absence of a decarbonylation process [28]
*5	$-C(O)CH_2Br$	I (45%, by stirring at 0 °C for several hours) [41] orange crystals, m.p. 112 to 115 °C [41] ^1H NMR (CS_2): 5.2 (CH_2), 5.6 (C_5H_5) [41] IR (KBr): 1645 (ν(C=O)), 1930, 1950, 2020 (ν(CO)) [41]
*6	$-C(O)CF_3$	I (31%, using $(CF_3CO)_2O$; the reaction mixture was stirred overnight at 25 °C; separation from $(C_5H_5Mo(CO)_3)_2$ by extraction with CH_2Cl_2) [3] also formed by proton acid-promoted hydrolysis of $C_5H_5Mo(CO)_3C_2F_5$ using HBF_4 in CH_3NO_2 for 12 h [38] yellow crystals (from pentane at −78 °C), m.p. 64 to 65 °C [3] ^1H NMR ($CHCl_3$): 5.55 (C_5H_5) [3] ^{19}F NMR ($CHCl_3$): −78.3 [3] IR (KBr): 1126, 1174, 1220 (ν(CF)), 3080 (ν(CH)); additional bands at 708, 828, 842, 1007, 1420 [3]

References on pp. 142/3

Table 3 (continued)

No.	^1L	method of preparation (yield) properties and remarks
		IR (halocarbon oil): 1627 (ν(C=O)) [3]; (cyclohexane): 1971 (A″, ν(CO)), 1977 (A′, ν(CO)), 2039 (A′, ν(CO)); k_1 (CO diagonal to CH_3) = 15.49, k_2 (CO diagonal to another) = 15.80, k_{id} = 0.51; for calculations the ratio of the diagonal interaction force constant (k_{id}) and the lateral interaction force constant (k_{is}) was assumed to be constant, k_{id}/k_{is} = 2 [10]; similar in halocarbon oil [3] IR (Ar matrix at −261 °C): 1148.5, 1195.0, 1240.5 (ν(CF)), 1657.0 (A′, ν(C=O)), 1965.0 (A″, ν(CO)), 1979.5 (A′, ν(CO)), 2049.0 (A′, ν(CO)) [46]; for analogous results in CH_4, CO, and N_2 matrices under these conditions, see [44] UV (Ar matrix at −261 °C, ε): 258 (10500), 304 (ca. 3700), 360 (ca. 1550); the absorption around 260 has been assigned to a metal-to-ligand charge transfer involving the acyl group, while the weaker absorptions have been assigned to Mo-centered transitions [44]; for cyclohexane as the solvent at 25 °C, compare also [3] crystallizes in the tetragonal space group $I4_1cd-C_{4v}^{12}$ (No. 110) with an Mo-C distance of 2.17(3) Å; no further details given [7]
*7	-C(O)C_3F_7	I (34%, by stirring overnight at 25 °C) [3]; already described in [1], but according to sublimation as the purification method and the given IR spectrum, this product has been proposed to consist mainly of $C_5H_5Mo(CO)_3C_3F_7$ [3] yellow crystals (from pentane at −78 °C) [1, 3], m.p. 46 to 47 °C [3] ^1H NMR ($CHCl_3$): 5.55 (C_5H_5) [3, 4] ^{13}C NMR (benzene): 94.6 (C_5H_5, d; J(C, H ≅ 175) [4] ^{19}F NMR ($CHCl_3$): −126.3 (-CF_2-), −109.6 (-C(O)CF_2-), −81.2 (-CF_3) [3] IR (KBr): 1110, 1136, 1180, 1200, 1220 (ν(CF)), 3080 (ν(CH)); additional bands at 696, 713, 730, 765, 825, 860, 945, 1008, 1270, 1336, 1425 [3] IR (cyclohexane): 1972 (A″, ν(CO)), 1976 (A′, ν(CO)), 2039 (A′, ν(CO)); k_1 (CO diagonal to CH_3) = 15.75, k_2 (CO diagonal to another) = 16.06, k_{id} = 0.53; for calculations the ratio of the diagonal interaction force constant (k_{id}) and the lateral interaction force constant (k_{is}) was assumed to be constant, k_{id}/k_{is} = 2 [10]; similar in halocarbon oil [3] UV (cyclohexane, ε): 262 (11600) [3]

References on pp. 142/3

Table 3 (continued)

No.	^1L	method of preparation (yield) properties and remarks
8	$-C(O)C{\equiv}CC_6H_5$	I (3%; before isolation the reaction mixture was warmed up to 0 °C; obtained along with mainly $C_5H_5Mo(CO)_3C{\equiv}CC_6H_5$ subsequently formed from No. 8; purification by TLC with benzene) [26] m.p. 84 to 85 °C [26] ^1H NMR ($CDCl_3$): 5.49 (s, C_5H_5), 7.22 to 7.40 (C_6H_5) [26] IR (KBr): 1580 (ν(C=O)), 2170 (ν(C≡C)); ($CHCl_3$): 1950, 2040 (ν(CO)) [26] fairly stable in the solid state, but in solution gradual decomposition occurred [26]
*9	$-C(O)C_6H_5$	I (18%; the reaction mixture was slowly warmed up to 25 °C followed by stirring for 0.5 h; isolated along with $(C_5H_5Mo(CO)_3)_2$, $C_6H_5C_5H_4Mo(CO)_3Mo(CO)_3C_5H_5$, and $(C_5H_5Mo(CO)_3)_2Hg$; purified by chromatography on alumina) [12] crystalline solid (from hexane at low temperatures), m.p. 82 to 83 °C [12] ^1H NMR ($CDCl_3$): 5.64 (s, C_5H_5), 7.29 to 7.45 (m, C_6H_5) [12] IR ($CHCl_3$): 1640 (ν(C=O)), 1934, 2022 (ν(CO)) [12]
10	$-C(O)C_5H_4Mn(CO)_3$	I (55%, for 1.5 h; precipitated from the mother liquid by addition of petroleum ether) [39] yellow crystals (from THF/pentane), m.p. 89.5 to 90 °C [39] ^1H NMR ($CHCl_3$, with hexamethyldisiloxane as an internal standard): 4.70, 4.97 (both t, C_5H_4), 5.53 (s, C_5H_5) [39] IR (CCl_4): 1629 (ν(C=O)), 1940, 1950 to 1970, 2030, 2040 (ν(CO)) [39] stable in air for extended periods of time, but although thermally stable above its melting point, the complex is easily decarbonylated on heating in THF to give $C_5H_5Mo(CO)_3C_5H_4Mn(CO)_3$ [39]; accordingly, the fragmentation pattern observed in the mass spectrum is nearly identical to the breakdown observed for the corresponding aryl compound $C_5H_5Mo(CO)_3C_5H_4Mn(CO)_3$ (Section 1.5.1.4.1.12.3.3) [32]
11	$-C(O)(CH_2)_3Fe(CO)_2C_5H_5$	from $C_5H_5Mo(CO)_3(CH_2)_3Fe(CO)_2C_5H_5$ (Section 1.5.1.4.1.12.2) in CH_3CN by bubbling CO through this solution for 105 min with stirring (47%) [47] dark yellow, very air-sensitive oil (from hot hexane) [47]

References on pp. 142/3

Table 3 (continued)

No.	^{1}L	method of preparation (yield) properties and remarks
		IR (CH_2Cl_2): 1645 (ν(C=O)), 1902, 1943, 2001, 2078 (ν(CO)) [47]
12	-C(O)C_6F_5	II (16%, purification by column chromatography on alumina with benzene/hexane as eluant) [5] yellow crystals (from hexane at −10°C), m.p. 77°C [5] ^{1}H NMR ($CDCl_3$): 5.50 (s, C_5H_5) [5] IR (CCl_4): 1615, 1645 (ν(C=O)), 1960, 2030 (ν(CO)), 3120 (ν(CH)); other bands at 475, 555, 580, 622, 695, 935, 980, 1100, 1300, 1400, 1425, 1492 [5]
*13	-C(O)$NHCH_3$	II (rapid decomposition prevented pure isolation) [13] IR (CCl_4): 1614 (ν(C=O)), 1935, 2020 (ν(CO)) [13]
*14	-C(O)N$(CH_3)_2$	II (11%; precipitation from the reaction mixture on cooling to −78°C) [13] orange-yellow crystals, m.p. 70°C (dec.) [13] ^{1}H NMR (acetone-d_6): 2.60 (CH_3), 5.63 (C_5H_5) [13] IR (hexane): 1594 (ν(C=O)), 1934, 1941, 2019 (ν(CO)) [13]
*15	-C(O)N$(CH_2)_5$-cyclo	II [13] coarse yellow crystals, m.p. 70°C (dec.) [13] IR (heptane): 1590 (ν(C=O)), 1931, 1937, 2018 (ν(CO)) [13]
*16	-C(O)C(O)C_6H_5	I (71%, stirring for 30 min at 30°C) [46] yellow solid (from benzene/hexane) [46] ^{1}H NMR ($CDCl_3$): 5.43 (s, C_5H_5), 7.43 (m, H-3,5), 7.53 (m, H-4), 7.81 (d, H-2,6; J = 7.1) [46] ^{13}C NMR ($CDCl_3$): 94.65 (C_5H_5), 129.00, 130.25, 134.07 (all C_6H_5), 190.41 (MoC(O)**C**(O)-), 225.57, 234.04 (both MoCO), 263.67 (Mo**C**(O)C(O)-) [46] IR (KBr): 1601, 1674 (ν(C=O)), 1934, 1945, 2040 (ν(CO)) [46]
17	-C(O)CO_2CH_3	I (77%, for 2 h) [20, 30] yellow crystals (from benzene at 0°C), m.p. 56 to 58°C [30] ^{1}H NMR (benzene-d_6): 3.31 (s, OCH_3), 4.65 (s, C_5H_5) [30] IR (THF): 1625, 1727 (ν(C=O)), 1942, 2047 (ν(CO)) [30] mass spectrum: $[M-CO]^+$, $[M-CO_2CH_3]^+$ [30] can be converted at −50°C to the kinetically labile anionic formyl derivative $[C_5H_5Mo(CO)_2(CHO)C(O)CO_2CH_3]^-$ by the addition of Li$[(C_2H_5)_3BH]$ dissolved in THF [25, 35]

References on pp. 142/3

Table 3 (continued)

No.	${}^{1}L$	method of preparation (yield) properties and remarks
*18		for preparation see "Further information" at the end of Table 3 isolated as a 1:1 mixture of two isomers which adopt the structures shown, as indicated by NMR studies; this isomerism arises because the CH_2 and $CHCH_3$ groups can adopt transoid or cisoid orientations with respect to the $C_6H_4CH_3$-4 group [42] yellow microcrystals [42] ${}^{1}H$ NMR (CD_2Cl_2): 1.42 (d, $CHC\mathbf{H}_3$; $J(CH_3, H^a) = 7$), 1.43 (d, $CHC\mathbf{H}_3$; $J(CH_3, H^a) = 7$), 2.30 (s, 2 CH_3-4), 2.33 (dd, 1 H^b; $J(H^a, H^b) = 3$, $J(H^b, H^c) = 2$), 2.46 (dd, 1 H^b; $J(H^a, H^b) = 3$, $J(H^b, H^c) = 1$), 2.56 (d, H^c), 3.26 (dq, 1 H^a), 3.42 (dq, 1 H^a), 5.12 (s, C_5H_5), 5.13 (s, C_5H_5), 7.05, 7.19 ("A_2B_2", 2 C_6H_4; $J(A,B) = 8$) [42] ${}^{13}C$ NMR (CD_2Cl_2/CH_2Cl_2, −40 °C): 14.3, 15.1 (each $C\mathbf{C}H_3$), 21.3 (4-CH_3), 46.3, 47.9 (both CH_2), 66.0, 67.3 (both $\mathbf{C}CH_3$), 94.1 (C_5H_5), 97.8, 98.2 (both $\mathbf{C}C_3$), 117.3, 117.5 (both $\mathbf{C}C_6H_4CH_3$-4), 128.3, 128.4, 129.4, 129.5, 130.5, 130.8, 130.9, 131.0, 134.3, 137.7 (all C_6H_4), 210.9, 211.5, 212.7, 213.2 (all FeCO), 229.1, 229.2, 229.3, 229.8, 238.6 (all MoCO), 248.1, 248.8 (each C=O) [42] IR (light petroleum): 1643 (ν(C=O)), 1936, 1953, 1987, 2000, 2022, 2058 (ν(CO)) [42]
19	$-C(CH_2C_6H_5)=NC_6H_{11}$-cyclo	III (51%) [11] yellow-orange crystals, isomeric mixture of the syn and anti forms, m.p. 170 °C (dec.) [11] ${}^{1}H$ NMR ($(CD_3)_2SO$): 0.8 to 2.1 (br, C_6H_{11}), 5.30 (s, CH_2 of syn form), 5.54 (s, C_5H_5), 5.72 (s, CH_2 of anti form), 7.0 to 7.8 (m, C_6H_5) [11] IR (KBr): 1594 (ν(CN)), 1880, 1954 (ν(CO)) [11] oxidative degradation with 3% aqueous H_2O_2 gave $C_6H_5CH_2C(O)NHC_6H_{11}$-cyclo [11]
20	$-C(CH_2C_6H_4OCH_3\text{-}4)=NC_6H_{11}$-cyclo	III (40%) [11] isomeric mixture of the syn and anti forms, m.p. 157 to 159 °C (dec.) [11] ${}^{1}H$ NMR ($(CD_3)_2SO$): 0.8 to 1.8 (br, C_6H_{11}), 3.71 (s, OCH_3), 5.39 (s, CH_2 of syn form), 5.51 (s, C_5H_5), 5.69 (s, CH_2 of anti form), 7.20 ("AB" type, C_6H_4; "J(A, B)" = 8.7) [11] IR (KBr): 1598 (ν(CN)), 1886, 1948 (ν(CO)) [11]
21	$-C(CH_2C_6H_4Cl\text{-}4)=NC_6H_{11}$-cyclo	III (72%) [11] isomeric mixture of the syn and anti forms, m.p. 210 °C (dec.) [11]

Table 3 (continued)

No.	^{1}L	method of preparation (yield) properties and remarks
		^{1}H NMR ($(CD_3)_2SO$): 0.8 to 2.1 (br, C_6H_{11}), 5.31 (s, CH_2 of syn form), 5.57 (s, C_5H_5), 5.74 (s, CH_2 of anti form), 7.48 ("AB" type, C_6H_4; "J(A, B)" = 8.4) [11] IR (KBr): 1593 (ν(CN)), 1892, 1954 (ν(CO)) [11]

*Further information:

$C_5H_5Mo(CO)_3C(O)CH_3$ (Table **3**, No. **1**) reacted with $(CH_3)_3P{=}CH_2$ in benzene at room temperature to yield $[P(CH_3)_4][C_5H_5Mo(CO)_3]$ and $(CH_3)_3P{=}CHC(O)CH_3$. A mechanism is discussed concerning a nucleophilic attack of the ylide at the acyl carbon to give $[C_5H_5Mo(CO)_3$-$CO^-(CH_3)CH_2P^+(CH_3)_3]$ followed by the rupture of the Mo-C σ bond. The produced $[(CH_3)_3PCH_2C(O)CH_3][C_5H_5Mo(CO)_3]$ then protonates a second mol of $(CH_3)_3P{=}CH_2$ to give the products. None of these intermediates could be detected during the reaction which revealed the first addition as the rate-determining step [34].

$C_5H_5Mo(CO)_3C(O)CH_2X$ (X = Cl, Br; Table **3**, Nos. **4**, **5**) are readily soluble in C_6H_6, CS_2, CH_2Cl_2, $CHCl_3$, and CCl_4, but only sparingly soluble in n-pentane or n-hexane. Both complexes are heat-sensitive and photo-sensitive. Thermal decomposition at 75 to 100°C in vacuum was shown to involve β-halogen elimination giving ketene and $C_5H_5Mo(CO)_3X$, while photodegradation in petroleum ether produced the latter halo complexes together with trace amounts of the chloromethyl and bromoethyl derivatives $C_5H_5Mo(CO)_3CH_2Cl$ [27, 28] and $C_5H_5Mo(CO)_3CH_2Br$ [41], respectively.

$C_5H_5Mo(CO)_3C(O)R$ (R = CF_3, C_3F_7; Table **3**, Nos. **6**, **7**) are somewhat air-sensitive turning blue on exposure to air for several hours, especially when impure [1, 3]. They exhibit essentially the same mass spectra corresponding to compounds of the type $C_5H_5Mo(CO)_3R$ which is in accordance with the ease of CO elimination undergone by the perfluoroacyls on heating; see [9].

Both complexes are appreciably volatile, but on vacuum sublimation partial thermal decarbonylation led to an approximate 1:1 mixture of $C_5H_5Mo(CO)_3CF_3$ with its acyl precursor and to 80% of $C_5H_5Mo(CO)_3C_3F_7$ along with 20% of the starting complex, respectively. Full conversion of the perfluoroacyls to the corresponding perfluoroalkyl derivatives accompanied by some decomposition may be achieved by heating at 120°C at atmospheric pressure until no further gas evolution occurs. Attempts at decarbonylation by UV irradiation at room temperature led to a black insoluble and nonvolatile material along with only small amounts of $C_5H_5Mo(CO)_3R$ [3]. Photolysis of $C_5H_5Mo(CO)_3C(O)CF_3$ isolated at high dilution in argon, CH_4, CO, and N_2 matrices at −261 °C was studied in detail and similar results were found for each matrix. Initial ejection of a terminal CO group followed by migration of the CF_3 moiety resulted in $C_5H_5Mo(CO)_3CF_3$ and free CO in a rapid reaction. Prolonged irradiation resulted in further degradation to trans-$C_5H_5Mo(CO)_2({=}CF_2)F$ as the final product with $C_5H_5Mo(CO)_2CF_3$ as a proved intermediate. Additionally in the N_2 matrix under long-wavelength irradiation, $C_5H_5Mo(CO)_2CF_3(N_2)$ was formed reversibly. Analogously in ^{13}CO doped CH_4 matrices, $C_5H_5Mo(CO)_2(^{13}CO)CF_3$ was generated [44].

Treatment of $C_5H_5Mo(CO)_3C(O)CF_3$ (No. 6) with $Rh(P(C_6H_5)_3)_3Cl$ in stirred CH_2Cl_2 at room temperature resulted in CO substitution to give $C_5H_5Mo(CO)_2(P(C_6H_5)_3)C(O)CF_3$ along

References on pp. 142/3

with $Rh(CO)(P(C_6H_5)_3)_2Cl$. No $C_5H_5Mo(CO)_3CF_3$ was formed because CO abstraction is more rapid than CF_3 migration under these conditions [8, 14].

Reaction of $C_5H_5Mo(CO)_3C(O)CF_3$ (No. 6) with an excess of $CH_3C{\equiv}CCH_3$ in hexane at 60 °C resulted in $C_5H_5Mo(CO)_2(\eta^2\text{-}C(CH_3){=}C(CH_3)C(O)CF_3\text{-cyclo})$ (Formula Ia) as the principle product. This complex partially reacted further with CO, formed during the reaction, or with $CH_3C{\equiv}CCH_3$ to give $C_5H_5Mo(CO)_2(\eta^3\text{-}C(CF_3)C(CH_3)C(CH_3)CO_2\text{-cyclo})$ (Formula Ib) and $C_5H_5Mo(CO)(\eta^5\text{-}OC(CF_3){=}C(CH_3)C(CH_3)C(CH_3)C(CH_3)C(O)\text{-cyclo})$ (Formula Ic) as by-products. When No. 6 was treated with $CH_3C{\equiv}CCH_3$ under UV irradiation, the π-alkyne-substituted analogue of compound Ic, $C_5H_5Mo(\eta^2\text{-}CH_3C{\equiv}CCH_3)(\eta^5\text{-}OC(CF_3){=}C(CH_3)C(CH_3)\text{-}C(CH_3)C(CH_3)C(O)\text{-cyclo})$, was isolated [17, 23].

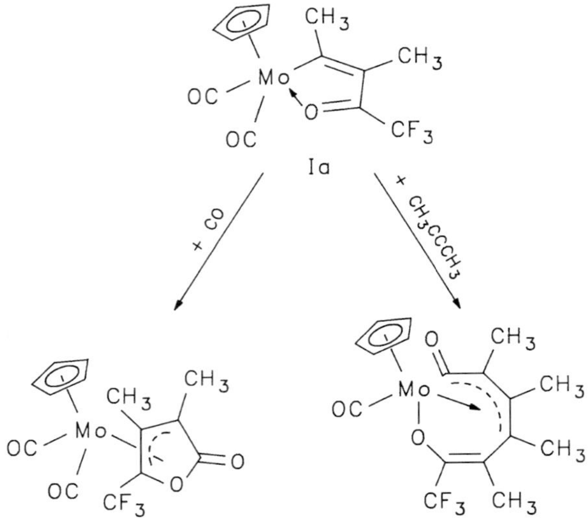

Treatment of No. 6 with $C_6H_5C{\equiv}CC_6H_5$ in refluxing hexane produced $C_5H_5Mo(CO)(\eta^2\text{-}C_6H_5C{\equiv}CC_6H_5)CF_3$. This reaction involved $C_5H_5Mo(CO)_3CF_3$ as an intermediate; compare also Section 1.5.1.4.1.12.2 where a related photo-assisted conversion of the alkyl complex by $CH_3C{\equiv}CCH_3$ is described. Although $C_5H_5Mo(CO)(\eta^2\text{-}C_6H_5C{\equiv}CC_6H_5)CF_3$ could not be converted to bis(alkyne) complexes, the reaction of $C_5H_5Mo(CO)_3C(O)CF_3$ (No. 6) with $CF_3C{\equiv}CCF_3$ in hexane under UV irradiation and with $CH_3O_2CC{\equiv}CCO_2CH_3$ in refluxing diethyl ether, respectively, gave $C_5H_5Mo(\eta^2\text{-}RC{\equiv}CR)_2CF_3$ without any evidence of a monocarbonyl species as an intermediate [21].

$C_5H_5Mo(CO)_3C(O)C_6H_5$ (Table **3**, No. **9**) is of limited stability in the solid state; in solution at room temperature, rapid decomposition, even in nonpolar solvents, resulted in $(C_6H_5\text{-}C_5H_4Mo(CO)_3)_2$ and $C_5H_5Mo(CO)_3Mo(CO)_3C_5H_4C_6H_5$. In refluxing benzene, No. 9 was mainly converted to $(C_6H_5C_5H_4Mo(CO)_3)_2$, while the latter complex was only isolated in traces. The observation that no biphenyl, dibenzyl, or diphenyl ketone was formed during the decomposition indicated that an intramolecular transfer of the phenyl radical to the cyclopentadienyl ring occurred.

$C_5H_5Mo(CO)_3C(O)C_6H_5$ was reacted with excess PR_3 (R = C_6H_5, OC_6H_5), either by standing in benzene solution at room temperature or by refluxing for 1 h in the same solvent. In

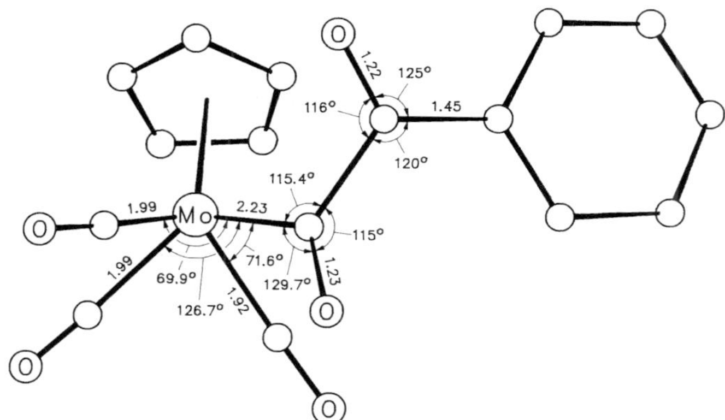

Fig. 9. The molecular structure of $C_5H_5Mo(CO)_3C(O)C(O)C_6H_5$ [46].

both cases $C_5H_5Mo(CO)_2C(O)C_6H_5(PR_3)$ was isolated. For R = C_6H_5 additionally a 6:1 mixture of cis- and trans-$Mo(CO)_4(P(C_6H_5)_3)_2$ was formed. In the case of R = OC_6H_5 noticeable quantities of the decarbonylation product $C_5H_5Mo(CO)_2C_6H_5(P(OC_6H_5)_3)$ and $Mo(CO)_3$-$(P(OC_6H_5)_3)_3$ were formed, but only traces of the tetracarbonyl compound [12].

$C_5H_5Mo(CO)_3C(O)NR_2$ (NR_2 = $NHCH_3$, $N(CH_3)_2$, $N(CH_2)_5$-cyclo; Table **3**, Nos. **13** to **15**). Only No. 14 and No. 15 were stable enough to be isolated in pure form. On the other hand, $C_5H_5Mo(CO)_3C(O)NHCH_3$ (No. 13) readily decomposed to $(C_5H_5Mo(CO)_3)_2$, and its presence was only established by the similarity of the CO stretching frequencies to those of the tungsten analogue and to the tertiary carboxamido complexes No. 14 and 15. Moreover, during the preparation of No. 13 no more than 2 equivalents of amine should be used to avoid formation of 1,3-dialkylurea and $[C_5H_5Mo(CO)_3]^-$, which were proved to be secondary products resulting from the reaction of the carboxamido complex with additional amine [13].

$C_5H_5Mo(CO)_3C(O)C(O)C_6H_5$ (Table **3**, No. **16**) crystallizes in the monoclinic space group $P2_1/c-C^5_{2h}$ (No. 14) with a = 6.674(3), b = 13.301(4), c = 16.903(6) Å, β = 90.82(5)°; Z = 4, D_{calc} = 1.594 g/cm³. The molecular structure shown in **Fig. 9** exhibits a four-legged piano-stool geometry. The s-trans oxalyl carbonyls constitute a nearly perpendicular torsional angle of 104(1)°. No interaction between the benzoylformyl α-carbonyl and the molybdenum atom could be estimated [46].

IIa IIb

References on pp. 142/3

$C_5H_5Mo(CO)_3(\mu\text{-}C(O)C(C_6H_4CH_3\text{-}4)(C(CH_2)CHCH_3))Fe(CO)_3$ (Table **3**, No. **18**) was prepared from an isomeric mixture of $C_5H_5Mo(CO)_3(\mu\text{-}C(C_6H_4CH_3\text{-}4)C(CHCH_3)CH_2)Fe(CO)_3$ (see Formula II) which was stirred under ca. 3 atm CO for 3 d at room temperature. The isolated crude residue was purified by chromatography on Florisil with CH_2Cl_2/light petroleum at 10 °C to give 72% of the pure product [42].

References:

[1] Kaesz, H. D.; King, R. B.; Stone, F. G. A. (Z. Naturforsch. **15b** [1960] 763/4).
[2] McCleverty, J. A.; Wilkinson, G. (J. Chem. Soc. **1963** 4096/9).
[3] King, R. B.; Bisnette, M. B. (J. Organomet. Chem. **2** [1964] 15/37).
[4] Lauterbur, P. C.; King, R. B. (J. Am. Chem. Soc. **87** [1965] 3266/7).
[5] Treichel, P. M.; Shubkin, R. L. (Inorg. Chem. **6** [1967] 1328/34).
[6] Barnett, K. W. (Diss. Univ. Wisconsin, Madison, Wisc., 1967; Diss. Abstr. Int. B **28** [1968] 3203).
[7] Fenessey, J. P. (Diss. Harvard Univ., Cambridge, Mass., 1968; Diss. Abstr. Int. B **29** [1968] 1594).
[8] Alexander, J. J.; Wojcicki, A. (J. Organomet. Chem. **15** [1968] P23/P24).
[9] King, R. B. (Appl. Spectrosc. **23** [1969] 137/47).
[10] King, R. B.; Houk, L. W. (Can. J. Chem. **47** [1969] 2959/64).

[11] Yamamoto, Y.; Yamazaki, H. (J. Organomet. Chem. **24** [1970] 717/24).
[12] Nesmeyanov, A. N.; Makarova, L. B.; Ustynyuk, N. A.; Bogatyreva, L. V. (J. Organomet. Chem. **46** [1972] 105/8).
[13] Jetz, W.; Angelici, R. J. (J. Am. Chem. Soc. **94** [1972] 3799/802).
[14] Alexander, J. J.; Wojcicki, A. (Inorg. Chem. **12** [1973] 74/6).
[15] Dean, W. K.; Treichel, P. M. (J. Organomet. Chem. **66** [1974] 87/93).
[16] Dean, W. K.; Heald, K. J.; Deming, S. N. (Science [Washington, D.C.] **189** [1975] 805/6).
[17] Davidson, J. L.; Green, M.; Nyathi, J. Z.; Scott, C.; Stone, F. G. A.; Welch, A. J.; Woodward, P. (J. Chem. Soc. Chem. Commun. **1976** 714/5).
[18] Pannell, K. H.; Cassias, J. B.; Crawford, G. M.; Flores, A. (Inorg. Chem. **15** [1976] 2671/5).
[19] Adams, R. D.; Chodosh, D. F.; Golembeski, N. M. (J. Organomet. Chem. **139** [1977] C39/C43).
[20] Gladysz, J. A.; Williams, G. M.; Tam, W.; Johnson, D. L. (J. Organomet. Chem. **140** [1977] C1/C6).

[21] Davidson, J. L.; Green, M.; Nyathi, J. Z.; Stone, F. G. A.; Welch, A. J. (J. Chem. Soc. Dalton Trans. **1977** 2246/55).
[22] Chodosh, D. F. (Diss. State Univ. New York, Buffalo, N.Y., 1977; Diss. Abstr. Int. B **38** [1978] 5359/60).
[23] Green, M.; Nyathi, J. Z.; Scott, C.; Stone, F. G. A.; Welch, A. J.; Woodward, P. (J. Chem. Soc. Dalton Trans. **1978** 1067/80).
[24] King, R. B.; King, A. D., Jr.; Iqbal, M. Z.; Frazier, C. C. (J. Am. Chem. Soc. **100** [1978] 1687/94).
[25] Gladysz, J. A.; Selover, J. C. (Tetrahedron Lett. **1978** 319/22).
[26] Nesmeyanov, A. N.; Makarova, L. G.; Vinogradova, V. N.; Ustynyuk, N. A. (Izv. Akad. Nauk SSSR Ser. Khim. **1978** 2375/9; Bull. Acad. Sci. USSR Div. Chem. Sci. [Engl. Transl.] **1978** 2108/12).
[27] Dilgassa, M. (Diss. Univ. Michigan, Ann Arbor, MI, 1978; Diss. Abstr. Int. B **39** [1979] 4879/80).

[28] Dilgassa, M.; Curtis, M. D. (J. Organomet. Chem. **172** [1979] 177/84).
[29] Butts, S. B.; Holt, E. M.; Strauss, S. H.; Alcock, N. W.; Stimson, R. E.; Shriver, D. F. (J. Am. Chem. Soc. **101** [1979] 5864/6).
[30] Gladysz, J. A.; Williams, G. M.; Tam, W.; Johnson, D. L.; Parker, D. W.; Selover, J. C. (Inorg. Chem. **18** [1979] 553/8).

[31] Butts, S. B. (Diss. Northwestern Univ., Evanston, IL, 1980; Diss. Abstr. Int. B **41** [1980] 2170/1).
[32] Sizoi, V. F.; Nekrasov, Yu. S.; Sukharev, Yu. N.; Leontyeva, L. I.; Eremin, S. A.; Kolobova, N. E.; Solodova, M. Ya.; Valueva, Z. P. (J. Organomet. Chem. **202** [1980] 83/90).
[33] Butts, S. B.; Strauss, S. H.; Holt, E. M.; Stimson, R. E.; Alcock, N. W.; Shriver, D. F. (J. Am. Chem. Soc. **102** [1980] 5093/100).
[34] Malisch, W.; Blau, H.; Haaf, F. J. (Chem. Ber. **114** [1981] 2956/70).
[35] Selover, J. C.; Marsi, M.; Parker, D. W.; Gladysz, J. A. (J. Organomet. Chem. **206** [1981] 317/29).
[36] LaCroce, S. J.; Cutler, A. R. (J. Am. Chem. Soc. **104** [1982] 2312/4).
[37] Cotton, J. D.; Dunstan, P. R. (Inorg. Chim. Acta **88** [1984] 223/7).
[38] Richmond, T. G.; Crespi, A. M.; Shriver, D. F. (Organometallics **3** [1984] 314/9).
[39] Leont'eva, L. I.; Perevalova, E. G. (Izv. Akad. Nauk SSSR Ser. Khim. **1984** 2352/8; Bull. Acad. Sci. USSR Div. Chem. Sci. [Engl. Transl.] **1984** 2149/54).
[40] Nolan, S. P.; López de la Vega, R.; Mukerjee, S. L.; Hoff, C. D. (Inorg. Chem. **25** [1986] 1160/5).

[41] Tafesse, F.; Dilgassa, M. (Bull. Chem. Soc. Ethiop. **1** [1987] 83/6).
[42] Garcia, M. E.; Jeffery, J. C.; Stone, F. G. A. (J. Chem. Soc. Dalton Trans. **1988** 2431/42).
[43] Wong, A.; Morgan, R. L. II; Golder, J. M.; Quimbita, G. E.; Pawlick, R. V. (Organometallics **8** [1989] 844/6).
[44] Campen, A. K.; Mahmoud, K. A.; Rest, A. J.; Willis, P. A. (J. Chem. Soc. Dalton Trans. **1990** 2817/23).
[45] Schenk, W. A.; Pfeffermann, J. (J. Organomet. Chem. **440** [1992] 341/51).
[46] Yeh, Y.-S.; Chen, J.-T; Lee, G.-H.; Wang, Y. (J. Chin. Chem. Soc. [Taipei] **39** [1992] 271/4).
[47] Friedrich, H. B.; Moss, J. R. (J. Chem. Soc. Dalton Trans. **1993** 2863/9).

1.5.1.4.1.12.3.3 Derivatives with Aryl and Heteroaryl Type Ligands

This chapter summarizes the limited number of stable aryl derivatives of the general type $C_5H_5Mo(CO)_3{}^1L$. In general, complexes of composition $C_5H_5Mo(CO)_3C_6H_{5-n}R_n$ are thermally unstable [3, 6, 8], although the molybdenum-aryl bond may be stabilized by chelation. Thus, treatment of $Na[C_5H_5Mo(CO)_3]$ with $[I(C_6H_5)_2]BF_4$ resulted only in the decomposition products $(C_5H_5Mo(CO)_3)_2$, $(C_6H_5C_5H_4Mo(CO)_3)_2$, and $C_6H_5C_5H_4Mo(CO)_3Mo(CO)_3C_5H_5$, arising from the spontaneous thermal rearrangement of $C_5H_5Mo(CO)_3C_6H_5$ formed as an intermediate [3]. On the other hand, reaction of $C_5H_5Mo(CO)_3Cl$ with $C_6H_5CH{=}NR$ or $Na[C_5H_5Mo(CO)_3]$ with 2-$BrC_6H_4CH{=}NR$ gave stable compounds of the type $C_5H_5Mo(CO)_2(C_6H_4CH{=}NR\text{-cyclo})$ (R = i-C_3H_7, $CH_2C_6H_5$, $CH(CH_3)C_6H_5$) due to the presence of chelating orthometalated Schiff base ligands [9].

Aryl halides, as a consequence of the low lability of the halogen atom in nucleophilic aromatic replacement reactions, do normally not react with $[C_5H_5Mo(CO)_3]^-$ and, in general, are unsuitable for the synthesis of σ-aryl derivatives $C_5H_5Mo(CO)_3{}^1L$. Thus, the molybdate

References on p. 147

ion is completely unreactive towards C_6H_5I [1] or even C_6F_6 [2, 5, 7, 13]. It only replaces fluoride ion from those polyfluoroheterocyclic systems which are known to be most reactive in nucleophilic aromatic substitution, e.g., 3,4,5,6-tetrafluoropyridazine, $C_4N_2F_4$, cyanuric fluoride, $C_3N_3F_3$ [4, 5], and metalated derivatives thereof, e.g., $(2\text{-}(CO)_5Mn)C_3N_3F_2$-4,6 [14], as described by the following method:

Method I: $C_5H_5Mo(CO)_3C_4N_2F_3$ and $C_5H_5Mo(CO)_3C_3N_3F_2$ were prepared by addition of $Na[C_5H_5Mo(CO)_3]$ to a stirred solution of perfluoropyridazine or cyanuric fluoride at −78 °C in THF. Subsequently the reaction mixture was warmed to room temperature overnight. Both crude products were purified by chromatography on Florisil [4, 5]. Similarly, $C_5H_5Mo(CO)_3C_3N_3NF(Mn(CO)_5)$ was obtained from $K[C_5H_5Mo(CO)_3]$ and $(2\text{-}(CO)_5Mn)C_3N_3F_2$-4,6 between −18 and −10 °C [14].

It was not possible to obtain a corresponding $C_5H_5Mo(CO)_3{}^1L$ derivative from the reaction between perfluoropyridine and $[C_5H_5Mo(CO)_3]^-$, although IR spectra taken of cyclohexane solutions of the reaction mixture presented some evidence that the desired complex **$C_5H_5Mo(CO)_3C_5NF_4$** was formed to some extent, but rapidly decomposed in solution: ν(CO) at 1973, 1977, and 2049; other bands at 658, 686, 726, 754, 777, 815, 915, 1190, 1550, 1615, 1625, and 3510 cm^{-1} [5, 13].

Table 4
$C_5H_5Mo(CO)_3{}^1L$ Derivatives Containing Aryl and Heteroaryl Ligands.
An asterisk indicates further information at the end of the table.
For explanations, abbreviations, and units, see p. X.

No.	^1L	method of preparation (yield) properties and remarks
*1	$\text{-}C_6H_4NO_2$-4	for the formation see "Further information" at the end of Table 4; no details given
2	F^2, N, N, F^6, F^5	I (20%) [5] yellow crystals (from benzene/hexane), m.p. 181 °C [4, 5] ^{19}F NMR ($CHCl_3$): −107.4 (F^5; $J(F^5, F^6) = J(F^2, F^5) = 31.0$); $\delta(F^2)$ and $\delta(F^6)$ not measured because of the insolubility of this complex [5] IR (cyclohexane): 1967, 1974, 2046 (ν(CO)) [4, 5] IR (CCl_4): unassigned bands at 664, 715, 942, 1164, 1310, 1340, 1355, 1410, 1430 [5] stable when pure and crystalline, but slowly decomposes in organic solvents [5]
3	N, F, N, N, F	I (27%) [5] yellow crystals (by sublimation at 70 °C and ca. 1.3×10^{-4} atm), m.p. 110 °C (dec.) [5] ^{19}F NMR ($CHCl_3$): −42.0 [5] IR (cyclohexane): 1964, 1979, 2050 (ν(CO)); unassigned bands at 840, 1232, 1340, 1557, 1562, 1579 [5] stable when pure and crystalline, but slowly decomposes in organic solvents [5]

Table 4 (continued)

No.	${}^1L$	method of preparation (yield) properties and remarks
4	N F N N $Mn(CO)_5$	I (88%; the kinetics of the preparation reaction was studied) [14] yellow crystals (from benzene/pentane), m.p. > 130 °C (dec.) [14] ${}^{19}F$ NMR (THF-d_8): −39.1 [14] IR (THF): 1950, 2010, 2031, 2070, 2128 (ν(CO)) [14]
*5	$-C_5H_4Mn(CO)_3$	prepared from $C_5H_5Mo(CO)_3C(O)C_5H_4Mn(CO)_3$ (Section 1.5.1.4.1.12.3.2) by refluxing in THF for 40 min; purification was accomplished by chromatography on alumina with benzene as eluant (71%) [12] yellow crystals (from THF/pentane), m.p. 146 to 147 °C (dec.) [12] ${}^1H$ NMR ($CHCl_3$, with hexamethyldisiloxane as an internal standard): 4.47, 4.87 (both t, C_5H_4), 5.47 (s, C_5H_5) [12] IR (CCl_4): 1925, 1940, 1967, 2019, 2048 (ν(CO)) [12] mass spectrum: $[M]^+$, $[M-nCO]^+$, n = 1 to 6, for further decarbonylation, compare [11]
6	$-C_5H_4FeC_5H_5$	from a 1,2-dimethoxyethane solution of $Na[C_5H_5Mo(CO)_3]$ to which chlorocarbonylferrocene in THF was added; the reaction mixture was subsequently stirred for 12 h (16%); similar to the corresponding Fe and W complexes, $C_5H_5Mo(CO)_3C(O)C_5H_4FeC_5H_5$ is suggested as an intermediate which decarbonylates so rapidly that no evidence of its existence could be obtained [10] m.p. 140 °C (dec.) [10] ${}^1H$ NMR (no solvent given): 3.88, 4.20 (both t (br), C_5H_4), 3.92 (s, C_5H_5Fe), 5.28 (s, C_5H_5Mo) [10] IR (no medium given): 1941, 1948, 2025 (ν(CO)), the carbonyl absorptions are at high frequencies compared to the corresponding methyl complex, indicating a mesomeric effect which removes electron density from Mo into the aromatic ferrocene C_5H_5 ring [10] thermal and oxidative stability more closely resembles the characteristics of the ferrocene than of the carbonyl metalate portion of the molecule [10]
*7	$-C_5H_4Fe(CO)_2C_5H_4Mn(CO)_3$	from $C_5H_5Fe(CO)_2C_5H_4Mn(CO)_3$ in THF which was stirred with LiC_4H_9-n/hexane for 1.5 h at −78 °C; after addition of $C_5H_5Mo(CO)_3Cl$, stirring was continued for 2 h at −78 °C and subsequently for 1 h at 0 to 5 °C; purification by chromatography with ethyl acetate/hexane as eluant gave No. 7 in modest yield [16] yellow crystals (from hexane/CH_2Cl_2 at low temperature), m.p. 132 to 134 °C [16]

References on p. 147

Table 4 (continued)

No.	^1L	method of preparation (yield) properties and remarks
*7 (continued)		^1H NMR (benzene-d_6): 4.11, 4.33, 4.47, 4.61 (each t, each 2 C_5H_4-H), 4.57 (s, C_5H_5) [16] IR (CH_2Cl_2): 1919, 1950, 1990, 2005, 2020, 2040 (ν(CO)) [16] mass spectrum: $[M]^+$, $[M-nCO]^+$ (n = 5, 7, 8) [16]

*Further information:

$C_5H_5Mo(CO)_3C_6H_4NO_2$-4 (Table **4**, No. **1**) was believed to be obtained in 30% yield along with $C_5H_5Mo(CO)_3CH_2CN$ (Section 1.5.1.4.1.12.2) by electrolysis of 4-$O_2NC_6H_4I$ in the presence of $[C_5H_5Mo(CO)_3]^-$ in a diaphragm electrolyzer at the potential for the reduction of the aryl iodide to the corresponding radical anion using CH_3CN as the solvent. ^1H NMR, IR, and mass spectroscopy as well as polarography were mentioned as being used for identification, but no data were given [15]. In a later work, no significant amount of the σ-aryl complex could be identified during this electrochemically induced nucleophilic aromatic substitution. Only $C_5H_5Mo(CO)_3I$ and nitrobenzene were recovered, regardless of whether the electrolysis was carried out with prior accumulation of the $[C_5H_5Mo(CO)_3]^-$ nucleophile or not [17].

$C_5H_5Mo(CO)_3C_5H_4Mn(CO)_3$ (Table **4**, No. **5**) is stable as a solid but will gradually decompose in solution, particularly in light [12].

Thermal decomposition in a melt at 169 to 171 °C produced $C_5H_5Mo(CO)_3Mo(CO)_3$-$C_5H_4C_5H_4Mn(CO)_3$ together with $(C_5H_5Mo(CO)_3)_2$ and $C_5H_5Mn(CO)_3$ as well as small amounts of $C_5H_5Mo(CO)_3Mo(CO)_3C_5H_4C(O)C_5H_4Mn(CO)_3$, but no binuclear fulvalene derivative, $(C_5H_4Mo(CO)_3)_2$ [12].

$C_5H_5Mo(CO)_3C_5H_4Mn(CO)_3$ reacted with I_2 in $CHCl_3$ at 20 °C to form $C_5H_5Mo(CO)_3I_3$ as the sole product. The compound is inert versus HCl in aqueous dioxane and also toward HgX_2 (X = Cl, Br), even when refluxed in benzene or acetone. Photolysis in THF solution in the presence of $P(C_6H_5)_3$ afforded trans-$C_5H_5Mo(CO)_2(P(C_6H_5)_3)C_5H_4Mn(CO)_3$. Substitution of CO ligands by the phosphine in the $Mn(CO)_3$ moiety of the complex does not take place [12].

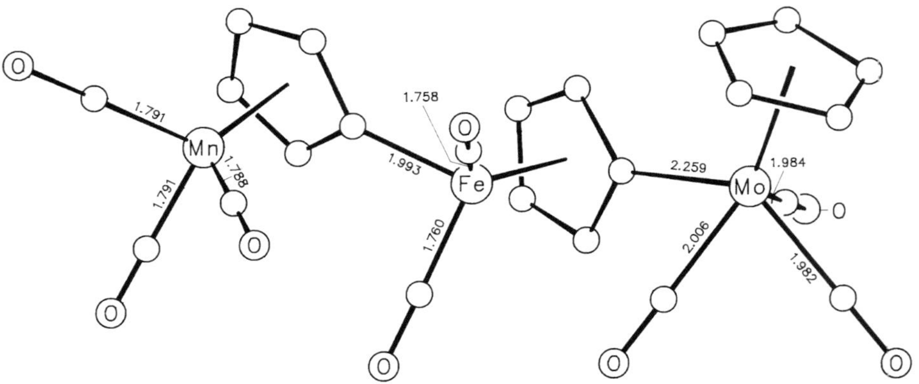

Fig. 10. The molecular structure of $C_5H_5Mo(CO)_3C_5H_4Fe(CO)_2C_5H_4Mn(CO)_3$ [16].

References on p. 147

$C_5H_5Mo(CO)_3C_5H_4Fe(CO)_2C_5H_4Mn(CO)_3$ (Table **4**, No. **7**) crystallizes in the triclinic space group $P\bar{1}-C_i^1$ (No. 2) with a = 7.600(1), b = 8.703(1), c = 18.204(2) Å, α = 92.00(1)°, β = 91.10(1)°, γ = 106.27(1)°; Z = 2, D_{calc} = 1.80 g/cm^3. The molecular structure is shown in **Fig. 10**. The C_5 rings π- and σ-bonded to the molybdenum atom are almost perpendicular (the dihedral angle between the two planes is 86.5°) while the C_5 rings linked to the iron center tend to adopt a coplanar arrangement (the dihedral angle is 22.9°) [16].

References:

[1] Piper, T. S.; Wilkinson, G. (Inorg. Nucl. Chem. Lett. **3** [1956] 104/24).
[2] King, R. B.; Bisnette, M. B. (J. Organomet. Chem. **2** [1964] 38/43).
[3] Nesmeyanov, A. N.; Chapovskii, Yu. A.; Lokshin, B. V.; Kisin, A. V.; Makarova, L. G. (Dokl. Akad. Nauk SSSR **171** [1966] 637/40; Dokl. Chem. Proc. [Engl. Transl.] **166/171** [1966] 1122/5).
[4] Cooke, J.; Green, M.; Stone, F. G. A. (Inorg. Nucl. Chem. Lett. **3** [1967] 47/9).
[5] Cooke, J.; Green, M.; Stone, F. G. A. (J. Chem. Soc. A **1968** 173/6).
[6] Nesmeyanov, A. N.; Makarova, L. G.; Ustynyuk, N. A. (J. Organomet. Chem. **23** [1970] 517/23).
[7] Bruce, M. I.; Sharrocks, D. N.; Stone, F. G. A. (J. Chem. Soc. A **1970** 680/3).
[8] Barnett, K. W.; Slocum, D. W. (J. Organomet. Chem. **44** [1972] 1/37).
[9] Brunner, H.; Wachter, J. (J. Organomet. Chem. **107** [1976] 307/14).
[10] Pannell, K. H.; Cassias, J. B.; Crawford, G. M.; Flores, A. (Inorg. Chem. **15** [1976] 2671/5).

[11] Sizoi, V. F.; Nekrasov, Yu. S.; Sukharev, Yu. N.; Leontyeva, L. I.; Eremin, S. A.; Kolobova, N. E.; Solodova, M. Ya.; Valueva, Z. P. (J. Organomet. Chem. **202** [1980] 83/90).
[12] Leont'eva, L. I.; Perevalova, E. G. (Izv. Akad. Nauk SSSR Ser. Khim. **1984** 2352/8; Bull. Acad. Sci. USSR Div. Chem. Sci. [Engl. Transl.] **1984** 2149/54).
[13] Artamkina, G. A.; Mil'chenko, A. Yu.; Beletskaya, I. P.; Reutov, O. A. (J. Organomet. Chem. **311** [1986] 199/206).
[14] Artamkina, G. A.; Mil'chenko, A. Yu.; Beletskaya, I. P.; Reutov, O. A. (J. Organomet. Chem. **321** [1986] 371/6).
[15] Denisovich, L. I.; Ustynyuk, N. A.; Peterleitner, M. G.; Vinogradova, V. N.; Kravtsov, D. N. (Izv. Akad. Nauk SSSR Ser. Khim. **1987** 2635/6; Bull. Acad. Sci. USSR Div. Chem. Sci. [Engl. Transl.] **1987** 2450).
[16] Orlova, T. Yu.; Setkina, V. N.; Batsanov, A. S.; Dzhafarov, M. Kh.; Struchkov, Yu, T.; Petrovskii, P. V. (Metallorg. Khim. **5** [1992] 1102/6; Organomet. Chem. USSR [Engl. Transl.] **5** [1992] 537/40).
[17] Magdesieva, T. V.; Kukhareva, I. I.; Artamkina, G. A.; Butin, K. P.; Beletskaya, I. P. (J. Organomet. Chem. **468** [1994] 213/21).

1.5.1.4.1.12.4 $C_5H_5Mo(CO)_3{}^1L$ Compounds Containing Mo-sp-C Bonds

The only compounds summarized in this section are the alkynyl derivatives $C_5H_5Mo(CO)_3C{\equiv}CR$ listed in Table 5. For the cyanide $C_5H_5Mo(CO)_3C{\equiv}N$ and complexes containing isonitrile-coordinated cations of composition $[C_5H_5Mo(CO)_3C{\equiv}NR]^+$, see the appropriate sections.

Based on the values of ν(CO) and $\delta(C_5H_5)$ (Table 5), which, in comparison to the corresponding parameters of acyl complexes $C_5H_5Mo(CO)_3C(O)R$ are significantly shifted toward higher wavenumbers and lower fields, respectively, the overall electron-acceptor influence

References on p. 153

of the alkynyl group in $C_5H_5Mo(CO)_3C{\equiv}CC_6H_5$ exceeds that of the acyl groups in $C_5H_5Mo(CO)_3C(O)R$ and is approximately equal to the acceptor influence of a perfluoroalkyl group or a halogen atom in $C_5H_5Mo(CO)_3R$ or $C_5H_5Mo(CO)_3X$ [2].

The following methods of synthesis are available:

Method I: From $[C_5H_5Mo(CO)_3]^-$.
$C_5H_5Mo(CO)_3C{\equiv}CC_6H_5$ was prepared by adding $[(C_6H_5)_3PC{\equiv}CC_6H_5]Br$ at room temperature to a solution of $K[C_5H_5Mo(CO)_3]$ in THF. Subsequently the reaction mixture was refluxed for 2 h. Separation from $C_5H_5Mo(CO)_3(P(C_6H_5)_3)Br$ and $(C_5H_5Mo(CO)_3)_2$ was accomplished by TLC with benzene/petroleum ether as eluant [3]. Similarly $C_5H_5Mo(CO)_3C{\equiv}CC_6H_4F$-4 was prepared using $BrC{\equiv}C$-C_6H_4F-4 for alkylation [7]. Addition of $ClC{\equiv}CCl$ dissolved in ether to $K[C_5H_5Mo(CO)_3]$ in THF at −70 °C and subsequent slow warming to room temperature gave $C_5H_5Mo(CO)_3C{\equiv}CCl$ [9].

Method II: From $C_5H_5Mo(CO)_3X$ (X = Cl, I).

a. $C_5H_5Mo(CO)_3C{\equiv}CR$ (R = H, $Si(CH_3)_3$, $C{\equiv}CFe(CO)_2C_5H_5$, $C{\equiv}CFe(CO)(P(C_6H_5)_3)C_5H_5$) were prepared from $C_5H_5Mo(CO)_3Cl$ and $LiC{\equiv}CR$, obtained in situ from $HC{\equiv}CR$ and LiC_4H_9-n [10, 15].

b. $C_5H_5Mo(CO)_3C{\equiv}CR$ (R = n-C_4H_9, C_6H_5) was prepared by the reaction of $C_5H_5Mo(CO)_3Cl$ with the respective Grignard reagent $BrMgC{\equiv}CR$ (obtained in situ from n-C_4H_9MgBr and $HC{\equiv}CR$) in THF at −60 °C. After warming to room temperature, the resulting solution was hydrolyzed with aqueous HCl. The crude product was purified by chromatography on alumina [1]. In a similar preparation, $C_5H_5Mo(CO)_3I$ reacted with $NaC{\equiv}CC_6H_5$ in THF between −30 to −50 °C [4]. $C_5H_5Mo(CO)_3C{\equiv}CC_6H_5$ was also obtained from $C_5H_5Mo(CO)_3I$ which reacted with $CuC{\equiv}CC_6H_5$ under ambient conditions. The latter compound was prepared in situ from $HC{\equiv}CC_6H_5$ and CuI in $N(C_2H_5)_3$ [8]. $C_5H_5Mo(CO)_3C{\equiv}CR$ with R = t-C_4H_9 and C_6H_4F-4 were apparently also prepared by the reactions of molybdenum halides $C_5H_5Mo(CO)_3X$ with the corresponding alk-1-ynes in organic amines in the presence of CuX as catalysts, as outlined for their tungsten homologues, $C_5H_5W(CO)_3C{\equiv}CC_6H_5$ [6], although no details were communicated [16].

Table 5
$C_5H_5Mo(CO)_3{}^1L$ Derivatives with ^1L = Alkyn-1-yl.
An asterisk indicates further information at the end of the table.
For explanations, abbreviations, and units, see p. X.

No.	^1L	method of preparation (yield) properties and remarks
1	$-C{\equiv}CH$	IIa (22%; by stirring at −78 °C for 20 min and subsequently at 25 °C for 2 h; chromatography and hydrolytic workup should be avoided due to considerable decomposition to $C_5H_5Mo(CO)_3CH_3$) [10] also in ca. 50% from desilylation of $C_5H_5Mo(CO)_3C{\equiv}CSi(CH_3)_3$ (No. 6) by reaction with $KF \cdot 2\,H_2O$ (CH_3OH, 40 °C), $[N(C_2H_5)_4]F \cdot 3\,HF$ (THF, 40 °C), calcined KF (CH_3CN, 40 °C), CsF (CH_3CN, 80 °C), or KOH (1 M in CH_3OH, 20 °C) [10]

Table 5 (continued)

No.	^{1}L	method of preparation (yield) properties and remarks
		yellowish crystals (from warm hexane), m.p. 98 to 100 °C [10] ^{1}H NMR (benzene-d_6): 2.85 (s, CH), 4.52 (s, C_5H_5) [10] ^{13}C NMR (benzene-d_6): 92.7 (C_5H_5), 108.7, 116.3 (both C≡C), 223.5 (CO cis MoC), 239.86 (CO trans MoC) [10] IR (hexane): 1970, 1985, 2057 (ν(CO)) [10] mass spectrum: $[M]^+$, $[M-CO]^+$, $[M-3CO]^+$, $[M-3CO-C_2H_2]^+$, $[M-3CO-2(C_2H_2)]^+$; additionally products of secondary ion-molecule reactions like $[M_2-2CO]^+$, $[M_2-3CO]^+$, $[M_2-4CO]^+$, $[M_2-5CO]^+$, $[M_2-6CO]^+$, $[M_2-6CO-C_2H_2]^+$, $[M_2-6CO-2(C_2H_2)]^+$ were observed; by chemical ionization in an NH_3 plasma only $[M]H^+$ and $[M]NH_4{}^+$ were formed [10]
2	$-C{\equiv}CC_4H_9$-n	IIb (<60%) [1] yellow-green solid, m.p. 55 °C (dec.) [1]; the value of 126 °C given in [1] is very probably confused with the value for $C_5H_5Mo(CO)_3C{\equiv}CC_6H_5$ (No. 4) ^{1}H NMR ($CDCl_3$): 0.75 to 2.34 (m, C_4H_9), 5.49 (s, C_5H_5) [1] IR (no medium given): 1960, 1970, 2045 (ν(CO)), 2100 (ν(C≡C)) [1] stable in air for prolonged periods and decomposes only slowly in $CHCl_3$ [1]
*3	$-C{\equiv}CC_4H_9$-t	IIb [16] no details given
*4	$-C{\equiv}CC_6H_5$	I (54%); similarly with $BrC{\equiv}CC_6H_5$ at −70 °C but with a decreased yield of 24%; furthermore, separation from simultaneously formed $C_5H_5Mo(CO)_3Br$ succeeded only partly by TLC and fractional crystallization, respectively [3] IIb (43%, with $NaC{\equiv}C_6H_5C$ [4]; up to 60%, with $C_6H_5C{\equiv}CMgBr$ [1]; 69%, with $CuI/C_6H_5C{\equiv}CH/N(C_2H_5)_3$) [8] also in 10% yield from $K[C_5H_5Mo(CO)_3]$ and $C_6H_5C{\equiv}C(O)Cl$ by decarbonylation of the initially formed phenylpropioloyl derivative $C_5H_5Mo(CO)_3C(O)C{\equiv}CC_6H_5$ [2]; see also Section 1.5.1.4.1.12.3.2 [2] yellow crystals (from benzene/heptane) [1, 3, 4], 126 °C (dec.) [4]; compare also [2, 3] ^{1}H NMR ($CDCl_3$): 5.51 (s, C_5H_5), 7.11 to 7.31 (C_6H_5) [2]; see also [1] ^{13}C NMR (CH_2Cl_2): 83.52 (MoC≡), 93.33 (C_5H_5), 126.14 (C-4), 126.9 (C-1), 128.15 (C-3,5), 129.39 (≡C),

References on p. 153

Table 5 (continued)

No.	^1L	method of preparation (yield) properties and remarks
*4 (continued)		130.88 (C-2,6), 223.35 (CO cis MoC), 238.8 (CO trans MoC) [5] IR (cyclohexane): 1970, 1982, 2050 (ν(CO)) [4]; similar in $CHCl_3$ [3]; see also [1, 2, 8]; (KBr): 2110 (ν(C≡C)) [1, 2] mass spectrum: $[M]^+$ [8] stable in air for prolonged periods and decomposes only slowly in $CHCl_3$ [1]
*5	-C≡CC_6H_4F-4	I (35%) [7]; IIb [16] ^{19}F NMR (toluene, relative to C_6H_5F): +1.57; the fluorine chemical shift was compared to other organometallic 4-fluoro-phenylethyne derivatives with respect to the shielding of fluorine and the polarity of the metal-carbon bond [11]
6	-C≡CSi$(CH_3)_3$	IIa (69%; the reactants were mixed at −78 °C; and subsequently warmed up to 25 °C for isolation; purification was accomplished by column chromatography on silica with petroleum ether/ether) [10] light yellow crystals (from hexane), m.p. 96 to 98 °C [10] ^1H NMR (benzene-d_6): 0.275 (s, CH_3), 4.65 (s, C_5H_5) [10] ^{13}C NMR (C_6D_6): 1.511 (CH_3), 92.854 (C_5H_5), 109.154, 138.956 (both C≡C), 210.5 (CO trans MoC), 223.354 (CO cis MoC) [10] IR (hexane): 1960, 1975, 2035, 2055 (ν(CO)) [10] possesses a C-Si bond which is more stable than those in simple silyl alkynes; it can be desilylated by reagents such as KF · 2 H_2O (CH_3OH, 40 °C), $[N(C_2H_5)_4]F$ · 3 HF (THF, 40 °C), calcined KF (CH_3CN, 40 °C), CsF (CH_3CN, 80 °C), and KOH (1 M in CH_3OH, 20 °C) to give $C_5H_5Mo(CO)_3$C≡CH (No. 1) in reasonable yields [10]
7	-C≡CCl	I (35%; the crude product was extracted with benzene and then precipitated from the concentrated solution by the addition of pentane) [9] orange crystals (from hexane), m.p. 72 to 74 °C (dec.) [9] IR (cyclohexane): 1960, 1980, 2055 (ν(CO)) [9] mass spectrum: $[M]^+$, $[M-nCO]^+$ (n = 1, 2, 3) [9]
8	-C≡CC≡CFe$(CO)_2C_5H_5$	IIa (58%) [15] yellow to orange solid [15] ^1H NMR ($CDCl_3$): 5.01 (s, C_5H_5), 5.50 (s, C_5H_5) [15] ^{13}C NMR ($CDCl_3$): 69.5, 79.7 (both C≡), 85.2, 92.9 (both C_5H_5), 101.2, 115.8 (both C≡), 210.8, 211.7, 230.9 (all CO) [15] IR (CH_2Cl_2): 1958, 1991, 2039 (ν(CO)), 2144 (ν(C≡C)) [15] moderately stable to air and heat in the solid state [15]

References on p. 153

Table 5 (continued)

No.	^{1}L	method of preparation (yield) properties and remarks
9	$-C{\equiv}CC{\equiv}CFe(CO)(P(C_6H_5)_3)C_5H_5$	IIa (58%) [15] yellow to orange solid [15] ^1H NMR ($CDCl_3$): 4.45 (d, C_5H_5Fe; J(P, H) = 1.2), 5.54 (s, C_5H_5Mo), 7.35 to 7.69 (m, C_6H_5) [15] ^{13}C NMR ($CDCl_3$): 57.8 (s, $MoC{\equiv}$), 84.5 (d, C_5H_5Fe; J(P, C) = 16.0), 92.9 (s, C_5H_5Mo), 96.7 (d, $FeC{\equiv}$; J(P, C) = 42.0), 105.3, 117.5 (both s, $\equiv CC\equiv$), 218.9 (d, FeCO; J(P, C) = 30.5), 221.7, 241.8 (both s, MoCO); $P(C_6H_5)_3$ at 127.9 (d; J(P, C) = 9.7), 129.7 (d; J(P, C) = 1.8), 133.4 (d; J(P, C) = 9.7), 135.9 (d; J(P, C) = 44.1) [15] IR (CH_2Cl_2): 1936, 1957, 2035 (ν(CO)), 2121 ($\nu(C{\equiv}C)$) [15] moderately stable to air and heat in the solid state; somewhat enhanced stability compared to the preceding No. 8 [15]

* Further information:

$C_5H_5Mo(CO)_3C{\equiv}CR$ (R = t-C_4H_9, C_6H_5, 4-FC_6H_4, Table **5**, Nos. **3** to **5**). Protonation of $C_5H_5Mo(CO)_3C{\equiv}CC_6H_5$ (No. 4) with HBF_4 in acetic anhydride at 0 °C in the presence of $C_6H_5C{\equiv}CH$ or $C_6H_5C{\equiv}CC_6H_5$ resulted in the formation of ionic bis(alkyne) complexes, $[C_5H_5Mo(CO)(C_6H_5C{\equiv}CH)_2]BF_4$ and $[C_5H_5Mo(CO)(C_6H_5C{\equiv}CC_6H_5)_2]BF_4$, respectively. With HBF_4 and $P(C_6H_5)_3$ as reactants at −10 °C in CH_2Cl_2, a closely related ionic bis(phosphane) derivative, $[C_5H_5Mo(CO)(P(C_6H_5)_3)_2]BF_4$, is produced after warming to room temperature [4].

Heating of $C_5H_5Mo(CO)_3C{\equiv}CC_6H_4X$-4 (X = H, F; Nos. 4, 5) in the solid state at 118 to 120 °C or in refluxing octane yielded $(C_5H_5Mo(CO)_2)_2(\mu$-4-$XC_6H_4C{\equiv}CC{\equiv}CC_6H_4X$-4) (Formula I). When Nos. 4 and 5 were heated together, additionally products of α,α-coupling of the σ-ethynyl groups, $(C_5H_5Mo(CO)_2)_2(\mu$-$HC_6H_4C{\equiv}CC{\equiv}CC_6H_4X$-4), were formed. The reaction mechanism discussed which starts with the dissociation of CO is indirectly confirmed by the formation of $C_5H_5Mo(CO)_2(C{\equiv}CC_6H_5)P(C_6H_5)_3$ from heating No. 4 in octane in the presence of $P(C_6H_5)_3$ [7].

```
                    Mo(CO)2C5H5
                    / | \
                   /  |  \                         X = X' = H, F
4-XC6H4—C————————C—C≡C—C6H4X'-4            X = F, X' = H
                   \  |  /                         X = H, X' = F
                    Mo(CO)2C5H5

                       I
```

Treatment of $C_5H_5Mo(CO)_3C{\equiv}CC_6H_5$ (No. 4) with $Fe_2(CO)_9$ in benzene at 40 °C yielded $C_5H_5Mo(CO)_2(\mu_3,\eta^2$-$C{\equiv}CC_6H_5)Fe_2(CO)_6$ (Formula IIa) containing a bridging σ,π-phenylethynyl ligand [5]. Formally related compounds of the type $C_5H_5Mo(CO)_2(\mu_3,\eta^2$-$C{\equiv}CR)$-$Ru_2(CO)_6$ (Formula IIb) were obtained from $C_5H_5Mo(CO)_3C{\equiv}CR$ (R = t-C_4H_9, C_6H_5, 4-FC_6H_4)

References on p. 153

and $Ru_3(CO)_{12}$ in refluxing toluene, but in these cases the alkynyl ligand preferred an asymmetric arrangement where it adopts an orthogonal orientation with respect to one of the Mo-Ru bonds, as shown in Formula IIb. Only in the case of the bulky t-C_4H_9 group (No. 3), did this asymmetric isomer exist in a mixture with small amounts of the symmetric isomer with the arrangement of the C-C bond orthogonal to the unique Ru-Ru bond as demonstrated in Formula IIa [16]. For R = C_6H_5 (No. 4), additionally small amounts of a pentanuclear alkynyl cluster $(C_5H_5Mo)_2(\mu\text{-}CO)_2(\mu_4,\eta^2\text{-}C{\equiv}CC_6H_5)_2Ru_3(CO)_8$ (see Formula III) were isolated [13].

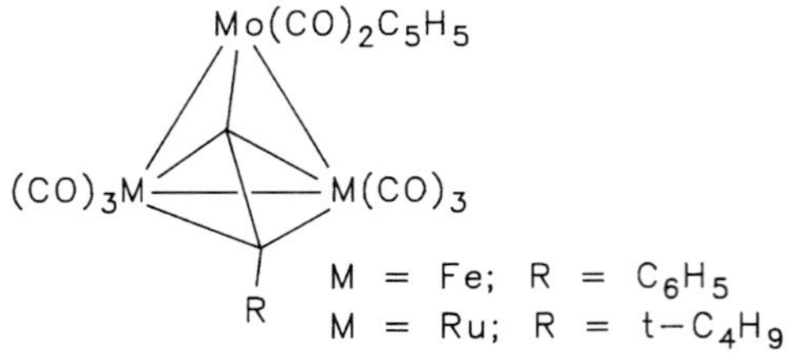

$Mo(CO)_2C_5H_5$, $(CO)_3M$, $M(CO)_3$, R

M = Fe; R = C_6H_5
M = Ru; R = t−C_4H_9

IIa

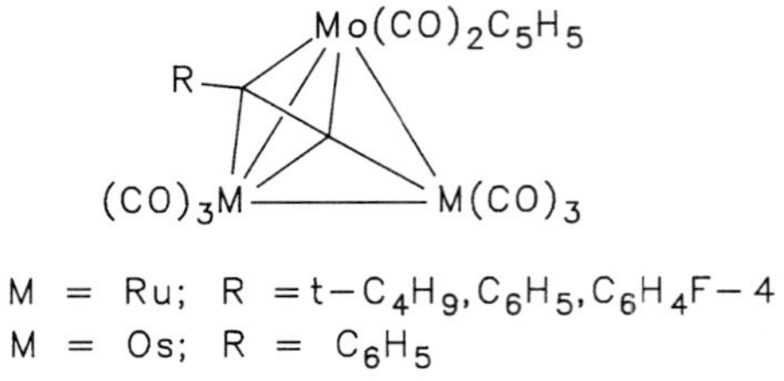

$Mo(CO)_2C_5H_5$, R, $(CO)_3M$, $M(CO)_3$

M = Ru; R = t−C_4H_9, C_6H_5, C_6H_4F−4
M = Os; R = C_6H_5

IIb

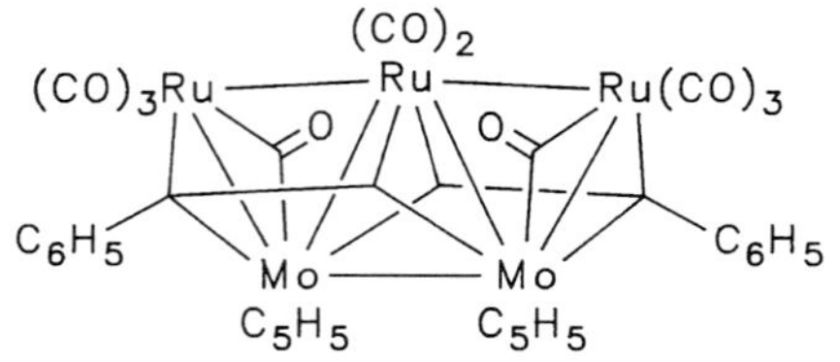

$(CO)_2$, $(CO)_3Ru$, Ru, $Ru(CO)_3$, O, O, C_6H_5, C_6H_5, Mo, Mo, C_5H_5, C_5H_5

III

Treatment of $C_5H_5Mo(CO)_3C{\equiv}CC_6H_5$ (No. 4) with $Os_3(CO)_{10}(NCCH_3)_2$ in refluxing toluene resulted in $C_5H_5Mo(CO)_2(\mu_4,\eta^2\text{-}C{\equiv}CC_6H_5)Os_3(CO)_9$ (Formula IV), probably along with $C_5H_5Mo(CO)_2(\mu_3,\eta^2\text{-}C{\equiv}CC_6H_5)Os_2(CO)_6$ (compare Formula IIb) as detailed for the tungsten analogue [12, 14]. Compounds of the type $C_5H_5M(CO)_2(\mu_4,\eta^2\text{-}C{\equiv}CC_6H_5)Os_3(CO)_9$ (M = W, Mo) reacted further with $C_5H_5Mo(CO)_3C{\equiv}CC_6H_5$ in a 1:2 molar ratio in refluxing toluene to give pentanuclear clusters of composition $(C_5H_5)_2MoM(CO)_3Os_3(CO)_8(\mu_5,\eta^4\text{-}C_2C_6H_5C_2\text{-}C_6H_5)$ which contain an $MoMOs_3$ skeleton defining a very flat double-butterfly geometry and a C_4 hydrocarbon fragment derived from head-to-tail coupling between the alkynyl ligands of the reactants and bonded to the metal core via two $(2\sigma + 1\pi)$ interactions as depicted in Formula V [14].

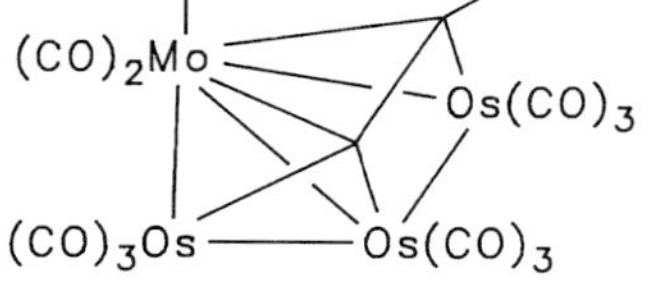

C_6H_5, $(CO)_2Mo$, $Os(CO)_3$, $(CO)_3Os$, $Os(CO)_3$

IV

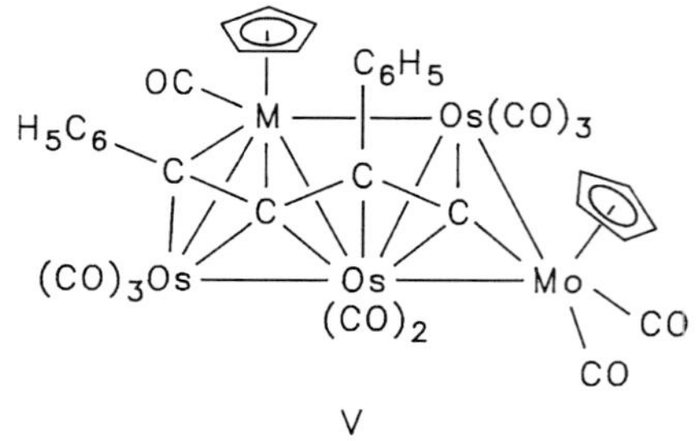

OC, C_6H_5, H_5C_6, M, $Os(CO)_3$, C, C, C, C, $(CO)_3Os$, Os, Mo, $(CO)_2$, CO, CO

V

References on p. 153

References:

[1] Green, M. L. H.; Mole, T. (J. Organomet. Chem. **12** [1968] 404/6).
[2] Nesmeyanov, A. N.; Makarova, L. G.; Vinogradova, V. N.; Ustynyuk, N. A. (Izv. Akad. Nauk SSSR Ser. Khim. **1978** 2375/9; Bull. Acad. Sci. USSR Div. Chem. Sci. [Engl. Transl.] **1978** 2108/12).
[3] Nesmeyanov, A. N.; Makarova, L. G.; Vinogradova, V. N.; Korneva, V. N.; Ustynyuk, N. A. (J. Organomet. Chem. **166** [1979] 217/22).
[4] Kolobova, N. E.; Skripkin, V. V.; Rozantseva, T. V.; Struchkov, Yu. T.; Aleksandrov, G. G.; Antonovich, V. A.; Bakhmutov, V. I. (Koord. Khim. **8** [1982] 1655/63; Sov. J. Coord. Chem. [Engl. Transl.] **8** [1982] 910/8).
[5] Ustynyuk, N. A.; Vinogradova, V. N.; Korneva, V. N.; Slovokhotov, Yu. L.; Struchkov, Yu. T. (Koord. Khim. **9** [1983] 631/7; Sov. J. Coord. Chem. [Engl. Transl.] **9** [1983] 363/9).
[6] Bruce, M. I.; Humphrey, M. G.; Matisons, J. G.; Roy, S. G.; Swincer, A. G. (Austral. J. Chem. **37** [1984] 1955/61).
[7] Ustynyuk, N. A.; Vinogradova, V. N.; Korneva, V. N.; Kravtsov, D. N.; Andrianov, V. G.; Struchkov, Yu. T. (J. Organomet. Chem. **277** [1984] 285/96).
[8] Villemin, D.; Schigeko, E. (J. Organomet. Chem. **346** [1988] C24/C26).
[9] Ustynyuk, N. A.; Vinogradova, V. N.; Kravtsov, D. N. (Metalloorg. Khim. **1** [1988] 85/8; Organomet. Chem. USSR [Engl. Transl.] **1** [1988] 45/7).
[10] Ustynyuk, N. A.; Vinogradova, V. N.; Kravtsov, D. N.; Oprunenko, Yu. F.; Piveń, V. A. (Metalloorg. Khim. **1** [1988] 884/8; Organomet. Chem. USSR [Engl. Transl.] **1** [1988] 488/91).
[11] Kravtsov, D. N. (Metalloorg. Khim. **2** [1989] 157/64; Organomet. Chem. USSR [Engl. Transl.] **2** [1989] 83/7).
[12] Chi, Y.; Lee, G.-H.; Peng, S.-M.; Wu, C.-H. (Organometallics **8** [1989] 1574/6).
[13] Hwang, D.-K.; Chi, Y.; Peng, S.-M.; Lee, G.-H. (J. Organomet. Chem. **389** [1990] C7/C11).
[14] Wu, C.-H.; Chi, Y.; Peng, S.-M.; Lee, G.-H. (J. Chem. Soc. Dalton Trans. **1990** 3025/31).
[15] Wong, A.; Kang, P. C. W.; Tagge, C. D.; Leon, D. R. (Organometallics **9** [1990] 1992/4).
[16] Hwang, D.-K.; Chi, Y.; Peng, S.-M.; Lee, G.-H. (Organometallics **9** [1990] 2709/18).

1.5.1.4.1.13 Compounds with $[C_5H_5Mo(CO)_3{=}CRR']^+$ Cations

Although no carbene intermediates could be observed directly, perfluorocarbene cations such as $[C_5H_5Mo(CO)_3{=}CF_2]^+$ and $[C_5H_5Mo(CO)_3{=}CF(C_2F_5)]^+$ are apparently implicated in the proton acid hydrolysis and the BF_3-induced cleavage reaction of $C_5H_5Mo(CO)_3CF_3$ (Section 1.5.1.4.1.12.2) affording $[C_5H_5Mo(CO)_4]^+$, as well as in the conversion of $C_5H_5Mo(CO)_3$-C_3F_7-n by BF_3 to a highly unstable red material that returns to the yellow starting compound above −20 °C [4, 5, 9]. Furthermore, the ability of electrophilic agents to promote the formation of well-defined cationic dihalocarbene complexes via fluoride ion abstraction from perfluoroalkyl precursors (see below) strongly suggests that such species are also reasonable intermediates in electrophilic halogen exchange reactions between $C_5H_5Mo(CO)_3R$ (Section 1.5.1.4.1.12.2, R = CF_3, C_2F_5, n-C_3F_7) and BX_3 (X = Cl, Br) [8].

$[C_5H_5Mo(CO)_3{=}CH_2]^+$ Salts. The $[C_5H_5Mo(CO)_3{=}CH_2]^+$ cation was first suggested as a reactive intermediate involved in the cleavage of the methoxy or acetoxy group from complexes of the type $C_5H_5Mo(CO)_3CH_2OCH_3$ and $C_5H_5Mo(CO)_3CH(R)OC(O)CH_3$ (R = H, CH_3)

References on p. 156

by acids such as HCl [2, 12], HBF_4 [1], or $R'CO_2H$ [12]. Although the carbene could not be detected, its intermediacy is proved from the reaction of the methoxy complex with tetrafluoroboric acid producing norcarane when carried out in the presence of cyclohexene [1]; nevertheless, in the case of the acetoxy group, no norcarane was found [12]. Additionally the ability of Mo to stabilize an intermediate carbene by formation of a metal-carbon double bond is discussed [2]. Further investigations showed $[C_5H_5Mo(CO)_3{=}CH_2]PF_6$ to be conveniently accessible by halide abstraction from $C_5H_5Mo(CO)_3CH_2Cl$ which was added to $AgPF_6$ in CH_2Cl_2 at −78 °C, but even at −80 °C, the $[C_5H_5Mo(CO)_3{=}CH_2]^+$ ion should best be regarded as a transient intermediate, prone to rapid disproportionation into $[C_5H_5Mo(CO)_3(CH_2{=}CH_2)]^+$ and the organometallic Lewis acid $[C_5H_5Mo(CO)_3]^+$ [10, 11].

Addition of a solution of $C_5H_5Mo(CO)_3H$ to in situ-generated $[C_5H_5Mo(CO)_3{=}CH_2]PF_6$ in CH_2Cl_2 at −78 °C and subsequent warming to room temperature gave a mixture of bimetallic μ-acetyl salts, $[(C_5H_5Mo(CO)_3)_2(\mu\text{-}CH_3CO)]PF_6$ and $[(C_5H_5Mo(CO)_2)_2(\mu,\eta^2\text{-}CH_3CO)]PF_6$, the methyl complex $C_5H_5Mo(CO)_3CH_3$, and the organometallic Lewis acid $[C_5H_5Mo(CO)_3\text{-}OC_4H_8]PF_6$ [10]. Under similar conditions $[C_5H_5Mo(CO)_3{=}CH_2]PF_6$ reacted with $C_5H_5Mo(CO)_3CH_2OCH_3$ to give approximately equal quantities of $[C_5H_5Mo(CO)_3{=}CH(OCH_3)]PF_6$ and $C_5H_5Mo(CO)_3CH_3$ as the major organometallic components [10].

Attempts to alkylate $C_5H_5Fe(CO)_2C(O)CH_3$ employing in situ-formed $[C_5H_5Mo(CO)_3{=}CH_2]PF_6$ in CH_2Cl_2 at −80 °C resulted in partial conversion of the starting acetyl complex to $[C_5H_5Mo(CO)_3OC(CH_3)Fe(CO)_2C_5H_5]PF_6$. In contrast, the more nucleophilic $C_5H_5Fe(CO)COCH_3(P(C_6H_5)_3)$ readily afforded the desired alkylation product, $[C_5H_5Mo(CO)_3CH_2OC(CH_3)Fe(CO)(P(C_6H_5)_3)C_5H_5]PF_6$ [11].

$[C_5H_5Mo(CO)_3{=}CH(N(CH_3)_2)][C_5H_5Mo(CO)_3]$ was prepared from $Na[C_5H_5Mo(CO)_3]$ and $[(CH_3)_2NCHCl]Cl$ by NaCl elimination and concomitant C-Cl oxidative addition when the two salts were stirred in 2:1 stoichiometry in THF for 12 h at room temperature. Diethyl ether was added to the filtered reaction mixture, and subsequent cooling down to −30 °C resulted in an 81% yield of yellow needles which decomposed at 140 °C [3, 6].

^1H NMR spectrum ($(CD_3)_2SO$): δ = 3.52 (s, NCH_3 trans CH), 3.61 (s, NCH_3 cis CH), 4.98 (s, C_5H_5 of anion), 6.08 (s, C_5H_5 of cation), 11.00 (s, CH) ppm. IR spectrum (CH_2Cl_2): 1753, 1773, 1889 (ν(CO) of anion), 1973, 2057 (ν(CO) of cation); (paraffin mull): 1588 (ν(CN)) cm^{-1} [6].

The reaction with $NaBF_4$ in CH_2Cl_2/CH_3CN (1:1) gave the tetrafluoroborate salt described below [6].

$[C_5H_5Mo(CO)_3{=}CH(N(CH_3)_2)]BF_4$ was obtained from $[C_5H_5Mo(CO)_3{=}CH(N(CH_3)_2)][C_5H_5Mo(CO)_3]$ (prepared in situ as described before) and excess anhydrous $NaBF_4$ by a metathetical reaction in CH_2Cl_2/CH_3CN (1:1). The resulting mixture was stirred at room temperature for 2 h and then evaporated to dryness. Extraction of the residue with CH_2Cl_2, followed by dilution with THF and cooling to −30 °C produced the title complex as yellow crystals in a 65% yield with a melting point of 150 °C [6].

^1H NMR spectrum ($(CD_3)_2SO$): δ = 3.49 (s, NCH_3 trans CH), 3.80 (s, NCH_3 cis CH), 6.05 (s, C_5H_5), 10.95 (s, CH) ppm. IR spectrum (CH_2Cl_2): 1601 (ν(BF)), 1060 (ν(CN)), 1905, 1975, 2045 (ν(CO)) cm^{-1} [6].

$[C_5H_5Mo(CO)_3{=}C(CH_3)N(C_2H_4)_2O\text{-}cyclo]BF_4$ resulted from protonation of the acryloyl complex $C_5H_5Mo(CO)_2(\eta^3\text{-}C(O)C(N(C_2H_4)_2O\text{-}cyclo){=}CH_2)$ with one equivalent of 54% ethereal HBF_4 in CH_2Cl_2 at −78 °C, followed by precipitation with diethyl ether [7].

^1H NMR spectrum (acetone-d_6): δ = 3.00 (s, CH_3), 3.8 to 4.5 (m, C_2H_4), 6.25 (s, C_5H_5) ppm. IR spectrum (CH_2Cl_2): 1532 (ν(CN)), 1964, 1986, 2056 (ν(CO)) cm^{-1} [7].

References on p. 156

$[C_5H_5Mo(CO)_3{=}C(CH_2CO_2C_2H_5)N(C_2H_5)_2]BF_4$ crystallized spontaneously on treating C_5H_5-$Mo(CO)_2(\eta^3$-$C(O)C(N(C_2H_5)_2){=}CHCO_2C_2H_5)$ with 54% ethereal HBF_4 in THF at room temperature [7].

^1H NMR spectrum (acetone-d_6): δ = 1.33, 1.55 (both t, amino CH_2; J = 7.0), 1.38 (t, ester CH_3; J = 7.0), 4.18, 4.30 (both q, NCH_2 and OCH_2), 4.43 (s, CCH_2), 6.18 (s, C_5H_5) ppm. IR spectrum (CH_2Cl_2): 1539 (ν(CN)), 1731 ($\nu(CO_2C_2H_5)$), 1962, 1990, 2059 (ν(CO)) cm^{-1} [7].

$[C_5H_5Mo(CO)_3{=}CHOCH_3]PF_6$ was prepared by addition of $[C(C_6H_5)_3]PF_6$ to a solution of $C_5H_5Mo(CO)_3CH_2OCH_3$ in CH_2Cl_2 at −25 °C. Yellow crystals of the methoxymethylidene salt appeared as the reaction mixture was warmed slowly to room temperature and were isolated in 69% yield. The title complex was also obtained from equimolar quantities of $C_5H_5Mo(CO)_3$-CH_2OCH_3 and $[C_5H_5Mo(CO)_3{=}CH_2]PF_6$ (in situ generated from $C_5H_5Mo(CO)_3CH_2Cl$ and $AgPF_6$) in CH_2Cl_2 combined at −78 °C. After being stirred at room temperature, the red-brown precipitate was extracted in nitromethane from which $[C_5H_5Mo(CO)_3{=}CHOCH_3]PF_6$ was deposited in 57% yield on dilution with excess ether. The product so obtained was contaminated by small amounts of $[C_5H_5Mo(CO)_4]PF_6$. $C_5H_5Mo(CO)_3CH_3$ was also found as a by-product. Additionally $[C_5H_5Mo(CO)_3{=}CHOCH_3]PF_6$ was prepared by treatment of $C_5H_5Mo(CO)_3FPF_5$ in CH_2Cl_2 at −20 °C with two equivalents of $C_5H_5Mo(CO)_3CH_2OCH_3$. Subsequently the reaction mixture was stirred at room temperature for 1.5 h to afford a yield of 46%. The orange precipitate was contaminated by trace amounts of $[(C_5H_5Mo(CO)_3)_2$-$(\mu$-$H)]PF_6$. When the stoichiometry of $C_5H_5Mo(CO)_3FPF_5$ to $C_5H_5Mo(CO)_3CH_2OCH_3$ was varied from 1:2 to 1:1 and 2:1, the yields of $[C_5H_5Mo(CO)_3{=}CHOCH_3]PF_6$ decreased [10].

^1H NMR spectrum (CD_3NO_2): δ = 4.81 (s, OCH_3), 6.10 (s, C_5H_5), 13.44 (s, CH) ppm. IR spectrum (CH_3NO_2): 1990, 2081 (ν(CO)) cm^{-1} [10].

Nitromethane solutions of the title complex were found to be stable at room temperature but reacted with $[N(C_4H_9\text{-}n)_4]I$ to quantitatively generate a 1:1 mixture of $C_5H_5Mo(CO)_3CH_2$-OCH_3 and $[C_5H_5Mo(CO)_4]PF_6$. The salt failed to abstract hydrogen from $C_5H_5Mo(CO)_3H$ [10].

$[C_5H_5Mo(CO)_3{=}CF_2]O_3SCF_3$ was prepared from $C_5H_5Mo(CO)_3CF_3$ which was stirred with two equivalents of $CF_3SO_2OSi(CH_3)_3$ in benzene for 1 h at room temperature. After 5 min yellow crystals of the title compound began to precipitate from the reaction mixture [13].

^1H NMR spectrum (SO_2, 0 °C): δ = 6.42 (s) ppm. ^{13}C NMR spectrum (SO_2, 0 °C): δ = 93.0 (dt, C_5H_5; ^1J(C, H) = 186, ^3J(C, H) = 6), 117.2 (q, CF_3; ^1J(F, C) = 319), 207.9 (s, CO trans Mo=C), 209.4 (t, 2 CO cis Mo=C; ^3J(F, C) = 9), 272.8 (t, CF_2; ^1J(F, C) = 472) ppm. ^{19}F NMR spectrum (SO_2, 0 °C; vs. $CFCl_3$): δ = −77.8 (s, CF_3), 163.7 (s, CF_2) ppm. IR spectrum (Nujol): 1032, 1155, 1195, 1270 ($\nu(CF_2)$, $\nu(CF_3)$), 2015, 2105 (ν(CO)) cm^{-1} [13].

The complex may be stored in the solid state at room temperature for several days without significant decomposition and may be stored indefinitely at −20 °C under nitrogen. The compound may also be kept in liquid SO_2 at 0 °C for several days without any spectroscopically noticeable changes but not in CH_2Cl_2 where rapid decomposition regenerated C_5H_5Mo-$(CO)_3CF_3$ as the major identifiable product [13].

$[C_5H_5Mo(CO)_3{=}CF_2]SbF_6$ was generated in SO_2 solution by treating $C_5H_5Mo(CO)_3CF_3$ with a 10% excess of SbF_5 in liquid sulfur dioxide at −78 °C. The following NMR spectra were recorded immediately [4, 5]:

^1H NMR spectrum (SO_2): δ = 5.6 (s) ppm. ^{13}C NMR spectrum (SO_2): δ = 99.8 (s, C_5H_5), 279.8 (t, CF_2; J(C, F) = 471) ppm. ^{19}F NMR spectrum (SO_2, vs. CF_3CO_2H): δ = 242.47 (s, CF_2) ppm [5].

References on p. 156

Attempts to isolate solid $[C_5H_5Mo(CO)_3{=}CF_2]SbF_6$ yielded only $[C_5H_5Mo(CO)_4]SbF_6$ resulting from disproportionation [4, 5].

$[C_5H_5Mo(CO)_3{=}CFC_2F_5]SbF_6$ resulted from treatment of $C_5H_5Mo(CO)_3C_3F_7$-n with one equivalent of SbF_5 in liquid SO_2 at −78 °C. After dilution with CH_2Cl_2, the sulfur dioxide was slowly evaporated by warming which caused the precipitation of the compound as a red solid in 92% yield [4, 5].

^1H NMR spectrum (SO_2): δ = 5.6 (s) ppm. ^{19}F NMR spectrum (SO_2; vs. CF_3CO_2H): δ = −28.35 (dq, CF_2; J(CF_2, CF) = 16.1, J(CF_2, CF_3) = 2.4), −1.62 (dt, CF_3; J(CF_3, CF) = 8.0), 185.03 (tq, CF) ppm. IR spectrum (Nujol): 640 to 680 (ν(SbF)), 1180 to 1210 (ν(CF)), 2035, 2120 (ν(CO)), 3080 (ν(C_5H_5)) cm^{-1} [5].

Solid samples decomposed completely after two days at room temperature under an inert atmosphere [5].

The corresponding **$[C_5H_5Mo(CO)_3{=}CFC_2F_5]BF_4$** salt was believed to be formed in the reaction of $C_5H_5Mo(CO)_3C_3F_7$-n with BF_3 in CH_2Cl_2 at −78 °C, but the red solid isolated at −78 °C decomposed to the starting material at temperatures above −20 °C [5].

References:

[1] Jolly, P. W.; Pettit, R. (J. Am. Chem. Soc. **88** [1966] 5044/5).
[2] Green, M. L. H.; Ishaq, M.; Whiteley, R. N. (J. Chem. Soc. A **1967** 1508/15).
[3] Hartshorn, A. J.; Lappert, M. F.; Turner, K. (J. Chem. Soc. Chem. Commun. **1975** 929/30).
[4] Dukes, M. D. (Diss. Univ. South Carolina, Columbia, S.C., 1977; Diss. Abstr. Int. B **38** [1978] 5925/6).
[5] Reger, D. L.; Dukes, M. D. (J. Organomet. Chem. **153** [1978] 67/72).
[6] Hartshorn, A. J.; Lappert, M. F.; Turner, K. (J. Chem. Soc. Dalton Trans. **1978** 348/56).
[7] Brix, H.; Beck, W. (J. Organomet. Chem. **234** [1982] 151/74).
[8] Richmond, T. G.; Shriver, D. F. (Organometallics **3** [1984] 305/14).
[9] Richmond, T. G.; Crespi, A. M.; Shriver, D. F. (Organometallics **3** [1984] 314/9).
[10] Markham, J.; Tolman, W.; Menard, K.; Cutler, A. (J. Organomet. Chem. **294** [1985] 45/58).
[11] Bodnar, T. W.; Cutler, A. R. (Organometallics **4** [1985] 1558/65).
[12] Himmel, S. E.; Young, G. B.; Fung, D. C. M.; Hollinshead, C. (Polyhedron **4** [1985] 349/56).
[13] Koola, J. D.; Roddick, D. M. (Organometallics **10** [1991] 591/7).

1.5.1.4.2 Compounds Containing the Fragment $^5LMo(CO)_3$ ($^5L \neq C_5H_5$)

1.5.1.4.2.1 Compounds Containing the Fragment $RC_5H_4Mo(CO)_3$

1.5.1.4.2.1.1 Mononuclear Complexes with Bonds between Molybdenum and Main Group Elements

The complexes dealt with under this heading are arranged in Table 6 as families of compounds: $M[RC_5H_4Mo(CO)_3]$ (obtained by Methods I to V), $RC_5H_4Mo(CO)_3H$ (obtained by Method VI), Mo−Ga-, Mo−In-, Mo−Sn-, Mo−P-, and Mo−Bi-bonded complexes $RC_5H_4Mo(CO)_3X$ (obtained by Methods VII and VIII), halides $RC_5H_4Mo(CO)_3X$ (obtained by Methods IX and X), M−C-bonded complexes $RC_5H_4Mo(CO)_3R'$ (obtained by Methods XI to XIII), $CH_3C_5H_4Mo(CO)_3{}^2D^{\cdot}$ (obtained by Method XIV), $[RC_5H_4Mo(CO)_3{}^2D]X$ (obtained by

References on pp. 253/8

Method XV), and polymer-attached $RC_5H_4Mo(CO)_3$ derivatives (obtained by Methods XVI and XVII). Within each family, the compounds are ordered according to increasing size and functionality of the cyclopentadienyl substituents R, the nature and position in the periodic table of the counterions M^+ or X^-, with the metal-bonded ligands X, R′ and ^2D being used as secondary criteria for sorting the various entries in Table 6. $[RC_5H_4Mo(CO)_3{}^2D]^+$ cations with accompanying $[RC_5H_4Mo(CO)_3]^-$ anions are dealt with under Nos. 11 to 14, 45, 47 to 51, and 54.

With regard to Method IV it should be noted that the $(CH_3C_5H_4Mo(CO)_3)_2$ dimer will disproportionate in photochemical reactions with amines (instead of phosphanes) only when the primary cationic product of this reaction, $[CH_3C_5H_4Mo(CO)_3NR_3]^+$, can undergo further reaction. By doing so, the cationic species is prevented from undergoing a fast back reaction with the anionic disproportionation product $[CH_3C_5H_4Mo(CO)_3]^-$, thereby reforming the parent homodimer. With pyridine, the initially formed $[CH_3C_5H_4Mo(CO)_3NC_5H_5]^+$ ion thus reacts to form $Mo(CO)_3(NC_5H_5)_3$ accompanied by $[C_5H_5NC_5H_4CH_3]^+$. With ethylene diamine and cyclohexylamine, photochemical net reactions also occur because the cationic disproportionation products, $[CH_3C_5H_4Mo(CO)_3NH_2R]^+$, react further to form carbamoyl complexes, $CH_3C_5H_4(CO)_2(NH_2R)C(O)NHR$, accompanied by $[NH_3R]^+$ (R = $C_2H_4NH_2$, C_6H_{11}-cyclo) [99]. The ratio of disproportionation to substitution products that formed was shown to be dependent (i) on the size and the electron-donating abilities of the entering ^2D ligands as well as on their concentration, (ii) on the polarity of the solvent used, and (iii) on the energy of the radiation employed. In nonpolar solvents, e.g., cyclohexane or benzene, phosphanes and phosphites with cone angles smaller than $P(C_6H_5)_3$ (θ = 145°) that are able to donate sufficient electron density, e.g., $P(OCH_3)_3$, $P(OC_2H_5)_3$, $P(OC_3H_7\text{-i})_3$, $P(OC_4H_9\text{-n})_3$, $P(OC_6H_5)_3$, $P(OC_6H_4CH_3\text{-4})_3$, $P(OC_6H_4OCH_3\text{-4})_3$, $P(C_2H_5)_3$, $P(C_4H_9\text{-n})_3$, $P(C_6H_5)_2CH_3$, $P(C_6H_5)_2C_4H_9$-n, and, to a lesser extent, also $P(C_4H_9\text{-i})_3$, will disproportionate the homodimer (especially at high concentrations) to produce the $[CH_3C_5H_4Mo(CO)_3]^-$ anion along with disubstituted cations, $[CH_3C_5H_4Mo(CO)_2(^2D)_2]^+$, irrespective of the radiation wavelength used (λ = 290, 366, 405, 505, or 525 nm) [44, 54, 69, 76].

With $P(C_6H_5)_3$ in benzene or cyclohexane, irradiation of $(CH_3C_5H_4Mo(CO)_3)_2$ at low energy ($\lambda > 366$ nm) resulted in substitution rather than disproportionation products. Disproportionation did, however, occur upon irradiation at higher energy (λ = 290 nm), $[CH_3C_5H_4Mo(CO)_2(P(C_6H_5)_3)_2][CH_3C_5H_4Mo(CO)_3]$ being formed under these conditions [35, 44, 76]. In the stronger polar CH_2Cl_2, on the other hand, disproportionation also came about for low-energy irradiation (λ = 525 nm), but then afforded $[CH_3C_5H_4Mo(CO)_3P(C_6H_5)_3]$-$[CH_3C_5H_4Mo(CO)_3]$ [76]. The dependence of the disproportionation reaction on solvent and radiation wavelength was shown to be a consequence of a facile back reaction, "$[CH_3C_5H_4Mo(CO)_3P(C_6H_5)_3]^+ + [CH_3C_5H_4Mo(CO)_3]^- \rightarrow (CH_3C_5H_4Mo(CO)_3)_2 + P(C_6H_5)_3$", proceeding too fast in nonpolar solvents to allow observation of the initially formed ions; i.e., no net disproportionation occurred under these conditions. Media of higher donicity, on the other hand, stabilized the primary disproportionation products with respect to recombination with the effect that net disproportionation was perceived in CH_2Cl_2 even for low-energy irradiation. Furthermore, high-energy irradiation at λ = 290 nm will also result in net disproportionation in nonpolar solvents such as benzene or cyclohexane because the disubstituted cation $[CH_3C_5H_4Mo(CO)_2(P(C_6H_5)_3)_2]^+$, produced under these conditions by a secondary photochemical ligand exchange reaction of the initially photogenerated $[CH_3C_5H_4Mo(CO)_3$-$P(C_6H_5)_3]^+$ cation, cannot back-react to give the starting homodimer [76].

Ligands with cone angles larger than $P(C_6H_5)_3$, e.g., $P(C_6H_5)_2C_3H_7$-i, $P(C_3H_7\text{-i})_3$, and $P(C_6H_{11}\text{-cyclo})_3$, are too bulky to take part in the secondary photosubstitution process induced by high-energy irradiation; instead, fast back reaction reforming $(CH_3C_5H_4Mo(CO)_3)_2$

References on pp. 253/8

will occur with the effect that no net disproportionation is encountered for these ligands at low- or high-energy irradiation in benzene or cyclohexane [44, 69, 76]. In a polar medium such as CH_2Cl_2, only the $[CH_3C_5H_4Mo(CO)_3]^-$ salts containing the monosubstituted cations $[CH_3C_5H_4Mo(CO)_3{}^2D]^+$ were obtained for these larger phosphanes upon irradiation at 290 or 525 nm, whereas salts with both mono- and disubstituted cations, $[CH_3C_5H_4Mo(CO)_3{}^2D][CH_3C_5H_4Mo(CO)_3]$ (favored under CO atmosphere) and $[CH_3C_5H_4Mo(CO)_2({}^2D)_2][CH_3C_5H_4Mo(CO)_3]$ (favored under N_2 atmosphere), were formed under these conditions for most of the aforementioned ligands smaller than $P(C_6H_5)_3$ [76]. Only disproportionation to $[CH_3C_5H_4Mo(CO)_2((C_6H_5)_2PC_2H_4P(C_6H_5)_2)][CH_3C_5H_4Mo(CO)_3]$ and not a competing substitution of $(CH_3C_5H_4Mo(CO)_3)_2$ occurred when the homodimer was irradiated at $\lambda = 405$ nm in benzene [44] or was exposed to room light in CH_3CN [98] in the presence of 1,2-diphenylphosphinoethane. Irradiation ($\lambda > 405$ nm) of $(CH_3C_5H_4Mo(CO)_3)_2$ in benzene containing added bis(2-diphenylphosphinoethyl)phenylphosphane similarly gave $[CH_3C_5H_4Mo(CO)_2P(C_2H_4P(C_6H_5)_2)_2C_6H_5][CH_3C_5H_4Mo(CO)_3]$ [52].

Irradiation at $\lambda > 525$ nm of the dicationic Mo_2 complexes $[(H_3NCH_2CH_2C_5H_4Mo(CO)_3)_2][X]_2$ in THF in the presence of $P(C_6H_5)_3$ ($X^- = PF_6^-$) or in aqueous solution containing the water-soluble phosphanes phosphatriazaadamantane, $PN_3(CH_2)_6$, or 2,3-bis(diphenylphosphanyl)maleic acid $((C_6H_5)_2)_2PC_2(CO_2H)_2$, together with an added pH 7 buffer ($X^- = NO_3^-$) likewise proceeded with the formation of $\{[H_3NCH_2CH_2C_5H_4Mo(CO)_{4-n}(P(C_6H_5)_3)_n][PF_6]\}^+$, $\{[H_3NCH_2CH_2C_5H_4Mo(CO)_{4-n}(PN_3(CH_2)_6)_n]NO_3\}^+$ (n = 1, 2), or $\{[H_3NCH_2CH_2C_5H_4Mo(CO)_2(P(C_6H_5)_2)_2C_2(CO_2H)_2]NO_3\}^+$, and $\{[H_3NCH_2CH_2C_5H_4Mo(CO)_3]X\}^-$ as products resulting from disproportionation chemistry [144, 154, 157]. In the case of the $P(C_6H_5)_3$-assisted photoreaction, the substituted binuclear derivative $[(H_3NCH_2CH_2C_5H_4Mo)_2(CO)_5P(C_6H_5)_3][PF_6]_2$ was observed as an accompanying species [144, 157]. Similarly, products containing $[RC_5H_4Mo(CO)_2(P(OC_2H_5)_3)_2]^+$ and $[RC_5H_4Mo(CO)_3]^-$ ions with functionalized cyclopentadienyl ligands were generated by irradiating the required homodimers $(RC_5H_4Mo(CO)_3)_2$ (R = $HOCH_2CH_2$, $CH_3(CH_2)_5NHC(O)OCH_2CH_2$, $HOCH_2C(O)$) in THF in the presence of triethyl phosphite [131]. Furthermore, irradiation of $(C_{14}H_7O_2{-}C(O)NHCH_2CH_2C_5H_4Mo(CO)_3)_2$ ($C_{14}H_7O_2$ = anthraquinon-2-yl) with low-energy light ($\lambda > 525$ nm) in THF in the presence of $P(C_6H_5)_3$ or $(C_6H_5)_2PC_2H_4P(C_6H_5)_2$ gave net disproportionation products, including $[C_{14}H_7O_2{-}C(O)NHCH_2CH_2C_5H_4Mo(CO)_3]^-$ and, as counterions, $[C_{14}H_7O_2{-}C(O)NHCH_2CH_2C_5H_4Mo(CO)_3P(C_6H_5)_3]^+$, $[C_{14}H_7O_2{-}C(O)NHCH_2CH_2C_5H_4Mo(CO)_2(P(C_6H_5)_3)_2]^+$, and $[C_{14}H_7O_2{-}C(O)NHCH_2CH_2C_5H_4Mo(CO)_2((C_6H_5)_2PC_2H_4P(C_6H_5)_2)]^+$, respectively [177]. Similar irradiations in acetone at room temperature did not yield net disproportionation because of the facile back reaction "$[CH_3C_5H_4Mo(CO)_3OC(CH_3)_2]^+ + [CH_3C_5H_4Mo(CO)_3]^- \rightarrow (CH_3C_5H_4Mo(CO)_3)_2 + (CH_3)_2CO$". This back reaction did not take place at low temperatures, and disproportionation of the homodimer with the formation of $[CH_3C_5H_4Mo(CO)_3OC(CH_3)_2][CH_3C_5H_4Mo(CO)_3]$ was observed upon 405 nm irradiation in neat acetone at −78 °C [62]. Photolysis with visible light ($\lambda > 525$ nm) of $(H_2NCH_2CH_2C_5H_4Mo(CO)_3)_2$ in THF yielded the internally amine-chelated cation $[C_5H_4CH_2CH_2NH_2Mo(CO)_3]^+$ together with $[H_2NCH_2CH_2C_5H_4Mo(CO)_3]^-$ as disproportionation products, which in the dark back-reacted to reform the parent Mo_2 complex within minutes [157]. Irradiation of the cationic homodimers $[(H_3NCH_2CH_2C_5H_4Mo(CO)_3)_2][X]_2$ in CH_3CN or THF ($X^- = PF_6^-$) or in water buffered at pH 7 ($X^- = NO_3^-$) similarly resulted in effective net disproportionation to give $\{[H_3NCH_2CH_2C_5H_4Mo(CO)_3{}^2D][PF_6]\}^+$ (${}^2D$ = CH_3CN, THF) or $\{[H_3NCH_2CH_2C_5H_4Mo(CO)_3OH_2]NO_3\}^+$ together with $\{[H_3NCH_2CH_2C_5H_4Mo(CO)_3]X\}^-$ as the first formed products [144, 157].

Method I: From $Mo(CO)_6$ and $[RC_5H_4]^-$ by substitution:

a. The hexacarbonyl was refluxed with the required lithium or sodium salt $M[RC_5H_4]$ in THF [1, 8, 17, 19, 22, 33, 34, 36, 38, 40, 42, 86, 91, 94, 130,

References on pp. 253/8

137, 144, 147, 151, 158 to 161, 171, 176], $CH_3OC_2H_4OCH_3$ [41, 104, 106], or $CH_3OC_2H_4OC_4H_4OCH_3$ [12, 83, 112, 135, 147] until no unreacted starting material was present. In general, the $[RC_5H_4Mo(CO)_3]^-$ anions so generated (R = CH_3 [1, 8, 12, 19, 83, 171], C_2H_5 [19, 171], i-C_3H_7 [86, 112], n-C_4H_9 [171], $C_6H_5CH_2$ [12], $C_6H_5C(CH_3)H$ [94], $RC_6H_4C(R')R''$ (R = H: R'/R'' = CH_3/CH_3, CH_3/C_2H_5, CH_3/n-C_3H_7, C_2H_5/C_2H_5, -$(CH_2)_5$-, -$(CH_2)_6$-; R = 3-CH_3: R'/R'' = CH_3/CH_3, CH_3/C_2H_5, -$(CH_2)_6$-; R = 4-CH_3, 4-OCH_3: R'/R'' = CH_3/CH_3, CH_3/C_2H_5) [160], CH_2=CH [34, 36], $CH_2=C(CH_3)$ [34, 42], $(CH_3)_2NC(CH_3)H$ [42], $H_2NCH_2CH_2$ [144], $(CH_3)_2NCH(R)CH_2$ (R = H, CH_3) [137], $CH_3OC(CH_3)_2OCH_2CH_2$ [130], HC(O) [17, 22, 38], $CH_3C(O)$ [17, 22, 38, 91, 135], $CH_3OC(CH_3)_2$-$OCH_2C(O)$ [95, 130], $CH_3OC(O)$ [17, 22, 38, 42, 135, 147, 176], $C_2H_5OC(O)$ [135, 147, 176], $(CH_3)_3Si$ [40], $(CH_3)_3Si(CH_3)_2Si$ [158, 159, 161], $C_6H_5N_2$ [151], $(C_6H_5)_2P$, $(4-CH_3C_6H_4)_2P$ [33]) were immediately employed for further syntheses without purification, either as solutions or, after evaporation of solvent, as solid residues. $Na[CH_3C_5H_4Mo(CO)_3]$ was isolated as a solvate with one molecule of 1,2-dimethoxyethane, $Na[CH_3C_5H_4Mo(CO)_3] \cdot CH_3OC_2H_4OCH_3$, by adding pentane to the reaction solution and cooling the mixture in dry ice as described in detail for the C_5H_5 analogue [41, 104, 106].

b. The ions $[RC_5H_4]^-$ (R = $CH_3C(O)$ or $CH_3OC(O)$) were generated by electroreduction on a mercury electrode of $(RC_5H_4)_2Fe$, either in THF or N,N-dimethylformamide containing added $[N(C_4H_9\text{-}n)_4]PF_6$ as the supporting electrolyte. The mixtures so obtained were filtered to remove metallic iron and then treated with $Mo(CO)_6$ at reflux temperature to furnish stock solutions of $[N(C_4H_9\text{-}n)_4][RC_5H_4Mo(CO)_3]$, suitable for further use [21].

Method II: The indenyl- and fluorenyl-functionalized cyclopentadienyl anions $[C_9H_7$-$C(CH_3)_2C_5H_4]^-$ and $[C_{13}H_9C(CH_3)_2C_5H_4]^-$ (generated as their lithium salts from the parent monosubstituted cyclopentadienes and one equivalent of $LiCH_3$ or LiC_4H_9-n in THF at 0 °C) were allowed to interact with $Mo(CO)_3$-$(NCCH_3)_3$ in refluxing THF to give the corresponding metalates, $Li[C_9H_7C(CH_3)_2C_5H_4Mo(CO)_3]$ and $Li[C_{13}H_9C(CH_3)_2C_5H_4Mo(CO)_3]$ (C_9H_7 = indenyl, $C_{13}H_9$ = fluorenyl), which were not isolated but immediately converted into neutral methyl derivatives [150]. When indenyl lithium was reacted with 6-(dimethylamino)fulvene in THF at 50 °C, the (benzopentafulven-1-yl)cyclopentadienide $Li[C_9H_6=CHC_5H_4]$ was formed, which was reacted further with one equivalent of $Mo(CO)_3(NCCH_3)_3$ in THF to produce the $[C_9H_6=CHC_5H_4Mo(CO)_3]^-$ anion. The latter was isolated as its tetrabutyl phosphonium salt, $[P(C_4H_9\text{-}n)_4][C_9H_6=CHC_5H_4Mo(CO)_3]$, by precipitation from the ethanolic solution of the evaporated mixture with $[P(C_4H_9\text{-}n)_4]Br$ in C_2H_5OH. For further purification, the crude precipitate was recrystallized from THF/diethyl ether [138]. Similarly, refluxing $Na[HC(O)C_5H_4]$ and Mo-$(CO)_3(NCCH_3)_3$ in 1 : 1 stoichiometry in THF and subsequent removal of solvent from the filtered mixture left crude $Na[HC(O)C_5H_4Mo(CO)_3]$ as an oil which solidified on stirring it vigorously with diethyl ether until all oily contaminants had dissolved [46]. Stirring $Mo(CO)_3(NCCH_3)_3$ in the presence of an equimolar amount of in situ-prepared $Na[C_6H_5OSi(CH_3)_2C_5H_4]$ in THF at room temperature formed the $[C_6H_5OSi(CH_3)_2C_5H_4Mo(CO)_3]^-$ anion, which was converted directly to $C_6H_5OSi(CH_3)_2C_5H_4Mo(CO)_3CH_3$ [124]. Preparing $Li[(C_6H_5)_2PC_5H_4Mo(CO)_3]$ from $Li[(C_6H_5)_2PC_5H_4]$ and $Mo(CO)_3(NCCH_3)_3$,

References on pp. 253/8

rather than from the lithium reagent and $Mo(CO)_6$ (see Method Ia), enabled the salt to be obtained at room temperature, rather than by prolonged reflux, thus giving a purer solution of the molybdate anion [139, 140].

Method III: From $(RC_5H_4Mo(CO)_3)_2$ derivatives by reduction:

a. The requisite homobimetallic $(RC_5H_4Mo(CO)_3)_2$ was added to sodium shot [2, 3, 119] or sodium amalgam [18, 28, 40, 53, 59], suspended in THF, and the mixture was stirred at room temperature [18, 28, 40, 119] or refluxed gently [2, 3] until the characteristic red color of the dimer had faded. The resulting pale yellow to virtually colorless solutions of $Na[RC_5H_4Mo(CO)_3]$ (R = CH_3 [18, 28, 40, 119], n-C_3H_7, i-C_3H_7, CH_3C-$(C_2H_5)H$, $CH_3C(C_3H_7$-n)H, $(C_2H_5)_2CH$, 4-$CH_3OC_6H_4CH_2$, $C_6H_5C(CH_3)H$, $(C_6H_5)_2CH$, cyclo-C_6H_{11} [2, 3]) obtained after decanting from the excess of sodium or sodium amalgam were subsequently used without further purification. The salts $Na[HO(CH_2)_nC_5H_4Mo(CO)_3]$ (n = 2, 3) were isolated from the evaporated mixtures by redissolving the residues in THF and precipitating the products with ether [53, 59].

b. Mixtures of $(CH_3C_5H_4Mo(CO)_3)_2$ or $[(H_3NCH_2CH_2C_5H_4Mo(CO)_3)_2][PF_6]_2$ and excess sodium/potassium alloy were stirred in THF, generating solutions of the potassium salts of the respective $[RC_5H_4Mo(CO)_3]^-$ anions (R = CH_3 [14], $H_2NCH_2CH_2$ [157]), useful for subsequent transformations.

Method IV: From $(RC_5H_4Mo(CO)_3)_2$ by photodisproportionation (see also introductory remarks):

a. Irradiation of $(CH_3C_5H_4Mo(CO)_3)_2$ in the presence of ^2D ligands such as phosphanes or phosphites resulted in disproportionation and formation of $[CH_3C_5H_4Mo(CO)_3{}^2D][CH_3C_5H_4Mo(CO)_3]$ and/or $[CH_3C_5H_4Mo(CO)_2$-$(^2D)_2][CH_3C_5H_4Mo(CO)_3]$, in addition to substitution giving $(CH_3C_5H_4Mo)_2$-$(CO)_5{}^2D$ and/or $(CH_3C_5H_4Mo)_2(CO)_4(^2D)_2$.

b. Irradiation (λ = 405 nm) of $(CH_3C_5H_4Mo(CO)_3)_2$ in neat CH_3CN or $(CH_3)_2SO$ formed $[CH_3C_5H_4Mo(CO)_3NCCH_3][CH_3C_5H_4Mo(CO)_3]$ and $[CH_3C_5H_4$-$Mo(CO)_3OS(CH_3)_2][CH_3C_5H_4Mo(CO)_3]$, respectively [62].

c. Solutions containing the salts $[ER_4][CH_3C_5H_4Mo(CO)_3]$ (E = N: R = n-C_4H_9; E = P: R = n-C_4H_9, C_6H_5) along with the halo complexes $CH_3C_5H_4Mo(CO)_3X$ (X = Cl, Br, I) were formed when $(CH_3C_5H_4Mo(CO)_3)_2$ was irradiated at λ = 405 and 505 nm in acetonitrile, acetone, dimethyl sulfoxide, or THF in the presence of halides X^-, added as $[N(C_4H_9$-n$)_4]X$ (X^- = Cl^-, Br^-, I^-), $[P(C_4H_9$-n$)_4]X$ (X^- = Cl^-, Br^-), or $[P(C_6H_5)_4]Br$ [62, 77]. For quantum yields and mechanisms of the various photodisproportionation processes, see the original literature [44, 52, 54, 62, 77, 131, 144, 157].

Method V: A saturated aqueous solution of $Na[HC(O)C_5H_4Mo(CO)_3]$ was precipitated by adding $[N(CH_3)_4]Cl$ or $[P(C_6H_5)_4]Br$. The precipitates were washed with water and ether and recrystallized from water or acetonitrile if necessary [46].

Method VI: The required lithium or sodium salt, $Li[C_{13}H_9C(CH_3)_2C_5H_4Mo(CO)_3]$ ($C_{13}H_9$ = fluorenyl) [111] or $Na[RC_5H_4Mo(CO)_3]$ (R = CH_3 [1, 24, 28], $HOCH_2CH_2$, $HOCH_2CH_2CH_2$ [53]), was treated with one equivalent of glacial acetic acid [1, 53], aqueous CH_3CO_2H [111], or p-toluenesulfonic acid [24, 28] in THF.

Representative procedures employed for isolation and purification of the hydrido products so generated were as follows:

R = CH_3: Solvent was removed and $CH_3C_5H_4Mo(CO)_3H$ separated by vacuum distillation as an oil [1, 24, 28];

R = $C_{13}H_9C(CH_3)_2$, where $C_{13}H_9$ = fluorenyl: the aqueous phase remaining after evaporation of volatile materials was extracted with pentane and then subjected to column chromatography on silica gel, eluting with pentane/toluene (5:1); final purification was achieved by crystallization from pentane at −78 °C [111];

R = $HOCH_2CH_2$ and $HOCH_2CH_2CH_2$: The residue remaining after evaporation was extracted with benzene; removal of volatile material from the extracts left the products as oils [53].

Solutions of $(C_6H_5)_2PC_5H_4Mo(CO)_3H$ were obtained similarly by adding one to three equivalents of acetic acid to pure $Li[(C_6H_5)_2PC_5H_4Mo(CO)_3]$ in toluene [152, 156]. Adding excess CF_3CO_2H to a THF solution of in situ-prepared $K[H_2NCH_2CH_2C_5H_4Mo(CO)_3]$ resulted in protonation of both the amino group and the molybdenum atom to form $[H_3NCH_2CH_2C_5H_4Mo(CO)_3H]$-$[CF_3CO_2]$ [157].

Method VII: $CH_3C_5H_4Mo(CO)_3I$ was treated in hot toluene with "GaI", generated by sonication of gallium metal with I_2 in the same solvent. Crystallization of the product from diethyl ether furnished crystalline $CH_3C_5H_4Mo(CO)_3GaI_2O$-$(C_2H_5)_2$ [113]. $CH_3C_5H_4Mo(CO)_3InCl_2$ was isolated from a similar reaction between equimolar amounts of InCl and $CH_3C_5H_4Mo(CO)_3Cl$ in toluene at 75 °C; the product was separated from the evaporated mixture in low yield by extraction into CH_2Cl_2, followed by solvent diffusion using hexane at −30 °C [128].

Method VIII: From $[RC_5H_4Mo(CO)_3]^-$ and $SnX_{4-n}R'_n$ or PCl_3 by substitution:

a. $CH_3C_5H_4Mo(CO)_3Sn(C_6H_5)_3$ was prepared by allowing $Na[CH_3C_5H_4Mo$-$(CO)_3]$ to react with $Sn(C_6H_5)_3Cl$ in THF at room temperature. Removal of solvent followed by extraction of the residue with pentane and renewed evaporation to dryness left an oil which solidified when treated with methanol. Further purification was achieved by recrystallization from CH_2Cl_2/light petroleum (1:1) [9, 57]. Dropwise addition of $Sn(CH_3)_2Cl_2$, dissolved in H_2O, to an aqueous solution containing an equimolar amount of Na-$[HC(O)C_5H_4Mo(CO)_3]$ caused the immediate precipitation of HC(O)-$C_5H_4Mo(CO)_3Sn(CH_3)_2Cl$, which was purified by crystallization from CH_2Cl_2/hexane [73].

b. Freshly prepared $Na[CH_3OC(O)C_5H_4Mo(CO)_3]$ was suspended in methylcyclohexane and allowed to interact with a slight excess of PCl_3 between −10 and −5 °C. After warming to room temperature, solvent was removed and the residue extracted into pentane. Cooling the concentrated extracts to −78 °C gave crystalline $CH_3OC(O)C_5H_4Mo(CO)_3PCl_2$ [135].

Method IX: From $[RC_5H_4Mo(CO)_3]^-$ or $RC_5H_4Mo(CO)_3H$ by halogenation:

a. Stirring equimolar amounts of $Na[RC_5H_4Mo(CO)_3]$ (R = $(CH_3)_2NCH_2CH_2$, $(CH_3)_2NCH(CH_3)CH_2$, HC(O), $(C_6H_5)_2P$) and I_2 in THF afforded the corresponding iodo complexes, $RC_5H_4Mo(CO)_3I$. The (dimethylamino)ethyl-substituted compounds $(CH_3)_2NCH(R)CH_2C_5H_4Mo(CO)_3I$ (R = H, CH_3)

References on pp. 253/8

were isolated by flash chromatography of solutions in CH_2Cl_2/CH_3OH (9:1) through silica gel, the acidity of which transformed the crude products into their ammonium salts. Deprotonation of the salts with formation of the pure iodides was achieved by washing CH_2Cl_2 solutions with aqueous K_2CO_3 [137]. $HC(O)C_5H_4Mo(CO)_3I$ was separated from the evaporated reaction mixture by extraction into CH_2Cl_2, addition of heptane, and cooling to −30 °C [73]. $(C_6H_5)_2PC_5H_4Mo(CO)_3I$ was purified by thin-layer chromatography on silica plates, eluting with CH_2Cl_2/hexane (1:3) [139]. $C_2H_5OC(O)C_5H_4Mo(CO)_3I$ was obtained similarly from $Na[C_2H_5OC(O)-C_5H_4Mo(CO)_3]$ and equimolar I_2 in $CH_3OC_2H_4OC_2H_4OCH_3$ [135].

b. The sodium salts $Na[RC_5H_4Mo(CO)_3]$ (R = $CH_3C(O)$, $CH_3OC(O)$, C_2H_5-OC(O)) were treated with phosphorus trihalides, phenyl dichlorophosphane, or methyl diiodoarsane (each reagent employed in slight excess in $CH_3OC_2H_4OC_2H_4OCH_3$ at ambient conditions) to give the corresponding halo complexes, $CH_3C(O)C_5H_4Mo(CO)_3X$ (X = Cl, Br, I), $CH_3OC(O)-C_5H_4Mo(CO)_3X$ (X = Cl, Br, I), and $C_2H_5OC(O)C_5H_4Mo(CO)_3X$ (X = Cl, Br) by thermal decomposition of the initially formed intermediates, $RC_5H_4Mo(CO)_3EX_2$ (E = P, As) [135].

c. Treatment of the lithium salts $Li[RC_5H_4Mo(CO)_3]$ (R = $3-CH_3C_6H_4-C(CH_3)_2$, $4-CH_3C_6H_4C(CH_3)_2$, $4-CH_3OC_6H_4C(CH_3)_2$, $C_6H_5C(C_2H_5)_2$, $(CH_3)_3-Si(CH_3)_2Si$) in THF first with acetic acid and then with CCl_4 or N-bromosuccinimide formed the halides $(CH_3)_3Si(CH_3)_2SiC_5H_4Mo(CO)_3X$ (X = Cl, Br) [148, 158, 159, 161], $R'C_6H_4C(CH_3)_2C_5H_4Mo(CO)_3Cl$ (R' = 3-CH_3, 4-CH_3), $R'C_6H_4C(CH_3)_2C_5H_4Mo(CO)_3Br$ (R' = 3-CH_3, 4-OCH_3), and $C_6H_5C(C_2H_5)_2C_5H_4Mo(CO)_3Br$, of which the latter five were isolated and purified by crystallization from $CHCl_3/CH_3OH$ (1:4) [160]. Prolonged stirring of THF solutions of $Li[(CH_3)_3Si(CH_3)_2SiC_5H_4Mo(CO)_3]$ with 1,2-dihaloethanes, XC_2H_4X, at ambient conditions likewise produced the halo compounds $(CH_3)_3Si(CH_3)_2SiC_5H_4Mo(CO)_3X$ (X = Cl, Br, I), which in this case were separated from the reaction mixtures by column chromatography on Al_2O_3, eluting with CH_2Cl_2/petroleum ether 1:19 [159]. (cyclo-$(t-C_4H_9)_3C_3)C_5H_4Mo(CO)_3Cl$ was made similarly by reacting (cyclo-$(t-C_4H_9)_3C_3)C_5H_4Mo(CO)_3H$ with CCl_4 [6].

Method X: From $(RC_5H_4Mo(CO)_3)_2$ by oxidative cleavage:

a. The ring-substituted dimer $(RC_5H_4Mo(CO)_3)_2$ in CH_2Cl_2 or $CHCl_3$ was treated dropwise with iodine in the same solvent. Upon completion of the reaction (monitored by IR spectroscopy), the mixture was shaken with aqueous $Na_2[S_2O_3]$ and the products, $RC_5H_4Mo(CO)_3I$, were isolated from the organic layer by removing the volatiles (R = CH_3, t-C_4H_9 [146]) or were crystallized from light petroleum [2, 3] or hexane/$CHCl_3$ (1:1) [21] (R = i-C_3H_7, $CH_3C(C_2H_5)H$, $CH_3C(C_3H_7-n)H$, $(C_2H_5)_2CH$, $4-CH_3OC_6H_4-CH_2$, $C_6H_5C(CH_3)H$ [2, 3], $CH_3C(O)$, $CH_3OC(O)$ [21]). The iodo complexes $RC_5H_4Mo(CO)_3I$ with R = $(CH_3)_3E$ (E = C, Si), $C_6H_5(CH_3)_2Si$, and $CH_3(C_6H_5)_2Si$ [5] or $(CH_3)_3Si(CH_3)_2Si$ [148, 159, 161] were prepared analogously.

b. Quantitative yields of $CH_3C_5H_4Mo(CO)_3Cl$ were attained when $(CH_3-C_5H_4Mo(CO)_3)_2$ was irradiated with visible light, either in CCl_4 ($\lambda >$ 490, 545, or 560 nm) or in solvent systems such as acetone/CCl_4 ($\lambda >$ 490 nm) or benzene/C_2Cl_6 ($\lambda >$ 540 nm) [48]. The chloro complexes

$RC_5H_4Mo(CO)_3Cl$ with R = $H_3NCH_2CH_2$ (PF_6^- salt) [144, 157], $C_{14}H_7O_2$–C(O)$NHCH_2CH_2$ ($C_{14}H_7O_2$ = anthraquinon-2-yl) [177], $HOCH_2CH_2$, CH_3-$(CH_2)_5NHC(O)OCH_2CH_2$, and $HOCH_2C(O)$ [131] as well as the iodo derivative $C_{14}H_7O_2$–$C(O)NHCH_2CH_2C_5H_4Mo(CO)_3I$ [177] were formed analogously by irradiation at $\lambda > 525$ nm of the requisite homodimers $(RC_5H_4Mo(CO)_3)_2$ in carbon tetrachloride [131] or in THF solutions containing added CCl_4 [144, 157, 177] or CH_3I [177]. Trapping of the primary photolysis product $\{[H_3NCH_2CH_2C_5H_4Mo(CO)_3][PF_6]\}^{\cdot}$ with $CHCl_3$ was markedly slower than with CCl_4, allowing the disproportionation of the parent homodimer to $[C_5H_4CH_2CH_2NH_2Mo(CO)_3]PF_6$ and $[H_3NCH_2CH_2$-$C_5H_4Mo(CO)_3H]PF_6$ to become competitive [157]. Similarly, irradiation of $[(H_3NCH_2CH_2C_5H_4Mo(CO)_3)_2](NO_3)_2$ and Cl_3CCH_2OH in aqueous solution at pH 7 or pH 2 resulted in the clean formation of $[H_3NCH_2CH_2$-$C_5H_4Mo(CO)_3Cl]NO_3$ [144].

Method XI: From $[RC_5H_4Mo(CO)_3]^-$ and R′X by nucleophilic substitution:

a. The requisite alkyl halide, R′Cl (R′ = $CH_2C_6H_5$ [58]), R′Br (R′ = CH_2-C_6H_5, $CD_2C_6D_5$ [18, 28]), or R′I (R′ = CH_3 [2, 3, 8, 17, 19, 22, 28, 38, 86, 111, 124, 150, 151, 161], CD_3 [18, 28], C_2H_5 [2, 3, 19]), was allowed to interact in THF solution with the required lithium or sodium salt $M[RC_5H_4Mo(CO)_3]$, either at ambient conditions [17, 18, 22, 28, 38, 58, 111, 124, 150, 151, 161], at 40 °C [8, 19], or at reflux temperature [2, 3, 86] to give the desired alkyl $RC_5H_4Mo(CO)_3R'$ (R = CH_3: R′ = CH_3 [8, 18, 19, 28], CD_3 [18, 28], C_2H_5 [19], $CH_2C_6H_5$ [18, 28, 58], $CD_2C_6D_5$ [18, 28]; R = C_2H_5: R′ = CH_3, C_2H_5 [19]; R = i-C_3H_7: R′ = CH_3 [2, 3, 86], C_2H_5 [2, 3]; R = $(C_2H_5)_2CH$: R′ = CH_3, C_2H_5 [2, 3]; R = $C_6H_5C(CH_3)H$: R′ = CH_3 [2, 3]; R = $C_9H_7C(CH_3)_2$ (C_9H_7 = indenyl), $C_{13}H_9C(CH_3)_2$ ($C_{13}H_9$ = fluorenyl): R′ = CH_3 [111, 150]; R = HC(O), $CH_3C(O)$, $CH_3OC(O)$: R′ = CH_3 [17, 22, 38]; R = $C_6H_5OSi(CH_3)_2$: R′ = CH_3 [124]; R = $(CH_3)_3Si(CH_3)_2Si$: R′ = CH_3 [161]; R = $C_6H_5N_2$: R′ = CH_3 [151]). Reported procedures used for separation and purification of the products included: high-vacuum sublimation or distillation of the residues left after removal of volatile material, giving pure $RC_5H_4Mo(CO)_3R'$ (R = CH_3: R′ = CH_3 [8, 18, 28], CD_3 [18, 28]; R = i-C_3H_7: R′ = CH_3 [2, 3, 86], C_2H_5 [2, 3]; R = $(C_2H_5)_2CH$: R′ = CH_3, C_2H_5 [2, 3]; R = $C_6H_5C(CH_3)H$: R′ = CH_3 [2, 3]) (also used after chromatography (see below) for obtaining analytical samples of $RC_5H_4Mo(CO)_3CH_3$ (R = HC(O), $CH_3C(O)$, and $CH_3OC(O)$ [38])); column chromatography on alumina, eluting with hexane or petroleum ether (used during workup of $CH_3C_5H_4MoCH_2C_6H_5$ [58] and $(CH_3)_3Si(CH_3)_2SiC_5H_4Mo(CO)_3CH_3$ [161]) or, respectively, with pentane/ether or hexane/ether mixtures (to afford the alkyls CH_3-$C_5H_4Mo(CO)_3R'$ and $C_2H_5C_5H_4Mo(CO)_3R'$ (R′ = CH_3, C_2H_5 [19]) as well as the functionally substituted methyl derivatives $RC_5H_4Mo(CO)_3CH_3$ with R = HC(O), $CH_3C(O)$, and $CH_3OC(O)$ [38]); column chromatography on silica gel, eluting with benzene (employed for $CH_3C_5H_4Mo(CO)_3R'$ with R′ = $CH_2C_6H_5$ and $CD_2C_6D_5$ [18, 28]) or hexane/toluene (2:1) (used to separate and purify $C_6H_5OSi(CH_3)_2C_5H_4Mo(CO)_3CH_3$ [124]); column chromatography (SiO_2, pentane/toluene (5:1)) with subsequent low-temperature crystallization from pentane or CH_2Cl_2/pentane (20:1), providing pure samples of the indenyl- and fluorenyl-functionalized derivatives $C_9H_7C(CH_3)_2C_5H_4Mo(CO)_3CH_3$ and $C_{13}H_9C(CH_3)_2C_5H_4Mo(CO)_3CH_3$

References on pp. 253/8

[111, 150]; chromatography, followed by recrystallization from petroleum ether, giving $C_6H_5N_2C_5H_4Mo(CO)_3CH_3$ [151]. In one of the syntheses leading to i-$C_3H_7C_5H_4Mo(CO)_3CH_3$, a solution of Na[i-$C_3H_7C_5H_4Mo(CO)_3$] in $CH_3OC_2H_4OC_2H_4OCH_3$ was treated with excess CH_3I at 30 °C. After removal of the volatiles, the residue was extracted with hot light petroleum. The filtered extracts were then reduced to a mobile oil and residual $CH_3OC_2H_4OC_2H_4OCH_3$ was removed in vacuum to leave the alkyl as an analytically pure product [112].

b. A THF solution of [N(C_4H_9-n)$_4$][$RC_5H_4Mo(CO)_3$] (R = $CH_3C(O)$, CH_3-OC(O)), prepared according to Method Ib, was allowed to react with excess methyl iodide at room temperature. Volatile materials were then removed and the residue was extracted with diethyl ether. Evaporation to dryness followed by crystallization of the residue from diethyl ether/hexane (1:1) furnished the pure methyl products $RC_5H_4Mo(CO)_3CH_3$ [21]. C_5H_4=$CHC_5H_4Mo(CO)_3CH_3$ resulted from stirring the phosphonium salt [P(C_6H_5)$_4$][C_5H_4=$CHC_5H_4Mo(CO)_3$] with CH_3I in THF. The product was extracted from the evaporated mixture with ether and then chromatographed on a silica gel column, eluting with hexane/ether (10:1). For further purification, the salt was crystallized from hexane at −30 °C [110]. The (benzopentafulvene-1-yl)cyclopentadienyl analogue C_9H_6=CHC_5H_4-$Mo(CO)_3CH_3$ was made similarly by stirring [P(C_4H_9-n)$_4$][C_9H_6=CH-$C_5H_4Mo(CO)_3$] with equimolar methyl iodide in THF. The methyl complex was separated by column chromatography on Al_2O_3, eluting with n-pentane/ether (95:5) and subsequently recrystallized from hexane [138].

Method XII: $Mo(CO)_3(CH_3OC_2H_4OC_2H_4OCH_3)$, prepared in situ by heating $Mo(CO)_6$ with $CH_3OC_2H_4OC_2H_4OCH_3$ in benzene, was suspended in hexane and treated with excess cyclo-$C_5H_4R_2$-5,5 (R/R = CH_3/CH_3, C_2H_5/C_2H_5, CH_3/C_2H_5) at 60 to 70 °C. Evaporation of the solvent and chromatography of the residues on alumina columns using hexane/ether mixtures (10:1) as eluants afforded $CH_3C_5H_4Mo(CO)_3CH_3$, $C_2H_5C_5H_4Mo(CO)_3C_2H_5$, and $C_2H_5C_5H_4Mo(CO)_3CH_3$, but none of the isomeric $CH_3C_5H_4Mo(CO)_3C_2H_5$ [19]. Only $CH_3C_5H_4Mo$-$(CO)_3CH_3$ was isolated from the competitive reaction of $Mo(CO)_3(CH_3OC_2H_4$-$OC_2H_4OCH_3)$ with a 1:1 mixture of cyclo-$C_5H_4(CH_3)_2$ and cyclo-$C_5H_4(C_2H_5)_2$ [19]. Similar procedures allowing $Mo(CO)_3(CH_3OC_2H_4OC_2H_4OCH_3)$ or Mo-$(CO)_3(NCCH_3)_3$ to interact with spirocyclic pentadienes, such as spiro[2,4]-hepta-4,6-diene (Formula I, R = H), or spiro[4,4]nona-1,3-diene (Formula II), in hexane at 40 °C gave alkane-α,ω-diyl-bridged products, $C_5H_4(CH_2)_n$-$Mo(CO)_3$ (n = 2, 4), isolated and purified by chromatography (Al_2O_3/hexane) and subsequent crystallization from hexane at −60 °C [8, 72]. Stirring Mo-$(CO)_3(CH_3OC_2H_4OC_2H_4OCH_3)$ with 1-methyl-spiro[2,4]hepta-4,6-diene (Formula I, R = CH_3), in THF between 0 °C and room temperature furnished the isomeric complexes $C_5H_4CH(CH_3)CH_2Mo(CO)_3$ and $C_5H_4CH_2CH(CH_3)$-$Mo(CO)_3$ as a 2:1 mixture, from which the first one could be separated by repeated column chromatography on Al_2O_3 eluting with hexane, followed by low-temperature crystallization from the same solvent [19]. Combination of $Mo(CO)_3(CH_3OC_2H_4OC_2H_4OCH_3)$ with 1-vinyl-spiro[2,4]hepta-4,6-diene (Formula I, R = $CH{=}CH_2$), in THF at room temperature and subsequent chromatography on alumina with hexane/ether (4:1) eluant yielded the butenyl-bridged species $C_5H_4CH_2CH{=}CHCH_2Mo(CO)_3$ together with the acyl $C_5H_4CH_2C({=}CHCH_3)C(O)Mo(CO)_3$, which could be separated and purified further by recrystallization from hexane [7, 19].

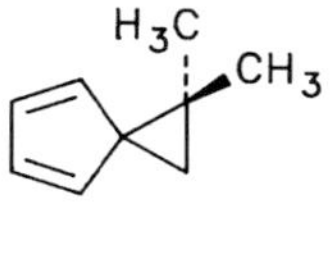

III 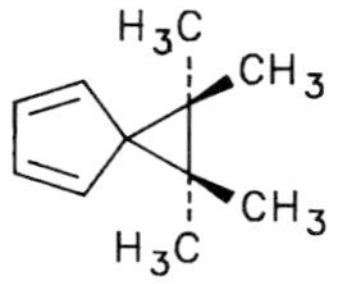 IV 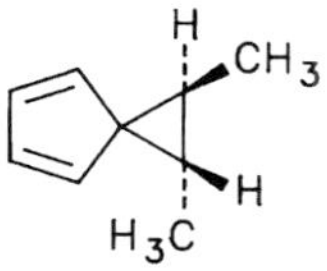 (E)-V 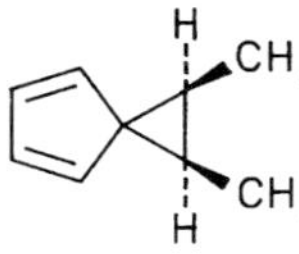 (Z)-V

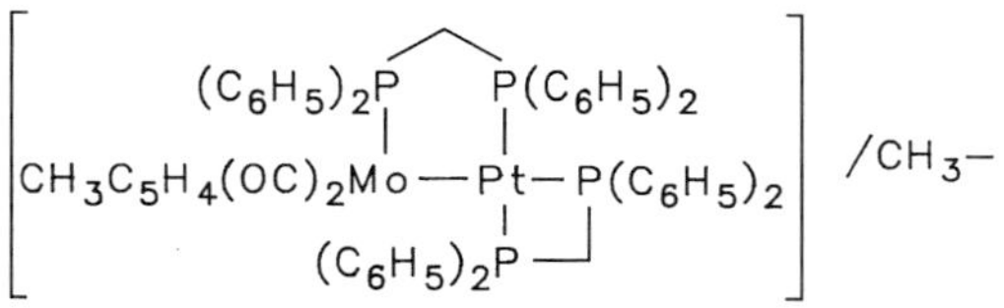

$$\left[\begin{array}{c}(C_6H_5)_2P\frown P(C_6H_5)_2 \\ CH_3C_5H_4(OC)_2Mo{-}Pt{-}P(C_6H_5)_2 \\ (C_6H_5)_2P\end{array}\right]/CH_3^-$$

Metal-induced ring opening was also investigated using $Mo(CO)_3(NCCH_3)_3$ in THF for 1,1-dimethyl-spiro[2,4]hepta-4,6-diene (Formula III), 1,1,2,2-tetramethyl-spiro[2,4]hepta-4,6-diene (Formula IV), and the E- and Z-isomers of 1,2-dimethyl-spiro[2,4]hepta-4,6-diene (Formula V). The reactions were carried out at room temperature; subsequent workup was accomplished by column chromatography on Al_2O_3, eluting with hexane (for $C_5H_4C(CH_3)_2CH_2Mo(CO)_3$ and (E)- and (Z)-$C_5H_4CH(CH_3)CH(CH_3)Mo(CO)_3$) or with a 10:1 hexane/ether mixture (for $C_5H_4C(CH_3)_2C(CH_3)_2Mo(CO)_3$). The products isolated from the reactions with (E)- and (Z)-isomers indicated that ring opening with the retention of configuration had occurred [55].

Method XIII: The same general reaction procedure outlined for $(C_6H_5)_3SnC_5H_4Mo(CO)_3$-$CH_3$ was used for all systems $C_5H_5Mo(CO)_3ER_3$ (E = Ge: R = CH_3; E = Sn, Pb: R = CH_3, C_6H_5) studied. Excess $Li[N(C_3H_7\text{-}i)_2]$ was added to the molybdenum-tin complex stirred in THF at 0 °C. As soon as IR spectroscopy had indicated complete formation of the $[(C_6H_5)_3SnC_5H_4Mo(CO)_3]^-$ anion, excess methyl iodide was added and the mixture was allowed to warm to room temperature. Removal of solvent followed by extraction of the residue with hexane/CH_2Cl_2 (7:3), renewed evaporation and dissolution of the remaining solid in CH_2Cl_2, and column chromatography on alumina, eluting with hexane/CH_2Cl_2 (9:1), afforded the product, which was purified further by recrystallization from the same solvent composition. Trimethylgermyl and triphenylplumbyl groups also readily migrated to the cyclopentadienyl ring on treatment of the respective $C_5H_5Mo(CO)_3ER_3$ derivatives with lithium diisopropylamide. With trimethylstannyl and trimethylplumbyl precursors, competition with Mo–Sn and Mo–Pb bond cleavage giving $[C_5H_5Mo(CO)_3]^-$ was observed as the predominant pathway [100].

References on pp. 253/8

Method XIV: The homodimer was irradiated at $\lambda > 310$ nm in toluene below $-40\,^{\circ}C$ in the presence of para- or ortho-quinones (1,4-benzoquinone, 2-methyl-1,4-benzoquinone, 2-phenyl-1,4-benzoquinone, 2,5-dimethyl-1,4-benzoquinone, 2,5-di-t-butyl-1,4-benzoquinone, 2,5-diphenyl-1,4-benzoquinone, 2,6-dimethyl-1,4-benzoquinone, 2,6-di-t-butyl-1,4-benzoquinone, 2,3,5,6-tetrafluoro-1,4-benzoquinone, 2,3,5,6-tetrachloro-1,4-benzoquinone, 9,10-phenanthrenequinone, 1,2-acenaphthenequinone) [105, 127] or at $\lambda > 520$ nm in THF or CH_2Cl_2 solutions containing added (2,3-bisdiphenylphosphanyl)-maleic anhydride ($T = 25\,^{\circ}C$) [118].

Method XV: To a CH_2Cl_2 solution of $RC_5H_4Mo(CO)_3H$ (R = $HOCH_2CH_2$, $HOCH_2CH_2CH_2$ [53]) was added one equivalent of $[C(C_6H_5)_3]BF_4$ at $-30\,^{\circ}C$ to generate the corresponding Lewis acidic intermediate, $RC_5H_4Mo(CO)_3FBF_3$. Subsequent treatment with an equimolar quantity of PR_3 (R = C_6H_5, $C_6H_5CH_3$-4), followed by warming to room temperature and adding diethyl ether, caused the ionic complexes $[HO(CH_2)_nC_5H_4Mo(CO)_3PR_3]BF_4$ (n = 2: R = C_6H_5, $C_6H_5CH_3$-4; n = 3: R = C_6H_5) to precipitate as crystalline materials [53]. For preparative approaches to $[RC_5H_4Mo(CO)_3{}^2D][RC_5H_4Mo(CO)_3]$ salts formed by photodisproportionation of $(RC_5H_4Mo(CO)_3)_2$ homodimers, see Methods IVa and IVb.

Method XVI: Introduction of the $Mo(CO)_3$ fragment to a macroreticular styrene-18%-divinyl benzene copolymer, "(PS)", was achieved by treatment of the polystyrylmethylenecyclopentadienyl anion, $(PS)-CH_2C_5H_4^-$ (Li^+ salt, containing about 1.5 mmol $-CH_2C_5H_4^-$ per g of polymer), with $Mo(CO)_6$ in N,N-dimethylformamide at $130\,^{\circ}C$, followed by washing with $(CH_3)_2NCHO$ to remove excess $Mo(CO)_6$, and with ether to remove the N,N-dimethylformamide [10]. The material was then treated with a 2 M solution of acetic acid in water/1,4-dioxane (1:1); crude $(PS)-CH_2C_5H_4Mo(CO)_3H$ was filtered off and washed neutral with water/1,4-dioxane, with 1,4-dioxane, and finally with diethyl ether [10].

Method XVII: Polystyryl-p-phenylene-cyclopentadiene, $(PS)-p-C_6H_4C_5H_5$ ("(PS)" = macroporous polystyrene, 3% cross-linked with divinyl benzene) was slurried in THF, combined with excess $Mo(CO)_3(NCCH_3)_3$, and the mixture stirred at $70\,^{\circ}C$. The resulting resin, red after extensive washing with THF and drying in vacuum, exhibited IR bands corresponding to a mixture of protonated and dimerized metal centers on the polymer, $(PS)-p-C_6H_4C_5H_5Mo(CO)_3H$ and $(PS)-p-C_6H_4C_5H_4Mo(CO)_3Mo(CO)_3C_5H_4C_6H_4-p-(PS)$, respectively. $Li[(C_2H_5)_3BH]$ in THF converted both the molybdenum hydride and the dimolybdenum groups to the lithium salt of polymer-attached molybdate anion, $[(PS)-p-C_6H_4C_5H_4Mo(CO)_3]^-$, and proved to be sufficiently basic to deprotonate residual nonmetalated resin-bound cyclopentadiene. Protonation of all anionic sites by means of p-toluenesulfonic acid in THF, followed by washing with THF, afforded polymer-bound molybdenum hydride $(PS)-p-C_6H_4C_5H_5Mo(CO)_3H$, which was then selectively deprotonated at its metal$-$H bonds with either lithium or sodium diethyl malonate, yielding $M[(PS)-p-C_6H_4C_5H_4Mo(CO)_3]$ ($M^+ = Li^+, Na^+$), the residual polymer-bound cyclopentadiene remaining neutral under these conditions [23, 25, 29].

References on pp. 253/8

Table 6
Mononuclear Complexes with Bonds between the $RC_5H_4Mo(CO)_3$ Fragment and Main Group Elements.
An asterisk indicates further information at the end of the table.
For explanations, abbreviations, and units, see p. X.

No.	compound	method of preparation (yield) properties and remarks
$M[RC_5H_4Mo(CO)_3]$ salts:		
M^+/R =		
1	Li^+/CH_3-	Ia (not isolated) [171]; also from 6-(dimethylamino)fulvene by consecutive treatment with $Li[AlH_4]$ and $Mo(CO)_6$ in THF, first at 0 to 25 °C, then at reflux temperature (not isolated) [42] combination with C_6H_5HgCl in THF cleanly produced $(CH_3C_5H_4Mo(CO)_3)_2Hg$ [171] $4\text{-}CH_3C_6H_4SO_2N(NO)CH_3$ reacted to form $CH_3C_5H_4Mo(CO)_2NO$ [42]
*2	Na^+/CH_3-	Ia [1, 8, 12, 19, 83, 104]; isolated as the solvate $Na[CH_3C_5H_4Mo(CO)_3] \cdot CH_3OC_2H_4OCH_3$ [104]; IIIa (not isolated) [18, 28, 40, 119]
*3	K^+/CH_3-	IIIb (not isolated) [14]
4	$[N(C_2H_5)_4]^+/CH_3-$	combination with $[CH_3C_5H_4Mo(CO)_3P(C_6H_5)_3]PF_6$ in CH_2Cl_2, followed by addition of benzene, gave $(CH_3C_5H_4Mo(CO)_3)_2$ [76]
5	$[N(C_4H_9\text{-}n)_4]^+/CH_3-$	IVc (not isolated) [62]; formed together with $CH_3C_5H_4Mo(CO)_3Br$ also from $[CH_3C_5H_4Mo(CO)_3OC(CH_3)_2][CH_3C_5H_4\text{-}Mo(CO)_3]$ and $[N(C_4H_9\text{-}n)_4]Br$ in acetone [62]
6	$[H_3NC_2H_4NH_2]^+/CH_3-$	by irradiation ($\lambda > 525$ nm) of $(CH_3C_5H_4Mo(CO)_3)_2$ in C_6H_6 containing added $H_2NC_2H_4NH_2$ (see introductory remarks; not isolated) [99] IR (benzene/ethylene diamine): 1776, 1892 (CO) [99]
7	$[NH_3C_6H_{11}\text{-cyclo}]^+/CH_3-$	by irradiating $(CH_3C_5H_4Mo(CO)_3)_2$ at $\lambda > 525$ nm in benzene in the presence of cyclohexylamine (see introductory remarks; not isolated) [99] IR (benzene/cyclohexylamine): 1762, 1782, 1892 (CO) [99]
8	$[C_5H_5NC_5H_4CH_3]^+/CH_3-$	from $CH_3C_5H_4Mo(CO)_3Cl$ by irradiation at $\lambda > 345$ nm in pyridine [88] or from $(CH_3C_5H_4Mo(CO)_3)_2$ by irradiation ($\lambda > 525$ nm) in benzene containing pyridine at concentrations $\geq 50\%$ (see introductory

References on pp. 253/8

Table 6 (continued)

No.	compound	method of preparation (yield) properties and remarks
8 (continued)		remarks; not isolated; cation not unambiguously identified) [99] IR (benzene/pyridine): 1775, 1892 (CO) [99]
9	$[PR_4]^+/CH_3-$ (R = n-C_4H_9, C_6H_5)	IVc (not isolated) [62, 77]
10	$[(C_6H_5)_3NPN(C_6H_5)_3]^+/CH_3-$	reacted with $C_5H_5Mo(CO)_3CH_3$ in THF to slowly form $CH_3C_5H_4Mo(CO)_3CH_3$ together with $[(C_6H_5)_3NPN(C_6H_5)_3][C_5H_5Mo(CO)_3]$ at equilibrium concentrations; the equilibration was found to be first-order in $C_5H_5Mo(CO)_3CH_3$ and independent of $[CH_3C_5H_4Mo(CO)_3]^-$ ($k_{obs} = 5.3 \times 10^{-6}\ s^{-1}$) [145]
11	$[CH_3C_5H_4Mo(CO)_3NCCH_3]^+/CH_3-$	IVb (not isolated) [62] IR (CH_3CN): 1773, 1892 (CO of anion), 1989, 2069 (CO of cation) [62]
12	$[CH_3C_5H_4Mo(CO)_3{}^2D]^+/CH_3-$ (^2D = $P(C_2H_5)_3$, $P(C_3H_7\text{-i})_3$, $P(C_4H_9\text{-n})_3$, $P(C_3H_7\text{-i})_3$, $P(C_4H_9\text{-n})_3$, $P(C_6H_{11}\text{-cyclo})_3$, $P(C_6H_5)_2CH_3$, $P(C_6H_5)_2C_3H_7\text{-i}$, $P(C_6H_5)_2C_4H_9\text{-n}$, $P(C_6H_5)_3$; $P(OCH_3)_3$, $P(OC_3H_7\text{-i})_3$, $P(OC_4H_9\text{-n})_3$)	IVa (not isolated) [76] IR (CH_2Cl_2; ^2D = $P(C_6H_5)_3$): 1774, 1894, 1975, 1995, 2055 (CO) [76] for photolytic CO elimination, see No. 15
13	$[CH_3C_5H_4Mo(CO)_3OC(CH_3)_2]^+/CH_3-$	IVb (not isolated) [62] IR (acetone, −78°C): 1779, 1898 (CO of anion), 1955, 2067 (CO of cation) [62] addition of $[N(C_4H_9\text{-n})_4]Br$ to acetone solutions produced $CH_3C_5H_4Mo(CO)_3Br$ along with $[N(C_4H_9\text{-n})_4][CH_3C_5H_4Mo(CO)_3]$ [62]
14	$[CH_3C_5H_4Mo(CO)_3OS(CH_3)_2]^+/CH_3-$	IVb (not isolated) [62] IR (dimethyl sulfoxide): 1770, 1891 (CO of anion), 1968, 2049 (CO of cation) [62]
15	$[CH_3C_5H_4Mo(CO)_2({}^2D)_2]^+/CH_3-$ (^2D = NC_5H_5, $P(C_2H_5)_3$, $P(C_4H_9\text{-n})_3$, $P(C_4H_9\text{-i})_3$, $P(C_6H_5)_2CH_3$, $P(C_6H_5)_2C_4H_9\text{-n}$, $P(C_6H_5)_3$; $P(OCH_3)_3$, $P(OC_2H_5)_3$, $P(OC_3H_7\text{-i})_3$, $P(OC_4H_9\text{-n})_3$, $P(OC_6H_5)_3$, $P(OC_6H_4CH_3\text{-4})_3$, $P(OC_6H_4OCH_3\text{-4})_3$)	IVa (not isolated) [35, 44, 54, 69, 76]; also formed by 290 nm irradiation of $[CH_3C_5H_4Mo(CO)_3{}^2D]$-$[CH_3C_5H_4Mo(CO)_3]P(C_6H_5)_2CH_3$, in CH_2Cl_2 in the presence of a ^2D ligand as demonstrated for ^2D = $P(C_6H_5)_3$ [76] IR (cyclohexane): 1755, 1775 (^2D = NC_5H_5) [44]; 1765 (^2D = $P(OCH_3)_3$) [54]; (C_6H_6; ^2D = $P(C_6H_5)_3$): 1776, 1894 (all CO of anion) [99]

References on pp. 253/8

Table 6 (continued)

No.	compound	method of preparation (yield) properties and remarks
16	$[CH_3C_5H_4Mo(CO)_2((C_6H_5)_2PC_2H_4P(C_6H_5)_2)]^+/CH_3-$	IVa (not isolated) [44, 98] IR (C_6H_6 (?)): 1904, 1970 (CO of cation) [52]
17	$[CH_3C_5H_4Mo(CO)_2P(C_2H_4P(C_6H_5)_2)_2C_6H_5]^+/CH_3-$	IVa (not isolated) [52] ^{31}P NMR (benzene): −10.8 ("dangling" $P(C_6H_5)_2$ group), 69.9, 84.6 (coordinated central and terminal P atoms) [52] IR (C_6H_6): 1901, 1966 (CO of cation) [52] for photolytically induced CO elimination, see No. 18
18	$[CH_3C_5H_4Mo(CO)P(C_2H_4P(C_6H_5)_2)_2C_6H_5]^+/CH_3-$	by irradiation at λ = 366 nm of No. 17 in benzene solution (not isolated) [52] ^{31}P NMR (benzene): 90.28 (PC_6H_5), 108.0 ($P(C_6H_5)_2$) [52]
19	$[$ $(C_6H_5)_2P$ ⌒ $P(C_6H_5)_2$; $CH_3C_5H_4(OC)_2Mo—Pt—P(C_6H_5)_2$; $(C_6H_5)_2P$ $]^+$ $/CH_3-$	from trans-$(CH_3C_5H_4Mo(CO)_3)_2Pt(NCC_6H_5)_2$ and two equivalents of $(C_6H_5)_2PCH_2P(C_6H_5)_2$ in THF [51] ^1H NMR ($CDCl_3$): 1.30 (s, CH_3 of cation), 2.06 (s, CH_3 of anion), 4.15 (m, CH_2), 4.40, 4.63 (both m, C_5H_4 of cation), 4.86 (m, CH_2), 5.05 (m, C_5H_4 of anion), 6.5 to 7.6 (m, C_6H_5) [51] IR (THF): 1778, 1893 (CO of anion), 1808, 1875 (CO of cation) [51] bubbling air into acetone solutions resulted in smooth oxidation of the anion to $[Mo_2O_7]^{2-}$ [51] metathetical exchange with NH_4PF_6 in CH_2Cl_2 gave the PF_6^- salt [51]
20	Li^+/C_2H_5-	Ia (not isolated) [171] C_6H_5HgCl in THF reacted to form $(C_2H_5C_5H_4Mo(CO)_3)_2Hg$ [171]
21	Na^+/C_2H_5-	Ia (not isolated) [19] nucleophilic substitution of alkyl iodides in THF at 40 °C gave $C_2H_5C_5H_4Mo(CO)_3R$ (R = CH_3, C_2H_5) [19]
22	$Na^+/n\text{-}C_3H_7-$	IIIa (not isolated) [2, 3]

References on pp. 253/8

Table 6 (continued)

No.	compound	method of preparation (yield) properties and remarks
23	Na^+/i-C_3H_7–	Ia (not isolated) [86, 112]; IIIa (not isolated) [2, 3] alkylation with RI in THF or $CH_3OC_2H_4OC_2H_4OCH_3$ formed the alkyls i-$C_3H_7C_5H_4Mo(CO)_3R$ (R = CH_3 [2, 3, 86, 112], C_2H_5 [2, 3])
24	Li^+/n-C_4H_9–	Ia (not isolated) [171] (n-$C_4H_9C_5H_4Mo(CO)_3)_2Hg$ resulted from combination with C_6H_5HgCl in THF [171]
25	K^+/t-C_4H_9–	reaction with $Cr(CO)_5PCl_3$ (0.33 equivalent) in THF (−78°C to +20°C) afforded t-$C_4H_9C_5H_4Mo(CO)_2P_3$-cyclo accompanied by (t-$C_4H_9C_5H_4Mo(CO)_2)_2(\mu$-$P_2)$, (t-$C_4H_9C_5H_4Mo(CO)_3)_2$, and $Cr(CO)_6$ [149]
26	Na^+/CH_3CHR– (R = C_2H_5, n-C_3H_7)	IIIa (not isolated) [2, 3]
27	$Na^+/(C_2H_5)_2CH$–	IIIa (not isolated) [2, 3] alkyl derivatives, $(C_2H_5)_2CHC_5H_5Mo(CO)_3R$ (R = CH_3, C_2H_5), were obtained by adding RI in THF [2, 3]
28	$Na^+/C_6H_5CH_2$–	Ia (not isolated) [12] $(C_6H_5CH_2C_5H_4Mo(CO)_3)_2$ resulted from oxidation with aqueous $Fe_2(SO_4)_3$ in $CH_3OC_2H_4OC_2H_4$-OCH_3 in the presence of added acetic acid [12]
29	Na^+/4-$CH_3OC_6H_4CH_2$–	IIIa (not isolated) [2, 3]
30	$Li^+/C_6H_5C(CH_3)H$–	Ia (not isolated) [94] $C_6H_5C(CH_3)HC_5H_4Mo(CO)_2NO$ was formed upon treatment in THF solution with 4-$CH_3C_6H_4SO_2N(NO)CH_3$ [94]
31	$Na^+/C_6H_5C(CH_3)H$–	IIIa (not isolated) [2, 3] treatment with CH_3I in THF gave $C_6H_5C(CH_3)HC_5H_4Mo(CO)_3CH_3$ [2, 3]
32	$Na^+/(C_6H_5)_2CH$–	IIIa (not isolated) [2, 3]
33	$Li^+/(C_6H_5)_2CH$– (R/R′ = CH_3/CH_3, CH_3/C_2H_5, CH_3/n-C_3H_7, C_2H_5/C_2H_5; -$(CH_2)_5$-, -$(CH_2)_6$-)	Ia (not isolated) [160] oxidation with $Fe_2(SO_4)_3/CH_3CO_2H$ in THF afforded the corresponding homodimers, $(C_6H_5C(R)R'C_5H_4Mo(CO)_3)_2$ [160] $C_6H_5C(C_2H_5)_2C_5H_4Mo(CO)_3Br$ resulted from treatment of an acidified THF solution of the requisite lithium salt with N-bromosuccinimide [160]

References on pp. 253/8

Table 6 (continued)

No.	compound	method of preparation (yield) properties and remarks
34	Li^+/3-$CH_3C_6H_4C(R)R'$– (R/R' = CH_3/CH_3, CH_3/C_2H_5; -$(CH_2)_6$-)	Ia (not isolated) [160] homobimetallic complexes, $(3\text{-}CH_3C_6H_4C(R)R'C_5H_4Mo(CO)_3)_2$, were obtained by oxidizing THF solutions of all three lithium salts with $Fe_2(SO_4)_3$ in the presence of aqueous CH_3CO_2H [160] combination of $Li[3\text{-}CH_3C_6H_4C(CH_3)_2C_5H_4Mo(CO)_3]$ with CCl_4 or N-bromosuccinimide in THF containing added acetic acid produced $3\text{-}CH_3C_6H_4C(CH_3)_2C_5H_4Mo(CO)_3X$ (X = Cl, Br) [160]
35	Li^+/4-$RC_6H_4C(R')R''$– (R = CH_3 or OCH_3, R'/R'' = CH_3/CH_3, CH_3/C_2H_5)	Ia (not isolated) [160] all four salts were cleanly oxidized by $Fe_2(SO)_4/CH_3CO_2H$ in aqueous THF to give the corresponding homodimers $(4\text{-}RC_6H_4CR'(R'')C_5H_4Mo(CO)_3)_2$ [160] treatment of CH_3CO_2H-containing THF solutions of the $[4\text{-}RC_6H_4C(CH_3)_2C_5H_4Mo(CO)_3]^-$ anions with CCl_4 or N-bromosuccinimide furnished $4\text{-}CH_3C_6H_4C(CH_3)_2C_5H_4Mo(CO)_3Cl$ and $4\text{-}CH_3OC_6H_4C(CH_3)_2C_5H_4Mo(CO)_3Br$, respectively [160]
36	Li^+/ 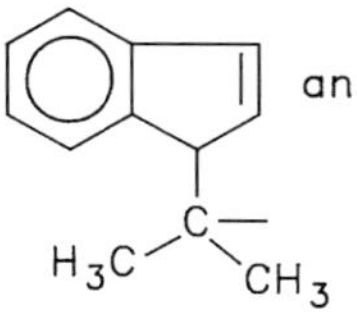 [indenyl–C(CH_3)$_2$– structure] and [isomeric indenyl–C(CH_3)$_2$– structure] H_3C C CH_3 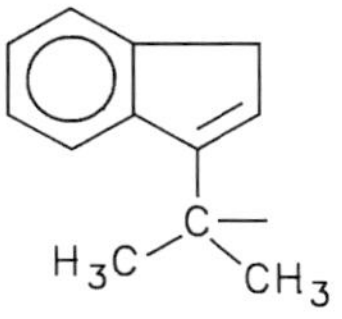	II (mixture of two isomers); neither isolated nor separated, but directly derivatized to $C_9H_7C(CH_3)_2C_5H_4Mo(CO)_3CH_3$ [150]
37	Li^+/ 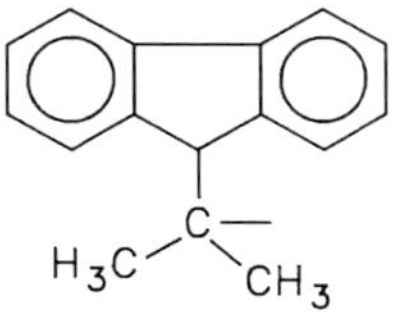 [fluorenyl–C(CH_3)$_2$– structure] H_3C C CH_3	II [150]; furthermore by refluxing $Mo(CO)_5OC_4H_8$ and $Li[C_{13}H_9C(CH_3)_2C_5H_4]$ in 1:1 stoichiometry in THF [111]; the salt was not isolated but treated in situ with aqueous CH_3CO_2H or CH_3I to give $C_{13}H_9C(CH_3)_2C_5H_4Mo(CO)_3H$ and $C_{13}H_9C(CH_3)_2C_5H_4Mo(CO)_3CH_3$, respectively [111, 150]
38	Na^+/cyclo-C_6H_{11}–	IIIa (not isolated) [2, 3]
39	Na^+/CH_2=CR– (R = H, CH_3)	Ia (not isolated) [34, 36, 42] in situ nitrosylation with N-methyl-N-nitroso-p-toluene sulfonamide led to CH_2=$C(R)C_5H_4Mo(CO)_2NO$ [34, 36, 42]

References on pp. 253/8

Table 6 (continued)

No.	compound	method of preparation (yield) properties and remarks
*40	$[P(C_6H_5)_4]^+$/ 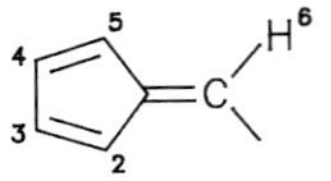	the fulvene complex $(CH_3)_2NCHC_5H_4Mo(CO)_3$ was allowed to react with a 10% excess of $Na[C_5H_5]$ in THF at room temperature; after dissolution of the evaporated mixture in ethanol, the salt was precipitated by adding excess $[P(C_6H_5)_4]Br$ in THF (88%) [110] purple needles [110] ^1H NMR (CD_2Cl_2): 5.25, 5.72 (both "t", C_5H_4=$CHC_5\mathbf{H}_4$), 6.20 (m, H-5; J(H-2,5) = 1.95, J(H-3,5) = 1.65, J(H-4,5) = 4.89), 6.25 (m, H-4; J(H-2,4) = 1.28, J(H-3,4) = 2.19), 6.44 (m, H-3; J(H-2,3) = 5.05), 6.75 (m, H-2), 6.96 ("s", H-6), 7.6 to 8.0 (m, C_6H_5) [110] ^{13}C NMR (THF-d_8): 88.12, 91.92, 99.41 (all C_5H_4=$CH\mathbf{C}_5H_4$); 118.65, 119.32, 119.55, 125.24, 127.08, 129.61 (all $\mathbf{C}_5H_4$=CHC_5H_4); 131.36 (d, J(C, P) = 12.87), 135.72 (d, J(C, P) = 11.03), 136.29 (d, J(C, P) = 2.76), 137.86 (all C_6H_5); 233.72 (CO) [110] IR (CH_2Cl_2): 1786, 1793, 1901 (CO) [110] C_5H_4=$CHC_5H_4Mo(CO)_3CH_3$ was derived from alkylation with CH_3I in THF [110]
41	$[P(C_4H_9\text{-}n)_4]^+$/ 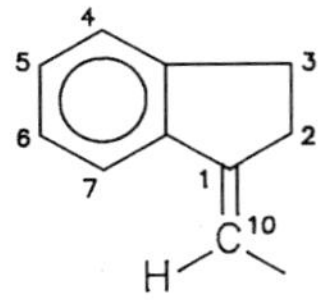	II (81%, isolated as a hydrate containing 1 molecule of H_2O) [138] purple crystals, m.p. 85°C [138] ^1H NMR (THF-d_8): 5.13, 5.66 (both "t", C_5H_4), 6.73 (d, H-3; J(H-2,3) = 5.5), 7.02 (m, 2H), 7.16 (d, H-2), 7.19 (m, 1H), 7.21 ("s", H-10), 7.54 (m, 1H); spectrum depicted in [138] ^{13}C NMR (THF-d_8): 88.7, 92.3, 100.8 (all C_5H_4), 119.5 (C-4), 121.6 (C-7), 124.8 (C-5), 126.1 (C-6), 127.6 (C-2), 129.3 (C-3), 129.4 (C-10), 131.6 (C-1), 139.8 (C-8), 142.6 (C-9), 234.7 (CO) [138] IR (CH_2Cl_2): 1796, 1909 (CO) [138] alkylation with CH_3I in THF produced C_9H_6=$CHC_5H_4Mo(CO)_3CH_3$ [138]
42	$Li^+/(CH_3)_2NC(CH_3)H-$	Ia (not isolated) [42] nitrosylation with N-methyl-N-nitroso-p-toluenesulfonamide in THF at 25°C furnished $(CH_3)_2NC(CH_3)HC_5H_4Mo(CO)_2NO$ together with a trace of CH_2=$CHC_5H_4Mo(CO)_2NO$ [42]

References on pp. 253/8

Table 6 (continued)

No.	compound	method of preparation (yield) properties and remarks
43	$Na^+/H_2NCH_2CH_2-$	Ia (not isolated); immediately oxidized to $[(H_3NCH_2CH_2C_5H_4Mo(CO)_3)_2][NO_3]_2$ using aqueous $Fe(NO_3)_3 \cdot 9\ H_2O$ [144]
44	$K^+/H_2NCH_2CH_2-$	IIIb (not isolated) [157] ^1H NMR (THF-d_8): 1.23 (t, NH_2; J = 6.9), 2.33 (m, $NCH_2C\mathbf{H}_2C$), 2.66 (qui, $NC\mathbf{H}_2CH_2C$), 4.85, 4.94 (both "d", C_5H_4; "J" = 2.1) [157] IR (THF): 1786, 1813, 1908 (CO) [157] protonation with excess CF_3CO_2H in THF furnished the cationic hydrido complex $[H_3NCH_2CH_2C_5H_4Mo(CO)_3H]^+$ as its trifluoroacetate [157]
45	$[C_5H_4CH_2CH_2NH_2Mo(CO_3)]^+/H_2NCH_2CH_2-$	IVb (not isolated; for spectroscopic properties of cation, compare No. 214) [157]
46	$[C_{14}H_{18}N_2H]^+/H_2NCH_2CH_2-$ ($C_{14}H_{18}N_2$ = 1,8-bis(dimethyl-amino)naphthalene)	from $[H_3NCH_2CH_2C_5H_4Mo(CO)_3H]PF_6$ by deprotonation of both the ammonium group and the metal atom, using 1,8-bis(dimethylamino)naphthalene in THF (not isolated) [157]
47	$[H_3NCH_2CH_2C_5H_4Mo(CO)_3NCCH_3]^{2+}/H_3N^+CH_2CH_2-PF_6^-$	IVb (not isolated) [157] IR (CH_3CN): 1993, 2077 (CO of cation) [157] facile proton transfer from the ammonium group of the cation to the anion with formation of the intramolecularly amine-coordinated cationic complex $[C_5H_4CH_2CH_2NH_2Mo(CO)_3]^+$ and the cationic hydride species $[H_3NCH_2CH_2C_5H_4Mo(CO)_3H]^+$ prevented the ions from back-reacting to the parent homo-dimer $[(H_3NCH_2CH_2C_5H_4Mo(CO)_3)_2][PF_6]_2$ [157]
48	$[H_3NCH_2CH_2C_5H_4Mo(CO)_3OH_2]^{2+}/H_3N^+CH_2CH_2-NO_3^-$	IVb (not isolated) [144] IR (H_2O with pH 7 buffer): 1729, 1783, 1905 (CO of anion), 1967, 1996, 2062 (CO of cation); −(CO) of cation in H_2O at pH 2: 1981, 1993, 2067 [144] in aqueous solution at pH 2, the anion underwent smooth protonation producing the cationic hydride $[H_3NCH_2CH_2C_5H_4Mo(CO)_3H]^+$ [144]
49	$[H_3NCH_2CH_2C_5H_4Mo(CO)_3OC_4H_8]^{2+}/H_3N^+CH_2CH_2-PF_6^-$	IVb (not isolated) [157] IR (THF): 1786, 1813, 1908 (CO of anion), 1979, 2072 (CO of cation) [157]

References on pp. 253/8

Table 6 (continued)

No.	compound	method of preparation (yield) properties and remarks
49 (continued)		prone to undergo cation → anion proton transfer, forming $[C_5H_4CH_2CH_2NH_2Mo(CO)_3]PF_6$ together with $[H_3NCH_2CH_2C_5H_4Mo(CO)_3H]PF_6$ [157]
50	$[H_3NCH_2CH_2C_5H_4Mo(CO)_{4-n}(P(C_6H_5)_3)_n]^{2+}/H_3N^+CH_2CH_2-PF_6^-$ (n = 1, 2)	IVa (not isolated) [144, 157] ^1H NMR (THF-d_8): 2.83 ($NC\mathbf{H}_2CH_2C$), 3.15 ($NCH_2C\mathbf{H}_2C$), 5.67, 6.18 (both C_5H_4), 7.3 (H_3N), 7.5 (C_6H_5) (all for $[H_3NCH_2CH_2C_5H_4Mo(CO)_3P(C_6H_5)_3]^{2+}$) [157] IR (THF): 1786, 1813, 1908 (CO of anion); ν(CO) of $[H_3NCH_2CH_2C_5H_4Mo(CO)_3P(C_6H_5)_3]^{2+}$ at 1965, 1987, 2059; ν(CO) of $[H_3NCH_2CH_2C_5H_4Mo(CO)_2(P(C_6H_5)_3)_2]^{2+}$ cation at 1900 and 1965 [144, 157] proton transfer from the ammonium group to the central metal of the anion yielded the cationic hydrido species $[H_3NCH_2CH_2C_5H_4Mo(CO)_3H]^+$ together with the amine-chelated cations cis- and trans- $[C_5H_4CH_2CH_2NH_2Mo(CO)_2P(C_6H_5)_3]^+$ [157]
51	$[H_3NCH_2CH_2C_5H_4Mo(CO)_{4-n}(PN_3(CH_2)_6)_n]^{2+}/H_3N^+CH_2CH_2-NO_3^-$ ($PN_3(CH_2)_6$ = phosphatri-azaadamantane; n = 1, 2)	IVa (not isolated) [144] IR (H_2O with pH 7 buffer): 1729, 1783, 1905 (CO of anion); ν(CO) of $[H_3NCH_2CH_2C_5H_4Mo(CO)_3PN_3(CH_2)_6]^{2+}$ at 1966, 1993, and 2066; ν(CO) of $[H_3NCH_2CH_2C_5H_4Mo(CO)_2(PN_3(CH_2)_6)_2]^{2+}$ at 1910 and 1957 [144]
52	$[H_3NCH_2CH_2C_5H_4Mo(CO)_2(P(C_6H_5)_2)_2C_2(CO_2H)_2]^{2+}/H_3N^+CH_2CH_2-NO_3^-$ ($(P(C_6H_5)_2)_2C_2(CO_2H)_2$ = 2,3-bis(diphenylphosphanyl)-maleic acid)	IVa (not isolated) [154] IR (water with pH 7 buffer; phosphane ligand deprotonated): 1780, 1905 (CO of anion); ν(CO) of cation at 1916 and 1956 [154]
53	$Li^+/(CH_3)_2NCH(R)CH_2-$ (R = H, CH_3)	Ia; not isolated but immediately oxidized (I_2/THF) to the corresponding iodides, $(CH_3)_2NCH(R)CH_2C_5H_4Mo(CO)_3I$ [137]
54	$[C_{14}H_7O_2-C(O)NHCH_2CH_2C_5H_4Mo(CO)_{4-n}{}^2D_n]^+/$ 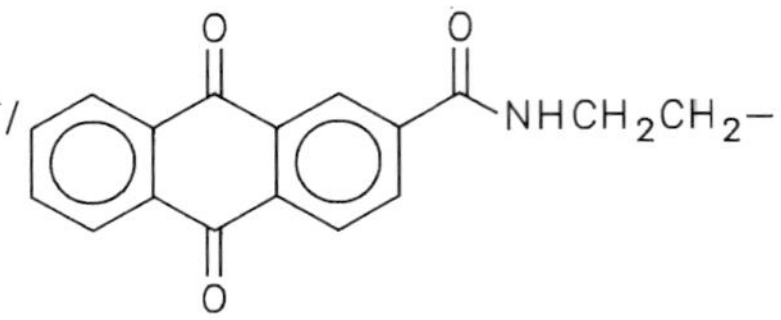 ($C_{14}H_7O_2$ = anthraquinon-2-yl; n = 1: $^2D = P(C_6H_5)_3$; n = 2: $^2D = P(C_6H_5)_3$, 0.5 $(C_6H_5)_2PC_2H_4P(C_6H_5)_2$)	O O NHCH$_2$CH$_2$– O

Table 6 (continued)

No.	compound	method of preparation (yield) properties and remarks
		IVa (not isolated) [177]; also by dark reactions of $(C_{14}H_7O_2-C(O)NHCH_2CH_2C_5H_4Mo(CO)_3)_2$ with 0.1 molar equivalent of cobaltocene or 1e-reduced anthraquinone in the presence of either $P(C_6H_5)_3$ or $(C_6H_5)_2PC_2H_4P(C_6H_5)_2$ in THF [177] IR (THF): 1678 (ν(C=O)), 1766, 1784, 1895 (CO of anion); ν(CO) of $[C_{14}H_7O_2-C(O)NHCH_2CH_2C_5H_4Mo(CO)_3$-$P(C_6H_5)_3]^+$ cation at 1956, 1987, 2059; ν(CO) of $[C_{14}H_7O_2-C(O)NHCH_2CH_2C_5H_4Mo$-$(CO)_2{}^2D_2]^+$ cations at 1900 and 1965 ($^2D = P(C_6H_5)_3$), and 1900 and 1979 ($^2D = 0.5\ (C_6H_5)_2PC_2H_4P(C_6H_5)_2$), respectively [177]
55	$[(C_5H_5)_2Co]^+$/ (anthraquinon-2-yl)-C(O)NHCH$_2$CH$_2$–	$(C_{14}H_7O_2-C(O)NHCH_2CH_2C_5H_4Mo(CO)_3)_2$ ($C_{14}H_7O_2$ = anthraquinon-2-yl) was reduced with $(C_5H_5)_2Co$ in THF (not isolated) [177] IR (THF): 1678 (ν(C=O)), 1766, 1784, 1895 (CO) [177]
56	$Na^+/HOCH_2CH_2-$	IIIa (75%) [53, 59] pale yellow solid [53, 59] ^{1}H NMR (acetone-d$_6$): 2.46 (t, C**H**$_2$C$_5$H$_4$), 3.55 (t, OCH$_2$), 4.90 (AA'BB' system, C$_5$H$_4$); HO variable [53, 59] IR (THF): 1750, 1795, 1900 (CO) [53, 59] protonation (CH_3CO_2H in THF) yielded $HOCH_2CH_2C_5H_4Mo(CO)_3H$ [53]
57	$[HOCH_2CH_2C_5H_4Mo(CO)_2(P(OC_2H_5)_3)_2]^+/HOCH_2CH_2-$	IVa (not isolated) [131] IR (THF): 1782, 1895, 1916, 1991 (CO) [131]
58	$Na^+/CH_3OC(CH_3)_2OCH_2CH_2-$	Ia (not isolated) [130] oxidation of THF solutions with $Fe(NO_3)_3 \cdot 9\ H_2O$ in the presence of aqueous CH_3CO_2H yielded $(HOCH_2CH_2C_5H_4Mo(CO)_3)_2$ [130]

References on pp. 253/8

Table 6 (continued)

No.	compound	method of preparation (yield) properties and remarks
59	$[CH_3(CH_2)_5NHC(O)OCH_2CH_2C_5H_4Mo(CO)_2(P(OC_2H_5)_3)_2]^+$/$CH_3(CH_2)_5$-NHC(O)$OCH_2CH_2$–	IVa (not isolated) [131] IR (THF): 1533 (NH, "amide II"), 1722 (ν(C=O)), 1783, 1895, 1917, 1991 (CO) [131]
60	Na^+/$HOCH_2CH_2CH_2$–	IIIa [53, 59] ^{1}H NMR (acetone-d_6): 1.75 (m, CCH_2C), 2.32 (t, $CH_2C_5H_4$), 3.57 (t, OCH_2), 4.88 (AA'BB' system, C_5H_4); HO variable [53, 59] IR (THF): 1742, 1796, 1900 (CO) [53, 59] the hydride $HOCH_2CH_2CH_2Mo(CO)_3H$ resulted from treating a THF solution with acetic acid [53]
*61	Na^+/HC(O)–	Ia [17, 22, 38]; II (89%) [46] isolated as an orange-yellow solvate, $Na[HC(O)C_5H_4Mo(CO)_3] \cdot 0.5\ O(C_2H_5)_2$, dec. >238 °C [46] ^{1}H NMR (acetone-d_6): 5.12, 5.57 (both "t", C_5H_4), 9.29 (s, HCO) [46] IR (KBr): 1609 (ν(C=O)), 1780, 1907 (CO) [46]
62	$[N(CH_3)_4]^+$/HC(O)–	V (51%) [46] fairly air-stable lustrous yellow platelets (from water) [46, 73], m.p. >180 °C (dec.) [46] ^{1}H NMR (acetone-d_6): 3.42 (s, CH_3), 5.13, 5.59 (both "t", C_5H_4), 9.34 (s, HCO) [46] IR (KBr): 1644 (ν(C=O)), 1778, 1910 (CO) [46]
63	$[P(C_6H_5)_4]^+$/HC(O)–	V (79%) [46] reasonably air-stable brownish yellow microcrystals [46, 73], m.p. 155 to 158 °C [46] ^{1}H NMR (acetone-d_6): 5.08, 5.55 (both "t", C_5H_4), 7.79 to 8.03 (m, C_6H_5), 9.32 (s, HCO) [46] ^{13}C NMR (acetone-d_6): 89.2, 90.4 (C-2,5 and C-3,4 of C_5H_4), 101.6 (C-1 of C_5H_4), 118.8, 131.3, 135.6 (C_6H_5), 183.0 (HCO), 233.0 (MoCO) [46] IR (KBr): 1642 (ν(C=O)), 1776, 1783, 1789, 1896 (CO) [46]
*64	Na^+/$CH_3C(O)$–	Ia (not isolated) [17, 22, 38, 91, 135]
65	$[N(C_4H_9\text{-}n)_4]^+$/$CH_3C(O)$–	Ib [21]; also formed by two-electron electroreduction of $(CH_3C(O)C_5H_4Mo(CO)_3)_2$ in THF with added $[N(C_4H_9\text{-}n)_4]PF_6$ at a glassy carbon electrode (E_{red} = −0.83 V vs. SCE) as well as by two-electron reduction of $CH_3C(O)C_5H_4Mo(CO)_3I$ or by one-electron reduction of

References on pp. 253/8

Table 6 (continued)

No.	compound	method of preparation (yield) properties and remarks
		$CH_3C(O)C_5H_4Mo(CO)_3CH_3$ (E_{red} = −0.82 and −1.62 V vs. SCE, respectively; same conditions as before) [21] IR (CH_2Cl_2): 1670 (ν(C=O)), 1800, 1900 (CO) [21] at a Pt electrode, one-electron oxidation in THF/$[N(C_4H_9\text{-}n)_4]PF_6$ occurred at E_{ox} = +0.22 V vs. SCE to afford $CH_3C(O)C_5H_4Mo(CO)_3^{\bullet}$, rapidly dimerizing to $(CH_3C(O)C_5H_4Mo(CO)_3)_2$; at a Hg electrode, $(CH_3C(O)C_5H_4Mo(CO)_3)_2Hg$ was formed [21] remarkably stable toward air and moisture [21] alkylation with CH_3I in THF produced $CH_3C(O)C_5H_4Mo(CO)_3CH_3$ [21]
66	$[HOCH_2C(O)C_5H_4Mo(CO)_2(P(OC_2H_5)_3)_2]^+/HOCH_2C(O)-$	IVa (not isolated) [131] IR (THF): 1644, 1706 (ν(C=O)), 1806, 1911, 1926, 1999 (CO) [131]
67	$Na^+/CH_3OC(CH_3)_2OCH_2C(O)-$	Ia [95, 130] ^1H NMR (D_2O): 1.44 (s, $C(CH_3)_2$), 3.26 (s, OCH_3), 4.59 (s, OCH_2), 5.41, 5.97 (both m, C_5H_4) [130] treatment in THF solution with $Fe(NO_3)_3 \cdot 9\,H_2O$/aqueous CH_3CO_2H afforded $(HOCH_2C(O)C_5H_4Mo(CO)_3)_2$ [95, 130]
*68	$Na^+/CH_3OC(O)-$	Ia [17, 22, 38, 42, 135, 147, 176] yellow powder [135]
69	$[N(C_4H_9\text{-}n)_4]^+/CH_3OC(O)-$	Ib [21]; also observed upon two-electron electroreduction, at a glassy carbon electrode, of $(CH_3OC(O)C_5H_4Mo(CO)_3)_2$ or $CH_3OC(O)C_5H_4Mo(CO)_3I$ in THF containing added $[N(C_4H_9\text{-}n)_4]PF_6$ (E_{red} = −0.80 and −0.87 V vs. SCE, respectively) or, under similar conditions, upon one-electron reduction of $CH_3OC(O)C_5H_4Mo(CO)_3CH_3$ (E_{red} = −1.80 V vs. SCE) [21] IR: 1722 (ν(C=O)), 1800, 1912 (CO) [21] one-electron oxidation at a Pt electrode in THF/$[N(C_4H_9\text{-}n)_4][PF_6]$ gave $(CH_3OC(O)C_5H_4Mo(CO)_3)_2$ via intermediately generated $CH_3OC(O)C_5H_4Mo(CO)_3^{\bullet}$ (E_{ox} = +0.22 V vs. SCE); electrolysis over mercury produced $(CH_3OC(O)C_5H_4Mo(CO)_3)_2Hg$ [21] air-stable and insensitive to moisture [21] $CH_3OC(O)C_5H_4Mo(CO)_3CH_3$ was obtained by treatment with CH_3I in THF [21]

References on pp. 253/8

Table 6 (continued)

No.	compound	method of preparation (yield) properties and remarks
*70	$Na^+/C_2H_5OC(O)-$	Ia (not isolated) [135, 147, 176]
71	$Na^+/(CH_3)_3Si-$	Ia (not isolated) [40] reaction with $I(CH_2)_3I$ in THF at room temperature afforded the 2-oxacyclopentylidene complex $(CH_3)_3SiC_5H_4Mo(CO)_2(=CO(CH_2)_3\text{-cyclo})I$, contaminated by some $C_5H_5Mo(CO)_2(=CO(CH_2)_3\text{-cyclo})I$ [40]
72	$Na^+/C_6H_5OSi(CH_3)_2-$	II; not isolated but directly converted to $C_6H_5OSi(CH_3)_2C_5H_4Mo(CO)_3CH_3$ by alkylation with CH_3I [124]
*73	$Li^+/(CH_3)_3Si(CH_3)_2Si-$	Ia (not isolated) [158, 159, 161]
74	$Li^+/(CH_3)_3Sn-$	by $Li[N(C_3H_7\text{-}i)_2]$-induced $(CH_3)_3Sn$ migration in $C_5H_5Mo(CO)_3Sn(CH_3)_3$; not isolated but quenched with CH_3I [100]; compare Method XIII IR (THF): 1718, 1783, 1807, 1906 (CO) [100]
75	$Li^+/(C_6H_5)_3Sn-$	observed as the primary intermediate in the preparation of $(C_6H_5)_3SnC_5H_4Mo(CO)_3CH_3$ from $C_5H_5Mo(CO)_3Sn(C_6H_5)_3/Li[N(C_3H_7\text{-}i)_2]$ [100]; compare Method XIII [100] IR (THF): 1716, 1782, 1806, 1905 (CO) [100]
76	$Li^+/C_6H_5N{=}N-$	Ia; not isolated but transformed in situ to $C_6H_5N_2C_5H_4Mo(CO)_3CH_3$ [151]
*77	$Li^+/(C_6H_5)_2P-$	Ia; II (on both routes isolated as an impure tan solid that appeared by ^1H NMR and IR spectroscopy to be, at best, an 85:15 mixture with $Li[Mo(CO)_5P(C_6H_5)_2C_5H_4]$ and contained ca. 1.6 molecules of THF per Mo [33, 139, 140]); pure samples of the salt were obtained by addition of $Li[(C_6H_5)_2PC_5H_4]$ to $(\text{cyclo-}C_7H_8)$-$Mo(CO)_3$ in THF at 70°C (90%) [152, 156] beige powder [156] ^1H NMR (C_6D_6): 5.50 (br, C_5H_4), 7.14 (m, meta- and para-H), 7.46, 7.70 (both m, both ortho-H) [152, 156]; similar in [33]; (THF-d_8): 4.94, 5.09 (both m, C_5H_4), 7.15 (m, meta- and para-H), 7.32 (m, ortho-H) [33] ^{13}C NMR (THF-d_8, 0.07 M $Cr(CH_3C(O)CH_2\text{-}C(O)CH_3)_3$): 88.8 (s, C-3,4 of C_5H_4), 94.3 (d, C-2,5 of C_5H_4; J(P, C) = 14.6), 96.4 (br, C-1 of C_5H_4), 128.1 (s, meta- and para-C), 133.9 (d, ortho-C; J(P, C) = 18.3), 141.3 (d, ipso-C; J(P, C) = 12.8), 235.4 (s, CO) [33]

References on pp. 253/8

Table 6 (continued)

No.	compound	method of preparation (yield) properties and remarks
		^{31}P NMR (C_6D_6): −18.2 [152, 156]; (THF): −16.3 [33] IR (THF): 1724, 1790, 1813, 1909 (CO) [152, 156]; similar in [33]
78	$Li^+/(4\text{-}CH_3C_6H_4)_2P-$	Ia; isolated as a nearly pure golden yellow solvate containing ca. 1.5 molecules of THF per Mo; the major impurity (ca. 5%) was $Li[Mo(CO)_5P(C_6H_4CH_3\text{-}4)_2C_5H_4]$ [33] ^{1}H NMR (C_6D_6): 2.09 (s, CH_3), 5.43 ("d", C_5H_4; "J" = 1.8), 5.46 ("t", C_5H_4), 7.04 (d, meta-H; J = 9.2), 7.61 ("t", ortho-H; $\Sigma\|J(H, H) + J(P, H)\| \approx 15$); THF at δ = 1.57 and 3.72 [33] ^{13}C NMR (THF-d_8, 0.07 M $Cr(CH_3C(O)CH_2\text{-}C(O)CH_3)_3$): 21.5 (s, CH_3), 88.8 (s, C-3,4 of C_5H_4), 94.1 (d, C-2,5 of C_5H_4; J(P, C) = 11.0), 97.7 (br, C-1 of C_5H_4), 128.7 (d, ipso-C), 129.1 (br, meta-C), 133.9 (d, ortho-C; J(P, C) = 18.3), 137.6 (s, para-C), 235.7 (s, CO) [33] IR (THF): 1723, 1790, 1813, 1912 (CO) [33] reactions of the molybdate anion with both the Mn and Re complexes $(M(CO)_4(\mu\text{-}Br))_2$ and with rhodium and iridium compounds such as $(Rh(R_2PC_2H_4PR_2)(\mu\text{-}Cl))_2$, $(Rh(CO)_2(\mu\text{-}Cl))_2$, $Ir(CO)_2(NH_2R)Cl$ (R = 4-$CH_3C_6H_4$), or $Ir(CO)(P(C_6H_5)_3)_2Cl$ in THF under varied conditions afforded metal−metal-bonded bridged heterobimetallics, $Mo(CO)_3(\mu\text{-}C_5H_4P(C_6H_4CH_3\text{-}4)_2)M(CO)_4$ (M = Mn, Re) [33], $Mo(CO)_3(\mu\text{-}C_5H_4\text{-}P(C_6H_4CH_3\text{-}4)_2)Rh(R_2PC_2H_4PR_2)$, $Mo(CO)_3(\mu\text{-}C_5H_4P(C_6H_4CH_3\text{-}4)_2)M'(CO)_2$ (M′ = Rh, Ir), and $Mo(CO)_3(\mu\text{-}C_5H_4\text{-}P(C_6H_4CH_3\text{-}4)_2)Ir(CO)P(C_6H_5)_3$ [45], respectively
hydrides $RC_5H_4Mo(CO)_3H$:		
R =		
79	CH_3-	VI [1, 24, 28] pale yellow, thermally unstable oil [1, 24, 28]; b.p. 55 to 60°C/0.01 Torr [24, 28] ^{1}H NMR (THF-d_8): −5.42 (s, MoH), 2.12 (s, CH_3), 5.46 (AA′BB′ quartets, C_5H_4; J(A, B) = 17, J(A, A′) = 2.1) [24, 28] IR (THF): 1930, 2015 (CO) [24, 28] air-oxidation of ether solutions afforded $(CH_3C_5H_4Mo(CO)_3)_2$ [1]

References on pp. 253/8

Table 6 (continued)

No.	compound	method of preparation (yield) properties and remarks
79 (continued)		$CH_3C_5H_4Mo(CO)_2(\mu\text{-}CO)Zr(C_5H_4CH_3)_2CH_3$ resulted from methane elimination between No. 79 and $(CH_3C_5H_4)_2Zr(CH_3)_2$ in THF [64] heating a mixture of $CH_3C_5H_4Mo(CO)_3H$ and $C_5H_5Mo(CO)_3CH_3$ (10% excess) in THF-d_8 at 50 °C to 100 °C gave CH_3CHO together with $CH_3C_5H_4Mo(CO)_nMo(CO)_nC_5H_5$, $(CH_3C_5H_4Mo(CO)_n)_2$, and $(C_5H_5Mo(CO)_n)_2$ (n = 3, 2); a similar reaction with $C_5H_5Mo(CO)_3C_2H_5$ at room temperature resulted in C_2H_5CHO, accompanied by the crossed dimers, $CH_3C_5H_4Mo(CO)_nMo\text{-}(CO)_nC_5H_5$ (n = 2, 3), as the predominant kinetic products [24, 26, 28]; see also [61] reaction with $Co_2(CO)_8$ yielded $Co(CO)_4H$ together with $CH_3C_5H_4MoCo(CO)_4$ [114]; see also [101]
80	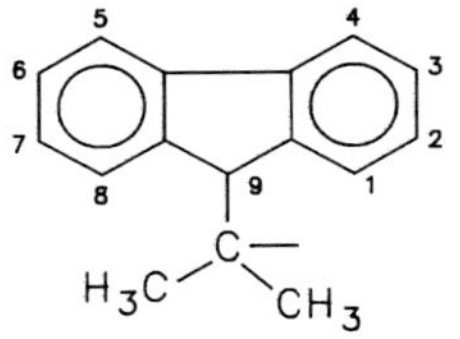 9-fluorenyl ring (positions 1–9) bearing at C-9 the group $C(CH_3)_2$– (H_3C, CH_3)	VI [111] m.p. 198 °C [111] ^1H NMR (CD_2Cl_2): −5.59 (s, MoH), 1.29 (s, CH_3), 3.92 (s, H-9), 4.95, 5.09 (both "t", C_5H_4), 7.28, 7.78 (both m, H-1 to H-8) [111] ^{13}C NMR (CD_2Cl_2): 28.8 ($C(\mathbf{C}H_3)_2$), 38.9 ($\mathbf{C}(CH_3)_2$), 59.4 (C-9); 87.6, 90.5, 127.2 (all C_5H_4); 119.7, 126.2, 126.8, 127.7, 142.3, 145.2 (all fluorenyl), 228.1 (CO) [111] IR (pentane): 1944, 2026 (CO) [111] thermal treatment with $P(CH_3)_3$ in toluene yielded cis- and trans-$C_{13}H_9C(CH_3)_2C_5H_4Mo\text{-}(CO)_2(P(CH_3)_3)H$; the corresponding photo-induced reaction employing excess phosphane gave trans-$C_{13}H_9C(CH_3)_2C_5H_4Mo(CO)\text{-}(P(CH_3)_3)_2H$ [111]
81	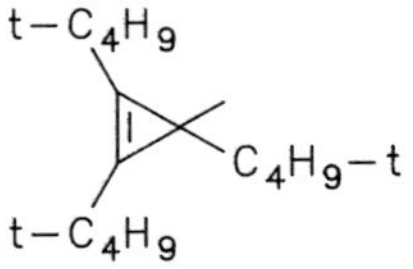 cyclopropenyl ring bearing three groups: $t\text{-}C_4H_9$, $C_4H_9\text{-}t$, $t\text{-}C_4H_9$	obtained in an almost quantitative yield by reaction of $Na[C_5H_5Mo(CO)_3]$ with $[cyclo\text{-}C_3(C_4H_9\text{-}t)_3][BF_4]$ in THF [6] air-sensitive, pale yellow crystals, m.p. 57 to 60 °C (dec.) [6] ^1H NMR (C_6D_6): −5.04 (s, MoH), 0.93 (s, $3CH_3$), 1.19 (s, $6CH_3$), 4.59, 4.97 (both "t", C_5H_4; "J" = 2) [6] IR (hexane): 1930, 2014 (CO) [6] reaction with CCl_4 occurred rapidly to afford $(cyclo\text{-}(t\text{-}C_4H_9)_3C_3)C_5H_4Mo(CO)_3Cl$ [6]

Table 6 (continued)

No.	compound	method of preparation (yield) properties and remarks
82	$H_3N^+CH_2CH_2-NO_3^-$	observed upon photodisproportionation of $[(H_3NCH_2CH_2C_5H_4Mo(CO)_3)_2][NO_3]_2$ in aqueous solution buffered at pH 2 (compare No. 48) [144] IR (H_2O with pH 2 buffer): 1929, 1943, 2024 (CO) [144]
83	$H_3N^+CH_2CH_2-PF_6^-$	readily formed from the initial photodisproportionation products of $[(H_3NCH_2CH_2C_5H_4Mo(CO)_3)_2][PF_6]_2$ in coordinating solvents or in the presence of added donor ligands, $[H_3NCH_2CH_2C_5H_4Mo(CO)_3{}^2D]^{2+}$ and $[H_3NCH_2CH_2C_5H_4Mo(CO)_3]^-$ (^2D = CH_3CN, THF, $P(C_6H_5)_3$), by proton transfer from the cation to the anion (not isolated or separated from the intramolecularly amine-coordinated by-products, $[C_5H_4CH_2CH_2NH_2Mo(CO)_3]PF_6$ and cis- and trans-$[C_5H_4CH_2CH_2NH_2Mo(CO)_2P(C_6H_5)_3]PF_6$, observed in the absence or presence of added phosphane) [157] ^1H NMR (THF-d_8): −5.48 (d (?), MoH), 2.65 (t, $NCH_2C\mathbf{H}_2C$; J = 6.3), 3.04 (t, $NC\mathbf{H}_2CH_2C$), 5.50, 5.64 ("s", C_5H_4), 7.27 (br, H_3N) [157] IR (THF): 1914, 1933, 2022 (CO) [157] addition of $P(C_6H_5)_3$ to photochemically generated THF solutions resulted in slow substitution with formation of the cis- and trans-substituted cations $[H_3NCH_2CH_2C_5H_4Mo(CO)_2(P(C_6H_5)_3)H]^+$ [157] 1,8-bis(dimethylamino)naphthalene caused deprotonation of both the ammonium group and the metal atom, forming the $[H_2NCH_2CH_2C_5H_4Mo(CO)_3]^-$ anion [157]
84	$H_3N^+CH_2CH_2-CF_3CO_2^-$	VI (quantitative; not isolated) [157] ^1H NMR (THF-d_8): −5.55 (s, MoH), 2.72 (t, $NCH_2C\mathbf{H}_2C$; J = 7.5), 3.10 (br, $NC\mathbf{H}_2CH_2$), 5.41, 5.62 ("s", C_5H_4), 7.79 (br, H_3N) [157] combination with $P(C_6H_5)_3$ in THF slowly produced cis- and trans-$[H_3NCH_2CH_2C_5H_4Mo(CO)_2(P(C_6H_5)_3)H]^+$ [157]
85	$HOCH_2CH_2-$	VI (quantitative) [53] thermally unstable golden oil [53] ^1H NMR (C_6D_6): −5.30 (s, MoH), 2.04 (t, $C\mathbf{H}_2C_5H_4$), 3.22 (t, OCH_2), 4.75 (AA′BB′ system, C_5H_4); HO variable [53]

References on pp. 253/8

Table 6 (continued)

No.	compound	method of preparation (yield) properties and remarks
85 (continued)		IR (C_6D_6): 1932, 1981, 2013 (CO) [53] hydride abstraction using one equivalent of $[C(C_6H_5)_3]BF_4$ in CH_2Cl_2 at −30 °C, followed by adding an equimolar quantity of phosphanes, PR_3, furnished $[HOCH_2CH_2C_5H_4Mo(CO)_3PR_3]BF_4$ (R = C_6H_5, $C_6H_4CH_3$-4) [53]
86	$HOCH_2CH_2CH_2-$	VI (quantitative) [53] thermally unstable golden oil [53] ^{1}H NMR (C_6D_6): −5.30 (s, MoH), 1.02 (m, CCH_2C), 2.00 (t, $C\mathbf{H}_2C_5H_4$), 3.15 (t, OCH_2), 4.63 (AA′BB′ system, C_5H_4); HO variable [53] IR (C_6D_6): 1930, 2020 (CO) [53] $[HOCH_2CH_2CH_2C_5H_4Mo(CO)_3P(C_6H_5)_3]BF_4$ was prepared by consecutive treatment with $[C(C_6H_5)_3]BF_4$ and $P(C_6H_5)_3$ in CH_2Cl_2 at −30 °C [53]
87	$(C_6H_5)_2P-$	VI (characterized in solution only) [156] ^{1}H NMR (C_6D_6): −5.26 (MoH), 4.74 (H-3,4 of C_5H_4), 4.98 (H-2,5 of C_5H_4), 7.20 (m, meta- and para-H), 7.45, 7.70 (both m, ortho-H) [152, 156] ^{31}P NMR (C_6D_6): −19.5 [152, 156] IR (toluene): 1937, 2024 (CO) [156]; same in THF [152] irradiation of toluene solutions with an Hg low-pressure lamp resulted in elimination of both H_2 and CO to give Mo−Mo single-bonded and doubly bridged $(\mu\text{-}(C_6H_5)_2PC_5H_4Mo(CO)_2)_2$ [152, 156]

Mo−Ga-, Mo−In-, Mo−Sn-, Mo−P-, and Mo−Bi-bonded complexes $RC_5H_4Mo(CO)_3X$:

R/X =

No.	compound	method of preparation (yield) properties and remarks
*88	$CH_3-/-GaI_2O(C_2H_5)_2$	VII [113] amber crystals [113] ^{1}H NMR (C_6D_6): 0.92 (t, $CH_2C\mathbf{H}_3$), 1.59 (s, $C\mathbf{H}_3C_5H_4$), 3.80 (q, CH_2), 4.72 ("d", C_5H_4) [113] the adduct readily lost some of the coordinated $O(C_2H_5)_2$ [113]; see also No. 89
89	$CH_3-/-GaI_2NC_5H_5$	from No. 88 by replacing diethyl ether for pyridine [113] ^{1}H NMR (C_6D_6): 1.69 (s, CH_3), 4.84, 4.91 (AA′BB′ system, C_5H_4), 6.16 (t, meta-H), 6.42 (t, para-H), 9.03 (d, ortho-H) [113]

Table 6 (continued)

No.	compound	method of preparation (yield) properties and remarks
90	CH_3–/–$InCl_2$	VII [128] yellow crystals [128] IR (THF): 1909, 1931, 2003 (CO); depicted in [128]
91	CH_3–/–$Sn(C_6H_5)_2I$	by treating $CH_3C_5H_4Mo(CO)_3Sn(C_6H_5)_3$ with equimolar I_2 in CCl_4 (competing Mo–Sn fission gave $CH_3C_5H_4Mo(CO)_3I$ as an accompanying product) [9]
92	CH_3–/–$Sn(C_6H_5)_3$	VIIIa (35%) [9, 57] yellow solid [57] ^{1}H NMR (CCl_4): 1.25 (CH_3), 5.03, 5.20 (both C_5H_4), 7.35 (C_6H_5) [9] IR (CCl_4): 1906, 1932, 2003 (CO) [9] treatment with one equivalent of I_2 in CCl_4 produced an equimolar mixture of $CH_3C_5H_4Mo$-$(CO)_3I$ and $CH_3C_5H_4Mo(CO)_3Sn(C_6H_5)_2I$ [9] electrophilic cleavage of the Mo–Sn bond by mercury(II) halides, HgX_2 (2 equivalents in refluxing CH_3OH), was used as a method of synthesis for $CH_3C_5H_4Mo(CO)_3HgX$ complexes (X = Cl, Br) [57]
93	HC(O)–/–$Sn(CH_3)_2Cl$	VIIIa (62%) [73] orange-brown microcrystals, m.p. 103 °C [73] ^{1}H NMR ($CDCl_3$): 0.93 (s, CH_3), 5.53, 5.92 (both m, C_5H_4), 9.64 (s, HCO) [73] IR (KBr): 1688 (ν(C=O)), 1884, 1905, 1992 (CO) [73] mass spectrum (70 eV, 250 °C): $[M]^+$, $[M-CH_3]^+$, $[M-nCO]^+$ (n = 2, 3), $[M-3CO-CH_3]^+$, $[HC(O)C_5H_4Mo]^+$, $[SnCl]^+$, $[Mo]^+$, $[HC(O)C_5H_4]^+$ [73]
94	$CH_3C(O)$–/–$SnCl_3$	formed on prolonged refluxing of acetone solutions containing in situ-generated $CH_3C(O)C_5H_4Mo(CO)_3HgCl$ together with excess $SnCl_2$; extraction of the evaporated mixture into CH_2Cl_2 with subsequent crystallization from light petroleum gave a 28% yield [91] brownish yellow crystals [91] IR (Nujol): 326, 338 (ν(SnCl)), 1689 (ν(C=O)), 1942, 1966, 1986, 2005, 2053 (CO) [91]; ($CHCl_3$): 1698 (ν(C=O)), 1973, 1998, 2055 (CO) [91] ^{1}H NMR ($CDCl_3$): 2.31 (s, CH_3), 5.76, 6.11 (both "t", C_5H_4) [91]

References on pp. 253/8

Table 6 (continued)

No.	compound	method of preparation (yield) properties and remarks
95	$CH_3OC(O)-/-PCl_2$	VIIIb (30%) [135] yellow crystals, m.p. 37 to 40°C [135] ^1H NMR ($CDCl_3$): 3.28 (s, CH_3), 5.44 to 5.56 (m, H-3,4 of C_5H_4), 6.26 to 6.38 (m, H-2,5 of C_5H_4) [135] ^{31}P NMR (C_6D_6): 358.51 [135] IR (KBr): 1729 (ν(C=O)), 1967, 2065 (CO) [135] mass spectrum: $[M-PCl_2]^+$ (4%), $[M-PCl_2-CO]^+$ (4%), $[M-PCl_2-3CO]^+$ (10%) [135] thermal decomposition in THF at room temperature gave $CH_3OC(O)C_5H_4Mo(CO)_3Cl$ [135]
96	$CH_3-/-BiCl_2$	formed by redistribution between $BiCl_3$ and one equivalent of $(CH_3C_5H_4Mo(CO)_3)_2BiCl$ in diethyl ether at room temperature; removal of solvent provided the complex in 85% yield [87] orange crystals [87] ^1H NMR (THF-d_8): 2.20 (s, CH_3), 5.58, 5.66 (both m, C_5H_4) [87]; (acetone-d_6): 2.37 (s, CH_3), 5.78, 5.89 (both m, C_5H_4) [87] ^{13}C NMR (THF-d_8): 14.0 (CH_3), 93.5, 95.8 (C-2,5 and C-3,4 of C_5H_4), 112.9 (C-1 of C_5H_4) [87]; (acetone-d_6): 14.4 (CH_3), 92.9, 95.6 (C-2,5 and C-3,4 of C_5H_4), 113.3 (C-1 of C_5H_4) [87] IR (THF): 1945, 2008 (CO) [87]
halides $RC_5H_4Mo(CO)_3X$:		
R/X =		
*97	CH_3-/Cl	Xb (quantitative) [48]; see also under "Further information" IR (CCl_4): 1957, 1981, 2051 (CO) [48]; same in THF [144, 157, 177]
*98	CH_3-/Br	see under "Further information" ^1H NMR ($CDCl_3$): 2.20 (s, CH_3), 5.26, 5.55 (both "t", C_5H_4) [32] IR (KBr): 1945, 1974, 2043 (CO) [32]
*99	CH_3-/I	Xa (72%) [146]; see also under "Further information" ^1H NMR (C_6D_6): 1.67 (CH_3), 4.23 (H-3,4 of C_5H_4), 4.47 (H-2,5 of C_5H_4) [146]; ($CDCl_3$): 2.28 (s, CH_3), 5.35, 5.46 (both "t", C_5H_4; "J" = 2.3) [125] ^{13}C NMR ($CDCl_3$): 14.88 (CH_3), 90.95, 95.33, 114.78 (all C_5H_4), 220.82 (CO) [125] IR (CCl_4): 1952, 1972, 2041 (CO) [9]; similar in [125]

Table 6 (continued)

No.	compound	method of preparation (yield) properties and remarks
		mass spectrum: $[M]^+$, $[M-nCO]^+$ (n = 1 to 3), $[C_6H_6]^+$ [125]
100	$i\text{-}C_3H_7-/Br$	formed by bromination of $(i\text{-}C_3H_7C_5H_4Mo(CO)_2)_2(\mu_3\text{-}Te_2)Fe(CO)_3$ (−78 °C to +25 °C, CH_2Cl_2 solution, CO atmosphere); the other product isolated was $i\text{-}C_3H_7C_5H_4Mo(CO)_2(\mu\text{-}Te_2Br)Fe(CO)_3$ [60]
101	$i\text{-}C_3H_7-/I$	Xa (69%) [2, 3] m.p. 92 to 94 °C [2, 3] IR (CCl_4 or CS_2): 1974, 2052 (CO) [2, 3]
102	$t\text{-}C_4H_9-/I$	Xa (72%) [146]; see also [5] red solid [146] ^1H NMR (C_6D_6): 0.84 (CH_3), 4.62, 4.63 (both C_5H_4) [146] $(C_5H_5Fe(CO)_2)_2$-catalyzed reactions with PR_3 and RNC, in refluxing benzene gave monosubstituted cis/trans isomeric derivatives, $CH_3C_5H_4Mo(CO)_2(PR_3)I$ (PR_3 = $P(CH_2PC_6H_5)_3$, $P(C_6H_5)_2CH_3$, $P(C_6H_5)_3$, $P(OCH_3)_3$, $P(OCH_2C(CH_3)_3)_3$) and $CH_3C_5H_4Mo(CO)_2(CNR)I$ (R = $t\text{-}C_4H_9$, $C_6H_3(CH_3)_2$-2,6), respectively [146]
103	$CH_3C(C_2H_5)H-/I$	Xa (70%) [2, 3] m.p. 75 °C [2, 3] IR (CCl_4 or CS_2): 1974, 2051 (CO) [2, 3]
104	$CH_3C(C_3H_7\text{-}n)H-/I$	Xa (40%) [2, 3] oil [2, 3] IR (CCl_4 or CS_2): 1973, 2050 (CO) [2, 3]
105	$(C_2H_5)_2CH-/I$	Xa (43%) [2, 3] m.p. 45 °C [2, 3] IR (CCl_4 or CS_2): 1976, 2050 (CO) [2, 3]
106	$4\text{-}CH_3OC_6H_4CH_2-/I$	Xa (70%) [2, 3] m.p. 78 °C [2, 3] IR (CCl_4 or CS_2): 1975, 2055 (CO) [2, 3]
107	$C_6H_5C(CH_3)H/-I$	Xa (59%) [2, 3] m.p. 76 °C [2, 3] IR (CCl_4 or CS_2): 1974, 2052 (CO) [2, 3]
108	$3\text{-}CH_3C_6H_4C(CH_3)_2-/Cl$	IXc (64%, using CCl_4) [160] brown crystals, m.p. 110 to 112 °C [160]

References on pp. 253/8

Table 6 (continued)

No.	compound	method of preparation (yield) properties and remarks
108 (continued)		^{1}H NMR ($CDCl_3$): 1.68 (s, $C(CH_3)_2$), 2.35 (s, 3-CH_3), 5.38, 5.58 (both "t", C_5H_4), 7.00 to 7.36 (m, C_6H_4) [160] IR (KBr): 1955, 2008 (CO) [160]
*109	3-$CH_3C_6H_4C(CH_3)_2$–/Br	IXc (60%, with N-bromosuccinimide) [160] orange-yellow crystals, m.p. 118 to 120°C [160] ^{1}H NMR ($CDCl_3$): 1.66 (s, $C(CH_3)_2$), 2.34 (s, 3-CH_3), 5.38, 5.56 (both "t", C_5H_4), 7.12 (m, C_6H_4) [160] IR (KBr): 1950, 2000 (CO) [160]
110	4-$CH_3C_6H_4C(CH_3)_2$–/Cl	IXc (53%, using CCl_4) [160] brown crystals, m.p. 119 to 121°C [160] ^{1}H NMR ($CDCl_3$): 1.68 (s, $C(CH_3)_2$), 2.32 (s, 4-CH_3), 5.38, 5.58 (both "t", C_5H_4), 7.04 to 7.32 (m, C_6H_4) [160] IR (KBr): 1950, 2000 (CO) [160]
111	4-$CH_3OC_6H_4C(CH_3)_2$–/Br	IXc (50%, with N-bromosuccinimide) [160] orange-yellow crystals, m.p. 90 to 92°C [160] ^{1}H NMR ($CDCl_3$): 1.64 (s, $C(CH_3)_2$), 3.80 (s, CH_3O), 5.34, 5.56 (both "t", C_5H_4), 6.72 to 7.36 (m, C_6H_4) [160] IR (KBr): 1955, 2010 (CO) [160]
112	$C_6H_5C(C_2H_5)_2$–/Br	IXc (76%, with N-bromosuccinimide) [160] golden yellow crystals, m.p. 92 to 94°C [160] ^{1}H NMR ($CDCl_3$): 0.78 (t, CH_3), 1.73 to 2.24 (m, CH_2), 5.08, 5.48 (both "t", C_5H_4), 7.10 to 7.48 (m, C_6H_4) [160] IR (KBr): 1950, 2005 (CO) [160]
113	t-C_4H_9, t-C_4H_9 (cyclopropene ring), C_4H_9-t /Cl	IXc [6] air-stable, orange crystals [6] IR (CCl_4): 1956, 1976, 2052 (CO) [6]
114	$NO_3^-H_3N^+CH_2CH_2$–/Cl	Xb (quantitative; not isolated; disappearance quantum yield at 550 nm for the reaction of $[(H_3NCH_2CH_2C_5H_4Mo(CO)_3)_2][NO_3]_2$ with 50 mM Cl_3CCH_2OH in aqueous solution buffered at pH 7: 0.78 ± 0.15) [144] IR (H_2O with pH 7 buffer): 1955, 1968, 2060 (CO); (H_2O with pH 2 buffer): 1981, 2060, 2067 (CO) [144]

References on pp. 253/8

Table 6 (continued)

No.	compound	method of preparation (yield) properties and remarks
115	$PF_6^-H_3N^+CH_2CH_2-/Cl$	Xb (quantitative; not isolated; disappearance quantum yield at 550 nm for the reaction of $[(H_3NCH_2CH_2C_5H_4Mo(CO)_3)_2][PF_6]_2$ with 1 M CCl_4 in THF: 0.36 ± 0.05 to 0.39 ± 0.04) [144, 157] ^{1}H NMR (THF-d_8): 2.61 (t, $NCH_2\mathbf{CH_2}C$; J = 7.2), 3.07 (br, $N\mathbf{CH_2}CH_2C$), 5.40, 5.99 (both "s", C_5H_4), 7.90 (br, H_3N) [157] IR (THF): 1956, 1971, 2050 (CO) [144, 157] in situ treatment with 1,8-bis(dimethylamino)-naphthalene resulted in elimination of HCl, producing the intramolecularly amine-coordinated $[C_5H_4CH_2CH_2NH_2Mo(CO)_3]PF_6$ [157]
116	$(CH_3)_2NCH_2CH_2-/I$	IXa (75%) [137] red liquid [137] ^{1}H NMR ($CDCl_3$): 2.24 (s, CH_3), 2.41 to 2.50, 2.59 to 2.68 (both m, CH_2CH_2), 5.41, 5.55 (both "t", C_5H_4; "J" = 2.2) [137] ^{13}C NMR ($CDCl_3$): 27.1 ($NCH_2\mathbf{C}H_2C$), 45.3 (CH_3), 60.2 ($N\mathbf{C}H_2CH_2C$), 91.3, 95.1 (C-2,5 and C-3,4 of C_5H_4), 116.9 (C-1 of C_5H_4), 220.9 (CO_{cis}), 237.3 (CO_{trans}) [137] IR (THF): 1956, 2036 (CO) [137] mass spectrum (70 eV): $[M]^+$ (4%), $[M-nCO]^+$ (n = 1 (17%), 2 (25%)) [137] thin-layer chromatography (SiO_2): R_f = 0.39 (10% CH_3OH in CH_2Cl_2) [137] UV irradiation of THF solutions afforded $C_5H_4CH_2CH_2N(CH_3)_2Mo(CO)_2I$ containing an intramolecularly coordinated amino group [137]
117	$(CH_3)_2NCH(CH_3)CH_2-/I$	IXa (70%) [137] red solid, m.p. 35 to 36°C [137] ^{1}H NMR ($CDCl_3$): 1.01 (d, CCH_3; J = 6.5), 2.26 (s, $N(CH_3)_2$), 2.31 to 2.38 (m, CH), 2.60 to 2.74 (m, CH_2), 5.39 to 5.43 (m, C_5H_4, 2H), 5.51 to 5.54, 5.56 to 5.59 (both m, C_5H_4, 1H each) [137] ^{13}C NMR ($CDCl_3$): 13.2 ($C\mathbf{C}H_3$), 33.1 (CH_2), 40.2 ($N(CH_3)_2$), 61.2 (NCH), 91.1 (C_5H_4, 2C), 95.5, 95.8 (C_5H_4, 1C each), 116.4 (C_5H_4, 1C), 220.8 (CO_{cis}), 237.3 (CO_{trans}) [137] photolysis in THF resulted in CO elimination and intramolecular amino group coordination, producing $C_5H_4CH_2CH(CH_3)N(CH_3)_2Mo(CO)_2I$ [137]

References on pp. 253/8

Table 6 (continued)

No.	compound	method of preparation (yield) properties and remarks
118	anthraquinon-2-yl–C(O)NHCH$_2$CH$_2$– /Cl	Xb (quantitative; not isolated) [177] IR (THF): 1956, 1971, 2050 (CO) [177]
119	anthraquinon-2-yl–C(O)NHCH$_2$CH$_2$– /I	Xb (quantitative; not isolated) [177]; together with $C_{14}H_7O_2-C(O)NHCH_2CH_2C_5H_4Mo(CO)_3CH_3$; also by reduction of $(C_{14}H_7O_2-C(O)NHCH_2CH_2C_5H_4Mo(CO)_3)_2$ ($C_{14}H_7O_2$ = anthraquinon-2-yl) with cobaltocene or 1e-reduced anthraquinone in THF in the presence of CH_3I (for mechanism, see [177]) IR (THF): 1954, 1960, 2037 (CO) [177]
120	$HOCH_2CH_2$–/Cl	Xb (quantum yield, Φ_{504} = 0.66 (2 M CCl_4 in THF)) [131] IR (THF): 1969, 2048 (CO) [131]
121	$CH_3C(O)OCH_2$–/Cl	from $Na[HC(O)C_5H_4Mo(CO)_3]$ by treatment with $CH_3C(O)Cl$ in THF; the evaporated mixture was chromatographed on silica gel, eluting with toluene (94%); further elution, first with CH_2Cl_2 and then with THF, afforded $(HC(O)C_5H_4Mo(CO)_3)_2$ followed by $CH_3C(O)OCH_2C_5H_4Mo(CO)_3C(O)CH_3$ [80] dark red crystals, m.p. 200°C (dec.) [80] ^{1}H NMR (C_6D_6): 1.64 (s, CH_3), 4.47 (s, CH_2), 4.29, 4.90 (both "t", C_5H_4) [80]; ($CDCl_3$): 2.11 (s, CH_3), 4.81 (s, CH_2), 5.47, 5.75 (both "t", C_5H_4) [80] ^{13}C NMR (C_6D_6): 20.1 (CH_3), 59.3 (CH_2), 92.3, 95.8 (C-2,5 and C-3,4 of C_5H_4), 115.9 (C-1 of C_5H_4), 169.6 (C=O), 224.3 (CO_{cis}), 242.8 (CO_{trans}) [80] IR (KBr): 1725 (ν(C=O)), 1950, 1973, 1983, 2050 (CO) [80]; ($CDCl_3$): 1743 (ν(C=O)), 1987, 2060 (CO) [80] freely soluble in toluene and CH_2Cl_2 [73]
122	$C_6H_5C(O)OCH_2$–/Cl	similarly to No. 121 from $Na[HC(O)C_5H_4Mo(CO)_3]$ and $C_6H_5C(O)Cl$ in THF (38%); Mo–Mo- or Mo–C-bonded homobinuclear and acyl by-products were not observed [80]

Table 6 (continued)

No.	compound	method of preparation (yield) properties and remarks
		red microcrystals, m.p. 200 °C (dec.) [80] IR (KBr): 1705 (ν(C=O)), 1905, 1950, 2040 (CO) [80] readily soluble in toluene and CH_2Cl_2 [80]
123	$CH_3(CH_2)_5NHC(O)OCH_2CH_2-$/Cl	Xb (quantum yield, Φ_{504} = 0.59 (2 M CCl_4 in THF) [131] IR (THF): 1539 (NH, "amide II"), 1728 (ν(C=O)), 1970, 2049 (CO) [131]
124	HC(O)−/I	IXa (33%) [73] red-brown crystals, m.p. 128 to 130 °C (dec.) [73] ^{1}H NMR ($CDCl_3$): 5.75, 6.02 (both "t", C_5H_4), 9.73 (s, HCO) [73] IR (KBr): 1690 (ν(C=O)), 1962, 1973, 2047 (CO) [73]
125	$CH_3C(O)-$/Cl	IXb (36%, using $P(C_6H_5)Cl_2$) [135] red solid, m.p. 133 °C [135] ^{1}H NMR ($CDCl_3$): 2.43 (s, CH_3), 5.73 ("t", H-3,4 of C_5H_4), 6.00 ("t", H-2,5 of C_5H_4) [135] IR (KBr): 1688 (ν(C=O)), 1971, 1999, 2039 (CO) [135] mass spectrum: $[M]^+$ (3.7%), $[M-nCO]^+$ (n = 1 (19.7%), 2 (41.7%), 3 (100%)) [135]
126	$CH_3C(O)-$/Br	IXb (47%, using PBr_3) [135] red crystals, m.p. 118 to 119 °C [135] ^{1}H NMR ($CDCl_3$): 2.46 (s, CH_3), 5.73 ("t", H-3,4 of C_5H_4), 6.03 ("t", H-2,5 of C_5H_4) [135] IR (KBr): 1688 (ν(C=O)), 1971, 1992, 2051 (CO) [135] mass spectrum: $[M]^+$ (10.5%), $[M-nCO]^+$ (n = 1 (32.1%), 2 (79.7%), 3 (100%)) [135]
*127	$CH_3C(O)-$/I	IXb (77%, with $As(CH_3)I_2$) [135]; Xa (57%) [21] red crystals [135], m.p. 88 °C [21], 91 to 93 °C [135] ^{1}H NMR ($CDCl_3$): 2.43 (s, CH_3), 5.73 ("t", H-3,4 of C_5H_4), 6.03 ("t", H-2,5 of C_5H_4) [135]; similar in [21] IR (KBr): 1688 (ν(C=O)), 1950, 1986, 2045 (CO) [135]; ($CHCl_3$): 1695 (ν(C=O)), 1984, 2059 (CO) [21] mass spectrum: $[M]^+$ (33.7%), $[M-nCO]^+$ (n = 1 (88.2%), 2 (88.4%), 3 (100%)), $[M-3CO-I]^+$ [21, 135]

References on pp. 253/8

Table 6 (continued)

No.	compound	method of preparation (yield) properties and remarks
*127 (continued)		in THF/[N(C_4H_9-n)$_4$]PF_6 solution, two-electron electroreduction with formation of [$CH_3C(O)C_5H_4Mo(CO)_3$]$^-$ and I^- occurred at E_{red} = −0.82 V vs. SCE [21]
128	$HOCH_2C(O)$−/Cl	Xb [131] IR (THF): 1703 (ν(C=O)), 1981, 2058 (CO) [131]
129	$CH_3OC(O)$−/Cl	IXb (20%, using PCl_3) [135]; also by controlled thermal decomposition of preformed $CH_3OC(O)C_5H_4Mo(CO)_3PCl_2$ in THF (isolated by column chromatography on silica gel eluting with CH_2Cl_2/petroleum ether (1:1), 24%) [135] red crystals, m.p. 97 to 98 °C [135] ^{1}H NMR ($CDCl_3$): 3.87 (s, CH_3), 5.80 ("t", H-3,4 of C_5H_4), 6.10 ("t", H-2,5 of C_5H_4) [135] IR (KBr): 1721 (ν(C=O)), 1967, 1983, 2008, 2065 (CO) [135] mass spectrum: [M]$^+$ (3%), [M−nCO]$^+$ (n = 1 (31%), 2 (50%), 3 (75%)) [135]
130	$CH_3OC(O)$−/Br	IXb (58%, using PBr_3) [135] red crystals, m.p. 103 to 104 °C [135] ^{1}H NMR ($CDCl_3$): 3.87 (s, CH_3), 5.77 ("t", H-3,4 of C_5H_4), 6.10 ("t", H-2,5 of C_5H_4) [135] IR (KBr): 1729 (ν(C=O)), 1973, 1991, 2054 (CO) [135] mass spectrum: [M]$^+$ (5%), [M−nCO]$^+$ (n = 1 (29%), 2 (64%), 3 (42%)) [135]
131	$CH_3OC(O)$−/I	IXb (84%, using PI_3) [135]; Xa (70%) [21] orange-red flakes [21], red crystals [135], m.p. 91 to 93 °C [135], 115 °C [21] ^{1}H NMR ($CDCl_3$): 3.87 (s, CH_3), 5.73 ("t", H-3,4 of C_5H_4), 6.07 ("t", H-2,5 of C_5H_4) [135]; similar in [21] IR (KBr): 1721 (ν(C=O)), 1962, 1988, 2050 (CO) [135]; (hexane): 1735 (ν(C=O)), 1975, 2042 (CO) [21] mass spectrum: [M]$^+$ (25%), [M−nCO]$^+$ (n = 1 (69%), 2 (61%), 3 (36%)) [135]; see also [21] two-electron electroreduction in THF/[N(C_4H_9-n)$_4$]PF_6 produced [$CH_3OC(O)C_5H_4Mo(CO)_3$]$^-$ accompanied by I^- (E_{red} = −0.87 V vs. SCE) [21]
132	$C_2H_5OC(O)$−/Cl	IXb (20%, using PCl_3) [135] red crystals, m.p. 73 to 74 °C [135]

References on pp. 253/8

Table 6 (continued)

No.	compound	method of preparation (yield) properties and remarks
		^{1}H NMR ($CDCl_3$): 1.35 (t, CH_3; J = 7.2), 4.32 (q, CH_2), 5.78 ("t", H-3,4 of C_5H_4), 6.05 ("t", H-2,5 of C_5H_4) [135] IR (KBr): 1713 (ν(C=O)), 1975, 2000, 2057 (CO) [135] mass spectrum: $[M]^+$ (6.1%), $[M-nCO]^+$ (n = 1 (36.3%), 2 (65.1%), 3 (100%)) [135]
133	$C_2H_5OC(O)-/Br$	IXb (50%, using PBr_3) [135] red crystals, m.p. 62 to 64°C [135] ^{1}H NMR ($CDCl_3$): 1.33 (t, CH_3; J = 7.2), 4.37 (q, CH_2), 5.80 ("t", H-3,4 of C_5H_4), 6.10 ("t", H-2,5 of C_5H_4) [135] IR (KBr): 1721 (ν(C=O)), 1974, 2056 (CO) [135] mass spectrum: $[M]^+$ (8.3%), $[M-nCO]^+$ (n = 1 (50.8%), 2 (100%), 3 (97.8%)) [135]
134	$C_2H_5OC(O)-/I$	IXa (74%) [135] red crystals, m.p. 79 to 81°C [135] ^{1}H NMR ($CDCl_3$): 1.35 (t, CH_3; J = 7.2), 4.32 (q, CH_2), 5.76 ("t", H-3,4 of C_5H_4), 6.04 ("t", H-2,5 of C_5H_4) [135] IR (KBr): 1721 (ν(C=O)), 1972, 2046 (CO) [135] mass spectrum: $[M]^+$ (20.4%), $[M-nCO]^+$ (n = 1 (55.0%), 2 (51.1%), 3 (41.9%)) [135]
135	$CH_3(R)R'Si-/I$ (R/R' = CH_3/CH_3, CH_3/C_6H_5, C_6H_5/C_6H_5)	Xa [5]
136	$(CH_3)_3Si(CH_3)_2Si-/Cl$	IXc (10%, using ClC_2H_4Cl) [159], (50%, using CCl_4) [148, 159, 161] red crystals, m.p. 83.5°C [159, 161] ^{1}H NMR ($CDCl_3$): −0.07 (s, $Si(CH_3)_3$), 0.08 (s, $Si(CH_3)_2$), 5.30, 5.76 (both "t", C_5H_4) [159, 161]; 0.05 (s, $Si(CH_3)_3$), 0.15 (s, $Si(CH_3)_2$), 5.37, 5.83 (both "t", C_5H_4) [148] IR (KBr): 1942.2, 1966.8, 2048.8 (CO) [148, 159, 161]
*137	$(CH_3)_3Si(CH_3)_2Si-/Br$	IXc (82%, using N-bromosuccinimide) [148, 158, 159, 161], (88%, using BrC_2H_4Br) [159] orange-red crystals, m.p. 76 to 77°C [158, 159, 161] ^{1}H NMR ($CDCl_3$): 0.15 (s, $Si(CH_3)_3$), 0.38 (s, $Si(CH_3)_2$), 5.36, 5.81 (both "t", C_5H_4) [158, 159, 161]; 0.19 (s, $Si(CH_3)_3$), 0.42 (s, $Si(CH_3)_2$), 5.40, 5.85 (both "t", C_5H_4) [148]

References on pp. 253/8

Table 6 (continued)

No.	compound	method of preparation (yield) properties and remarks
*137 (continued)		IR (KBr): 1942.2, 1975.0, 2040.6 (CO) [148, 158, 159, 161] prolonged treatment with excess $P(C_6H_5)_3$ in CH_2Cl_2/hexane gave $(CH_3)_3Si(CH_3)_2SiC_5H_4Mo(CO)(P(C_6H_5)_3)_2Br$ as a mixture of cis and trans isomers [158]
138	$(CH_3)_3Si(CH_3)_2Si-/I$	IXc (95%, using ICH_2CH_2I) [159]; Xa (91%) [148, 159, 161] purple-red crystals, m.p. 68 °C [159, 161] ^1H NMR ($CDCl_3$): 0.08 (s, $Si(CH_3)_3$), 0.31 (s, $Si(CH_3)_2$), 5.39, 5.78 (both "t", C_5H_4) [148, 159, 161] IR (KBr): 1937.6, 1960.7, 2032.4 (CO) [148, 159, 161]
139	$(C_6H_5)_2P-/I$	IXa (44%) [139]; also by iodination of $(C_6H_5)_2PC_5H_4Mo(CO)_3Mn(CO)_5$ (quantitative) [178] slightly oily red material [139] ^1H NMR ($CDCl_3$): 5.40, 5.83 (both "t", C_5H_4; J = 2.2), 7.3 to 7.8 (m C_6H_5) [139]; see also [178] ^{13}C NMR ($CDCl_3$): 97.7 (d, C-2,5 of C_5H_4; J(P, C) = 12.1), 98.8 (s, C-3,4 of C_5H_4), 100.8 (br, C-1 of C_5H_4), 128 to 136 (m, C_6H_5), 217.9, 219.5, 232.0 (s, 1CO each) [139]; see also [178] ^{31}P NMR ($CDCl_3$): 32.39 [178]; −159.4 (vs. $P(OCH_3)_3$) [139] IR (hexane): 1950, 1970, 2044 (CO) [139]; (CH_2Cl_2): 1968.5, 2004.7, 2017.8, 2044.4, 2085.9 (CO) [178] combination with $Li[(C_6H_5)_2PC_5H_4Mo(CO)_3]$ in 1:1 stoichiometry in refluxing THF afforded $(\mu\text{-}(C_6H_5)_2PC_5H_4Mo(CO)_2)_2$ accompanied by minor amounts of $(C_6H_5)_2PC_5H_4Mo(CO)_2$-$(\mu\text{-}(C_6H_5)_2PC_5H_4)Mo(CO)_3$ [139]; treatment with $Pd(P(C_6H_5)_3)_4$ or with Pd(dibenzylindeneacetone)$_2$/$P(C_6H_5)_3$ in THF gave $Mo(CO)_3(\mu\text{-}C_5H_4P(C_6H_5)_2)Pd(P(C_6H_5)_3)I$ [178]
Mo−C-bonded complexes $RC_5H_4Mo(CO)_3R'$:		
R/R′ =		
*140	$CH_3-/-CH_3$	XIa (45 to 65%) [8, 18, 19, 28]; XII (7%) [19] bright yellow oil, m.p. ca. 10 °C [8, 19]; b.p. 50 °C/10^{-4} Torr [18, 28], 65 °C/10^{-3} Torr [8]

Table 6 (continued)

No.	compound	method of preparation (yield) properties and remarks
		^1H NMR (THF-d_8): 0.29 (s, $MoCH_3$), 1.93 (s, CCH_3), 5.25 (m, C_5H_4) [18, 28]; only slightly different in [8] ^{13}C NMR (C_6D_6): −18.6 (q, $MoCH_3$; $^1J(C, H)$ = 137.1), 12.9 (q, C**C**H_3; $^1J(C, H)$ = 128.5), 89.2 (dm, C-3,4 of C_5H_4; $^1J(C, H)$ = 177.5), 93.2 (dm, C-2,5 of C_5H_4; $^1J(C, H)$ = 175.0), 111.3 (s, C-1 of C_5H_4), 227.7 (CO_{cis}), 241.0 (s, CO_{trans}) [11] IR (KBr): 1925, 2015 (CO) [8] UV (n-hexane, log ε): 222 (4.18), 258 (3.82), 314 (3.33), 353 (3.07) [8]
141	CH_3−/−CD_3	XIa (60%) [18, 28] ^1H NMR (THF-d_8): 1.93 (s, CH_3), 5.25 (m, C_5H_4) [18, 28] heating an equimolar mixture of $CH_3C_5H_4Mo(CO)_3CD_3$ and $C_5H_5Mo(CO)_3CH_3$ in THF-d_8 to 50°C gave none of the expected crossover products, $CH_3C_5H_4Mo(CO)_3CH_3$ or $C_5H_5Mo(CO)_3CD_3$; adding $C_5H_5Mo(CO)_3H$ to the mixture induced the smooth formation of acetaldehydes, CH_3CHO and CD_3CHO, still without any sign of crossover [16, 18, 28]
142	CH_3−/−C_2H_5	XIa (35%) [19] yellow oil [19] ^1H NMR (C_6D_6): 1.48 (s, CH_3), 1.53 (m, C_2H_5), 4.52 ("s", C_5H_4) [19] IR (film): 1905, 2010 (CO) [19] UV (n-hexane, log ε): 254 (3.92), 315 (3.36), 357 (3.04) [19] for m/e values of principal fragments in mass spectrum, see [19]
143	CH_3−/−$CH_2C_6H_5$	XIa (64%) [18, 28, 58] ^1H NMR (THF-d_8): 1.96 (s, CH_3), 2.82 (s, CH_2), 5.20 (m, C_5H_4), 7.12 (m, C_6H_5) [18, 28] ^{95}Mo NMR ($CHCl_3$): T_1 = 5.84 ± 0.07 ms [65] $P(CH_3)_2C_6H_5$-induced CO insertion in CH_3CN at 30°C afforded cis- and trans-$CH_3C_5H_4$-$Mo(CO)_2(P(CH_3)_2C_6H_5)C(O)CH_2C_6H_5$ (rate constant, k_1, for the formation of the initial solvent-stabilized acyl intermediate: 4.7×10^{-4} s^{-1}); under the same conditions, t-C_4H_9NC gave the expected $CH_3C_5H_4Mo(CO)_2(CNC_4H_9$-t$)C(O)CH_2C_6H_5$ with

References on pp. 253/8

Table 6 (continued)

No.	compound	method of preparation (yield) properties and remarks
143 (continued)		rate constants k_1 of $5.3 \times 10^{-4}\ s^{-1}$ and k_3 (for acyl formation by direct attack of the entering nucleophile) of $5.2 \times 10^{-3}\ L \cdot mol^{-1} \cdot s^{-1}$ [58]
144	CH_3–/–$CD_2C_6D_5$	XIa [18, 28] m.p. 49 to 50°C [18, 28] ^{1}H NMR (THF-d_8): 1.96 (s, CH_3), 5.20 (m, C_5H_4) [18, 28] heating at 50°C in THF-d_8 in the presence of $C_5H_5Mo(CO)_3CH_2C_6H_5$ gave traces of $CH_3C_5H_4Mo(CO)_3CH_2C_6H_5$; with added $C_5H_5Mo(CO)_3H$, competing homolytic Mo–C bond cleavage and CO insertion resulted in the formation of toluene in addition to phenylacetaldehyde [16, 18, 28]
145	C_2H_5–/–CH_3	XIa (60%) [19]; XII (7%) [19] yellow oil [19] ^{1}H NMR (C_6D_6): 0.48 (s, $MoCH_3$), 0.88 (t, CCH_3), 1.90 (q, CH_2), 4.58 ("s", C_5H_4) [19] IR (film): 1935, 2020 (CO) [19] UV (n-hexane, log ε): 316 (3.30), 360 (2.97) [19] for most abundant ions in mass spectrum (m/e only), see [19]
146	C_2H_5–/–C_2H_5	XIa (65%) [19]; XII (5%) [19] bright yellow oil [19] ^{1}H NMR (C_6D_6): 0.82 (t, CH_3), 1.55 ("s", C_2H_5), 1.85 (q, CH_2), 4.50 ("s", C_5H_4) [19] IR (CCl_4): 1910, 2015 (CO) [19] UV (n-hexane, log ε): 256 (3.94), 312 (3.34), 363 (2.99) [19] for m/e values of mass spectroscopic fragments, see [19]
147	i-C_3H_7–/–CH_3	XIa (54 to 70%) [2, 3, 86], (80 to 90%) [112] orange-yellow oil [112], m.p. ca. −25°C [2, 3]; b.p. 105 to 110°C (bath temperature) [86] ^{1}H NMR (C_6D_6): 0.44 (s, $MoCH_3$), 0.83 (d, $CH(C\mathbf{H}_3)_2$; J = 7), 2.10 (sept, $C\mathbf{H}(CH_3)_2$), 4.44, 4.48 (both "t", C_5H_4; "J" = 2.5) [112]; similar in [86] ^{13}C NMR (C_6D_6): −20.7 (q, $MoCH_3$; J(C, H) = 137), 23.7 (q, $CH(\mathbf{C}H_3)_2$; J(C, H) = 126), 27.1 (d, $\mathbf{C}H(CH_3)_2$; J(C, H) = 128), 90.8, 91.0 (both d, C-2,5 and C-3,4 of C_5H_4; J(C, H) = 172 and 173), 122.8 (s, C-1 of C_5H_4), 227.2 (CO_{cis}), 240.6 (CO_{trans}) [86]; similar in [112]

Table 6 (continued)

No.	compound	method of preparation (yield) properties and remarks
		IR (film): 1925, 2017 (CO) [86]; (heptane): 1940, 2018 (CO) [86]; (CCl_4 or CS_2): 1938, 2030 (CO) [2, 3] treatment with PCl_5 (2.5 equivalents) in CH_2Cl_2, followed by refluxing, formed i-$C_3H_7C_5H_4MoCl_4$ [112]
148	i-C_3H_7–/–C_2H_5	XIa (60%) [2, 3] m.p. ca. −15°C [2, 3] IR (CCl_4 or CS_2): 1931, 2025 (CO) [2, 3]
149	$(C_2H_5)_2CH$–/–CH_3	XIa (60%) [2, 3] oil [2, 3] IR (CCl_4 or CS_2): 1931, 2028 (CO) [2, 3]
150	$(C_2H_5)_2CH$–/–C_2H_5	XIa [2, 3] oil [2, 3] IR (CCl_4 or CS_2): 1927, 2021 (CO) [2, 3]
151	$CH_3C(C_6H_5)H$–/–CH_3	XIa [2, 3] oil [2, 3] IR (CCl_4 or CS_2): 1931, 2021 (CO) [2, 3]
*152	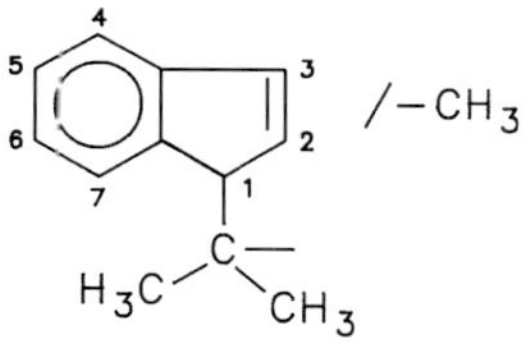 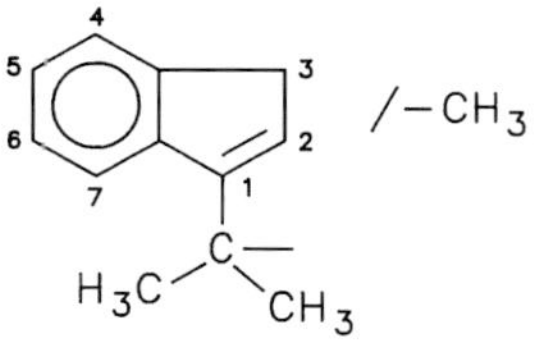 /–CH_3 a /–CH_3 b	XIa (27%); isolated as a 1:1 mixture of double bond isomers a and b [150] yellow solid, m.p. 83°C (dec.) (isomeric mixture) [150] isomer a: ^{1}H NMR (CD_2Cl_2): 0.45 (s, $MoCH_3$), 1.67 (s, $C(CH_3)_2$), 5.54, 5.60 (both "t", C_5H_4) [150] (indenyl resonances probably erroneous) ^{13}C NMR (CD_2Cl_2): −21.0 ($MoCH_3$), 29.9 ($C(\mathbf{C}H_3)_2$), 37.5 ($\mathbf{C}(CH_3)_2$), 62.6 (C-1), 92.0, 93.1, 145.8 (all C_5H_4), 124.7, 126.0, 127.3, 137.3, 132.8, 144.7, 152.2 (all indenyl), 227.8, 241.4 (both CO) [150] IR (pentane): 1936, 2020 (CO) [150] UV photolysis of the isomeric mixture in pentane solution resulted in CO elimination and intramolecular coordination of the five-membered indenyl ring producing $C_5H_4C(CH_3)_2C_9H_7Mo(CO)_2CH_3$ as a mixture of two diastereoisomers [150]; upon photolysis in toluene, the mixture of complexes No. 152a and 152b underwent dealkylation to give CH_4 together with the inden-1,2-diyl-bridged tricarbonyl $C_5H_4C(CH_3)_2C_9H_6Mo(CO)_3$ (No. 179) [153]

Table 6 (continued)

No.	compound	method of preparation (yield) properties and remarks
*152 (continued)		isomer b: yellow rods [150] ^1H NMR (CD_2Cl_2): 0.40 (s, $MoCH_3$), 1.25, 1.39 (both s, $1CCH_3$ each), 5.30, 5.40, 5.43, 5.48 (all m, C_5H_4) [150] (indenyl resonances probably erroneous) ^{13}C NMR (CD_2Cl_2): −21.1 ($MoCH_3$), 26.8, 29.2 (1C**C**H_3 each), 37.0 (CH_2), 38.1 (**C**CH_3), 90.6, 91.9, 92.7, 92.9, 146.1 (all C_5H_4), 124.5, 124.8, 125.4, 128.8, 137.5, 143.6, 152.2 (all indenyl), 227.8, 241.4 (both CO) [150] IR (pentane): 1936, 2020 (CO) [150]
*153	9-fluorenyl-C(CH_3)$_2$— (positions 1–9 numbered) /−CH_3	XIa [111] yellow rods [150], m.p. 119 °C [111, 126] ^1H NMR (CD_2Cl_2): 0.35 (s, $MoCH_3$), 1.32 (s, $C(CH_3)_2$), 3.88 (s, H-9), 4.68, 4.99 (both "t", C_5H_4), 7.28, 7.64 (both m, H-1 to 8) [111, 126] ^{13}C NMR (CD_2Cl_2): −21.0 ($MoCH_3$), 28.1 (C(**C**H_3)$_2$), 39.4 (**C**(CH_3)$_2$), 59.8 (C-9); 90.8, 93.1, 124.8 (all C_5H_4); 119.7, 126.3, 126.9, 127.8, 142.4, 145.2 (all fluorenyl); 227.8, 241.3 (both CO) [111, 126] IR (pentane): 1937, 2019 (CO) [111, 126] photolysis in toluene or THF resulted in elimination of methane and formation of the fluorene-1,9-diyl-bridged product $C_5H_4C(CH_3)_2C_{13}H_8Mo(CO)_3$ (No. 180) [111, 153] attempts to deprotonate the fluorenyl group at C-9 by treatment with either Na or NaH or LiC_4H_9-n were unsuccessful [111] UV irradiation in toluene in the presence of alkynes, RC≡CR, led to alkyne-substituted fluorene-1,9-diyl-bridged derivatives, $C_5H_4C(CH_3)_2C_{13}H_8Mo(RC{\equiv}CR)CO$ (R = H, CH_3) [126]
154	CH_2=CH−/−CH_3	synthesized by treating $HC(O)C_5H_4Mo(CO)_3CH_3$ with one equivalent of $(C_6H_5)_3P{=}CH_2$ in diethyl ether at 25 °C; isolated by column chromatography on Florisil, eluting with hexane (11%) [17, 22, 38]; in even lower yield (2%) also by dehydration of $CH_3CH(OH)C_5H_4Mo(CO)_3CH_3$ with catalytic amounts of 4-$CH_3C_6H_4SO_3H$ in refluxing benzene [38] distillable yellow liquid [38]

Table 6 (continued)

No.	compound	method of preparation (yield) properties and remarks
		^1H NMR ($CDCl_3$): 0.32 (s, CH_3), 5.1 to 5.5 (m, C_5H_4 and CH_2=), 5.9 to 6.5 (m, =CH) [22]; similar in [38] IR (KBr): 1920, 2005 (CO) [22, 38]
155	CH_2=C(CH_3)−/−CH_3	by analogy to No. 154 above from $CH_3C(O)C_5H_4Mo(CO)_3CH_3$ and $(C_6H_5)_3P$=CH_2 in ether at 25°C; yield after column chromatography (Florisil/hexane): 26% [22, 38] yellow solid subliming at 40°C/0.01 Torr [38], m.p. 54.5 to 56°C [38] ^1H NMR ($CDCl_3$): 0.33 (s, $MoCH_3$), 1.91 (d, =C(CH_3)), 5.0 to 5.2 (m, CH_2=), 5.28, 5.38 (both m, H-3,4 and H-2,5 of C_5H_4) [22, 38] IR (KBr): 1915, 2005 [38] (CO); similar in [22] attempts at cationic polymerization using $SnCl_4$ or $(C_2H_5)_2O \cdot BF_3$ as initiators were unsuccessful, the monomer being recovered largely unreacted [66]
156	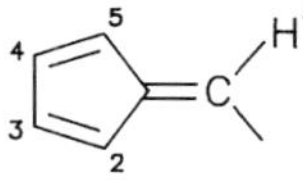 [structure: fulvene C_5H_4=C(H[6])– with ring positions 2–5] /−CH_3	XIb (68%) [110] yellow-orange needles [110] ^1H NMR (CD_2Cl_2): 0.39 (s, CH_3), 5.42, 5.62 (both "t", C_5H_4=CHC_5**H**$_4$), 6.23 (m, H-5; J(H-2,5) = 2.02, J(H-3,5) = 1.46, J(H-4,5) = 5.13), 6.45 (m, H-3; J(H-2,3) = 5.22, J(H-3,4) = 2.02), 6.51 (m, H-4; J(H-2,4) = 1.28), 6.61 (m, H-2), 6.67 ("s", H-6) [110] ^{13}C NMR (THF-d_8): −17.75 (CH_3), 93.42, 94.80, 107.84 (all C_5H_4=CH**C**$_5H_4$), 119.54, 127.23, 129.98, 131.38, 135.93, 145.17 (all **C**$_5H_4$=CHC_5H_4); 226.32, 239.30 (both CO) [110] IR (CH_2Cl_2): 1932, 2019 (CO) [110]
157	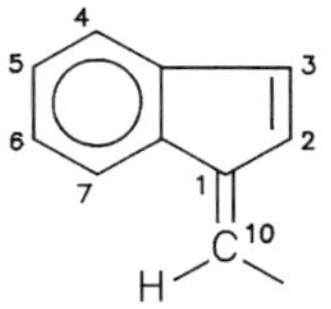 [structure: benzofulvene with positions 1–7, exocyclic =C[10](H)–] /−CH_3	XIb (66%) [138] amber needles, m.p. 123°C (dec.) [138] ^1H NMR ($CDCl_3$): 0.69 (s, CH_3), 5.36, 5.60 (both "t", C_5H_4), 6.79 (d, H-2; J(H-2,3) = 5.5), 6.85 ("s", H-10), 6.97 (dd, H-3; J(H-3,10) = 1.22), 7.20 to 7.29 (m, 3H), 7.56 (dd, 1H; J = 1.22 and 7.32) [138]; data correspond only partly with the spectrum shown in [138] ^{13}C NMR (THF-d_8): −19.5 (CH_3), 93.4, 95.2, 110.3 (all C_5H_4), 119.8 (C-4), 121.3 (C-10), 121.8 (C-7), 126.0 (C-2), 126.1 (C-5), 128.4 (C-6), 135.5 (C-3), 138.4 (C-8), 141.2 (C-1), 143.0 (C-9), 221.8, 240.3 (both CO) [138] IR (CH_2Cl_2): 1934, 2016 (CO) [138]

References on pp. 253/8

Table 6 (continued)

No.	compound	method of preparation (yield) properties and remarks
158	a: 3-cyclopentadienyl (C₅H₅ ring, 1,3-diene isomer a) /−CH₃ b: cyclopentadienyl isomer b /−CH₃	two isomers were obtained by dehydration of cyclo-3-$HOC_5H_6C_5H_4Mo(CO)_3CH_3$ (No. 162) using catalytic amounts of p-toluenesulfonic acid in benzene at 55 °C for short periods of time; total yield: 84% [78] yellow crystals, m.p. 105 to 107 °C (dec.) (isomeric mixture) [78] isomer a: ^1H NMR ($CDCl_3$): 0.21 (s, CH_3), 3.15 (ddd, CH_2; $J_1 = J_2 = J_3 = 1.3$); 5.24, 5.47 (both "t", C_5H_4; "J" = 2.3), 6.35 (m, 1 =CH), 6.48 (m, 2 =CH) [78] IR (CCl_4): 1941, 2019 (CO) [78] mass spectrum: $[M]^+$ (12%), $[C_{10}H_8Mo]^+$ (100%) [78] treatment of the isomeric mixture with $M(CO)_3(NCCH_3)_3$ (M = Cr, W) in refluxing THF produced the fulvalene-bridged metal−metal-bonded heterobimetallics $Mo(CO)_3(\mu\text{-}C_5H_4C_5H_4)M(CO)_3$ along with acetaldehyde [78] deprotonation of the diene using NaH in THF, followed by addition of one equivalent of $(Rh(CO)_2(\mu\text{-}Cl))_2$, yielded the Mo−Rh-bonded acyl $Mo(CO)_3(\mu\text{-}C_5H_4C_5H_4)Rh(CO)C(O)CH_3$ [166] isomer b: ^1H NMR ($CDCl_3$): 0.21 (s, CH_3), 3.08 (ddd, CH_2; $J_1 = J_2 = J_3 = 1.3$); 5.21, 5.40 (both "t", C_5H_4; J = 2.2), 6.32 (m, 1 =CH), 6.51 (m, 2 =CH) [78] further properties and reactions as for isomer a
159	HC≡C−/−CH_3	equimolar amounts of $(n\text{-}C_4H_9)_3SnC-CH$ and $IC_5H_4Mo(CO)_3CH_3$ were stirred in N,N-dimethylformamide in the presence of $Pd(NCCH_3)_2Cl_2$; after dilution with ether, addition of 50% aqueous KF, and separation of the organic phase, the mixture was chromatographed over silica using hexane as the eluant to give the product in trace quantities [115]
160	$C(CH_3)_2CH_2$− /−CH_3 (2-pyridyl substituent)	IR (hexane): 1940, 2020 (CO) [96] neither thermolysis in refluxing octane nor UV irradiation caused chelation of the pyridine moiety as a consequence of decarbonylation or CO insertion into the Mo−CH_3 bond [96]

Table 6 (continued)

No.	compound	method of preparation (yield) properties and remarks
161	(anthraquinone-2-yl)–C(O)NHCH$_2$CH$_2$– /–CH$_3$	together with $C_{14}H_7O_2-C(O)NHCH_2CH_2C_5H_4$-$Mo(CO)_3I$ by reduction of $(C_{14}H_7O_2-C(O)NHCH_2CH_2C_5H_4Mo(CO)_3)_2$ ($C_{14}H_7O_2$ = anthraquinon-2-yl) with cobaltocene or 1e-reduced anthraquinone in THF in the presence of CH_3I (not isolated; for the mechanism, see [177]) IR (THF): 1927, 2016 (CO) [177]
162	HO–(3-methylcyclopent-2-enyl) /–CH$_3$	obtained in 94% yield by reduction of No. 163, using Li[AlH_4] in diethyl ether at 0°C [78] yellow crystals, m.p. 74 to 76°C [78] ^{1}H NMR ($CDCl_3$): 0.27 (s, CH_3); 1.41, 1.81, 2.34, 2.58 (all br or m, CH_2CH_2 and C**H**OH), 4.94 (br, OH), 5.25 ("t", 2H of C_5H_4; "J" = 2.2), 5.33, 5.55 (both m, both 1H of C_5H_4), 5.85 (ddd, =CH; $J_1 = J_2 = J_3 = 2.0$) [78] IR (CCl_4): 1939, 2018 (CO) [78] mass spectrum: $[M]^+$ (0.7%), $[M-H_2O]^+$ (4.8%), $[C_{10}H_8Mo]^+$ (100%) [78] dehydration (catalytic 4-$CH_3C_6H_4SO_3H$ in C_6H_6 at 55°C) furnished an isomeric mixture of No. 158a, b [78]
163	O=(3-methylcyclopent-2-enone) /–CH$_3$	the following sequence of synthetic steps was adopted: treatment of Na[C_5H_5] · $CH_3OC_2H_4$-OCH_3 with a fourfold excess of dimethyl succinate in THF led to the precipitation of Na[$C_5H_4C(O)C_2H_4C(O)OCH_3$]; addition of $LiCH_2P(O)(OCH_3)_2$ at −78°C followed by dilution with hexane caused Na[$C_5H_4C(O)$-$C_2H_4C(O)CH_2P(O)(OCH_3)_2$] to separate; exposure of the latter to $Mo(CO)_3(NCC_2H_5)_3$ in refluxing $CH_3OC_2H_4OCH_3$ followed by methylation (CH_3I, 23°C) permitted cyclization to occur, producing No. 163 (27%) [78] yellow crystals, m.p. 72 to 73°C [78] ^{1}H NMR ($CDCl_3$): 0.35 (s, CH_3); 2.51, 2.73 (both m, CH_2CH_2), 5.43, 5.52 (both "t", C_5H_4; "J" = 2.3), 6.09 (t, =CH; J = 1.7) [78] IR (CCl_4): 1610, 1717 (ν(C=CC=O)), 1940, 2021 (CO) [78] mass spectrum: $[M]^+$ (16%), $[C_4H_9]^+$ (100%) [78]

References on pp. 253/8

Table 6 (continued)

No.	compound	method of preparation (yield) properties and remarks
163 (continued)		treatment with $Li[AlH_4]$ in ether at 0 °C resulted in reduction of the 3-oxocyclopent-1-enyl ring to the corresponding 3-hydroxy unit; see No. 162 [78]
164	$CH_3C(OH)H-/-CH_3$	$HC(O)C_5H_4Mo(CO)_3CH_3$ was combined with excess CH_3MgI in ether at 25 °C; after hydrolysis with aqueous NH_4Cl and removal of the volatiles, the product was separated by chromatography on alumina using pentane/ether as the eluant (32%) [38] distillable yellow liquid [38] ^1H NMR ($CDCl_3$): 0.38 (s, $MoCH_3$), 5.27 (C_5H_4) [38] IR (neat): 1910, 2005 (CO) [38] dehydration with p-toluenesulfonic acid resulted in low yields of $CH_2{=}CHC_5H_4Mo(CO)_3CH_3$ [38]
165	$HC(O)-/-CH_3$	XIa (63%) [17, 22, 38] air-stable, yellow crystals subliming at 50 °C/0.01 Torr, m.p. 75 °C (dec.) [38] ^1H NMR ($CDCl_3$): 0.45 (s, CH_3), 5.52 ("t", H-3,4 of C_5H_4), 5.72 ("t", H-2,5 of C_5H_4), 9.58 (s, HCO) [22, 38] IR (KBr): 1680 (ν(C=O)), 1920, 2020 (both CO) [22, 38] carbonyl olefination by means of $(C_6H_5)_3P{=}CH_2$ in diethyl ether at 25 °C led to $CH_2{=}CHC_5H_4Mo(CO)_3CH_3$ [17, 22, 38]; treatment with ethereal CH_3MgI, followed by hydrolysis, provided $CH_3CH(OH)C_5H_4Mo(CO)_3CH_3$ [38]
*166	$CH_3C(O)-/-CH_3$	XIa (91%) [17, 22, 38]; XIb (61%) [21] air-stable, yellow crystals subliming at 50 °C/0.01 Torr [21, 38], m.p. 61.5 to 63 °C [38], 66 °C [21] ^1H NMR ($CDCl_3$): 0.05 (s, $MoCH_3$), 2.40 (s, $CH_3C(O)$), 5.80, 6.00 (both "t", C_5H_4) [21]; 0.44 (s, $MoCH_3$), 2.35 (s, $CH_3C(O)$), 5.47 (t, H-3,4 of C_5H_4), 5.73 ("t", H-2,5 of C_5H_4) [22, 38] IR (film): 1665 (ν(C=O)), 1920, 2020 (CO) [38]; similar in [22]; (hexane): 1704 (ν(C=O)), 1947, 1955, 2030 (CO) [21] mass spectrum: $[M]^+$, $[M-nCO]^+$ (n = 1 to 3), $[M-3CO-CH_3]^+$ [21] in THF/$[N(C_4H_9\text{-}n)_4]PF_6$, irreversible one-electron reduction with formation of $[CH_3C(O)C_5H_4$-

Table 6 (continued)

No.	compound	method of preparation (yield) properties and remarks
		$Mo(CO)_3]^-$ and, presumably, $CH_3^{\bullet}$ occurred at $E_{red} = -1.62$ V vs. SCE [21] conversion under Wittig conditions as described for No. 165 above resulted in smooth production of $CH_2{=}C(CH_3)C_5H_4Mo(CO)_3CH_3$ [22]
167	$HO_2C-/-CH_3$	$C_5H_5Mo(CO)_3CH_3$ was reacted consecutively with excess LiC_4H_9-s and solid CO_2 in THF at −78°C; acidification (10% aqueous HCl, −30°C) followed by CH_2Cl_2 extraction afforded a 61% yield [167] yellow microcrystalline solid [167] ^{1}H NMR (acetone-d_6): 0.44 (s, CH_3), 5.68, 5.90 (both "t", C_5H_4; "J" = 2.5) [167] ^{13}C NMR (acetone-d_6): −19.60 (CH_3), 95.99, 96.15, 99.33 (all C_5H_4), 165.40 (CO_2H), 227, 239 (both CO) [167] IR (KBr): 1680 (ν(C=O)), 1917, 1950, 2025 (CO) [167] pK_a (C_2H_5OH/H_2O (1:1)): 4.2 [167] decomposed after short standing in base (pH 11 to 12) [167] treatment with equimolar N-hydroxy succinimide in the presence of dicyclohexyl carbodiimide in THF or with disuccinimidyl carbonate in CH_3CN in the presence of pyridine produced the activated ester derivative cyclo-$C_2H_4(CO)_2NO$-$C(O)C_5H_4Mo(CO)_3CH_3$ (No. 169) [167]
168	$CH_3OC(O)-/-CH_3$	XIa (36%) [17, 22, 38]; XIb (62%) [21] air-stable, lemon-yellow crystals subliming at 45°C/0.01 Torr [38], m.p. 61 to 62.5°C [22, 38], 77°C [21] ^{1}H NMR ($CDCl_3$): 0.04 (s, $MoCH_3$), 3.75 (s, CH_3O), 5.60, 5.85 (both "t", C_5H_4) [21]; 0.45 (s, CH_3), 3.83 (s, CH_3O), 5.39 ("t", H-3,4 of C_5H_4), 5.78 ("t", H-2,5 of C_5H_4) [22, 38] IR (KBr): 1710 (ν(C=O)), 1915, 2010 (CO) [38]; similar in [22]; (hexane): 1743 (ν(C=O)), 1950, 2030 (CO) [21] mass spectrum: $[M]^+$, $[M-nCO]^+$ (n = 1 to 3), $[M-3CO-CH_3]^+$ [21] in THF/$[N(C_4H_9\text{-}n)_4]PF_6$, irreversible one-electron reduction at $E_{red} = -1.80$ V vs. SCE generated the $[CH_3OC(O)C_5H_4Mo(CO)_3]^-$ anion and, presumably, also the $CH_3^{\bullet}$ radical [21]

References on pp. 253/8

Table 6 (continued)

No.	compound	method of preparation (yield) properties and remarks
169	succinimido-$NOC(O)-$ /$-CH_3$ (O=C ring C=O)	the metallocarboxylic acid $HO_2CC_5H_4Mo(CO)_3CH_3$ was treated with N-hydroxy succinimide in THF in the presence of equimolar dicyclohexyl carbodiimide; higher yields (57%) were attained by reacting the organometallic acid with disuccinimidyl carbonate in 2:1 stoichiometry in CH_3CN containing one equivalent of pyridine [167] ^1H NMR (acetone-d_6): 0.55 (s, CH_3), 2.92 (s, CH_2CH_2), 5.83, 6.10 (both "t", C_5H_4; "J" = 2.5) [167] ^{13}C NMR (acetone-d_6): −19.36 (CH_3), 22.50 (CH_2CH_2), 90.47, 94.60, 95.94 (all C_5H_4), 160.38, 166.76 (both C=O), 223.15, 235.71 (both CO) [167] IR (KBr): 1736, 1769, 1802 (ν(C=O)), 1936, 2028 (CO) [167]
*170	$-CH_2CH_2-$	XII (17%, using $Mo(CO)_3(CH_3OC_2H_4OC_2H_4OCH_3)_3)$ [8], (27%, using $Mo(CO)_3(NCCH_3)_3)$ [72] slightly air-sensitive orange needles, m.p. 104 to 106°C [8] ^1H NMR (CCl_4): −0.36 (t, $MoCH_2$; J = 8), 2.94 (t, CCH_2), 5.2 (m, C_5H_4) [8] ^{13}C NMR (C_6D_6): −49.6 (tt, $MoCH_2$; ^1J(C, H) = 151.9, ^3J(C, H) = 4.3), 21.9 (tm, **C**H_2; ^1J(C, H) = 135.1), 67.0 (s, C-1 of C_5H_4), 88.0 (dm, C-2,5 of C_5H_4; ^1J(C, H) = 177.5), 89.6 (dm, C-3,4; ^1J(C, H) = 178.3); 231 (CO) [11] IR (KBr): 1895, 2000 (CO) [8] UV (n-hexane, log ε): 220 (4.35), 262 (3.94), 296 (3.60), 426 (2.59) [8]
171	$-CH(CH_3)CH_2-$	XII [19] orange platelets, m.p. 57 to 58°C [19] ^1H NMR (C_6D_6): −0.92 ("t", 1 $MoCH_2$-H), −0.05 (dd, 1 $MoCH_2$-H; J = 6.2 and 9.8), 0.75 (d, CH_3; J = 7.0), 2.1 to 2.6 (m, CH), 4.6 (m, C_5H_4) [19] ^{13}C NMR (C_6D_6): −35.8 ($MoCH_2$), 21.7, 28.0 (CH and CH_3), 75.6, 85.3, 87.6, 89.7 (all C_5H_4) [19] IR ($CHCl_3$): 1912, 1940, 2005 (CO) [8] UV (n-hexane, log ε): 265 (3.9), 304 (3.5), 423 (2.5) [8] mass spectrum (70 eV): $[M]^+$ (18%), $[M-nCO]^+$ (n = 1 to 3) [19]

References on pp. 253/8

Table 6 (continued)

No.	compound	method of preparation (yield) properties and remarks
172	$-CH_2CH(CH_3)-$	XII (only obtained as a mixture with No. 171) [19] ^1H NMR (C_6D_6): 0.17 (m, MoCH), 1.39 (d, CH_3; J = 6.8), 1.94 (dd, $1CH_2$-H; J = 6.5 and 12.7), 2.74 (dd, $1CH_2$-H; J = 10.0 and 12.7), 4.4 to 4.6 (m, C_5H_4) [19] ^{13}C NMR (C_6D_6): −32.2 (MoCH), 26.8, 31.7 (CH_2 and CH_3), 69.1, 88.1, 89.3, 89.8 (all C_5H_4) [19]
173	$-C(CH_3)_2CH_2-$	XII (79%) [55] orange crystals, m.p. 61 °C [55] ^1H NMR (C_6D_6): −0.43 (s, $MoCH_2$), 0.77 (s, CH_3), 4.59 ("s", C_5H_4) [55] IR (KBr): 1892, 1962, 2006 (CO) [55] mass spectrum (70 eV): $[M]^+$ (19%), $[M-nCO]^+$ (n = 1 (18%), 2 (21%), 3 (100%)), $[Mo]^+$ (42%) [55]
174	H_3C H H—┼—┼—CH_3	XII (59%) [55] orange crystals, m.p. 39°C [55] ^1H NMR (C_6D_6): −0.18 (dq, MoCH; ^3J(HCCH) = $^3J(HCCH_3)$ = 7), 0.72, 1.41 (both d, both CH_3), 1.91 (dq, CCH; $^3J(HCCH_3)$ = 7), 4.39, 4.49 (both m, C_5H_4, 1H each), 4.54 (m, C_5H_4, 2H) [55] IR (KBr): 1890, 1905, 1930, 2010 (CO) [55] mass spectrum (70 eV): $[M]^+$ (9%), $[M-nCO]^+$ (n = 1 (8%), 2 (15%), 3 (100%)) [55]
175	H_3C CH_3 H—┼—┼—H	XII (56%) [55] orange crystals; m.p. 49°C [55] ^1H NMR (C_6D_6): 0.68 (dq, MoCH; ^3J(HCCH) = 10, $^3J(HCCH_3)$ = 7.25), 0.79, 1.35 (both d, CH_3), 2.70 (dq, CCH; $^3J(HCCH_3)$ = 7.25), 4.4 to 4.5 (m, C_5H_4, 3H), 4.62 (m, C_5H_4, 1H) [55] IR (KBr): 1953, 1995 (CO) [55] mass spectrum (70 eV): $[M]^+$ (9%), $[M-nCO]^+$ (n = 1 (8%), 2 (15%), 3 (100%)) [55]
176	$-C(CH_3)_2C(CH_3)_2-$	XII (61%) [55] orange crystals, m.p. 69°C [55] ^1H NMR (C_6D_6): 0.82 (s, $MoC(CH_3)_2$), 1.51 (s, $CC(CH_3)_2$), 4.46 (m, C_5H_4) [55] IR ($CHCl_3$): 1900, 1930, 2010 (CO) [55] mass spectrum (70 eV): $[M]^+$ (19%), $[M-CO-2H]^+$ (19%), [M−nCO−4H] (n = 2 (37%), 3 (100%)), $[Mo]^+$ (31%) [55]

References on pp. 253/8

Table 6 (continued)

No.	compound	method of preparation (yield) properties and remarks
177	$-CH_2CH_2CH_2CH_2-$	XII (21%) [8] relatively air-stable, bright yellow platelets, m.p. 54 °C [8] ^1H NMR ($CDCl_3$): 1.7 (m, 1CH_2), 2.3 (m, 3CH_2), 5.14 (m, C_5H_4) [8] IR (KBr): 1904, 2000 (CO) [8] UV (n-hexane, log ε): 221 (4.27), 254 (3.95), 313 (3.38), 361 (2.58) [8]
178	H H $-CH_2$ CH_2-	XII (13%) [7, 19] bright yellow needles, m.p. 91 °C [7, 19] ^1H NMR ($CDCl_3$): 2.32 (d, CH_2; J = 7.5), 2.74 (d, CH_2; J = 5.0), 5.03, 5.54 (both "t", C_5H_4; "J" = 2.2), 6.4 (m, CH=CH) [7, 19] IR (KBr): 1910, 2005 (CO) [7, 19] mass spectrum (70 eV): $[M]^+$ (4%), $[M-nCO]^+$ (n = 1 to 3) [19]
*179	4 5 3 6 2 7 1 C— H_3C CH_3	$C_9H_7C(CH_3)_2C_5H_4Mo(CO)_3CH_3$ (No. 152) was photochemically dealkylated in toluene; column chromatography on silica gel eluting with pentane/toluene followed by crystallization from CH_2Cl_2/hexane gave the crystalline product (40%) [153] orange rods, m.p. 103 °C (dec.) [153] ^1H NMR (acetone-d_6): 1.64 (s, CH_3), 3.32 (s, H-3), 5.37, 6.57 (both "t", C_5H_4; "J" = 4.6), 6.90, 7.10 (both "t", H-5,6; "J" = 5.3 each), 7.22, 7.28 (both d, H-4,7; J = 7.8 and 7.7) [153] ^{13}C NMR (acetone-d_6): 21.4 (C(**C**$H_3)_2$), 40.6 (**C**$(CH_3)_2$), 50.3 (C-3), 86.6, 91.9, 129.7 (all C_5H_4), 116.5, 122.3, 123.0, 126.1, 144.3, 146.1, 149.1, 151.1 (all indenyl), 227.5, 239.4 (both CO) [153] IR (pentane): 1941, 1956, 2028 (CO) [153]
180	5 4 6 3 7 2 8 9 1 C— H_3C CH_3	by UV irradiation of $C_{13}H_9C(CH_3)_2C_5H_4Mo(CO)_3CH_3$ (No. 153) in toluene or THF; isolated by chromatographing the evaporated mixture on silica gel with pentane/toluene (2:1) as the eluant [111] m.p. 122 °C (dec.) [111] ^1H NMR (CD_2Cl_2): 0.34, 1.81 (both s, CH_3), 3.84 (s, H-9), 5.15, 5.27, 5.57, 6.06 (all m, C_5H_4), 7.01 to 7.71 (m, H-2 to H-8); spectrum depicted in [111]

References on pp. 253/8

Table 6 (continued)

No.	compound	method of preparation (yield) properties and remarks
		^{13}C NMR (CD_2Cl_2): 19.1, 31.6, 35.8 (all $C(CH_3)_2$), 65.4 (C-9); 84.7, 86.0, 89.2, 91.1, 134.7 (all C_5H_4), 116.1, 120.0, 126.2, 127.8, 146.4 (all fluorenyl CH), 140.5, 142.9, 143.2, 144.1, 154.0 (all fluorenyl C_{quart}), 227.0, 230.3, 239.8 (all CO) [111] IR (pentane): 1945, 1952, 2019, 2027 (CO) [111]
*181	SCH_3 CH_3S CH_3S CH_3S	see under "Further information" yellow crystals, m.p. 71 °C (dec.) [31] ^1H NMR ($CDCl_3$): 2.11, 2.19, 2.21, 2.27 (all s, all SCH_3), 4.22 (d, 1H), 4.80 to 5.36 (m, 6H), 6.42 (m, 1H) [31] IR (hexane): 1934, 1940, 2014 (CO) [31] solutions underwent rapid decomposition at room temperature [31]
182	$C_6H_5OSi(CH_3)_2$–/–CH_3	XIa (69%) [124] air-stable, yellow-green crystals, m.p. 53 °C [124] ^1H NMR ($CDCl_3$): 0.39 (s, $MoCH_3$), 0.49 (s, $Si(CH_3)_2$; $^3J(Si, H)$ = 6.9), 5.24, 5.51 (both "t", C_5H_4; "J" = 2.1), 6.72 (d, ortho-H; J = 7.4), 6.90 (t, para-H; J = 7.4), 7.13 ("t", meta-H) [124] IR (Nujol): 1934, 2020 (CO) [124]
183	$(CH_3)_3Si(CH_3)_2Si$–/–CH_3	XIa (32%) [161] yellow solid [161] ^1H NMR ($CDCl_3$): 0.1 (s, $Si(CH_3)_3$), 0.30 (s, $Si(CH_3)_2$), 0.36 (s, $MoCH_3$), 5.1, 5.5 (both "t", C_5H_4) [161] IR (mull): 1911.8, 2005.0 (CO) [161]
184	$(CH_3)_3Ge$–/–CH_3	XIII (30%) [100] m.p. 61 to 62 °C [100] ^1H NMR ($CDCl_3$): 0.33 ($Ge(CH_3)_3$), 0.57 ($MoCH_3$), 4.68, 4.89 (both C_5H_4) [100] ^{13}C NMR ($CDCl_3$): −21.7 ($MoCH_3$), −0.71 ($Ge(CH_3)_3$), 92.5, 96.6, 97.2 (all C_5H_4), 227.8 (CO) [100] IR (hexane): 1904, 1936, 2021 (CO) [100]
185	$(CH_3)_3Sn$–/–CH_3	XIII (competitive Mo–Sn bond cleavage in $C_5H_5Mo(CO)_3Sn(CH_3)_3$ resulted in the formation of a 1:9 mixture with $C_5H_5Mo(CO)_3CH_3$) [100] ^1H NMR (C_6D_6): 0.14 ($Sn(CH_3)_3$), 0.43 ($MoCH_3$) [100]

References on pp. 253/8

Table 6 (continued)

No.	compound	method of preparation (yield) properties and remarks
185 (continued)		^{13}C NMR ($CDCl_3$): −21.9 ($MoCH_3$), −8.57 ($Sn(CH_3)_3$), 94.0, 97.3, 99.0 (all C_5H_4), 227.8 (CO) [100] IR (hexane): 1934, 2021 (CO) [100]
186	$(C_6H_5)_3Sn$−/−CH_3	XIII (40%) [100] yellow solid, m.p. 110 to 111 °C [100] ^{1}H NMR ($CDCl_3$): 0.26 ($MoCH_3$), 5.2, 5.6 (both C_5H_4), 7.3 to 7.6 (C_6H_5) [100] ^{13}C NMR ($CDCl_3$): −21.7 ($MoCH_3$), 92.5, 96.9, 99.6 (all C_5H_4), 128.7, 129.5, 136.9 (all C_6H_5), 226.2 (CO) [100] IR (hexane): 1929, 2019 (CO) [100]
187	$(CH_3)_3Pb$−/−CH_3	XIII (isolated as a 1:1 mixture with $C_5H_5Mo(CO)_3CH_3$, the latter resulting from competitive Mo−Pb bond cleavage in $C_5H_5Mo(CO)_3Pb(CH_3)_3$) [100] ^{1}H NMR ($CDCl_3$): 0.3 ($MoCH_3$), 0.9 ($Pb(CH_3)_3$), 5.0, 5.2 (both C_5H_4) [100] ^{13}C NMR ($CDCl_3$): −21.7 ($MoCH_3$), 0.31 ($Pb(CH_3)_3$), 92.8, 99.8, 100.3 (all C_5H_4), 227.1 (CO) [100] ^{207}Pb NMR (C_6D_6): −4.1 (vs. $Pb(CH_3)_4$) [116, 120] IR (hexane): 1932, 2018 (CO) [100]
188	$(C_6H_5)_3Pb$−/−CH_3	XIII (68%) [100] m.p. 92 to 94 °C [100] ^{1}H NMR ($CDCl_3$): 0.29 ($MoCH_3$), 5.3, 5.5 (both C_5H_4), 7.3 to 7.7 (C_6H_5) [100] ^{13}C NMR ($CDCl_3$): −21.1 ($MoCH_3$), 97.4, 100.4 (both C_5H_4), 129.0, 130.0, 138.0, 149.0 (all C_6H_5), 227 (CO) [100] ^{207}Pb NMR (C_6D_6): −174.6 (vs. $Pb(CH_3)_4$) [116, 120] IR (hexane): 1938, 2021 (CO) [100]
189	$C_6H_5N{=}N$−/−CH_3	XIa (11%; biphenyl and $C_6H_5N_2C_5H_4Mo(CO)_3C(O)CH_3$ were observed as by-products) [151] orange-red crystals, m.p. 84 °C [151] ^{1}H NMR ($CDCl_3$): 0.49 (s, CH_3), 5.40 (m, H-3,4 of C_5H_4), 6.01 (m, H-2,5 of C_5H_4), 7.55 to 7.85 (m, C_6H_5) [151] ^{13}C NMR ($CDCl_3$): 1.03 (q, CH_3; $^1J(C, H)$ = 168.4), 88.11 (dm, C-3,4 of C_5H_4; $^1J(C, H)$ = 150.4), 90.91 (dm, C-2,5 of C_5H_4; $^1J(C, H)$ = 164.1), 122.66 (m, ortho-C), 129.145

Table 6 (continued)

No.	compound	method of preparation (yield) properties and remarks
		(m, meta-C), 131.365 (m, para-C), 152.42 (s, ipso-C), 225.31 (CO) [151] IR (petroleum ether): 1945, 2025 (CO) [151] mass spectrum: $[M]^+$, $[M-nCO]^+$ (n = 1 to 3), $[M-3CO-CH_3]^+$ [151]
190	I−/−CH_3	$C_5H_5Mo(CO)_3CH_3$, dissolved in THF, was treated at −78°C with an equimolar quantity of LiC_4H_9-s/hexane, followed by one equivalent of IC_2H_4I; chromatographic separation over silica using benzene/hexane (1:1) as the eluant afforded a 74% yield [103] green solid, m.p. 30 to 31°C [103] ^1H NMR ($CDCl_3$): 0.47 (s, $MoCH_3$), 5.20, 5.43 (both "t", C_5H_4; J = 2.3) [103] ^{13}C NMR ($CDCl_3$): −16.00 ($MoCH_3$), 52.22, 92.47, 100.76 (all C_5H_4), 225.50 (MoCO) [103] IR (CCl_4): 1955, 2000 (CO) [103] mass spectrum: $[M-2CO]^+$, $[M-3CO]^+$ [103] $Pd(NCCH_3)_2Cl_2$-catalyzed coupling with $(n-C_4H_9)_3SnC{\equiv}CH$ in N,N-dimethylformamide yielded $HC{\equiv}CC_5H_4Mo(CO)_3CH_3$ [115]; with $(n-C_4H_9)_3SnC{\equiv}CSn(C_4H_9-n)_3$, $CH_3Mo(CO)_3C_5H_4C{\equiv}CC_5H_4Mo(CO)_3CH_3$ was obtained [103]; methyl/iodide scrambling with the formation of $CH_3C_5H_4Mo(CO)_3I$ occurred in the presence of $Pd(NCCH_3)_2Cl_2/SnR_3R'$ (R = R' = CH_3, n-C_4H_9; R = CH_3, R' = $C_5H_4Ti(C_5H_5)Cl_2$) [125] heterobimetallic complexes $CH_3Mo(CO)_3C_5H_4C-CC_5H_4M$ were derived from palladium-catalyzed coupling reactions (5 mol% of $Pd(NCCH_3)_2Cl_2$ in $HC(O)N(CH_3)_2$) with equimolar amounts of $R_3SnC{\equiv}CC_5H_4M$ (R = CH_3, n-C_4H_9; M = $W(CO)_3CH_3$, $Mn(CO)_3$, $Re(CO)_3$, $Fe(CO)_2CH_3$) [115]
191	$-CH_2CH_2C(O)-$	formed by irreversible CO insertion on heating $C_5H_4CH_2CH_2Mo(CO)_3$ in benzene under 10 bars of carbon monoxide at 120°C; isolated in 29% yield by column chromatography on alumina with hexane/ether (1:1) as the eluant [19] yellow needles, m.p. 132 to 133°C [19] ^1H NMR ($CDCl_3$): 2.22, 3.43 (both t, CH_2; J = 7.5), 5.12, 5.92 (both "t", C_5H_4; J = 2.0) [19] IR (KBr): 1645 (ν(C=O)), 1910, 1935, 2015 (CO) [19]

References on pp. 253/8

Table 6 (continued)

No.	compound	method of preparation (yield) properties and remarks
191 (continued)		UV (CH_3OH, log ε): 214 (4.3), 261 (4.0), 307 (3.6) [19] mass spectrum (70 eV): $[M]^+$ (4%), $[M-nCO]^+$ (n = 1 to 4) [19]
192	$-CH_2C(=CHCH_3)C(O)-$	XII (6%; by-product of No. 178) [7, 19] yellow needles, m.p. 104 °C [7, 19] ^1H NMR ($CDCl_3$): 1.68 (dt, CH_3; J = 1.5 and 8.0), 3.04 (m CH_2), 5.12 ("t", $2C_5H_4$-H; "J" = 2.0), 5.78 (m, CH), 5.85 ("t", $2C_5H_4$-H) [7, 19] IR (KBr): 1645 (ν(C=CC=O)), 1915, 2015 (CO) [7, 19] mass spectrum (70 eV): $[M]^+$ (12%), $[M-nCO]^+$ (n = 1 to 4) [19]
193	$CH_3C(O)OCH_2-/-C(O)CH_3$	by-product of No. 121 (see there); not isolated in pure form [80]
194	$C_6H_5N_2-/-C(O)CH_3$	by-product of No. 189; no characterizing data available [151]

$CH_3C_5H_4Mo(CO)_3{}^2D^{\bullet}$ radicals:

^2D =

No.	compound	method of preparation (yield) properties and remarks
195	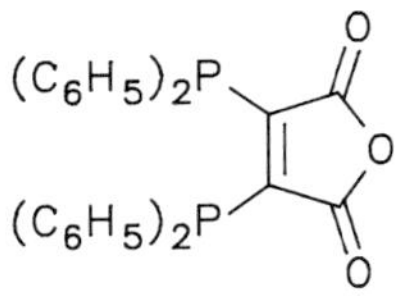 $(C_6H_5)_2P$, $(C_6H_5)_2P$ on maleic anhydride ring (O, O, O)	XIV [118] ESR (THF, 25 °C): g = 2.0039; a(Mo) = 0.52, $a(P_{coord})$ = 9.02, $a(P_{free})$ = 2.45 G [118]; same in CH_2Cl_2 [129] IR (CH_2Cl_2): 1653, 1737 (ν(C=O)), 1958, 1981, 2049 (CO) [118] E_{ox} (CH_2Cl_2/0.1 M $[N(C_4H_9\text{-}n)_4]BF_4$) = −0.09 V vs. SCE (→ $[CH_3C_5H_4Mo(CO)_3P(C_6H_5)_2\text{-}C_2P(C_6H_5)_2(CO)_2O]^+$) [118] ring-closure with formation of $CH_3C_5H_4Mo(CO)_2\text{-}(P(C_6H_5)_2)_2C_2(CO)_2O^{\bullet}$ proceeded slowly at room temperature; back reaction by radical recombination restored the starting $(CH_3C_5H_4Mo(CO)_3)_2$ dimer [118]
196	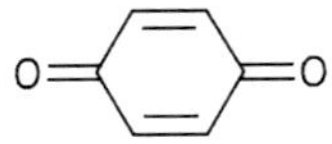 O=⟨ring⟩=O (p-benzoquinone)	XIV [105, 127] ESR (toluene, −40 °C): g = 2.0048; a(Mo) = 0.3, a(H-2) = a(H-6) = 4.65, $a(^{13}CO_{cis})$ = 0.78, $a(^{13}CO_{trans})$ = 0.35 G; (THF, −40 °C): g = 2.0048; a(Mo) = 0.4, a(H-2) = a(H-6) = 4.53, a(H-3) = a(H-5) = 0.22, $a(^{13}CO_{cis})$ = 0.73, $a(^{13}CO_{trans})$ = 0.25 G; (2-methyltetrahydrofuran, −40 °C): g = 2.0048; a(Mo) = 0.4, a(H-2) = a(H-6) = 4.57, a(H-3) = a(H-5) =

Table 6 (continued)

No.	compound	method of preparation (yield) properties and remarks
		0.19, a($^{13}CO_{cis}$) = 0.73, a($^{13}CO_{trans}$) = 0.25 G [105, 127] activation energy for restricted rotation of the quinone ring: 25 ± 2 kJ/mol (temperature-dependent ESR spectra shown in [105, 127])
197	2-methyl-1,4-benzoquinone (H_3C, O=, =O)	XIV (bonded through O of C-4) [127] ESR (toluene, −40 °C): g = 2.0046; a(H-3) = a(H-5) = 0.3, a(H-6) = 4.7, a(CH_3-2) = 4.2 G [127]
198	2-phenyl-1,4-benzoquinone (C_6H_5, O=, =O)	XIV [127] ESR (toluene, −40 °C): g = 2.0047; a(H-6) = 4.5 (bonded through O of C-4, major isomer); g = 2.0046; a(H-3) = 4.1, a(H-5) = 5.1 G (bonded through O of C-1, minor isomer) [127]
199	2,5-dimethyl-1,4-benzoquinone (H_3C, CH_3)	XIV [127] ESR (toluene, −40 °C): g = 2.0044; a(H-6) = 3.84, a(CH_3-2) = 5.23 G [127]
200	2,5-di-t-butyl-1,4-benzoquinone (t-C_4H_9, t-C_4H_9)	XIV [127] ESR (toluene, −40 °C): g = 2.0043; a(H-6) = 4.18, a(^{13}CO) = 1.13 G [127]
201	2,5-diphenyl-1,4-benzoquinone (C_6H_5, C_6H_5)	XIV [127] ESR (toluene, −40 °C): g = 2.0045; a(H-6) = 3.87 G [127]
202	2,6-dimethyl-1,4-benzoquinone (H_3C, H_3C)	XIV (bonded through O of C-4) [105, 127] ESR (toluene, −40 °C): g = 2.0045; a(Mo) = 0.1, a(CH_3-2) = a(CH_3-6) = 4.20, a($^{13}CO_{cis}$) = 0.65, a($^{13}CO_{trans}$) = 0.41 G; (THF, −40 °C): g = 2.0045; a(Mo) = 0.2, a(H-3) = a(H-5) = 0.13, a(CH_3-2) = a(CH_3-6) = 4.00 G; (2-methyltetrahydrofuran, −40 °C): g = 2.0045; a(Mo) = 0.2, a(CH_3-2) = a(CH_3-6) = 4.05 G [105, 127] activation energy for restricted rotation of the quinone ring: 29 ± 2 kJ/mol (temperature-dependent ESR spectra shown in [105, 127])

References on pp. 253/8

Table 6 (continued)

No.	compound	method of preparation (yield) properties and remarks
203	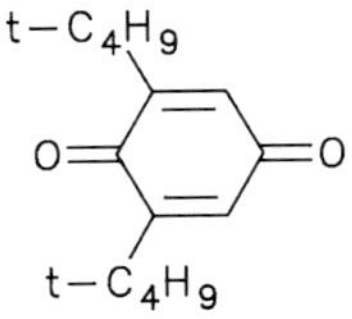	XIV (bonded through O of C-4) [127] ESR (toluene, −40 °C): g = 2.0044; a(Mo) = 0.3, $a(^{13}CO)$ = 0.5 G [127]
204	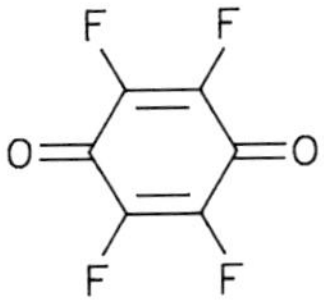	XIV [127] ESR (toluene, −40 °C): g = 2.0052; a(F-2) = a(F-6) = 10.0, a(F-3) = a(F-5) = 1.6 G [127]
205	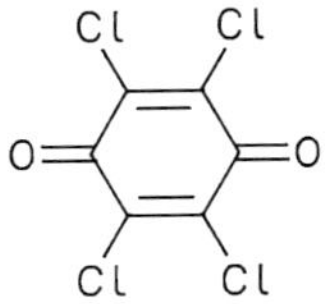	XIV [127] ESR (toluene, −40 °C): g = 2.0062; a(Mo) = 2.8 G [127]
206	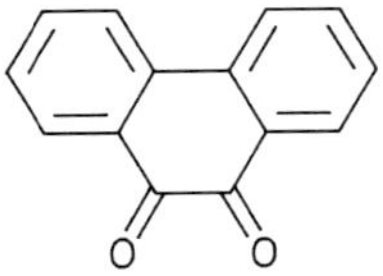	XIV [127] ESR (toluene, −40 °C): g = 2.0051; a(H) = 2.5 (2H) and 0.8 (4H) G [127] thermal decarbonylation above −40 °C gave chelated $CH_3C_5H_4Mo(CO)_2O_2C_{14}H_6^{\cdot}$ [127]
207	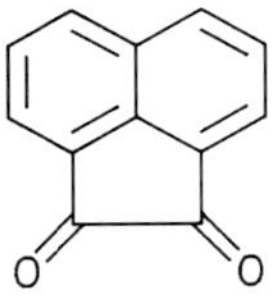	XIV [127] ESR (toluene, −40 °C): g = 2.0058; a(H) = 1.1 (4H) G [127] the more stable chelated radical $CH_3C_5H_4Mo(CO)_2O_2C_{12}H_6^{\cdot}$ resulted from facile CO ejection above −40 °C [127]
$[RC_5H_4Mo(CO)_3{}^2D]X$ salts:		
$R/{}^2D/X^-$ =		
208	$CH_3-/NC_5H_5/[B(C_6H_5)_4]^-$	by treatment of $[(CH_3C_5H_4Mo(CO)_3)_2(\mu\text{-}I)][B(C_6H_5)_4]$ with pyridine in acetone or CH_2Cl_2 at room temperature [99] as detailed for $[C_5H_5Mo(CO)_3NC_5H_5][B(C_6H_5)_4]$ [15] IR (pyridine): 1962, 1989, 2058 (CO) [99]; (CH_2Cl_2): 1972, 2000, 2064 (CO) [99]
209	$CH_3-/P(C_6H_5)_3/PF_6^-$	by analogy to $[C_5H_5Mo(CO)_3P(C_6H_5)_3]PF_6$ [13] from $[CH_3C_5H_4Mo(CO)_3N_2H_4]PF_6$ and $P(C_6H_5)_3$ [76]

Table 6 (continued)

No.	compound	method of preparation (yield) properties and remarks
		IR (CH_2Cl_2): 1973, 1995, 2057 (CO) [76] 290 nm irradiation in CH_2Cl_2 in the presence of excess $P(C_6H_5)_3$ furnished $[CH_3C_5H_4Mo(CO)_2(P(C_6H_5)_3)_2]PF_6$ [76] treatment with $[N(C_2H_5)_4][CH_3C_5H_4Mo(CO)_3]$ in CH_2Cl_2 followed by addition of benzene produced $(CH_3C_5H_4Mo(CO)_3)_2$ [76]
210	$PF_6^-H_3N^+CH_2CH_2$–/OC_4H_8/[1,4-$((NC)_2C)_2C_6H_4$]$^-$	from $[(H_3NCH_2CH_2C_5H_4Mo(CO)_3)_2][PF_6]_2$ and excess 1,4-bis(dicyanomethylene)cyclohexa-2,5-diene in THF solution (not isolated) [157] IR (THF): 1979, 2072 (CO) [157]
211	$HOCH_2CH_2$–/$P(C_6H_5)_3$/BF_4^-	XV [53] yellow crystals, m.p. 200 to 206°C (dec.) [53] ^{1}H NMR (acetone-d_6): 2.64 (t, $CH_2C_5H_4$), 3.71 (t, OCH_2), 6.00 (AA′BB′ system, C_5H_4), 7.31 to 7.81 (m, C_6H_5); HO variable [53] ^{31}P NMR (acetone-d_6): 50.18 (vs. free $P(C_6H_5)_3$ in the same solvent) [53] IR (CH_2Cl_2): 1969, 1995, 2056 (CO) [53] reaction with $Na[OCH_3]$ in acetone afforded a cyclized intramolecular metalloester product, $C_5H_4CH_2CH_2O(O)CMo(CO)_2P(C_6H_5)_3$ [53, 70]
212	$HOCH_2CH_2$–/$P(C_6H_4CH_3\text{-}4)_3$/BF_4^-	XV [53] yellow crystals, m.p. 230 to 235°C (dec.) [53] ^{1}H NMR (acetone-d_6): 2.43 (s, CH_3), 2.63 (t, $CH_2C_5H_4$), 3.71 (t, OCH_2), 5.95 (AA′BB′ system, C_5H_4), 7.15 to 7.89 (m, C_6H_4); HO not observed [53] ^{31}P NMR (acetone-d_6): 50.01 (vs. free $P(C_6H_4CH_3\text{-}4)_3$ in the same solvent) [53] IR (CH_2Cl_2): 1971, 1980, 2055 (CO) [53] the intramolecular metalloester $C_5H_4CH_2CH_2O(O)CMo(CO)_2P(C_6H_4CH_3\text{-}4)_3$ was obtained from the reaction with Na[OMe] in acetone [53]
213	$HOCH_2CH_2CH_2$–/$P(C_6H_5)_3$/BF_4^-	XV [53] yellow crystals, m.p. 189 to 191°C (dec.) [53] ^{1}H NMR (acetone-d_6): 1.77 (m, CCH_2C), 2.57 (t, $CH_2C_5H_4$), 3.56 (t, OCH_2), 6.01 (AA′BB′ system, C_5H_4), 7.28 to 7.82 (m, C_6H_5); HO variable [53] ^{31}P NMR (acetone-d_6): 49.86 (vs. free $P(C_6H_5)_3$ in the same solvent) [53]

References on pp. 253/8

Table 6 (continued)

No.	compound	method of preparation (yield) properties and remarks
213 (continued)		IR (CH_2Cl_2): 1970, 1996, 2050 (CO) [53] treatment with sodium methoxide formed $HOCH_2CH_2CH_2C_5H_4Mo(CO)_2(P(C_6H_5)_3)$-$C(O)OCH_3$ rather than a cyclized metalloester compound such as was observed for Nos. 211 and 212 above [53]
214	$-CH_2CH_2NH_2\rightarrow$ (intramolecularly coordinated)/PF_6^-	$[H_3NCH_2CH_2C_5H_4Mo(CO)_3Cl]PF_6$ was photogenerated from $[(H_3NCH_2CH_2C_5H_4Mo(CO)_3)_2][PF_6]_2$ and CCl_4 in THF and subsequently dehydrochlorinated by stirring with 1,8-bis(dimethylamino)naphthalene (not isolated) [157]; accompanied by the hydride $[H_3NCH_2CH_2C_5H_4Mo(CO)_3H]PF_6$ also from $[(H_3NCH_2CH_2C_5H_4Mo(CO)_3)_2][PF_6]_2$ by photochemical disproportionation in CH_3CN or THF in the absence of donor ligands [157]; for the initial disproportionation products, see Nos. 47 and 49 ^1H NMR (THF-d_8): 2.49 (t, $CC\mathbf{H}_2CH_2N$; J = 6.4), 3.59 (m, $CCH_2C\mathbf{H}_2N$), 4.65 (br, NH_2), 5.41, 6.41 (both "d", C_5H_4; "J" = 1.8) [157] IR (THF): 1958, 1987, 2060 (CO) [157] addition of $P(C_6H_5)_3$ to THF solutions caused the slow formation of $[C_5H_4CH_2CH_2NH_2Mo(CO)_2$-$P(C_6H_5)_3]PF_6$ as a mixture of cis- and trans-substituted isomers; treatment with $P(C_6H_5)_3$ followed by the base 1,8-bis(dimethylamino)-naphthalene gave the chelated carbamoyl $C_5H_4CH_2CH_2NHC(O)Mo(CO)_2P(C_6H_5)_3$ [157]
polymer-attached $RC_5H_4Mo(CO)_3$ derivatives:		
215	$Li[(PS)-CH_2C_5H_4Mo(CO)_3]$ ("(PS)" = polystyrene, 18% crosslinked with divinyl benzene)	XVI [10] acidification with CH_3CO_2H in water/1,4-dioxane yielded $(PS)-CH_2C_5H_4Mo(CO)_3H$ [10]
216	$M[(PS)-p\text{-}C_6H_4C_5H_4Mo(CO)_3]$ ("(PS)" = styrene-3%-divinyl benzene copolymer; $M^+ = Li^+, Na^+$)	XVII [23, 25, 29] light beige resins containing 0.67 mmol of Mo and 0.49 mmol of Li and, respectively, 0.74 mmol of Mo and 0.64 mmol of Na per g of polymer [23, 25] IR (KBr, lithium salt): 1710, 1775, 1803, 1905 (CO); (KBr, sodium salt): 1730, 1840, 1900 (CO) [23, 25]

References on pp. 253/8

Table 6 (continued)

No.	compound	method of preparation (yield) properties and remarks
		in proton-transfer and alkylation reactions, producing (PS)−p-$C_6H_4C_5H_4Mo(CO)_3R$ (R = H, CH_3), the Li^+ and Na^+ salts displayed a nearly identical behavior arising from anion-inherent properties rather than being controlled by ion-pairing energetics [23, 25, 29]; compare No. 218
217	(PS)−$CH_2C_5H_4Mo(CO)_3H$ ("(PS)" = polystyrene, 18% crosslinked with divinyl benzene)	XVI [10] yellowish material containing about one −$CH_2C_5H_4Mo(CO)_3$ group per 20 styryl units [10] IR: 1930, 2030 (CO) [10] underwent endothermic decomposition with elimination of H_2 and CO at 200 to 210°C; loss of C_5H_6 was observed at 170 to 230°C; because of site-site isolation, no binuclear Mo−Mo-bonded species were detectable after partial thermal decomposition [10]
218	(PS)−p-$C_6H_4C_5H_4Mo(CO)_3H$ ("(PS)" = styrene-3%-divinyl benzene copolymer)	XVII [23, 25, 29] beige resin [23, 25] IR (KBr): 1930, 2010 [23, 25] equilibration experiments in THF using combinations of No. 218 and $M[C_5H_5Mo(CO)_3]$ (M^+ = Li^+, Na^+) or the Li^+ or Na^+ salt of polymer-attached anion No. 216 together with $C_5H_5Mo(CO)_3H$ demonstrated that little change in pK_a was induced by binding the molybdenum hydride to the polymer [23, 25, 29] treatment with the enolate anions of diethyl malonate, ethyl acetoacetate, and 2-carboethoxycyclopentanone or 2-carboethoxycyclohexanone in THF led to complete deprotonation [23, 25]
219	(PS)−p-$C_6H_4C_5H_4Mo(CO)_3CH_3$ ("(PS)" = styrene-3%-divinyl benzene copolymer)	from $[(PS)-CH_2C_5H_4Mo(CO)_3]^-$ and CH_3I [23, 25] IR (KBr): 1935, 2012 (CO) [23, 25]

*Further information:

$M[CH_3C_5H_4Mo(CO)_3]$ (M^+ = Na^+, K^+; Table **6**, Nos. **2** and **3**). Protonation of the $[CH_3C_5H_4Mo(CO)_3]^-$ anion (sodium salt) with glacial acetic acid in THF produced $CH_3C_5H_4Mo(CO)_3H$ [1].

Reaction of the In−Fe-bonded complex $InCl_2(\mu\text{-}(C_6H_5)_2PCH_2P(C_6H_5)_2)Fe(CO)_3Si(OCH_3)_3$ with a slight excess of $Na[CH_3C_5H_4Mo(CO)_3] \cdot CH_3OC_2H_4OCH_3$ in THF at 0°C gave

References on pp. 253/8

the heterotrimetallic Mo−In−Fe chain compound $CH_3C_5H_4Mo(CO)_3InCl(\mu\text{-}(C_6H_5)_2PCH_2\text{-}P(C_6H_5)_2)Fe(CO)_3Si(OCH_3)_3$ [169]. Salt elimination between $CF_3SO_2OSiH_2SiH_2OSO_2CF_3$ (generated in situ from $C_6H_5SiH_2SiH_2C_6H_5$ and CF_3SO_3H in pentane at −40 °C) and two equivalents of $K[CH_3C_5H_4Mo(CO)_3]$ in 1,2-dimethoxyethane at −60 °C provided $CH_3C_5H_4Mo\text{-}(CO)_3SiH_2SiH_2Mo(CO)_3C_5H_4CH_3$ [168]. $CH_3C_5H_4Mo(CO)_3Sn(C_6H_5)_3$ was obtained by allowing $Na[CH_3C_5H_4Mo(CO)_3]$ to react with $Sn(C_6H_5)_3Cl$ in THF at ambient conditions [9, 57]. The reaction between $BiCl_3$ and three equivalents of $Na[CH_3C_5H_4Mo(CO)_3]$ in THF at room temperature afforded $(CH_3C_5H_4Mo(CO)_3)_3Bi$; the corresponding reaction involving two equivalents of the sodium salt gave the disubstituted product, $(CH_3C_5H_4Mo(CO)_3)_2BiCl$ [87].

Oxidation of $Na[CH_3C_5H_4Mo(CO)_3]$ with aqueous ferric sulfate in $CH_3OC_2H_4OC_2H_4OCH_3$ solution in the presence of acetic acid yielded $(CH_3C_5H_4Mo(CO)_3)_2$ [12]. The homo- and heterobimetallics $(CH_3C_5H_4Mo(CO)_3)_2$, $(C_5H_5Fe(CO)_2)_2$, and $CH_3C_5H_4Mo(CO)_3Fe(CO)_2C_5H_5$ resulted from the treatment of the sodium salt with $C_5H_5Fe(CO)_2I$ in $CH_3OC_2H_4OC_2H_4OCH_3$ at ambient temperature [83]. Metal exchange with $Na[CH_3C_5H_4Mo(CO)_3]$ in $(Co(CO)_3)_2\text{-}(\mu\text{-}RC{\equiv}CH)$ (THF, reflux temperature) yielded $CH_3C_5H_4Mo(CO)_2(\mu\text{-}RC{\equiv}CH)Co(CO)_3$ (R = CH_2OH, CH_2O-menthyl, CH_2O-bornyl) [119]. Single- and double-exchanged $MoCo_2$ and Mo_2Co heteroclusters, $CH_3C_5H_4Mo(CO)_2(\mu_3\text{-}RC)(Co(CO)_3)_2$ and $(CH_3C_5H_4Mo(CO)_2)_2(\mu_3\text{-}RC)Co(CO)_3$, resulted from similar reactions between the sodium salt and the alkylidyne-bridged tricobalt precursors $(Co(CO)_3)_3(\mu_3\text{-}RC)$ (R = C_6H_5, $C_2H_5OC(O)$) [175]. The reaction of two equivalents of $Na[CH_3C_5H_4Mo(CO)_3] \cdot CH_3OC_2H_4OCH_3$ with trans-$Pt(NCC_6H_5)_2Cl_2$ in THF gave trans-$(CH_3C_5H_4Mo(CO)_3)_2Pt(NCC_6H_5)_2$ [51]; with $(Pt(P(C_6H_5)_2H)Cl(\mu\text{-}P(C_6H_5)_2))_2$, a heterotetrametallic Mo_2Pt_2 product (Formula VI, R = C_6H_5) was formed [104]. The Mo−Cd−Fe heterotrimetallic complex $CH_3C_5H_4Mo(CO)_3Cd(\mu\text{-}(C_6H_5)_2PCH_2P(C_6H_5)_2)Fe\text{-}(CO)_3Si(OCH_3)_3$ was synthesized from the Cd−Fe-bonded tetranuclear precursor $((CH_3O)_3\text{-}SiFe(CO)_3(\mu\text{-}P(C_6H_5)_2CH_2P(C_6H_5)_2)Cd(\mu\text{-}Cl))_2$ and one equivalent of $Na[CH_3C_5H_4Mo(CO)_3] \cdot CH_3OC_2H_4OCH_3$ in THF [169]. Adding an aqueous solution of $Hg(CN)_2$ to a THF solution of $Na[CH_3C_5H_4Mo(CO)_3]$ resulted in the immediate precipitation of $(CH_3C_5H_4Mo(CO)_3)_2Hg$ [57].

```
                          R2   H
              R2P\      /P\   /    /Mo(CO)2C5H4CH3
CH3C5H4(OC)2Mo----Pt      Pt------
                   /  \P/    \PR2
                  H    R2
```

VI

The alkyls $CH_3C_5H_4Mo(CO)_3R$ with R = CH_3 [8, 18, 19, 28], CD_3 [18, 28], C_2H_5 [8, 19], $CH_2C_6H_5$ [18, 28], and $CD_2C_6D_5$ [18, 28] resulted from reactions of the sodium salt with excess RBr (R = $CH_2C_6H_5$, $CD_2C_6D_5$) or RI (R = CH_3, CD_3, C_2H_5) in THF at ambient conditions or 40 °C. Combination with a 10% excess of 1,3-diiodopropane in THF at reflux temperature provided the 2-oxacyclopentylidene dicarbonyl $CH_3C_5H_4Mo(CO)_2(=CO(CH_2)_3\text{-}cyclo)I$ [39, 40, 74].

Equimolar quantities of $Na[CH_3C_5H_4Mo(CO)_3]$ and $[4\text{-}FC_6H_4N_2]BF_4$, reacted together in THF at −60 °C and worked up by extracting the evaporated mixture with CH_2Cl_2, yielded $CH_3C_5H_4Mo(CO)_2N_2C_6H_4F\text{-}4$, accompanied by minor quantities of $CH_3C_5H_4Mo(N_2C_6H_4\text{-}F\text{-}4)_2Cl$. The latter was not formed if CH_2Cl_2 was omitted from the workup [27]. Reaction of the $[CH_3C_5H_4Mo(CO)_3]^-$ anion (from $(CH_3C_5H_4Mo(CO)_3)_2$ by Na/K reduction in THF) with 4,4′-dimethoxythiobenzophenone in THF afforded a 1:1 isomeric mixture of 2- and 3-substituted fulvenes, $(4\text{-}CH_3OC_6H_4)_2C{=}C_5H_3CH_3\text{-}2$ and $(4\text{-}CH_3OC_6H_4)_2C{=}C_5H_3CH_3\text{-}3$, respectively [14].

References on pp. 253/8

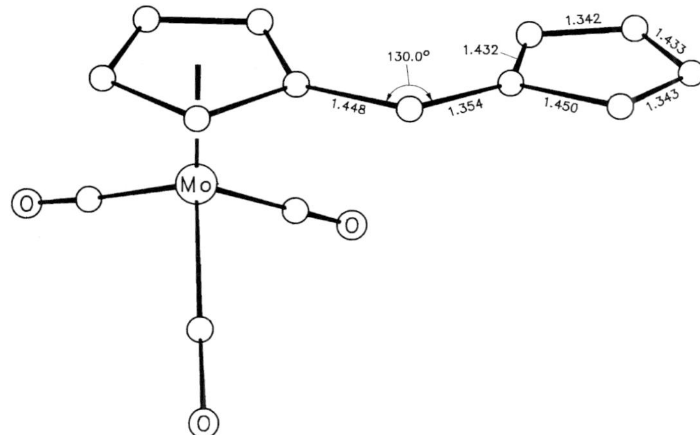

Fig. 11. The molecular structure of $[C_5H_4{=}CHC_5H_4Mo(CO)_3]^-$ [110].

$[P(C_6H_5)_4][C_5H_4{=}CHC_5H_4Mo(CO)_3]$ (Table **6**, No. **40**). Crystals grown from THF/hexane at −30 °C belonged to the monoclinic space group $P2_1/n$ (nonconventional setting of $P2_1/c$)−C^5_{2h} (No. 14) with a = 1131.89(3), b = 1726.86(6), c = 1692.67(10) pm, β = 105.09(1)°; Z = 4 molecules per unit cell, D_{calc} = 2.445 g/cm³. The structural model of the anion is represented in **Fig. 11** [110].

$Na[HC(O)C_5H_4Mo(CO)_3]$ (Table **6**, No. **61**) was described as soluble in H_2O, THF, and acetone and very sensitive toward air [46, 73].

$HC(O)C_5H_4Mo(CO)_3I$ resulted from the reaction with equimolar I_2 in THF at ambient conditions [73].

Treatment with $Sn(CH_3)_2Cl_2$ in water afforded $HC(O)C_5H_4Mo(CO)_3Sn(CH_3)_2Cl$ [73]. Precipitation of aqueous solutions with $[N(CH_3)_4]Cl$ or $[P(C_6H_5)_4]Br$ gave $M[HC(O)C_5H_4Mo(CO)_3]$ (M^+ = $[N(CH_3)_4]^+$, $[P(C_6H_5)_4]^+$) [46].

Oxidation with aqueous $Fe(NO_3)_3$ in the presence of acetic acid led to $(HC(O)C_5H_4Mo(CO)_3)_2$ [73]. Metal exchange reactions with $(Co(CO)_3)_3(\mu_3\text{-}CC_6H_5)$ or $Fe(CO)_3(\mu_3\text{-}S)(Co(CO)_3)_2$ in boiling THF yielded $HC(O)C_5H_4Mo(CO)_2(\mu_3\text{-}C_6H_5C)(Co(CO)_3)_2$ [173] and $HC(O)C_5H_4Mo(CO)_2(\mu_3\text{-}S)(Fe(CO)_3)Co(CO)_3$ [174], respectively. $HC(O)C_5H_4Mo(CO)_3AuP(C_6H_5)_3$ was formed in the reaction with $Au(P(C_6H_5)_3)Cl$ in THF at 40 °C [73]. Addition of $Hg(CN)_2$ to aqueous solutions of the title compound precipitated $(HC(O)C_5H_4Mo(CO)_3)_2Hg$ [46].

Methylation with CH_3I in THF at 25 °C provided $HC(O)C_5H_4Mo(CO)_3CH_3$ [17, 22, 38]. $CH_3C(O)Cl$ in THF reacted to produce $CH_3C(O)OCH_2C_5H_4Mo(CO)_3Cl$ together with $CH_3C(O)OCH_2C_5H_4Mo(CO)_3C(O)CH_3$ and $(HC(O)C_5H_4Mo(CO)_3)_2$ [80]. $C_6H_5C(O)Cl$ only yielded $C_6H_5C(O)OCH_2C_5H_4Mo(CO)_3Cl$ [80]. Attempts to prepare $HC(O)C_5H_4Mo(CO)_2NO$ by reacting the salt with $4\text{-}CH_3C_6H_4SO_2N(NO)CH_3$ in the presence or absence of acid were unsuccessful [42].

$Na[CH_3C(O)C_5H_4Mo(CO)_3]$ (Table **6**, No. **64**) was reacted with $SnCl_2$ in acetone to give $(CH_3C(O)C_5H_4Mo(CO)_3)_2SnCl_2$ accompanied by elemental tin. Reaction with $SnCl_4$ in 3:1 stoichiometry in THF also led to $(CH_3C(O)C_5H_4Mo(CO)_3)_2SnCl_2$ [91]. Treatment with $P(C_6H_5)Cl_2$, PBr_3, or $As(CH_3)I_2$ in $CH_3OC_2H_4OC_2H_4OCH_3$ at ambient conditions gave the

 References on pp. 253/8

halo derivatives $CH_3C(O)C_5H_4Mo(CO)_3X$ (X = Cl, Br, I) [135]. Metal exchange with $(Co(CO)_3)_3(\mu_3\text{-}CC_6H_5)$ in refluxing THF produced $CH_3C(O)C_5H_4Mo(CO)_2(\mu_3\text{-}C_6H_5C)(Co(CO)_3)_2$ [173]. With $Fe(CO)_3(\mu_3\text{-}S)(Co(CO)_3)_2$, $CH_3C(O)C_5H_4Mo(CO)_2(\mu_3\text{-}S)(Fe(CO)_3)Co(CO)_3$ was obtained [174]. Adding aqueous $Hg(CN)_2$ to THF solutions caused the precipitation of $(CH_3C(O)C_5H_4Mo(CO)_3)_2Hg$ [91]. $CH_3C(O)C_5H_4Mo(CO)_3CH_3$ was formed by reaction with CH_3I in THF at 25°C [17, 22, 38]. Combination with $I(CH_2)_3I$ (10% excess in refluxing THF) gave a cyclic carbene derivative, $CH_3C(O)C_5H_4Mo(CO)(=CO(CH_2)_3\text{-cyclo})I$ [40]. No $CH_3C(O)C_5H_4Mo(CO)_3(CO)_2NO$ could be obtained from reactions with N-methyl-N-nitroso-p-toluenesulfonamide in the presence or absence of acid [42].

$Na[CH_3OC(O)C_5H_4Mo(CO)_3]$ (Table **6**, No. **68**). Combination with phosphorus trihalides in $CH_3OC_2H_4OC_2H_4OCH_3$ at ambient conditions resulted in clean formation of halo products, $CH_3OC(O)C_5H_4Mo(CO)_3X$ (X = Cl, Br, I), via condensation and successive decomposition of the $CH_3OC(O)C_5H_4Mo(CO)_3PX_2$ intermediates, so formed [135]. $CH_3OC(O)C_5H_4Mo(CO)_3PCl_2$ could be isolated from the reaction with PCl_3 in methylcyclohexane at −10 to −5°C [135]. Metal exchange with $Fe(CO)_3(\mu_3\text{-}S)(Co(CO)_3)_2$ in refluxing THF provided the functional cluster $CH_3OC(O)Mo(CO)_2(\mu_3\text{-}S)(Fe(CO)_3)Co(CO)_3$ [176]. Metal−metal bond formation to give $(CH_3OC(O)C_5H_4Mo(CO)_3)_2Hg$ took place between $Na[CH_3OC(O)C_5H_4Mo(CO)_3]$ and equimolar C_2H_5HgCl in $CH_3OC_2H_4OC_2H_4OCH_3$ at 80°C [147]. $CH_3OC(O)C_5H_4Mo(CO)_3CH_3$ resulted from combination with CH_3I in THF at ambient conditions [17, 22, 38]. Acidification of THF solutions with acetic acid and subsequent nitrosylation with N-methyl-N-nitroso-p-toluenesulfonamide afforded $CH_3OC(O)C_5H_4Mo(CO)_2NO$ [17, 22, 42].

$Na[C_2H_5OC(O)C_5H_4Mo(CO)_3]$ (Table **6**, No. **70**). Iodination in $CH_3OC_2H_4OC_2H_4OCH_3$ at room temperature produced $C_2H_5OC(O)C_5H_4Mo(CO)_3I$ [135]. Reactions with PX_3 (X = Cl, Br; $CH_3OC_2H_4OC_2H_4OCH_3$, ambient) led to the corresponding halo complexes, $C_2H_5OC(O)C_5H_4Mo(CO)_3X$ [135]. Combination with $(Co(CO)_3)_3(\mu_3\text{-}CC_6H_5)$ in THF at reflux temperature resulted in metal exchange with formation of $C_2H_5OC(O)C_5H_4Mo(CO)_2(\mu_3\text{-}C_6H_5C)(Co(CO)_3)_2$ [173]. $C_2H_5OC(O)C_5H_4Mo(CO)_2(\mu_3\text{-}S)(Fe(CO)_3)Co(CO)_3$ was isolated from the exchange reaction with $Fe(CO)_3(\mu_3\text{-}S)(Co(CO)_3)_2$ under similar conditions [174, 176]. Treatment with equimolar C_6H_5HgCl in THF for short periods at ice/salt bath temperature gave $C_2H_5OC(O)C_5H_4Mo(CO)_3HgC_6H_5$; prolonged reaction and/or warming resulted in symmetrization with the formation of $(C_2H_5OC(O)C_5H_4Mo(CO)_3)_2Hg$ [147].

$Li[(CH_3)_3Si(CH_3)_2SiC_5H_4Mo(CO)_3]$ (Table **6**, No. **73**). On oxidation with $Fe_2(SO_4)_3/CH_3CO_2H$ in THF, the salt produced homobimetallic $((CH_3)_3Si(CH_3)_2SiC_5H_4Mo(CO)_3)_2$ [148, 159, 161]. Combination with excess CH_3I in THF gave $(CH_3)_3Si(CH_3)_2SiC_5H_4Mo(CO)_3CH_3$ [161]. The halo complexes $(CH_3)_3Si(CH_3)_2SiC_5H_4Mo(CO)_3X$ (X = Cl, Br, I) resulted from treatment with the requisite 1,2-dihaloethane in THF at room temperature [159]. Similar reactions with either 1,3-dibromo- or 1,3-diiodopropane yielded the cyclic carbene complexes $(CH_3)_3Si(CH_3)_2SiC_5H_4Mo(CO)_2(=C(CH_2)_3O\text{-cyclo})X$ (X = Br, I) [148]; treatment of THF solutions with acetic acid, followed by reaction with CCl_4 or N-bromosuccinimide, afforded the halo derivatives $(CH_3)_3Si(CH_3)_2SiC_5H_4Mo(CO)_3X$ (X = Cl, Br) [148, 158, 159, 161].

$Li[(C_6H_5)_2PC_5H_4Mo(CO)_3]$ (Table **6**, No. **77**). Oxidative electrolysis at +100 mV vs. SCE, on a platinum-gauze electrode, in acetone containing 0.1 M $[N(C_2H_5)_4][BF_4]$, was used to prepare $((C_6H_5)_2PC_5H_4Mo(CO)_3)_2$ [156].

Reaction with equimolar I_2 in THF at room temperature gave $(C_6H_5)_2PC_5H_4Mo(CO)_3I$ [139]. Nitrosylation of the anion using $NOBF_4$ or N-methyl-N-nitroso-p-toluenesulfonamide in THF produced $(C_6H_5)_2PC_5H_4Mo(CO)_2NO$ [139]. Homo- and heterobimetallics, $Mo(CO)_3(\mu\text{-}C_5H_4P(C_6H_5)_2)M(CO)_2C_5H_5$ were formed in reactions with $C_5H_5M(CO)_3I$ (M = Mo, W) in refluxing THF [139]; combination with $(C_6H_5)_2PC_5H_4Mo(CO)_3I$ in 1:1 stoichiometry under

References on pp. 253/8

similar conditions yielded (μ-$(C_6H_5)_2PC_5H_4Mo(CO)_2)_2$ together with minor amounts of $(C_6H_5)_2PC_5H_4Mo(CO)_2(\mu$-$(C_6H_5)_2PC_5H_4)Mo(CO)_3$ [139]. The reaction with $Mn(CO)_5Br$ formed $(C_6H_5)_2PC_5H_4Mo(CO)_3Mn(CO)_5$ [178]; treatment with $(Mn(CO)_4(\mu$-$Br))_2$ in THF at 0 °C gave $Mo(CO)_3(\mu$-$C_5H_4P(C_6H_5)_2)Mn(CO)_4$ [33]. Oxidation with $Fe_2(SO_4)_3$ in the presence of CH_3CO_2H produced (μ-$(C_6H_5)_2PC_5H_4Mo(CO)_2)_2$ [139]. Adding one equivalent of solid $AgBF_4$ to a toluene solution of the salt resulted in oxidation of the anion to heterotetrametallic $(Mo(CO)_3C_5H_4P(C_6H_5)_2Ag)_2$ containing two phosphane-coordinated silver atoms bonded to two molybdenum centers [156].

Attempts to prepare protonated, alkylated, halogenated, or cation-exchanged derivatives by treatment of the rather impure salt obtained from $Li[(C_6H_5)_2PC_5H_4]$ and $Mo(CO)_3(NCCH_3)_3$ with CH_3CO_2H, CH_3I, CCl_4, BrC_2H_4Br, N-bromosuccinimide, and $[NR_4]Cl$ (R = CH_3, C_2H_5), respectively, were unsuccessful [33]. The hydrido complex $(C_6H_5)_2PC_5H_4Mo(CO)_3H$ however was readily prepared by adding three equivalents of acetic acid to toluene solutions of pure $Li[(C_6H_5)_2PC_5H_4Mo(CO)_3]$ (from $Li[(C_6H_5)_2PC_5H_4]$ and (cyclo-$C_7H_8)Mo(CO)_3$) [156].

$CH_3C_5H_4Mo(CO)_3GaI_2O(C_2H_5)_2$ (Table **6**, No. **88**) formed triclinic crystals, space group $P\bar{1}-C_i^1$ (No. 2), with a = 12.570(5), b = 10.049(14), c = 11.919(18) Å, α = 138.47(7), β = 91.12(7), γ = 94.31(7)°; Z = 2 molecules per unit cell, D_{calc} = 2.209 g/cm^3. The molecular structure is shown in **Fig. 12** [113].

$CH_3C_5H_4Mo(CO)_3X$ (X = Cl, Br, I; Table **6**, Nos. **97** to **99**). All three complexes resulted from photochemical disproportionation of $(CH_3C_5H_4Mo(CO)_3)_2$ by halides X^- in solvents such as acetonitrile, acetone, dimethyl sulfoxide, and THF; the other organometallic complex formed was $[CH_3C_5H_4Mo(CO)_3]^-$ [62, 77]; see Method IVc.

$CH_3C_5H_4Mo(CO)_3Br$ was also isolated from reactions of the triple-bonded dimer $(CH_3C_5H_4Mo(CO)_2)_2$ with α-bromo ketones such as 4-$RC_6H_4C(O)CH_2Br$ (R = H, Br, CH_3O, C_6H_5) or α-bromocamphor in toluene at 60 to 70 °C, which gave the corresponding methyl ketones, 4-$RC_6H_4C(O)CH_3$ and camphor, together with $CH_3C_5H_4Mo(CO)_3Br$ as the ac-

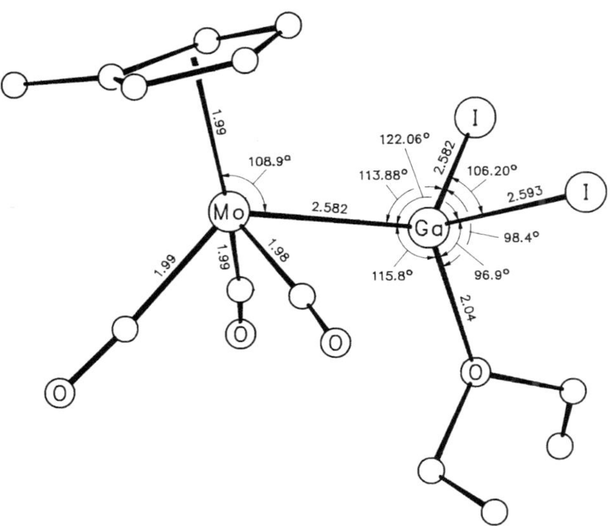

Fig. 12. The molecular structure of $CH_3C_5H_4Mo(CO)_3GaI_2O(C_2H_5)_2$ [113].

References on pp. 253/8

companying organometallic compound. The products were separated from the concentrated mixtures by preparative silica gel thin-layer chromatography using CH_2Cl_2/hexane (4:1) as the solvent mixture for development of the plate. Yields of $CH_3C_5H_4Mo(CO)_3Br$ ranged from 25 to 38% [32]. Treatment of acetone solutions containing in situ-generated $[CH_3C_5H_4Mo(CO)_3OC(CH_3)_2][CH_3C_5H_4Mo(CO)_3]$ with $[N(C_4H_9\text{-n})_4]Br$ formed $CH_3C_5H_4Mo(CO)_3Br$ together with $[N(C_4H_9\text{-n})_4][CH_3C_5H_4Mo(CO)_3]$ [62]. Along with $CH_3C_5H_4Mo(CO)_2(\mu\text{-}Te_2Br)\text{-}Fe(CO)_3$, the bromo complex also resulted from bromination of $(CH_3C_5H_4Mo(CO)_2)_2(\mu_3\text{-}Te_2)\text{-}Fe(CO)_3$ in CH_2Cl_2 solution, under a blanket of CO, between −78 and +25 °C [60].

$IC_5H_4Mo(CO)_3CH_3$ underwent rearrangement to $CH_3C_5H_4Mo(CO)_3I$ when stirred in N,N-dimethylformamide in the presence of a catalytic amount of $Pd(NCCH_3)_2Cl_2$ and two equivalents (relative to Pd) of a reducing tin reagent ($Sn(CH_3)_4$, $Sn(C_4H_9\text{-n})_4$, or $(CH_3)_3SnC_5H_4Ti(C_5H_5)Cl_2$) [125]. Accompanied by $CH_3C_5H_4Mo(CO)_3Sn(C_6H_5)_2I$, the iodo complex was also formed, when $CH_3C_5H_4Mo(CO)_3Sn(C_6H_5)_3$ was treated with one equivalent of iodine in CCl_4 (rate constant of formation at 30 °C: $k = 0.58\ L \cdot mol^{-1} \cdot s^{-1}$) [9]. The crossover reaction between $(CH_3C_5H_4Mo(CO)_3)_2$ and $C_5H_5Mo(CO)_3I$ generated $CH_3C_5H_4Mo(CO)_3I$ together with $(C_5H_5Mo(CO)_3)_2$ [79].

Insertion of "GaI" (generated in situ by ultrasonic irradiation of gallium metal with I_2 in toluene) across the Mo−I bond of $CH_3C_5H_4Mo(CO)_3I$, followed by isolation of the product from diethyl ether, afforded $CH_3C_5H_4Mo(CO)_3GaI_2O(C_2H_5)_2$ [113]. $CH_3C_5H_4Mo(CO)_3InCl_2$ was obtained similarly by combination of $CH_3C_5H_4Mo(CO)_3Cl$ with InCl in 1:1 stoichiometry in toluene at 75 °C [128]. Reactions of $CH_3C_5H_4Mo(CO)_3I$ with various PR_3 and RNC ligands in refluxing benzene in the presence of catalysts (PdO, $(C_5H_5Mo(CO)_3)_2$, or, preferentially, $(C_5H_5Fe(CO)_2)_2$), yielded monosubstituted derivatives, $CH_3C_5H_4Mo(CO)_2(PR_3)I$ (PR_3 = $P(CH_2PC_6H_5)_3$, $P(C_6H_5)_2CH_3$, $P(C_6H_5)_3$, $P(OCH_3)_3$, $P(OC_2H_5)_3$) and $CH_3C_5H_4Mo(CO)_2(CNR)I$ (R = t-C_4H_9, $C_6H_3(CH_3)_2$-2,6), respectively, generally as mixtures of cis and trans isomers [146]. Reaction of the chloro complex with equimolar $Na[Fe(CO)_3NO]$ in CH_3OH/THF, at room temperature under CO atmosphere, furnished $CH_3C_5H_4Mo(\mu\text{-}CO)(\mu\text{-}NO)(\mu_3\text{-}NH)\text{-}(Fe(CO)_3)_2$, accompanied by $(CH_3C_5H_4Mo(CO)_3)_2$ and $CH_3C_5H_4Mo(CO)_2NO$ [134, 136, 141, 162]. Treatment of $(\mu\text{-}H)(Fe(CO)_3)_2(\mu_3\text{-}S)Co(CO)_3$ with $CH_3C_5H_4Mo(CO)_3Cl$ in refluxing THF produced single- and double-exchanged heteroclusters, $CH_3C_5H_4Mo(CO)_2(\mu_3\text{-}S)(Fe(CO)_3)\text{-}Co(CO)_3$ and $(CH_3C_5H_4Mo(CO)_2)_2(\mu_3\text{-}S)Fe(CO)_3$ [165, 170]. The bis(dimethylglyoximato)-cobalt(II) complex $Co(CH_3C(NO)C(NOH)CH_3)_2NC_5H_5$ (solution in CH_3OH) and $CH_3C_5H_4Mo(CO)_3Cl$ reacted via chlorine atom abstraction to produce $CoCl(CH_3C(NO)C(NOH)CH_3)_2\text{-}NC_5H_5$ together with $(CH_3C_5H_4Mo(CO)_3)_2$ [90].

Irradiation at $\lambda > 490$ nm of $CH_3C_5H_4Mo(CO)_3Cl$ in pyridine yielded $CH_3C_5H_4Mo(CO)_2\text{-}(NC_5H_5)Cl$. At higher energy ($\lambda > 345$ nm), $CH_3C_5H_4Mo(CO)(NC_5H_5)_2Cl$ was also formed. Prolonged irradiation at $\lambda > 345$ nm resulted in the appearance of $(CH_3C_5H_4Mo)_2(CO)_5\text{-}NC_5H_5$, $[CH_3C_5H_4Mo(CO)_3NC_5H_5]^+$, and $[CH_3C_5H_4Mo(CO)_3]^-$, arising from intermediately generated $(CH_3C_5H_4Mo(CO)_3)_2$ (via Mo−Cl homolysis and radical combination) by substitution and disproportionation (compare Methods IVa and IVb). These transformations were ultimately followed by conversion of the $[CH_3C_5H_4Mo(CO)_3NC_5H_5]^+$ cation to $[C_5H_5N\text{-}C_5H_4CH_3]^+$ and $Mo(CO)_3(NC_5H_5)_3$ and further substitution of the latter by CO to $Mo(CO)_4\text{-}(NC_5H_5)_2$ [88]; see also [99]. Photolysis of $CH_3C_5H_4Mo(CO)_3Cl$ in dimethyl sulfoxide initially yielded $CH_3C_5H_4Mo(CO)_{3-n}(OS(CH_3)_2)_nCl$ (n = 1, 2). Continued irradiation then gave $[CH_3C_5H_4Mo(CO)_2(OS(CH_3)_2)_2]^+$ and Cl^- via Mo−Cl heterolysis, but no $[CH_3C_5H_4Mo(CO)_3]^-$ [88]. Irradiation of the chloro complex in benzene solution containing added cyclo-C_6H_{12}-N=CHCH=NC_6H_{12}-cyclo produced the cationic dicarbonyl $[CH_3C_5H_4Mo(CO)_2(C_6H_{12}N{=}CH\text{-}CH{=}NC_6H_{12})]^+$ [71].

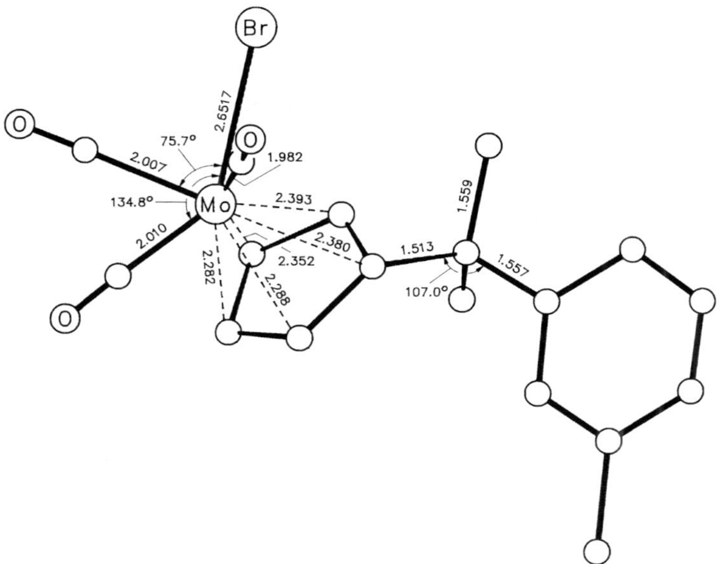

Fig. 13. The molecular structure of $3\text{-}CH_3C_6H_4C(CH_3)_2C_5H_4Mo(CO)_3Br$ [160].

$3\text{-}CH_3C_6H_4C(CH_3)_2C_5H_4Mo(CO)_3Br$ (Table **6**, No. **109**). Crystals belonged to the monoclinic system, space group $P2_1/a$ (nonstandard setting of $P2_1/c)-C^5_{2h}$ (No. 14), with a = 1.175(3), b = 1.4196(2), c = 1.1894(4) nm, β = 118.03(2)°; Z = 4 molecules per unit cell, D_{calc} = 1.733 g/cm³. A drawing of the molecular structure is given in **Fig. 13** [160].

$CH_3C(O)C_5H_4Mo(CO)_3I$ (Table **6**, No. **127**) crystallized in the monoclinic system, space group $P2_1/c-C^5_{2h}$ (No. 14), with a = 0.6663(2), b = 1.2364(6), c = 1.5464(5) nm, β = 99.02(3)°; Z = 4 molecules per unit cell, D_{calc} = 2.186 g/cm³. **Fig. 14** gives a representation of the structural model [135].

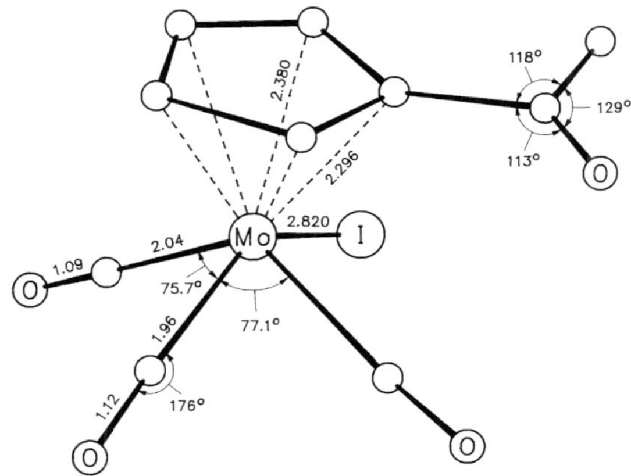

Fig. 14. The molecular structure of $CH_3C(O)C_5H_4Mo(CO)_3I$ [135].

References on pp. 253/8

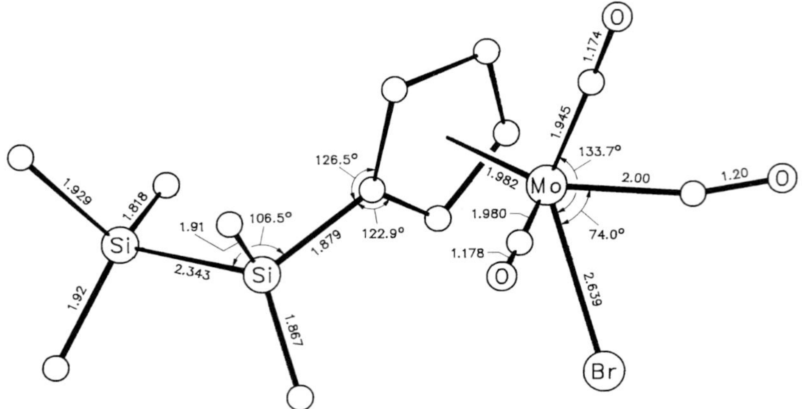

Fig. 15. The molecular structure of $(CH_3)_3Si(CH_3)_2SiC_5H_4Mo(CO)_3Br$ [148, 161].

$(CH_3)_3Si(CH_3)_2SiC_5H_4Mo(CO)_3Br$ (Table **6**, No. **137**). Single crystals were orthorhombic, space group Pbca$-D_{2h}^{15}$ (No. 61), with a = 1.3378(6), b = 1.5884(5), c = 1.7417(6) nm; Z = 8 molecules per unit cell, D_{calc} = 1.63 g/cm^3. The molecule is shown in **Fig. 15** [148, 161].

$CH_3C_5H_4Mo(CO)_3CH_3$ (Table **6**, No. **140**). $CH_3C_5H_4MoCl_4$ resulted from treatment with 2.5 equivalents of PCl_5 in CH_2Cl_2, followed by refluxing [112].

A thermal reaction with one equivalent of $C_5H_5Mo(CO)_3H$ in THF-d_8 at 50°C gave CH_3CHO together with the crossed dimers $CH_3C_5H_4Mo(CO)_nMo(CO)_nC_5H_5$ (n = 2, 3) as the predominant kinetic products; because of the competing thermal decomposition of the hydride, a significant amount of $(C_5H_5Mo(CO)_3)_2$ was observed as well [24, 26, 28].

The photoinduced reaction with C_2H_2 in pentane solution yielded the metallacyclic alkenyl ketone derivative (Formula VII) [56].

$CH_3C_5H_4(OC)_2Mo$ — O, CH_3, H, H

VII

$C_9H_7C(CH_3)_2C_5H_4Mo(CO)_3CH_3$ (Table **6**, No. **152**). Single crystals belonged to the orthorhombic space group Pnma$-D_{2h}^{16}$ (No. 62) with a = 13.461(3), b = 9.748(3), c = 14.037(4) Å; Z = 4 molecules per unit cell, D_{calc} = 1.50 g/cm^3. **Fig. 16** gives a view of the molecule residing on a crystallographic mirror plane that encompasses the indenyl system, the two quaternary carbon atoms of the $-C(CH_3)_2C_5H_4$ moiety, and the trans-OC$-$Mo$-CH_3$ fragment in which the methyl group and the carbonyl ligand were found to be mutually disordered [150].

$C_{13}H_9C(CH_3)_2C_5H_4Mo(CO)_3CH_3$ (Table **6**, No. **153**). Single crystals were orthorhombic, space group Pbca$-D_{2h}^{15}$ (No. 61), with a = 13.059(9), b = 13.089(6), c = 25.014(9) Å; Z = 8 molecules per unit cell, D_{calc} = 1.44 g/cm^3 [150]. The structure of the molecule is given in **Fig. 17**.

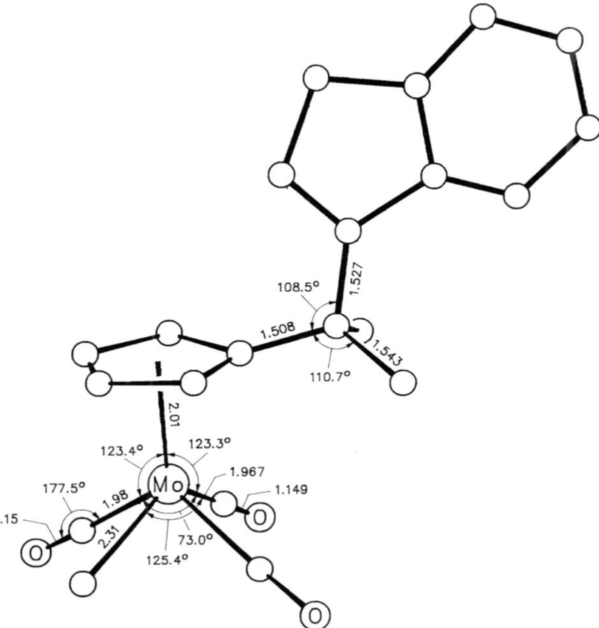

Fig. 16. The molecular structure of $C_9H_7C(CH_3)_2C_5H_4Mo(CO)_3CH_3$ [150].

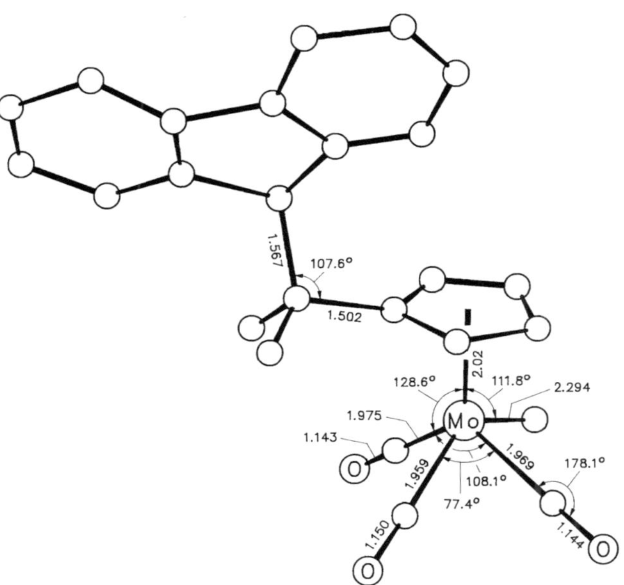

Fig. 17. The molecular structure of $C_{13}H_9C(CH_3)_2C_5H_4Mo(CO)_3CH_3$ [150].

References on pp. 253/8

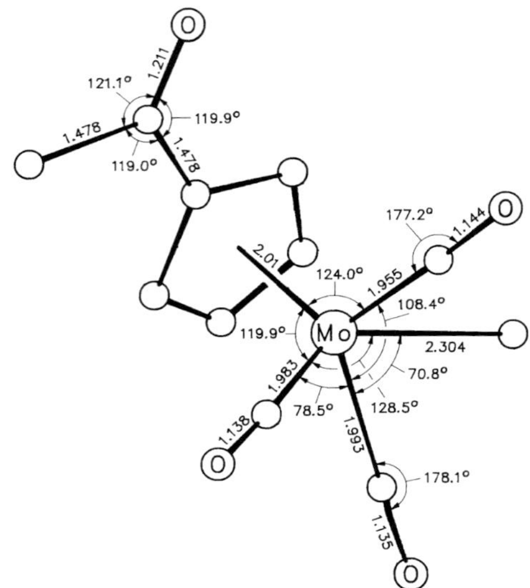

Fig. 18. The molecular structure of $CH_3C(O)C_5H_4Mo(CO)_3CH_3$ [122].

$CH_3C(O)C_5H_4Mo(CO)_3CH_3$ (Table **6**, No. **166**) crystallized in the monoclinic space group $P2_1/c-C^5_{2h}$ (No. 14) with a = 10.205(6), b = 14.192(8), c = 8.135(6) Å, β = 93.43(4)°; Z = 4 molecules per unit cell, D_{calc} = 1.71 g/cm³. In the molecule (**Fig. 18**), the substituted ring carbon atom is positioned such that it bisects one of the OC–Mo–CO angles giving it the maximally staggered position. The acetyl substituent is twisted slightly (twist angle: 4.6°) with respect to the C_5 plane; the CO group is bent away from the central metal, and the methyl residue inclined toward the Mo atom [122].

$C_5H_4CH_2CH_2Mo(CO)_3$ (Table **6**, No. **170**). Medium-pressure carbonylation in benzene at 120 °C gave the bridged acyl $C_5H_4CH_2CH_2C(O)Mo(CO)_3$ [19].

Photochemical treatment between 203 and 243 K of the title complex in pentane or petroleum ether solution with various ^{2}L ligands (ethene, propene, but-1-ene, E- and Z-but-2-ene, 2-methylpropene, 2-methylbut-2-ene, styrene, and allene) resulted in smooth CO substitution with formation of dicarbonyl(π-alkene) derivatives, $C_5H_4CH_2CH_2Mo(CO)_2{}^2L$. The complex $C_5H_4CH_2CH_2Mo(CO)_2((CH_3)_2C{=}CHCH_3)$, which could not be isolated but was only detected spectroscopically, appears to represent a limit for such compounds with sterically demanding ^{2}L ligands as there was no indication for the conversion of $C_5H_4CH_2CH_2Mo(CO)_3$ to $C_5H_4CH_2CH_2Mo(CO)_2((CH_3)_2C{=}C(CH_3)_2)$ in its reaction with 2,3-dimethylbut-2-ene [93]. UV irradiation in n-pentane or n-hexane solution at 218 to 253 K in the presence of buta-1,3-diene, (E)-penta-1,3-diene, or 2-methylpenta-1,3-diene produced the corresponding dicarbonyl-^{2}L and monocarbonyl-^{4}L complexes, $C_5H_4CH_2CH_2Mo(CO)_2(\eta^2\text{-}RCH{=}C(R')\text{-}CH{=}CH_2)$ and $C_5H_4CH_2CH_2Mo(CO)(\eta^4\text{-}RCH{=}C(R')CH{=}CH_2)$ (R = R' = H; R = CH_3, R' = H; R = H, R' = CH_3), by stepwise substitution of CO. Competitive insertion of buta-1,3-diene and (E)-penta-1,3-diene into the Mo–C σ bonds of the initially formed dicarbonyls gave bridged allyls of type VIII (R = H, CH_3) as by-products [72]. Photolysis at 223 K in pentane solution containing added cycloolefins (cyclopenta-1,3-diene, cyclohexa-1,3-diene, norbornadiene, cycloocta-1,5-diene, 2,3-dihydrofurane, and cyclohepta-1,3,5-triene) proceeded analogously affording dicarbonyls of the $C_5H_4CH_2CH_2Mo(CO)_2{}^2L$ type as the first-formed products. With cyclo-C_5H_6, cyclo-C_6H_8, cyclo-C_7H_8, and cyclo-C_8H_{12} as reactants, the respective monocarbonyls, $C_5H_4CH_2CH_2Mo(CO)^4L$, were also detected [108]. Cycloocta-1,3,5,7-tetraene inserted into the Mo–C bond to give an allyl product (Formula IX) [108]. Attempts to photochemically react the title complex with furane, spiro[2,4]hepta-4,6-diene, or acetylene were unsuccessful [93, 108].

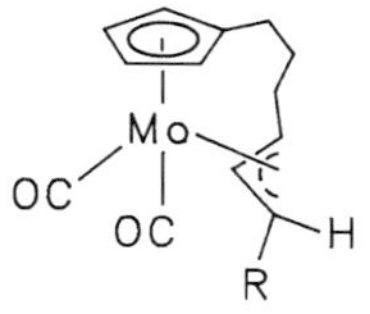

VIII 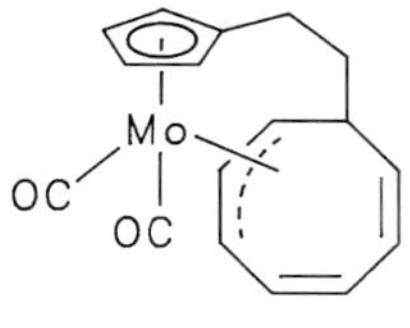 IX

$C_5H_4C(CH_3)_2C_9H_6Mo(CO)_3$ (Table **6**, No. **179**). Single crystals of the toluene solvate $C_5H_4C(CH_3)_2C_9H_6Mo(CO)_3 \cdot 0.5$ toluene belonged to the triclinic space group $P\bar{1}-C_i^1$ (No. 2) with a = 9.923(8), b = 10.307(4), c = 10.577(4) Å, α = 85.46(3), β = 73.65(5), γ = 76.14(4)°; Z = 2 molecules per unit cell, D_{calc} = 1.47 g/cm^3. **Fig. 19** presents a model of the molecular structure indicating that the indenyl double bond is localized within the cyclometalated ring [153].

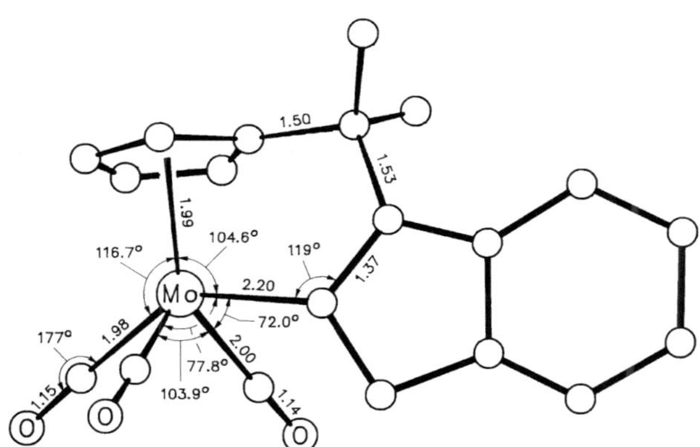

Fig. 19. The molecular structure of $C_5H_4C(CH_3)_2C_9H_6Mo(CO)_3$ [153].

$(C_{16}H_{20}S_4)Mo(CO)_3$ (Table **6**, No. **181**) was formed in the reaction between $Mo(CO)_3$-$(NCCH_3)_3$ and 6,6-bis(methylthio)fulvene in refluxing pentane by the addition of a second $(CH_3S)_2C{=}C_5H_4$ molecule to $(CH_3S)_2CC_5H_4Mo(CO)_3$, produced as an unstable primary intermediate. On cooling the mixture to −35 °C, the title compound crystallized together with its allylic decarbonyation product, $(C_{16}H_{20}S_4)Mo(CO)_2$ (Formula X), from which it was separated by repeated crystallization from pentane; yield: ca. 2% [31].

Single crystals grown from pentane were triclinic, space group $P\bar{1}-C_i^1$ (No. 2) with a = 1008.7(3), b = 1073.3(3), c = 1201.4(3) pm, α = 92.41(2), β = 110.73(2), γ =

References on pp. 253/8

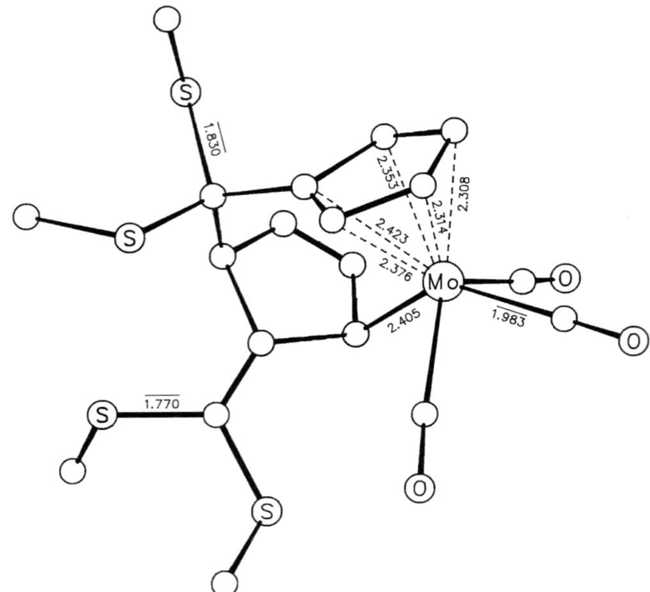

Fig. 20. The molecular structure of $(C_{16}H_{20}S_4)Mo(CO)_3$ [31].

115.79(3)°; Z = 2 molecules per unit cell, D_{calc} = 1.62 g/cm^3 [31]. A perspective drawing of the molecular model is shown in **Fig. 20**.

1.5.1.4.2.1.2 Heterobimetallic Complexes

1.5.1.4.2.1.2.1 RC_5H_4-Bridged Complexes without Metal–Metal Bonds

$Mn(CO)_3C_5H_4CH_2C_5H_4Mo(CO)_3Br$. A THF solution of $Li[C_5H_4CH_2C_5H_4Mn(CO)_3]$ (obtained in situ from $C_5H_4{=}CHC_5H_4Mn(CO)_3$ and $Li[(C_2H_5)_3BH]$) was stirred with $Mo(CO)_3(NCCH_3)_3$ (one equivalent) at room temperature to generate the $[Mn(CO)_3C_5H_4CH_2C_5H_4Mo(CO)_3]^-$ anion. Oxidation with Br_2 at 0 °C followed by filtration over silica gel, eluting with diethyl ether, afforded a 60% yield of the heterobinuclear product as red microcrystals [92].

^{1}H NMR spectrum (C_6D_6): δ = 2.80 (s, CH_2), 3.87, 3.95 (both "t", C_5H_4Mn), 4.35, 4.61 (both "t", C_5H_4Mo) ppm. ^{13}C NMR spectrum ($CDCl_3$): δ = 28.02 (CH_2), 82.03, 83.44, 91.39, 95.05, 102.62, 116.95 (all C_5H_4), 224.2, 227.0, 241.2 (CO) ppm. IR spectrum (CH_2Cl_2): 1936, 1966, 2022, 2042 (CO) cm^{-1} [92].

$Mn(CO)_3C_6H_6C_5H_4Mo(CO)_3CH_3$ (Formula I). The lithiated derivative made in situ by treating tricarbonyl(exo-(cyclopentadienyl)η^5-cyclohexadienyl)manganese, $C_5H_5C_6H_6Mn(CO)_3$, with LiC_4H_9-n/hexane in THF was refluxed with $Mo(CO)_3(NCCH_3)_3$ in THF solution and then treated with CH_3I at room temperature. Extraction of the evaporated mixture into diethyl ether, followed by column chromatography on alumina with ether/hexane (1:10), gave the title complex in 60% yield [143].

Yellow solid, m.p. 114 °C (dec.). ^{1}H NMR spectrum ($CDCl_3$): δ = 0.31 (s, CH_3), 3.26 (m, 3H), 4.88 ("t", 2H; "J" = 2.25 Hz), 4.94 ("t", 2H; "J" = 5.40 Hz), 5.09 ("t", 2H; J = 2.28 Hz),

5.77 (t, 1H; J = 5.26 Hz) ppm. IR spectrum (NaCl): 1903, 1928, 1946, 2015 (CO) cm^{-1} [143].

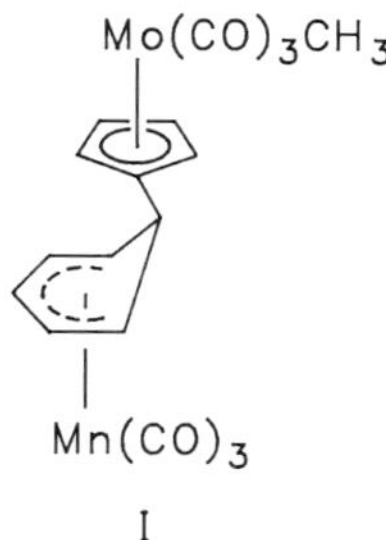

$C_5H_5Fe(CO)_2CH_2CH_2C_5H_4Mo(CO)_3CH_3$. To a suspension of $Mo(CO)_3(NCCH_3)_3$ in THF was added a THF solution of excess $Na[C_5H_5Fe(CO)_2CH_2CH_2C_5H_4]$ (from spiro[2,4]hepta-4,6-diene by nucleophilic cyclopropane ring opening with $Na[C_5H_5Fe(CO)_2]$). The mixture was stirred at room temperature and then treated with CH_3I in THF. Chromatography on an Al_2O_3 column, eluting with hexane/ether (20:1), afforded some $C_5H_5Fe(CO)_2CH_2CH_2C_5H_5$, followed by the title compound, which was purified further by recrystallization from hexane/ether; yield: 36% [30].

Yellow needles, m.p. 83 to 84 °C. ^1H NMR spectrum (C_6D_6): δ = 0.50 (s, CH_3), 1.24 to 1.50 (m, $FeCH_2$), 2.05 to 2.13 (m, $CH_2C_5H_4$), 4.03 (s, C_5H_5), 4.50, 4.65 (both "t", C_5H_4; "J" = 2.0 Hz) ppm. IR (KBr): 1895, 1945, 2000 (CO) cm^{-1}. Mass spectrum (70 eV): $[M]^+$ (10%), $[M-nCO]^+$ (n = 2 (3%), 3 (7%), 4 (6%), 5 (16%)), $[M-5CO-CH_4]^+$ (63%); for additional fragments, see [30].

$MC_5H_4C{\equiv}CC_5H_4Mo(CO)_3CH_3$ (M = $CH_3W(CO)_3$, $Mn(CO)_3$, $Re(CO)_3$, $CH_3Fe(CO)_2$). The following general method of synthesis was adopted: The iodocyclopentadienyl complex IC_5H_4-$Mo(CO)_3CH_3$ was coupled with the required (trialkylstannyl)ethynylcyclopentadienyl compound $R_3SnC{\equiv}CC_5H_4M$ (R = CH_3, n-C_4H_9) using equimolar quantities of the two coupling partners in the presence of 5 mol% of $Pd(NCCH_3)_2Cl_2$ in N,N-dimethylformamide at room temperature. The use of the $(CH_3)_3SnC{\equiv}CC_5H_4M$ derivatives was more convenient because the reaction was faster, and the $Sn(CH_3)_3I$ formed as a by-product was removed by extraction with water after diethyl ether had been added. Removal of the corresponding $Sn(C_4H_9\text{-}n)_3I$ required transformation to the insoluble fluoride by adding 50% aqueous KF. Evaporation of the ethereal phases gave the crude products which were purified further by chromatographic separation over silica using hexane/benzene (5:1), followed by sublimation at 100 °C/10^{-5} Torr [115]; for yields and properties, see the following table:

M	yield and properties
$W(CO)_3CH_3$	(76%) yellow crystals, m.p. > 150 °C (dec.) ^1H NMR ($CDCl_3$): 0.49 ($MoCH_3$), 0.53 (s, WCH_3), 5.23, 5.44 (both "t", C_5H_4; "J" = 2.3), 5.29, 5.51 (both "t", C_5H_4; "J" = 2.4) ^{13}C NMR ($CDCl_3$): −28.69 (WCH_3), −15.87 ($MoCH_3$), 82.86, 89.35, 90.53, 91.69, 93.05, 94.60, 95.49, 96.02 (all $C_5H_4C{\equiv}CC_5H_4$), 214.81, 225.31 (both CO) IR (CCl_4): 1940, 2015 (CO)

References on pp. 253/8

M	yield and properties
$Mn(CO)_3$	(70%) yellow crystals, m.p. 92 to 93°C (dec.) ^1H NMR ($CDCl_3$): 0.49 (s, CH_3), 4.69, 5.03 (both "t", C_5H_4; "J" = 2.1), 5.22, 5.44 (both "t", C_5H_4; "J" = 2.2) ^{13}C NMR ($CDCl_3$): −15.27 (CH_3), 80.73, 82.03, 83.23, 86.97, 88.10, 90.28, 93.70, 95.81 (all $C_5H_4C{\equiv}CC_5H_4$), 223.90, 225.37 (both CO) IR (CCl_4): 1950, 2018, 2023 (CO)
$Re(CO)_3$	(68%) yellow crystals, m.p. 118 to 120°C (dec.) ^1H NMR ($CDCl_3$): 0.49 (s, CH_3), 5.22, 5.44 (both "t", C_5H_4; "J" = 2.3), 5.28, 5.64 (both "t", C_5H_4; "J" = 2.1) ^{13}C NMR ($CDCl_3$): −15.56 (CH_3), 81.40, 83.59, 83.90, 87.83, 88.39, 90.44, 93.21, 95.88 (all $C_5H_4C{\equiv}CC_5H_4$), 192.83, 225.30 (both CO) IR (CCl_4): 1945, 2018, 2022 (CO)
$Fe(CO)_2CH_3$	(82%) orange crystals, m.p. > 150°C (dec.) ^1H NMR ($CDCl_3$): 0.33 (s, $FeCH_3$), 0.49 (s, $MoCH_3$), 4.75, 4.90 (both "t", C_5H_4; "J" = 2.0), 5.22, 5.44 (both "t", C_5H_4; "J" = 2.1) ^{13}C NMR ($CDCl_3$): −19.14 ($FeCH_3$), −15.07 ($MoCH_3$), 81.01, 82.38, 83.18, 83.97, 89.86, 90.16, 94.17, 95.81 (all $C_5H_4C{\equiv}CC_5H_4$), 216.00, 225.42 (both CO) IR (CCl_4): 1945, 1968, 2008, 2018 (CO)

$Mo(CO)_3C_5H_4C_5H_4Ru(CO)(P(CH_3)_3)_2$ (Formula II) was first made by adding a THF solution of $Mo(CO)_3(NCCH_3)_3$ (10% excess) to the fulvalene precursor $C_5H_4{=}C_5H_4Ru(CO)(P(CH_3)_3)_2$ in THF without stirring. The zwitterionic title complex precipitated in 97% yield [78]. An alternative approach involved the $C_5H_5FeC_6(CH_3)_6$-initiated intramolecular electron transfer reaction of the parent Mo−Ru-bonded fulvalene complex $Mo(CO)_3(\mu\text{-}C_5H_4C_5H_4)Ru(CO)_2$ in the presence of excess $P(CH_3)_3$, which was complete after stirring the reactants for a few minutes in THF/1,2-dimethoxyethane mixtures at room temperature. The product was isolated from the evaporated mixture by extraction into CH_3CN, followed by crystallization from acetone/ether at low temperature [163].

Yellow needles, m.p. 280 to 282°C (dec.). ^1H NMR spectrum (CD_3CN): δ = 1.52 ("d", CH_3; "J" = 10.1 Hz), 5.08, 5.17, 5.36, 5.55 (all "t", $C_5H_4C_5H_4$; "J" = 2.0 (2nd and 3rd "t") and 2.3 (1st and 4th "t") Hz) ppm. ^{31}P NMR spectrum (CD_3CN): δ = 7.3 ppm. IR spectrum (CH_3CN): 1792, 1908, 1972 (all CO) cm^{-1}. Mass spectrum: $[M-CO]^+$ (18%), $[M-P(CH_3)_3]^+$ (100%) [78].

$Mo(CO)_3C_5H_4C_5H_4Rh(P(CH_3)_3)_2C(O)CH_3$ (Formula III) was synthesized by treating the W−Rh-bonded fulvalene complex $W(CO)_3(\mu\text{-}C_5H_4C_5H_4)Rh(CO)C(O)CH_3$ initially with a

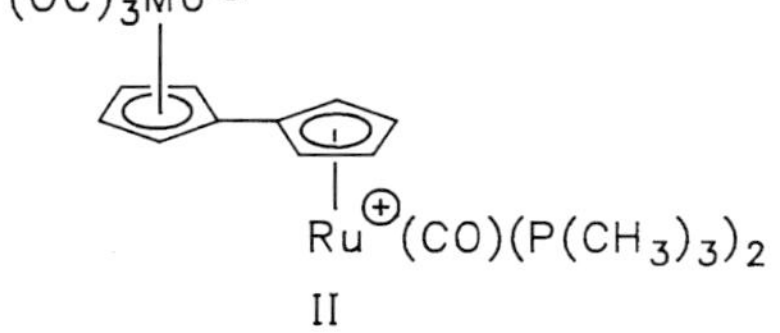

$(OC)_3Mo^{\ominus}$ … $Ru^{\oplus}(CO)(P(CH_3)_3)_2$

II

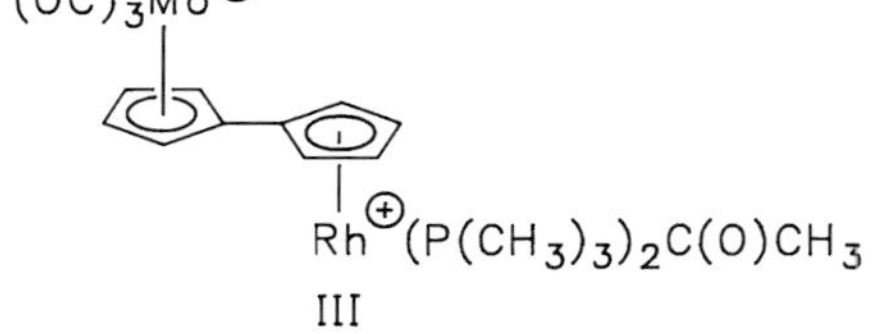

$(OC)_3Mo^{\ominus}$ … $Rh^{\oplus}(P(CH_3)_3)_2C(O)CH_3$

III

References on pp. 253/8

large excess of trimethyl phosphane in CH_3CN at room temperature. The residue obtained after evaporation of the solvent was redissolved in THF to give a solution of the zwitterionic intermediate $C_5H_4^{(-)}C_5H_4Rh^{(+)}(P(CH_3)_3)_2C(O)CH_3$ admixed with $W(CO)_3(P(CH_3)_3)_3$. Exposure of this solution to $Mo(CO)_3(NCC_2H_5)_3$ resulted in re-complexation of the free cyclopentadienyl ring, depositing the title complex as a precipitate that was crystallized by diffusion of benzene into a CH_3CN solution (13%) [166].

Orange crystals, m.p. 215 to 217 °C. ^1H NMR spectrum (CD_3CN): δ = 1.47 ("dd", $P(CH_3)_3$; $\Sigma|^2J(P, H) + {}^4J(P, H)|$ = 9.9, J(Rh, H) = 0.7 Hz), 2.41 (s, $C(O)CH_3$), 5.15 ("t", 2 fulvalene-H; "J"(H, H) = 2.3 Hz), 5.56 (m, 4 fulvalene-H), 5.64 ("t", 2 fulvalene-H; "J"(H, H) = 2.3 Hz) ppm. ^{13}C NMR spectrum (CD_3CN): δ = 18.88 ("t", $P(CH_3)_3$; $\Sigma|^1J(P, C) + {}^3J(P, C)|$ = 34.7 Hz), 50.10 (C(O)**C**H_3), 86.95 (br), 88.60, 88.64, 92.91 (br), 94.08 (all fulvalene-C), 235.22 (MoCO) ppm (quaternary carbon nuclei attached to Rh not detected). ^{31}P NMR spectrum (CD_3CN): δ = 4.77 (d, J(Rh, P) = 157.2 Hz) ppm. IR spectrum (KBr): 1636 (ν(C=O)), 1773, 1892 (CO) cm^{-1}. FAB mass spectrum (sulfolane): $[M]^+$, $[M-Mo(CO)_3]^+$ [166].

1.5.1.4.2.1.2.2 Metal–Metal-Bonded Complexes without Bridging RC_5H_4 Ligands

$(C_6H_5)_2PC_5H_4Mo(CO)_3Mn(CO)_5$ was isolated in 46% yield from the reaction between $Li[(C_6H_5)_2PC_5H_4Mo(CO)_3]$ and $Mn(CO)_5Br$ [178].

^1H NMR spectrum ($CDCl_3$): δ = 4.49 ("q", H-3,4 of C_5H_4; "J" = 2.1 Hz), 5.43 ("q", H-2,5 of C_5H_4), 7.43 to 7.46 (m, H-3 to 5 of C_6H_5), 7.90 to 7.97 (m, H-2,6 of C_6H_5) ppm. ^{13}C NMR spectrum ($CDCl_3$): δ = 56.68 (d, C-1 of C_5H_4; J(P,C) = 50.6 Hz), 87.89 (d, C-2,5 of C_5H_4; J(P,C) = 6.2 Hz), 129.27 (d, J(P,C) = 10.6 Hz), 131.42 (d; J(P,C) = 12.4 Hz), 131.65 (s), 133.32 (d; J(P,C) = 42.5 Hz), (all C_6H_5), 215.34 (br), 225.73, 231.03 (all CO) ppm. ^{31}P NMR spectrum ($CDCl_3$): δ = 58.74 ppm. IR spectrum (CH_2Cl_2): 1881.7, 1908.7, 1938.2, 1971.8, 1988.0, 2057.5 (CO) cm^{-1} [178].

Iodination gave $(C_6H_5)_2PC_5H_4Mo(CO)_3I$ in quantitative yield [178].

$CH_3C_5H_4Mo(CO)_3Fe(CO)_2C_5H_5$ resulted from the reaction between $C_5H_5Fe(CO)_2I$ and $Na[CH_3C_5H_4Mo(CO)_3]$ in $CH_3OC_2H_4OC_2H_4OCH_3$ at ambient conditions. After evaporation of the solvent at 70 °C in vacuum, the mixture was chromatographed on a silica gel column. Elution with ether/pentane (1 : 1) provided a red fraction containing the title complex together with $(CH_3C_5H_4Mo(CO)_3)_2$. The two compounds were separated by renewed chromatography on SiO_2, eluting with n-pentane. Further purification of the Mo–Fe-bonded heterodimer was accomplished by recrystallization from n-pentane/diethyl ether at −80 °C; yield: 37% [83].

IR spectrum (Nujol): 1870, 1890, 1903, 1920, 1945, 1960 (CO); additional bands at 460, 490, 535, 570, 589, 625, 820, 837, 1005, 1045, and 1062 cm^{-1} [83].

$RC_5H_4Mo(CO)_3Co(CO)_4$ (R = CH_3, $CH_3C(O)$, $C_2H_5OC(O)$). $CH_3C_5H_4Mo(CO)_3Co(CO)_4$ was isolated from the mixture of products resulting from the slow addition at room temperature of $CH_3C_5H_4Mo(CO)_3Ni(CO)C_5H_5$, dissolved in toluene, to an equimolar quantity of $Co_2(CO)_8$ in the same solvent. Column chromatography on silica gel with hexane/ether mixtures (2 : 1) afforded a dark brown band, characterized as containing unreacted $Co_2(CO)_8$ in addition to $(CH_3C_5H_4Mo(CO)_3)_2$ and $CH_3C_5H_4Mo(CO)_3Co(CO)_4$. This mixture was subsequently separated by renewed chromatography, using pure hexane as the eluting solvent. Multiple crystallizations from hexane yielded the product in 41% yield. When $Co_2(CO)_8$ was slowly added to $CH_3C_5H_4Mo(CO)_3Ni(CO)C_5H_5$ in toluene, chromatography under the conditions outlined above provided a 35% yield of $CH_3C_5H_4Mo(CO)_3Co(CO)_4$ [109]; for other products separated from these metal/metal metathesis reactions, see under $CH_3C_5H_4Mo(CO)_3Ni(CO)C_5H_5$ below. An alternative synthesis was based on the reaction between $CH_3C_5H_4Mo(CO)_3H$ and $Co_2(CO)_8$ [114], analogous to the method employed for the preparation of the related species $C_5H_5Mo(CO)_3Co(CO)_4$ [101].

In the syntheses of $CH_3C(O)C_5H_4Mo(CO)_3Co(CO)_4$ and $C_2H_5OC(O)C_5H_4Mo(CO)_3Co(CO)_4$, the triple-bonded homodimers $(RC_5H_4Mo(CO)_2)_2$ (R = $CH_3C(O)$, $C_2H_5OC(O)$) were reacted with excess $Co_2(CO)_8$ in toluene at room temperature until analysis using thin-layer chromatography showed no starting material was left. Column chromatography of the evaporated mixtures on silica gel, eluting with petroleum ether/CH_2Cl_2 (2:3), gave a black band corresponding to $Co_4(CO)_{12}$, followed by a pink band arising from $CH_3C(O)C_5H_4Mo(CO)_3Co(CO)_4$ (43%) and $C_2H_5OC(O)C_5H_4Mo(CO)_3Co(CO)_4$ (46%), respectively. A third red band was shown to contain the single-bonded dimers $(RC_5H_4Mo(CO)_3)_2$, formed as additional by-products [172].

Some physical properties are given in the following table [172]:

R	properties
CH_3	purple solid [114] ^1H NMR ($CDCl_3$): 2.12 (s, CH_3), 5.39, 5.44 (AA′BB′ m, C_5H_4) [114] IR (hexane): 1913, 1918, 1941, 1944, 1959, 1977, 1987, 2021, 2044, 2058, 2075 (CO) [114]
$CH_3C(O)$	pink solid, m.p. 75 to 76°C [172] ^1H NMR ($CDCl_3$): 2.37 (s, CH_3), 5.77 ("s", H-3,4 of C_5H_4), 5.93 ("s", H-2,5 of C_5H_4) [172] IR (KBr): 1690 (ν(C=O)), 1924, 1949, 1981, 2000, 2013, 2075 (CO) [172] mass spectrum: $[M-nCO]^+$ (n = 2 (7%), 3 (5%), 4 (2%), 5 (2%), 6 (5%), 7 (4%)), $[CH_3C(O)C_5H_4Mo(CO)_n]^+$ (n = 3 (15%), 2 (31%), 1 (19%), 0 (76%)), $[Co(CO)_n]^+$ (n = 4 (21%), 3 (12%), 2 (5%), 1 (5%)), $[Mo]^+$ (24%), $[Co]^+$ (11%) [172]
$C_2H_5OC(O)$	pink solid, m.p. 68 to 69°C [172] ^1H NMR ($CDCl_3$): 1.33 (t, CH_3; J = 7.0), 4.29 (q, CH_2), 5.63 ("s", H-3,4 of C_5H_4), 5.92 ("s", H-2,5 of C_5H_4) [172] IR (KBr): 1725 (ν(C=O)), 1931, 1974, 2028, 2075 (CO) [172] mass spectrum: $[M]^+$ (12%), $[M-nCO]^+$ (n = 1 (3%), 2 (3%), 4 (1%), 5 (1%), 6 (9%), 7 (6%)), $[C_2H_5OC(O)C_5H_4Mo(CO)_n]^+$ (n = 3 (25%), 2 (19%), 1 (40%), 0 (36%)), $[Co(CO)_4]^+$ (16%), $[Mo]^+$ (23%) [172]

Crystals of $CH_3C(O)C_5H_4Mo(CO)_3Co(CO)_4$ grown from CH_2Cl_2/petroleum ether were shown to be monoclinic, space group $P2_1/n$ (nonstandard setting of $P2_1/c)-C^5_{2h}$ (No. 14), with a = 7.461(2), b = 13.863(2), c = 15.919(5) Å, β = 96.43(2)°; Z = 4 molecules per unit cell, D_{calc} = 1.86 g/cm^3. The molecular structure is shown in **Fig. 21** [172].

Interaction of $CH_3C_5H_4Mo(CO)_3Co(CO)_4$ with $(CH_3C_5H_4W(CO)_2)_2$ in toluene at 55°C gave a mixture of products from which the bimetallic species $CH_3C_5H_4M(CO)_3Co(CO)_4$, $(CH_3C_5H_4M(CO)_3)_2$ (M = Mo, W), and $CH_3C_5H_4Mo(CO)_3W(CO)_3C_5H_4CH_3$ were recovered in addition to the tetrahedral clusters $(CH_3C_5H_4W(CO)_2)_2(Co(CO)_3)_2$, $(CH_3C_5H_4W(CO)_2)_3Co(CO)_3$, and $CH_3C_5H_4Mo(CO)_2(CH_3C_5H_4W(CO)_2)_2Co(CO)_3$ [114]. A reaction with $Co_2(CO)_8$ in refluxing toluene caused the formation of $(CH_3C_5H_4Mo(CO)_3)_2$, $(CH_3C_5H_4Mo(CO)_2)_2(Co(CO)_3)_2$, $CH_3C_5H_4Mo(CO)_2(Co(CO)_3)_3$, and a homonuclear cobalt cluster of unknown identity [114].

References on pp. 253/8

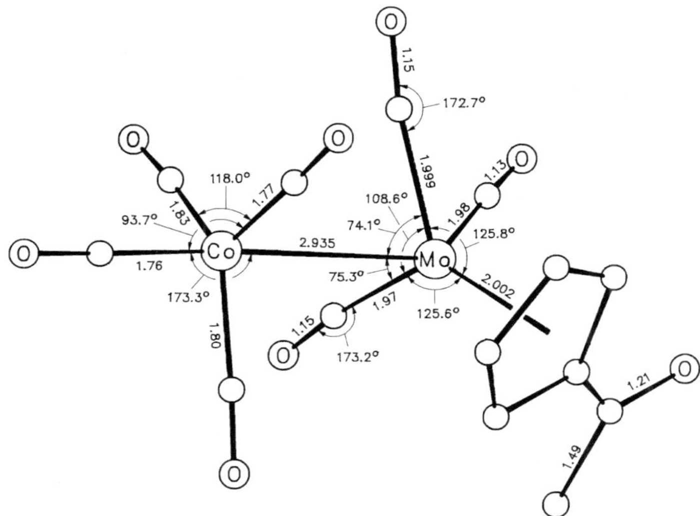

Fig. 21. The molecular structure of $CH_3C(O)C_5H_4Mo(CO)_3Co(CO)_4$ [172].

$CH_3C_5H_4Mo(CO)_3NiC_5(CH_3)_5$. The product of mild thermolysis of **$CH_3C_5H_4Mo(CO)_3Ni(CO)$-$C_5(CH_3)_5$** in hexane [117, 142, 155] reacted with $P(C_6H_5)_2H$ in hexane between −78 and −50 °C to form $CH_3C_5H_4Mo(CO)(P(C_6H_5)_2H)(\mu\text{-}CO)_2NiC_5(CH_3)_5$ together with $((CH_3)_5C_5Ni$-$CO)_2$ and much decomposed material [142]. Treatment with one equivalent of diphenyl disulfide in CH_2Cl_2/ether at dry-ice temperature led to $CH_3C_5H_4Mo(CO)(SC_6H_5)(\mu\text{-}CO)$-$(\mu\text{-}SC_6H_5)NiC_5(CH_3)_5$ [155].

$CH_3C_5H_4Mo(CO)_3Ni(CO)C_5H_5$ was originally made by analogy to $C_5H_5Mo(CO)_3Ni(CO)C_5H_5$ [63], the latter being formed at high equilibrium concentrations when $(C_5H_5Mo(CO)_3)_2$ and $(C_5H_5NiCO)_2$ were combined in benzene [20]. In a slight modification of that procedure, an equimolar ratio of $(CH_3C_5H_4Mo(CO)_3)_2$ and $(C_5H_5NiCO)_2$ was irradiated in toluene at ambient conditions [81, 89].

^1H NMR spectrum (C_6D_6): δ = 1.52 (s, CH_3), 4.54, 4.57 (both m, C_5H_4), 5.24 (s, C_5H_5) ppm [89].

Refluxing in hexane caused partial decarbonylation with the formation of $CH_3C_5H_4Mo$-$(CO)_3NiC_5H_5$ [117, 142, 155]. Thermolysis in cycloocta-1,5-diene at 70 °C afforded $(CH_3$-$C_5H_4Mo(CO)_3)_2$ together with a tetranuclear paramagnetic $MoNi_3$ complex, $CH_3C_5H_4Mo$-$(\mu_3\text{-}CO)_3(NiC_5H_5)_3$, consisting of an approximately isosceles nickel atom triangle capped by a $CH_3C_5H_4Mo$ unit [97].

Slow addition of toluene solutions of the title complex to equimolar solutions of $Co_2(CO)_8$ in the same solvent caused the formation of a mixture of mixed-metal products, from which the dinuclear species $(CH_3C_5H_4Mo(CO)_3)_2$ and $CH_3C_5H_4Mo(CO)_3Co(CO)_4$ as well as the tetrahedral clusters $CH_3C_5H_4Mo(CO)_2(Co(CO)_3)_2NiC_5H_5$ and $CH_3C_5H_4Mo(CO)_2(Co(CO)_3)_3$ could be separated. When the order of addition was reversed, and a solution of $Co_2(CO)_8$ was added slowly to $CH_3C_5H_4Mo(CO)_3Ni(CO)C_5H_5$, all species reported before except $(CH_3C_5H_4Mo(CO)_3)_2$ were obtained, albeit in slightly different relative ratios [109].

Metal/metal metathesis with formation of a 1 : 1 : 1 : 1 mixture of $C_5H_5Mo(CO)_3Ni(CO)C_5H_5$, $CH_3C_5H_4Mo(CO)_3Ni(CO)C_5H_5$, $C_5H_5Mo(CO)_3Ni(CO)C_5H_4CH_3$, and $CH_3C_5H_4Mo(CO)_3Ni$-

References on pp. 253/8

$(CO)C_5H_4CH_3$ took place when the title complex was allowed to equilibrate with C_5H_5Mo-$(CO)_3Ni(CO)C_5H_4CH_3$ in C_6D_6 [89]. In the presence of phenylacetylene, the metalatetrahedrane and nickelacyclobutenone compounds I and II (each with R = H and R′ = CH_3 or R = CH_3 and R′ = H) were observed to form without any evidence of crossover products, implying that the alkyne reacted with the intact heterobimetallic complexes and not with monometallic radicals derived therefrom by Mo–Ni bond fission [89].

$RC_5H_4(OC)Mo(\mu\text{-}C_6H_5C_2H)(\mu\text{-}CO)NiC_5H_4R'$

I

$RC_5H_4(OC)_2Mo[\mu\text{-}C(H)=C(C_6H_5)C(=O)]NiC_5H_4R'$

II

$CH_3C_5H_4(OC)Mo(\mu\text{-}RC_2R')(\mu\text{-}CO)NiC_5H_5$

III

$CH_3C_5H_4(OC)_2Mo[\mu\text{-}C(R)=C(R')C(=O)]NiC_5H_5$

IV

Treatment with alkynes such as but-2-yne, pent-2-yne, phenylacetylene, or diphenylacetylene in toluene at 25 °C resulted in mixtures of dimetalatetrahedrane compounds (Formula III: R = R′ = CH_3, C_6H_5; R = CH_3, R′ = C_2H_5; R = H, R′ = C_6H_5) and nickelacyclobutenone derivatives (Formula IV: R = R′ = CH_3, C_6H_5; R/R′= CH_3/C_2H_5 (mixture of two isomers); R = H, R′ = C_6H_5) [63, 81, 89]. The rate of the reaction with phenylacetylene showed a first-order dependence on both the title compound and the alkyne for moderate concentrations of title complex and $C_6H_5C{\equiv}CH$. At 20.0 ± 0.5 °C in benzene, the second-order rate constant was measured as $(2.9 \pm 0.4) \times 10^{-3}$ L · mol^{-1} · s^{-1} [89].

The reaction with excess allene in hexane at ambient conditions resulted in the formation of the heterobimetallic species V [132, 133]. Combination with 1,1-dimethylallene in hexane/toluene (1:1) afforded the 2-nickel-substituted allylic product (Formula VI) (and not the dimethylallene-bridged species $CH_3C_5H_4Mo(CO)_2(\mu\text{-}(CH_3)_2C{=}C{=}CH_2)NiC_5H_5$ proposed originally [102]), which could be isolated by cooling the concentrated reaction mixture to −20 °C. On attempted chromatography on silica gel, the product underwent isomerization with the formation of a 1-nickel-substituted allyl complex (Formula VII) via an effective 1,4-proton shift (rather than a 1,2-methyl group migration) as demonstrated by labeling studies using 1,1-dimethylallene-3-d_1 [132, 133].

Combination with t-C_4H_9NC in diethyl ether/hexane at −78 °C, followed by warming to room temperature, produced $CH_3C_5H_4Mo(CO)_2(\mu\text{-}CNC_4H_9\text{-}t)(\mu\text{-}CO)NiC_5H_5$ [142].

$CH_3C_5H_4(OC)_2Mo$—(allyl)—CH_2—C(=CHH)—(allyl)—NiC_5H_5

V

$CH_3C_5H_4(OC)Mo[\mu\text{-}(H_3C)_2C\text{-}C\text{-}CH_2](\mu\text{-}CO)NiC_5H_5$

VI

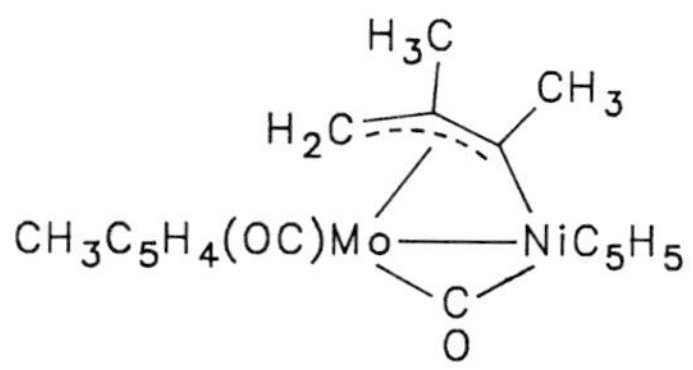

VII

$CH_3C_5H_4Mo(CO)_3Ni(CO)C_5H_4CH_3$ was prepared from $(CH_3C_5H_4Mo(CO)_3)_2$ and $(CH_3C_5H_4$-$NiCO)_2$ as outlined for $CH_3C_5H_4Mo(CO)_3Ni(CO)C_5H_5$ [89]; for its formation in the metal/metal metathesis reaction occurring between $C_5H_5Mo(CO)_3Ni(CO)C_5H_4CH_3$ and $CH_3C_5H_4MoNi$-$(CO)C_5H_5$ [89], see above.

^1H NMR spectrum (C_6D_6): δ = 1.55, 1.87 (both s, CH_3), 4.53, 4.55, 5.12, 5.16, 5.18 (all m, C_5H_4) ppm [89].

$HC(O)C_5H_4Mo(CO)_3AuP(C_6H_5)_3$ was briefly cited from a conference report [67] in "Organomolybdenum Compounds" Pt. 6, 1990, p. 383. The complex was first prepared by mixing the 6-(dimethylamino)fulvene derivative $Me_2NCHC_5H_4Mo(CO)_3$ with a deficiency of $[O(AuP(C_6H_5)_3)_3]BF_4$ between −5 and 0 °C in THF in the presence of H_2O and K_2CO_3, which afforded the product in 67% yield [68]. The complex was also obtained from $Na[HC(O)C_5H_4$-$Mo(CO)_3]$ and equimolar $Au(P(C_6H_5)_3)Cl$ in THF at 40 °C. Crystallization of the evaporated reaction mixture from toluene/hexane provided a 60% yield [73].

Yellow orange crystals, m.p. 132 °C (from benzene/pentane) [68]; ivory-colored microcrystals, m.p. 143 °C (from toluene/hexane) [73]. ^1H NMR spectrum ($CDCl_3$): δ = 5.45, 5.80 (both "t", C_5H_4; $\Sigma|^3J(H, H) + {}^4J(H, H)|$ = 4.6 Hz), 7.48 (m, C_6H_5), 9.62 (s, HCO) ppm [68]; similar in [73]. ^{13}C NMR spectrum ($CDCl_3$): δ = 89.71, 90.30, 102.97 (all s, all C_5H_4), 129.15 (d; J(P, C) = 11.8 Hz), 131.36 (s), 134.02 (d; J(P, C) = 14.7 Hz) (all C_6H_5), 184.57 (s, HCO), 230.06 (CO) ppm [68]. IR (KBr): 1850, 1883, 1960 (CO) cm^{-1} [68]; 1679 (ν(C=O)), 1840, 1881, 1955 (CO) cm^{-1} [73].

Single crystals were shown to belong to the monoclinic system, space group $P2_1/c-C^5_{2h}$ (No. 14), with a = 13.072(9), b = 11.403(8), c = 16.451(12) Å, β = 94.62(5)°; Z = 4 molecules per unit cell, D_{calc} = 1.99 g/cm^3. The molecular structure is shown in **Fig. 22**. The formyl group extrudes from the plane of the cyclopentadienyl ring toward the molybdenum atom (dihedral angle: 4.2(2)°). Two of the three CO ligands are not purely terminal but rather act as semibridging ligands, as indicated by their comparatively short Au···C intramolecular contacts and by the marked deviation from linearity of the fragments [68].

The mass spectrum (70 eV, 250 °C) contained an intense molecular ion peak. Competitive fragmentation via one-stage loss of three carbonyl groups and cleavage of the Mo−Au bond generated $[HC(O)C_5H_4Mo(CO)_3]^+$, $[AuP(C_6H_5)_3]^+$, and the $[M-3CO]^+$ ion, which then eliminated the formyl CO as well, giving $[C_5H_5MoAuP(C_6H_5)_3]^+$. Other characteristics of the fragmentation process were the destruction of the cyclopentadienyl ligand via predominant elimination of C_2H_2 from the ions $[HC(O)C_5H_4Mo(CO)_n]^+$ (n = 0 to 3) and the degradation of the CO-free $[C_5H_5MoAuP(C_6H_5)_3]^+$ fragment via competitive elimination of Au and $P(C_6H_5)_3$. For further details, see [82].

Photolysis (500 < λ < 680 nm) in toluene at −70 °C in the presence of 3,6-di-t-butyl-o-benzoquinone gave the radicals $3,6\text{-}(t\text{-}C_4H_9)_2C_6H_2O_2AuP(C_6H_5)_3^{\bullet}$ and $(3,6\text{-}(t\text{-}C_4H_9)_2\text{-}C_6H_2O_2)_2MoC_5H_5^{\bullet}$ [82].

References on pp. 253/8

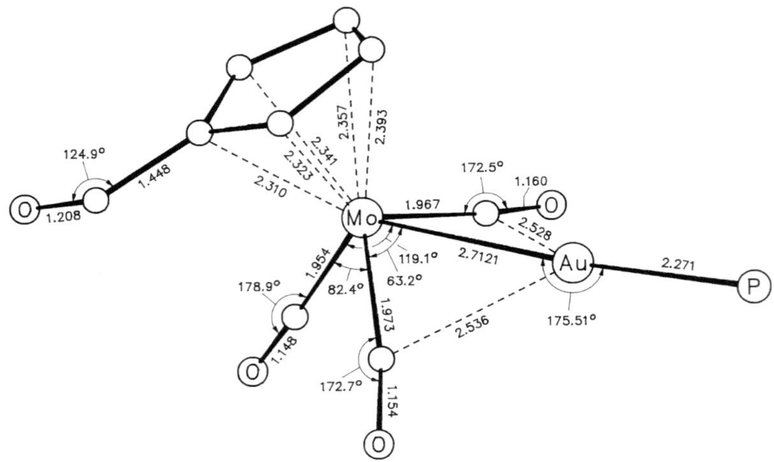

Fig. 22. The molecular structure of $HC(O)C_5H_4Mo(CO)_3AuP(C_6H_5)_3$ (phenyl rings omitted) [68].

$CH_3C_5H_4Mo(CO)_3CdBr$ was prepared from $(CH_3C_5H_4Mo(CO)_3)_2Cd$ and $CdBr_2$ by redistribution in methanol as described for the unsubstituted cyclopentadienyl analogue in [4]. It was then reacted with the anionic iron silyl complex $K[(CH_3O)_3SiFe(CO)_3P(C_6H_5)_2CH_2$-$P(C_6H_5)_2]$ in THF to form a Mo–Cd–Fe-bonded chain compound, $CH_3C_5H_4Mo(CO)_3$-$Cd(\mu$-$(C_6H_5)_2PCH_2P(C_6H_5)_2)Fe(CO)_3Si(OCH_3)_3$ [169].

$RC_5H_4Mo(CO)_3HgX$ (R = CH_3: X = Cl, Br, I, SCN; R = $CH_3C(O)$: X = Cl, Br, I). These Mo–Hg-bonded complexes were prepared in two ways: (I) by cleavage of the Mo–Sn bond in the reactions of $CH_3C_5H_4Mo(CO)_3Sn(C_6H_5)_3$ with HgX_2 and (II) by exchange between $(RC_5H_4Mo(CO)_3)_2Hg$ (R = CH_3, $CH_3C(O)$) and mercury(II) halides or pseudohalides. In (I) a methanolic solution of $CH_3C_5H_4Mo(CO)_3Sn(C_6H_5)_3$ and HgX_2 (X = Cl, Br; 2 equivalents) was refluxed and then kept at 0°C, which caused $CH_3C_5H_4Mo(CO)_3HgCl$ and $CH_3C_5H_4Mo$-$(CO)_3HgBr$ to separate as crystals [57]. In (II) solutions of $(CH_3C_5H_4Mo(CO)_3)_2Hg$ or $(CH_3C(O)C_5H_4Mo(CO)_3)_2Hg$ in acetone were mixed with an equimolar quantity of the required HgX_2 reagent in the same solvent. Volatile materials were removed and the crude products, $CH_3C_5H_4Mo(CO)_3HgX$ (X = Cl, Br, I, SCN) [57] and $CH_3C(O)C_5H_4Mo(CO)_3HgX$ (X = Cl, Br, I) [91], crystallized from methanol and ethanol, respectively. Yields and some properties of the compounds so prepared are listed in the following table:

R/X	method of preparation (yield), properties, and remarks
CH_3/Cl	I, II (60%) [57] yellow crystals, m.p. 132 to 134°C [57] conductivity (ca. 10^{-3} M in $C_6H_5NO_2$ at 22°C): $\Lambda = 0.60\ \Omega^{-1} \cdot cm^2 \cdot mol^{-1}$ [57] ^1H NMR ($(CD_3)_2SO$): 2.05 (s, CH_3), 5.65, 5.90 (both m, C_5H_4) [57, 85] ^{95}Mo NMR (CH_2Cl_2/C_6D_6): −1793 (vs. aqueous $Na_2[MoO_4]$ at pH 11); $\Delta\nu_{1/2}$ = 50 Hz [85] ^{199}Hg NMR (CH_2Cl_2/C_6D_6): −577 (vs. $Hg(CH_3)_2$ neat); $\Delta\nu_{1/2}$ = 6 Hz [85] IR (Nujol): 266 (ν(HgCl)); (KBr): 1916, 1930, 2000 (CO) [57, 85]

R/X	method of preparation (yield), properties, and remarks
	UV (CCl_4, log ε; assignment): 255 (4.24; CT Mo → π*(CO)), 280 (4.13; CT Mo → π*(CO)), 365 (3.83; Mo–Hg(σ → σ*) + Mo(d → d) + CT Mo → π*(CO)) [57]
CH_3/Br	I, II (49%) [57] yellow crystals, m.p. 142 °C [57] ^{1}H NMR ($(CD_3)_2SO$): 2.01 (s, CH_3), 5.50, 5.78 (both m, C_5H_4) [57, 85] ^{95}Mo NMR (CH_2Cl_2/C_6D_6): −1775 (vs. aqueous $Na_2[MoO_4]$ at pH 11); $\Delta\nu_{1/2}$ = 80 Hz [85] ^{199}Hg NMR (CH_2Cl_2/C_6D_6): −765 (vs. $Hg(CH_3)_2$ neat); $\Delta\nu_{1/2}$ = 10 Hz [85] IR (KBr): 1890, 1922, 1996 (CO) [57] UV (CCl_4, log ε; assignment): 265 (4.16; CT Mo → π*(CO)), 280 (4.12; CT Mo → π*(CO)), 365 (3.40; Mo–Hg(σ → σ*) + Mo(d → d) + CT Mo → π*(CO)) [57]
CH_3/I	II (50%) [57] yellow-orange crystals, m.p. 135 °C [57] conductivity (ca. 10^{-3} M in $C_6H_5NO_2$ at 22 °C): $\Lambda = 0.24\ \Omega^{-1} \cdot cm^2 \cdot mol^{-1}$ [57] ^{1}H NMR ($(CD_3)_2SO$): 2.00 (s, CH_3), 5.43, 5.66 (both m, C_5H_4) [57, 85] ^{95}Mo NMR (CH_2Cl_2/C_6D_6): −1756 (vs. aqueous $Na_2[MoO_4]$ at pH 11); $\Delta\nu_{1/2}$ = 100 Hz [85] ^{199}Hg NMR (CH_2Cl_2/C_6D_6): −1135 (vs. $Hg(CH_3)_2$ neat); $\Delta\nu_{1/2}$ = 40 Hz [85] IR (KBr): 1884, 1916, 1992 (CO) [57] UV (CCl_4, log ε; assignment): 265 (4.17; CT Mo → π*(CO)), 305 (4.23; CT Mo → π*(CO)), 380 (2.93; Mo–Hg(σ → σ*) + Mo(d → d) + CT Mo → π*(CO)) [57]
CH_3/SCN	II (52%) [57] yellow crystals, m.p. 122 °C [57] conductivity (ca. 10^{-3} M in $C_6H_5NO_2$ at 22 °C): $\Lambda = 0.43\ \Omega^{-1} \cdot cm^2 \cdot mol^{-1}$ [57] ^{1}H NMR ($(CD_3)_2SO$): 2.0 (s, CH_3), 5.43, 5.71 (both m, C_5H_4) [57] IR (KBr): 240 (ν(HgS)), 1892, 1926, 2000 (CO), 2110 (ν(SCN)) [57] UV (CCl_4, log ε; assignment): 275 (4.05; CT Mo → π*(CO)), 290 (4.01; CT Mo → π*(CO)), 365 (4.38; Mo–Hg (σ → σ*) + Mo(d → d) + CT Mo → π*(CO)) [57]
$CH_3C(O)$/Cl	II (69%) [91] yellow-brown solid [91] ^{1}H NMR ($CDCl_3$): 2.31 (s, CH_3), 5.62, 5.95 (both "t", C_5H_4) [91] IR ($CHCl_3$): 1692 (ν(C=O)), 1941, 1958, 2025 (CO) [91] $CH_3C(O)C_5H_4Mo(CO)_3SnCl_3$ resulted from insertion across the Mo–Hg bond of $SnCl_2$ in refluxing acetone [91]
$CH_3C(O)$/Br	II (66%) [91] yellow-brown solid [91] ^{1}H NMR ($CDCl_3$): 2.31 (s, CH_3), 5.61, 5.95 (both "t", C_5H_4) [91] IR ($CHCl_3$): 1692 (ν(C=O)), 1939, 1958, 2023 (CO) [91]

References on pp. 253/8

R/X	method of preparation (yield), properties, and remarks
$CH_3C(O)/I$	II (65%) [91] yellow-brown solid [91] ^{1}H NMR ($CDCl_3$): 2.30 (s, CH_3), 5.59, 5.94 (both "t", C_5H_4) [91] IR ($CHCl_3$): 1692 (ν(C=O)), 1934, 1952, 2019 (CO) [91]

$CH_3C_5H_4Mo(CO)_3HgCl$ crystallized in the monoclinic space group $P2_1/c-C^5_{2h}$ (No. 14) with a = 6.613(2), b = 13.647(4), c = 13.257(4) Å, β = 101.85(3)°; Z = 4 molecules per unit cell, D_{calc} = 2.81 g/cm³. **Fig. 23** shows the geometry of the molecule. In addition to the Hg–Cl bond, there is an interaction between the mercury atom and one chlorine atom of a neighboring molecule (d(Hg···Cl), 3.053 Å) [57].

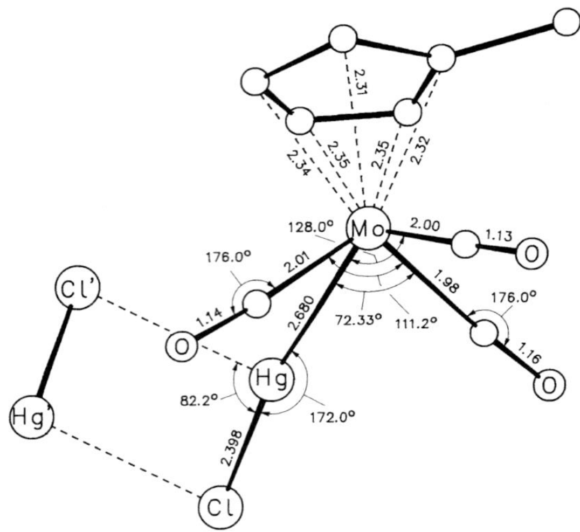

Fig. 23. The molecular structure of $CH_3C_5H_4Mo(CO)_3HgCl$ [57].

All four compounds $CH_3C_5H_4Mo(CO)_3HgX$ proved to be moderately soluble in polar solvents, in which gradual decomposition with formation of metallic Hg occurred; they are stable toward air [57].

Reactions of the acetyl-substituted title complexes $CH_3C(O)C_5H_4Mo(CO)_3HgX$ with an excess of $P(C_6H_5)_3$ in boiling ethanol or with $P(OCH_3)_3$ in methylcyclohexane at reflux temperature gave $CH_3C(O)C_5H_4Mo(CO)_2(P(C_6H_5)_3)HgX$ and $CH_3C(O)C_5H_4Mo(CO)_2(P(OCH_3)_3)$-HgX, respectively [121].

$C_2H_5OC(O)C_5H_4Mo(CO)_3HgC_6H_5$ was isolated from a short-time reaction (1 min) between equimolar quantities of $Na[C_2H_5OC(O)C_5H_4Mo(CO)_3]$ and C_6H_5HgCl in THF at ice/salt bath temperature. Chromatographic workup on a silica gel column eluting with CH_2Cl_2/petroleum ether (1:5) gave a 52% yield [147].

Yellow crystals, m.p. 78 to 80°C. ^{1}H NMR spectrum: δ = 1.24 (t, CH_3; J = 7.2 Hz), 4.20 (q, CH_2), 5.40 ("t", H-3,4 of C_5H_4), 5.92 ("t", H-2,5 of C_5H_4), 7.12 to 7.52 (m, C_6H_5) ppm. IR spectrum: 1713 (ν(C=O)), 1877, 1926, 1967 (CO) cm^{-1}. Mass spectrum: $[M-C_6H_5]^+$ (2%), $[M-C_6H_5-nCO]^+$ (n = 1 (4%), 2 (6%), 3 (3%)), $[M-C_6H_5-Hg]^+$ (100%), $[M-C_6H_5-$

References on pp. 253/8

Hg−nCO]$^+$ (n = 1 (58%), 2 (35%), 3 (28%)), [HgC_6H_5]$^+$ (4%), [C_6H_5]$^+$ (2%), [Mo]$^+$ (3%), [Hg]$^+$ (20%) [147].

Attempted crystallization from CH_2Cl_2/petroleum ether (1:1) resulted in facile symmetrization yielding ($C_2H_5OC(O)C_5H_4Mo(CO)_3)_2Hg$ [147].

1.5.1.4.2.1.2.3 Metal−Metal-Bonded Complexes with Bridging RC_5H_4 Ligands

$Mo(CO)_3(\mu\text{-}C_5H_4C_5H_4)Cr(CO)_3$ (Formula I; M = Cr, n = 3) was prepared in 28% yield by reacting the isomeric mixture of dicyclopentadienyl methyl complexes II with $Cr(CO)_3$-$(NCCH_3)_3$ in THF at 66 °C [78]; for the reaction pathway and intermediates, see below under $Mo(CO)_3(\mu\text{-}C_5H_4C_5H_4)W(CO)_3$.

Dark green crystals, m.p. 250 °C (dec.). ^{1}H NMR spectrum (acetone-d_6): δ = 4.34, 4.88, 5.28, 5.59 (all "t", 2H each; "J" = 2.2 Hz each) ppm. IR spectrum (THF): 1894, 1916, 1937, 1962, 2015 (CO) cm^{-1}. UV spectrum (THF, ε): 416 (10.900), 610 (230) nm. Mass spectrum: [M]$^+$ (7.2%), [$C_{10}H_8Mo$]$^+$ (100%) [78].

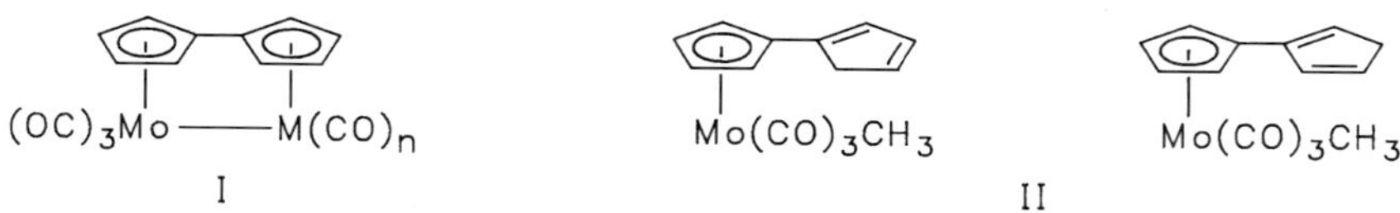

$Mo(CO)_3(\mu\text{-}C_5H_4C_5H_4)W(CO)_3$ (Formula I; M = W, n = 3) was generated by oxidative C−H addition and subsequent elimination of acetaldehyde when the mixture of isomers (Formula II) was treated with a slight excess of $W(CO)_3(NCCH_3)_3$ in THF. At 23 °C, **CH_3Mo-$(CO)_3(\mu\text{-}C_5H_4C_5H_4)W(CO)_3H$** was identified by ^{1}H NMR (THF-d_8: δ(WH) = −7.0 ppm) as the first-formed intermediate. Heating the THF solution at 66 °C caused the formation of the metal−metal bond along with CH_3CHO; isolated yield: 34% [78].

Purple crystals, m.p. 250 °C (dec.). ^{1}H NMR spectrum (acetone-d_6): δ = 4.83, 4.94, 5.57, 5.59 (all "t", 2H each; J = 2.3 Hz each) ppm. IR spectrum (THF): 1888, 1913, 1934, 1969, 2015 (CO) cm^{-1}. UV spectrum (THF): λ_{max} (ε) = 355 (16.000), 552 (540) nm. Mass spectrum: [M]$^+$ (16%) [78].

$Mo(CO)_3(\mu\text{-}C_5H_4C_5H_4)Fe(CO)_2$ (Formula I; M = Fe, n = 2) resulted in 8% yield from the reaction of $Fe_2(CO)_9$ with the fulvalene compound $C_5H_4{=}C_5H_4Mo(CO)_2(P(CH_3)_3)_2$ in THF, first at ambient conditions and then at reflux temperature. $Mo(CO)_2(P(CH_3)_3)(\mu\text{-}C_5H_4C_5H_4)$-$Fe(CO)_2$ was formed as the major product [78].

Brown powder. ^{1}H NMR (acetone-d_6): δ = 4.18, 4.96, 5.36, 5.54 (all "t", 2H each; J = 2.2 to 2.3 Hz) ppm. IR spectrum (THF): 1891, 1959, 2014 (CO) cm^{-1}. Mass spectrum: [M]$^+$ (5.3%), [$C_{10}H_8Mo$]$^+$ (100%) [78].

$Mo(CO)_3(\mu\text{-}C_5H_4C_5H_4)Ru(CO)_2$ (Formula I; M = Ru, n = 2). A 3:1 molar mixture of $Mo(CO)_6$ and $Ru_3(CO)_{12}$ in $CH_3OC_2H_4OCH_3$ was heated to reflux. A cold heptane solution of purified dihydrofulvalene (50% excess; freshly prepared from Na[C_5H_5] · $CH_3OC_2H_4OCH_3$ by oxidation with I_2 in THF and subsequent low-temperature THF/H_2O-hydrocarbon extraction) was slowly added over 2 h and the solution heated to reflux for an additional 20 h. Filtration of the cooled mixture through a CH_2Cl_2-packed alumina column followed by washing with CH_2Cl_2 and renewed column chromatography on Al_2O_3, eluting first with hexane and then with dichloromethane, gave a crude product which was analyzed by NMR to contain (i) $Mo(CO)_3(\mu\text{-}C_5H_4C_5H_4)Ru(CO)_2$ (58%), (ii) $Mo(CO)_3(\mu\text{-}C_5H_4C_5H_4)Mo(CO)_3$ (24%), and (iii)

References on pp. 253/8

$Ru(CO)_2(\mu\text{-}C_5H_4C_5H_4)Ru(CO)_2$ (18%). The mixture was dissolved in THF and separated by preparative high-pressure liquid chromatography using silica gel columns and 10% THF in hexane as the eluant. The compounds separated in the order (iii), (i), and (ii) and were purified further by crystallizing from CH_2Cl_2; yield of title complex: 42%, in addition to 7% of (ii) and 9% of (iii) [47]. Adding freshly prepared THF solutions of crude dihydrofulvalene (also containing NaI and other minor impurities) to boiling $CH_3OC_2H_4OCH_3$ solutions of the metal carbonyls led to much lower yields of the desired fulvalene-bridged products (title compound: 18%) and large amounts of insoluble precipitates [43]. Combination of the monometallic fulvalene complex $C_5H_4{=}C_5H_4Mo(CO)_2(P(CH_3)_3)_2$ with $Ru_3(CO)_{12}$ in refluxing 1,2-dimethoxyethane furnished $Ru_3(CO)_9(P(CH_3)_3)_3$ together with the Mo–Ru bonded systems $Mo(CO)_3(\mu\text{-}C_5H_4C_5H_4)Ru(CO)_2$ and $Mo(CO)_2(P(CH_3)_3)(\mu\text{-}C_5H_4C_5H_4)Ru(CO)_2$ in an approximate 1:1 ratio (70% overall yield) [78]. When $MoO(\mu\text{-}C_6H_5C{\equiv}CC_6H_5)(\mu\text{-}C_5H_4C_5H_4)$-$Ru(CO)_2$ (the product of photolysis of the title complex in THF in the presence of diphenylacetylene and dioxygen; see below) was hydrogenated in the presence of CO, $Mo(CO)_3$-$(\mu\text{-}C_5H_4C_5H_4)Ru(CO)_2$ was regenerated quantitatively [49].

Metallic orange flakes, m.p. 256 to 258°C [43, 47]. ^1H NMR spectrum (acetone-d_6): δ = 4.42, 4.79, 5.53, 5.88 (all dd, 2H each; J = 2.1 to 2.4 Hz) ppm [47]. IR spectrum (KBr): 1868, 1935, 1942, 2020 (CO) cm^{-1} [47]. UV spectrum: λ_{max} (ε) = 309 (15000), 378 (9100), 470 (880) nm (intense absorption at 309 nm assigned to metal–metal $\sigma \rightarrow \sigma^*$ transition) [47, 75]. Mass spectrum (70 ev): $[M]^+$ (19.6%), $[M-nCO]^+$ (n = 1 (15.9%), 2 (7.0%), 3 (71.0%), 4 (16.2%), 5 (100%)) [47].

$Mo(CO)_3(\mu\text{-}C_5H_4C_5H_4)Ru(CO)_2$ proved to be soluble and stable in a wide variety of solvents, including THF, acetone, propylene carbonate, dimethyl sulfoxide, 1,2-dimethoxyethane, butyronitrile, and SO_2, but was insoluble in NH_3 and water [75]. Prolonged heating to 120°C or irradiating in solution did not lead to any detectable formation of $Mo(CO)_3(\mu\text{-}C_5H_4C_5H_4)Mo(CO)_3$ and $Ru(CO)_2(\mu\text{-}C_5H_4C_5H_4)Ru(CO)_2$, respectively [47]. The complex was completely air-stable as a solid and oxidized only slowly in solutions protected from light [47].

1.15×10^{-3} M solutions in 1 M $[N(C_4H_9\text{-}n)_4]BF_4$/acetone were irreversibly oxidized in a two-electron process at +1.03 V vs. NHE. Two-electron reduction with cleavage of the metal–metal bond occurred at −1.25 V vs. NHE to give the $[Mo(CO)_3(\mu\text{-}C_5H_4C_5H_4)Ru(CO)_2]^{2-}$ dianion, which was re-oxidized to the starting material in another two-electron step at −0.54 V vs. NHE [75]. $C_5H_5FeC_6(CH_3)_6$-initiated intramolecular electron transfer in THF/1,2-dimethoxyethane mixtures in the presence of excess $P(CH_3)_3$ resulted in cleavage of the metal–metal bond, smoothly producing the heterobimetallic zwitterion $Mo^{(-)}(CO)_3(\mu\text{-}C_5H_4C_5H_4)Ru^{(+)}(CO)(P(CH_3)_3)_2$ [163].

UV photolysis in THF in the presence of alk-1-ynes, $RC{\equiv}CH$, yielded metal–metal-bonded products containing side-on bridging four-electron alkenylidene ligands, $Mo(CO)_2(\mu\text{-}C_5H_4C_5H_4)({=}C{=}CHR\text{-}\mu,\eta^2)Ru(CO)_2$ (R = C_6H_5, $Si(CH_3)_3$). Similar phototransformations with disubstituted alkynes, $RC{\equiv}CR'$, gave tri- and tetracarbonyls exchanged at the molybdenum center, $Mo(CO)(RC{\equiv}CR')(\mu\text{-}C_5H_4C_5H_4)Ru(CO)_2$ (R/R' = C_6H_5/C_6H_5, 4-$CH_3C_6H_4/C_6H_4CH_3$-4, $C_6H_5/Si(CH_3)_3$; major products) and $Mo(CO)_2(RC{\equiv}CR')(\mu\text{-}C_5H_4C_5H_4)Ru(CO)_2$ (R/R' = C_6H_5/C_6H_5, $C_6H_5/Si(CH_3)_3$; minor products), respectively [123]. Irradiation at 300 nm in THF in the presence of diphenylacetylene and traces of adventitious dioxygen produced a reactive dinuclear alkyne-bridged oxo complex, $MoO(\mu\text{-}C_6H_5C{\equiv}CC_6H_5)(\mu\text{-}C_5H_4C_5H_4)Ru(CO)_2$ [49].

$Mo(CO)_3(\mu\text{-}C_5H_4C_5H_4)Rh(CO)CH_3$ (Formula III; R = CH_3) was synthesized by thermolyzing the acyl $Mo(CO)_3(\mu\text{-}C_5H_4C_5H_4)Rh(CO)C(O)CH_3$ (Formula III; R = $CH_3C(O)$) in toluene or $CH_3OC_2H_4OC_2H_4OCH_3$ at reflux temperature. The product was isolated from the reaction

$$(OC)_3Mo—Rh(CO)R \quad \text{(fulvalene-bridged)}$$

III

$$(OC)_3Mo—W(CO)_2C_5H_5 \quad (C_5H_4\text{-}P(C_6H_5)_2 \text{ bridged})$$

IV

mixture by column chromatography on silica gel eluting with ether/pentane and was further purified by recrystallization of the evaporated eluates from THF/hexane (95%) [166].

Brown crystals, m.p. 170°C (dec.). ^{1}H NMR spectrum (C_6D_6): δ = 0.94 (d, CH_3; J(Rh, H) = 2.6 Hz), 3.21, 3.28, 3.43, 3.83, 4.38, 4.64, 4.87, 5.02 (all m, 1 fulvalene-H each) ppm; (THF-d_8): δ = 0.69 (d, CH_3; J(Rh, H) = 2.6 Hz), 4.30, 4.47, 4.58, 4.91, 5.38, 5.62, 5.80, 6.04 (all m, 1 fulvalene-H each) ppm. ^{13}C NMR spectrum (THF-d_8): δ = −20.31 (d, CH_3; J(Rh, C) = 24.3 Hz), 84.96 (d, RhC_5H_4; J(Rh, C) = 4.1 Hz), 85.47 (d, RhC_5H_4; J(Rh, C) = 2.5 Hz), 86.42 (s, MoC_5H_4), 87.75 (d, RhC_5H_4; J(Rh, C) = 3.0 Hz), 93.96 (s, MoC_5H_4), 95.09 (d, RhC_5H_4; J(Rh, C) = 3.8 Hz), 97.32 (s, MoC_5H_4), 98.61 (d, RhC_5H_4; J(Rh, C) = 2.3 Hz), 100.53 (s, MoC_5H_4) ppm. IR spectrum (KBr): 1878, 1896, 1955, 2005 (CO) cm^{-1}. UV spectrum (THF): λ_{max} (log ε) = 237 (4.57), 307 (4.41), 426 (4.08), 532 (3.48) nm. Mass spectrum: $[M]^+$ (18.4%), $[M-nCO]^+$ (n = 1 (40.2%), 3 (100%)), $[M-nCO-CH_3]^+$ (n = 1 (5.3%), 2 (7.7%)) [166].

$Mo(CO)_3(\mu\text{-}C_5H_4C_5H_4)Rh(CO)C(O)CH_3$ (Formula III; R = $C(O)CH_3$). The isomeric mixture of dicyclopentadienyl methyl complexes $C_5H_5C_5H_4Mo(CO)_3CH_3$ (Formula II), was deprotonated by treating a THF solution with a slight excess of sodium hydride. Excess NaH was filtered off prior to the addition of the stoichiometrically required amount of $(Rh(CO)_2\text{-}(\mu\text{-}Cl))_2$. After stirring at room temperature and removal of volatile material, the residue was chromatographed on a column packed with silica gel using a 1:1 ether/pentane mixture as the eluant. Crystallization by diffusion of hexane into a THF solution of the product gave an 18% yield [166].

Red crystals, m.p. 162°C (dec.). ^{1}H NMR (C_6D_6): δ = 2.66 (s, CH_3), 2.87, 3.74, 3.76, 3.81, 4.24, 4.63, 5.05, 5.37 (all m, 1 fulvalene-H each) ppm; (THF-d_8): = 2.53 (s, CH_3), 3.98, 4.25, 4.62, 4.97, 5.44, 5.72, 6.08, 6.12 (all m, 1 fulvalene-H each) ppm. ^{13}C NMR spectrum (THF-d_8): δ = 50.48 (d, CH_3; J(Rh, C) = 3.8 Hz), 85.33 (d, RhC_5H_4; J(Rh, C) = 1.94 Hz), 87.11, 87.50, 88.08, 88.26 (all s, all $C_5H_4C_5H_4$), 89.38 (d, RhC_5H_4; J(Rh, C) = 4.8 Hz), 92.90 (d, RhC_5H_4; J(Rh, C) = 4.8 Hz), 95.81, 100.25, 101.46 (all s, all $C_5H_4C_5H_4$), 197.43 (d, RhCO; J(Rh, C) = 93.4 Hz), 226.83 (d, C=O; J(Rh, C) = 30.4 Hz) ppm. IR spectrum (KBr): 1631 (ν(C=O)), 1887, 1891, 1967, 2015 (CO) cm^{-1}. UV spectrum (THF): λ_{max} (log ε) = 275 (4.23), 288 (4.21), 402 (3.99), 530 (3.04) nm. Mass spectrum: $[M]^+$ (30.0%), $[M-nCO]^+$ (n = 1 (23.4%), 2 (23.6%), 3 (5.0%)), $[M-2CO-CH_3]^+$ (4.1%), $[M-Mo(CO)_3]^+$ (12.0%), $[C(O)CH_3]^+$ (100%) [166].

Thermolysis in refluxing toluene or $CH_3OC_2H_4OC_2H_4OCH_3$ resulted in decarbonylation of the acetyl group to furnish the methyl complex $Mo(CO)_3(\mu\text{-}C_5H_4C_5H_4)Rh(CO)CH_3$ (Formula III; R = CH_3) [166].

$Mo(CO)_3(\mu\text{-}C_5H_4P(C_6H_5)_2)W(CO)_2C_5H_5$ (Formula IV) resulted from combination of Li-$[(C_6H_5)_2PC_5H_4Mo(CO)_3]$ with an equimolar quantity of $C_5H_5W(CO)_3I$ in refluxing THF. The evaporated mixture was purified by thin-layer chromatography on silica with CH_2Cl_2/hexane (1:3) as the eluant. A red and a brown band were eluted and yielded, respectively, unreacted $C_5H_5W(CO)_3I$ and $Mo(CO)_3(\mu\text{-}C_5H_4P(C_6H_5)_2)W(CO)_2C_5H_5$ (31%) [139].

^{1}H NMR spectrum ($CDCl_3$): δ = 3.30, 4.52 (both m, 1C_5H_4-H each), 5.10 (s, C_5H_5), 5.34, 5.40 (both m, 1C_5H_4-H each), 7.0 to 7.7 (m, C_6H_5) ppm. ^{31}P NMR spectrum ($CDCl_3$): δ =

References on pp. 253/8

−109.6 (vs. $P(OCH_3)_3$) ppm; J(W, P) = 295 Hz. IR spectrum (hexane): 1867, 1887, 1939, 1993 (CO) cm^{-1}. Mass spectrum: $[M]^+$, $[M-nCO]^+$ (n = 1 to 5) [139].

Reaction with excess HBF_4 in CH_2Cl_2 resulted in protonation of the Mo−W bond, giving $[Mo(CO)_3(\mu\text{-}C_5H_4P(C_6H_5)_2)(\mu\text{-}H)W(CO)_2C_5H_5]BF_4$. The phosphane-substituted derivative $Mo(CO)_2(P(C_6H_5)_2CH_3)(\mu\text{-}C_5H_4P(C_6H_5)_2)W(CO)_2C_5H_5$ was generated when the title complex was reacted with $P(C_6H_5)_2CH_3$ in refluxing toluene [139].

$[(OC)_3Mo(\mu\text{-}C_5H_4P(C_6H_5)_2)(\mu\text{-}H)W(CO)_2C_5H_5]BF_4$

V

$(OC)_3Mo(\mu\text{-}C_5H_4PR_2)M(CO)_4$

VI

$[Mo(CO)_3(\mu\text{-}C_5H_4P(C_6H_5)_2)(\mu\text{-}H)W(CO)_2C_5H_5]BF_4$ (Formula V) was obtained by adding an excess of HBF_4 to a CH_2Cl_2 solution of $Mo(CO)_3(\mu\text{-}C_5H_4P(C_6H_5)_2)W(CO)_2C_5H_5$ cooled to 195 K. After warming to room temperature, solvent was removed to leave the protonated derivative as a brown air-sensitive powder [139].

^1H NMR spectrum ($CDCl_3$): δ = −18.62 (d with W satellites, μ-H; J(P, H) = 11.3, J(W, H) = 36.1 Hz), 3.82 (br, $1C_5H_4$-H), 5.58 (s, C_5H_5), 5.70, 6.22, 6.25 (all br, $1C_5H_4$-H each), 7.3 to 7.7 (m, C_6H_5) ppm. IR (CH_2Cl_2): 1908, 1984, 2061 (CO) cm^{-1} [139].

$Mo(CO)_3(\mu\text{-}C_5H_4PR_2)M(CO)_4$ (Formula VI; M = Mn: R = C_6H_5, $C_6H_4CH_3$-4; M = Re: R = C_6H_{11}-cyclo, $C_6H_4CH_3$-4). The manganese and rhenium complexes with R = C_6H_5 and $C_6H_4CH_3$-4 were obtained by reacting the appropriate lithium molybdate $Li[R_2PC_5H_4\text{-}Mo(CO)_3]$ with the bromo-bridged dimanganese and dirhenium compounds $(M(CO)_4(\mu\text{-}Br)_2)_2$ in approximate 2:1 stoichiometry in THF at 0°C (M = Mn) or between −78°C and room temperature (M = Re). The crudes remaining after removal of solvent were purified by col-

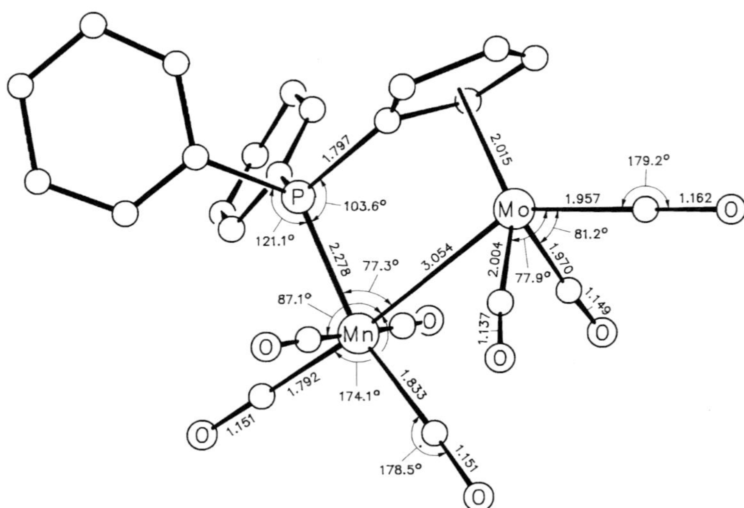

Fig. 24. The molecular structure of $Mo(CO)_3(\mu\text{-}C_5H_4P(C_6H_5)_2)Mn(CO)_4$ [33].

umn chromatography (silica gel, hexane/ether (9:1); M = Mn) or by preparative thin-layer chromatography (silica gel, hexane/ether (1:1); M = Re) [33]. Yields and physical properties are given in the table below. In addition, $Mo(CO)_3(\mu\text{-}C_5H_4P(C_6H_{11}\text{-cyclo})_2)Re(CO)_4$ was briefly mentioned as one of the products resulting from treatment of an equimolar mixture of $Re_2(CO)_{10}$ and $(C_5H_5Mo(CO)_3)_2$ with dicyclohexylphosphane in toluene at 180 °C in a sealed vessel, but no characterizing data were reported [164].

Single crystals of $Mo(CO)_3(\mu\text{-}C_5H_4P(C_6H_5)_2)Mn(CO)_4$ grown by slow cooling of hot toluene solutions belonged to the orthorhombic system, space group Pbca$-D^{15}_{2h}$ (No. 61); crystal data (24 °C): a = 11.782(2), b = 17.215(3), c = 23.129(6) Å; Z = 8 molecules per unit cell, D_{calc} = 1.6921(2) g/cm^3. The molecular model derived from the structure analysis revealed a nearly planar, slightly strained ring system made up of the two metal atoms, the phosphorus atom, and the cyclopentadienyl centroid [33]; see **Fig. 24**.

M/R	yield, properties, and remarks
Mn/C_6H_5	(45%) [33] orange solid, m.p. 184 to 187 °C (dec.) [33] ^{1}H NMR (C_6D_6): 3.80, 4.53 (both "q", C_5H_4; "J" = 2.1), 6.84 (m, meta- and para-H), 7.63 to 7.72 (m, ortho-H) [33]; (acetone-d_6): 4.70, 5.77 (both "q", C_5H_4; "J" = 2.1), 7.57 (m, meta- and para-H), 8.1 (m, ortho-H) [33] ^{13}C NMR (acetone-d_6, 0.07 M $Cr(CH_3C(O)CH_2C(O)CH_3)_3$; −40 °C): 57.8 (d, C-1 of C_5H_4; J(P, C) = 51.3), 89.2 (d, C-2,5 (?) of C_5H_4; J(P, C) = 9.2), 91.0 (br, C-3,4 (?) of C_5H_4), 130.3 (d, meta-C; J(P, C) = 10.9), 132.3 (d, ortho-C; J(P, C) = 12.8), 132.7 (s, para-C), 134.2 (d, ipso-C; J(P, C) = 42.2), 216.2 (2MnCO), 222.9, 224.8 (1MnCO each), 227.0 (s, MoCO cis to Mn), 236.8 (s, MoCO trans to Mn); at ambient temperature, the $Mn(CO)_4$ group appeared as a quadrupole-broadened signal centered at δ = 215.9 [33] IR (cyclohexane): 1901, 1921, 1941, 1976, 1980, 1994, 2056 (CO) [33] for EHMO calculations, see [84]
Mn/$C_6H_4CH_3$-4	(32%) [33] orange solid, m.p. 169 to 170 °C (dec.) [33] ^{1}H NMR (C_6D_6): 1.84 (s, CH_3), 3.93, 4.55 (both m, C_5H_4), 6.72 (br d, meta-H), 7.65 (dd, ortho-H; J(H, H) = 7.9, J(P, H) = 11.6) [33] ^{13}C NMR (CD_2Cl_2, 0.07 M $Cr(CH_3C(O)CH_2C(O)CH_3)_3$): 21.0 (s, CH_3), 57.3 (d, C-1 of C_5H_4; J(P, C) = 50.5), 88.0 (d, C-2,5 of C_5H_4; J(P, C) = 18.1), 89.4 (d, C-3,4 of C_5H_4; J(P, C) = 9.1), 129.9 (d, meta-C; J(P, C) = 10.7), 131.2 (d, ortho-C; J(P, C) = 12.2), 142.3 (s, para-C), 215.8 (quadrupole-broadened $Mn(CO)_4$), 225.0 (s, MoCO cis to Mn), 231.4 (s, MoCO trans to Mn); ipso-C partially obscured by either the δ = 129.9 or the δ = 131.2 peak [33] IR (cyclohexane): 1900, 1919, 1939, 1973, 1978, 1996, 2059 (CO) [33]
Re/$C_6H_4CH_3$-4	(31%) [33] yellow solid, m.p. 167 to 169 °C (dec.) [33] ^{1}H NMR (C_6D_6): 1.84 (s, CH_3), 4.02, 4.66 (both "q", C_5H_4; "J" = 2.1), 6.70 (dd, meta-H; J(H, H) = 8.0, J(P, H) = 2.0), 7.58 (dd, ortho-H; J(H, H) = 8.4, J(P, H) = 12.0) [33]

References on pp. 253/8

M/R	yield, properties, and remarks
	^{13}C NMR (CD_2Cl_2, 0.07 M $Cr(CH_3C(O)CH_2C(O)CH_3)_3$; −50 °C): 21.1 (s, CH_3), 61.6 (d, C-1 of C_5H_4; J(P, C) = 59.7), 89.2 (d, C-2,5 (?) of C_5H_4; J(P, C) = 9.2), 89.7 (d, C-3,4 (?) of C_5H_4; J(P, C) = 7.6), 128.0 (d, ipso-C; J(P, C) = 49.0), 130.0 (d, meta-C; J(P, C) = 12.2), 131.8 (d, ortho-C; J(P, C) = 13.8), 142.7 (s, para-C), 187.4 (d, ReCO trans to Mo; J(P, C) = 4.6), 189.7 (d, 2ReCO cis to P; J(P, C) = 9.2), 192.4 (d, ReCO trans to P; J(P, C) = 47.7), 223.9 (s, MoCO cis to Re), 230.7 (s, MoCO trans to Re); at +25 °C, the $Mo(CO)_3$ group appeared as a broad resonance centered at δ = 224.2 [33] IR (cyclohexane): 1886, 1910, 1939, 1974, 1992, 2006, 2081 (CO) [33]

Both $Mo(CO)_3(\mu\text{-}C_5H_4P(C_6H_5)_2)Mn(CO)_4$ and $Mo(CO)_3(\mu\text{-}C_5H_4P(C_6H_4CH_3\text{-}4)_2)Re(CO)_4$ thermally decomposed when heated in cyclohexane or THF solution at 120 to 170 °C under 1200 to 1400 psi of H_2, but no evidence was obtained for the formation of any dihydrido derivative resulting from the addition of H_2 across the metal–metal bond. Small amounts of thermal decomposition were also noted when the two complexes were heated in toluene at 180 to 205 °C under a high pressure of synthesis gas (1200 to 1400 psi of 1:1 H_2/CO); no CH_4 or CH_3OH was formed under these conditions. Similarly, when toluene solutions containing added cyclohexene were heated at 165 to 170 °C under 1200 to 1400 psi of H_2, no cyclohexane could be detected [33].

UV irradiation with a medium-pressure mercury lamp of a toluene solution of $Mo(CO)_3(\mu\text{-}C_5H_4P(C_6H_4CH_3\text{-}4)_2)Mn(CO)_4$ under 1 atm of H_2 resulted in photochemical hydrogenolysis of the C_5H_4–P bond and formation of the phosphido/hydrido-bridged cyclopentadienyl complex $C_5H_5Mo(CO)_2(\mu\text{-}H)(\mu\text{-}P(C_6H_4CH_3\text{-}4)_2)Mn(CO)_4$ [50]. Photolysis in acetone under 1 atm of 90% ^{13}CO led to random incorporation of about 0.56 ^{13}CO per molecule of $Mo(CO)_3(\mu\text{-}C_5H_4P(C_6H_4CH_3\text{-}4)_2)Mn(CO)_4$ [50].

Thermal reactions of $Mo(CO)_3(\mu\text{-}C_5H_4P(C_6H_5)_2)Mn(CO)_4$ with one equivalent of the less bulky phosphanes $P(C_6H_5)_2H$ or $P(CH_3)_2C_6H_5$ in refluxing toluene (T = 383 K), or reactions with one equivalent of phosphanes of intermediate bulk, $P(C_6H_5)_2R$ (R = CH_3, C_2H_5, $CH{=}CH_2$), in refluxing hexane/toluene (5:1) at 353 K, resulted in monosubstitution by the entering nucleophile, ^2D, at the manganese atom to give derivatives $Mo(CO)_3(\mu\text{-}C_5H_4\text{-}P(C_6H_5)_2)Mn(CO)_3{}^2D$-mer bearing the substituent trans to the Mn–P bond (Formula VII). In the case of $P(C_6H_5)_2H$, the hydrido/phosphido-bridged complex $Mo(CO)_2(\mu\text{-}C_5H_4\text{-}P(C_6H_5)_2)(\mu\text{-}H)(\mu\text{-}P(C_6H_5)_2)Mn(CO)_3$ was isolated as a second product, and was shown to originate from thermal decarbonylation of the initially formed phosphane derivative (Formula VII; ^2D = $P(C_6H_5)_2H$). An analogous compound, $Mo(CO)_2(\mu\text{-}C_5H_4P(C_6H_5)_2)(\mu\text{-}H)(\mu\text{-}P(C_6H_5)H)Mn(CO)_3$, was obtained as the only product from thermal treatment of $Mo(CO)_3(\mu\text{-}C_5H_4P(C_6H_5)_2)Mn(CO)_4$ with $PH_2C_6H_5$ at 383 K in toluene. Combination of $Mo(CO)_3(\mu\text{-}C_5H_4P(C_6H_5)_2)Mn(CO)_4$ with $P(C_6H_5)_2C_2H_5$ in 1:1 stoichiometry in refluxing toluene rather than in hexane/toluene (5:1) at 353 K afforded a product that was monosubstituted at the molybdenum atom, trans-$Mo(CO)_2(P(C_6H_5)_2C_2H_5)(\mu\text{-}C_5H_4P(C_6H_5)_2)Mn(CO)_4$, in a 1:2 ratio with the Mn-substituted complex (Formula VII; ^2D = $P(C_6H_5)_2C_2H_5$) as an inseparable mixture. When equimolar quantities of $Mo(CO)_3(\mu\text{-}C_5H_4P(C_6H_5)_2)Mn(CO)_4$ and $P(C_6H_5)_2CH{=}CH_2$ were heated in toluene at 383 K rather than in 5:1 hexane/toluene at 353 K, the derivatives isolated included the Mo-monosubstituted cis and trans isomers of $Mo(CO)_2(P(C_6H_5)_2CH{=}CH_2)(\mu\text{-}C_5H_4P(C_6H_5)_2)Mn(CO)_4$, together with the Mo/Mn-disubstituted product trans-$Mo(CO)_2(P(C_6H_5)_2CH=CH_2)(\mu-C_5H_4P(C_6H_5)_2)Mn(CO)_3P(C_6H_5)_2CH=CH_2$-mer,trans and a complex of composition $Mo(CO)_2(\mu-C_5H_4P(C_6H_5)_2)(\mu-CH_2=CHP(C_6H_5)_2)$-$Mn(CO)_3$, in which the bridging vinylphosphane ligand is bonded through phosphorus to manganese and through the vinyl group to the molybdenum atom. The stereochemically demanding triphenylphosphane did not interact with $Mo(CO)_3(\mu-C_5H_4P(C_6H_5)_2)Mn(CO)_4$ in refluxing hexane/toluene (5:1) to give a manganese-substituted product of type VII, but afforded the Mo-monosubstituted derivative trans-$Mo(CO)_2(P(C_6H_5)_3)(\mu-C_5H_4P(C_6H_5)_2)$-$Mn(CO)_4$ when combined with the parent tricarbonyl in equimolar ratio in boiling toluene. Trimethylphosphite, which has a cone angle even smaller than $P(C_6H_5)_2H$ or $P(CH_3)_2C_6H_5$, gave a mixture of two Mn-monosubstituted meridional isomers (Formulas VII and VIII; ^2D = $P(OCH_3)_3$), on reacting with equimolar $Mo(CO)_3(\mu-C_5H_4P(C_6H_5)_2)Mn(CO)_4$ in hexane at 343 K. Disubstitution with formation of trans-$Mo(CO)_2(P(C_6H_5)_2CH_3)(\mu-C_5H_4P(C_6H_5)_2)$-$Mn(CO)_3P(C_6H_5)_2CH_3$-mer,trans and trans-$Mo(CO)_2(P(OCH_3)_3)(\mu-C_5H_4P(C_6H_5)_2)Mn(CO)_3$-$P(OCH_3)_3$ (mixture of mer,trans and mer,cis isomers) occurred in reactions of $Mo(CO)_3(\mu$-$C_5H_4P(C_6H_5)_2)Mn(CO)_4$ with excess $P(C_6H_5)_2CH_3$ or $P(OCH_3)_3$ in boiling toluene [140].

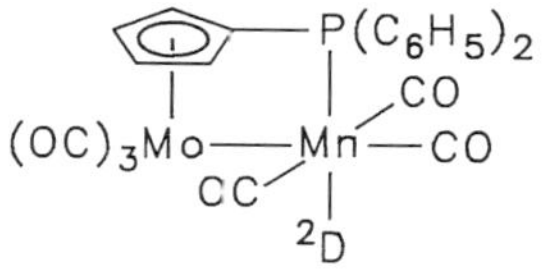

VII

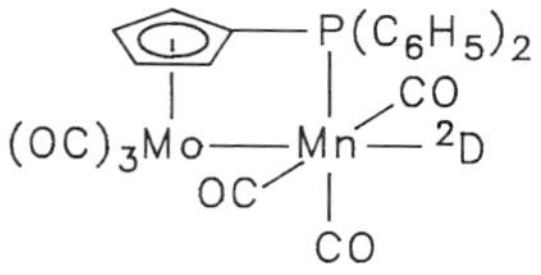

VIII

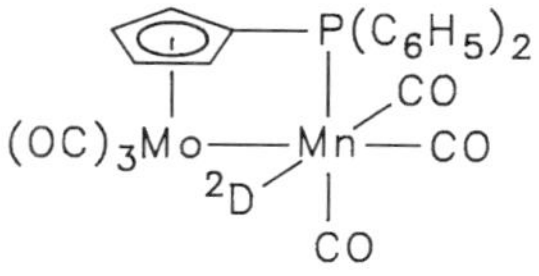

IX

Reactions of $Mo(CO)_3(\mu-C_5H_4P(C_6H_5)_2)Mn(CO)_4$ with equimolar amounts of isonitriles (which are less hindered at their donor atom than either phosphanes or phosphites) in boiling hexane/toluene (5:1) or in toluene at reflux temperature led to monosubstitution at the manganese atom trans to a carbonyl group yielding $Mo(CO)_3(\mu-C_5H_4P(C_6H_5)_2)Mn(CO)_3$-$^2D$-fac derivatives of type IX (^2D = t-C_4H_9NC, $C_6H_5CH_2NC$) [140].

Addition of trifluoromethane sulfonic or trifluoroacetic acids to each of the Mo–Mn and Mo–Re heterobimetallics $Mo(CO)_3(\mu-C_5H_4P(C_6H_4CH_3\text{-}4)_2)M(CO)_4$ in CD_2Cl_2 caused the protonation of the metal–metal bond giving cationic compounds with bridging hydrides, [Mo-$(CO)_3(\mu-C_5H_4P(C_6H_4CH_3\text{-}4)_2)(\mu\text{-}H)Mn(CO)_4]^+$ and $[Mo(CO)_3(\mu-C_5H_4P(C_6H_4CH_3\text{-}4)_2)(\mu\text{-}H)$-$Re(CO)_4]^+$ (see below), from which the starting materials were regenerated in high yield on treatment with base ($N(C_2H_5)_3$, $Li[(C_2H_5)_3BH]$, or even acetone). The basicity of the Mo–Re bond was shown to be greater than the basicity of the Mo–Mn bond, as only the $[Mo(CO)_3$-$(\mu-C_5H_4P(C_6H_4CH_3\text{-}4)_2)(\mu\text{-}H)Re(CO)_4]^+$ cation was observed to result from a competition experiment using equimolar $Mo(CO)_3(\mu-C_5H_4P(C_6H_4CH_3\text{-}4)_2)Mn(CO)_4$ and $Mo(CO)_3(\mu$-$C_5H_4P(C_6H_4CH_3\text{-}4)_2)Re(CO)_4$ together with 1.5 equivalents of CF_3SO_3H in CD_2Cl_2 [37].

$Mo(CO)_3(\mu-C_5H_4P(C_6H_5)_2)Mn(CO)_3{}^2D$ (Formula VII; ^2D = $P(CH_3)_2C_6H_5$, $P(C_6H_5)_2H$, $P(C_6H_5)_2CH_3$, $P(C_6H_5)_2C_2H_5$, $P(C_6H_5)_2CH=CH_2$, $P(OCH_3)_3$. Formula VIII; ^2D = $P(OCH_3)_3$. Formula IX; ^2D = t-C_4H_9NC, $C_6H_5CH_2NC$) were obtained by thermal treatment of Mo-$(CO)_3(\mu-C_5H_4P(C_6H_5)_2)Mn(CO)_4$ with one equivalent of the required nucleophile in refluxing hexane (T = 343 K; ^2D = $P(OCH_3)_3$), hexane/toluene (5:1) (T = 353 K; ^2D = $P(C_6H_5)_2CH_3$, $P(C_6H_5)_2C_2H_5$, $P(C_6H_5)_2CH=CH_2$, $C_6H_5CH_2NC$), or toluene (T = 383 K; ^2D = $P(CH_3)_2C_6H_5$, $P(C_6H_5)_2H$, t-C_4H_9NC). Isolation from the evaporated mixtures was accomplished either by column chromatography on SiO_2 eluting with CH_2Cl_2/hexane (1:4) (^2D = $P(C_6H_5)_2H$) or by preparative thin-layer chromatography on silica gel using CH_2Cl_2/hexane (3:7) (^2D = $P(CH_3)_2C_6H_5$, $P(C_6H_5)_2CH_3$, $P(C_6H_5)_2C_2H_5$, $P(C_6H_5)_2CH=CH_2$) or CH_2Cl_2/hexane (1:4) (^2D = $P(OCH_3)_3$, t-C_4H_9NC, $C_6H_5CH_2NC$) as the eluants. An impurity of $Mo(CO)_2(\mu-C_5H_4$-

References on pp. 253/8

$P(C_6H_5)_2)(\mu\text{-}H)(\mu\text{-}P(C_6H_5)_2)Mn(CO)_3$ formed during the synthesis of $Mo(CO)_3(\mu\text{-}C_5H_4\text{-}P(C_6H_5)_2)Mn(CO)_3P(C_6H_5)_2H$ was removed by recrystallization of the latter from CH_2Cl_2/hexane (1:1). Yields and properties are listed in the following table (the ^{31}P NMR shifts are given versus $P(OCH_3)_3$) [140]:

^2D/isomer	isomer yield, properties, and remarks
$P(CH_3)_2C_6H_5$/VII	20%, orange powder ^1H NMR ($CDCl_3$): 1.99 (d, CH_3; J(P, H) = 8.6), 4.64, 5.36 (both "q", C_5H_4; "J" = 2), 7.3 to 8.1 (m, C_6H_5) IR (CH_2Cl_2): 1847, 1879, 1939, 2014 (CO)
$P(C_6H_5)_2H$/VII	35%, orange crystals ^1H NMR ($CDCl_3$): 4.69, 5.36 (both "q", C_5H_4; "J" = 2), 7.14 (dd, PH; ^1J(P, H) = 360, ^3J(P, H) = 3.8), 7.4 to 7.6 (m, meta- and para-H), 7.9 to 8.2 (ortho-H) ^{13}C NMR (CH_2Cl_2, 234 K): 56.4 (C-1 of C_5H_4; J(P, C) = 50.1), 88.5 (C-2,5 (?) of C_5H_4; J(P, C) = 10.4), 89.1 (br, C-3,4 (?) of C_5H_4), 127 to 134 (m, C_6H_5), 217.2 (br, MnCO), 228.5 (s, MoCO trans to Mn), 232.2 (MoCO cis to Mn) ^{31}P NMR ($CDCl_3$): −80.1 (s, $P(C_6H_5)_2H$), −72.3 (s, $C_5H_4P(C_6H_5)_2$) IR (hexane): 1861, 1881, 1942, 1959, 2109 (CO) thermolysis in toluene at 383 K gave $Mo(CO)_2(\mu\text{-}C_5H_4P(C_6H_5)_2)(\mu\text{-}H)(\mu\text{-}P(C_6H_5)_2)Mn(CO)_3$
$P(C_6H_5)_2CH_3$/VII	34%, orange powder ^1H NMR ($CDCl_3$): 2.24 (d, CH_3; J(P, H) = 8.4), 4.69, 5.35 (both "q", C_5H_4; "J" = 2), 7.3 to 7.7 (m, meta- and para-H), 7.7 to 8.0 (m, ortho-H) ^{31}P NMR (CD_2Cl_2, 225 K): −95.1 (d, $P(C_6H_5)_2CH_3$; J(P, P) = 10), −71.6 (d, $C_5H_4P(C_6H_5)_2$); ($CDCl_3$, 298 K): −95.4 (s, $P(C_6H_5)_2CH_3$), −71.4 (s, $C_5H_4P(C_6H_5)_2$) IR (CH_2Cl_2): 1852, 1880, 1938, 2012 (CO) treatment with excess $P(C_6H_5)_2CH_3$ in refluxing toluene resulted in further CO substitution at the Mo atom with formation of trans-$Mo(CO)_2(P(C_6H_5)_2CH_3)(\mu\text{-}C_5H_4P(C_6H_5)_2)Mn(CO)_3P(C_6H_5)_2CH_3$-mer,trans; combination with equimolar $P(OCH_3)_3$ under the same conditions afforded $Mo(CO)_3(\mu\text{-}C_5H_4P(C_6H_5)_2)Mn(CO)_3P(OCH_3)_3$ as a 3:1 mixture of isomers VII and VIII
$P(C_6H_5)_2C_2H_5$/VII	48%, orange solid ^1H NMR (CD_2Cl_2): 0.94 (dt, CH_3; J(P, H) = 15.1, J(H, H) = 6.3), 2.63 (qui, CH_2; J(P, H) = J(H, H) = 6.3), 4.67, 5.32 (both br, C_5H_4), 7.3 to 7.6 (m, meta- and para-H), 7.6 to 8.0 (m, ortho-H) IR (CH_2Cl_2): 1853, 1880, 1934, 1946, 2009 (CO)
$P(C_6H_5)_2CH{=}CH_2$/VII	42%, orange solid ^1H NMR ($CDCl_3$): 4.69 ("q", C_5H_4; J = 2), 5.06 (t, $=CH_2$ (H cis to P); cis-^3J(P, H) = trans-^3J(H, H) = 18.1), 5.34 ("q", C_5H_4), 5.91 (dd, $=CH_2$ (H trans to P); trans-^3J(P, H) = 34.4, cis-^3J(H, H) = 12.6), 6.76 (ddd, CH=; gem-^2J(P, H) = 28.7), 7.1 to 7.5 (m, meta- and para-H), 7.6 to 8.1 (m, ortho-H)

References on pp. 253/8

^{2}D/isomer	isomer yield, properties, and remarks
	^{13}C NMR ($CDCl_3$): 60.5 (d, C-1 of C_5H_4; J(P, C) = 46.8), 88.8 (d, C-2,5 (?) of C_5H_4; J(P, C) = 6.4), 89.3 (d, C-3,4 (?) of C_5H_4; J(P, C) = 10.1), 128.2 (d, meta-C; J(P, C) = 9.5), 128.5 (s, =CH_2), 128.8 (d, meta-C; J(P, C) = 10.1), 129.8, 130.7 (both s, both para-C), 131.5, 133.3 (both d, both ortho-C; J(P, C) = 11.5 and 9.7), 134.9 (d, PCH=; J(P, C) = 30), 135.2, 135.4 (both d, both ipso-C; J(P, C) = 42.1 and 40.0), 220 (br, MnCO), 227 (s, MoCO trans to Mn), 229.1 (s, MoCO cis to Mn) ^{31}P NMR ($CDCl_3$): −84.4 (s, $P(C_6H_5)_2CH_3$), −70.6 (s, $C_5H_4P(C_6H_5)_2$) IR (CH_2Cl_2): 1858, 1886, 1936, 1949, 2011 (CO) FAB mass spectrum: $[M]^+$ observed heating in toluene at 383 K produced $Mo(CO)_2(\mu\text{-}C_5H_4P(C_6H_5)_2)(\mu\text{-}CH_2{=}CHP(C_6H_5)_2)Mn(CO)_3$, containing a vinylphosphane bridge linked to manganese through phosphorus and to molybdenum through the vinyl group
$P(OCH_3)_3$/VII, VIII	34% (isolated as a 3:1 mixture of noninterconverting isomers VII and VIII by crystallization from CH_2Cl_2/hexane (1:1)); also resulting from thermal treatment of $Mo(CO)_3(\mu\text{-}C_5H_4P(C_6H_5)_2)Mn(CO)_3P(C_6H_5)_2CH_3$ with one equivalent of $P(OCH_3)_3$ in toluene at reflux temperature (59%) orange solid ^{1}H NMR ($CDCl_3$): 3.82 (d, OCH_3 of VIII; J(P, H) = 11.0), 3.83 (d, OCH_3 of VII; J(P, H) = 10.9), 4.29 ("q", C_5H_4 of VIII; "J" = 2), 4.60 ("q", C_5H_4 of VII; "J" = 2), 5.33 (m, C_5H_4 of VII and VIII), 7.3 to 7.6 (m, meta and para-H), 7.6 to 8.0 (m, ortho-H) ^{13}C NMR ($CDCl_3$; isomer VII): 53.0 (d, OCH_3; J(P, C) = 7.8), 59.0 (d, C-1 of C_5H_4; J(P, C) = 48.1), 88 (m, C-2,5 and C-3,4 of C_5H_4), 128.8 (d, meta-C; J(P, C) = 10.2), 130.9 (s, para-C), 131.5 (d, ortho-C; J(P, C) = 11.9), 135.2 (d, ipso-C; J(P, C) = 39.6), 218 (br, MnCO), 227.9 (s, MoCO cis to Mn), 233.8 (s, MoCO trans to Mn); ($CDCl_3$; isomer VIII): 52.7 (d, OCH_3; J(P, C) = 6.3), 55 (d, C-1 of C_5H_4; J(P, C) = 40), 88 (m, C-2,5 and C-3,4 of C_5H_4), 128.5 (d, meta-C; J(P, C) = 10.0), 130.7 (s, para-C), 132.2 (d, ortho-C; J(P, C) = 12.1), 134.2 (d, ipso-C; J(P, C) = 37.8), 218 (br, MnCO), 227.2 (s, MoCO cis to Mn); (CD_2Cl_2, 218 K; isomer VII): 217 (br, MnCO), 227.7 (s, MoCO cis to Mn), 233.7 (s, MoCO trans to Mn); (CD_2Cl_2, 218 K; isomer VIII): 217 (br, MnCO), 227.2 (s, MoCO cis to Mn), 231.7 (s, MoCO trans to Mn) ^{31}P NMR ($CDCl_3$; isomer VII): −73.5 (s, $C_5H_4P(C_6H_5)_2$), 40.8 (s (br), $P(OCH_3)_3$); ($CDCl_3$; isomer VIII): −85.7 (s, $C_5H_4P(C_6H_5)_2$), 57.1 (s (br), $P(OCH_3)_3$) IR (hexane): 1899, 1949, 1961, 1999, 2025 (CO) mass spectrum: $[M]^+$, $[M-nCO]^+$ (n = 1 to 3)
t-C_4H_9NC/IX	37%, orange powder ^{1}H NMR ($CDCl_3$): 0.95 (s, CH_3), 4.25, 4.34 (both "q", 1C_5H_4-H each); "J" = 1.5), 5.40 (m, 2C_5H_4-H), 7.2 to 7.5 (m, meta- and para-H), 7.9 to 8.1 (m, ortho-H)

References on pp. 253/8

^2D/isomer	isomer yield, properties, and remarks
	IR (hexane): 1882, 1903, 1919, 1960, 1968, 2021 (CO), 2148 (ν(NC)) mass spectrum: $[M]^+$, $[M-nCO]^+$ (n = 3 to 6)
$C_6H_5CH_2NC$/IX	22%, orange powder ^1H NMR ($CDCl_3$): 3.50 (m, $1C_5H_4$-H), 4.21 (d, $1CH_2$-H; J(H, H) = 3.5), 4.30 (m, $1C_5H_4$-H), 4.39 (dd, $1CH_2$-H; J(P, H) = 15.9), 5.17, 5.30 (both m, $1C_5H_4$-H each), 6.64 (d, ortho-H of $C_6H_5CH_2NC$; J(P, H) = 7.1), 7.1 to 7.5 (m, meta- and para-H), 7.8 to 8.0 (m, ortho-H of $P(C_6H_5)_2$) ^{13}C NMR ($CDCl_3$): 48.3 (s, CH_2), 53.3 (d, C-1 of C_5H_4; J(P, C) = 48.9), 87.6, 87.7, 88.5, 88.6 (all d, C-2,5 of C_5H_4; J(P, C) = 9.9, 9.8, 6.2, and 6.2), 127 to 136 (m, C_6H_5), 167 (br, NC), 217 (br, MnCO), 225.4, 228.0, 232.4 (all MoCO) ^{31}P NMR ($CDCl_3$): −77.6 IR (hexane): 1879, 1904, 1921, 1960, 1969, 2022 (CO), 2148 (ν(NC)) mass spectrum: $[M]^+$, $[M-nCO]^+$ (n = 3 to 6)

$[(OC)_3Mo(\mu\text{-}C_5H_4P(C_6H_4CH_3\text{-}4)_2)(\mu\text{-}H)M(CO)_4]X$

X

$(C_5H_4)P(C_6H_4CH_3\text{-}4)_2$; $(OC)_3Mo—M—{}^2D$; ${}^2D'$

XI

$[Mo(CO)_3(\mu\text{-}C_5H_4P(C_6H_4CH_3\text{-}4)_2)(\mu\text{-}H)M(CO)_4]X$ (Formula X; M = Mn, Re; X^- = $CF_3SO_3^-$, $CF_3CO_2^-$). The cationic complexes formed in solution when CF_3SO_3H (1.5 to 2.0 equivalents) or excess CF_3CO_2H was added to the respective Mo−Mn or Mo−Re precursor $Mo(CO)_3(\mu\text{-}C_5H_4P(C_6H_4CH_3\text{-}4)_2)M(CO)_4$ in CD_2Cl_2 [37]; spectroscopic data reported for the in situ-generated $CF_3SO_3^-$ salts are summarized in the following table:

M	properties
Mn	^1H NMR (CD_2Cl_2): −19.21 (d, μ-H; J(P, H) = 6.2), 2.42 (s, CH_3), 5.32, 6.02 (both "s", C_5H_4), 7.46 (br d, meta-H), 7.78 (dd, ortho-H; J(H, H) = 7.9, J(P, H) = 14.0) ^{13}C NMR (CD_2Cl_2): 21.6 (s, CH_3), 78.6 (d, C-1 of C_5H_4; J(P, C) = 41.3), 93.4 (d, C-2,5 of C_5H_4; J(P, C) = 9.2), 94.9 (d, C-3,4 of C_5H_4; J(P, C) = 4.6), 119.5 (q, CF_3; J(F, C) = 316.6), 126.4 (d, ipso-C; J(P, C) = 50.5), 131.8 (meta-C; J(P, C) = 12.2), 132.4 (d, ortho-C; J(P, C) = 12.2), 145.8 (s, para-C), 207 (quadrupole-broadened $Mn(CO)_4$), 219.6 (s, 2MoCO cis to Mn), 225.5 (s, MoCO trans to Mn) IR (CH_2Cl_2): 1978, 1999, 2029, 2040, 2062, 2106 (CO) [33]
Re	^1H NMR (CD_2Cl_2): −16.87 (s, μ-H), 2.42 (s, CH_3), 5.37, 6.08 (both br, C_5H_4), 7.43 (dd, meta-H; J(H, H) = 7.9, J(P, H) = 2.3), 7.72 (dd, ortho-H; J(H, H) = 8.2, J(P, H) = 13.2) ^{13}C NMR (CD_2Cl_2): 21.6 (s, CH_3), 84.4 (d, C-1 of C_5H_4; J(P, C) = 47.4), 94.1 (d, C-2,5 of C_5H_4; J(P, C) = 9.2), 95.4 (d, C-3,4 of C_5H_4; J(P, C) = 6.1),

M	properties
	120.2 (q, CF_3; J(F, C) = 320.0), 124.9 (d, ipso-C; J(P, C) = 55.1), 131.6 (d, meta-C; J(P, C) = 12.2), 132.8 (d, ortho-C; J(P, C) = 13.8), 145.5 (s, para-C), 178.9 (d, ReCO trans to P; J(P, C) = 39.7), 180.2 (d, ReCO trans to Mo; J(P, C) = 7.6), 180.8 (d, 2 ReCO cis to P; J(P, C) = 9.2), 220.0 (s, 2 MoCO cis to Re), 226.7 (MoCO trans to Re)

Deprotonation using bases such as $N(C_2H_5)_3$ or $Li[(C_2H_5)_3BH]$, or even acetone converted the cations to their parent heterobimetallics in almost quantitative yields [37].

$Mo(CO)_3(\mu\text{-}C_5H_4P(C_6H_4CH_3\text{-}4)_2)M(^2D)^2D'$ (Formula XI; M = Rh: ^2D = $^2D'$ = CO; ^2D = CO, $^2D'$ = $P(CH_3)_3$, $P(C_6H_4CH_3\text{-}4)_3$; ^2D, $^2D'$ = $(C_6H_4CH_3\text{-}4)_2PC_2H_4P(C_6H_4CH_3\text{-}4)_2$. M = Ir: ^2D = $^2D'$ = CO; ^2D = CO, $^2D'$ = $P(CH_3)_3$, $P(C_6H_5)_3$). For this family of Mo–Rh and Mo–Ir heterobimetallics, which are listed in the table below, two general methods of synthesis are available (for yields, see table). (I) Two equivalents of $Li[(4\text{-}CH_3C_6H_4)_2PC_5H_4Mo(CO)_3]$, dissolved in toluene or THF, were added below −72°C to $(Rh(CO)_2(\mu\text{-}Cl))_2$ (solution in toluene) and $(Rh(R_2PC_2H_4PR_2)(\mu\text{-}Cl))_2$ (R = $C_6H_4CH_3$-4; solution in THF), respectively. The products remaining after warming to room temperature and evaporation to dryness were separated and purified as follows: Column chromatography (alumina, toluene) afforded pure $Mo(CO)_3(\mu\text{-}C_5H_4P(C_6H_4CH_3\text{-}4)_2)Rh(CO)_2$; $Mo(CO)_3(\mu\text{-}C_5H_4P(C_6H_4CH_3\text{-}4)_2)Rh(R_2PC_2H_4PR_2)$ (R = $C_6H_4CH_3$-4) was isolated by preparative thin-layer chromatography (silica gel, toluene). In the preparation of $Mo(CO)_3(\mu\text{-}C_5H_4P(C_6H_4CH_3\text{-}4)_2)Ir(CO)_2$, a 1:1 molar mixture of $Li[(4\text{-}CH_3C_6H_4)_2PC_5H_4Mo(CO)_3]$ and $Ir(CO)_2(NH_2C_6H_4CH_3\text{-}4)Cl$ was refluxed in THF, and column chromatography on alumina, eluting with ether, was employed to isolate the product. All three compounds so obtained were purified further by recrystallization from toluene/pentane [45]. (II) THF solutions of the Mo–Rh and Mo–Ir heterobimetallics $Mo(CO)_3(\mu\text{-}C_5H_4\text{-}P(C_6H_4CH_3\text{-}4)_2)M(CO)_2$, prepared according to (I), were stirred at room temperature with a slight excess of a tertiary phosphane, PR_3 (M = Rh: R = CH_3, $C_6H_4CH_3$-4; M = Ir: R = CH_3, C_6H_5). The crude substitution products remaining after removal of solvent, $Mo(CO)_3(\mu\text{-}C_5H_4P(C_6H_4CH_3\text{-}4)_2)M(CO)PR_3$, were purified either by recrystallization from toluene/pentane (M = Rh, R = CH_3), or by preparative thin-layer chromatography (silica gel, diethyl ether/hexane (1:1)) with subsequent crystallization from CH_2Cl_2/hexane (M = Rh, R = $C_6H_4CH_3$-4) or toluene/hexane (M = Ir, R = C_6H_5), or by column chromatography (alumina, toluene) followed by recrystallization from toluene/hexane (M = Ir, R = CH_3) [45].

$M/^2D/^2D'$	method of synthesis (yield), properties, and remarks
Rh/CO/CO	I (41%) olive-green microcrystals, m.p. 168 to 171 °C (dec.) ^1H NMR (C_6D_6): 1.91 (s, CH_3), 4.02, 4.79 (both "q", C_5H_4; "J" = 2), 7.94 (?) (dd, meta-H; J(H, H) = 8.2, J(P, H) = 1.9), 7.69 (dd, ortho-H; J(H, H) = 7.9, J(P, H) = 12.5) ^{13}C NMR (acetone-d_6, 0.07 M $Cr(CH_3C(O)CH_2C(O)CH_3)_3$): 21.5 (s, CH_3), 55.0 (d, C-1 of C_5H_4; J(P, C) = 60.3), 93.2 (m, C-2 to 5 of C_5H_4), 126.4 (d, ipso-C; J(P, C) = 51.3), 130.9 (d, meta-C; J(P, C) = 11.0), 134.7 (d, ortho-C; J(P, C) = 11.0), 143.9 (s, para-C), 184 (br, RhCO), 222.4 (s, MoCO cis to Rh), 235.3 (s, MoCO trans to Rh); ^{13}CO-enriched samples (acetone-d_6, −70 °C) showed RhCO trans to Mo at

References on pp. 253/8

$M/^2D/^2D'$	method of synthesis (yield), properties, and remarks
	δ = 183.2 (d; J(Rh, C) = 75.1) and RhCO trans to P at δ = 200.8 (dd; J(P, C) = 62.3, J(Rh, C) = 108.1) ^{31}P NMR (acetone-d_6, 0.07 M $Cr(CH_3C(O)CH_2C(O)CH_3)_3$): 42.7 (d; J(Rh, P) = 118) IR ($CH_3OC_2H_4OCH_3$): 1867, 1884, 1959, 1988, 2062 (CO) samples enriched in ^{13}CO at all carbonyl sites were obtained by a combination of heating to 50 °C under ^{13}CO in benzene and photolysis
$Rh/CO/P(CH_3)_3$	II (71%) yellow solid, m.p. 248 to 257 °C (dec.) ^1H NMR (C_6D_6): 1.48 (dm, $P(CH_3)_3$; ^2J(P, H) = 8.6), 1.92 (s, CH_3), 4.52, 5.01 (both "q", C_5H_4; "J" = 2), 6.86 (d, meta-H), 7.95 (dd, ortho-H; J(H, H) = 8.1, J(P, H) = 11.8) ^{13}C NMR (CD_2Cl_2, 0.07 M $Cr(CH_3C(O)CH_2C(O)CH_3)_3$): 16.6 (d, $P(CH_3)_3$; J(P, C) = 27.5), 21.1 (s, CH_3), 62.2 (d, C-1 of C_5H_4; J(P, C) = 45.9), 92.1 (br, C-3,4 (?) of C_5H_4), 93.9 (d, C-2,5 (?) of C_5H_4; J(P, C) = 10.7), 128.4 (d, ipso-C; J(P, C) = 42.8), 129.4 (d, meta-C; J(P, C) = 10.7), 133.6 (d, ortho-C; J(P, C) = 13.8), 141.7 (s, para-C), 225.7 (s, MoCO cis to Rh), 235.0 (s, MoCO trans to Rh); RhCO not observed ^{31}P NMR (CD_2Cl_2, 0.07 M $Cr(CH_3C(O)CH_2C(O)CH_3)_3$): −1.7 (dd, $P(CH_3)_3$; J(P, P) = 348, J(Rh, P) = 127), 45.4 (dd, $P(C_6H_4CH_3\text{-}4)_2$; J(Rh, P) = 112) IR (THF): 1828, 1851, 1932, 1957 (CO)
$Rh/CO/P(C_6H_4CH_3\text{-}4)_3$	II (56%) yellow powder, m.p. 236 to 242 °C ^1H NMR (C_6D_6): 1.99 (s, $2CH_3$), 2.05 (s, $3CH_3$), 4.56, 4.93 (both "q", C_5H_4; "J" = 2), 6.92 (d, 4 meta-H; J(H, H) = 6.7), 7.12 (d, 6 meta-H; J(H, H) = 7.1), 8.08 (m, 10 ortho-H) ^{13}C NMR (CD_2Cl_2, 0.07 M $Cr(CH_3C(O)CH_2C(O)CH_3)_3$): 20.9 (s, CH_3), 60.1 (d, C-1 of C_5H_4; J(P, C) = 50.5), 91.75 (br, C-3,4 (?) of C_5H_4), 93.5 (d, C-2,5 (?) of C_5H_4; J(P, C) = 10.7), 128.4 (d, 6 meta-C; J(P, C) = 7.7), 129.1 (d, 4 meta-C; J(P, C) = 9.2), 130.2 (d, ipso-C; J(P, C) = 42.8), 133.4 (d, 4 ortho-C; J(P, C) = 13.8), 134.3 (d, 6 ortho-C; J(P, C) = 10.7), 139.7 (s, 3 para-C), 141.6 (s, 2 para-C), 225.5 (s, MoCO cis to Rh), 234.6 (s, MoCO trans to Rh) ^{31}P NMR (CD_2Cl_2, 0.07 M $Cr(CH_3C(O)CH_2C(O)CH_3)_3$): 32.3 (dd, $P(C_6H_4CH_3\text{-}4)_3$; J(P, P) = 332, J(Rh, P) = 130), 47.2 (dd, $P(C_6H_4CH_3\text{-}4)_2$; J(Rh, P) = 120) IR (THF): 1832, 1859, 1938, 1966 (CO)
$Rh/(4\text{-}CH_3C_6H_4)_2PC_2H_4P(C_6H_4CH_3\text{-}4)_2$	I (25%) orange solid, m.p. 240 to 244 °C (dec.) ^1H NMR (C_6D_6): 0.89, 1.36 (both m, CH_2), 1.99, 2.00, 2.14 (all s, all CH_3), 4.62, 5.04 ("q" and m, C_5H_4; "J" = 2), 6.78 (d, 8 meta-H; J(H, H) = 7.3), 7.20, (d, 4 meta-H; J(H, H) = 6.6),

References on pp. 253/8

$M/^2D/^2D'$	method of synthesis (yield), properties, and remarks
	7.46, 7.58, 8.00 (all dd, all 4 ortho-H; J(H, H) = 8.1, J(P, H) = 9.9, 10.7, and 9.6) ^{13}C NMR (C_6D_6, 0.07 M $Cr(CH_3C(O)CH_2C(O)CH_3)_3$): 229.7 (s, MoCO cis to Rh), 236.9 (s, MoCO trans to Rh), aromatic carbons omitted; other resonances reported in THF-d_8: 21.2, 21.4, 21.5 (all s, all CH_3), 28.3, 30.0 (both m, both CH_2), 90.5 (d, C-3,4 of C_5H_4; J(P, C) = 6.1), 92.9 (d, C-2,5 of C_5H_4; J(P, C) = 12.2) ^{31}P NMR (C_6D_6; "P_A" = C_5H_4P; "P_B" cis to "P_A", "P_C" trans to "P_A"): 38.6 (ddd, P_A; J(P_A, P_B) = 26, J(P_A, P_C) = 345, J(Rh, P_A) = 129), 54.0 (ddd, P_B; J(P_B, P_C) = 32, J(Rh, P_B) = 183), 65.4 (ddd, P_C; J(Rh, P_C) = 146); spectrum reproduced in the original literature IR (toluene): 1801, 1841, 1922 (CO)
Ir/CO/CO	I (28%) orange microcrystalline solid, m.p. 178 to 181°C (dec.) ^{1}H NMR (C_6D_6): 1.91 (s, CH_3), 4.16, 4.76 (both "q", C_5H_4; "J" = 2.2), 6.79 (dd, meta-H; J(H, H) = 7.7, J(P, H) = 2.2), 7.70 (dd, ortho-H; J(H, H) = 8.3, J(P, H) = 12.7) ^{13}C NMR (acetone-d_6, 0.07 M $Cr(CH_3C(O)CH_2C(O)CH_3)_3$): 21.6 (s, CH_3), 94.1 (d, C-2,5 (?) of C_5H_4; J(P, C) = 9.2), 94.9 (d, C-3,4 (?) of C_5H_4; J(P, C) = 3.0), 125.2 (d, ipso-C; J(P, C) = 59.7), 130.9 (d, meta-C; J(P, C) = 12.2), 134.5 (d, ortho-C; J(P, C) = 12.2), 144.0 (s, para-C), 220.7 (s, MoCO cis to Ir), 234.2 (s, MoCO trans to Ir); C-1 and IrCO not observed ^{31}P NMR (acetone-d_6, 0.07 M $Cr(CH_3C(O)CH_2C(O)CH_3)_3$): 32.2 (s) IR (THF): 1887, 1896, 1971, 2047 (CO)
$Ir/CO/P(CH_3)_3$	II (77%) orange crystals, m.p. 250 to 254°C (dec.) ^{1}H NMR (C_6D_6): 1.60 (dd, $P(CH_3)_3$; ^{2}J(P, H) = 9.5, ^{4}J(P, H) = 2.3), 1.92 (s, CH_3), 4.68, 4.99 (both "q", C_5H_4; "J" = 2), 6.86 (dm, meta-H), 7.97 (dd, ortho-H; J(H, H) = 8.1, J(P, H) = 11.9) ^{13}C NMR (CD_2Cl_2, 0.07 M $Cr(CH_3C(O)CH_2C(O)CH_3)_3$): 16.9 (d, $P(CH_3)_3$; J(P, C) = 33.6), 21.2 (s, CH_3), 65.9 (d, C-1 of C_5H_4; J(P, C) = 55.1), 93.1 (d, C-3,4 of C_5H_4; J(P, C) = 6.1), 94.9 (d, C-2,5 of C_5H_4; J(P, C) = 12.2), 127.8 (d, ipso-C; J(P, C) = 52.0), 129.4 (d, meta-C; J(P, C) = 12.2), 133.6 (d, ortho-C; J(P, C) = 12.2), 141.9 (s, para-C), 175.1 (m, IrCO), 224.4 (s, MoCO cis to Ir), 234.5 (s, MoCO trans to Ir) ^{31}P NMR (CD_2Cl_2, 0.07 M $Cr(CH_3C(O)CH_2C(O)CH_3)_3$): −14.9 (d, $P(CH_3)_3$; J(P, P) = 334), 33.2 (d, $P(C_6H_4CH_3\text{-}4)_2$) IR (THF): 1830, 1854, 1929, 1959 (CO)
$Ir/CO/P(C_6H_5)_3$	II (74%); also isolated from the reaction of $Li[(4\text{-}CH_3C_6H_4)_2PC_5H_4Mo(CO)_3]$ with $Ir(CO)(P(C_6H_5)_3)_2Cl$, albeit in a less purer form orange crystals, m.p. 218 to 224°C ^{1}H NMR (C_6D_6): 1.98 (s, CH_3), 4.71, 4.90 (both "q", C_5H_4; "J" = 2), 6.90 (dd, meta-H of C_6H_4; J(H, H) = 7.9, J(P, H) = 1.8),

References on pp. 253/8

$M/^2D/^2D'$	method of synthesis (yield), properties, and remarks
	7.2 (m, meta- and para-H of C_6H_5), 8.04 (dd, ortho-H of C_6H_4; J(H, H) = 8.1, J(P, H) = 11.6), 8.16 (m, ortho-H of C_6H_5) ^{13}C NMR (CD_2Cl_2, 0.07 M $Cr(CH_3C(O)CH_2C(O)CH_3)_3$): 20.9 (s, CH_3), 93.8 (d, C-3,4 of C_5H_4; J(P, C) = 7.6), 94.6 (d, C-2,5 of C_5H_4; J(P, C) = 12.2), 127.7 (d, meta-C of C_6H_5; J(P, C) = 9.2), 129.2 (d, meta-C of C_6H_4; J(P, C) = 10.7), 129.9 (s, para-C of C_6H_5), 133.3 (d, ortho-C of C_6H_4; J(P, C) = 12.2), 134.7 (d, ortho-C of C_6H_5; J(P, C) = 10.7), 141.8 (s, para-C of C_6H_4), 224.1 (s, MoCO cis to Ir), 233.6 (s, MoCO trans to Ir); additional resonances observed in THF-d_8: 64.8 (d, C-1 of C_5H_4; J(P, C) = 54.0), 174.4 (s, IrCO); ipso carbons not detected ^{31}P NMR (C_6D_6): 21.1 (d, $P(C_6H_5)_3$; J(P, P) = 336), 31.0 (d, $P(C_6H_4CH_3\text{-}4)_2$) IR (THF): 1834, 1863, 1938, 1958 (CO)

Even under forced conditions, i.e., in toluene solutions heated at 70 to 115 °C under 1000 to 1200 psi of H_2, none of the Mo–Rh complexes described above reacted with dihydrogen to form observable metal hydrides. In contrast, all three Mo–Ir compounds interacted readily and reversibly with H_2 or D_2 to afford iridium-centered dihydrides, XII (^2D = CO, $P(CH_3)_3$, $P(C_6H_5)_3$), under very mild conditions (≤1 atm H_2, room temperature) [45].

Attempted CO hydrogenation experiments were unsuccessful; thus, no CH_4 or CH_3OH was observed when toluene solutions of $Mo(CO)_3(\mu\text{-}C_5H_4P(C_6H_4CH_3\text{-}4)_2)Rh(CO)P(CH_3)_3$ were heated at 120 °C under 600 psi CO and 600 psi H_2 [45]. Only traces of cyclohexene were hydrogenated when $Mo(CO)_3(\mu\text{-}C_5H_4P(C_6H_4CH_3\text{-}4)_2)Rh(CO)P(C_6H_4CH_3\text{-}4)_3$ was exposed in toluene at 140 °C, in the presence of cyclo-C_6H_{10}, to 1000 psi of H_2 [45].

C_5H_4–$P(C_6H_4CH_3-4)_2$; $(OC)_3Mo$–Ir; Ir: CO, H, H, ^2D

XII

$Mo(CO)_3(\mu\text{-}C_5H_4P(C_6H_4CH_3\text{-}4)_2)Ir(CO)(^2D)H_2$ (XII; ^2D = CO, $P(CH_3)_3$, $P(C_6H_5)_3$). The complexes were formed readily at room temperature when solutions of the parent $Mo(CO)_3(\mu\text{-}C_5H_4P(C_6H_4CH_3\text{-}4)_2)Ir(CO)^2D$ compound were stirred under H_2 or D_2 (≤1 atm). Due to steric retardation, the rate of formation of $Mo(CO)_3(\mu\text{-}C_5H_4P(C_6H_4CH_3\text{-}4)_2)Ir(CO)(P(C_6H_5)_3)H_2$ (half-life of H_2-free precursor in 12 mM C_6D_6 under 400 Torr H_2: 30 min) was significantly slower than that of either $Mo(CO)_3(\mu\text{-}C_5H_4P(C_6H_4CH_3\text{-}4)_2)Ir(CO)(P(CH_3)_3)H_2$ (reaction in 28 mM C_6D_6 under 396 Torr H_2 complete in less than 6 min) or $Mo(CO)_3(\mu\text{-}C_5H_4P(C_6H_4\text{-}CH_3\text{-}4)_2)Ir(CO)_2H_2$ (44 mM under 565 Torr H_2: formation complete within 7 min). Because of the reversibility of the addition of H_2 to the coordinatively unsaturated starting complexes (vide infra), $Mo(CO)_3(\mu\text{-}C_5H_4P(C_6H_4CH_3\text{-}4)_2)Ir(CO)_2H_2$ could be characterized only in solutions kept under H_2. The more stable phosphane-substituted H_2 adducts $Mo(CO)_3(\mu\text{-}C_5H_4\text{-}P(C_6H_4CH_3\text{-}4)_2)Ir(CO)(P(CH_3)_3)H_2$ and $Mo(CO)_3(\mu\text{-}C_5H_4P(C_6H_4CH_3\text{-}4)_2)Ir(CO)(P(C_6H_5)_3)H_2$, on the other hand, were isolated by precipitation from toluene using hydrogen-saturated hexane, which afforded the former in an analytically pure state but the latter as a 93:7 mixture with its H_2-free precursor [45].

References on pp. 253/8

In C_6D_6 solutions kept at 22 °C under ca. 60 Torr H_2 pressure, the equilibrium constants $K_{eq} = [XII][XII - H_2]^{-1}[H_2]^{-1}$ for "(XII − H_2) + $H_2 \rightleftarrows$ XII" were estimated by NMR as $K_{eq} = 37\ atm^{-1}$ for ^2D = CO, $K_{eq} = 440\ atm^{-1}$ for $^2D = P(C_6H_5)_3$, and $K_{eq} > 6500\ atm^{-1}$ for $^2D = P(CH_3)_3$, respectively [45]. Standing of the title complexes in solution at ambient conditions resulted in reductive elimination of H_2, with the degree of conversion back to their parent compounds, (XII − H_2), decreasing in the order ^2D = CO ≫ $P(C_6H_5)_3 > P(CH_3)_3$ [45].

^2D	(yield), properties
CO	only stable under a H_2 atmosphere ^1H NMR (C_6D_6, 500 Torr H_2): −11.58 (dd, IrH; J(H, H) = 2.6, J(P, H) = 17.2), −11.37 (dd, IrH; J(P, H) = 14.8), 1.89, 1.91 (both s, $1CH_3$ each), 3.89, 4.0, 4.74, 4.79 (all m, C_5H_4), 6.50 (dd, meta-H; J(H, H) = 8.2, J(P, H) = 2.1), 6.80 (dd, meta-H; J(H, H) = 8.4, J(P, H) = 2.1), 7.55 (dd, ortho-H; J(H, H) = 8.0, J(P, H) = 12.7), 7.66 (dd, ortho-H; J(H, H) = 8.4, J(P, H) = 13.5) ^{13}C NMR (CD_2Cl_2, 0.07 M $Cr(CH_3C(O)CH_2C(O)CH_3)_3$, 500 Torr H_2): 20.9, 21.1 (both s, both CH_3), 56.5 (d, C-1 of C_5H_4; J(P, C) = 67.3), 90.2 (m, C_5H_4, 3C), 91.0 (d, C_5H_4, 1C; J(P, C) = 6.1), 126.2 (d, ipso-C; J(P, C) = 67), 126.3 (d, ipso-C; J(P, C) = 55), 129.3, 129.7 (both d, meta-C; J(P, C) = 12.2 each), 131.9, 133.4 (both d, ortho-C; J(P, C) = 15.3 each), 142.6, 142.9 (both s, para-C), 172.1, 172.4 (both s, IrCO), 225 (br, MoCO) ^{31}P NMR (CD_2Cl_2, 0.07 M $Cr(CH_3C(O)CH_2C(O)CH_3)_3$): 8.5 (s) IR (THF, 1 atm H_2): 1877, 1896, 1956 (MoCO), 2008 (IrCO trans to H), 2059 (IrCO trans to P); 2120 (ν(IrH)); (THF, 1 atm D_2): 1877, 1897, 1961 (MoCO), 2029 (IrCO trans to D), 2071 (IrCO trans to P); ν(IrD) obscured by other bands
$P(CH_3)_3$	(50%) yellow solid, m.p. 246 to 248 °C (dec.) ^1H NMR (C_6D_6): −13.81 (ddd, IrH; J(H, H) = 3.4, J(P, H) = 11.0 and 14.6), −10.43 (ddd, IrH; J(P, H) = 15.6 and 21.6), 1.60 (dd, $P(CH_3)_3$; ^2J(P, H) = 10.3, ^4J(P, H) = 2.6), 1.90, 1.95 (both s, both CH_3), 4.26, 4.59, 4.92 (all m, C_5H_4), 6.83 (dd, meta-H; J(H, H) = 8.4, J(P, H) = 1.9), 6.92 (dd, meta-H; J(H, H) = 8.2, J(P, H) = 1.7), 7.82 (dd, ortho-H; J(H, H) = 8.2, J(P, H) = 12.0), 7.88 (dd, ortho-H; J(H, H) = 8.2, J(P, H) = 12.5) ^{13}C NMR (CD_2Cl_2, 0.07 M $Cr(CH_3C(O)CH_2C(O)CH_3)_3$): 21.1 (d, $P(CH_3)_3$; J(P, C) = 6.1), 22.2, 22.9 (both s, both CH_3), 61.6 (d, C-1 of C_5H_4; J(P, C) = 41.9), 90.0, 90.7 (both d, C_5H_4, 1C; J(P, C) = 6.1), 91.3 (d, C_5H_4, 1C; J(P, C) = 9.2), 92.3 (d, C_5H_4, 1C; J(P, C) = 12.2), 127.7 ("s", tentatively assigned as one resonance of an ipso tolyl doublet), 129.0, 129.4 (both d, both meta-C; J(P, C) = 12.2 each), 130.0 ("s", tentatively assigned as one resonance of an ipso tolyl doublet), 133.1, 133.7 (both d, ortho-C; J(P, C) = 12.2 each), 141.8 (br, para-C), 177.6 (m, IrCO), 229 (br, MoCO) ^{31}P NMR (CD_2Cl_2, 0.07 M $Cr(CH_3C(O)CH_2C(O)CH_3)_3$): −48.9 (d, $P(CH_3)_3$; J(P, P) = 308), 11.6 (d, $P(C_6H_4CH_3\text{-}4)_2$) IR (THF): 1844, 1871, 1938 (MoCO); 1981 (IrCO), 2093 (ν(IrH)); IrD_2 analogue in THF: 1492 (ν(IrD)), 1846, 1870, 1943 (MoCO), 2006 (IrCO)
$P(C_6H_5)_3$	(71%) yellow powder decomposing at 175 to 180 °C (contaminated by small amounts of $Mo(CO)_3(\mu\text{-}C_5H_4P(C_6H_4CH_3\text{-}4)_2Ir(CO)P(C_6H_5)_3)$ ^1H NMR (C_6D_6): −12.93 (ddd, IrH; J(H, H) = 3.7, J(P, H) = 11.3 and 15.6), −10.49 (td, IrH; J(P, H) = 17.4 each), 1.93, 1.95 (both s, CH_3), 4.40, 4.68,

References on pp. 253/8

^{2}D	(yield), properties
	4.85 (all m, C_5H_4), 6.84 (d, meta-H of C_6H_4; J(H, H) = 7.7), 7.02 to 7.18 (m, meta- and para-H of C_6H_5), 7.83 (dd, ortho-H of one C_6H_4; J(H, H) = 8.2, J(P, H) = 11.2), 7.98 (m, ortho-H of C_6H_5 and one C_6H_4)
	^{13}C NMR (CD_2Cl_2, 0.07 M $Cr(CH_3C(O)CH_2C(O)CH_3)_3$, −30 °C): 21.0 (s, CH_3), 59.6 (d, C-1 of C_5H_4; J(P, C) = 64.2), 89.6, 90.7 (both d, C_5H_4, 1C; J(P, C) = 6.1 each), 91.8 (d, C_5H_4, 1C; J(P, C) = 12.2), 92.1 (d, C_5H_4, 1C; J(P, C) = 9.2), 127.7 (d, meta-C of C_6H_5; J(P, C) = 6.1), 129.0, 129.3 (both d, meta-C of C_6H_4; J(P, C) = 9.2 each), 130.0 (s, para-C of C_6H_5), 131.8, 133.4 (both d, ortho-C of C_6H_4; J(P, C) = 12.2 each), 134.0 (d, ortho-C of C_6H_5; J(P, C) = 9.2), 135.0 (dd, ipso-C of C_6H_5; ^{1}J(P, C) = 47.4, ^{3}J(P, C) ≈ 7), 141.6, 141.8 (both s, para-C of C_6H_4), 177.4 (s, IrCO), 224.9, 227.6, 233.9 (all s, all MoCO); ipso-carbons of $P(C_6H_4CH_3\text{-}4)$ not detected; at +30 °C, the $Mo(CO)_3$ group appeared as a broad multiplet centered at δ = 229.5 ppm
	^{31}P NMR (CD_2Cl_2, 0.07 M $Cr(CH_3C(O)CH_2C(O)CH_3)_3$): 9.0 (d, $P(C_6H_4CH_3\text{-}4)_2$; J(P, P) = 312), 14.0 (d, $P(C_6H_5)_3$)
	IR (THF): 1853, 1874, 1941 (MoCO); 1992 (IrCO); 2100 (ν(IrH)); IrD_2 analogue in THF: 1497 (ν(IrD)); 1852, 1876, 1946 (MoCO); 2015 (IrCO)

$Mo(CO)_3(\mu\text{-}C_5H_4P(C_6H_5)_2)Pd(P(C_6H_5)_3)I$ was synthesized by stirring $(C_6H_5)_2PC_5H_4Mo(CO)_3I$ with an equimolar mixture of Pd(benzylideneacetone)$_2$ and $P(C_6H_5)_3$ in THF. Chromatography on silica with CH_2Cl_2 gave 32% yield. The same result was obtained by reacting the iodo complex with $Pd(P(C_6H_5)_3)_4$ under the same conditions [178].

Red solid. ^{1}H NMR spectrum ($CDCl_3$): δ = 4.45 ("q", H-3,4 of C_5H_4; "J" = 2.1 Hz), 5.47 ("q", H-2,5 of C_5H_4), 7.45 to 7.55 (m, H-3 to 5 of C_6H_5), 8.09 to 8.16 (H-2,6 of C_6H_5) ppm. ^{13}C NMR spectrum ($CDCl_3$): δ = 60.24 (d, C-1 of C_5H_4; J(P,C) = 41.3 Hz), 92.46 (d, C-3,4 of C_5H_4; J(P,C) = 6.9 Hz), 93.33 (d, C-2,5 of C_5H_4; J(P,C) = 11.7 Hz), 127.92 (d; J(P,C) = 9.5 Hz), 128.53 (d; J(P,C) = 11.5 Hz), 130.00 (s), 131.76 (s), 132.20 (d; J(P,C) = 38.6 Hz), 135.58 (d; J(P,C) = 11.7 Hz) (all C_6H_5), 224.93, 235.75 (both CO) ppm. ^{31}P NMR spectrum ($CDCl_3$): δ = 17.00, 33.50 (both d; trans-^{2}J(P,P) = 459 Hz) ppm. IR spectrum (CH_2Cl_2): 1867.7, 1891.4, 1966.1 (CO) cm^{-1} [178].

1.5.1.4.2.1.3 Heterotrimetallic Complexes

$CH_3W(CO)_3(\mu\text{-}C_5H_4C_5H_3C_5H_4Mo(CO)_3H)Re(CO)_3$ (Formula I; R = H) resulted from sequential reactions of II with $Al(C_4H_9\text{-}i)_2H$ (CH_2Cl_2, 0 °C), $4\text{-}CH_3C_6H_4SO_3H$ (C_6H_6, 60 °C), and $Mo(CO)_3(NCC_2H_5)_3$ (THF, 23 °C); yield: 83% [107]. Deprotonation and alkylation gives the methyl derivative (Formula I; R = CH_3).

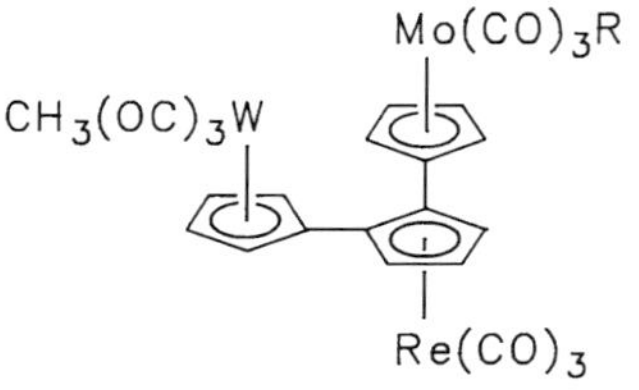

$Mo(CO)_3R$
$CH_3(OC)_3W$
$Re(CO)_3$
I

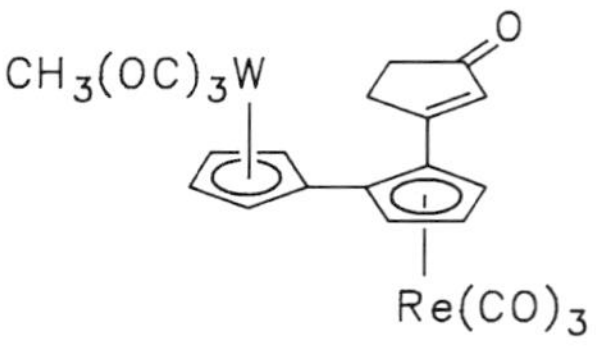

$CH_3(OC)_3W$
O
$Re(CO)_3$
II

References on pp. 253/8

$CH_3W(CO)_3(\mu-C_5H_4C_5H_3C_5H_4Mo(CO)_3CH_3)Re(CO)_3$ (Formula I; R = CH_3) was made in 45% yield by deprotonating the hydride (Formula I; R = H) with $K[OC_4H_9-t]$ in THF at room temperature, followed by alkylation of the ionic intermediate $K[CH_3W(CO)_3(\mu-C_5H_4C_5H_3C_5H_4-Mo(CO)_3)Re(CO)_3]$ so generated with methyl iodide (THF, 23 °C) [107].

Yellow crystals, m.p. 142 to 145 °C. ^1H NMR spectrum (acetone-d_6): δ = 0.44 (s, $MoCH_3$), 0.48 (s with W satellites, WCH_3; J(^{183}W, H) = 3.5 Hz), 5.43, 5.50 (both m, 1CH each), 5.62 (m, 2CH), 5.67 (dd, 1CH; J_1 = J_2 = 2.9 Hz), 5.72 (m, 4CH), 5.82 (dd, 1CH; J_1 = 2.0, J_2 = 2.9 Hz), 5.85 (dd, 1CH; J_1 = 2.0, J_2 = 2.7 Hz) ppm. IR spectrum (ether): 1937, 2016, 2027 (all CO) cm^{-1}. Mass spectrum: molecular ion observed; for m/e values of additional fragments, see [107].

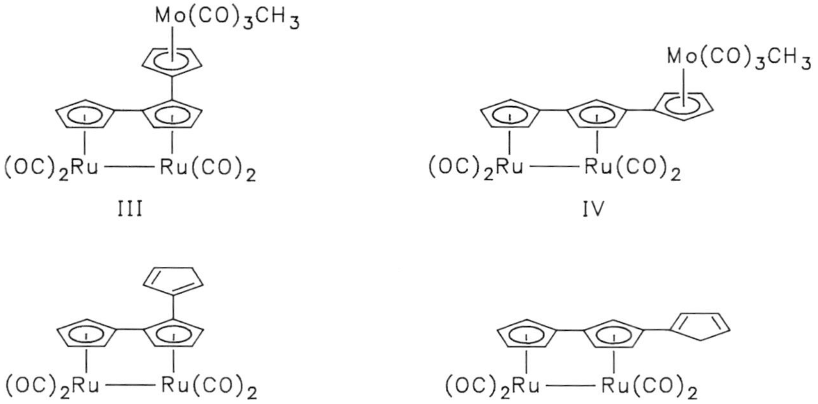

Mo(CO)$_3$CH$_3$

(OC)$_2$Ru——Ru(CO)$_2$

III

Mo(CO)$_3$CH$_3$

(OC)$_2$Ru——Ru(CO)$_2$

IV

(OC)$_2$Ru——Ru(CO)$_2$

V

(OC)$_2$Ru——Ru(CO)$_2$

VI

$Ru(CO)_2(\mu-C_5H_4C_5H_3C_5H_4Mo(CO)_3CH_3)Ru(CO)_2$ (regioisomers Formulas III and IV). The two isomers resulted from consecutive treatment of the diruthenium precursors V and VI with $NaNH_2$ and $Mo(CO)_3(NCCH_3)_3$ in THF at 23 °C, giving the respective regioisomers of $Na[Ru(CO)_2(\mu-C_5H_4C_5H_3C_5H_4Mo(CO)_3)Ru(CO)_2]$, followed by alkylation with methyl iodide at ambient conditions; yields: 65% (isomer III) and 44% (isomer IV) [107].

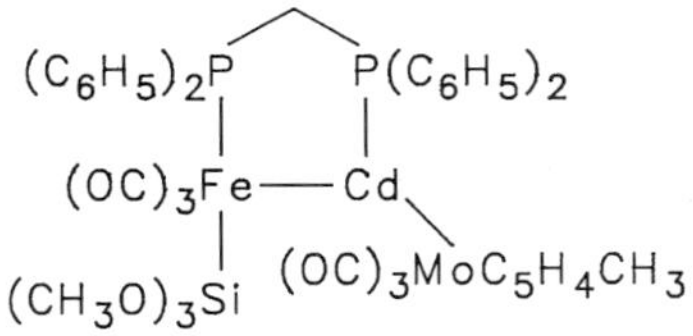

(C$_6$H$_5$)$_2$P P(C$_6$H$_5$)$_2$

(OC)$_3$Fe——Cd

(CH$_3$O)$_3$Si (OC)$_3$MoC$_5$H$_4$CH$_3$

VII

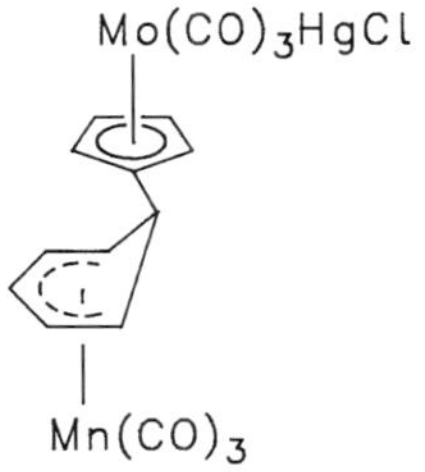

Mo(CO)$_3$HgCl

Mn(CO)$_3$

VIII

$CH_3C_5H_4Mo(CO)_3Cd(\mu-(C_6H_5)_2PCH_2P(C_6H_5)_2)Fe(CO)_3Si(OCH_3)_3$ (Formula VII) was synthesized by stirring solid $Na[CH_3C_5H_4Mo(CO)_3] \cdot CH_3OC_2H_4OCH_3$ with the Cd–Fe-bonded heterotetranuclear complex $((CH_3O)_3SiFe(CO)_3(\mu-(C_6H_5)_2PCH_2P(C_6H_5)_2)Cd(\mu-Cl))_2$ in the molar ratio 2:1 in THF. The same product was obtained when a THF solution of the anionic silyl iron complex $K[(CH_3O)_3SiFe(CO)_3P(C_6H_5)_2CH_2P(C_6H_5)_2]$ was treated with $CH_3C_5H_4Mo-(CO)_3CdBr$. The resulting filtered mixtures were used for ^{31}P NMR and IR characterization:

References on pp. 253/8

^{31}P NMR spectrum (THF/C_6D_6): δ = −15.4 (Cd-flanked d, CdP; $\Sigma|^2J(P, P) + {}^3J(P, P)|$ = 113, $^1J(^{111, 113}Cd, P)$ = 398 Hz), 64.3 (Cd-flanked d, FeP; $\Sigma|^2J(^{111, 113}Cd, P) + {}^3J(^{111, 113}Cd, P)|$ = 61 Hz) ppm. IR spectrum (THF): 1836, 1871, 1955 (MoCO); 1899, 1919, 1983 (FeCO) cm^{-1} [169].

$CH_3C_5H_4Mo(CO)_3InCl(\mu\text{-}(C_6H_5)_2PCH_2P(C_6H_5)_2)Fe(CO)_3Si(OCH_3)_3$ was obtained by treating the Fe−In-bonded complex $InCl_2(\mu\text{-}(C_6H_5)_2PCH_2P(C_6H_5)_2)Fe(CO)_3Si(OCH_3)_3$ with a slight excess of $Na[CH_3C_5H_4Mo(CO)_3] \cdot CH_3OC_2H_4OCH_3$ in THF at 0°C. Slow diffusion of hexane into the filtered and concentrated mixture afforded the product as yellow crystals (70%) [169].

^1H NMR spectrum (C_6D_6): δ = 1.37 (s, CCH_3), 3.31 (m, CH_2), 3.82 (s, OCH_3), 5.13, 5.29 (both m, C_5H_4), 6.72 to 7.71 (m, C_6H_5) ppm; ($CDCl_3$): δ = 2.04 (s, CCH_3), 3.52 (m, CH_2), 3.75 (s, OCH_3), 5.4 (m, C_5H_4), 6.96 to 7.69 (m, C_6H_5) ppm. ^{31}P NMR spectrum (THF/C_6D_6): −17.6 (d, InP; $\Sigma|^2J(P, P) + {}^3J(P, P)|$ = 122 Hz), 59.7 (d, FeP). IR spectrum (THF): 1861, 1917, 1980 (MoCO); 1944, 1950, 2011 (FeCO) cm^{-1}. FIR spectrum (polyethylene): 165 (unassigned), 261 (ν(InCl)) cm^{-1} [169].

Single crystals grown by diffusion of hexane into a chlorobenzene solution at 5°C contained one molecule of C_6H_5Cl per formula unit and were triclinic, space group $P\bar{1}-C_i^1$ (No. 2), with a = 10.785(3), b = 12.141(3), c = 21.028(6) Å, α = 98.17(2), β = 96.09(2), γ = 114.42(2)°; Z = 2 molecules per unit cell, D_{calc} = 1.589 g/cm^3. The molecule contains a tetrahedrally coordinated indium center; the iron-bound carbonyl groups occupy meridional positions (**Fig. 25**) [169].

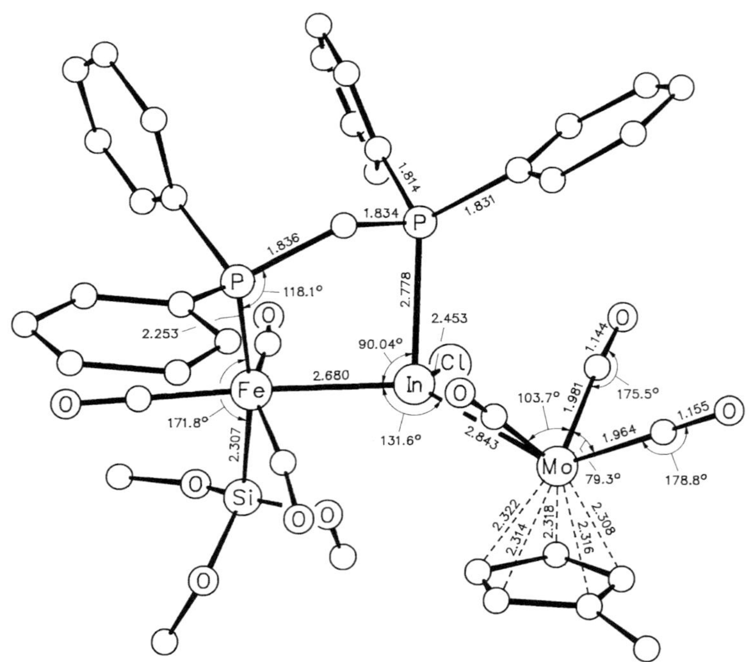

Fig. 25. The molecular structure of $CH_3C_5H_4Mo(CO)_3InCl(\mu\text{-}(C_6H_5)_2PCH_2P(C_6H_5)_2)Fe(CO)_3Si(OCH_3)_3$ [169].

References on pp. 253/8

$Mn(CO)_3C_6H_6C_5H_4Mo(CO)_3HgCl$ (Formula VIII). The lithio compound generated from tricarbonyl(exo-cyclopentadienyl-η^5-cyclohexadienyl)manganese, $C_5H_5C_6H_6Mn(CO)_3$, and LiC_4H_9-n/hexane in THF was refluxed with $Mo(CO)_3(NCCH_3)_3$ in the same solvent and subsequently treated with $HgCl_2$. Column chromatographic workup on alumina eluting with ether/hexane (1:10) gave a 72% yield [143].

Air-stable yellow crystals, m.p. 82°C (dec.). ^1H NMR spectrum ($CDCl_3$): δ = 3.22 (m, 3H), 4.96 ("t", 2H; "J" = 5.59 Hz), 5.24 ("s", C_5H_4), 5.80 (t, 1H; J = 5.37 Hz) ppm. IR spectrum (NaCl): 1911, 2000, 2007 (CO) cm^{-1} [143].

References:

[1] Reynolds, L. T.; Wilkinson, G. (J. Inorg. Nucl. Chem. **9** [1959] 86/92).
[2] Abel, E. W.; Singh, A.; Wilkinson, G. (J. Chem. Soc. **1960** 1321/4).
[3] Wilkinson, G.; Ethyl Corp. (U.S. 3109010 [1960/63] 4 pp.; C.A. **60** [1964] 14538).
[4] Mays, M. J.; Robb, J. D. (J. Chem. Soc. A **1968** 561/4).
[5] Conville, J. J., Jr. (Diss. Lehigh University, Bethlehem, Pa., 1972; Diss. Abstr. Int. B **33** [1973] 3530/1).
[6] Green, M.; Hughes, R. P. (J. Chem. Soc. Chem. Commun. **1975** 862/3).
[7] Eilbracht, P. (J. Organomet. Chem. **120** [1976] C37/C38).
[8] Eilbracht, P. (Chem. Ber. **109** [1976] 1429/35).
[9] Chipperfield, J. R.; Ford, J.; Hayter, A. C.; Webster, D. E. (J. Chem. Soc. Dalton Trans. **1976** 360/6).
[10] Gubitosa, G.; Brintzinger, H. H. (J. Organomet. Chem. **140** [1977] 187/93).

[11] Braun, S.; Dahler, P.; Eilbracht, P. (J. Organomet. Chem. **146** [1978] 135/41).
[12] Birdwhistell, R.; Hackett, P.; Manning, A. R. (J. Organomet. Chem. **157** [1978] 239/41).
[13] Nolte, M. J.; Reimann, R. H. (J. Chem. Soc. Dalton Trans. **1978** 932/7).
[14] Alper, H.; Paik, H.-N. (J. Am. Chem. Soc. **100** [1978] 508/12).
[15] Burckett-St. Laurent, J. C. T. R.; Field, J. S.; Haines, R. J.; McMahon, M. (J. Organomet. Chem. **181** [1979] 117/30).
[16] Jones, W. D.; Bergman, R. G. (J. Am. Chem. Soc. **101** [1979] 5447/9).
[17] Hart, W. P.; Macomber, D. W.; Rausch, M. D. (Org. Coat. Plast. Chem. **41** [1979] 47/51).
[18] Jones, W. D. (Report LBL-10214 [1979] 1/178; Energy Res. Abstr. **5** [1980] Abstr. No. 32658; C.A. **95** [1981] No. 6002).
[19] Eilbracht, P.; Dahler, P.; Mayser, U.; Henkes, E. (Chem. Ber. **113** [1980] 1033/46).
[20] Madach, T.; Vahrenkamp, H. (Chem. Ber. **113** [1980] 1675/85).

[21] Chaloyard, A.; El Murr, N. (Inorg. Chem. **19** [1980] 3217/20).
[22] Hart, W. P.; Macomber, D. W.; Rausch, M. D. (J. Am. Chem. Soc. **102** [1980] 1196/8).
[23] Frommer, J. E.; Bergman, R. G. (J. Am. Chem. Soc. **102** [1980] 5227/34).
[24] Huggins, J. M. (Report LBL-11697 [1980] 1/206; Energy Res. Abstr. **6** [1981] Abstr. No. 5132; C.A. **95** [1981] No. 41785).
[25] Frommer, J. E. (Report LBL-11698 [1980] 1/158; Energy Res. Abstr. **6** [1981] Abstr. No. 2654; C.A. **95** [1981] No. 115730).
[26] Huggins, J. M. (Diss. California Inst. Technol., Pasadena, CA, 1981; Diss. Abstr. Int. B **41** [1980] 2172).
[27] Lalor, F. J.; Condon, D.; Ferguson, G.; Khan, M. A. (Inorg. Chem. **20** [1981] 2178/81).
[28] Jones, W. D.; Huggins, J. M.; Bergman, R. G. (J. Am. Chem. Soc. **103** [1981] 4415/23).

[29] Frommer, J. E. (Diss. California Inst. Technol., Pasadena, CA, 1981; Diss. Abstr. Int. B **41** [1981] 3448/9).
[30] Eilbracht, P.; Totzauer, W. (Chem. Ber. **115** [1982] 1669/81).

[31] Koch, O.; Edelmann, F.; Lubke, B.; Behrens, U. (Chem. Ber. **115** [1982] 3049/62).
[32] Petrignani, J.-F.; Alper, H. (Organometallics **1** [1982] 1095/7).
[33] Casey, C. P.; Bullock, R. M.; Fultz, W. C.; Rheingold, A. L. (Organometallics **1** [1982] 1591/6).
[34] Macomber, D. W.; Hart, W. P.; Rausch, M. D.; Priester, R. D.; Pittman, C. U., Jr. (J. Am. Chem. Soc. **104** [1982] 884/6).
[35] Stiegman, A. E.; Tyler, D. R. (J. Am. Chem. Soc. **104** [1982] 2944/5).
[36] Macomber, D. W.; Spink, W. C.; Rausch, M. D. (J. Organomet. Chem. **250** [1983] 311/8).
[37] Casey, C. P.; Bullock, R. M. (J. Organomet. Chem. **251** [1983] 245/8).
[38] Macomber, D. W.; Rausch, M. D. (J. Organomet. Chem. **258** [1983] 331/41).
[39] Adams, H.; Bailey, N. A.; Cahill, P.; Rogers, D.; Winter, M. J. (J. Chem. Soc. Chem. Commun. **1983** 831/3).
[40] Bailey, N. A.; Chell, P. L.; Manuel, C. P.; Mukhopadhyay, A.; Rogers, D.; Tabbron, H. E.; Winter, M. J. (J. Chem. Soc. Dalton Trans. **1983** 2397/403).

[41] Bender, R.; Braunstein, P.; Jud, J. M.; Dusausoy, Y. (Inorg. Chem. **22** [1983] 3394/407).
[42] Macomber, D. W.; Rausch, M. D. (Organometallics **2** [1983] 1523/9).
[43] Vollhardt, K. P. C.; Weidman, T. W. (J. Am. Chem. Soc. **105** [1983] 1676/7).
[44] Stiegman, A. E.; Stieglitz, M.; Tyler, D. R. (J. Am. Chem. Soc. **105** [1983] 6032/7).
[45] Casey, C. P.; Bullock, R. M.; Nief, F. (J. Am. Chem. Soc. **105** [1983] 7574/80).
[46] Edelmann, F.; Behrens, U. (Chem. Ber. **117** [1984] 3463/72).
[47] Vollhardt, K. P. C.; Weidman, T. W. (Organometallics **3** [1984] 82/6).
[48] Goldman, A. S.; Tyler, D. R. (Organometallics **3** [1984] 449/56).
[49] Drage, J. S.; Tilset, M.; Vollhardt, K. P. C.; Weidman, T. W. (Organometallics **3** [1984] 812/4).
[50] Casey, C. P.; Bullock, R. M. (Organometallics **3** [1984] 1100/4).

[51] Braunstein, P.; De Meric de Bellefon, C.; Lanfranchi, M.; Tiripicchio, A. (Organometallics **3** [1984] 1772/4).
[52] Goldman, A. S.; Tyler, D. R. (J. Am. Chem. Soc. **106** [1984] 4066/7).
[53] Coolbaugh, T. S.; Santarsiero, B. D.; Grubbs, R. H. (J. Am. Chem. Soc. **106** [1984] 6310/8).
[54] Stiegman, A. E.; Tyler, D. R. (J. Photochem. **24** [1984] 311/4).
[55] Eilbracht, P.; Fassmann, W.; Diehl, W. (Chem. Ber. **118** [1985] 2314/29).
[56] Alt, H. G.; Engelhardt, H. E.; Thewalt, U.; Riede, J. (J. Organomet. Chem. **288** [1985] 165/77).
[57] Cano, M.; Criado, R.; Gutierrez-Puebla, E.; Monge, A.; Pardo, M. P. (J. Organomet. Chem. **292** [1985] 375/83).
[58] Cotton, J. D.; Kimlin, H. A. (J. Organomet. Chem. **294** [1985] 213/7).
[59] Coolbaugh, T. S.; Coots, R. J.; Santarsiero, B. D.; Grubbs, R. H. (Inorg. Chim. Acta **98** [1985] 99/105).
[60] Bogan, L. E., Jr.; Rauchfuss, T. B.; Rheingold, A. L. (Inorg. Chem. **24** [1985] 3720/2).

[61] Turaki, N. N.; Huggins, J. M. (Organometallics **4** [1985] 1766/9).
[62] Stiegman, A. E.; Tyler, D. R. (J. Am. Chem. Soc. **107** [1985] 967/71).
[63] Azar, M. C.; Chetcuti, M. J.; Eigenbrot, C.; Green, K. A. (J. Am. Chem. Soc. **107** [1985] 7209/10).

[64] Martin, B. D.; Matchett, S. A.; Norton, J. R.; Anderson, O. P. (J. Am. Chem. Soc. **107** [1985] 7952/9).
[65] Brownlee, R. T. C.; O'Connor, M. J.; Shehan, B. P.; Wedd, A. G. (J. Magn. Reson. **61** [1985] 516/20).
[66] Gonsalves, K.; Lin, Z.-R.; Lenz, R. W.; Rausch, M. D. (J. Polym. Sci.: Polym. Chem. Ed. **23** [1985] 1702/22).
[67] Perevalova, E.; Grandberg, K.; Smyslova, E.; Vasyukova, N.; Strunin, B. (12th Int. Conf. Organomet. Chem., Vienna 1985, Abstr. p. 246).
[68] Strunin, B. N.; Grandberg, K. I.; Andrianov, V. G.; Setkina, V. N.; Perevalova, E. G.; Struchkov, Yu. T.; Kursanov, D. N. (Dokl. Akad. Nauk SSSR **281** [1985] 599/602; Dokl. Chem. [Engl. Transl.] **280/285** [1985] 106/9).
[69] Stiegman, A. E.; Tyler, D. R. (Coord. Chem. Rev. **63** [1985] 217/40).
[70] Coolbaugh, T. S. (Diss. California Inst. Technol., Pasadena, CA, 1985; Diss. Abstr. Int. B **45** [1985] 3499).

[71] Bruce, A. E. (Diss. Columbia Univ., New York, NY, 1985; Diss. Abstr. Int. B **47** [1986] 191).
[72] Kreiter, C. G.; Michels, W.; Wenz, M. (Chem. Ber. **119** [1986] 1994/2005).
[73] Edelmann, F.; Töfke, S.; Behrens, U. (J. Organomet. Chem. **309** [1986] 87/108).
[74] Adams, H.; Bailey, N. A.; Cahill, P.; Rogers, D.; Winter, M. J. (J. Chem. Soc. Dalton Trans. **1986** 2119/26).
[75] Moulton, R.; Weidman, T. W.; Vollhardt, K. P. C.; Bard, A. J. (Inorg. Chem. **25** [1986] 1846/51).
[76] Philbin, C. E.; Goldman, A. S.; Tyler, D. R. (Inorg. Chem. **25** [1986] 4434/6).
[77] Philbin, C. E.; Granatir, C. A.; Tyler, D. R. (Inorg. Chem. **25** [1986] 4806/7).
[78] Huffman, M. A.; Newman, D. A.; Tilset, M.; Tolman, W. B.; Vollhardt, K. P. C. (Organometallics **5** [1986] 1926/8).
[79] Stiegman, A. E.; Tyler, D. R. (Comments Inorg. Chem. **5** [1986] 215/45).
[80] Töfke, S.; Behrens, U. (J. Organomet. Chem. **331** [1987] 229/38).

[81] Chetcuti, M. J.; Eigenbrot, C.; Green, K. A. (Organometallics **6** [1987] 2298/306).
[82] Vasyukova, N. I.; Tumanskii, B. L.; Strunin, B. N.; Grandberg, K. I.; Nekrasov, Yu. S.; Bubnov, N. N.; Solodovnikov, S. P.; Prokof'ev, A. I. (Izv. Akad. Nauk. SSSR Ser. Khim. **1987** 2702/8; Bull. Acad. Sci. USSR Div. Chem. Sci. [Engl. Transl.] **1987** 2504/9).
[83] Caballero, C.; Oberdorfer, F.; Guggolz, E.; Nuber, B.; Korswagen, R. P.; Ziegler, M. L. (Bol. Soc. Quim. Peru **53** [1987] 135/49).
[84] Wang, X.; Zhang, Z. (Jiegou Huaxue **6** [1987] 1/6; C.A. **107** [1987] No. 141448).
[85] Kubicki, M. M.; Le Gall, J. Y.; Pichon, R.; Salaun, J. Y.; Cano, M.; Campo, J. A. (J. Organomet. Chem. **348** [1988] 349/56).
[86] Sitzmann, H. (J. Organomet. Chem. **354** [1988] 203/14).
[87] Clegg, W.; Compton, N. A.; Errington, R. J.; Norman, N. C.; Tucker, A. J.; Winter, M. J. (J. Chem. Soc. Dalton Trans. **1988** 2941/51).
[88] Pan, X.; Philbin, C. E.; Castellani, M. P.; Tyler, D. R. (Inorg. Chem. **27** [1988] 671/6).
[89] Chetcuti, M. J.; Green, K. A. (Organometallics **7** [1988] 2450/7).
[90] Baker, K. B.; Bury, A. (Synth. React. Inorg. Met.-Org. Chem. **18** [1988] 529/34).

[91] Medina, R. M.; Masaguer, J. R. (An. Quim. Ser. B **84** [1988] 223/6).
[92] Härter, P.; Boguth, G.; Herdtweck, E.; Riede, J. (Angew. Chem. **101** [1989] 1058/9; Angew. Chem. Int. Ed. Engl. **28** [1989] 1008).
[93] Kreiter, C. G.; Wenz, M.; Michels, W. (Z. Naturforsch. **44b** [1989] 1247/59).
[94] Moriarty, K. J.; Rausch, M. D. (J. Organomet. Chem. **370** [1989] 75/80).
[95] Tenhaeff, S. C.; Tyler, D. R. (J. Chem. Soc. Chem. Commun. **1989** 1459/61).

[96] Clark, T. J.; Nile, T. A.; McPhail, D.; McPhail, A. D. (Polyhedron **8** [1989] 1804/6).
[97] Chetcuti, M. J.; Huffman, J. C.; McDonald, S. R. (Inorg. Chem. **28** [1989] 238/42).
[98] Silavwe, N. D.; Goldman, A. S.; Ritter, A. S.; Tyler, D. R. (Inorg. Chem. **28** [1989] 1231/6).
[99] Philbin, C. E.; Stiegman, A. E.; Tenhaeff, S. C.; Tyler, D. R. (Inorg. Chem. **28** [1989] 4414/7).
[100] Cervantes, J.; Vincenti, S. P.; Kapoor, R. N.; Pannell, K. H. (Organometallics **8** [1989] 744/8).

[101] Kovacs, I.; Sisak, A.; Ungvary, F.; Marko, L. (Organometallics **8** [1989] 1873/7).
[102] Chetcuti, M. J.; McDonald, S. R.; Rath, N. P. (Organometallics **8** [1989] 2077/9).
[103] Lo Sterzo, C.; Miller, M. M.; Stille, J. K. (Organometallics **8** [1989] 2331/7).
[104] Blum, T.; Braunstein, P.; Tiripicchio, A.; Tiripicchio Camellini, M. (Organometallics **8** [1989] 2504/13).
[105] Hanaya, M.; Iwaizumi, M. (Chem. Lett. **1989** 1381/4).
[106] Braunstein, P.; Bender, R.; Jud, J.-M. (Inorg. Synth. **26** [1989] 341/50).
[107] Boese, R.; Myrabo, R. L.; Newman, D. A.; Vollhardt, K. P. C. (Angew. Chem. **102** [1990] 589/92; Angew. Chem. Int. Ed. Engl. **29** [1990] 549).
[108] Kreiter, C. G.; Wenz, M.; Bell, P. (J. Organomet. Chem. **394** [1990] 195/211).
[109] Chetcuti, M. J.; Cunningham, P. N.; Gordon, J. C.; Grant, B. E.; Klaiss, J. (J. Organomet. Chem. **394** [1990] 765/75).
[110] Härter, P.; Kiprof, P. (J. Organomet. Chem. **397** [1990] C1/C5).

[111] Alt, H. G.; Maisel, H. E.; Han, J. S.; Wrackmeyer, B.; Razawi, A. (J. Organomet. Chem. **399** [1990] 131/9).
[112] Green, M. L. H.; Hubert, J. D.; Mountford, P. (J. Chem. Soc. Dalton Trans. **1990** 3793/800).
[113] Green, M. L. H.; Mountford, P.; Smout, G. J.; Speel, S. R. (Polyhedron **9** [1990] 2763/5).
[114] Chetcuti, M. J.; Gordon, J. C.; Fanwick, P. E. (Inorg. Chem. **29** [1990] 3781/7).
[115] Lo Sterzo, C.; Stille, J. K. (Organometallics **9** [1990] 687/94).
[116] Pannell, K. H.; Rozell, J. M.; Cortez, S.; Kapoor, R. (Organometallics **9** [1990] 1332/4).
[117] Chetcuti, M. J.; Grant, B. E.; Fanwick, P. E.; Geselbracht, M. J.; Stacy, A. M. (Organometallics **9** [1990] 1343/5).
[118] Mao, F.; Philbin, C. E.; Weakley, T. J. R.; Tyler, D. R. (Organometallics **9** [1990] 1510/6).
[119] D'Agostino, M. F.; Frampton, C. S.; McGlinchey, M. J. (Organometallics **9** [1990] 2972/84).
[120] Pannell, K. H.; Rozell, J. M.; Cortez, S.; Kapoor, R. (Organometallics **9** [1990] 3006).

[121] Medina, R. M.; Masaguer, J. R. (An. Quim. Ser. B **86** [1990] 733/6).
[122] Rogers, R. D.; Atwood, J. L.; Rausch, M. D.; Macomber, D. W. (J. Crystallogr. Spectrosc. Res. **20** [1990] 555/60).
[123] Boese, R.; Huffman, M. A.; Vollhard, K. P. C. (Angew. Chem. **103** [1991] 1542/4; Angew. Chem. Int. Ed. Engl. **30** [1991] 1463).
[124] Plenio, H. (Chem. Ber. **124** [1991] 2185/90).
[125] Lo Sterzo, C. (J. Organomet. Chem. **408** [1991] 253/9).
[126] Alt, H. G.; Han, J. S.; Maisel, H. E. (J. Organomet. Chem. **409** [1991] 197/205).
[127] Hanaya, H.; Iwaizumi, M. (J. Organomet. Chem. **417** [1991] 407/20).
[128] Clarkson, L. M.; Clegg, W.; Hockless, D. C. R.; Norman, N. C.; Farrugia, L. J.; Bott, S. G.; Atwood, J. L. (J. Chem. Soc. Dalton Trans. **1991** 2241/52).

[129] Mao, F.; Sur, S. K.; Tyler, D. R. (Organometallics **10** [1991] 419/23).
[130] Tenhaeff, S. C.; Tyler, D. R. (Organometallics **10** [1991] 473/82).

[131] Tenhaeff, S. C.; Tyler, D. R. (Organometallics **10** [1991] 1116/23).
[132] Chetcuti, M. J.; Fanwick, E. P.; McDonald, S. R.; Rath, N. N. (Organometallics **10** [1991] 1551/60).
[133] Chetcuti, M. J.; Fanwick, E. P.; McDonald, S. R.; Rath, N. N. (Organometallics **10** [1991] 3411).
[134] Sun, W.-H.; Yang, S.-Y.; Yin, Y.-Q.; Zhou, Z.-Y.; Yu, K.-B. (Youji Huaxue **11** [1991] 87/90; C.A. **114** [1991] No. 185686).
[135] Song, L.-C.; Dong, Q.; Hu, Q.-M. (Huaxue Xuebao **49** [1991] 1129/35; C.A. **116** [1992] No. 129168).
[136] Sun, W.-H.; Yang, S.-Y.; Yin, Y.-Q.; Yu, K.-B.; Zhou, Z.-Y. (Chin. Chem. Lett. **2** [1991] 263/6).
[137] Wang, T.-F.; Lee, T.-Y.; Chou, J.-W.; Ong, C.-W. (J. Organomet. Chem. **423** [1992] 31/8).
[138] Härter, P.; Behm, J.; Burkert, K. J. (J. Organomet. Chem. **438** [1992] 297/307).
[139] Duckworth, T. J.; Mays, M. J.; Conole, G.; McPartlin, M. M. (J. Organomet. Chem. **439** [1992] 327/39).
[140] Doyle, M. J.; Duckworth, T. J.; Manojlovic-Muir, Lj.; Mays, M. J.; Raithby, P. R.; Robertson, F. J. (J. Chem. Soc. Dalton Trans. **1992** 2703/14).

[141] Sun, W.-H.; Yang, S.-Y.; Wang, H.-Q.; Yin, Y.-Q.; Yu, K.-B. (Polyhedron **11** [1992] 1143/4).
[142] Chetcuti, M. J.; Deck, K. J.; Gordon, J. C.; Grant, B. E.; Fanwick, P. E. (Organometallics **11** [1992] 2128/39).
[143] Chung, T.-M.; Chung, Y. K. (Organometallics **11** [1992] 2822/6).
[144] Avey, A.; Tyler, D. R. (Organometallics **11** [1992] 3856/63).
[145] Wang, P.; Atwood, J. D. (J. Am. Chem. Soc. **114** [1992] 6424/7).
[146] Coville, N. J.; Loonat, M. S.; Carlton, L. (S. Afr. J. Chem. **45** [1992] 63/72).
[147] Song, L.-C.; Dong, Q.; Hu, Q.-M. (Huaxue Xuebao **50** [1992] 193/9; C.A. **116** [1992] No. 214648).
[148] Zhou, X.-Z.; Xu, S.-S.; Kuang, D.-Z. (Chin. Chem. Lett. **3** [1992] 1015/6).
[149] Scheer, M.; Friedrich, G.; Schuster, K. (Angew. Chem. **105** [1993] 641/3; Angew. Chem. Int. Ed. Engl. **32** [1993] 593).
[150] Alt, H. G.; Han, J. S.; Rogers, R. D. (J. Organomet. Chem. **445** [1993] 115/24).

[151] Knox, G. R.; Pauson, P. L.; Willison, D. (J. Organomet. Chem. **450** [1993] 177/84).
[152] Brumas, B.; Dahan, F.; de Montauzon, D.; Poilblanc, P. (J. Organomet. Chem. **453** [1993] C13/C14).
[153] Alt, H. G.; Han, J. S.; Rogers, R. D. (J. Organomet. Chem. **454** [1993] 165/72).
[154] Avey, A.; Schut, D. M.; Weakley, T. J. R.; Tyler, D. R. (Inorg. Chem. **32** [1993] 233/6).
[155] Bartlone, A. F.; Chetcuti, M. J.; Fanwick, P. E.; Haller, K. J. (Inorg. Chem. **32** [1993] 1435/41).
[156] Brumas, B.; de Caro, D.; Dahan, F.; de Montauzon, D.; Poilblanc, R. (Organometallics **12** [1993] 1503/5).
[157] Avey, A.; Weakley, T. J. R.; Tyler, D. R. (J. Am. Chem. Soc. **115** [1993] 7706/15).
[158] Kuang, D.-Z.; Zhou, X.-Z. (Huaxue Tongbao **1993** 30/1; C.A. **119** [1993] No. 226118).
[159] Kuang, D.-Z.; Zhou, X.-Z. (Huaxue Tongbao **1993** 37/9; C.A. **121** [1994] No. 35772).
[160] Wang, J.-X.; Chen, S.-S.; Wang, X.-K.; Wang, H.-G. (Huaxue Xuebao **51** [1993] 720/6; C.A. **120** [1994] No. 30867).

[161] Kuang, D.-Z.; Zhou, X.-Z.; Xu, S.-S. (Huaxue Xuebao **51** [1993] 1035/40; C.A. **120** [1994] No. 298836).
[162] Sun, W.-H.; Yang, S.-Y.; Yin, Y.-Q.; Wang, H.-Q.; Yu, K.-B. (Chin. J. Chem. **11** [1993] 151/8).
[163] Brown, D. S.; Delville-Desbois, M.-H.; Boese, R.; Vollhardt, K. P. C.; Astruc, D. (Angew. Chem. **106** [1994] 715/7; Angew. Chem. Int. Ed. Engl. **33** [1994] 661).
[164] Haupt, H.-J.; Merla, A.; Flörke, U. (Z. Anorg. Allg. Chem. **620** [1994] 999/1005).
[165] Sun, W.-H.; Yang, S.-Y.; Wang, H.-Q.; Zhou, Q.-F.; Yu, K.-B. (J. Organomet. Chem. **465** [1994] 263/5).
[166] Kahn, P. A.; Boese, R.; Blümel, J.; Vollhardt, K. P. C. (J. Organomet. Chem. **472** [1994] 149/62).
[167] El Mouatassim, B.; El Amouri, H.; Salmain, M.; Jaouen, G. (J. Organomet. Chem. **479** [1994] C18/C20).
[168] Stadelmann, B.; Lassacher, P.; Stüger, H.; Hengge, E. (J. Organomet. Chem. **482** [1994] 201/6).
[169] Braunstein, P.; Knorr, M.; Strampfer, M.; DeCian, A.; Fischer, J. (J. Chem. Soc. Dalton Trans. **1994** 117/34).
[170] Sun, W.-H.; Wang, H.-Q.; Yang, S.-Y.; Zhou, Q.-F.; Yu, K.-B. (Polyhedron **13** [1994] 389/94).
[171] Song, L.-C.; Tao, Z.-F.; Hu, Q.-M.; Wang, R.-J.; Wang, H.-G. (Polyhedron **13** [1994] 2179/84).
[172] Song, L.-C.; Shen, J.-Y.; Wang, J.-Q.; Hu, Q.-M.; Wang, R.-J.; Wang, H.-G. (Polyhedron **13** [1994] 3235/41).
[173] Wu, H.-P.; Yin, Y.-Q.; Jin, D.-S. (Chin. Chem. Lett. **5** [1994] 889/92).
[174] Wu, H.-P.; Yin, Y.-Q.; Huang, X.-Y. (Chin. Chem. Lett. **5** [1994] 987/90).
[175] Wu, H.-P.; Zhao, Z.-Y.; Yin, Y.-Q.; Jin, D.-S.; Huang, X.-Y. (Polyhedron **14** [1995] 1543/6).
[176] Song, L.-C.; Shen, J.-Y.; Hu, Q.-H.; Huang, X.-Y. (Organometallics **14** [1995] 98/106).
[177] Avey, A.; Nieckarz, G. F.; Keana, K.; Tyler, D. R. (Organometallics **14** [1995] 2790/98).
[178] Spadoni, L.; Lo Sterzo, C.; Crescenzi, P.; Frachey, G. (Organometallics **14** [1995] 3149/51).

1.5.1.4.2.2 Compounds Containing the Fragment $R_2C_5H_3Mo(CO)_3$

1.5.1.4.2.2.1 Compounds with Indenyl Ligands

The most common methods of synthesis used in preparations of $Mo(CO)_3$ complexes possessing unsubstituted or substituted indenyl ligands, C_9H_7 (Formula I) and $R_nC_9H_{7-n}$, respectively, are given in Methods I to VII. Known derivatives are listed in Table 7. In addition, the salt **$[C_2(N(CH_3)_2)_4][C_9H_7Mo(CO)_3]_2$** was briefly mentioned. The complex was detected in a small quantity in the insoluble brown material resulting from the treatment of $(C_9H_7Mo(CO)_3)_2$ with tetrakis(dimethylamino)ethene in THF solution, but no further information was given [3].

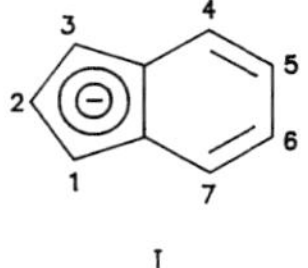

I

References on pp. 268/9

Method I: Equimolar quantities of $Li[C_9H_7]$ and $Mo(CO)_6$ were heated in N,N-dimethylformamide at reflux temperature until the evolution of carbon monoxide had ceased. Solvent was then removed and the red oily residue dissolved in THF to give a solution of crude $Li[C_9H_7Mo(CO)_3]$ suitable for further high-yield preparations without purification [14]. In a modified version, refluxing THF was used as the reaction medium from the beginning [32]. In contrast, none of the hoped-for $Na[C_9H_7Mo(CO)_3]$ was formed when $Na[C_9H_7]$ and $Mo(CO)_6$ were combined in boiling THF [2], and the potassium salt $K[C_9H_7Mo(CO)_3]$ was only obtained in low purity from the hexacarbonyl and $K[C_9H_7]$ in refluxing 1,2-dimethoxyethane [15, 19]. For that reason, this method was only used to prepare the mercury derivative $(C_9H_7Mo(CO)_3)_2Hg$ to serve as a starting material for procedure III by precipitating the in situ-generated potassium salt with aqueous $Hg(CN)_2$ [15, 19].

Method II: From $(C_9H_7Mo(CO)_3)_2$ or $(5,6\text{-}R_2C_9H_5Mo(CO)_3)_2$ by reduction:

a. Reduction of $(C_9H_7Mo(CO)_3)_2$ with sodium amalgam in THF at room temperature gave a dark brown solution containing the $[C_9H_7Mo(CO)_3]^-$ anion as demonstrated by the reaction with CH_3I to produce $C_9H_7Mo(CO)_3CH_3$ (see below) [2, 4, 5]. $(5,6\text{-}(CH_3O)_2C_9H_5Mo(CO)_3)_2$ behaved similarly when stirred with sodium sand in THF at ambient conditions [7].

b. The homodimer was added to a suspension of Na/K alloy in THF and the mixture was stirred until the characteristic red-violet color of $(C_9H_7Mo(CO)_3)_2$ had disappeared. The green-yellow solution of $K[C_9H_7Mo(CO)_3]$ was then filtered to remove excess alloy. No further purification was required for further use in subsequent syntheses [15].

Method III: $K[C_9H_7Mo(CO)_3]$ could be generated in high yield and purity in solution by reducing $(C_9H_7Mo(CO)_3)_2Hg$ with Na/K alloy in THF as summarized above for $(C_9H_7Mo(CO)_3)_2$. The process was monitored by following the IR spectrum [19].

Method IV: $Ge(C_2H_5)_3Br$ or $Sn(CH_3)_3Cl$ were added in equimolar portions to THF solutions of $K[C_9H_7Mo(CO)_3]$, prepared by Na/K reduction of $(C_9H_7Mo(CO)_3)_2Hg$. While the reaction with $Ge(C_2H_5)_3Br$ proceeded at 25 °C, the reaction with $Sn(CH_3)_3Cl$ required heating at 40 to 50 °C for completion. After removal of solvent, $C_9H_7Mo(CO)_3Ge(C_2H_5)_3$ was obtained as an impure, thermally unstable oil by extraction of the residue with hexane and evaporation of volatile material. $C_9H_7Mo(CO)_3Sn(CH_3)_3$ could by crystallized from the hexane extract at reduced temperature [19]. $C_9H_7Mo(CO)_3Sn(C_6H_5)_3$ was prepared by treating in situ-generated $Li[C_9H_7Mo(CO)_3]$ in THF with $Sn(C_6H_5)_3Cl$ in 1:1 stoichiometry in THF at ambient conditions. The reaction mixture was repeatedly filtered through alumina before removing the solvent. Dissolution of the resulting oil in light petroleum/CH_2Cl_2, followed by evaporation of the volatiles, yielded the product as a microcrystalline powder [32].

Method V: The reaction of $(C_9H_7Mo(CO)_3)_2$ with iodine had been originally reported to unexpectedly give the 16e complex $C_9H_7Mo(CO)_2I$ [2, 5, 6]. Repeating that preparation, later authors [10, 34] showed, however, that the expected tricarbonyl species, $C_9H_7Mo(CO)_3I$, was actually formed: A mixture of $(C_9H_7Mo(CO)_3)_2$ and I_2 (molar ratio 1:1 to 1:2) was stirred in CH_2Cl_2 at room temperature. After completion of the reaction, monitored by IR spectroscopy, unreacted iodine was removed by shaking the organic phase with aqueous $Na_2S_2O_3 \cdot H_2O$ as reported previously [2, 34] or by chromatography on silica gel of the evaporated mixture, using CH_2Cl_2/petroleum as the eluant, as recommended later [10]. Further puri-

References on pp. 268/9

fication was achieved by recrystallization from CH_2Cl_2/hexane [2, 10]. Using the same technique described above, 5,6-$(CH_3O)_2C_9H_5Mo(CO)_3I$ was obtained from $(5,6\text{-}(CH_3O)_2C_9H_5Mo(CO)_3)_2$ by reacting with iodine [11].

Method VI: $(C_9H_7Mo(CO)_3)_2$ was irradiated in $CHCl_3$ or $CHBr_3$ under CO at 30 °C and volatile material was then removed. The residues were chromatographed on silica gel eluting with CH_2Cl_2/petroleum. Further purification by recrystallization from CH_2Cl_2/hexane afforded the halo complexes $C_9H_7Mo(CO)_3X$ (X = Cl, Br) in an analytically pure state [10].

Method VII: Excess CH_3I was added to the THF solutions of $M[C_9H_7Mo(CO)_3]$ ($M^+ = Li^+$ [14], Na^+ [2], K^+ [15]) generated as outlined above (see Methods I and II). After stirring at ambient conditions and removal of volatile material, the product, $C_9H_7Mo(CO)_3CH_3$, was isolated from the residues by extraction with CH_2Cl_2 [2] or benzene [14, 15], followed by renewed evaporation of the filtered extracts, extraction of the residuals with pentane or hexane, and low-temperature crystallization [2, 14, 15]. With a decreased yield, the methyl complex could also be purified by vacuum sublimation at 70 to 100 °C/0.1 Torr [2]. 5,6-$(CH_3O)_2C_9H_5Mo(CO)_3CH_3$ was made similarly by adding excess CH_3I to the THF solution of $Na[5,6\text{-}(CH_3O)_2C_9H_5Mo(CO)_3]$, in situ-generated from the corresponding homodimer by reduction. The complex was separated from the evaporated reaction mixture by extraction with petroleum ether and subsequent vacuum sublimation at 100 to 110 °C/0.1 Torr [7].

Table 7
Compounds with Indenyl Ligands.
An asterisk indicates further information at the end of the table.
For explanations, abbreviations, and units see p. X.

No.	compound	method of preparation (yield) properties and remarks
$M[C_9H_7Mo(CO)_3]$ and $M[5,6\text{-}R_2C_9H_5Mo(CO)_3]$ salts:		
M^+/R =		
*1	Li^+/H	I (not isolated) [14, 32]; for other methods used to generate solutions of the lithium salt, see under "Further information"
*2	Na^+/H	IIa (>20%; not isolated) [2, 4, 5]; see also under "Further information" IR (ether): 1738, 1790, 1920 (CO) [21]
*3	K^+/H	I [15, 19]; IIb [15]; III [19] (the latter two representing the preferred procedures [15]; not isolated) IR (THF): 1766, 1804, 1908 (CO) [15]
4	Na^+/CH_3O	IIa (not isolated) [7] treatment with CH_3I gave 5,6-$(CH_3O)_2C_9H_5Mo(CO)_3CH_3$ [7]
$C_9H_7Mo(CO)_3ER_n$ (E = Ge, Sn):		
ER_3 =		
5	$Ge(C_2H_5)_3$	IV (ca. 40%) [19] oil (not obtained in pure form) [19]

References on pp. 268/9

Table 7 (continued)

No.	compound	method of preparation (yield) properties and remarks
		^{1}H NMR ($CDCl_3$): 5.38 (t, H-2; J = 3), 5.78 (d, H-1,3), 7.07, 7.52 (AA'BB' system, H-4,7, H-5,6) [19] IR (hexane): 1911, 1936, 2006 (CO) [19]
6	$Sn(CH_3)_3$	IV (52%) [19] m.p. 92 to 93 °C; sublimed at 90 °C/0.1 Torr [19] ^{1}H NMR ($CDCl_3$): 0.48 (s, CH_3), 5.26 (t, H-2; J = 3), 5.75 (d, H-1,3), 7.06, 7.47 (AA'BB' system, H-4,7, H-5,6) [19] IR (hexane): 1910, 1937, 2008 (CO) [19]
7	$Sn(C_6H_5)_3$	IV (71%) [32] yellow solid; m.p. 143 to 145 °C [32] ^{1}H NMR ($CDCl_3$): 5.12 (t, H-2; J = 3), 5.85 (d, H-1,3), 7.04, 7.24 (both m, H-4,7, H-5,6), 7.38 (m, meta- and para-H of C_6H_5), 7.61 (m, ortho-H of C_6H_5) [32] ^{13}C NMR ($CDCl_3$): 79.7, 93.1, 108.8, 123.6, 125.6 (all C_9H_7), 128.3 (meta-C of C_6H_5), 128.6 (para-C of C_6H_5), 136.5 (ortho-C of C_6H_5), 141.9 (ipso-C of C_6H_5), 223.6 (CO_{cis}), 227.8 (CO_{trans}) [32] IR: 1909, 1937, 2005 (CO) [32] treatment with LiC_6H_5 in ether, followed by hydrolysis and alkylation with $[O(C_2H_5)_3][BF_4]$, resulted in formation of the carbene product trans-$C_9H_7Mo(CO)_2(=C(OC_2H_5)C_6H_5)Sn(C_6H_5)_3$ [30, 32]

halo complexes $C_9H_7Mo(CO)_3X$ and 5,6-$R_2C_9H_5Mo(CO)_3X$:

No.	compound	method of preparation (yield) properties and remarks
*8	Cl/H	VI (45%) [10] orange crystals, m.p. 98 °C (dec.) [10] ^{1}H NMR (acetone-d_6): 5.82 (t, H-2), 6.19 (d, H-1,3) [10] IR ($CHCl_3$): 1963, 1995, 2063 (CO) [10]
*9	Br/H	VI (45%) [10] red crystals, m.p. 98 °C (dec.) [10] ^{1}H NMR (acetone-d_6): 5.82 (t, H-2), 6.28 (d, H-1,3) [10] IR (pentane): 1957, 1988, 2052 [10]
*10	I/H	V (19%) [2, 10], (72%) [34]; erroneously formulated as "$C_9H_7Mo(CO)_2I$" in [2, 5, 6]; see [10] dark reddish brown plates [2, 10, 13], m.p. 141 to 144 °C [13], 142 to 145 °C (dec.) [2], 143 to 144 °C (dec.) [10] ^{1}H NMR (C_6D_6): 4.16 (H-2), 4.95 (H-1,3) [34]; (acetone-d_6): 5.79 (t, H-2), 6.32 (d, H-1,3), 7.36 (H-4,7), 7.68 (H-5,6) [10]; similar in [2], additionally giving J(H-4,5) = J(H-6,7) = 7 and J(H-4,6) = J(H-5,7) = 3 IR (KBr): 3020 (ν(CH)); fingerprint bands at 751, 820, 845, 854, 950, 1029, 1040, 1200, 1240, 1330, 1380, 1435, 1477, 1520, and 1600 [2]; (halocarbon mull): 1958, 2028 (CO) [10]; almost identical in [2]; (n-pentane): 1957, 1981, 2043 (CO) [10]; similar in [13]; (CH_2Cl_2): 1967, 2020 (CO) [34] UV (1,4-dioxane, ε) = 213 (55500) [2]

References on pp. 268/9

Table 7 (continued)

No.	compound	method of preparation (yield) properties and remarks
11	I/CH_3O	V [11] the difference in rate at 30°C between the reactions of $C_9H_7Mo(CO)_3I$ and No. 11 with PR_3 ligands to give $C_9H_7Mo(CO)_2(PR_3)I$ and 5,6-$(CH_3O)_2C_9H_5Mo(CO)_2(PR_3)I$, respectively, was only a factor of 3, indicating that no substantial charge migration occurred during CO substitution [11]

$C_9H_7Mo(CO)_3R$, 1-R′$C_9H_6Mo(CO)_3R$, and 5,6-$R''_2C_9H_5Mo(CO)_3R$:

R/R′/R″ =

No.	compound	method of preparation (yield) properties and remarks
*12	CH_3/-/-	VII (19%) [2], (57%) [15], (80 to 85%) [14] yellow-orange volatile crystals [2, 4, 5, 15] subliming at 70 to 100°C/0.1 Torr [2], m.p. 91 to 93°C (dec.) [2, 4] ^1H NMR (CS_2): −0.43 (CH_3), 5.35 (H-2), 5.69 (H-1,3), 7.18 (H-4 to H-7) [2]; (acetone-d_6): −0.40 ($MoCH_3$), 5.57 (t, H-2), 5.97 (d, H-1,3), 7.27 (m, H-4 to H-7) [14] IR (KBr): 2860, 2950, 3050 (ν(CH)); fingerprint bands at 750, 815, 895, 1027, 1038, 1150, 1205, 1335, 1380, 1445, 1475, 1520, and 1605 [2]; (halocarbon oil mull): 1911, 1945, 2024 (CO) [2]; (hexane): 1935, 1942, 2020 (CO) [14]; (cyclohexane): 1941 (A″), 1949 (A′), 2016 (A′); for values of the Cotton-Kraihanzel stretching force constants, see [9]; (THF): 1910, 1944, 2023 (CO) [7] UV (cyclohexane, ε) = 222 (43400) [2]
13	CH_3/-/CH_3O	VII (25%) [7] yellow needles subliming at 100 to 110°C/ 0.1 Torr [7] in the reaction with $P(OCH_3)_3$ in THF, methyl migration with formation of the acetyl derivative 5,6-$(CH_3O)_2C_9H_5Mo(CO)_2(P(OCH_3)_3)C(O)CH_3$ was only slightly slower than with the unsubstituted indenyl complex $C_9H_7Mo(CO)_3CH_3$, thus ruling out the possibility that appreciable charge separation in the indenyl system occurred during this migratory insertion [7]
14	$-CH_2CH_2-$/-	a mixture of spirocyclopropane-1,1-indene (excess), $Mo(CO)_3(NC_5H_5)_3$, and $(C_2H_5)_2O \cdot BF_3$ in ether was stirred at room temperature, then diluted with hexane, and decomposed with water; the organic layer was chromatographed on a Florisil column eluting with hexane; low-temperature crystallization afforded No. 14 in 6% yield [27] orange-red crystals, m.p. 70 to 71°C [27] ^1H NMR (C_6D_6): 0.6 (m, $MoCH_2$), 2.78, 3.31 (both m, both CCH_2C), 4.70, 5.12 (AB pattern, H-2,3; J = 2), 6.96 (m, C_6H_4) [27] IR (pentane): 1930, 1957, 2022 (CO) [27]

*Further information:

$M[C_9H_7Mo(CO)_3]$ (M^+ = Li^+, Na^+, K^+; Table **7**, Nos. **1** to **3**). Solutions of the lithium salt were also generated by deprotonation with LiC_4H_9-n/hexane of the ^6L indene complex

$C_9H_8Mo(CO)_3$ in diethyl ether at −70 °C. The initial formation of the ^{6}L anion was accompanied by the deposition of a red precipitate. When the temperature was raised gradually to −15 °C, the precipitate dissolved and CO stretching bands appeared in the IR spectrum which were assigned to the Li^+ salt of the rearranged ^{5}L anion. These absorptions disappeared rapidly, even when the solution was stored below 0 °C [21].

The enthalpy of reaction of 1,4-$(CH_3)_2C_6H_4Mo(CO)_3$ with $Na[C_9H_7]$ forming $Na[C_9H_7Mo(CO)_3]$ was measured in THF at 30 °C as $\Delta H = -17.3$ kcal/mol, approximately 10 kcal/mol less negative than the corresponding reaction with $Na[C_5H_5]$ [33].

The results of comparative studies of the reactions of the salts $K[C_9H_7M(CO)_3]$ (M = Cr, Mo, W) with electrophiles suggest that with chromium as the metal center, ^{6}L indene structures of the products are, as a rule, thermodynamically favored over the ^{5}L indenyl structures, while stabilities of the ^{5}L derivatives increase on going first to molybdenum and then to tungsten. Thus, the tungsten salt $K[C_9H_7W(CO)_3]$ was observed to react with acetic acid in THF forming a stable hydride, $C_9H_7W(CO)_3H$, whereas the corresponding molybdate, $K[C_9H_7Mo(CO)_3]$, behaved like its chromium analogue giving, on protonation, an exceedingly unstable ^{6}L indene compound, $C_9H_8Mo(CO)_3$ [15]; see also [33].

Reactions of $K[C_9H_7Mo(CO)_3]$ with $Ge(C_2H_5)_3Br$ and $Sn(CH_3)_3Cl$ in 1:1 stoichiometry in THF afforded the expected Mo−Ge- and Mo−Sn-bonded heterobimetallics, $C_9H_7Mo(CO)_3Ge(C_2H_5)_3$ and $C_9H_7Mo(CO)_3Sn(CH_3)_3$, respectively [19]. $C_9H_7Mo(CO)_3Sn(C_6H_5)_3$ was obtained analogously by treating a THF solution of $Li[C_9H_7Mo(CO)_3]$ with one equivalent of $Sn(C_6H_5)_3Cl$ [32].

In contrast to the anions $[RC_5H_4Mo(CO)_3]^-$ (R = H, CH_3, $CH_2C_6H_5$) that underwent smooth oxidation to $(RC_5H_4Mo(CO)_3)_2$ on treatment with aqueous $Fe_2(SO_4)_2$, the indenyl-substituted anion $[C_9H_7Mo(CO)_3]^-$ did not give the expected homodimer $(C_9H_7Mo(CO)_3)_2$ on oxidation with Fe^{3+} ion, but instead afforded small amounts of $Mo(CO)_6$ as the only isolated product [16]. On the other hand, oxidation of $K[C_9H_7Mo(CO)_3]$ using $[(C_5H_5)_2Fe]BF_4$ as a one-electron oxidant did afford the homodimer, presumably via initial oxidation of the anion to the $C_9H_7Mo(CO)_3^{\bullet}$ radical, followed by radical pairing [19].

The reaction of $K[C_9H_7Mo(CO)_3]$ with $Hg(CN)_2$ in THF or 1,2-dimethoxyethane led to $(C_9H_7Mo(CO)_3)_2Hg$ [15, 19].

The THF solutions of the lithium, sodium, and potassium salts of the $[C_9H_7Mo(CO)_3]^-$ anion, prepared either from $Mo(CO)_6$ by substitution, or from $(C_9H_7Mo(CO)_3)_2$ by reduction, reacted with CH_3I to form the expected $C_9H_7Mo(CO)_3CH_3$ [2, 4, 5, 14, 15]. However, combination of CH_2=$CHCH_2Cl$, CH_2=CHCH=$CHCH_2Cl$, and CH_3SCH_2Cl with the $[C_9H_7Mo(CO)_3]^-$ anion under similar conditions gave dicarbonyls, $C_9H_7Mo(CO)_2(\eta^3\text{-}CH_2CHCH_2)$, $C_9H_7Mo(CO)_2(\eta^3\text{-}CH_2CHCHCH{=}CH_2)$, and $C_9H_7Mo(CO)_2(\eta^2\text{-}CH_2SCH_3)$, in unusually facile decarbonylation reactions [2, 4, 5, 29, 31].

$(C_9H_7Mo(CO)_2)_2$ was readily synthesized by oxidation of $Li[C_9H_7Mo(CO)_3]$ with $[C(C_6H_5)_3]Cl$ in THF and subsequent thermolysis (toluene, 111 °C) of the $(C_9H_7Mo(CO)_3)_2$ intermediate so generated [24]. $[C_6H_5N_2][BF_4]$ reacted smoothly with equimolar $K[C_9H_7Mo(CO)_3]$ in THF to form a phenyldiazonium derivative of the ^{5}L indenyl series, $C_9H_7Mo(CO)_2N_2C_6H_5$, even at temperatures as low as −60 °C [19]. While $C_9H_7W(CO)_3C_6H_5$ could be isolated from reactions of the tungstate $K[C_9H_7W(CO)_3]$ with $[I(C_6H_5)_2]X$ ($X^- = I^-$, $[BF_4]^-$) in THF, no products could be obtained from similar reactions of $K[C_9H_7M(CO)_3]$ (M = Cr, Mo) with diphenyliodonium salts [19].

$C_9H_7Mo(CO)_3X$ (X = Cl, Br, I; Table **7**; Nos. **8** to **10**). The red-brown plates of $C_9H_7Mo(CO)_3I$ grown from CH_2Cl_2/hexane were triclinic, space group $P\bar{1}-C_i^1$ (No. 2) with a = 8.61, b =

References on pp. 268/9

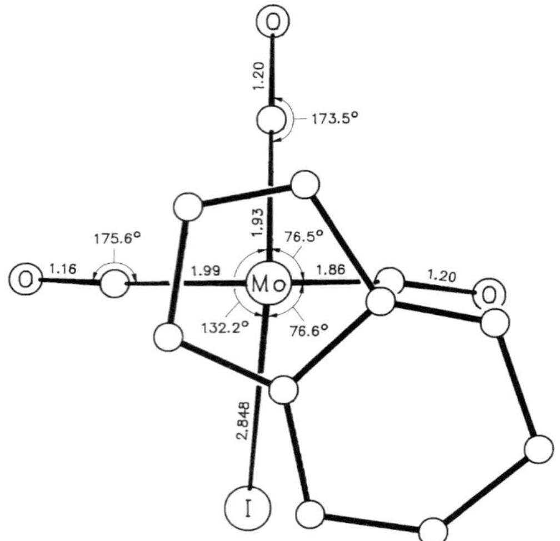

Fig. 26. The molecular structure of $C_9H_7Mo(CO)_3I$ [13].

9.61, c = 8.06 Å, α = 105.80°, β = 102.69°, γ = 86.58°; Z = 2 molecules per unit cell, D_{meas} = 2.18, D_{calc} = 2.21 g/cm^3. A view of the molecule including bond lengths and angles is shown in **Fig. 26**. The indenyl ligand does not deviate significantly from planarity, its best mean plane being approximately parallel to the plane defined by the carbon atoms in the three carbonyl groups [13]. For ring-slip parameters for this compound and related indenyl complexes, see [23].

Heating the three halides under high vacuum at 100°C converted them to new compounds believed to be $(C_9H_7Mo(CO)_2(\mu\text{-}X))_2$ (X = Cl, Br, I) [10]. Accordingly, the mass spectra of all three complexes were generally complicated by the formation of binuclear fragments caused by thermal decomposition of the halo complexes under the conditions necessary to obtain the spectra [10]. Also the fragmentation pattern reported in [8] for "$C_9H_7Mo(CO)_2I$" was shown to actually be for $C_9H_7Mo(CO)_3I$ [10].

All three compounds $C_9H_7Mo(CO)_3X$ reacted with PR_3 ligands in solvents such as benzene, CH_3CN, acetone, THF, CH_2Cl_2, or $CHCl_3$ to form monosubstituted complexes, cis- and trans-$C_9H_7Mo(CO)_2(PR_3)X$ (PR_3 = $P(C_6H_5)_2H$, $P(C_4H_9\text{-}n)_3$, $P(CH_2C_6H_5)_3$, $P(C_6H_5)_2CH_3$, $P(C_6H_5)_3$, $P(OCH_3)_3$, $P(OC_2H_5)_3$, $P(OC_3H_7\text{-}i)_3$, $P(OC_6H_5)_3$) [10, 11, 34]. Reactions of the bromo and iodo compound with $P(C_4H_9\text{-}n)_3$ additionally yielded small amounts of cis-$Mo(CO)_4(P(C_4H_9\text{-}n)_3)_2$ and fac-$Mo(CO)_3(P(C_4H_9\text{-}n)_3)_3$ [10]. Observed rate constants for the reactions of the bromo and iodo compound could be fitted to the expression $k_{obs} = k_A + k_B[PR_3]$, where, for a given complex, solvent, and temperature, k_A was found to be independent of the nature and the concentration of the PR_3 ligand, but the values of k_B increased in the order $PR_3 = P(OC_6H_5)_3 < P(C_6H_5)_3 < P(OCH_3)_3 < P(C_4H_9\text{-}n)_3$. $C_9H_7Mo(CO)_3Br$ reacted faster than $C_9H_7Mo(CO)_3I$ (the sluggish reactions of which were catalyzed by $(C_5H_5Fe(CO)_2)_2$ [34]) and the reactions of $C_9H_7Mo(CO)_3Cl$ were too rapid at room temperature to be studied kinetically. The rate expression was attributed to two competing mechanisms being operative, one (S_N1, k_A term) involving a rate-determining loss of carbon monoxide followed by rapid uptake of PR_3, and the other (S_N2, k_B term) involving simultaneous attack by PR_3 and the loss of carbon monoxide. The values of k_A for the indenyl complexes, C_9H_7-

References on pp. 268/9

$Mo(CO)_3X$, were several thousand times greater than those of the corresponding cyclopentadienyl complexes, $C_5H_5Mo(CO)_3X$ (fitting the simpler expression $k_{obs} = k_A$), due to a drop in $\Delta H^‡$ by several kcal/mol [11]. Pertinent activation parameters are given in the following table:

X/PR_3	first-order (k_A)		second-order (k_B)	
	$\Delta H^‡$ (kcal/mol)	$\Delta S^‡$ ($cal \cdot mol^{-1} \cdot K^{-1}$)	$\Delta H^‡$ (kcal/mol)	$\Delta S^‡$ ($cal \cdot mol^{-1} \cdot K^{-1}$)
$Br/P(C_6H_5)_2$	23.2 ± 0.5	$+7.6 \pm 0.5$	13.5 ± 0.6	-26.7 ± 0.5
$Br/P(OCH_3)_3$			12.8 ± 0.6	-24.8 ± 0.7
$I/P(OCH_3)_3$			13.7 ± 0.8	-25 ± 1
$I/P(OC_6H_5)_3$	25.5 ± 0.4	$+8.4 \pm 0.6$	13.3 ± 0.7	-35.7 ± 0.8

The reaction of $[Re(CO)_5]^-$ with $C_9H_7Mo(CO)_3Br$ giving $[C_9H_7Mo(CO)_3]^-$ together with $Re_2(CO)_{10}$ and $(C_9H_7Mo(CO)_3)_2$ occurred approximately three times faster than with the cyclopentadienyl analogue [37].

$C_9H_7Mo(CO)_3CH_3$ (Table **7**, No. **12**). In the 70 eV mass spectrum, the parent ion exhibited stepwise elimination of its CO groups; fragmentation of the CH_3 substituent competed with this successive loss of carbonyl ligands. Loss of the last CO group from $[C_9H_7Mo(CO)CH_3]^+$ was accompanied by dehydrogenation giving carbonyl-free $[C_{10}H_8Mo]^+$. Carbonyl-free $[C_9H_7Mo]^+$ was formed by CO-loss from methyl-free $[C_9H_7MoCO]^+$ without accompanying dehydrogenation. Both $[C_{10}H_8Mo]^+$ and $[C_9H_7Mo]^+$ underwent C_2H_2 elimination with the formation of $[C_8H_6Mo]^+$ and $[C_7H_5Mo]^+$, respectively. The usual metal-free ions of π-indenyl derivatives, including, e.g. $[C_9H_8]^+$, $[C_9H_7]^+$, $[C_7H_5]^+$, and $[C_5H_3]^+$, as well as the highly abundant acetyl ion, $[CH_3CO]^+$, were also observed. For further details, see [8].

Thermal reactions with tertiary phosphanes or phosphites in THF or n-hexane yielded acetyl derivatives, $C_9H_7Mo(CO)_2(PR_3)C(O)CH_3$ ($PR_3 = P(C_4H_9\text{-}n)_3$, $P(C_6H_5)_3$, $P(OCH_3)_3$, $P(OC_6H_5)_3$). The initially formed species, assumed to possess the trans configuration, underwent a further transformation to secondary products believed to be a trans/cis isomerization, except for $PR_3 = P(C_4H_9\text{-}n)_3$, where the nature of the subsequent reaction remained unclear [7]. Possibly due to a $^5L \rightarrow {}^3L$ ring-slippage of the indenyl system during methyl migration, these PR_3-assisted CO insertion reactions were considerably faster (by a factor of 10) than those of the cyclopentadienyl complex $C_5H_5Mo(CO)_3CH_3$. All the reactions were first-order in $C_9H_7Mo(CO)_3CH_3$. The dependence of the reaction rate on ligand concentration for the reaction with $P(C_4H_9\text{-}n)_3$ was of the type $k_{obs} = k_A + k_B[P(C_4H_9\text{-}n)_3]$, indicating two competing mechanisms, an essentially unassisted methyl migration (k_A term), and an associative process in which attack by the incoming ligand takes place simultaneously with methyl migration (k_B term). Activation data derived from values of k_{obs} measured over the temperature range 33 to 50°C (see the original literature) are as follows: $\Delta H^‡ = 14.7 \pm 0.5$ kcal/mol, $\Delta S^‡ = -24.2 \pm 0.5$ $cal \cdot mol^{-1} \cdot K^{-1}$ for the reaction in n-hexane and $\Delta H^‡ = 13.4 \pm 0.1$ kcal/mol, $\Delta S^‡ = -27.7 \pm 0.2$ $cal \cdot mol^{-1} \cdot K^{-1}$ for the reaction in THF, respectively [7].

Protolytic cleavage of the Mo–C bond using $(C_2H_5)_2O \cdot HBF_4$ in CH_2Cl_2 at −78°C followed by reaction of the $C_9H_7Mo(CO)_3FBF_3$ intermediate so generated with acetonitrile (−78°C → +20°C) afforded $[C_9H_7Mo(CO)_2(NCCH_3)_2]BF_4$ [25].

Photolysis in pentane in the presence of $P(CH_3)_3$ yielded $C_9H_7Mo(CO)_2(P(CH_3)_3)CH_3$ and $C_9H_7Mo(CO)(P(CH_3)_3)_2CH_3$. Trace amounts of a binuclear, asymmetrically substituted species, $C_9H_7Mo(CO)_3{-}Mo(CO)(P(CH_3)_3)_2C_9H_7$, were also formed [17].

References on pp. 268/9

The compounds $C_9H_7Mo(CO)_3CH_3$ and $C_5H_5W(CO)_2(\equiv CC_6H_4CH_3\text{-}4)$ reacted readily in toluene at 60 °C to give the alkyne-bridged dimetalatetrahedrane complex $1\text{-}CH_3C(O)C_9H_6Mo(CO)_2(\mu\text{-}C_2(C_6H_4CH_3\text{-}4)_2)W(CO)_2C_5H_5$ [22].

Methylenecyclopropane (neat, sealed tube) interacted slowly with $C_9H_7Mo(CO)_3CH_3$ to produce the ring-opened η^3-allyl product $C_9H_7Mo(CO)_2(\eta^3\text{-anti-}CH_2C(C(O)CH_3)CHCH_3)$. A corresponding reaction with buta-1,3-diene likewise led to an allyl compound, $C_9H_7Mo(CO)_2(\eta^3\text{-}CH_2CHCHR)$ ($R = CH_2C(O)CH_3$). The mechanism of formation of these products is discussed in detail in [18]. The reaction of $C_9H_7Mo(CO)_3CH_3$ with neat but-2-yne at room temperature in a closed vessel yielded a vinyl ketone complex (Formula II, $R = R' = CH_3$). In contrast, 3,3-dimethylbut-1-yne reacted under similar conditions to give consecutively a vinyl ketone derivative (Formula II; $R = H$, $R' = t\text{-}C_4H_9$) and a π-pyranyl compound (Formula III, $R = t\text{-}C_4H_9$). For possible pathways of these reactions, see [18]. Vinyl ketone complexes (Formula II; $R = R' = H$; $R = CH_3$, $R' = H$; $R = C_6H_5$, $R' = H$) were also formed in photoreactions of $C_9H_7Mo(CO)_3CH_3$ with alk-1-ynes, $RC\equiv CH$ ($R = H$, CH_3, C_6H_5), in pentane [14].

Mo OC OC R R' O CH3

II

Mo R OC H CH3 OC H O R

III

1.5.1.4.2.2.2 Compounds with 1,2 and 1,3-Disubstituted $R_2C_5H_3$ Ligands

$M[1,2\text{-}(HC(O))_2C_5H_3Mo(CO)_3]$ ($M^+ = Na^+$, $[N(CH_3)_4]^+$, $[P(C_6H_5)_4]^+$, $[(C_6H_5)_3PNP(C_6H_5)_3]^+$). The sodium salt was obtained in 85% yield from the reaction between equimolar quantities of $Mo(CO)_3(NCCH_3)_3$ and $Na[1,2\text{-}(HC(O))_2C_5H_3]$ in refluxing THF. The oily residue remaining after removal of the volatiles in vacuum solidified on prolonged stirring with ether to give the complex as brick red air-sensitive microcrystals, m.p. 255 °C (dec.) [20].

An alternative approach involved the hydrolysis of the 1-formyl(6-dimethylamino)fulvene precursor $((CH_3)_2NCHC_5H_3C(O)H)Mo(CO)_3$ in 5% aqueous NaOH at 50 °C through which was bubbled a vigorous stream of nitrogen to remove the dimethyl amine released during reaction. The solution so-obtained was precipitated by adding $[N(CH_3)_4]Cl$, $[P(C_6H_5)_4]Cl$, or $[(C_6H_5)_3PNP(C_6H_5)_3]Cl$, which caused the corresponding $M[1,2\text{-}(HC(O))_2C_5H_3Mo(CO)_3]$ salts to separate as sparingly soluble materials [20].

The $[N(CH_3)_4]^+$ and $[P(C_6H_5)_4]^+$ salts proved to be thermally highly unstable, decomposing rapidly with the formation of $Mo(CO)_6$. The more stable $[(C_6H_5)_3PNP(C_6H_5)_3][1,2\text{-}(HC(O))_2C_5H_3Mo(CO)_3]$ was isolated in 54% yield as orange-red crystals, m.p. 138 to 140 °C [20].

$M^+ = Na^+$: ^1H NMR spectrum (acetone-d_6): $\delta = 5.15$ (t, H-4; J = 3.0 Hz), 5.85 (d, H-3,5), 9.81 (s, HCO) ppm. IR spectrum (KBr): 1623, 1655; 1791, 1823, 1927 (ν(CO)) cm^{-1} [20].

$M^+ = [(C_6H_5)_3PNP(C_6H_5)_3]^+$: ^1H NMR spectrum (acetone-d_6): $\delta = 5.12$ (t, H-4; J = 3.1 Hz), 5.82 (d, H-3,5), 7.51 to 7.84 (m, C_6H_5), 9.82 (s, HCO) ppm. IR spectrum (KBr): 1648, 1658; 1815, 1917 (ν(CO)) cm^{-1} [20].

$C_9H_{11}Mo(CO)_3Cl$ (C_9H_{11} = 4,5,6,7-tetrahydroindenyl) was made from the product of the reaction between $Mo(CO)_6$ and spiro-[4,4]nona-1,3-diene in refluxing light petroleum (believed to be $(C_9H_{11}Mo(CO)_3)_2$), which in turn reacted with chloroform at 30 °C under mild irradiation to give the title complex. Chromatography of the crude product on silica gel, eluting with CH_2Cl_2/light petroleum (1:1) and subsequent crystallization from CH_2Cl_2/hexane, afforded red crystals, m.p. 86 to 88 °C, in ca. 60% yield [12].

^1H NMR spectrum ($CDCl_3$): δ = 1.6, 2.2 (each m, each $2CH_2$), 4.14 (t, H-2), 4.47 (d, H-1,3) ppm. IR spectrum (pentane): 1961, 1988, 2058 (ν(CO)) cm^{-1} [12].

Substitution reactions with phosphanes and phosphites in THF, acetone, or benzene afforded dicarbonyls, $C_9H_{11}Mo(CO)_2(PR_3)Cl$ (PR_3 = $P(C_6H_5)_3$, $P(OCH_3)_3$, $P(OC_6H_5)_3$). These reactions were found to be first-order in $C_9H_{11}Mo(CO)_3Cl$, essentially independent of the nature and the concentration of the incoming ligand PR_3, and not markedly sensitive to the solvent used, suggesting an S_N1 mechanism in which the rate-determining step involves the loss of CO. Mechanism, rate, and activation parameters were markedly different from those of the indenyl complex $C_9H_7Mo(CO)_3Cl$, but closely similar to those of the cyclopentadienyl compound $C_5H_5Mo(CO)_3Cl$: k_{obs} (THF, 40 °C) = 1.5×10^{-4} s^{-1}, $\Delta H^{\ddagger}$ = 27.0 ± 0.2 kcal/mol, $\Delta S^{\ddagger}$ = +9.9 ± 0.5 cal · mol^{-1} · K^{-1} [12].

$K[1,3\text{-}(t\text{-}C_4H_9)_2C_5H_3Mo(CO)_3]$ (preparation not described) reacted with 0.33 equivalent of $Cr(CO)_5PCl_3$ in THF between −78 °C and room temperature to give $1,3\text{-}(t\text{-}C_4H_9)_2C_5H_3\text{-}Mo(CO)_2P_3$-cyclo together with $(1,3\text{-}(t\text{-}C_4H_9)_2C_5H_3Mo(CO)_2)_2(\mu\text{-}P_2)$, $(1,3\text{-}(t\text{-}C_4H_9)_2C_5H_3Mo\text{-}(CO)_3)_2$, and $Cr(CO)_6$ [35].

$1,3\text{-}((CH_3)_3Si)_2C_5H_3Mo(CO)_3H$ was prepared by protonation of the $[1,3\text{-}((CH_3)_3Si)_2C_5H_3\text{-}Mo(CO)_3]^-$ anion using the procedure described for $C_5H_5Mo(CO)_3H$ in [1] but replacing $Na[C_5H_5]$ by $Li[1,3\text{-}((CH_3)_3Si)_2C_5H_3]$ [26].

Reaction with half an equivalent of $Pb(N(Si(CH_3)_3)_2)_2$ in THF at 0 °C afforded $(1,3\text{-}((CH_3)_3Si)_2C_5H_3Mo(CO)_3)_2Pb \cdot OC_4H_8$; the same reaction in pentane at 25 °C gave the solvent-free analogue, $(1,3\text{-}((CH_3)_3Si)_2C_5H_3Mo(CO)_3)_2Pb$ [26].

$1,3\text{-}(i\text{-}C_3H_7)_2C_5H_3Mo(CO)_3CH_3$. The sodium salt $Na[(i\text{-}C_3H_7)_2C_5H_3]$ (mixture of 1,3 (main) and 1,2 (minor) isomers) was refluxed with $Mo(CO)_6$ (5% deficiency) in THF to generate the $[1,3\text{-}(i\text{-}C_3H_7)_2C_5H_3Mo(CO)_3]^-$ anion. Subsequent treatment of the solution with CH_3I in the molar ratio 1:1 (20 °C → 65 °C) and crystallization of the evaporated mixture from petroleum ether at −80 °C furnished the isomerically pure title complex as the first fraction (57%) [28].

Yellow crystals, m.p. 53 °C, soluble in all common organic solvents. ^1H NMR spectrum (C_6D_6): δ = 0.43 (s, $MoCH_3$), 0.83, 0.96 (both d, $CH(C\mathit{H}_3)_2$; J = 6.8 Hz), 2.12 (sept, $C\mathit{H}(CH_3)_2$), 4.28 (t, H-2; J = 3.0 Hz), 4.61 (d, H-4,5) ppm. ^{13}C NMR spectrum (C_6D_6): δ = −20.7 (q, $MoCH_3$; J(C,H) = 137 Hz), 23.9, 24.9 (both q, both $CH(\mathit{C}H_3)_2$; J(C,H) = 126 and 130 Hz), 25.3 (d, $\mathit{C}H(CH_3)_2$; J(C,H) = 131 Hz), 87.5 (d, C-4,5; J(C,H) = 180 Hz), 95.9 (d, C-2; J(C,H) = 176 Hz), 119.0 (s, C-1,3), 228.0 (s, $MoCO_{cis}$), 241.8 (s, $MoCO_{trans}$) ppm. IR spectrum (heptane): 1933, 2012 (ν(CO)) cm^{-1} [28].

$1,3\text{-}(cyclo\text{-}C_6H_{10}CH_3\text{-}1')_2C_5H_3Mo(CO)_3CH_3$. To $Li[1,3\text{-}(cyclo\text{-}C_6H_{10}CH_3\text{-}1')_2C_5H_3]$, in situ-generated by combining equimolar quantities of LiC_4H_9-n/hexane and bis-(1′-methylcyclohexyl)cyclopenta-1,3-diene in THF at 0 °C, was added one equivalent of $Mo(CO)_3(NCC_2H_5)_3$ and the resulting mixture was refluxed to form the $[1,3\text{-}(cyclo\text{-}C_6H_{10}CH_3\text{-}1')_2C_5H_3Mo(CO)_3]^-$ anion. The stoichiometrically required amount of CH_3I was then added and the solution was stirred at room temperature. The residue remaining after removal of all volatile material was purified by chromatography on silica using petroleum ether as the eluant to provide pure product in 21% yield [36].

References on pp. 268/9

Yellow solid, m.p. 96 to 98 °C. ^{1}H NMR spectrum ($CDCl_3$): δ = 0.34 (s, $MoCH_3$), 1.06 (s, cyclo-$C_6H_{10}CH_3$), 1.35 to 1.60 (m, cyclo-$C_6H_{10}CH_3$), 4.93 ("s", C_5H_3) ppm [36].

References:

[1] King, R. B.; Stone, F. G. A. (Inorg. Synth. **7** [1963] 99/115).
[2] King, R. B.; Bisnette, M. B. (Inorg. Chem. **4** [1965] 475/81).
[3] King, R. B. (Inorg. Chem. **4** [1965] 1518/20).
[4] King, R. B.; Bisnette, M. B. (Abstr. Pap. 149th Meet. Am. Chem. Soc., Detroit 1965; Abstr. Pap. No. 110).
[5] King, R. B. (U.S. Clearinghouse Fed. Sci. Tech. Inform. AD-665429 [1967] 36 pp.; U.S. Gov. Res. Dev. Rep. **68** [1968] 48; C.A. **69** [1968] No. 56601).
[6] Hart-Davis, A. J.; Jones, J. D.; Mawby, R. J.; White, C. (Proc. 1st Int. Symp. Aspects Chem. Met. Carbonyls Deriv., Padua, Italy, 1968, pp. C3/C9).
[7] Hart-Davis, A. J.; Mawby, R. J. (J. Chem. Soc. A **1969** 2403/7).
[8] King, R. B. (Can. J. Chem. **47** [1969] 559/68).
[9] King, R. B.; Houk, L. W. (Can. J. Chem. **47** [1969] 2958/64).
[10] Hart-Davis, A. J.; White, C.; Mawby, R. J. (Inorg. Chim. Acta **4** [1970] 431/4).

[11] Hart-Davis, A. J.; White, C.; Mawby, R. J. (Inorg. Chim. Acta **4** [1970] 441/6).
[12] White, C.; Mawby, R. J. (J. Chem. Soc. A **1971** 940/2).
[13] Mawby, A.; Pringle, G. E. (J. Inorg. Nucl. Chem. **34** [1972] 525/30).
[14] Alt, H. G. (Z. Naturforsch. **32b** [1977] 1139/44).
[15] Nesmeyanov, A. N.; Ustynyuk, N. A.; Makarova, L. G.; Andre, S.; Ustynyuk, Yu. A.; Novikova, L. N.; Luzikov, Yu. N. (J. Organomet. Chem. **154** [1978] 45/63).
[16] Birdwhistell, R.; Hackett, P.; Manning, A. R. (J. Organomet. Chem. **157** [1978] 239/41).
[17] Alt, H. G.; Schwärzle, J. A. (J. Organomet. Chem. **162** [1978] 45/56).
[18] Bottrill, M.; Green, M. (J. Chem. Soc. Dalton Trans. **1979** 820/5).
[19] Nesmeyanov, A. N.; Ustynyuk, N. A.; Novikova, L. N.; Rybina, T. N.; Ustynyuk, Yu. A.; Oprunenko, Yu. F.; Trifonova, O. I. (J. Organomet. Chem. **184** [1980] 63/75).
[20] Edelmann, F.; Behrens, U. (Chem. Ber. **117** [1984] 3463/72).

[21] Novikova, L. N.; Ustynyuk, N. A.; Zvorykin, V. E.; Dneprovskaya, L. S.; Ustynyuk, Yu. A. (J. Organomet. Chem. **292** [1985] 237/43).
[22] Jeffery, J. C.; Laurie, J. C. V.; Stone, F. G. A. (Polyhedron **4** [1985] 1135/9).
[23] Faller, J. W.; Crabtree, R. H.; Habib, A. (Organometallics **4** [1985] 929/35).
[24] Mercer, J. R.; Green, M.; Orpen, A. G. (J. Chem. Soc. Chem. Commun. **1986** 567/9).
[25] Allen, S. R.; Beevor, R. G.; Green, M.; Orpen, A. G.; Paddick, K. E.; Williams, D. (J. Chem. Soc. Dalton Trans. **1987** 591/604).
[26] Hitchcock, P. B.; Lappert, M. F.; Michalczyk, M. J. (J. Chem. Soc. Dalton Trans. **1987** 2635/42).
[27] Ustynyuk, N. A.; Novikova, L. N.; Zvorykin, V. E.; Kravtsov, D. N.; Ustynyuk, Yu. (J. Organomet. Chem. **338** [1988] 19/28).
[28] Sitzmann, H. (J. Organomet. Chem. **354** [1988] 203/14).
[29] Lee, G.-H.; Peng, S.-M.; Liu, F.-C.; Liu, R.-S. (J. Organomet. Chem. **377** [1989] 123/32).
[30] Bentley, G. W.; Hough, G.; Winter, M. J.; Woodward, S. (Polyhedron **8** [1989] 1861/2).

[31] Lee, G.-H.; Peng, S.-M.; Liu, F.-C.; Mu, D.; Liu, R.-S. (Organometallics **8** [1989] 402/7).
[32] Adams, H.; Bailey, N. A.; Bentley, G. W.; Hough, G.; Winter, M. J.; Woodward, S. (J. Chem. Soc. Dalton Trans. **1991** 749/58).

[33] Kubas, G. J.; Kiss, G.; Hoff, C. D. (Organometallics **10** [1991] 2870/6).
[34] Coville, N. J.; Loonat, M. S.; Carlton, L. (S. Afr. J. Chem. **45** [1992] 63/72).
[35] Scheer, M.; Friedrich, G.; Schuster, K. (Angew. Chem. **105** [1993] 641/3; Angew. Chem. Int. Ed. Engl. **32** [1993] 593).
[36] Clark, T. J.; Killian, C. M.; Luthra, S.; Nile, T. A. (J. Organomet. Chem. **462** [1993] 247/57).
[37] Striejewske, W. S.; See, R. F.; Churchill, M. R.; Atwood, J. D. (Organometallics **12** [1993] 4413/9).

1.5.1.4.2.3 Compounds Containing the Fragment $R_3C_5H_2Mo(CO)_3$

$Na[(CH_3)_3C_5H_2Mo(CO)_3]$ (substitution pattern, 1,2,3-$(CH_3)_3C_5H_2$ or 1,2,4-$(CH_3)_3C_5H_2$, not specified) was obtained in solution by reacting $Mo(CO)_6$ with $Na[(CH_3)_3C_5H_2]$ in THF [3].

^{13}C NMR spectrum (THF): δ = 14.01 ($2CH_3$), 15.92 ($1CH_3$), 87.48 (ring-CH), 103.20 (2 ring-CCH_3), 103.67 (1 ring-CCH_3), 238.77 (CO) ppm. ^{17}O NMR spectrum (THF): δ = 353.8 ppm (vs. H_2O). ^{95}Mo NMR spectrum (THF): δ = −1985 ppm (vs. $Na_2[MoO_4]$) [3].

$(C_2H_5)_3C_5H_2Mo(CO)_3CH_3$ (5:6 mixture of 1,2,3- and 1,2,4-isomers). An isomeric mixture of the lithio derivatives Li[1,2,3-$(C_2H_5)_3C_5H_2$] and Li[1,2,4-$(C_2H_5)_3C_5H_2$] was heated with 0.95 equivalent of $Mo(CO)_6$ in THF at 65 °C to form the respective isomers of the $[(C_2H_5)_3C_5H_2Mo(CO)_3]^-$ anion, which was subsequently alkylated by treatment with equimolar CH_3I (25 °C → 65 °C → 25 °C). The methyl derivative was separated from the evaporated mixture by column chromatography (silica gel/petroleum ether) and further purified by crystallization from diethyl ether at −78 °C (88%) [5].

M.p. < 20 °C. ^1H NMR spectrum (C_6D_6): δ = 0.28, 0.29 (both s, $MoCH_3$ of isomers), 0.77 to 0.90 (m, CH_2CH_3), 1.70 to 1.95 (m, CH_2CH_3), 4.50, 4.58 (both s, C_5H_2 of isomers) ppm. ^{13}C NMR spectrum (C_6D_6): δ = −15.0, −14.7 ($MoCH_3$ of isomers; J(C,H) = 136 Hz each), 14.6, 14.9, 15.0, 18.1, 19.5, 21.5 (all C_2H_5), 86.5, 89.3 (ring-CH of isomers; J(C,H) = 174 and 171 Hz), 112.7, 113.8 (ring-CC_2H_5 of isomers), 228.9, 242.2, 242.4 (CO of isomers) ppm. IR spectrum (hexane): 1932, 2014 (ν(CO)) cm^{-1} [5].

1,2,4-$(i\text{-}C_3H_7)_3C_5H_2Mo(CO)_3CH_3$. $Na[(i\text{-}C_3H_7)_3C_5H_2Mo(CO)_3]$ was formed by combining a 5% deficiency of $Mo(CO)_6$ with a 1:4 isomeric mixture of Na[1,2,3-i-$(C_3H_7)_3C_5H_2$] and Na[1,2,4-i-$(C_3H_7)_3C_5H_2$] in boiling THF. It was allowed to react with CH_3I in 1:1 stoichiometry in the same solvent, first at room temperature and subsequently under reflux. Evaporation and crystallization of the residue from petroleum ether at −80 °C afforded the isomerically pure title compound as yellow crystals, which sublimed above 110 °C and proved to be soluble in all common organic solvents; yield: 51% [1].

^1H NMR spectrum (C_6D_6): δ = 0.43 (s, $MoCH_3$), 0.88 ("t", $2CH(CH_3)_2$; "J" = 6.7 Hz), 1.12 (d, $1CH(CH_3)_2$; J = 7.2 Hz), 2.04 (sept, $1CH(CH_3)_2$), 2.24 (sept, $2CH(CH_3)_2$), 4.73 (s, C_5H_2) ppm. ^{13}C NMR spectrum (C_6D_6): δ = −15.0 (q, $MoCH_3$; J(C,H) = 136 Hz), 23.3, 24.3, 25.8 (all q, $CH(CH_3)_2$), 25.8 (d, $2CH(CH_3)_2$), 26.9 (d, $1CH(CH_3)_2$), 86.3 (d, ring-CH; J(C,H) = 171 Hz), 118.0 (s, 2 ring-CC_3H_7-i), 121.5 (s, 1 ring-CC_3H_7-i), 228.5 ($MoCO_{cis}$), 242.8 ($MoCO_{trans}$) ppm. IR spectrum (heptane): 1933, 2012 (ν(CO)) cm^{-1} [1].

1,2,4-$(t\text{-}C_4H_9)_3C_5H_2Mo(CO)_3CH_3$. $Mo(CO)_6$ was reacted with an equimolar quantity of Li[1,2,4-$(t\text{-}C_4H_9)_3C_5H_2$] in THF at reflux temperature to generate the $[1,2,4\text{-}(t\text{-}C_4H_9)_3C_5H_2Mo(CO)_3]^-$ anion, which was then methylated using an equimolar quantity of CH_3I in the same solvent at 65 °C. The evaporated mixture was extracted into petroleum ether, solvent was removed, and the residue crystallized from diethyl ether at −30 °C to furnish the product in 59% yield as bright yellow crystals decomposing above 110 °C [4].

References on p. 270

^1H NMR spectrum (C_6D_6): δ = 0.65 ($MoCH_3$), 1.05 ($1C(CH_3)_3$), 1.22 ($2C(CH_3)_3$), 4.95 (C_5H_2) ppm. ^{13}C NMR spectrum (C_6D_6): δ = −19.1 ($MoCH_3$; J(C,H) = 137 Hz), 31.4 ($1C(\mathit{C}H_3)_3$), 31.6 ($1\mathit{C}(CH_3)_3$), 33.5 ($2\mathit{C}(CH_3)_3$), 34.2 ($2C(\mathit{C}H_3)_3$), 93.5 (ring-CH; J(C,H) = 170 Hz), 120.6 (2 ring-$\mathit{C}C_4H_9$-t), 122.8 (1 ring-$\mathit{C}C_4H_9$-t), 229.0, 243.4 (both CO) ppm. IR spectrum (petroleum ether): 1938, 2012 (ν(CO)) cm^{-1}. Mass spectrum (70 eV, 60 °C): $[M]^+$ (51%), $[M-CO]^+$ (55%), $[M-2\ CO]^+$ or $[M-i\text{-butene}]^+$ (59%) [4].

I

$Mo(CO)_3(\mu\text{-}1,3\text{-}(t\text{-}C_4H_9)_2C_5H_2C_5H_2(C_4H_9\text{-}t)_2\text{-}1',3')Fe(CO)_2$ (Formula I; M = Mo, m = 3; M′ = Fe, n = 2, R = t-C_4H_9) was synthesized by heating an equimolar mixture of 1,1′,3,3′-tetra-t-butyl-5,5′-dihydropentafulvalene, $Mo(CO)_6$, and $Fe(CO)_5$ in refluxing 1,2-dimethoxyethane for several days. Solvent and unreacted metal carbonyls were then removed in vacuum and the residue was extracted into CH_2Cl_2. Fractional crystallization from the concentrated extracts furnished a first fraction consisting of $Fe(CO)_2(\mu\text{-}1,3\text{-}(t\text{-}C_4H_9)_2C_5H_2C_5H_2(C_4H_9\text{-}t)_2\text{-}1',3')Fe(CO)_2$ (Formula I; M = M′ = Fe; m = n = 2; 14%), followed by the title complex (11%) and $Mo(CO)_3(\mu\text{-}1,3\text{-}(t\text{-}C_4H_9)_2C_5H_2C_5H_2(C_4H_9\text{-}t)_2\text{-}1',3')Mo(CO)_3$ (Formula I; M = M′ = Mo; m = n = 3; 17%) [2].

Brown powder; m.p. 107 °C (dec.). ^1H NMR spectrum ($CDCl_3$): δ = 1.06, 1.16, 1.29, 1.41 (all s, CH_3), 3.11, 4.06, 5.05, 5.69 (all d, H-2,2′,4,4′; J = 2.1 Hz) ppm. IR spectrum (CsI): 1878, 1937, 1951, 1995 (ν(CO)) cm^{-1}. Mass spectrum: $[M]^+$ (16%), $[M-n\ CO]^+$ (n = 2 (77%), 3 (37%), 4 (15%), 5 (42%)). The compound is air-stable and insensitive to light [2].

References:

[1] Sitzmann, H. (J. Organomet. Chem. **354** [1988] 203/14).
[2] Jutzi, P.; Schnittger, J. (Chem. Ber. **122** [1989] 629/32).
[3] Lyatifov, I. R.; Gasanov, T. Kh.; Petrovski, P. V.; Lutsenko, A. I. (J. Organomet. Chem. **361** [1989] 181/6).
[4] Sitzmann, H. (Chem. Ber. **123** [1990] 2311/5).
[5] Stein, D.; Sitzmann, H. (J. Organomet. Chem. **402** [1991] 249/57).

1.5.1.4.2.4 Compounds Containing the Fragment $R_4C_5HMo(CO)_3$

$M[C_{13}H_9Mo(CO)_3]$ ($C_{13}H_9$: Formula I; M^+ = Li^+, Na^+, K^+, $[N(C_2H_5)_4]^+$). The lithium salts $Li[C_{13}H_9M(CO)_3]$ (M = Cr, Mo, W) were first mentioned as products resulting from the reaction of the respective hexacarbonyls with lithium fluorenide under unspecified conditions. The $[C_{13}H_9M(CO)_3]^-$ anions were isolated from these reaction mixtures as their tetraethyl ammonium salts, $[N(C_2H_5)_4][C_{13}H_9M(CO)_3]$. No characterizing data were given [2]. Generation of $Li[C_{13}H_9Mo(CO)_3]$ in solution was also accomplished by the deprotonation with LiC_4H_9-n/hexane of the ^6L fluorene complex $C_{13}H_{10}Mo(CO)_3$ in diethyl ether at −70 °C. The initial formation of the ^6L anion, $[C_{13}H_9Mo(CO)_3]^-$, was accompanied by the deposition of a red precipitate from the highly colored solution. When the temperature was raised gradually to −15 °C, the precipitate dissolved and CO stretching bands appeared in the IR spectrum; the bands were assigned to the Li^+ salt of the rearranged ^5L anion, $[C_{13}H_9Mo(CO)_3]^-$ (ν(CO):

1732, 1788, and 1922 cm^{-1}). These absorptions disappeared rapidly when the solution was stored, even below 0 °C. On treatment of the ethereal reaction mixture with $Hg(CN)_2$ and CH_3I at −15 °C, the corresponding mercury and methyl derivatives, $(C_{13}H_9Mo(CO)_3)_2Hg$ and $\mathbf{C_{13}H_9Mo(CO)_3CH_3}$ (not isolated; ν(CO): 1946, 2024 cm^{-1}), were formed [3]. Protonation of the lithium salt restored the ^{6}L fluorene complex, $C_{13}H_{10}Mo(CO)_3$ [2].

Amber yellow solutions containing $Na[C_{13}H_9Mo(CO)_3]$ or $K[C_{13}H_9Mo(CO)_3]$ were obtained by refluxing equimolar quantities of $Mo(CO)_3(NCCH_3)_3$ and $M[C_{13}H_9]$ ($M^+ = Na^+, K^+$) in THF or 1,4-dioxane. A typical ^{1}H NMR spectrum obtained for $K[C_{13}H_9M(CO)_3]$ in 1,4-dioxane is shown in [1] (δ(H-9) ≅ 3.6 ppm). Treatment of the sodium salt with 1-chloropenta-2,4-diene in THF at 0 °C produced the ^{3}L pentadienyl dicarbonyl $C_{13}H_9Mo(CO)_2(\eta^3\text{-}CH_2CHCHCH{=}CH_2)$ [7, 8].

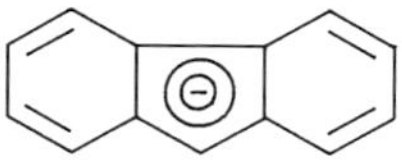

I

$\mathbf{M[R_4C_5HMo(CO)_3]}$ ($M^+ = Li^+, Na^+$; R = CH_3, C_2H_5, i-C_3H_7, C_6H_5). Solutions of these carbonyl molybdates were obtained by treating $Mo(CO)_6$ with the requisite lithium or sodium cyclopentadienides, $Li[R_4C_5H]$ (R = CH_3 [4], C_2H_5 [11], C_6H_5 [4]) and $Na[R_4C_5H]$ (R = CH_3 [6], i-C_3H_7 [9]) in THF at reflux temperature. In general, the products were not isolated, but were employed for further synthesis without purification.

For $Na[(CH_3)_4C_5HMo(CO)_3]$, the following NMR data were reported: ^{13}C NMR spectrum (THF): δ = 12.38, 14.34 (both CH_3), 85.43 (ring-CH), 101.80, 102.04 (both ring-$\mathit{C}CH_3$), 239.42 (CO) ppm. ^{17}O NMR spectrum (THF): δ = 354.6 ppm (vs. H_2O). ^{95}Mo NMR spectrum (THF): δ = −1946 ppm (vs. $Na_2[MoO_4]$) [6].

Spectroscopic data for $Na[(i\text{-}C_3H_7)_4C_5HMo(CO)_3]$: ^{1}H NMR spectrum (toluene-d_8): δ = 1.30, 1.46, 1.52, 1.58 (all d, $CH(C\mathit{H}_3)_2$; J = 6.7, 6.8, 7.1, and 7.2 Hz), 2.95, 3.04 (both sept, $C\mathit{H}(CH_3)_2$), 5.22 (s, C_5H) ppm. ^{13}C NMR spectrum (toluene-d_8): δ = 26.0, 26.8, 26.9, 27.2, 27.3 (all $CH(\mathit{C}H_3)_2$ and $\mathit{C}H(CH_3)_2$), 80.6 (ring-CH; J(C,H) = 166 Hz), 116.4, 117.3 (2 ring-$\mathit{C}C_3H_7$-i each), 240.0 (CO) ppm [9].

For reactions of the salts with $SiCl(CH_3)_2H$, $CuCl_2$, $Hg(CN)_2/HgX_2$ (X = Cl, Br, I), and CH_3I, see under $(i\text{-}C_3H_7)_4C_5HMo(CO)_3Si(CH_3)_2H$, $(i\text{-}C_3H_7)_4C_5HMo(CO)_3Cl$, $R_4C_5HMo(CO)_3$-HgX (R = CH_3, C_6H_5; X = Cl, Br, I), and $R_4C_5HMo(CO)_3CH_3$ (R = C_2H_5, i-C_3H_7).

$Na[(i\text{-}C_3H_7)_4C_5HMo(CO)_3]$ formed $(i\text{-}C_3H_7)_4C_5HMo(CO)_2{\equiv}Mo(CO)_2C_5(CH_3)_5$ when combined with an equimolar quantity of $(CH_3)_5C_5Mo(CO)_3Cl$ in THF, followed by heating in toluene at reflux temperature [9].

$\mathbf{(i\text{-}C_3H_7)_4C_5HMo(CO)_3H}$ resulted from the reaction of 1,2,3,4-$(i\text{-}C_3H_7)_4C_5H_2$ with $Mo(CO)_6$ in refluxing heptane in the molar ratio 1:1. The evaporated mixture was subjected to column chromatography on silica gel eluting with petroleum ether. Removal of solvent and crystallization of the residue from diethyl ether at −30 °C furnished the hydride in 81% yield [9].

Highly air-sensitive, colorless crystals; m.p. 95 °C (dec.). ^{1}H NMR spectrum (C_6D_6): δ = −4.98 (s, MoH), 0.92, 1.06, 1.13, 1.22 (all d, $CH(C\mathit{H}_3)_2$; J = 6.7, 6.8, 7.2, and 7.2 Hz), 2.50 (2 superimposed sept, $C\mathit{H}(CH_3)_2$), 4.96 (s, C_5H) ppm. ^{13}C NMR spectrum (C_6D_6): δ = 24.8, 25.1, 25.4, 25.5, 25.6 (all $CH(\mathit{C}H_3)_2$ and $\mathit{C}H(CH_3)_2$), 82.9 (ring-CH; J(C,H) = 169 Hz), 116.8, 119.3 (2 ring-$\mathit{C}C_3H_7$-i each), 229.7 (CO) ppm. IR spectrum (heptane): 1783 (ν(MoH)), 1898, 1930, 1979, 2011 (ν(CO)) cm^{-1} [9].

References on pp. 273/4

Refluxing the hydride in toluene in the presence of norborna-2,5-diene resulted in elimination of both H_2 and CO to give $((i\text{-}C_3H_7)_4C_5HMo(CO)_2)_2$ [9].

$(i\text{-}C_3H_7)_4C_5HMo(CO)_3Si(CH_3)_2H$. A suspension of $Na[(i\text{-}C_3H_7)_4C_5HMo(CO)_3]$ in petroleum ether was stirred with a slight excess of $SiCl(CH_3)_2H$ at ambient conditions and then centrifuged. Cooling to −80 °C gave the product in 53% yield [9].

Very air-sensitive, brownish yellow crystals, m.p. 90 °C (dec.). ^1H NMR spectrum (C_6D_6): δ = 0.79 (d, $Si(CH_3)_2$; J = 3.8 Hz), 0.94, 1.12, 1.13, 1.27 (all d, $CH(CH_3)_2$; J = 6.8, 7.2, 6.9, and 7.2 Hz), 2.59, 2.61 (both sept, both $CH(CH_3)_2$), 4.77 (s, C_5H), 5.07 (sept, SiH) ppm. ^{13}C NMR spectrum (C_6D_6): δ = 3.0 ($Si(CH_3)_2$; J(C,H) = 123 Hz), 23.9, 24.9, 25.4, 26.0, 26.1, 26.4 (all $CH(CH_3)_2$ and $CH(CH_3)_2$), 86.5 (ring-CH; J(C,H) = 169 Hz), 115.5, 117.2 (2 ring-CC_3H_7-i each), 240.2, 241.2 (both CO) ppm. IR spectrum (heptane): 1897, 1916, 1991 (ν(CO)), 2097 (ν(SiH)) cm^{-1} [9].

$(i\text{-}C_3H_7)_4C_5HMo(CO)_3Cl$ was prepared by oxidizing $Na[(i\text{-}C_3H_7)_4C_5HMo(CO)_3]$ with excess $CuCl_2$ in THF at ambient conditions. Chromatographic workup on silica gel eluting with petroleum ether/toluene (1:1), followed by low-temperature crystallization at −80 °C, afforced the chloro complex in 13% yield [9].

Brick red crystals. ^1H NMR spectrum (C_6D_6): δ = 0.89, 0.97, 1.01, 1.21 (all d, $CH(CH_3)_2$; J = 6.9, 6.8, 7.2, and 7.1 Hz), 2.36, 2.58 (both sept, $CH(CH_3)_2$), 4.71 (s, C_5H) ppm. ^{13}C NMR spectrum (CH_2Cl_2): δ = 23.1, 23.4, 23.5, 25.4, 25.8, 26.1 (all $CH(CH_3)_2$ and $CH(CH_3)_2$), 88.8 (ring-CH), 120.1, 123.3 (2 ring-CC_3H_7-i each), 201.4, 228.0 (both CO) ppm. IR spectrum (heptane): 1942, 1948, 1965, 1972, 2038 (ν(CO)) cm^{-1} [9].

Combination with $Na[(CH_3)_5C_5Mo(CO)_3]$ in 1:1 stoichiometry in THF and subsequent heating in refluxing toluene provided $(i\text{-}C_3H_7)_4C_5HMo(CO)_2{\equiv}Mo(CO)_2C_5(CH_3)_5$ [9].

$(C_2H_5)_4C_5HMo(CO)_3CH_3$ resulted from consecutive treatment of $Mo(CO)_6$, first with $Li[(C_2H_5)_4C_5H]$ (5% excess) in THF at reflux temperature, and then with CH_3I (1 equivalent) in the same solvent between 25 and 65 °C. Removal of volatile material, followed by column chromatography (silica gel/petroleum ether) and crystallization from diethyl ether at −78 °C, gave the alkyl in 66% yield [11].

M.p. < 20 °C. ^1H NMR spectrum (C_6D_6): δ = 0.31 (s, $MoCH_3$), 0.85 (m, CH_2CH_3), 1.89 (m, CH_2CH_3), 4.58 (s, C_5H) ppm. ^{13}C NMR spectrum (C_6D_6): δ = −13.5 ($MoCH_3$; J(C,H) = 136 Hz), 14.7, 16.1 (both CH_2CH_3), 18.6, 19.7 (both CH_2CH_3), 87.1 (ring-CH; J(C,H) = 172 Hz), 111.2, 111.6 (2 ring-CC_2H_5 each), 229.3, 243.0 (both CO) ppm. IR spectrum (hexane): 1928, 2012 (ν(CO)) cm^{-1} [11].

$(i\text{-}C_3H_7)_4C_5HMo(CO)_3CH_3$ was made by refluxing a THF solution of $Na[(i\text{-}C_3H_7)_4C_5H]$ with 0.95 equivalent of $Mo(CO)_6$ followed by alkylating the $[(i\text{-}C_3H_7)_4C_5HMo(CO)_3]^-$ anion so-generated with equimolar CH_3I, first at room temperature and then at 65 °C. The product was isolated from the evaporated mixture by extraction into petroleum ether and low-temperature crystallization at −80 °C; yield: 49% as yellow volatile crystals (subliming at >110 °C), soluble in common organic solvents [5].

^1H NMR (C_6D_6): δ = 0.54 (s, $MoCH_3$), 0.94, 1.05, 1.06, 1.22 (all d, $CH(CH_3)_2$; J = 6.7, 7.2, 6.8, and 7.2 Hz), 2.30, 2.58 (both sept, $CH(CH_3)_2$; J = 7.2 and 6.7 Hz), 4.71 (s, C_5H) ppm. ^{13}C NMR spectrum (C_6D_6): δ = −17.9 (q, $MoCH_3$; J(C,H) = 136 Hz), 23.3, 23.8, 24.5, 25.2 (all q, $CH(CH_3)_2$), 25.7, 26.2 (both d, $CH(CH_3)_2$), 90.3 (d, ring-CH; J(C,H) = 169 Hz), 114.3, 118.2 (both s, 2 ring-CC_3H_7-i each), 229.7 (CO_{cis}), 243.4 (CO_{trans}) ppm. IR spectrum (heptane): 1930, 2005 (ν(CO)) cm^{-1} [5].

$R_4C_5HMo(CO)_3HgX$ (R = CH_3, C_6H_5; X = Cl, Br, I). The complexes were prepared by analogous methods starting from $Mo(CO)_6$ and the corresponding lithium cyclopenta-

References on pp. 273/4

dienides, followed by reactions of the $Li[R_4C_5HMo(CO)_3]$ intermediates with $Hg(CN)_2$ and treatment of the resulting heterotrimetallics, $(R_4C_5HMo(CO)_3)_2Hg$, with a molar proportion of the requisite mercury(II) halide. The products were recrystallized from acetone or from CH_2Cl_2/hexane [4]. The complexes and some of their physical properties are listed in the following table (the NMR spectra were recorded in CH_2Cl_2/C_6D_6; the ^{95}Mo chemical shifts are given versus aqueous $Na_2[MoO_4]$, the ^{199}Hg shifts versus neat $Hg(CH_3)_2$):

R/X	properties
CH_3/Cl	^1H NMR: 2.00, 2.11 (both CH_3), 5.59 (CH) ^{95}Mo NMR: −1635 ($\Delta\nu_{1/2}$ = 150 Hz) ^{199}Hg NMR: −614 ($\Delta\nu_{1/2}$ = 10 Hz) IR (Nujol): 1910, 1924, 1993 (CO)
CH_3/Br	^1H NMR: 2.00, 2.10 (both CH_3), 5.57 (CH) ^{95}Mo NMR: −1615 ($\Delta\nu_{1/2}$ = 130 Hz) ^{199}Hg NMR: −776 ($\Delta\nu_{1/2}$ = 40 Hz)
CH_3/I	^1H NMR: 1.99, 2.07 (both CH_3), 5.50 (CH) ^{95}Mo NMR: −1600 ($\Delta\nu_{1/2}$ = 100 Hz) ^{199}Hg NMR: −1059 ($\Delta\nu_{1/2}$ = 30 Hz)
C_6H_5/Cl	^1H NMR: 6.60 (CH), 7.21, 7.25 (both C_6H_5) ^{95}Mo NMR: −1547 ($\Delta\nu_{1/2}$ = 150 Hz) ^{199}Hg NMR: −685 ($\Delta\nu_{1/2}$ = 7 Hz) IR (Nujol): 1916, 1937, 2000 (CO)
C_6H_5/Br	^1H NMR: 6.56 (CH), 7.20, 7.24 (both C_6H_5) ^{95}Mo NMR: −1536 ($\Delta\nu_{1/2}$ = 270 Hz) ^{199}Hg NMR: −905 ($\Delta\nu_{1/2}$ = 10 Hz)
C_6H_5/I	^1H NMR: 6.50 (CH), 7.21, 7.25 (both C_6H_5) ^{95}Mo NMR: −1539 ($\Delta\nu_{1/2}$ = 150 Hz) ^{199}Hg NMR: −1309 ($\Delta\nu_{1/2}$ = 7 Hz)

$(C_6H_5)_4C_5HMo(CO)_3P(C_6H_5)_2C_2P(C_6H_5)_2(CO)_2O^{\bullet}$ ($P(C_6H_5)_2C_2P(C_6H_5)_2(CO)_2O$ = 2,3-bis(diphenylphosphanyl)maleic anhydride). The radical was observed spectroscopically when a highly dilute CH_2Cl_2 solution of $((C_6H_5)_4C_5HMo(CO)_3)_2$ was irradiated ($\lambda > 520$ nm) in the presence of the chelating bis(phosphane) at −20 °C [10].

ESR spectrum (CH_2Cl_2, −20 °C): g = 2.0039; $a(P_{coord})$ = 8.86, $a(P_{free})$ = 3.46 G [10, 12].

Ring-closure with the formation of $(C_6H_5)_4C_5HMo(CO)_2((C_6H_5)_2P)_2C_2(CO)_2O$ commenced slowly even at −20 °C and was rapid at room temperature [10].

References:

[1] Nicholas, K. M.; Kerber, R. C.; Stiefel, E. I. (Inorg. Chem. **10** [1971] 1519/21).

[2] Slocum, D. W.; Conway, B.; Kuchel, K.; Moronski, M.; Noble, R.; Duraj, S.; Siegel, A.; Owen, D. A. (Org. Coat. Plast. Chem. **41** [1979] 44/6).

[3] Novikova, L. N.; Ustynyuk, N. A.; Zvorykin, V. E.; Dneprovskaya, L. S.; Ustynyuk, Yu. A. (J. Organomet. Chem. **292** [1985] 237/43).

[4] Kubicki, M. M.; Le Gall, J. Y.; Pichon, R.; Salaun, J. Y.; Cano, M.; Campo, J. A. (J. Organomet. Chem. **348** [1988] 349/56).

[5] Sitzmann, H. (J. Organomet. Chem. **354** [1988] 203/14).
[6] Lyatifov, I. R.; Gasanov, T. Kh.; Petrovski, P. V.; Lutsenko, A. I. (J. Organomet. Chem. **361** [1989] 181/6).
[7] Lee, G.-H.; Peng, S.-M.; Liu, F.-C.; Liu, R.-S. (J. Organomet. Chem. **377** [1989] 123/32).
[8] Lee, G.-H.; Peng, S.-M.; Liu, F.-C.; Mu, D.; Liu, R.-S. (Organometallics **8** [1989] 402/7).
[9] Sitzmann, H. (Chem. Ber. **123** [1990] 2311/5).
[10] Mao, F.; Philbin, C. E.; Weakley, T. J. R.; Tyler, D. R. (Organometallics **9** [1990] 1510/6).
[11] Stein, D.; Sitzmann, H. (J. Organomet. Chem. **402** [1991] 249/57).
[12] Mao, F.; Sur, S. K.; Tyler, D. R. (Organometallics **10** [1991] 419/23).

1.5.1.4.2.5 Compounds Containing the Fragment $R_5C_5Mo(CO)_3$ or $R'R_4C_5Mo(CO)_3$

1.5.1.4.2.5.1 The Radical $(C_6H_5)_5C_5Mo(CO)_3^{\bullet}$

$(C_6H_5)_5C_5Mo(CO)_3^{\bullet}$. IR spectroscopy showed that the $((C_6H_5)_5C_5Mo(CO)_3)_2$ homodimer was in equilibrium with its 17e monomer in solution: "$((C_6H_5)_5C_5Mo(CO)_3)_2 \rightleftarrows 2\ (C_6H_5)_5C_5$-$Mo(CO)_3^{\bullet}$"; K_{eq}(toluene, 25 °C) = $(8.7 \pm 5.1) \times 10^{-5}$ mol/L (by electronic absorption spectroscopy) [94].

IR spectrum (C_6H_6): 1897, 1907, 1996 (ν(CO)) cm^{-1} [79, 94].

Trapping with 2,3-bis(diphenylphosphino)maleic anhydride in THF gave the 19e radical $(C_6H_5)_5C_5Mo(CO)_2((C_6H_5)_2P)_2C_2(CO)_2O^{\bullet}$ [94].

1.5.1.4.2.5.2 Complexes with Bonds between Molybdenum and Main Group Elements

The various compounds hitherto known are listed in Table 8. In addition, an unstable molybdoiminophosphane, **$(CH_3)_5C_5Mo(CO)_3P{=}NC_4H_9$-t**, was proposed as a key intermediate in the formation of the spirocyclic compound (Formula I) from $Mo(CO)_3(NCCH_3)_3$ and $(CH_3)_5C_5P{=}NC_4H_9$-t [40]. Monitoring the thermal reaction between $((CH_3)_5C_5Mo(CO)_2)_2$ and CH_3SSCH_3 by IR spectroscopy indicated that the μ-sulfido product of this reaction, $((CH_3)_5C_5MoCO)_2(\mu\text{-}SCH_3)_2(\mu\text{-}S)$, arises from decarbonylative dimerization of the intermediately formed **$(CH_3)_5C_5Mo(CO)_3SCH_3$** [118].

O
$(CH_3)_5C_5(CO)_2Mo$ — P — NC_4H_9-t
$(CH_3)_5C_5P$ — NC_4H_9-t

I

$^8CH_3$ $^9CH_3$ $^{10}CH_3$ CH_3 CH_3 CH_3 CH_3
1 2 3 4 5 6 7

II

Most of the complexes described in this section were obtained by the following methods of synthesis. Methods I to III describe the preparation of $[R_5C_5Mo(CO)_3]^-$, IV to VI of $R_5C_5Mo(CO)_3H$, VII of $R_5C_5Mo(CO)_3SnR'_3$, VIII and IX of $R_5C_5Mo(CO)_3ER_2$ (E = P, As), X of $R_5C_5Mo(CO)_3X$ (X^- is O- or S-bonded), XI to XVI of $R_5C_5Mo(CO)_3X$ (X = Cl, Br, I), XVII of $R_5C_5Mo(CO)_3X$ (X^- is a weakly coordinated anion), XVIII and XIX of $R_5C_5Mo(CO)_3{}^1L$, and XX to XXII of $[R_5C_5Mo(CO)_3{}^1L]^+$ and $[R_5C_5Mo(CO)_3{}^2D]^+$.

References on pp. 313/6

Method I: A solution of $Li[(CH_3)_5C_5]$ in THF or $CH_3OC_2H_4OCH_3$ was refluxed with $Mo(CO)_6$ (equimolar or small excess) until IR spectroscopy indicated complete formation of the $[(CH_3)_5C_5Mo(CO)_3]^-$ anion, which is suitable for further synthesis without isolation of the crude lithium salt so formed [1, 12, 35, 55, 56, 59, 92, 95]. Similar in situ-methods for the lithium and sodium salts $Li[(C_2H_5)_5C_5Mo(CO)_3]$ [86] and $Na[(i\text{-}C_3H_7)_5C_5Mo(CO)_3]$ [74] involved treatment of the hexacarbonyl with $Li[(C_2H_5)_5C_5]$ or $Na[(i\text{-}C_3H_7)_5C_5]$ in approximate 1:1 stoichiometry in boiling THF. By analogy, the reactions of the lithium and potassium salts $M[(C_6H_5)_5C_5]$ with $Mo(CO)_6$ in $CH_3OC_2H_4OCH_3$ at 85 °C (M = Li) or in refluxing $CH_3OC_2H_4OC_2H_4OCH_3$ (M = K) gave $Li[(C_6H_5)_5C_5Mo(CO)_3]$ [19, 22, 33] and $K[(C_6H_5)_5C_5Mo(CO)_3]$ [94], respectively, which were either isolated as the tetraethyl ammonium derivative, $[N(C_2H_5)_4][(C_6H_5)_5C_5Mo(CO)_3]$ (see there), or were converted directly into other complexes containing the $(C_6H_5)_5C_5Mo(CO)_3$ moiety. The sodium salts $Na[R'R_4C_5Mo(CO)_3]$ (R = CH_3: R' = CH_3 [53, 62], C_2H_5, $CH_2{=}CH(CH_2)_2$ [53]; R = C_6H_5: R' = 2,5-$(CH_3O)_2C_6H_3$ [109]) were generated similarly by refluxing $Mo(CO)_6$ in THF with the required cyclopentadienide, $Na[R'R_4C_5]$. In a similar way, the lithio derivatives of cyclo-(−)-$C_2H_5(C_6H_5)CH(CH_3)_4C_5H$, pinan-3-yl tetramethylcyclopentadienide (Formula II) and cyclo-(4-CH_3-$C_6H_4)_2P(CH_3)_4C_5H$ formed $Li[C_2H_5(C_6H_5)CH(CH_3)_4C_5Mo(CO)_3]$, Li[pinane-3-yl$(CH_3)_4C_5Mo(CO)_3$], and $Li[(4\text{-}CH_3C_6H_4)_2P(CH_3)_4C_5Mo(CO)_3]$, when combined with $Mo(CO)_6$ in THF at reflux temperature. The salts were not isolated but transformed immediately into (−)-$(C_2H_5(C_6H_5)CH(CH_3)_4C_5Mo(CO)_3)_2$ [18], (−)-pinane-3-yl$(CH_3)_4C_5Mo(CO)_3CH_3$ [108], and $Mo(CO)_3(\mu\text{-}(CH_3)_4\text{-}C_5PR_2)Rh(R_2PC_2H_4PR_2)$ (R = $C_6H_4CH_3$-4) [16], respectively.

Method II: THF solutions of $Na[(CH_3)_5C_5Mo(CO)_3]$ are conveniently accessible by stirring a slurry of $((CH_3)_5C_5Mo(CO)_3)_2$ in dry THF with an excess of sodium (mineral oil dispersion) at ambient conditions [31]. An alternative procedure uses activated potassium in THF as a stoichiometric reducing agent: Ultrasound irradiation (20 kHz) of clean potassium metal with a titanium horn directly immersed into the liquid produced an activated colloidal potassium dispersion in THF to which was added half an equivalent of the $((CH_3)_5C_5Mo(CO)_3)_2$ dimer with continued sonication (15 min) before addition of a stoichiometric amount of an electrophile RX (see Method XVIII) [63].

Method III: A solution of $((CH_3)_5C_5Mo(CO)_2)_2$ in THF was stirred with excess sodium amalgam. Unreacted amalgam was then removed and the resulting solution used without isolation of the sodium salt, $Na[(CH_3)_5C_5Mo(CO)_3]$, so generated [6].

Method IV: Treatment of the crude $Li[(CH_3)_5C_5Mo(CO)_3]$ obtained in THF solution according to procedure I with glacial acetic acid at 0 °C or at room temperature gave $(CH_3)_5C_5Mo(CO)_3H$, which was separated from the evaporated mixture either by sublimation at 90 °C [12] or by extracting the residue with pentane and subliming off any unreacted $Mo(CO)_6$ from the evaporated extract at 40 °C [56]; see also [1, 51].

Method V: $(CH_3)_5C_5Mo(CO)_3H$ was prepared in excellent yield using a single-flask procedure in which $Mo(CO)_3(NCC_2H_5)_3$, preformed from $Mo(CO)_6$ and propionitrile, was treated with cyclo-$(CH_3)_5C_5H$ in toluene at 50 to 60 °C directly upon removal of excess C_2H_5CN in vacuum. When all of the solid had dissolved,

References on pp. 313/6

the mixture was stripped of all liquid volatiles and the product was sublimed at 55 °C in vacuum in a sublimation apparatus covered with aluminium foil to prevent photolytic decomposition [97]; see also [31]. Starting with Mo-$(CO)_3(NCCH_3)_3$, the above procedure was somewhat less facile and gave reduced yields [97]. Keeping an equimolar mixture of cyclo-$(CH_3)_5C_5H$ and fac-$Mo(CO)_3(OCHN(CH_3)_2)_3$ in methanol at room temperature in the dark similarly resulted in high yields of $(CH_3)_5C_5Mo(CO)_3H$ precipitating from the filtered and concentrated solution upon cooling to −78 °C [104].

Method VI: The cycloheptatriene complex $C_7H_8Mo(CO)_3$ and an equimolar amount of cyclo-$(CH_3)_5C_5H$ in anhydrous methanol were kept at room temperature in the dark under hydrogen (to avoid the formation of considerable amounts of H_2-loss products, $((CH_3)_5C_5Mo(CO)_3)_2$ and $((CH_3)_5C_5Mo(CO)_2)_2$, respectively). Concentration of the mixture by cooling afforded the hydride $(CH_3)_5$-$C_5Mo(CO)_3H$ as a crystalline precipitate which did not need further purification [38]. High yields of $(CH_3)_5C_5Mo(CO)_3H$ were attained from reactions between $CH_3C_6H_5Mo(CO)_3$ or 1,4-$(CH_3)_2C_6H_4Mo(CO)_3$ and pentamethylcyclopentadiene in THF. In these procedures, the product was isolated from the evaporated mixtures by slow sublimation at 60 °C and 10^{-5} Torr onto a water-cooled probe [23, 60].

Method VII: The THF solution of $Na[(CH_3)_5C_5Mo(CO)_3]$, obtained by the reduction with Na/Hg of the triple-bonded dimer $((CH_3)_5C_5Mo(CO)_2)_2$ (see Method III), was treated with a deficiency of $Sn(C_6H_5)_3Cl$ in THF at room temperature. Column chromatography of the evaporated reaction mixture on Florisil with a 1:3 mixture of CH_2Cl_2 and hexane as the eluent yielded $(CH_3)_5C_5Mo(CO)_3$-$Sn(C_6H_5)_3$ in low yield [6]. Higher yields of Mo−Sn derivatives, $(CH_3)_5C_5Mo$-$(CO)_3SnR_3$ (R = C_6H_5, $C_6H_4CH_3$-4), were attained by stirring the lithium salt of the $[(CH_3)_5C_5Mo(CO)_3]^-$ anion (cf. Method I) with the respective triaryl iodo tin compound SnR_3I in THF at ambient conditions. After removal of solvent, extraction of the residues into light petroleum/CH_2Cl_2 (1:1), and filtration of the extracts over Al_2O_3, the products were crystallized from the same solvent mixture [92].

Method VIII: Metalation of PCl_3 to give $(CH_3)_5C_5Mo(CO)_3PCl_2$ was achieved by reacting a suspension of $Na[(CH_3)_5C_5Mo(CO)_3]$ in methylcyclohexane with phosphorus(III) chloride in a 1:1 molar ratio at −78 °C. After removal of solvent and extraction of the residue with pentane, the product was crystallized at −78 °C [20]. The metallophosphane $(CH_3)_5C_5Mo(CO)_3P(C_6H_5)_2$ was prepared by nucleophilic metalation of $P(C_6H_5)_2Cl$ with $Na[(CH_3)_5C_5Mo(CO)_3]$ in benzene [17]. The heterogeneous metalation of $ClPCH(Si(CH_3)_3)C(Si$-$(CH_3)_3)_2$-cyclo with one equivalent of $K[(CH_3)_5C_5Mo(CO)_3]$ at −60 °C, followed by warming to 25 °C and heating at 40 °C, afforded $(CH_3)_5C_5Mo(CO)_3$-$PCH(Si(CH_3)_3)C(Si(CH_3)_3)_2$-cyclo, which remained as an impure viscous oil after evaporation of all volatile material [98]. Treatment of $XP{=}C(Si(CH_3)_3)R$ with $Na[(CH_3)_5C_5Mo(CO)_3]$ in toluene/THF (20:1) between 0 and 20 °C (X = Cl, R = $Si(CH_3)_3$), or in THF between −78 °C and room temperature (X = Br, R = C_6H_5), gave the corresponding molybdenum-substituted phosphaalkenes, $(CH_3)_5C_5Mo(CO)_3P{=}C(Si(CH_3)_3)R$, which were isolated from the evaporated reaction mixtures by crystallization from hexane at −80 °C (R = $Si(CH_3)_3$) [26, 107] or from pentane at −25 °C (R = C_6H_5) [107].

References on pp. 313/6

Method IX: The reaction of $(CH_3)_5C_5PCH(Si(CH_3)_3)C(Si(CH_3)_3)_2$-cyclo with $Mo(CO)_3(NCCH_3)_3$ in THF at 40°C produced $(CH_3)_5C_5Mo(CO)_3PCH(Si(CH_3)_3)C(Si(CH_3)_3)_2$-cyclo, which could be isolated after removing the volatiles by crystallization of the residue from pentane [98]. UV irradiation (Pyrex filter) of an equimolar mixture of $(CH_3)_5C_5P{=}C(Si(CH_3)_3)_2$ and $Mo(CO)_3(NCCH_3)_3$ in pentane, followed by cooling the concentrated solution to −80°C, provided crystalline $(CH_3)_5C_5Mo(CO)_3P{=}C(Si(CH_3)_3)_2$ [32, 107]. Prolonged stirring of $(CH_3)_5C_5As{=}PC_6H_2(C_4H_9\text{-t})_3$-2,4,6 with excess $Mo(CO)_3(NCCH_3)_3$ in toluene with subsequent evaporation of solvent and extraction of unreacted $Mo(CO)_3(NCCH_3)_3$ into hexane gave $(CH_3)_5C_5Mo(CO)_3As{=}PC_6H_2(C_4H_9\text{-t})_3$-2,4,6 [52].

Method X: $(CH_3)_5C_5Mo(CO)_3SO_2H$ resulted from treating a concentrated solution of $(CH_3)_5C_5Mo(CO)_3H$ in diethyl ether at 0°C with a stream of SO_2. The mixture immediately turned red, soon followed by the precipitation of the sulfinato complex [31, 36]. $(CH_3)_5C_5Mo(CO)_3SO_2CH_3$ was obtained from the reaction of $(CH_3)_5C_5Mo(CO)_3CH_3$ with liquid sulfur dioxide at −50 to −40°C. After allowing the SO_2 to evaporate at room temperature, the residue was redissolved in CH_2Cl_2/hexane (1:3), then chromatographed on Florisil eluting with acetone, and finally recrystallized from CH_2Cl_2/hexane [6].

Method XI: $Li[(CH_3)_5C_5Mo(CO)_3]$, in situ-generated in THF solution as summarized by procedure I, was treated at 0°C with anhydrous $FeCl_3$ in the molar ratio 1:2. After removing the solvent in vacuum, the solid residue was extracted with ether and chromatographed on an alumina column. Initial elution with hexane removed by-products, and $(CH_3)_5C_5Mo(CO)_3Cl$ was then eluted with CH_2Cl_2 and was purified further by recrystallization from CH_2Cl_2/hexane mixtures [56]. Iodination of $Na[(CH_3)_5C_5Mo(CO)_3]$ using the stoichiometric quantity of I_2 in THF (as detailed in the synthesis of $C_5H_5Mo(CO)_3I$ [65]) was described as a high-yield route to $(CH_3)_5C_5Mo(CO)_3I$ [88]. A THF solution of crude $Na[2,5\text{-}(CH_3O)_2C_6H_3(C_6H_5)_4C_5Mo(CO)_3]$ was acidified with glacial acetic acid, cooled to −10°C, and then treated with an equimolar amount of N-bromosuccinimide. Removal of the volatiles left the crude bromo product, $2,5\text{-}(CH_3O)_2C_6H_3(C_6H_5)_4C_5Mo(CO)_3Br$, which was extracted into CH_2Cl_2 and then recrystallized from CH_2Cl_2/CH_3OH [109].

Method XII: $(CH_3)_5C_5Mo(CO)_3H$ was slowly added at −20°C to CCl_4 which induced the precipitation of $(CH_3)_5C_5Mo(CO)_3Cl$ within minutes [38]; see also [96]. A modification of this approach involved the addition of 2.5 molar equivalents of CCl_4 to a pentane solution of the hydride at 20°C. Removal of solvent and crystallization of the residue from CH_2Cl_2/hexane provided the chloro complex as crystals in high yields [56].

Method XIII: $(CH_3)_5C_5Mo(CO)_3CH_3$, dissolved in CH_2Cl_2, was treated dropwise with a CH_2Cl_2 solution of a near equimolar amount of iodine in the same solvent. The mixture was then refluxed, solvent was removed, and the residue was chromatographed on a Florisil column. Elution with CH_2Cl_2/hexane (1:1) followed by recrystallization from the same solvent mixture afforded the iodo complex, $(CH_3)_5C_5Mo(CO)_3I$, in reasonable yield [6].

Method XIV: The interaction of $((CH_3)_5C_5Mo(CO)_3)_2$ with X_2 ($C_6H_5I \cdot Cl_2$, Br_2, I_2) in a 1:3 molar ratio at room temperature in CH_2Cl_2 (X = Cl, Br) or THF (X = I) was monitored by IR spectroscopy. The initially formed halo complexes,

References on pp. 313/6

$(CH_3)_5C_5Mo(CO)_3X$, were identified but not isolated, as the ultimately stable $(CH_3)_5C_5Mo(CO)_2X_3$ products precipitated directly out of solution [90]; see also [115, 116].

Method XV: The homodimer was irradiated with near-UV light in CCl_4 solution until an IR spectrum showed no starting material remaining. The CCl_4 was removed and $(CH_3)_5C_5Mo(CO)_3Cl$ purified by recrystallization from CH_2Cl_2/i-octane [7, 8].

Method XVI: Addition of a slight deficiency of I_2 to $((CH_3)_5C_5Mo(CO)_2)_2$ in dichloromethane solution, followed by stirring at ambient conditions and column chromatography using Florisil as the stationary phase and CH_2Cl_2/hexane (1:3) as the eluent, gave modest yields of $(CH_3)_5C_5Mo(CO)_3I$ [6].

Method XVII: $(CH_3)_5C_5Mo(CO)_3FBF_3$ was prepared by reacting $(CH_3)_5C_5Mo(CO)_3H$ with equimolar $[C(C_6H_5)_3]BF_4$ in CH_2Cl_2 at −30 °C. Dilution of the mixture with n-hexane and storage at −70 °C caused the weakly coordinated complex to precipitate [57]. The synthesis of $(CH_3)_5C_5Mo(CO)_3FPF_5$ was carried out analogously using $[C(C_6H_5)_3]PF_6$ in CH_2Cl_2 at 0 °C. The product separated from the concentrated solution after adding pentane at −40 °C [56].

Method XVIII: From $[R_5C_5Mo(CO)_3]^-$ and R′X or $(R'C(O))_2O$ by substitution:

a. A THF solution of $M[R_5C_5Mo(CO)_3]$ (see Method I or II) was treated with R′X; see "Remarks on Preparation".

b. For preparing the acyl $(CH_3)_5C_5Mo(CO)_3C(O)CF_3$, a twofold excess of trifluoroacetic anhydride was added to a $CH_3OC_2H_4OCH_3$ solution of in situ-generated $Li[(CH_3)_5C_5Mo(CO)_3]$ chilled to −78 °C. The residue, remaining after warming slowly to room temperature and removing the volatiles under vacuum, was extracted with several portions of petroleum ether. Concentrating the combined extracts and cooling to −78 °C caused the product to crystallize [95].

Method XIX: A mixture of excess $Mo(CO)_6$ and 5-acetyl-pentamethylcyclopentadiene in 2,2,5-trimethylhexane was boiled under reflux for a prolonged period. The solvent was removed at 50 °C in vacuum and the residue was extracted with boiling hexane; workup (see "Remarks on Preparation") yielded crystalline $(CH_3)_5C_5Mo(CO)_3CH_3$ [3, 5]. The methyl complex was also obtained in comparable yield by reacting $Mo(CO)_3(NCCH_3)_3$ with a near-equimolar amount of $(CH_3)_5C_5C(O)CH_3$ in refluxing methylcyclohexane. In this case, the residue remaining after the removal of solvent was extracted with pentane and then chromatographed on a Florisil column eluting with pentane [5]. While $((CH_3)_5C_5Mo(CO)_2)_2$ was isolated as an additional product from the reaction between $(CH_3)_5C_5C(O)CH_3$ and $Mo(CO)_6$ in boiling 2,2,5-trimethylhexane [3, 4, 5], this Mo≡Mo triple-bonded dimer was absent from the reaction mixture formed with $Mo(CO)_3(NCCH_3)_3$ in refluxing methylcyclohexane [5].

Method XX: $(CH_3)_5C_5Mo(CO)_3H$ was allowed to react with $[C(C_6H_5)_3]BF_4$ in 1:1 stoichiometry in CH_2Cl_2 at −78 °C to generate a solution of the highly reactive organometallic Lewis acid $(CH_3)_5C_5Mo(CO)_3FBF_3$. Addition of one equivalent of a phosphite ligand, $P(OC_2H_5)_3$ or $P(OC_6H_5)_3$, followed by warming to room temperature and diluting the concentrated solutions with diethyl ether, resulted in the precipitation of the crude tetrafluoroborates, $[(CH_3)_5C_5Mo(CO)_3P(OR)_3]BF_4$, which were purified by recrystallization from CH_2Cl_2/ether (1:9) (R = C_2H_5) or methanol/ether (R = C_6H_5) [73]. The complexes

$[(CH_3)_5C_5Mo(CO)_3{}^2D]PF_6$ (^2D = CH_3CN, $P(CH_3)_3$, $P(C_4H_9\text{-}n)_3$, $P(C_6H_5)_2CH_3$, $P(C_6H_5)_3$, $P(OC_6H_5)_3$) were obtained by the following standard procedure: To a CH_2Cl_2 solution of $[C(C_6H_5)_3]PF_6$ cooled to 0 °C was added one equivalent of $(CH_3)_5C_5Mo(CO)_3H$. When the reactive intermediate, $(CH_3)_5C_5Mo(CO)_3FPF_5$, had formed, as indicated by the purple color of the mixture, the requisite ^2D ligand was added in 25% excess, and the temperature raised to 20 °C. After concentration of the solution, ether was added to precipitate the desired salt, $[(CH_3)_5C_5Mo(CO)_3{}^2D]PF_6$, which was subsequently chromatographed on alumina with $CHCl_3$ as the eluent and then crystallized by diluting the eluate with ether [56].

Method XXI: $(CH_3)_5C_5Mo(CO)_3CH_3$ was treated with one equivalent of $[HO(C_2H_5)_2]BF_4$ in CH_2Cl_2 at −78 °C. The reddish violet solution of $(CH_3)_5C_5Mo(CO)_3FBF_3$ so-generated was allowed to warm to room temperature and recooled to −78 °C before acetonitrile was added. The oil obtained after warming to ambient temperature and removing insoluble and volatile materials was dissolved in CH_2Cl_2 and then combined with ether to deposit crystals of $[(CH_3)_5C_5Mo(CO)_3NCCH_3]BF_4$ [93].

Method XXII: $CNC_6H_4CH_3\text{-}4$ was added to an equimolar quantity of $(CH_3)_5C_5Mo(CO)_3FBF_3$ in CH_2Cl_2. After stirring at room temperature overnight and evaporation of the solvent, the residue was redissolved in THF. Cooling to −70 °C caused the precipitation of crystalline $[(CH_3)_5C_5Mo(CO)_3CNC_6H_4CH_3\text{-}4]BF_4$ [57]. In the synthesis of the salts $[(CH_3)_5C_5Mo(CO)_3{}^2D]BF_4$ (^2D = $P(C_6H_5)_3$, $P(OCH_3)_3$), the Lewis acid complex $(CH_3)_5C_5Mo(CO)_3FBF_3$ was allowed to interact with the respective ^2D ligand in 1:1 stoichiometry in CH_2Cl_2 at −30 °C. Warming to room temperature and dilution of the concentrated mixtures with diethyl ether resulted in the deposition of the saltlike products as oily precipitates, which solidified on stirring the mixtures vigorously. Further purification was accomplished by recrystallization from methanol/ether (1:1.5) [57].

Remarks on Preparation (Method XVIIIa): A THF solution of crude $Li[(CH_3)_5C_5Mo(CO)_3]$ (Method I) was stirred with excess CH_3I or $C_6H_5CH_2Cl$. For the isolation of $(CH_3)_5C_5Mo(CO)_3CH_3$, the residue left behind after removal of volatile material was extracted with CH_2Cl_2. The solvent was evaporated again and the resulting solid was treated with boiling hexane. Cooling the extracts to −78 °C afforded crystals of the methyl complex which were purified further by repeated sublimation in vacuum [1]. $(CH_3)_5C_5Mo(CO)_3CH_2C_6H_5$ was separated from the reaction mixture by column chromatography on alumina using a hexane eluent [25]. $(C_2H_5)_5C_5Mo(CO)_3CH_3$ was obtained from in situ-prepared $Li[(C_2H_5)_5C_5Mo(CO)_3]$ and equimolar CH_3I in THF (25 °C → 65 °C → 25 °C) and was isolated and purified by column chromatography (silica gel/petroleum ether), followed by crystallization from diethyl ether at −78 °C [86]. The analogous reaction between equimolar quantities of $Na[(i\text{-}C_3H_7)_5C_5Mo(CO)_3]$ (in situ in THF) and CH_3I produced $(i\text{-}C_3H_7)_5C_5Mo(CO)_3CH_3$, which was extracted from the evaporated solution with petroleum ether and then crystallized at −80 °C [74]. The complex $(C_6H_5)_5C_5Mo(CO)_3CH_3$ was synthesized similarly by reacting $Li[(C_6H_5)_5C_5Mo(CO)_3]$ with iodomethane in refluxing 1,2-dimethoxyethane [19, 22, 33].

The in situ alkylation of $K[(CH_3)_5C_5Mo(CO)_3]$ (Method II) using stoichiometric quantities of CH_3I or CH_3OCH_2Cl between −80 °C and room temperature provided a facile and high-yield method of synthesis for the corresponding alkyl derivatives $(CH_3)_5C_5Mo(CO)_3R$ (R = CH_3, CH_2OCH_3) [63]. In the syntheses of the halo- and methoxymethyl derivatives $(CH_3)_5C_5Mo(CO)_3CH_2X$ (X = Cl, Br, OCH_3), solutions of $Li[(CH_3)_5C_5Mo(CO)_3]$ in THF were

References on pp. 313/6

slowly added to THF solutions of XCH_2Cl (X = Br, I), CH_2Br_2, and $ClCH_2OCH_3$, respectively. After stirring at room temperature and removing volatile material, the residues were extracted with hexane and the respective $(CH_3)_5C_5Mo(CO)_3CH_2X$ products were isolated from the extracts as follows. (i) X = Cl: The extract was cooled to 0°C, which caused some $(CH_3)_5C_5Mo(CO)_3Cl$, formed as a by-product, to precipitate; further cooling to −78°C afforded the chloromethyl complex as crystals. (ii) X = Br: The hexane extract was cooled to −78°C to give oily crystals of the bromomethyl product, contaminated with some cyclo-$(CH_3)_5C_5H$; because of its instability, the complex was not purified further. (iii) X = OCH_3: The extract was chromatographed on a silica gel column; elution with ether/hexane (1:5) gave the crystalline methoxymethyl compound; further elution with CH_2Cl_2 afforded some additionally formed $(CH_3)_5C_5Mo(CO)_3Cl$ [35]. Stirring $Li[(CH_3)_5C_5Mo(CO)_3]$ with excess $Br(CH_2)_3Br$ in THF, followed by extraction of the evaporated mixture into light petroleum and filtration over Al_2O_3, gave a solution from which $(CH_3)_5C_5Mo(CO)_3(CH_2)_3Br$ could be isolated by crystallization [59].

(−)-Pinane-3-yl$(CH_3)_4C_5Mo(CO)_3CH_3$ (for ^5L ligand, see Formula II) was obtained by in situ treatment of Li[pinane-3-yl$(CH_3)_4C_5Mo(CO)_3$] with a slight excess of CH_3I in refluxing THF. The residue remaining after pumping off the solvent was extracted with hexane and then subjected to column chromatography (SiO_2/hexane) [108].

Properties: For a series of $(CH_3)_5C_5Mo(CO)_3X$ complexes, the following summary of experimental Mo−X bond dissociation energies was reported (X given parenthetically): 66.0 (H), 72.4 (Cl), 60.6 (Br), 51.8 (I), 47.0 (CH_3), 37.0 (C_2H_5), and 32.0 ($CH_2C_6H_5$) kcal/mol. A molecular interpretation based on a model considering the electrostatic and covalent bond forming tendencies, as well as the electron-transfer and electron-receptor tendencies, of the atoms forming the positive and negative ends of the Mo−X dipoles was offered in [103].

Table 8
Complexes with Bonds between Molybdenum and Main Group Elements.
An asterisk indicates further information at the end of the table.
For explanations, abbreviations, and units see p. X.

No.	compound	method of preparation (yield) properties and remarks
$M[R_5C_5Mo(CO)_3]$ salts:		
M^+/R =		
*1	Li^+/CH_3	I (not isolated) [1, 12, 35, 55, 56, 92, 95] IR (THF): 1709, 1791, 1889 (CO) [92]
2	$[Li((CH_3)_2NC_2H_4N(CH_3)_2)_2]^+/CH_3$	$(CH_3)_5C_5Mo(CO)_3H$ in toluene was deprotonated using LiC_4H_9-n in the presence of $(CH_3)_2C_2H_4N(CH_3)_2$ (1:1:2 ratio); the salt precipitated from the reaction mixture in 70% yield [38] yellow crystalline solid, highly sensitive to air and moisture [38] ^1H NMR (C_6D_6): 0.5 (s, $(CH_3)_5C_5$), 1.9 (s, C_2H_4), 2.3 (s, $N(CH_3)_2$) [38] IR (Nujol): 1745, 1755, 1885 (all CO) [38]

References on pp. 313/6

Table 8 (continued)

No.	compound	method of preparation (yield) properties and remarks
*3	Na^+/CH_3	I [53, 62]; II [31]; III [6] (not isolated); stock solutions in THF were also prepared by deprotonation with NaH of the parent hydride, $(CH_3)_5C_5Mo(CO)_3H$ [23] ^{13}C NMR (THF): 12.25 ($(CH_3)_5C_5$), 100.11 ($(CH_3)_5C_5$), 239.97 (CO) [62] ^{17}O NMR (THF): 354.9 (vs. H_2O) [62] ^{95}Mo NMR (THF): −1900 (vs. $Na_2[MoO_4]$) [62]
4	K^+/CH_3	II (not isolated) [63] in situ alkylation using stoichiometric quantities of CH_3I or CH_3OCH_2Cl between −80°C and room temperature gave high yields of the corresponding alkyls, $(CH_3)_5C_5Mo(CO)_3R$ (R = CH_3, CH_2OCH_3) [63]
*5	$[N(C_4H_9\text{-}n)_4]^+/CH_3$	for generation in solution, see "Further information" IR ($CH_2Cl_2/[N(C_4H_9\text{-}n)_4]BF_4$): 1740, 1870 (CO) [58]
6	$[H_2NC(N(CH_3)_2)_2]^+/CH_3$	generated in solution by deprotonating $(CH_3)_5C_5Mo(CO)_3H$ with tetramethyl guanidine in CH_3CN [42] IR (CH_3CN; molar absorptivities ($M^{-1} \cdot cm^{-1}$) in parentheses): 1762.0 (2152), 1881.1 (1915) (CO) [42]
7	$[P(C_6H_5)_4]^+/CH_3$	reaction with $Ru_3(CO)_{12}$ in refluxing THF, followed by addition of excess CF_3CO_2H in CH_2Cl_2 at room temperature led to an $MoRu_3$ cluster compound containing a quadruply bridging CO ligand, $(CH_3)_5C_5Mo(CO)_2(\mu_4\text{-}CO)(Ru(CO)_3)_3(\mu\text{-}H)$ [114]
8	$[(C_5H_5)_3UOC_4H_8]^+/CH_3$	observed in THF solutions of the isocarbonyl-bridged heterobimetallic $(CH_3)_5C_5Mo(CO)_2(\mu\text{-}CO)U(C_5H_5)_3$ [27] IR (THF): 1731, 1785, 1888 (CO) [27]
9	Li^+/C_2H_5	I (not isolated) [86] $(C_2H_5)_5C_5Mo(CO)_3CH_3$ resulted from the reaction with equimolar CH_3I in boiling THF [86]
10	$Na^+/i\text{-}C_3H_7$	I (not isolated) [74] methylation (CH_3I, THF, 65°C) gave $(i\text{-}C_3H_7)_5C_5Mo(CO)_3CH_3$ [74]
11	Li^+/C_6H_5	I [19, 22, 33] yellow to orange solid [19, 33] ^{13}C NMR (C_6D_6/THF-d_8): 111.79 ($(C_6H_5)_5C_5$), 125.6, 127.0, 134.1, 137.3 (all C_6H_5), 236.6 (CO) [19, 33] combination with $[N(C_2H_5)_4]Br$ in CH_2Cl_2 resulted in cation exchange with the formation of $[N(C_2H_5)_4][(C_6H_5)_5C_5Mo(CO)_3]$ [19, 33]

References on pp. 313/6

Table 8 (continued)

No.	compound	method of preparation (yield) properties and remarks
11 (continued)		interaction with CH_3I in $CH_3OC_2H_4OCH_3$ at reflux temperature produced $(C_6H_5)_5C_5Mo(CO)_3CH_3$ [19, 22, 33]
12	K^+/C_6H_5	I (not isolated) [94] chemical oxidation with $Fe(NO_3)_3 \cdot 9\,H_2O$ in THF in the presence of acetic acid gave $((C_6H_5)_5C_5Mo(CO)_3)_2$ [94]
13	$[N(C_2H_5)_4]^+/C_6H_5$	obtained by salt metathesis of No. 11 with $[N(C_2H_5)_4]Br$ in CH_2Cl_2 [19, 33]
$M[R'R_4C_5Mo(CO)_3]$ salts:		
$M^+/R'/R$ =		
14	$Na^+/C_2H_5/CH_3$	I (not isolated) [53] consecutive treatments with glacial acetic acid and 4-$CH_3C_6H_4SO_2N(NO)CH_3$ in THF at ambient conditions provided $C_2H_5(CH_3)_4C_5Mo(CO)_2NO$ [53]
15	$Na^+/CH_2{=}CH(CH_2)_2/CH_3$	I (not isolated) [53] $CH_2{=}CH(CH_2)_2(CH_3)_4C_5Mo(CO)_2NO$ was obtained by reacting the salt in THF first with CH_3CO_2H and then with 4-$CH_3C_6H_4SO_2N(NO)CH_3$ [53]
16	$Li^+/C_2H_5(C_6H_5)CH/CH_3$	I (using the lithium salt of enantiopure cyclo-(−)-$C_2H_5(C_6H_5)CH(CH_3)_4C_5H$; not isolated) [18] IR (THF): 1735, 1785, 1887 (CO) [18] oxidation with $Fe_2(SO_4)_3$ yielded optically pure (−)-$(C_2H_5(C_6H_5)CH(CH_3)_4C_5Mo(CO)_3)_2$ [18]
17	$Li^+/(4\text{-}CH_3C_6H_4)_2P/CH_3$	I (not isolated; formed as an approximately equimolar mixture with $Li[Mo(CO)_5P(C_6H_4CH_3\text{-}4)_2C_5(CH_3)_4\text{-cyclo}])$ [16] IR (THF): 1715, 1782, 1801, 1895 (CO) [16] reaction with half an equivalent of $(Rh(R_2PC_2H_4PR_2)(\mu\text{-}Cl))_2$ in THF gave metal−metal-bonded $Mo(CO)_3(\mu\text{-}(CH_3)_4C_5PR_2)Rh(R_2PC_2H_4PR_2)$ $(R = C_6H_4CH_3\text{-}4)_2$ [16]
18	Li^+/pinane-3-yl (see Formula II)/CH_3	I; not isolated but immediately transformed into (−)-pinane-3-yl$(CH_3)_4C_5Mo(CO)_3CH_3$ [108]
19	$Na^+/2,5\text{-}(CH_3O)_2C_6H_3/C_6H_5$	I [109]; also by Na/Hg reduction of 2,5-$(CH_3O)_2C_6H_3$-$(C_6H_5)_4C_5Mo(CO)_3Br$ in $CH_3OC_2H_4OCH_3$ [113] yellow powder [109] cation metathesis with $[P(C_6H_5)_4]Br$ in CH_2Cl_2 afforded $[P(C_6H_5)_4][2,5\text{-}(CH_3O)_2C_6H_3(C_6H_5)_4C_5Mo(CO)_3]$ [109]

References on pp. 313/6

Table 8 (continued)

No.	compound	method of preparation (yield) properties and remarks
		bromination using N-bromosuccinimide in THF in the presence of glacial acetic acid led to $2,5\text{-}(CH_3O)_2C_6H_3(C_6H_5)_4C_5Mo(CO)_3Br$ [109]
20	$[N(C_4H_9\text{-}n)_4]^+/2,5\text{-}(CH_3O)_2C_6H_3/C_6H_5$	by bulk electrolysis of $2,5\text{-}(CH_3O)_2C_6H_3(C_6H_5)_4C_5Mo(CO)_3Br$ in $C_2H_4Cl_2$ containing 0.4 M $[N(C_4H_9\text{-}n)_4]PF_6$ at $E_{red} < -1.70$ V vs. $(C_5H_5)_2Fe/[(C_5H_5)_2Fe]^+$ (not isolated) [113] IR ($C_2H_4Cl_2/[N(C_4H_9\text{-}n)_4]PF_6$): 1780, 1893 (CO) [113] cyclovoltammetry in $C_2H_4Cl_2/[N(C_4H_9\text{-}n)_4]PF_6$ revealed an electrochemically quasi-reversible $[2,5\text{-}(CH_3O)_2C_6H_3(C_6H_5)_4C_5Mo(CO)_3]^-/$ $2,5\text{-}(CH_3O)_2C_6H_3(C_6H_5)_4C_5Mo(CO)_3^{\bullet}$ couple at $E_{ox} = -0.54$ V vs. $(C_5H_5)_2Fe/[(C_5H_5)_2Fe]^+$ [113]
21	$[P(C_6H_5)_4]^+/2,5\text{-}(CH_3O)_2C_6H_3/C_6H_5$	crude $Na[2,5\text{-}(CH_3O)_2C_6H_3Mo(CO)_3]$ was dissolved in CH_2Cl_2 and then treated with equimolar $[P(C_6H_5)_4]Br$; the precipitate was recrystallized from hot CH_3OH/CH_2Cl_2 (86%) [109] yellow crystals, m.p. 130 to 132°C [109, 113] ^1H NMR ($CDCl_3$): 2.99, 3.46 (both s, CH_3O), 6.48 (d, C_6H_3, 1H; ^3J = 8.9), 6.54 (dd, C_6H_3, 1H; ^3J = 8.9, ^4J = 3.0), 6.82 to 7.16 (m, $(C_6H_5)_4C_5$ and C_6H_3, 1H), 7.53 to 7.82 (m, C_6H_5 of cation) [109, 113] ^{13}C NMR ($CDCl_3$): 55.23, 55.59 (both OCH_3), 109.244, 111.37, 112.46, 113.94, 116.88, 118.067, 123.51, 124.58, 124.70, 126.21, 126.38, 130.635, 130.81, 131.75, 132.47, 133.46, 134.34, 134.47, 135.74, 136.78, 137.17, 152.09 (all unassigned), 236.04 (CO) [113] IR (paraffin mull): 1770, 1894 (CO); additional bands at 531, 695, 726, 1112, 1224, 1381 [109]; (CH_2Cl_2): 1781, 1896 (CO) [109]; similar in [113]
hydrides:		
*22	$(CH_3)_5C_5Mo(CO)_3H$	IV (44%) [12], (90%) [56]; see also [1, 51]; V (90%, starting with $Mo(CO)_3(OCHN(CH_3)_2)_3$ [104]), (88% employing $Mo(CO)_3(NCC_2H_5)_3$) [97], (60%, starting with $Mo(CO)_3(NCCH_3)_3$) [97]; see also [31]; VI (95%, using $CH_3C_6H_5Mo(CO)_3$) [23], (87%, using $1,4\text{-}(CH_3)_2C_6H_4Mo(CO)_3$) [60], (60%, starting with $C_7H_8Mo(CO)_3$) [38] air-sensitive white crystals subliming at 90°C [1, 12]; pale yellow crystals subliming at 55 to 60°C/10^{-5} Torr [23, 60, 97] (the orange or pink-tinged colors reported

References on pp. 313/6

Table 8 (continued)

No.	compound	method of preparation (yield) properties and remarks
*22 (continued)		in [12, 97] are due to small amounts of $((CH_3)_5C_5Mo(CO)_3)_2$ present in the product) ^{1}H NMR (C_6D_6): −2.9 (s, MoH), 1.5 (s, CH_3) [23, 60]; (acetone-d_6, −30 °C): −5.41 (s, MoH), 2.09 (s, CH_3) [48]; ($CDCl_3$): −5.41 (s, MoH), 2.11 (s, CH_3) [12] ^{13}C NMR (acetone-d_6, −20 °C): 10.6 (($\mathit{C}H_3)_5C_5$), 105.7 ($(CH_3)_5\mathit{C}_5$), 231.4 (CO) [48] IR (pentane): 1929, 2015 (CO) [48]; (heptane): 1936, 2015 (CO) [23]; (cyclohexane): 1930, 1960, 2010, 2030 (CO) [12]; (CH_3CN; molar absorptivities ($M^{-1} \cdot cm^{-1}$) in parentheses): 1918.3 (3167), 2009.3 (2086) (CO) [42]; (THF): 1921, 2010 (O) [60]; (CS_2): 1763 (ν(MoH)) [23]; (Ar/H_2 matrix at 12 K): 1938.7, 2021.2 (CO; force and interaction constants: k_1 (CO_{cis}) = 15.601, k_2 (CO_{trans}) = 15.615, k_{cis} = 0.437, k_{trans} = 0.440 mdyn/Å) [43] mass spectrum: $[M]^+$, $[M-H-nCO]^+$ (n = 1 to 3) [12] solutions in CH_3CN were found to gradually decompose to $Mo(CO)_3(NCCH_3)_3$ by reductive elimination of cyclo-$(CH_3)_5C_5H$ [31, 36]
23	$(CH_3)_5C_5Mo(CO)_3D$	matrix-isolated from $(CH_3)_5C_5Mo(CO)_3H$ at high dilution by prolonged photolysis with a low-pressure Hg lamp at 12 K in Ar matrices containing 20 mol% of D_2 in addition to 7 mol% of CO [39, 43] IR (Ar/D_2 matrix at 12 K): 1938.6, 2020.6 (CO) [43]
$(CH_3)_5C_5Mo(CO)_3SnR_3$:		
SnR_3 =		
24	$Sn(C_6H_5)_3$	VII (5%) [6], (51%) [92] yellow crystals [6, 92], m.p. 154 to 156 °C [6], 171 to 173 °C [92] ^{1}H NMR ($CDCl_3$): 1.93 (s, CH_3), 7.27 to 7.33 (m, meta- and para-H), 7.54 to 7.61 (m, ortho-H) [92]; similar in [6] ^{13}C NMR ($CDCl_3$): 11.2 (($\mathit{C}H_3)_5C_5$), 103.8 ($(CH_3)_5\mathit{C}_5$), 127.9 (br, meta- and para-C), 136.9 (br, ortho-C), 227.6 (CO_{cis}), 232.9 (CO_{trans}) [92] IR (CH_2Cl_2): 1891, 1915, 1988 (CO) [6]; similar in [92] treatment with LiC_6H_5 in ether followed by hydrolysis and addition of $[O(C_2H_5)_3]BF_4$ formed trans-$(CH_3)_5$-$C_5Mo(CO)_2(=C(OC_2H_5)C_6H_5)Sn(C_6H_5)_3$ [66, 92]
25	$Sn(C_6H_4CH_3\text{-}4)_3$	VII (73%) [92] yellow crystals, m.p. 195 to 197 °C (dec.) [92]

References on pp. 313/6

Table 8 (continued)

No.	compound	method of preparation (yield) properties and remarks
		^{1}H NMR ($CDCl_3$): 1.92 (s, $(CH_3)_5C_5$), 2.33 (s, para-CH_3), 7.13 (d, meta-H; J = 8), 7.46 (d, ortho-H; $J(^{117}Sn,H)$ = 42, $J(^{119}Sn,H)$ = 44) [92] ^{13}C NMR ($CDCl_3$): 11.2 ($(\mathit{C}H_3)_5C_5$), 21.5 (para-CH_3), 103.7 ($(CH_3)_5\mathit{C}_5$), 128.8 (ortho- and meta-C; J(Sn,C) = 44), 137.8 (para-C), 227.7 (CO_{cis}), 233.1 (CO_{trans}) [92] IR: 1897, 1911, 1985 (CO) [92] trans-$(CH_3)_5C_5Mo(CO)_2(=C(OC_2H_5)C_6H_4CH_3$-4)-$Sn(C_6H_4CH_3-4)_3$ was formed on reaction with $LiC_6H_4CH_3$-4 in ether followed by alkylation with aqueous $[O(C_2H_5)_3]BF_4$ [92]

Mo–P- and Mo–As-bonded complexes $(CH_3)_5C_5Mo(CO)_3X$:

X =

No.	compound	method of preparation (yield) properties and remarks
26	PCl_2	VIII (45%) [20] orange-red crystals, m.p. 85 °C [20] ^{1}H NMR (C_6D_6): 1.52 (d; J(P,H) = 0.8) [20] ^{31}P NMR (C_6D_6): 403.3 [20] IR (methylcyclohexane): 1938, 1958, 2014 (CO) [20] indefinitely stable as a solid below −15 °C; solutions gradually decomposed forming $(CH_3)_5C_5Mo(CO)_3Cl$ by elimination of "PCl" [20] extremely air- and moisture-sensitive [20] reaction with cyclo-S_8 in hydrocarbon solution gave $(CH_3)_5C_5Mo(CO)_3P(S)Cl_2$ [20] $(CH_3)_5C_5Mo(CO)_3PCl_2 \cdot BH_3$ resulted from treatment with $H_3B \cdot OC_4H_8$ in C_6H_6 [20] $P(CH_3)_3$ reacted with the formation of $(CH_3)_5C_5Mo(CO)_2(P(CH_3)_3)Cl$ via CO/$P(CH_3)_3$ exchange and subsequent elimination of "PCl" [20]
27	$P(C_6H_5)_2$	VIII [17]
28	$P(CF_3)CF_2H$	treatment of a 20% solution of $(CH_3)_5C_5Mo(CO)_3H$ in toluene-d_8 with a 5 to 10% excess of the perfluorophosphaalkene CF_2=PCF_3 at 20 °C resulted in rapid and quantitative formation of No. 28, which was isolated by evaporation of solvent and volatile materials; attempted sublimation at 40 °C in vacuum was accompanied by partial decomposition with the formation of the CO-loss dimer $((CH_3)_5C_5Mo(CO)_2(\mu$-$P(CF_3)CF_2H))_2$ [45] ^{1}H NMR (toluene-d_8): 1.7 (s, CH_3), 7.2 (dt, PCF_2H; $^{2}J(P,H)$ = 12.0, $^{2}J(F,H)$ = 53.0) [45] ^{19}F NMR (toluene-d_8): −109.0, −103.4 (each ddq, F_A and F_B of PCF_2; $^{2}J(P,F_A)$ = 85.0, $^{2}J(P,F_B)$ = 152.0,

References on pp. 313/6

Table 8 (continued)

No.	compound	method of preparation (yield) properties and remarks
28 (continued)		$^2J(F_A,F_B)$ = 315.0, $^4J(CF_2,CF_3)$ = 6.5), −41.6 (dt, CF_3; $^2J(P,F)$ = 53.0); PCF_2 spectrum depicted in [45] ^{31}P NMR (toluene-d_8): 28.9 (ddq) [45] IR (cyclohexane): 1945, 1960, 2020 (CO) [45] mass spectrum (70 eV): $[M]^+$ (12.0%), $[M-3CO]^+$ (0.4%), $[M-P(CF_3)CF_2H]^+$ (0.2%), $[(CH_3)_5C_5Mo(CO)_nF]^+$ (n = 2 (13.3%), 1 (12.6%), 0 (62.0%)), $[(CH_3)_5C_5Mo]^+$ (13.3%), ($[P(CF_3)CF_2H]^+$ (4.8%)), ($[(CH_3)_5C_5]^+$ (24.0%)), $[Mo]^+$ (100.0%) [45]
29	H–C(Si(CH$_3$)$_3$)(P)–C(Si(CH$_3$)$_3$)(Si(CH$_3$)$_3$) (three-membered P–C–C ring)	VIII (impure product) [98]; IX (55 to 60%) [98] m.p. 132 to 135°C [98] ^{1}H NMR (C_6D_6): 0.32, 0.36, 0.39 (all s, 1Si$(CH_3)_3$) each), 1.55 (d, PCH; J(P,H) = 1.8), 1.60 (s, $(CH_3)_5C_5$) [98] ^{13}C NMR (C_6D_6): 1.5 (s, 1Si$(CH_3)_3$), 1.6 (d, 1Si$(CH_3)_3$; J(P,C) = 3.5), 2.9 (d, 1Si$(CH_3)_3$; J(P,C) = 10.7), 10.0 (d, $(CH_3)_5C_5$; J(P,C) = 10.2), 13.1, 19.5 (both d, PC; J(P,C) = 70.7 and 67.2), 105.3 (s, $(CH_3)_5C_5$), 229.6, 230.8, 239.5 (all d, 1CO each; J(P,C) = 13.5, 7.5, and 10.5) [98] ^{31}P NMR (C_6D_6): −64.9 [98] IR (pentane): 1860, 1900, 1980 (CO) [98] mass spectrum (70 eV): $[M]^+$ (<1%), $[M-nCO]^+$ (n = 1 (9%), 2 (2%), 3 (4%)), $[PC_2H(Si(CH_3)_3)_3]^+$ (3%), $[Si(CH_3)_3]^+$ (100%) [98]
30	$P{=}C(Si(CH_3)_3)C_6H_5$	VIII (65%) [107] orange crystals, m.p. 86°C [107] ^{1}H NMR (C_6D_6): 0.18 (d, Si$(CH_3)_3$; J(P,H) = 1.0), 1.40 (s, $(CH_3)_5C_5$), 6.95 (br, C_6H_5) [107] ^{13}C NMR (C_6D_6): 0.5 (d, Si$(CH_3)_3$; J(P,C) = 10.1), 10.2 (s, $(CH_3)_5C_5$), 105.5 (s, $(CH_3)_5C_5$), 125.2 (s, para-C), 126.8 (s, meta-C), 127.6 (s, ortho-C), 150.0 (d, ipso-C; J(P,C) = 14.6), 217.9 (d, P=C; J(P,C) = 91.1), 227.5 (d, CO_{cis}; J(P,C) = 12.9), 237.6 (d, CO_{trans}; J(P,C) = 2) [107] ^{31}P NMR (C_6D_6): 471.3 [107] IR (pentane): 1910, 1938, 1991 (CO) [107] mass spectrum: $[M]^+$ (2%), $[M-nCO]^+$ (n = 1 (9%), 3 (100%)), $[M-3CO-Si(CH_3)_3]^+$ (49%), $[C_6H_5]^+$ (12%), $[Si(CH_3)_3]^+$ (46%) [107] thermolysis resulted in decomposition rather than the formation of $(CH_3)_5C_5Mo(CO)_2{=}P{=}C(Si(CH_3)_3)C_6H_5$ [107]
31	$P{=}C(Si(CH_3)_3)_2$	VIII (76%) [26, 107]; IX (40%) [32, 107] m.p. 74 to 76°C (dec.) [26, 107]

References on pp. 313/6

Table 8 (continued)

No.	compound	method of preparation (yield) properties and remarks
		^{13}C NMR (C_6D_6): 2.8 (d, $Si(CH_3)_3$; J(P,C) = 2.5), 3.4 (d, $Si(CH_3)_3$; J(P,C) = 15.1), 9.9 (d, $(CH_3)_5C_5$; J(P,C) = 8.7), 106.5 (s, $(CH_3)_5C_5$), 218.0 (d, P=C; J(P,C) = 106.9), 226.3 (d, CO_{cis}; J(P,C) = 14.1), 237.1 (d, CO_{trans}; J(P,C) = 7.9) [26, 107] ^{31}P NMR (C_6D_6): 588.8 [26, 107] IR (hexane): 1062 (ν(P=C)), 1914, 1924, 1992 (CO) [26, 107] UV (heptane, ε) = 252 (12400), 315 (18500), 385 (2200), 482 (240; n $\rightarrow$ π^*) [107] mass spectrum: $[M-nCO]^+$ (n = 1 (2%), 3 (9%)), $[Si(CH_3)_3]^+$ (100%) [107] on gentle heating in nonpolar solvents, intramolecular substitution of one CO ligand occurred with the formation of $(CH_3)_5C_5Mo(CO)_2$=P=C$(Si(CH_3)_3)_2$ [26, 32, 107]
32	$PCl_2 \cdot BH_3$	$H_3B \cdot OC_4H_8$ (3 equivalents, THF solution) was added to $(CH_3)_5C_5Mo(CO)_3PCl_2$ in benzene at room temperature; the product remained after removing the solvent and washing the residue with pentane (31%) [20] ocher crystals, m.p. 108°C (dec.) [20] ^1H NMR (C_6D_6): 1.63 (d, CH_3; J(P,H) = 0.84), 2.20 (q, BH_3; J(B,H) = 90) [20] $^{11}B\{^1H\}$ NMR (C_6D_6): −16.1 [20] ^{31}P NMR (C_6D_6): 296.4 [20] IR (C_6H_6): 1970, 2036 (CO); BH_3 at 802, 1071, 2340, 2410 [20] mass spectrum: $[M-BH_3]^+$ [20] stable toward air but very moisture-sensitive [20]
33	$P(S)Cl_2$	$(CH_3)_5C_5Mo(CO)_3PCl_2$ was stirred with equimolar cyclo-S_8 in benzene or methylcyclohexane; No. 33 was isolated from the evaporated mixture by extraction with pentane and crystallization from the extracts at −78°C (38%) [20] orange crystals, m.p. 78°C (dec.) [20] ^1H NMR (C_6D_6): 1.98 (d; J(P,H) = 1.8) [20] ^{31}P NMR (C_6D_6): 179.3 [20] IR (methylcyclohexane): 1962, 1977, 2020 (CO) [20] mass spectrum: $[M-3CO]^+$ [20] air-stable, but highly sensitive toward moisture [20]
34	As=$PC_6H_2(C_4H_9\text{-}t)_3$-2,4,6	IX (44%) [52] dark red powder, m.p. 183°C (dec.) [52] ^1H NMR ($CDCl_3$): 1.35 (s, para-$C(CH_3)_3$), 1.41 (s, $(CH_3)_5C_5$), 1.95 (s, ortho-$C(CH_3)_3$), 7.39 (s, meta-H) [52] ^{13}C NMR ($CDCl_3$): 10.8 (s, $(CH_3)_5C_5$), 31.5, 34.1 (both s, $C(CH_3)_3$), 34.8, 38.6 (both s, $C(CH_3)_3$), 105.6

References on pp. 313/6

Table 8 (continued)

No.	compound	method of preparation (yield) properties and remarks
34 (continued)		$((CH_3)_5C_5)$, 121.9 (s, aryl-CH), 148.8, 151.0 (both s, aryl-C), 226.2 (d, CO; $^3J(P,C)$ = 16.9), 236.2 (s, CO) [52] ^{31}P NMR ($CDCl_3$): 586.9 [52] IR (KBr): 1915, 1925, 1992 (CO) [52] mass spectrum: $[M-nCO]^+$ (n = 1 (12%), 3 (10%)), $[M-3CO-C_4H_9]^+$ (45%) [52]

Mo–O- and Mo–S-bonded complexes $R_5C_5Mo(CO)_3X$:

R/X =

No.	compound	method of preparation (yield) properties and remarks
35	$OC(O)CF_3/OC(O)CF_3$	by reaction of $C_5H_5Mo(CO)_3Sn(C_6H_5)_3$ with $Hg(OC(O)CF_3)_2$ in 1:1 or 1:5 molar ratios in CCl_4 [91] yellow solid [91] IR (KBr): 1420 ($\nu(CO_2)_{sym}$), 1685 ($\nu(CO_2)_{asym}$), 1945, 1992, 2040 (CO) [91]
36	$CH_3/SCH_2C_6H_5$	by prolonged irradiation of a toluene solution containing $((CH_3)_5C_5Mo(CO)_2)_2$ and excess benzyl mercaptan; $((CH_3)_5C_5Mo(CO)_2(\mu\text{-}SCH_2C_6H_5))_2$ precipitated; No. 36 was isolated from the filtrate by column chromatography (SiO_2, CH_2Cl_2/hexane (2:3) eluent) (10%) [118] orange colored [118] IR (CH_2Cl_2): 1930, 2015 (CO) [118] ^1H NMR ($CDCl_3$): 2.21 and 2.26 (both s, CH_3), 2.63 and 2.57 (both s, CH_2), 7.00 and 7.11 (both m, C_6H_5), the two sets of signals arise from the existence of two isomers (relative percentage: 44 and 56%) that differ by the orientation of the benzyl group relative to the CO plane [118]
37	CH_3/SO_2H	X (75 to 80%) [31, 36] ^1H NMR (SO_2, −40 °C): 2.03 (s, CH_3), 3.89 (br s, SO_2OH) [31] IR (Nujol): 753, 1003 (SO_2), 1317 (δ(OH)), 1924, 1948, 2016 (CO), 2540 (ν(OH)) [31] considerably more thermally stable than $C_5H_5Mo(CO)_3SO_2H$; heating to 70 °C gave $((CH_3)_5C_5MoO(\mu\text{-}S))_2$, a species of composition $((CH_3)_5C_5Mo)_3S_2O_4$, and $((CH_3)_5C_5Mo(CO)_3)_2$ [31, 36] air-stable for short periods [31]
38	CH_3/SO_2CH_3	X (58%) [6] golden yellow crystals, m.p. 125 to 127 °C [6] ^1H NMR ($CDCl_3$): 2.07 (s, $(CH_3)_5C_5$), 2.93 (s, SO_2CH_3) [6] IR (CH_2Cl_2): 1959, 2041 (CO) [6]

References on pp. 313/6

Table 8 (continued)

No.	compound	method of preparation (yield) properties and remarks
halo complexes $R_5C_5Mo(CO)_3X$:		
R/X =		
*39	CH_3/Cl	XI (60%) [56]; XII (80%) [38], (86%) [56]; XIV (not isolated but directly converted to $(CH_3)_5C_5Mo(CO)_2Cl_3$) [90, 115]; XV (80%) [7, 8] orange crystals [38]; red needles [56] ^1H NMR ($CDCl_3$): 1.9 [38] IR (Nujol): 1945, 1965, 2030 (CO) [38]; (CH_2Cl_2): 1938, 1958, 2041 (CO) [90]; (CCl_4): 1925, 1960, 2030 (CO) [7, 8]
40	CH_3/Br	XIV (not isolated but directly transformed to $(CH_3)_5C_5Mo(CO)_2Br_3$) [90, 116]; also formed in the reaction between $((CH_3)_5C_5Mo(CO)_2)_2$ and 4-$BrC_6H_4C(O)CH_2Br$ (toluene, 60 to 70°C) [13]; refluxing $(CH_3)_5C_5Mo(CO)_3CH_2Br$ in CH_3OH gave No. 40 (10%) together with $(CH_3)_5C_5Mo(CO)_3CH_2OCH_3$ (90%) [35]
*41	$CH_3/CH_3/I$	XI ("high") [88]; XIII (44%) [6]; XIV (not isolated; reaction with excess I_2 provided $(CH_3)_5C_5Mo(CO)_2I_3$) [90, 116]; XVI (26%) [6] dark red crystals, m.p. 208 to 210°C (dec.) [6] ^1H NMR ($CDCl_3$): 2.05 [6] IR (cyclohexane): 1939, 1960, 2036 (CO) [6]; (C_6H_6): 1945, 2028 (CO) [6]; (THF): 1934, 1955, 2045 (CO) [90]
halo complexes $R'R_4C_5Mo(CO)_3X$:		
R'/R/X =		
42	$CF_3/CH_3/I$	$CH_3C_6H_5Mo(CO)_3$ was consecutively reacted with cyclo-$CF_3(CH_3)_4C_5H$ (5% excess) and CF_3I (6 equivalents) in THF at room temperature; trituration of the evaporated mixture with hexane, followed by filtration over Celite and removal of solvent gave the product in 72% yield [106] red solid, m.p. 164 to 166°C [106] ^1H NMR ($CDCl_3$): 2.17, 2.21 (both s, $2CH_3$) [106] ^{13}C NMR ($CDCl_3$): 11.48, 11.59 (both s, $2CH_3$), 96.37 (q, $\mathit{C}CF_3$; J(F,C) = 36.4), 105.81, 112.66 (both s, $2\mathit{C}CH_3$), 124.38 (q, CF_3; J(F,C) = 272.1), 220.09, 237.32 (both s, CO) [106] IR (CH_2Cl_2): 1264 (CF_3), 1971, 2042 (CO) [106]
43	2,5-$(CH_3O)_2C_6H_3/C_6H_5/Br$	XI (82%) [109] orange crystals, m.p. 170 to 172°C [109, 113] ^1H NMR ($CDCl_3$): 3.19, 3.63 (both br s, CH_3O), 6.57 (d, C_6H_3, 1H; ^3J = 9.2), 6.75 (dd, C_6H_3, 1H;

References on pp. 313/6

Table 8 (continued)

No.	compound	method of preparation (yield) properties and remarks
43 (continued)		$^3J = 9.2$, $^4J = 3.1$), 6.93 to 7.19 (m, $(C_6H_5)_4C_5$ and C_6H_3, 1H) [109, 113]; two sets of CH_3O signals (δ = 3.02, 3.15 and 3.13, 3.66 ppm) were observed on cooling and attributed to rotamers arising from the CH_3O groups lying either between two CO ligands or between CO and Br [109] IR (paraffin mull): 1950, 1987, 2046 (CO); further bands at 387, 412, 566, 704, 729, 1019, 1042, 1218, 1274 [109, 113]; (CH_2Cl_2): 1960, 1979, 2047 (CO) [109, 113] mass spectrum: $[M-nCO]^+$ (n = 1 (12%), 2 (12%), 3 (100%)), $[M-3CO-Br]^+$ (43%) [113] thermolysis in refluxing o-xylene with slow N_2 sparing gave $(2,5\text{-}(CH_3O)_2C_6H_3(C_6H_5)_4C_5Mo(\mu\text{-}CO)(\mu\text{-}Br))_2$ [109] in $C_2H_4Cl_2$/0.1 M $[N(C_4H_9\text{-}n)_4][PF_6]$, electrochemical reduction occurred at −1.70 V vs. $(C_5H_5)_2Fe/[(C_5H_5)_2Fe]^+$ to produce the $[2,5\text{-}(CH_3O)_2C_6H_3(C_6H_5)_4C_5Mo(CO)_3]^-$ anion together with Br^- via intermediately formed 19e and 17e radicals $(2,5\text{-}(CH_3O)_2C_6H_3(C_6H_5)_4C_5Mo(CO)_3Br^{\bullet}$ and $2,5\text{-}(CH_3O)_2C_6H_3(C_6H_5)_4C_5Mo(CO)_3^{\bullet})$ [113] chemical reduction with Na/Hg in $CH_3OC_2H_4OCH_3$ cleanly gave $Na[2,5\text{-}(CH_3O)_2C_6H_3(C_6H_5)_4C_5Mo(CO)_3]$ [113] treatment with excess allyl bromide in boiling octane yielded $((2,5\text{-}(CH_3O)_2C_6H_3(C_6H_5)_4C_5Mo)(\mu\text{-}Br)_2)_2$ [109]

complexes of the type $(CH_3)_5C_5Mo(CO)_3X$ with weakly bonded ligands X^-:

X^- =

No.	compound	method of preparation (yield) properties and remarks
*44	FBF_3^-	XVII (70%) [57]; see also "Further information" deep violet microcrystalline, thermally stable solid [57] ^1H NMR ($CDCl_3$): 2.30 (s) [57] IR (Nujol): 1945, 1970, 2050 (CO) [57]; (CH_2Cl_2 with added $[N(C_4H_9\text{-}n)_4]BF_4$): 1960, 2010 (CO) [58]
*45	FPF_5^-	XVII (90%) [56]; purple solvate with 1 molecule of CH_2Cl_2 [56] ^1H NMR (CD_2Cl_2): 1.92 (s) [56] IR (Nujol): 1975, 2070 (CO) [56] thermally stable, but very air-sensitive [56] attempted reactions with chloride salts failed to give $(CH_3)_5C_5Mo(CO)_3Cl$ [56]

$R_5C_5Mo(CO)_3{}^1L$:

$R/{}^1L$ =

No.	compound	method of preparation (yield) properties and remarks
*46	CH_3/CH_3	XVIIIa (33%) [1], (90%, using $K[(CH_3)_5C_5Mo(CO)_3]$) [63]; XIX (37 to 43%) [3, 5]

References on pp. 313/6

Table 8 (continued)

No.	compound	method of preparation (yield) properties and remarks
		orange crystals subliming at 70 to 100°C/0.01 to 0.1 Torr [1, 5, 6], m.p. 140°C (dec.) [3, 5], 141 to 145°C (dec.) [1] ^{1}H NMR (CS_2): −0.03 ($MoCH_3$), 1.87 ($(CH_3)_5C_5$) [1] ^{13}C NMR ($CDCl_3$): −12.2 ($MoCH_3$), 10.4 ($(\mathit{C}H_3)_5C_5$), 104.1 ($(CH_3)_5\mathit{C}_5$), 232.2, 243.1 (both CO) [63] IR (KBr): 2860, 2915 (ν(CH)); fingerprint bands at 761, 792, 874, 903, 1028, 1066, 1150, 1374, 1418, 1442, 1471 [1]; (cyclohexane): 1931 (A′ and A″), 2004 (A′) (CO; force and interaction constants calculated on the basis k_{trans}/k_{cis} = 2: k_1 (CO_{cis}) = 15.57, k_2 (CO_{trans}) = 15.19, k_{trans} = 0.51 mdyn/Å) [2]; similar in [1] UV (cyclohexane, ε): 323 (2300) [1]
*47	$CH_3/CH_2C_6H_5$	XVIIIa [25] ^{95}Mo NMR ($CHCl_3$, 20 ± 2°C): T_1 = 3.39 ± 0.06 ms [34]
48	CH_3/C_6H_5	isolated in yields up to 50% from the iodination reaction of the carbene species $(CH_3)_5C_5Mo(CO)_2(=C(OC_2H_5)C_6H_5)Sn(C_6H_5)_3$ using I_2 in CH_2Cl_2 at −78°C; the initially formed iodo derivative, $(CH_3)_5C_5Mo(CO)_2(=C(OC_2H_5)C_6H_5)I$, decomposed with the loss of C_2H_5I on attempted chromatography of the reaction mixture on Al_2O_3, even at −60°C [92] yellow crystals, m.p. 133 to 135°C [92] ^{1}H NMR ($CDCl_3$): 1.78 (s, CH_3), 7.00 to 7.16 (m, meta- and para-H), 7.62 to 7.68 (m, ortho-H) [92] ^{13}C NMR ($CDCl_3$): 10.6 ($(\mathit{C}H_3)_5C_5$), 105.5 ($(CH_3)_5\mathit{C}_5$), 124.3 (meta-C), 128.3 (para-C), 144.6 (ortho-C), 150.7 (ipso-C), 230.0 (CO_{trans}), 242.2 (CO_{cis}) [92] IR: 1917, 2009 (CO) [92]
49	CH_3/CH_2OCH_3	XVIIIa (19%) [35] (95%, using $K[(CH_3)_5C_5Mo(CO)_3]$ [63]); in yields up to 90% also by Mo−C bond fission of $(CH_3)_5C_5Mo(CO)_3CH_2X$ (X = Cl, Br) in CH_3OH, preferentially at reflux temperature [35] yellow crystals, m.p. 57 to 65°C [35] ^{1}H NMR (C_6D_6): 1.62 (s, $(CH_3)_5C_5$), 3.24 (s, OCH_3), 4.29 (s, CH_2) [35] ^{13}C NMR (C_6D_6): 10.0 ($(\mathit{C}H_3)_5C_5$), 64.5 (OCH_3), 68.3 (CH_2), 105.0 ($(CH_3)_5\mathit{C}_5$), 230, 243 (both CO) [63] IR (n-hexane): 1921, 1937, 2016 (CO) [35] mass spectrum (70 eV): $[M-CO]^+$, $[M-2CO-2H]^+$, $[M-2CO-H_2O]^+$, $[M-3CO-2H]^+$, $[M-2CO-CH_2OCH_3-2H]^+$ [35] electrophilic cleavage with dry HX gas in a hydrocarbon solution led to $(CH_3)_5C_5Mo(CO)_3CH_2X$ (X = Cl, Br, I); $(CH_3)_5C_5Mo(CO)_3I$ was also formed [35]

References on pp. 313/6

Table 8 (continued)

No.	compound	method of preparation (yield) properties and remarks
50	CH_3/CH_2Cl	XVIIIa (40%, using $BrCH_2Cl$; 24%, using ICH_2Cl) [35]; treating $(CH_3)_5C_5Mo(CO)_3CH_2OCH_3$ with dry HCl gas in hexane at room temperature and cooling the resulting solution to −78°C gave a 78% yield of the chloromethyl complex [35] yellow to yellow-orange crystals, m.p. 87 to 94°C (dec.) [35] ^1H NMR (C_6D_6): 1.54 (s, CH_3), 3.91 (s, CH_2) [35] IR (n-hexane): 1929, 1946, 2024 (CO) [35] mass spectrum (70 eV): $[M-CO]^+$, $[M-CO-Cl]^+$, $[M-3CO]^+$ [35] with $P(C_6H_5)_3$ in CH_3CN at ambient conditions in the dark, $(CH_3)_5C_5Mo(CO)_2(P(C_6H_5)_3)Cl$ was formed; the same reaction in boiling CH_3OH afforded the substitution product together with $(CH_3)_5C_5Mo(CO)_3CH_2OCH_3$ [35]
51	CH_3/CH_2Br	XVIIIa (30%; contaminated with cyclo-$(CH_3)_5C_5H$) [35]; a 58% yield of the pure complex was attained by cleaving the Mo−C bond of $(CH_3)_5C_5Mo(CO)_3CH_2OCH_3$ with dry HBr in hexane and chilling the solution to −78°C [35] yellow-orange crystals, m.p. 75 to 85°C [35] ^1H NMR (C_6D_6): 1.35 (s, CH_3), 3.16 (s, CH_2) [35] IR (n-hexane): 1929, 1947, 2023 (CO) [35] mass spectrum (70 eV): $[M-CO]^+$, $[M-nCO-CH_2]^+$ (n = 1, 2), $[M-3CO]^+$ [35] in refluxing methanol, conversion into a 9:1 mixture of $(CH_3)_5C_5Mo(CO)_3CH_2OCH_3$ and $(CH_3)_5C_5Mo(CO)_3Br$ was observed [35]
52	CH_3/CH_2I	not isolated in pure form due to its instability; treatment of $(CH_3)_5C_5Mo(CO)_3CH_2OCH_3$ with dry HI in hexane at 0°C followed by removing the volatiles and crystallizing the residue from hexane at −78°C gave an unseparable mixture with $(CH_3)_5C_5Mo(CO)_3I$ [35] IR (n-hexane): 1929, 1949, 2027 (CO) [35] decomposed easily to $(CH_3)_5C_5Mo(CO)_3I$ [35]
53	$CH_3/CH_2CH_2CH_2Br$	XVIIIa (64%) [59] m.p. 81 to 83°C [59] ^1H NMR ($CDCl_3$): 0.86 (m, $MoCH_2$), 1.92 (s, CH_3), 2.15 (m, CCH_2C), 3.43 (t, CH_2Br; J = 7) [59] IR (CH_2Cl_2): 1909, 2003 (CO) [59] mass spectrum: molecular ion observed [59] reactions with LiX (X^- = Br^-, I^-, NCS^-) in THF or with KCN in CH_3OH provided 2-oxacyclopentylidene derivatives, $(CH_3)_5C_5Mo(CO)_2(=C(CH_2)_3O\text{-cyclo})X$, as cis and/or trans isomers [59]

References on pp. 313/6

Table 8 (continued)

No.	compound	method of preparation (yield) properties and remarks
		treatment with $Na[C_5H_5Fe(CO)_2]$ in THF resulted in elimination of Br^- and formation of $(CH_3)_5C_5Mo(CO)_3(CH_2)_3Fe(CO)_2C_5H_5$ [59]
*54	CH_3/CF_3	obtained by solid-state thermal decarbonylation of $(CH_3)_5C_5Mo(CO)_3C(O)CF_3$ at 110°C; analytical samples were recrystallized from petroleum ether [95]; also formed by thermal decomposition (CD_2Cl_2, 20°C) of the carbene triflate salt $[(CH_3)_5C_5Mo(CO)_3{=}CF_2]O_3SCF_3$ [95] yellow crystals, m.p. 165 to 167°C (dec.) [95] ^1H NMR (C_6D_6): 1.65 (s) [95] ^{13}C NMR (C_6D_6): 11.40 (q, $(CH_3)_5C_5$; J(C,H) = 128), 108.49 (s, $(CH_3)_5C_5$), 154.31 (q, CF_3; J(F,C) = 368), 232.26 (q, CO_{cis}; $^3J(F,C)$ = 6), 242.47 (s, CO_{trans}) [95] ^{19}F NMR (C_6D_6): 0.7 (s) [95] IR (Nujol): 1926, 1946, 2032 (CO); further characteristic bands at 981, 993, and 1047 [95] photosensitive but air-stable for brief periods [95] unreactive toward $(CH_3)_3SiOC(O)CF_3$ and $(CH_3)_3SiOSO_2C_6H_5$; reactions with $(CH_3)_3SiX$ (X = Cl, I) gave uncharacterized insoluble precipitates together with $(CH_3)_3SiF$; treatment with $(CH_3)_3SiOSO_2CF_3$ in benzene at ambient conditions provided high yields of the carbene triflate salt $[(CH_3)_5C_5Mo(CO)_3{=}CF_2]O_3SCF_3$ [95] reactions with excess $BF_3 \cdot O(C_2H_5)_2$, $[C(C_6H_5)_3]BF_4$, or $[C(C_6H_5)_3]PF_6$ resulted primarily in the formation of $[(CH_3)_5C_5Mo(CO)_4]^+$ salts [95]
55	C_2H_5/CH_3	XVIIIa (72%) [86] m.p. 78°C [86] ^1H NMR (C_6D_6): 0.28 (s, $MoCH_3$), 0.91 (t, CH_2CH_3; J = 7.6), 1.99 (q, CH_2CH_3) [86] ^{13}C NMR (C_6D_6): −13.2 (q, $MoCH_3$; J(C,H) = 136), 15.8 (q, CH_2CH_3; J(C,H) = 128), 17.9 (t, CH_2CH_3; J(C,H) = 128), 109.3 (s, $(C_2H_5)_5C_5$), 228.5, 242.2 (both CO) [86] IR (hexane): 1940, 2020 (CO) [86]
56	$i\text{-}C_3H_7/CH_3$	XVIIIa (47%) [74] bright yellow crystals decomposing above 110°C [74] ^1H NMR (toluene-d_8, 373 K): 0.60 (s, $MoCH_3$), 1.22 (d, $CH(CH_3)_2$; J = 7.2), 2.81 (sept, $CH(CH_3)_2$) [74] ^{13}C NMR (toluene-d_8, 373 K): −12.2 ($MoCH_3$; J(C,H) = 137), 26.9 ($CH(CH_3)_2$), 28.1 ($CH(CH_3)_2$),

References on pp. 313/6

Table 8 (continued)

No.	compound	method of preparation (yield) properties and remarks
56 (continued)		120.2 ((i-$C_3H_7)_5$*C_5*), 231.4, 245.7 (both CO); ^{13}C and ^{1}H NMR spectra taken at 193 K (^{13}C spectrum reproduced in the original literature) indicated the presence of two conformers differing in the orientation of the isopropyl groups with respect to the five-membered ring ($\Delta G^{\ddagger}$ of isopropyl rotation: 55.6 ± 4 kJ/mol) [74] IR (cyclohexane): 1927, 2007 (CO) [74]
57	C_6H_5/CH_3	XVIIIa (63%) [19, 22, 33] yellow [19, 33] ^{13}C NMR ($CDCl_3$): −5.26 (CH_3), 113.0 (($C_6H_5)_5$*C_5*), 127.3, 127.6, 131.8, 132.2 (all C_6H_5), 227.25 (CO) [19, 33]
58	$CH_3/C{\equiv}CR$ (R = t-C_4H_9, C_6H_5)	no details reported, but apparently prepared by reacting $(CH_3)_5C_5Mo(CO)_3Cl$ with the corresponding alk-1-ynes in diethylamine in the presence of catalytic CuCl [80], as outlined for $C_5H_5W(CO)_3C{\equiv}CC_6H_5$ in [21] combination with $Ru_3(CO)_{12}$ in refluxing toluene afforded alkynyl-bridged heterotrimetallics, $(CH_3)_5C_5Mo(CO)_2(\mu_3\text{-}C{\equiv}CR)(Ru(CO)_3)_2$ [80]
59	$CH_3/C(O)CF_3$	XVIIIb (71%) [95] relatively air-stable orange-yellow crystals, m.p. 64 to 65°C [95] ^{1}H NMR ($CDCl_3$): 1.93 (s) [95] ^{19}F NMR ($CDCl_3$): −80.5 (s) [95] IR (Nujol): 1637 (ν(C=O)), 1949, 2016 (CO); additional fingerprint bands at 710, 833, 1128, 1172, 1221 [95] thermal decarbonylation at 110°C produced $(CH_3)_5C_5Mo(CO)_3CF_3$ [95]

R′$R_4C_5Mo(CO)_3$^{1}L salts:

R′/R/^{1}L =

No.	compound	method of preparation (yield) properties and remarks
*60	pinane-3-yl (see Formula II)/CH_3/CH_3	XVIIIa (70%) [108] yellow crystals [108] $[\alpha]_D$ = −41.91° ± 0.3° (c = 10.61, CH_2Cl_2) [108] ^{1}H NMR ($CDCl_3$): 0.07 (s, $MoCH_3$), 0.98 (d, $C^{(10)}H_3$; J = 7.05), 1.16, 1.26 (both s, CH_3-8,9), 1.81, 1.89, 1.92, 1.99 (all s, $(CH_3)_4C_5$), 1.17 (d, H-4,exo; J ≈ 8), 1.83 to 1.96 (m, H-1,6-exo), 2.02 to 2.09 (m, H-3,5), 2.21 (dddd, H-6,endo; J = 2.1, 4.06, 10.1, 14), 2.35 (dtd, H-4,endo; J = 2.1, 6.2, 9.7), 2.93 (q, H-2; J = 9.4) [108] ^{13}C NMR ($CDCl_3$): −10.91 ($MoCH_3$), 10.10, 10.62, 11.96, 12.01 (all (*CH_3*)$_4C_5$), 21.61 (C-10), 23.54 (C-9), 28.85 (C-8), 34.2 (C-2), 34.53 (C-4), 37.81 (C-6), 39.54 (C-7),

References on pp. 313/6

Table 8 (continued)

No.	compound	method of preparation (yield) properties and remarks
		43.31 (C-5), 44.15 (C-3), 49.36 (C-1), 102.78, 103.18, 104.22, 107, 114.08 (all $R'(CH_3)_4C_5$), 228.93, 229.08, 243.26 (all CO) [108] IR (KBr): 1908, 1930, 2005 (CO) [108]
$[(CH_3)_5C_5Mo(CO)_3{}^1L]X$:		
$^1L/X^-$ =		
61	$CNC_6H_4CH_3$-4/BF_4	XXII (60%) [57] yellow microcrystals [57] ^1H NMR ($CDCl_3$): 2.2 (s, $(CH_3)_5C_5$), 2.5 (s, para-CH_3), 7.4 to 7.8 (AA'BB' q, C_6H_4) [57] IR (Nujol): 1940, 2020, 2075 (CO), 2165 (ν(CN)) [57] $((CH_3)_5C_5Mo(CO)_3)_2$ was generated by irreversible one-electron reduction in CH_2Cl_2 containing 0.1 M $[N(C_4H_9\text{-n})_4]BF_4$, $E_p = -1.17$ V vs. SCE (cyclic voltammogram depicted in [58]); exhaustive cathodic electrolysis allowed the isolation of the dimer in pure form on a preparative scale [58]
*62	$=CF_2/CF_3SO_3^-$	prepared by fluorine abstraction from $(CH_3)_5C_5Mo(CO)_3CF_3$ using two equivalents of $(CH_3)_3SiOSO_2CF_3$ in benzene at room temperature; the salt precipitated from the reaction mixture (87%) [95] orange prisms (from CH_2Cl_2 at $-55\,^\circ C$) [95] ^1H NMR (SO_2, 0 °C): 2.53 (s) [95] ^{13}C NMR (SO_2, 0 °C): 11.2 (q, $(\mathit{C}H_3)_5C_5$; J(C,H) = 130), 112.2 (s, $(CH_3)_5\mathit{C}_5$), 121.1 (q, $CF_3SO_3^-$; J(F,C) = 320), 213.3 (s, CO_{trans}), 216.0 (q, CO_{cis}; ^3J(F,C) = 9), 270.5 (t, $=CF_2$; J(F,C) = 465) [95] ^{19}F NMR (SO_2, 0 °C): −77.8 (s, $CF_3SO_3^-$), 143.3 (s, CF_2) [95] IR (Nujol): 2040, 2099 (CO) [95]; additional absorptions at 1032, 1133, 1142, 1225, 1272 [95] while solutions in SO_2 or CH_2Cl_2 stored at −20 to 0 °C could be kept for longer periods, rapid decomposition with the formation of $(CH_3)_5C_5Mo(CO)_3CF_3$ as the major product occurred in CH_2Cl_2 at 20 °C [95] attempts to form CF_2-bridged heterobimetallics via reactions with $CH_3C_5H_4Mn(CO)_2OC_4H_8$, $C_2H_4Pt(P(C_6H_5)_3)_2$, or $Pt(P(C_6H_{11}\text{-cyclo})_3)_2$ remained unsuccessful [95] no identifiable products were observed in reactions with potential nucleophiles and/or methathesis reagents, such as $Li[(C_2H_5)_3BH]$, $Na[AlH(C_4H_9\text{-i})_2C_4H_9\text{-n}]$, or LiR (R = CH_3, C_6H_5) [95]

References on pp. 313/6

Table 8 (continued)

No.	compound	method of preparation (yield) properties and remarks
*62 (continued)		excess HOC_2H_4OH in CH_2Cl_2 provided a 1:3 mixture of $[(CH_3)_5C_5Mo(CO)_4]O_3SCF_3$ and $[(CH_3)_5C_5Mo(CO)_3{=}COC_2H_4O\text{-cyclo}]O_3SCF_3$ [95]
63	(cyclic carbene: =C with O–CH₂CH₂–O ring) $/CF_3SO_3^-$	formed by adding excess HOC_2H_4OH to $[(CH_3)_5C_5Mo(CO)_3{=}CF_2]O_3SCF_3$ in CH_2Cl_2 at −78°C; after warming to room temperature, ^{13}C NMR analysis of the crude product revealed a 1:3 mixture of $[(CH_3)_5C_5Mo(CO)_4]O_3SCF_3$ and an ionic complex, identified as the cyclic carbene No. 63 [95] ^{13}C NMR (CD_2Cl_2): 10.88 (q, ($\mathit{C}H_3)_5C_5$; J(C,H) = 131), 74.27 (t, C_2H_4; J(C,H) = 161), 109.41 (s $(CH_3)_5\mathit{C}_5$), 226.65, 232.73 (both s, CO), 245.37 (s, =C) [95]

$[(CH_3)_5C_5Mo(CO)_3{}^2D]X$ salts:

$^2D/X^-$ =

No.	compound	method of preparation (yield) properties and remarks
64	CH_3CN/BF_4^-	XXI [93]; also by exhaustive constant-potential oxidation of $(CH_3)_5C_5Mo(CO)_3H$ (E_{ox} = +0.561 V vs. $(C_5H_5)_2Fe/[(C_5H_5)_2Fe]^+$) in CH_3CN containing 0.05 M $[N(CH_3)_4]BF_4$ together with 2×10^{-3} M hydride (isolated by extraction of the evaporated mixture into CH_2Cl_2 and subsequent crystallization from CH_2Cl_2/ether; 58%) [78] brick red crystals [93] IR (CH_2Cl_2): 1977, 2059 (CO) [93] refluxing in CH_3CN led to cis-$[(CH_3)_5C_5Mo(CO)_2(NCCH_3)_2][BF_4]$ [93]
*65	CH_3CN/PF_6^-	XX (70%) [56] yellow crystals [56] ^1H NMR (CD_2Cl_2): 2.05 (s, $(CH_3)_5C_5$), 2.65 (s, CH_3CN) [56] ^{13}C NMR (CD_2Cl_2): 6.2 ($\mathit{C}H_3CN$), 11.0 (s, $(\mathit{C}H_3)_5C_5$), 111.3 ($CH_3\mathit{C}N$), 118.8 ($(CH_3)_5\mathit{C}_5$), 228.8 (CO_{cis}), 238.6 (CO_{trans}) [56] ^{31}P NMR (CD_2Cl_2): −143.7 (sept, J(P,F) = 711) [56] IR (CH_2Cl_2): 1990, 2075 (CO) [56]
*66	$P(CH_3)_3/PF_6^-$	XX (96%) [56] yellow crystals [56] ^1H NMR ($CDCl_3$): 1.73 (d, $P(CH_3)_3$; J(P,H) = 10.1), 2.11 (s, $(CH_3)_5C_5$) [56] ^{13}C NMR ($CDCl_3$): 10.8 (s, $(\mathit{C}H_3)_5C_5$), 18.1 (d, $P(CH_3)_3$; J(P,C) = 34.0), 108.7 (s, $(CH_3)_5\mathit{C}_5$), 228.0 (d, CO_{cis}; J(P,C) = 31), 230.0 (d, CO_{trans}; J(P,C) = 4.0) [56] ^{31}P NMR ($CDCl_3$): −143.7 (sept, PF_6^-; J(P,F) = 712), 6.2 (s, $P(CH_3)_3$) [56]

References on pp. 313/6

Table 8 (continued)

No.	compound	method of preparation (yield) properties and remarks
		IR (Nujol): 1960, 1990, 2075 (CO) [56]; (CH_2Cl_2): 1965, 2055 (CO) [56] prolonged heating in 1,2-$Cl_2C_6H_4$ at 135 °C in the presence of excess $P(C_6H_5)_2CH_3$ led to trans-[$(CH_3)_5$-$C_5Mo(CO)_2(P(CH_3)_3)P(C_6H_5)_2CH_3]PF_6$ [56]
*67	$P(C_4H_9\text{-n})_3/PF_6^-$	XX (60%) [56] yellow crystals [56] ^1H NMR ($CDCl_3$): 0.9 to 2.0 (m, n-C_4H_9), 2.08 (s, $(CH_3)_5C_5$) [56] ^{13}C NMR ($CDCl_3$): 11.0 (s, $(\mathit{C}H_3)_5C_5$), 22.0 to 29.0 (m, n-C_4H_9), 108.8 (s, $(CH_3)_5\mathit{C}_5$), 229.3 (d, CO_{cis}; J(P,C) = 29.0), 231.7 (d, CO_{trans}; J(P,C) = 3.0) [56] ^{31}P NMR ($CDCl_3$): −142.7 (sept, PF_6^-; J(P,F) = 708), 23.1 (s, $P(C_4H_9\text{-n})_3$) [56] IR (CH_2Cl_2): 1962, 2055 (CO) [56]; ν(CO) values given in [56] for Nujol mulls are probably erroneous
*68	$P(C_6H_5)_2CH_3/PF_6^-$	XX (70%) [56] yellow crystals [56] ^1H NMR ($CDCl_3$): 1.90 (s, $(CH_3)_5C_5$), 2.30 (d, PCH_3; J(P,H) = 9.0), 7 to 8 (m, C_6H_5) [56] ^{13}C NMR ($CDCl_3$): 10.4 (s, $(\mathit{C}H_3)_5C_5$), 18.1 (d, PCH_3; J(P,C) = 31.6), 109.1 (s, $(CH_3)_5\mathit{C}_5$), 130 to 134 (m, C_6H_5), 228.6 (d, CO_{cis}; J(P,C) = 28.8), 231.7 (s, CO_{trans}) [56] ^{31}P NMR ($CDCl_3$): −143.6 (sept, PF_6^-; J(P,F) = 707), 26.4 (s, $P(C_6H_5)_2CH_3$) [56] IR (Nujol): 1935, 1960, 2040 (CO) [56]; (CH_2Cl_2): 1955, 2080 (CO) [56]
*69	$P(C_6H_5)_3/BF_4^-$	XXII (81%) [57] yellow crystals [57] ^1H NMR ($CDCl_3$): 2.0 (s, CH_3), 7.0 to 7.8 (m, C_6H_5) [57] IR (Nujol): 1960, 1990, 2050 (CO) [57] reaction with equimolar NaN_3 in methanol at ambient conditions furnished a mixture of cis-$(CH_3)_5C_5Mo(CO)_2$-$(P(C_6H_5)_3)N_3$ and cis-$(CH_3)_5C_5Mo(CO)_2(P(C_6H_5)_3)NCO$; the latter was also obtained by treating a CH_3OH solution of No. 69 with KNCO at 40 °C [68]
*70	$P(C_6H_5)_3/PF_6^-$	XX (70%) [56] yellow crystals [56] ^1H NMR ($CDCl_3$): 1.91 (s, CH_3), 7 to 7.8 (m, C_6H_5) [56] ^{13}C NMR ($CDCl_3$): 10.9 (s, $(\mathit{C}H_3)_5C_5$), 110.3 (s, $(CH_3)_5\mathit{C}_5$), 130.6 (d, para-C; J(P,C) = 11), 130.9 (d, ipso-C; J(P,C) = 47.9), 133.3 (d, meta-C; J(P,C) = 21), 134.2 (d, ortho-C; J(P,C) = 11), 229.5 (d, CO_{cis}; J(P,C) = 27.8), 232.9 (d, CO_{trans}; J(P,C) = 2.6) [56]

References on pp. 313/6

Table 8 (continued)

No.	compound	method of preparation (yield) properties and remarks
*70 (continued)		^{31}P NMR ($CDCl_3$): −143.4 (sept, PF_6^-; J(P,F) = 707), 44.4 (s, $P(C_6H_5)_3$) [56] IR (Nujol): 1955, 1995, 2070 (CO) [56]; (CH_2Cl_2): 1975, 2060 (CO) [56]
*71	$P(OCH_3)_3/BF_4^-$	XXII (66%) [57]; also formed from $[(CH_3)_5C_5Mo(CO)_2(P(OCH_3)_3)CH_3OH]BF_4$ by substitution of the CH_3OH ligand by CO [77] yellow crystals [57] ^1H NMR (CD_3OD): 2.15 (s, $(CH_3)_5C_5$), 4.00 (d, OCH_3; J(P,H) = 12) [57] IR (Nujol): 1950, 1995, 2040 (CO) [57] laser irradiation at 488 nm of a methanol solution gave cis- and trans-$[(CH_3)_5C_5Mo(CO)_2(P(OCH_3)_3)CH_3OH]$-$[BF_4]$ [77] $NaOCH_3$ (one equivalent in CH_3OH between −60 and 0 °C) caused smooth conversion to a mixture of cis- and trans-$(CH_3)_5C_5Mo(CO)_2(P(OCH_3)_3)C(O)OCH_3$ [67]
72	$P(OCH_3)_3/CF_3SO_3^-$	equimolar amounts of $(CH_3)_5C_5Mo(CO)_2(P(OCH_3)_3)CHO$ and CF_3SO_2OR (R = H, CF_3) were allowed to react in $CHCl_3$ at −60 °C; the organometallic products, $[(CH_3)_5C_5Mo(CO)_3P(OCH_3)_3]O_3SCF_3$ and $(CH_3)_5C_5Mo(CO)_2(P(OCH_3)_3)OSO_2CF_3$, were separated by column chromatography on silica gel eluting with THF/$CHCl_3$ (10:1); No. 72 was isolated from the second fraction by adding diethyl ether and cooling to −78 °C (41 to 44%) [83, 84] yellow solid [83, 84] ^1H NMR: 2.10 (s, $(CH_3)_5C_5$), 3.90 (d, OCH_3; J(P,H) = 12) [83, 84] IR (Nujol): 1960, 1980, 2040 (CO) [83, 84]
*73	$P(OC_2H_5)_3/BF_4^-$	XX (72%) [73] yellow crystals [73] ^1H NMR (CD_2Cl_2): 1.38 (t, CH_2CH_3; J (H,H) = 7.1), 2.07 (d, $(CH_3)_5C_5$; J(P,H) = 1.2), 4.14 ("qui", CH_2) [73] ^{13}C NMR (CD_2Cl_2): 10.72 (s, $(CH_3)_5C_5$), 15.89 (d, $POCH_2CH_3$; J(P,C) = 7.2), 65.68 (d, $POCH_2CH_3$; J(P,C) = 9.0), 109.13 (s, $(CH_3)_5C_5$), 226.92 (d, CO; J(P,C) = 41.2), 228.79 (d, CO; J(P,C) = 1.7) [73] IR (Nujol): 1957, 1974, 2047 (CO) [73]
*74	$P(OC_6H_5)_3/BF_4^-$	XX (64%) [73] yellow crystals [73] ^1H NMR (CD_2Cl_2): 2.18 (d, CH_3; J(P,H) = 1.1), 7.20 (m, C_6H_5) [73]

References on pp. 313/6

Table 8 (continued)

No.	compound	method of preparation (yield) properties and remarks
		^{13}C NMR (CD_2Cl_2): 11.00 (s, ($\mathit{C}H_3)_5C_5$), 109.89 (s, $(CH_3)_5\mathit{C}_5$), 121.09 (d, ortho-C; J(P,C) = 4.0), 127.14, 130.76 (both s, para and meta-C), 150.85 (d, ipso-C; J(P,C) = 12.5), 225.32 (s, CO), 226.88 (d, CO; J(P,C) = 40.5) [73] IR (Nujol): 1970, 1990, 2058 (CO) [73]
*75	$P(OC_6H_5)_3/PF_6^-$	XX (59%) [56] yellow crystals [56] ^1H NMR (CD_2Cl_2): 2.18 (d, CH_3; J(P,H) = 1.2), 7 to 7.5 (m, C_6H_5) [56] ^{13}C NMR (CD_2Cl_2): 10.8 (s, ($\mathit{C}H_3)_5C_5$), 109.8 (s, $(CH_3)_5\mathit{C}_5$), 120.8 (d, ortho-C; J(P,C) = 4.0), 126.9 (s, para-C), 130.5 (s, meta-C), 150.6 (d, ipso-C; J(P,C) = 13.0), 225.3 (s, CO_{trans}), 226.9 (d, CO_{cis}; J(P,C) = 40.0) [56] ^{31}P NMR ($CDCl_3$): −144.3 (sept, PF_6^-; J(P,F) = 708), 156.2 (s, $P(OC_6H_5)_3$) [56] IR (Nujol): 1955, 1965, 2050 (CO) [56]; (CH_2Cl_2): 1990, 2080 (CO) [56]

*Further information:

M[$(CH_3)_5C_5Mo(CO)_3$] (M^+ = Li^+, Na^+; Table **8**, Nos. **1** and **3**). The ^{13}C and ^{17}O NMR spectra of Na[$(CH_3)_5C_5Mo(CO)_3$] (see table) and its lower methyl homologues Na[$(CH_3)_nC_5H_{5-n}$-$Mo(CO)_3$] (n = 0, 1, 3, 4; see there) indicated that the data for the carbonyl groups tend to shift downfield with an increasing degree of ring substitution [62].

Oxidation of the sodium salt by I_2 (THF, molar ratio 1:1) provided high yields of $(CH_3)_5$-$C_5Mo(CO)_3I$ [88].

Protonation of Li[$(CH_3)_5C_5Mo(CO)_3$] with glacial acetic acid in THF is a frequently used standard method of synthesis for $(CH_3)_5C_5Mo(CO)_3H$ [1, 12, 51, 56] (see Method IV).

Combination of M[$(CH_3)_5C_5Mo(CO)_3$] (M^+ = Li^+, Na^+) with SnR_3X (R = C_6H_5, X = Cl [6], I [92]; R = $C_6H_4CH_3$-4, X = I [92]) in THF yielded $(CH_3)_5C_5Mo(CO)_3SnR_3$. Treatment of the sodium salt with PCl_3 in methylcyclohexane, with $P(C_6H_5)_2Cl$ in benzene, with ClP=C(Si-$(CH_3)_3)_2$ in toluene/THF (20:1), or with BrP=C($Si(CH_3)_3$)C_6H_5 in THF at varied conditions (see Method VIII) provided $(CH_3)_5C_5Mo(CO)_3PCl_2$ [20], $(CH_3)_5C_5Mo(CO)_3P(C_6H_5)_2$ [17], $(CH_3)_5C_5Mo(CO)_3P{=}C(Si(CH_3)_3)_2$ [26, 107], and $(CH_3)_5C_5Mo(CO)_3P{=}C(Si(CH_3)_3)C_6H_5$ [107], respectively. K[$(CH_3)_5C_5Mo(CO)_3$] and ClPCH($Si(CH_3)_3$)C($Si(CH_3)_3)_2$-cyclo similarly furnished $(CH_3)_5Mo(CO)_3PCH(Si(CH_3)_3)C(Si(CH_3)_3)_2$-cyclo, albeit in an impure form [98]. The reaction of Li[$(CH_3)_5C_5Mo(CO)_3$] with stoichiometric amounts of 2,6-(t-$C_4H_9)_2$-4-RC_6H_2O-P($C{\equiv}CC_6H_5$)Cl (R = CH_3, t-C_4H_9) or 2,4,6-(t-$C_4H_9)_3C_6H_2OP(CH{=}CHC_6H_5)Cl$ in THF at 25 °C resulted in partial decarbonylation with formation of the Mo=P double-bonded complexes $(CH_3)_5C_5Mo(CO)_2{=}P(C{\equiv}CC_6H_5)OC_6H_2R$-4-($C_4H_9$-t$)_2$-2,6 (R = CH_3, t-C_4H_9) and $(CH_3)_5C_5Mo$-$(CO)_2{=}P(CH{=}CHC_6H_5)OC_6H_2(C_4H_9\text{-t})_3$-2,4,6 [75, 76]. Combination of cyclo-P(Cl)N(C_4H_9-t)-P(Cl)NC_4H_9-t with 2 equivalents of Na[$(CH_3)_5C_5Mo(CO)_3$] in THF likewise led to CO elimi-

References on pp. 313/6

nation and production of the diazadiphosphetidine-bridged complex $((CH_3)_5C_5Mo(CO)_2)_2$-(μ-PN(C_4H_9-t)PN(C_4H_9-t)-cyclo) [15]. Reaction of Na[$(CH_3)_5C_5Mo(CO)_3$] in THF with excess SO_2 at 0 °C proceeded immediately yielding a mixture of $((CH_3)_5C_5Mo(CO)_3)_2$ and $Na_2[S_2O_4]$ [31].

NaCl elimination between the sodium salt and (i-C_3H_7)$_4$$C_5HMo(CO)_3Cl$ in THF with subsequent heating in refluxing toluene resulted in the formation of $(CH_3)_5C_5Mo(CO)_2{\equiv}Mo(CO)_2$-$C_5H(C_3H_7\text{-i})_4$ [74]. Addition of Li[$(CH_3)_5C_5Mo(CO)_3$] to [$C_7H_7Mo(CO)_2(NCCH_3)$]BF_4 in THF rapidly produced $(CH_3)_5C_5Mo(CO)_3Mo(CO)_2C_7H_7$ [46, 61]. Oxidation of the lithium salt in THF solution at 0 °C, using either anhydrous $FeCl_3$ or $Fe_2(SO_4)_3$ in water/aqueous acetic acid, provided high yields of $(CH_3)_5C_5Mo(CO)_3Cl$ and $((CH_3)_5C_5Mo(CO)_3)_2$, respectively [56]. The heterobinuclear complex $(CH_3)_5C_5M(CO)_2(\mu\text{-}OC(CH_3)O)_2CuNC_5H_4CH_3$-4 was obtained by treating a benzene suspension of the dimeric copper(II) acetate adduct (4-$CH_3C_5H_4N$-Cu)$_2$(μ-O_2CCH_3)$_4$ with Li[$(CH_3)_5C_5Mo(CO)_3$] at room temperature [44]. Reaction of the lithium salt with equimolar $Cu(P(C_6H_5)_3)_3Cl$ in THF yielded impure $(CH_3)_5C_5Mo(CO)(\mu\text{-}CO)_2$-$CuP(C_6H_5)_3$ [117]. Reaction of the $[(CH_3)_5C_5Mo(CO)_3]^-$ anion (Li^+ salt) with $Hg(CN)_2$ from THF yielded $((CH_3)_5C_5Mo(CO)_3)_2Hg$ [55].

$(CH_3)_5C_5Mo(CO)_3CH_3$ was made by reactions of the lithium salt with excess CH_3I in THF at ambient temperature [1]. Reactions of Li[$(CH_3)_5C_5Mo(CO)_3$] with XCH_2Cl (X = Br, I) or $ClCH_2OCH_3$ under similar conditions afforded $(CH_3)_5C_5Mo(CO)_3CH_2X'$ (X' = Cl, OCH_3) together with some $(CH_3)_5C_5Mo(CO)_3Cl$. Treatment of the lithium salt with CH_2Br_2 accordingly gave $(CH_3)_5C_5Mo(CO)_3CH_2Br$. In contrast, only $(CH_3)_5C_5Mo(CO)_3I$ could be isolated from the corresponding reaction with CH_2I_2 [35]. 1,3-Dibromopropane reacted to form $(CH_3)_5C_5Mo$-$(CO)_3(CH_2)_3Br$ [59].

$(CH_3)_5C_5Mo(CO)_3Cl$ rather than the transition metal-substituted carboxylic acid R_5C_5Mo-$(CO)_3CH_2CO_2H$ (which was observed for R = H) was formed on the combination of Li[$(CH_3)_5C_5Mo(CO)_3$] with $ClCH_2C(O)OSi(CH_3)_3$ followed by deprotection of the trimethylsilyl group on silica gel surfaces [54]. With excess trifluoroacetic anhydride in $CH_3OC_2H_4OCH_3$, the acyl $(CH_3)_5C_5Mo(CO)_3C(O)CF_3$ was obtained [95]. Reactions of Li[$(CH_3)_5C_5Mo(CO)_3$] with substituted 4-chlorobuta-1,2-dienes, $ClCH_2C(R){=}C{=}CH_2$ (excess, THF), resulted in an unusual cyclization leading to the formation of cyclobutenone derivatives (Formula III; R = CH_3, C_2H_5) [101].

Low-temperature reactions in THF of Na[$(CH_3)_5C_5Mo(CO)_3$] with symmetrically and unsymmetrically substituted cyclopropenyl salts, [cyclo-$C_3(CH_3)_3$]BF_4, [cyclo-$C_3(C_6H_5)_3$]PF_6, [cyclo-$C_2(C_6H_5)_2CH$]ClO_4, [cyclo-$C_2(C_6H_5)_2CCH_3$]BF_4, or [cyclo-$C_2(C_6H_5)_2CC_4H_9$-t]PF_6, cleanly produced ring-expanded oxocyclobutenyl products (Formula IV; R = R' = CH_3 [30], C_6H_5 [14, 30]; R = C_6H_5: R' = H, CH_3, t-C_4H_9 [30]).

Protonation (glacial acetic acid in THF) of Na[$(CH_3)_5C_5Mo(CO)_3$] followed by nitrosylation (4-$CH_3C_6H_4SO_2N(NO)CH_3$/THF) afforded $(CH_3)_5C_5Mo(CO)_2NO$ [53]. Fair yields of the latter

$(CH_3)_5C_5Mo(CO)_2$ — R, =O, H H

III

$(CH_3)_5C_5Mo(CO)_2$ — O, R', R, R

IV

References on pp. 313/6

were also provided by direct combination of 4-$CH_3C_6H_4SO_2N(NO)CH_3$ with the sodium salt in THF [11].

$[N(C_4H_9\text{-}n)_4][(CH_3)_5C_5Mo(CO)_3]$ (Table **8**, No. **5**) was formed in CH_2Cl_2 solution by disproportionation of $((CH_3)_5C_5Mo(CO)_3)_2$, proceeding readily in the presence of an excess of $[N(C_4H_9\text{-}n)_4]BF_4$; the other species observed was $(CH_3)_5C_5Mo(CO)_3FBF_3$ [58]. The complex was also electrogenerated via two-electron transfer to $((CH_3)_5C_5Mo(CO)_3)_2$ in $CH_2Cl_2/[N(C_4H_9\text{-}n)_4]X$ ($X^- = BF_4^-$, PF_6^-) by keeping the electrode rest potential negative enough (< -1.58 V [58] or < -1.71 V [56] vs. SCE) to cause reduction of the homodimer [58, 72]. Electrochemical reduction of several $[(CH_3)_5C_5Mo(CO)_3{}^2D]PF_6$ salts in $CH_2Cl_2/[N(C_4H_9\text{-}n)_4]PF_6$ also provided dilute CH_2Cl_2 solutions of the salt by specific loss of the ^2D ligand after electron transfer [56]. For pertinent reduction potentials, see "Further information" given for ionic $[(CH_3)_5C_5Mo(CO)_3{}^2D]X$ complexes.

The following peak potentials were reported for oxidation to $((CH_3)_5C_5Mo(CO)_3)_2$ in $CH_2Cl_2/[N(C_4H_9\text{-}n)_4]X$ ($X^- = BF_4^-$, PF_6^-): E_{ox} (SCE) $= -0.34$ V [56] or -0.22 V [58], E_{ox} ($(C_5H_5)_2Fe/[(C_5H_5)_2Fe]^+$) $= -0.709$ V (not corrected for kinetic potential shifts) and -0.603 V (reversible potential calculated after addition of an estimated kinetic potential shift of $+0.106$ V), and E_{ox} (NHE) $= -0.554$ V (reversible potential) [72, 78, 81, 100, 105].

$(CH_3)_5C_5Mo(CO)_3H$ (Table **8**, No. **22**). Accompanied by $P(C_6H_5)_3$, the hydride was observed as a product resulting from thermal decomposition of the formyl species $(CH_3)_5C_5Mo(CO)_2(PC_6H_5)_3CHO$ occurring in solution above $-40\,^\circ C$ [57]. Treatment of the "isocarbonyl"-bridged complex $(CH_3)_5C_5Mo(CO)_2(\mu\text{-}CO)U(C_5H_5)_3$ with anhydrous HCl in toluene at $-70\,^\circ C$ gave No. 22 together with $(C_5H_5)_3UCl$ [27].

The enthalpy of formation from $CH_3C_6H_5Mo(CO)_3$ and cyclo-$(CH_3)_5C_5H$ (Method VI) in THF was determined as -5.5 ± 0.2 kcal/mol. Comparison with the value measured for the reaction between $CH_3C_6H_5Mo(CO)_3$ and cyclo-C_5H_6 forming $C_5H_5Mo(CO)_3H$ ($\Delta H = -7.3 \pm 0.5$ kcal/mol) indicated that in THF, substitution of pentamethylcyclopentadiene for cyclopentadiene according to "$C_5H_5Mo(CO)_3H$ + cyclo-$(CH_3)_5C_5H \rightleftarrows (CH_3)_5C_5Mo(CO)_3H$ + cyclo-C_5H_6" is actually slightly endothermic by about 1.8 kcal/mol [23], in good agreement with an estimate of $+2.3$ kcal/mol made from calculated enthalpies for the exchange reactions between $C_6H_6Mo(CO)_3$ and cyclo-R_5C_5H (R = H, CH_3) in solution [41]. Thermochemical measurements of the ligand exchange between $R_5C_5Mo(CO)_3H$ (R = H, CH_3) and pyridine, which at 50 °C proceeded quantitatively with the formation of $Mo(CO)_3(NC_5H_5)_3$ and cyclo-R_5C_5H, also indicated a small difference in the $(CH_3)_5C_5{-}Mo$ and $C_5H_5{-}Mo$ bond strengths; the enthalpies of the two reactions are -16.7 ± 0.7 and -16.2 ± 0.3 kcal/mol for $C_5H_5Mo(CO)_3H$ and $(CH_3)_5C_5Mo(CO)_3H$, respectively [23]. For the addition of pentamethylcyclopentadiene to $Mo(CO)_3(NCCH_3)_3$ (Method V), an enthalpy of $+11.3$ kcal/mol was estimated from the enthalpies calculated for the exchange reactions between $C_6H_6Mo(CO)_3$ and cyclo-$(CH_3)_5C_5H$ or CH_3CN [41], in excellent agreement with a study of the temperature variation of the equilibrium constant for "$Mo(CO)_3(NCCH_3)_3$ + cyclo-$(CH_3)_5C_5H \rightleftarrows (CH_3)_5C_5Mo(CO)_3H$ + 3 CH_3CN", which yielded $+10.8$ kcal/mol [37]. The heat of hydrogenation of $(R_5C_5Mo(CO)_3)_2$ to yield two moles of $R_5C_5Mo(CO)_3H$ was found to be more favorable by 1.8 kcal/mol for R = CH_3 ($\Delta H = +5.5 \pm 0.6$ kcal/mol) compared to R = H ($\Delta H = +7.3 \pm 0.8$ kcal/mol), probably due to steric repulsion in the more crowded $((CH_3)_5C_5Mo(CO)_3)_2$ dimer [23]. By use of a thermochemical cycle that requires knowledge of the metal hydride pK_a (see below) and the oxidation potential of its conjugate base, the homolytic bond dissociation energy D(Mo−H) was estimated in acetonitrile solution as $+68.5$ to $+69$ kcal/mol [72, 81, 85, 100, 105], which compares well to an Mo−H bond strength of $+66.0$ kcal/mol derived from solution calorimetry data [41].

References on pp. 313/6

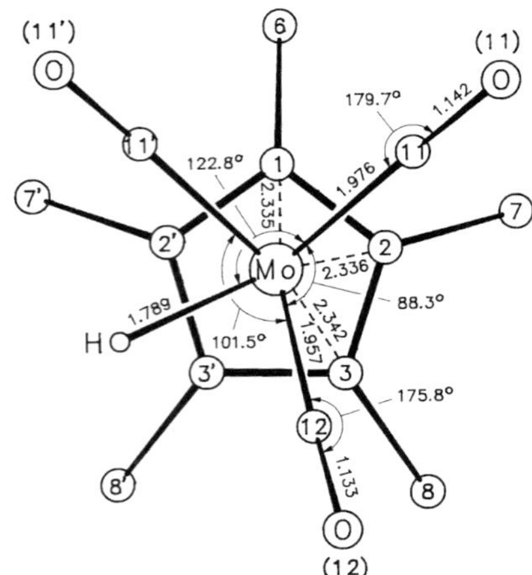

Fig. 27. The molecular structure of $(CH_3)_5C_5Mo(CO)_3H$ as determined by neutron diffraction [111].

No. 22 crystallizes in the monoclinic system. The X-ray structure determination reported originally (crystal data: space group $P2_1-C_2^2$ (No. 4), a = 8.003(2), b = 12.874(3), c = 7.087(2) Å, β = 103.06(2)°; Z = 2 molecules per unit cell, D_{calc} = 1.471 g/cm^3 [38]) was later shown to involve an incorrectly assigned space group, with a resultant inaccurate molecular geometry. Refinement of the structural model with mirror constraints in the centrosymmetric space group $P2_1/m-C_{2h}^2$ (No. 11) against X-ray data collected at 163(3) K (a = 6.969(2), b = 12.651(2), c = 7.954(2) Å, β = 103.36(2)°; Z = 2 molecules per unit cell, D_{calc} = 1.539 g/cm^3) converged at identical fit indices as in $P2_1$ but gave the more accurate and precise molecular parameters without unreasonable large variations among chemically equivalent bond lengths, such as noted in the $P2_1$ model. Neutron diffraction results obtained at 163.0(5) K (a = 6.968(2), b = 12.658(2), c = 7.953(2) Å, β = 103.34(2)°; Z = 2 molecules per unit cell, D_{calc} = 1.538 g/cm^3) confirmed the choice of $P2_1/m$ as the correct space group as refinement in $P2_1$ against the neutron data failed. The molecular geometry is shown in **Fig. 27**. The crystallographic mirror plane (passing through atoms Mo, C-1, C-6, and bisecting the C-3−C-3′ bond of the ring) required statistical disorder of both the hydride ligand and the carbonyl group C(12)−O(12). In addition, the $(CH_3)_5C_5$ ring exhibited rotational disorder [111].

Photolysis with a low-pressure mercury lamp in dihydrogen- or deuterium-containing argon matrices at 12 K generated dicarbonyl adducts of molecular hydrogen, $(CH_3)_5C_5$-$Mo(CO)_2(X_2)X$ (X = H and/or D), shown to arise from $(CH_3)_5C_5Mo(CO)_2X$ (the products of photolysis at $\lambda < 400$ nm) and X_2 with irradiation at $\lambda > 400$ nm. Prolonged photolysis in Ar matrices containing 20 mol% of D_2 in addition to 7 mol% of CO produced $(CH_3)_5C_5Mo(CO)_3D$ [39, 43]; see also "Organomolybdenum Compounds" Pt. 7, 1991, pp. 1/5.

Constant-current coulometry with derivative cyclic voltammetry monitoring of the oxidative consumption of the hydride in CH_3CN/0.1 M $[N(C_4H_9\text{-}n)_4][PF_6]$ required the passage of 2.5 ± 0.1 F/mol for the substrate oxidation peak (E_{ox} = +0.561 V versus $(C_5H_5)_2Fe/[(C_5H_5)_2Fe]^+$) to disappear completely. Preparative electrolysis (2×10^{-3} M hydride in CH_3CN with 0.05 M $[N(CH_3)_4]BF_4$) at a constant potential some 100 mV more anodic than E_{ox} yielded $[(CH_3)_5C_5Mo(CO)_3NCCH_3]BF_4$ [78]. A thermodynamic cycle was used to estimate

References on pp. 313/6

the acidity of the cation radical intermediate, $[(CH_3)_5C_5Mo(CO)_3H]^{+\cdot}$: $pK_a(CH_3CN) = -2.5$ ($pK_a = +17.1$ for the neutral hydride; see below) [78, 105].

The reaction of elemental sulfur with $(CH_3)_5C_5Mo(CO)_3H$ in a 1:1.5 molar ratio in refluxing THF produced insoluble polymeric material together with two soluble complexes, $((CH_3)_5C_5MoS(\mu\text{-}S))_2$ and $((CH_3)_5C_5MoS_5)_2$ (Formula V) containing both bridging and nonbridging S_2 ligands [12].

S—S S—S
$(CH_3)_5C_5Mo$ S S $MoC_5(CH_3)_5$
S—S S—S

V

In order to determine the relative acidities of $(CH_3)_5C_5Mo(CO)_3H$ and $C_5H_5Mo(CO)_3H$, the acid-base equilibrium "$(CH_3)_5C_5Mo(CO)_3H + [C_5H_5Mo(CO)_3]^- \rightleftarrows C_5H_5Mo(CO)_3H + [(CH_3)_5C_5Mo(CO)_3]^-$" was studied qualitatively in THF solution by IR spectrometry. The equilibrium was approached from both sides and lies far to the left, $K_{eq} \leq 10^{-3}$, suggesting that the pK_a of $(CH_3)_5C_5Mo(CO)_3H$ would exceed that of $C_5H_5Mo(CO)_3H$ by at least 3.0 units [23]. This prediction was confirmed by independent quantitative IR measurements of the position of deprotonation equilibria in acetonitrile between $R_5C_5Mo(CO)_3H$ (R = H, CH_3) and nitrogen bases such as pyridine and morpholine ($K_{eq} = 0.31$ for R = CH_3 at 25 °C), which led to pK_a values of 17.1 for the $(CH_3)_5C_5$ compound but 13.9 for the C_5H_5 complex [42]. $(CH_3)_5C_5Mo(CO)_3H$ was deprotonated quantitatively by LiC_4H_9-n in toluene in the presence of tetramethylethylenediamine to give $[Li((CH_3)_2NC_2H_4N(CH_3)_2)_2][(CH_3)_5C_5Mo(CO)_3]$ in high yield [38]. In a similar fashion, stock solutions of $Na[(CH_3)_5C_5Mo(CO)_3]$ were shown to be readily obtainable by reaction of the parent hydride with NaH in THF [23]. Quantitative deprotonation with the formation of $[H_2NC(N(CH_3)_2)_2][(CH_3)_5C_5Mo(CO)_3]$ was also accomplished using tetramethyl guanidine, $HN{=}C(N(CH_3)_2)_2$, in slight excess in CH_3CN [42].

Treatment in $CHCl_3$ or toluene solution with CO (1 atm), CO_2 (1 atm), or with other CO_2-like molecules (CS_2 or $C({=}NC_6H_4CH_3\text{-}4)_2$) resulted in no reaction with any of these ligands; the only detectable transformation was a slow conversion into the homodimer $((CH_3)_5C_5Mo(CO)_3)_2$ [38]. The molecule could be enriched with ^{13}CO by dissolving the ^{12}CO isotopomer in hexane and photolyzing the solution in a quartz vessel with a medium-pressure mercury lamp under an atmosphere of ^{13}CO [43]. In THF at 0 °C, the reaction of $Pb(N(Si(CH_3)_3)_2)_2$ with two equivalents of No. 22 produced $((CH_3)_5C_5Mo(CO)_3)_2Pb \cdot OC_4H_8$. When the same reaction was carried out in n-pentane/n-hexane (1:1), the bis(isocarbonyl)-bridged hexametallic complex $((CH_3)_5C_5Mo(CO)_3PbMo(CO)_2C_5(CH_3)_5(\mu\text{-}CO))_2$ containing a pair of $Mo{=}C{=}O \rightarrow Pb$ linkages was formed [51].

Thermal reaction with phosphanes, PR_3 (1.0 to 1.2 equivalents), in pentane at room temperature ($PR_3 = P(CH_3)_3$ [48]) or in toluene at 20 °C ($PR_3 = P(CH_3)_3$, $P(CH_3)_2C_6H_5$ [102]) or 65 °C ($PR_3 = P(C_6H_5)_3$ [102]) yielded monosubstituted products, $(CH_3)_5C_5Mo(CO)_2(PR_3)H$, either as a mixture of cis and trans isomers (CO substitution in pentane [48]) or as single cis isomers (reactions in toluene [102]). No reaction occurred with $P(C_6H_{11}\text{-cyclo})_3$, probably due to steric hindrance [97]. Irradiation in pentane solution in the presence of excess trimethylphosphane provided a disubstituted derivative, trans-$(CH_3)_5C_5Mo(CO)(P(CH_3)_3)_2H$, along with minor quantities of $(CH_3)_5C_5Mo(CO)_3Mo(CO)_2(P(CH_3)_3)C_5(CH_3)_5$ [48]. $(CH_3)_5C_5Mo(CO)_3H$ also reacted readily with phosphites, $P(OR)_3$, in benzene or toluene at 20 to 25 °C to give CO substitution with the formation of cis-$(CH_3)_5C_5Mo(CO)_2(P(OR)_3)H$ (R = CH_3,

References on pp. 313/6

C_6H_5) [97, 102]. Addition of the Mo–H bond to the phosphaalkene CF_2=PCF_3 (toluene-d_8, 20 °C) resulted in rapid and quantitative conversion to $(CH_3)_5C_5Mo(CO)_3P(CF_3)CF_2H$ [45].

Treatment of the hydride in concentrated diethyl ether solution at 0 °C with a stream of SO_2 gas afforded $(CH_3)_5C_5Mo(CO)_3SO_2H$ [31, 36]. In contrast, cooling dilute CH_3CN solutions containing excess SO_2 to −20 °C furnished $((CH_3)_5C_5Mo(CO)_3)_2(\mu\text{-}S_2O_4)$ [31, 36]. Prolonged reactions in solution of $(CH_3)_5C_5Mo(CO)_3H$ with SO_2 in the molar ratio 4:1 gave $((CH_3)_5C_5MoO(\mu\text{-}S))_2$, a species of composition $((CH_3)_5C_5Mo)_3S_2O_4$, and $((CH_3)_5C_5Mo(CO)_3)_2$, along with water [31, 36].

Heterotetrametallic μ-phosphido and μ-phosphinidene $MoRu_3$ clusters, Formulas VI and VII, resulted from the reaction of $Ru_3(CO)_{10}(\mu\text{-}H)(\mu\text{-}P(C_6H_5)_2)$ with excess $(CH_3)_5C_5Mo(CO)_3H$ in refluxing toluene [112].

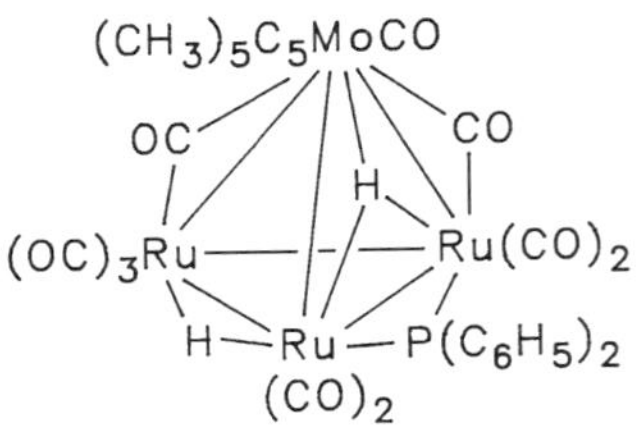

$(CH_3)_5C_5MoCO$; OC; CO; H; $(OC)_3Ru$; $Ru(CO)_2$; H; Ru; $P(C_6H_5)_2$; $(CO)_2$

VI

$(CH_3)_5C_5MoCO$; OC; PC_6H_5; H; $(OC)_2Ru$; $Ru(CO)_3$; OC; Ru; $(CO)_2$

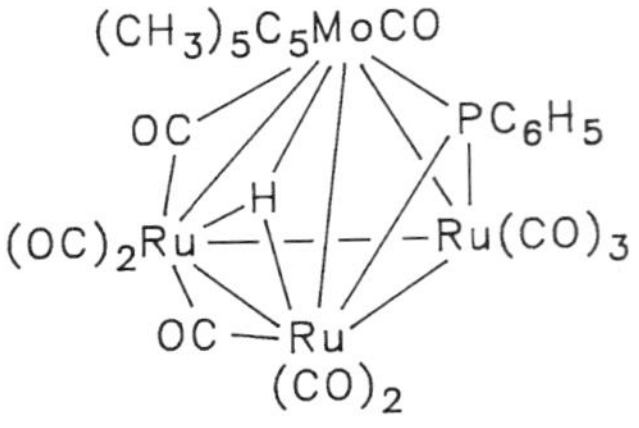

VII

The "isocarbonyl"-linked Mo=CO→U complex $(CH_3)_5C_5Mo(CO)_2(\mu\text{-}CO)U(C_5H_5)_3$ was quantitatively obtained by a methane elimination reaction between $(CH_3)_5C_5Mo(CO)_3H$ and $(C_5H_5)_3UCH_3$ in heptane at −50 °C [27].

The hydrogenation of α-cyclopropylstyrene by a series of transition metal hydrides including $(CH_3)_5C_5Mo(CO)_3H$ gave a mixture of unrearranged and rearranged hydrogenation products, $C_6H_5CH(C_3H_5\text{-cyclo})CH_3$, and (E)-$C_6H_5C(CH_3)$=$CHCH_2CH_3$, respectively. Generation of the carbon-centered radical $C_6H_5C^{\cdot}(C_3H_5\text{-cyclo})CH_3$ by H atom transfer from the metal hydride was shown to be the rate-determining step. In toluene at 60 °C, the second-order rate constant for hydrogen abstraction by the olefin from $(CH_3)_5C_5Mo(CO)_3H$ was measured as $6.3(4)\times10^{-5}$ $M^{-1}\cdot s^{-1}$; the relative rates, k_{rel}, ranged from k_{rel} = 1 for cis-$Mn(CO)_4(P(C_6H_5)_3)H$ to k_{rel} = 4 for $(CH_3)_5C_5Mo(CO)_3H$, and were k_{rel} = 93 for $C_5H_5Mo(CO)_3H$, and k_{rel} = 94 for $(CH_3)_5C_5Fe(CO)_2H$ [82]. A reaction with cyclopentadiene in THF did not occur at room temperature. At 50 °C, slow conversion to $((CH_3)_5C_5Mo(CO)_3)_2$ was observed [23]. Cyclohexa-1,3-diene in heptane at 50 °C reacted quantitatively according to "2 $(CH_3)_5C_5Mo(CO)_3H$ + cyclo-C_6H_8 → $((CH_3)_5C_5Mo(CO)_3)_2$ + cyclo-C_6H_{10}"; ΔH = −32.0 ± 0.6 kcal/mol [23, 60].

$(CH_3)_5C_5Mo(CO)_3H$ reacted with t-C_4H_9NC in pentane at room temperature forming trans-$(CH_3)_5C_5Mo(CO)_2(CNC_4H_9\text{-t})H$ [64]. Combination with neat pyridine at room temperature produced the anion $[(CH_3)_5C_5Mo(CO)_3]^-$ and also $Mo(CO)_3(NC_5H_5)_3$, the latter forming quantitatively at 50 °C [23]. Reductive elimination of cyclo-$(CH_3)_5C_5H$ with a clean formation of $Mo(CO)_3(NCCD_3)_3$ was complete in CD_3CN at room temperature within 2 h. For other strongly coordinating solvents, such as pyridine (see above) or dimethyl sulfoxide, the solvent-induced elimination was even faster than for nitriles (ca. 20 min). However, no evidence for a reaction in THF-d_8 was found even after 2 d at 25 °C [97].

Dissolution in an excess of CCl_4 caused the chloro complex $(CH_3)_5C_5Mo(CO)_3Cl$ to precipitate in high yields [38, 56]. If $CHCl_3$ or CH_2Cl_2 were used instead of CCl_4, the halogenation reaction was much slower and considerable amounts of $((CH_3)_5C_5Mo(CO)_3)_2$ and $((CH_3)_5C_5Mo(CO)_2)_2$ were formed, together with $(CH_3)_5C_5Mo(CO)_3Cl$ [38].

Addition of $[C(C_6H_5)_3]X$ ($X^- = BF_4^-$ or PF_6^-) to CH_2Cl_2 solutions of the title complex at low temperature resulted in hydride abstraction and generation of the corresponding weakly coordinated Lewis acid species, $(CH_3)_5C_5Mo(CO)_3FBF_3$ and $(CH_3)_5C_5Mo(CO)_3FPF_5$. These were either isolated [56, 57] or directly converted into $[(CH_3)_5C_5Mo(CO)_3{}^2D]X$ (see there), into $[(CH_3)_5C_5Mo(CO)_2((C_6H_5)_2PC_2H_4P(C_6H_5)_2)]PF_6$ [56], or into $[(CH_3)_5C_5Mo(CO)_4]PF_6$ [56] by adding CO, $(C_6H_5)_2PC_2H_4P(C_6H_5)_2$, or suitable ^2D ligands, e.g., nitriles [56], phosphanes [56], or phosphites [56, 73]; see also Method XX.

The tris(p-t-butylphenyl)methyl radical (made by treating a toluene solution of $(4\text{-t-}C_4H_9\text{-}C_6H_4)_3CBr$ with copper powder abstracted the hydrogen atom from $(CH_3)_5C_5Mo(CO)_3H$ (and various other transition metal hydrides) forming $(4\text{-t-}C_4H_9C_6H_4)_3CH$ together with the rapidly dimerizing $(CH_3)_5C_5Mo(CO)_3^{\cdot}$ radical. At 25 °C in toluene, the transfer rate constant for H atom transfer, k_H, was determined as 13.9(5) M/s. The activation parameters were $\Delta H^‡ = 7.7(3)$ kcal/mol and $\Delta S^‡ = -27.4(8)$ cal · mol^{-1} · K^{-1} [99]. The Arrhenius A factor for H atom abstraction from $(CH_3)_5C_5Mo(CO)_3H$ by the primary hex-5-enyl radical was estimated as $\log(A/M^{-1} \cdot s^{-1}) = 9.27 \pm 0.04-(1.35 \pm 0.11)/(2.3\ RT\ kcal/mol)$ [70].

$(CH_3)_5C_5Mo(CO)_3Cl$ (Table **8**, No. **39**). Adding $Li[(CH_3)_5C_5Mo(CO)_3]$ to XCH_2Cl (X = Br, I) or $ClCH_2OCH_3$ in THF afforded the title compound (minor product) in addition to $(CH_3)_5C_5Mo(CO)_3CH_2X'$ (X' = Cl, OCH_3; desired major products) [35]. Reactions between the lithium salt and trimethylsilyl chloroacetate, $ClCH_2C(O)OSi(CH_3)_3$, followed by deprotection of the silyl group during workup by column chromatography on silica gel, provided a 21% yield of the chloro complex rather than the transition metal-substituted carboxylic acid $R_5C_5Mo(CO)_3CH_2CO_2H$ (R = CH_3), which was obtained for R = H [54]. $(CH_3)_5C_5Mo(CO)_3Cl$ was also observed as the product resulting from thermal decomposition of $(CH_3)_5C_5Mo(CO)_3PCl_2$, which proceeded quantitatively in solution above −15 °C [20]. The reaction of $(CH_3)_5C_5Mo(CO)_2(\mu\text{-}CO)U(C_5H_5)_3$ with CCl_4 in toluene at low temperature provided a quantitative yield of $(CH_3)_5C_5Mo(CO)_3Cl$ accompanied by $(C_5H_5)_3UCl$ [27].

UV irradiation (medium-pressure Hg lamp) in toluene in the presence of oxygen yielded $(CH_3)_5C_5MoO_2Cl$ [96].

$AlCl_3$-promoted chloride expulsion in heptane under 100 atm of CO, followed by hydrolysis and treatment with aqueous HPF_6, gave $[(CH_3)_5C_5Mo(CO)_4]PF_6$; analogous reactions using phosphanes and arsanes in place of CO failed [56]. Irradiation in the presence of equimolar $P(C_6H_5)_3$ in toluene provided high yields of $(CH_3)_5C_5Mo(CO)_2(P(C_6H_5)_3)Cl$ [56].

Combination with an equimolar quantity of $Na[(i\text{-}C_3H_7)_4C_5HMo(CO)_3]$ in THF, followed by heating in toluene at reflux temperature, led to $(CH_3)_5C_5Mo(CO)_2{\equiv}Mo(CO)_2C_5H(C_3H_7\text{-i})_4$ [74].

$(CH_3)_5C_5Mo(CO)_3I$ (Table **8**, No. **41**). The reaction of $Li[(CH_3)_5C_5Mo(CO)_3]$ with CH_2I_2 in THF resulted only in the isolation of $(CH_3)_5C_5Mo(CO)_3I$ (yield: 25%) and gave none of the hoped-for $(CH_3)_5C_5Mo(CO)_3CH_2I$ [35]. Accompanied by the latter, No. 41 was also formed by the treatment of $(CH_3)_5C_5Mo(CO)_3OCH_3$ with dry HI gas in hexane solution [35].

The sodium amalgam reduction of $(CH_3)_5C_5Mo(CO)_3I$ in THF solution, followed by reaction of the presumably generated $[(CH_3)_5C_5Mo(CO)_3]^-$ anion with additional iodo complex, provided the heterotrimetallic mercury derivative $((CH_3)_5C_5Mo(CO)_3)_2Hg$, rather than the desired homobimetallic $((CH_3)_5C_5Mo(CO)_3)_2$ [6].

The cis and trans isomers of $(CH_3)_5C_5Mo(CO)_2(CNC_2H_5)I$ were formed quantitatively when the iodo complex was allowed to interact with one equivalent of ethyl isocyanide in CH_2Cl_2 between −50 °C and room temperature [88].

References on pp. 313/6

$(CH_3)_5C_5Mo(CO)_3FBF_3$ and **$(CH_3)_5C_5Mo(CO)_3FPF_5$** (Table **8**, Nos. **44** and **45**) were frequently employed for further synthesis as in situ intermediates generated in CH_2Cl_2 solution, either in a similar fashion as outlined by Method XVII [56, 73], or by protolytic cleavage of the Mo–C bond in $(CH_3)_5C_5Mo(CO)_3CH_3$ using $[HO(C_2H_5)_2]BF_4$ in stoichiometric quantity at −78 °C [93]. Redox cleavage of $((CH_3)_5C_5Mo(CO)_3)_2$ with AgX (X^- = BF_4^- [10], PF_6^- [28]) or two-electron electrooxidation of the latter homodimer in the presence of added $[N(C_4H_9\text{-}n)_4]BF_4$ (E_{ox} = +0.85 V vs. SCE) [58] or $[N(C_4H_9\text{-}n)_4]PF_6$ (E_{ox} = +0.67 V versus SCE) [56] were also used for the in situ-generation of the two complexes. Furthermore, solutions containing $(CH_3)_5C_5Mo(CO)_3FBF_3$ along with $[N(C_4H_9\text{-}n)_4][(CH_3)_5C_5Mo(CO)_3]$ were smoothly formed by BF_4^--assisted disproportionation of $((CH_3)_5C_5Mo(CO)_3)_2$ in CH_2Cl_2/$[N(C_4H_9\text{-}n)_4]BF_4$ [58].

The peak potentials for two-electron electroreduction to $((CH_3)_5C_5Mo(CO)_3)_2$ of −0.51 V (No. 44 [58]) and −0.62 V (No. 45 [56]) were measured versus SCE in CH_2Cl_2/0.1 M $[N(C_4H_9\text{-}n)_4]BF_4$ and CH_2Cl_2/0.1 M $[N(C_4H_9\text{-}n)_4][PF_6]$, respectively.

Exposure of CH_2Cl_2 solutions to CO or the addition of $CNC_6H_4CH_3$-4 or ^2D ligands (nitriles, phosphanes, and phosphites) to such solutions produced salt-like products, $[(CH_3)_5C_5Mo(CO)_4]PF_6$ [56]; $[(CH_3)_5C_5Mo(CO)_3{}^1L]BF_4$ (^1L = CO, $CNC_6H_4CH_3$-4) [57]; $[(CH_3)_5C_5Mo(CO)_3{}^2D]BF_4$ (^2D = CH_3CN [93], $P(C_6H_5)_3$ [57], $P(OCH_3)_3$ [57], $P(OC_2H_5)_3$, $P(OC_6H_5)_3$ [73]); $[(CH_3)_5C_5Mo(CO)_3{}^2D]PF_6$ (^2D = CH_3CN, PMe_3, $P(C_4H_9\text{-}n)_3$, $P(C_6H_5)_2CH_3$, $P(C_6H_5)_3$, $P(OC_6H_5)_3$) [56]; and $[(CH_3)_5C_5Mo(CO)_2((C_6H_5)_2PC_2H_4P(C_6H_5)_2)]PF_6$ [56], respectively. See also Methods XX and XXII.

An insoluble polynuclear precipitate, $((CH_3)_5C_5Mo(CO)_2OCH_3)_n$, was deposited from solution when $(CH_3)_5C_5Mo(CO)_3FBF_3$ reacted with $Na[OCH_3]$ in a 1:1 molar ratio in methanol between −40 and +20 °C [67].

$(CH_3)_5C_5Mo(CO)_3FBF_3$, generated in situ from $((CH_3)_5C_5Mo(CO)_3)_2$ and $AgBF_4$ in CH_2Cl_2 solution by redox cleavage of the Mo–Mo single bond, was readily captured by added methylenecyclopropane and cis- and trans-2,3-dimethylenecyclopropanes to form the dicarbonyl(π-trimethylenemethane) salts, $[(CH_3)_5C_5Mo(CO)_2(\mu\text{-}CH_2C(CHR)_2)]BF_4$ (R = H, CH_3); the stereochemistry of the substituted products implicates disrotatory ring-opening reactions [10]. Addition at −80 °C of $(CH_3)_5C_5Mo(CO)_3FPF_5$ (from the homodimer by oxidative cleavage with $AgPF_6$ in CH_2Cl_2) to a THF slurry of sodium 2,3-diphenyl-2-cyclopropene-1-carboxylate, $Na[cyclo\text{-}C_2(C_6H_5)_2CCO_2]$, resulted in displacement of CO and formation of a (cyclopropene)carboxylato complex (Formula VIII; R = C_6H_5) [28].

$(CH_3)_5C_5(OC)_2Mo$ … VIII | $(CH_3)_5C_5(OC)_2Mo$ … IX | $(CH_3)_5C_5(CO)Mo$ … X

$(CH_3)_5C_5Mo(CO)_3CH_3$ (Table **8**, No. **46**). UV irradiation in THF yielded $((CH_3)_5C_5Mo(CO)_2)_2$ instead of the hoped-for $((CH_3)_5C_5Mo(CO)_3)_2$ [6].

Combination with iodine (slight deficiency in CH_2Cl_2) led to $(CH_3)_5C_5Mo(CO)_3I$ [6].

Protolytic cleavage of the Mo–C bond using $[HO(C_2H_5)_2]BF_4$ in 1:1 stoichiometry in CH_2Cl_2 at −78 °C was adopted as an in situ method for generating solutions of $(CH_3)_5C_5Mo$-$(CO)_3FBF_3$ [93]. Heating a solution in tetradecane under 300 atm of CO at 137 °C resulted in the formation of $Mo(CO)_6$ and liberation of the CH_3 and $(CH_3)_5C_5$ ligands as the ketone $(CH_3)_5C_5C(O)CH_3$. This is the reverse of the reaction used for the preparation of the title compound by heating $Mo(CO)_6$ with 5-acetyl-pentamethyl-cyclopentadiene according to Method XIX [9]. The reaction between $(CH_3)_5C_5Mo(CO)_3CH_3$ and PCl_5 in refluxing CH_2Cl_2 furnished high yields of $(CH_3)_5C_5MoCl_4$ [29]. Stirring the methyl derivative in liquid SO_2 provided the sulfinato-S compound $(CH_3)_5C_5Mo(CO)_3SO_2CH_3$ [6]. The reaction with one equivalent of a polysulfide solution generated from a solid of nominal composition K_2S_3 in $CH_3CN/(CH_3)_2NCHO$ (5:1) resulted in oxidation of the metal center to Mo(IV) with concomitant dealkylation and decarbonylation to give an anionic (dithiocarbonato)polychalcogeno complex, $[(CH_3)_5C_5Mo(S_2CO)S_4]^-$, which was isolated as its $[P(C_6H_5)_4]^+$ salt [89].

UV irradiation of $(CH_3)_5C_5Mo(CO)_3CH_3$ in the presence of buta-1,3-diene in hexane at 253 K yielded a mixture of four complexes which could be separated by low-temperature column chromatography: the diene-substituted alkyl $(CH_3)_5C_5Mo(CO)(\eta^4\text{-}CH_2{=}CHCH{=}CH_2)\text{-}CH_3$, two π-enyl and π-enyl ketone derivatives (Formulas IX and X), as well as the triple-bonded dimer $((CH_3)_5C_5Mo(CO)_2)_2$ [50]. A similar experiment in a mixture of cycloocta-1,5-diene and cyclohexane gave only the triple-bonded dimer and none of a hoped-for cyclooctadiene derivative [6]. The metalacyclic alkenyl ketone derivative (Formula XI; R = R′ = H) was produced by the photoinduced reaction of the title complex with acetylene in pentane [24]. With phenylacetylene, the metalacycle (Formula XI) was formed as a 2:1 mixture of isomers, differing in the relative position of hydrogen and phenyl substituents: R = C_6H_5, R′ = H (66%) and R = H, R′ = C_6H_5 (33%), respectively [49].

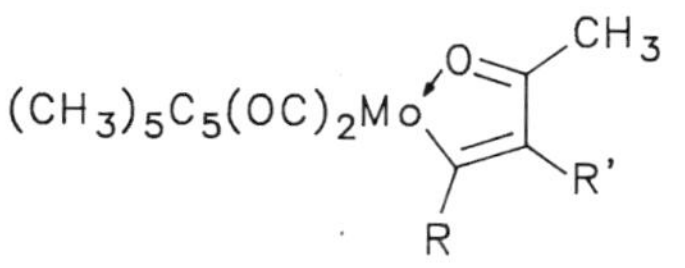

$(CH_3)_5C_5(OC)_2Mo$ — XI

$(CH_3)_5C_5(OC)_2Mo$ — XII (C_6H_4R-4)

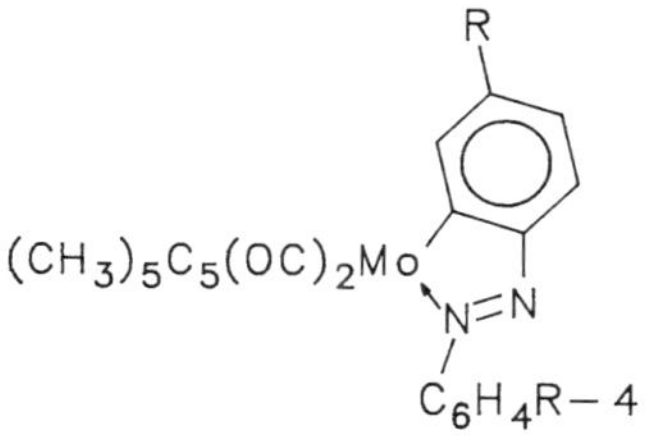

Heating the methyl complex in toluene in the presence of equimolar quantities of aromatic diazenes, $4\text{-}RC_6H_4N{=}NC_6H_4R\text{-}4'$, afforded cyclometalated products (Formula XII; R = H, CH_3) [87].

$(CH_3)_5C_5Mo(CO)_3CH_2C_6H_5$ (Table **8**, No. **47**). Reactions with tertiary phosphanes in dimethyl sulfoxide or acetonitrile to give substituted acyl complexes, $(CH_3)_5C_5Mo(CO)_2(PR_3)C(O)\text{-}CH_2C_6H_5$, as mixtures of cis and trans isomers were shown to be restricted to the smaller ligands, $P(C_6H_5)_3$, e.g., being unreactive. These CO insertions as well as analogous reactions induced by alkyl isocyanides in polar solvents yielding $(CH_3)_5C_5Mo(CO)_2(CNR)C(O)\text{-}CH_2C_6H_5$ were kinetically analyzed in terms of a two-stage process involving (i) a k_1k_2 path via a short-lived solvent-stabilized acyl intermediate, $(CH_3)_5C_5Mo(CO)_2C(O)CH_2C_6H_5\cdot solv$, which further reacts with the inducing nucleophile, ^2D, to give the final acyl complex ($k_2[^2D]$ step), and (ii) a direct second-order k_3 path from starting material to product: $k_{obs} = \{k_1k_2[^2D]/(k_{-1} + k_2[^2D]) + k_3[^2D]\}$ for reactions going to completion. With $P(CH_3)_2C_6H_5$ as the entering nucleophile, the k_1k_2 path predominated under the conditions used (CH_3CN solutions 8×10^{-3} M in benzyl complex, phosphane concentrations ranging from 0.22 to 0.25 mol · l^{-1}); the direct attack of the nucleophile (k_3 pathway) came into play in reactions with t-butyl isocyanide at very high concentrations. Rate constants obtained in CH_3CN at +30 °C including values for the corresponding reactions of $RC_5H_4Mo(CO)_3CH_2C_6H_5$ com-

References on pp. 313/6

plexes (R = H, CH_3) are listed in the following table. The marked increase in k_{-1}/k_2 observed for the reactions of the pentamethylcyclopentadiene compound highlights its substantially greater steric congestion [25]:

^5L	^2D	$10^4\,k_1$ (s^{-1})	$10^3\,k_{-1}/k_2$ (mol/L)	$10^3\,k_3$ ($L \cdot mol^{-1} \cdot s^1$)
C_5H_5	$P(CH_3)_2C_6H_5$	3.4	4.6	
$CH_3C_5H_4$		4.7	4.3	
$(CH_3)_5C_5$		9.2	120	
C_5H_5	$t\text{-}C_4H_9NC$	2.7	5.2	3.3
$CH_3C_5H_4$		5.3	23	5.2
$(CH_3)_5C_5$		7.9	160	11
C_5H_5	$cyclo\text{-}C_6H_{12}NC$	2.8	28	
$(CH_3)_5C_5$		12	150	

$(CH_3)_5C_5Mo(CO)_3CF_3$ (Table **8**, No. **54**). Crystals grown from benzene by slow evaporation of solvent belonged to the orthorhombic space group $P2_1cn$ (nonstandard setting of $Pna2_1$)–C_{2v}^9 (No. 33) with a = 9.2083(11), b = 9.8220(13), c = 17.176(2) Å; Z = 4 molecules per unit cell, D_{calc} = 1.643 g/cm^3 (at 24 °C). The molecule adopts the conformation commonly observed for $R_5C_5Mo(CO)_3X$ complexes, having the CF_3 ligand eclipsed with one ring carbon and the trans carbonyl bisecting the opposite bond. The molecular structure is shown in **Fig. 28** [95].

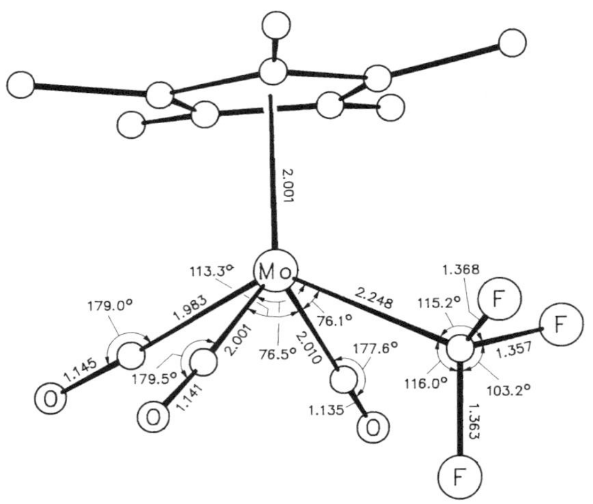

Fig. 28. The molecular structure of $(CH_3)_5C_5Mo(CO)_3CF_3$ [95].

(−)-Pinan-3-yl$(CH_3)_4C_5Mo(CO)_3CH_3$ (Table **8**, No. **60**) crystallized in the orthorhombic space group $P2_12_12_1$–D_2^4 (No. 19) with a = 7.439(3), b = 12.685(5), c = 24.04(1) Å; Z = 4 molecules per unit cell, D_{calc} = 1.313 g/cm^3. The structure exhibits the four-legged piano-stool geometry typical of $^5LMo(CO)_3R$ complexes; **Fig. 29** [108].

References on pp. 313/6

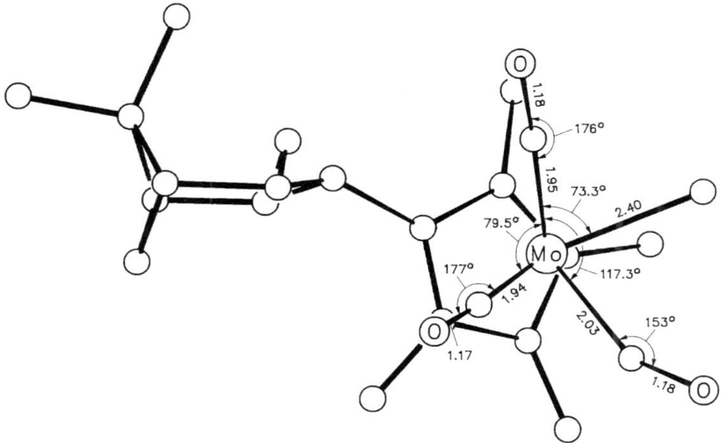

Fig. 29. The molecular structure of (–)-pinan-3-yl$(CH_3)_4C_5Mo(CO)_3CH_3$ [108].

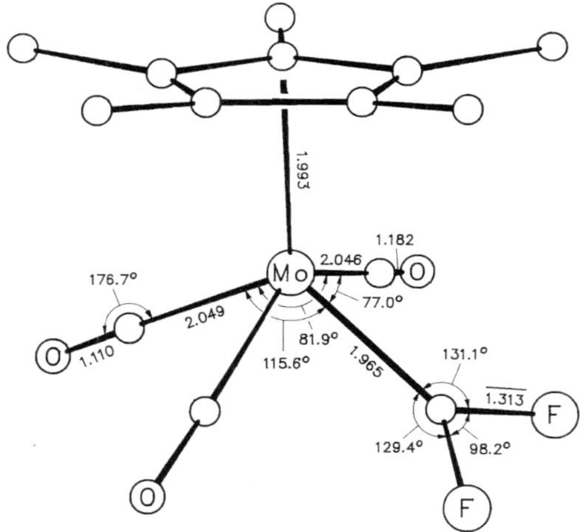

Fig. 30. The molecular structure of $[(CH_3)_5C_5Mo(CO)_3{=}CF_2]^+$ [95].

$[(CH_3)_5C_5Mo(CO)_3CF_2]O_3SCF_3$ (Table **8**, No. **62**). The following crystal data were obtained at −100 °C for prisms grown from CH_2Cl_2 at −55 °C: tetragonal, space group $P4_2/mbc-D_{4h}^{13}$ (No. 135), a = 17.895(2), c = 12.040(2) Å; Z = 8 molecules per unit cell, D_{calc} = 1.772 g/cm^3. The four-legged piano-stool structure of the cation is shown in **Fig. 30**. In contrast to $(CH_3)_5C_5Mo(CO)_3CF_3$ (see above), the complex has the carbonyl group trans to $=CF_2$ eclipsed with one ring carbon, the carbene ligand itself residing in the bisecting position. The cation adopts a lateral orientation of the planar $=CF_2$ carbene moiety with respect to the basal plane (angle between carbene plane and the plane defined by carbene carbon, central metal, and ring centroid: 80°). This is different from all other structurally characterized examples of four-legged piano-stool complexes with single-faced π-acceptor ligands, which prefer an upright ligand orientation [95].

References on pp. 313/6

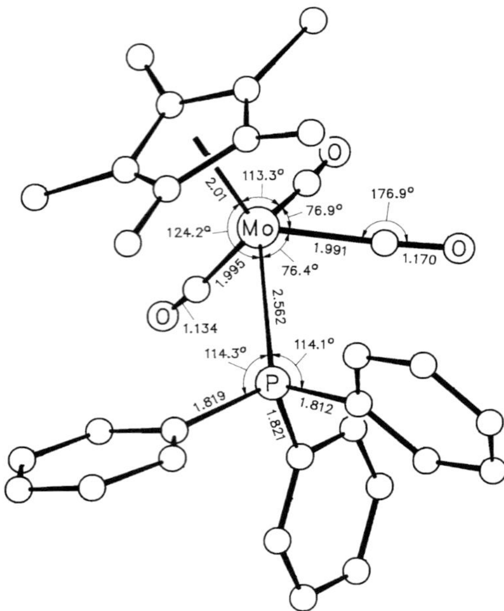

Fig. 31. The molecular structure of $[(CH_3)_5C_5Mo(CO)_3P(C_6H_5)_3]^+$ [57].

$[(CH_3)_5C_5Mo(CO)_3{}^2D]X$ ($X^- = PF_6^-$, $^2D = CH_3CN$, $P(CH_3)_3$, $P(C_4H_9\text{-}n)_3$, $P(C_6H_5)_2CH_3$, $P(C_6H_5)_3$, $P(OC_6H_5)_3$; Table **8**, Nos. **65** to **68**, **70**, and **75**; $X^- = BF_4^-$, $^2D = P(C_6H_5)_3$, $P(OCH_3)_3$, $P(OC_2H_5)_3$, $P(OC_6H_5)_3$; Table **8**, Nos. **69**, **71**, **73**, and **74**). The compounds were described as diamagnetic, thermally stable, crystalline solids [57].

Single crystals of $[(CH_3)_5C_5Mo(CO)_3P(C_6H_5)_3]BF_4 \cdot 0.5\ CH_3OH$, obtained by crystallization of No. 69 from $CH_3OH/(C_2H_5)_2O$ mixtures, belonged to the triclinic system, space group $P\bar{1}-C_i^1$ (No. 2), a = 13.548(3), b = 11.507(2), c = 11.141(2) Å, α = 108.34(1), β = 92.34(1), γ = 100.12(1)°; Z = 2 molecules per unit cell, D_{calc} = 1.40 g/cm^3. The structure of the cation is represented by **Fig. 31**. The essentially planar five-membered ring is slightly tilted from the plane perpendicular to the Mo···centroid axis as indicated by an angle of 3.3° between this axis and the normal to the $(CH_3)_5C_5$ ring plane. The methyl groups are shifted 0.18 to 0.27 Å, with respect to the ring plane, away from the metal atom [57].

Cyclic voltammograms of the hexafluorophosphates $[(CH_3)_5C_5Mo(CO)_3{}^2D]PF_6$ showed an irreversible wave for the initial oxidation process. The potential of the first reduction wave decreased with increasing electron-donating ability of the ^2D ligand (see the following table). The product waves were identical for all the salts and corresponded to the formation of the $[(CH_3)_5C_5Mo(CO)_3]^-$ anion with specific loss of the respective ^2D ligand: "$[(CH_3)_5C_5Mo(CO)_3{}^2D]^+ + 2\ e^- \rightarrow [(CH_3)_5C_5Mo(CO)_3]^- + {}^2D$" [56]. In contrast, electroreduction of $[(CH_3)_5C_5Mo(CO)_3P(OCH_3)_3]BF_4$ in CH_2Cl_2 containing 0.1 M $[N(C_4H_9\text{-}n)_4]BF_4$ produced $((CH_3)_5C_5Mo(CO)_2P(OCH_3)_3)_2$ by irreversible one-electron transfer occurring at E_p = −1.31 V versus SCE [58]. For selected cyclic voltammograms, see [56, 58]. The peak potentials (given in V versus SCE, Pt electrode) for oxidation and reduction of $[(CH_3)_5C_5Mo(CO)_3{}^2D]PF_6$ complexes in CH_2Cl_2/0.1 M $[N(C_4H_9\text{-}n)_4]PF_6$ are given in the following table [56].

^2D	E_{ox}	E_{red}
CH_3CN	1.45	−0.84
$P(OC_6H_5)_3$	1.96	−1.08
$P(C_6H_5)_2CH_3$	1.78	−1.12
$P(C_4H_9\text{-n})_3$	1.74	−1.28
$P(CH_3)_3$	1.72	−1.29

Cis- and trans-isomeric formyl complexes, $(CH_3)_5C_5Mo(CO)_2(^2D)CHO$ (^2D = $P(CH_3)_3$ [69], $P(C_6H_5)_2CH_3$ [69], $P(C_6H_5)_3$ [47, 57, 69], $P(OCH_3)_3$ [57], $P(OC_2H_5)_3$ [73], $P(OC_6H_5)_3$ [69, 73]), were isolated from reactions of the corresponding $[(CH_3)_5C_5Mo(CO)_3{}^2D]X$ salts (X = BF_4, PF_6) with equimolar quantities of $NaBH_4$ [47, 57, 69] or $[N(C_2H_5)_4]BH_4$ [73] in methanol at low temperatures. $LiAlH_4$ (THF, −20 °C) reduced the $[(CH_3)_5C_5Mo(CO)_3{}^2D]^+$ cations of the title complexes to cis-$(CH_3)_5C_5Mo(CO)_2(^2D)H$ (^2D = $P(C_6H_5)_3$, $P(OCH_3)_3$, $P(OC_6H_5)_3$) or, in the presence of added phosphane, to cis-$(CH_3)_5C_5Mo(CO)(^2D)_2H$ (^2D = $P(CH_3)_3$, $P(C_6H_5)_2CH_3$), depending on the steric demand and the electronic properties of the ^2D ligand coordinated. Treatment of $[(CH_3)_5C_5Mo(CO)_3P(C_6H_5)_2CH_3]PF_6$ with $LiAlH_4$ in the presence of free $P(CH_3)_3$ gave $(CH_3)_5C_5Mo(CO)_2(P(CH_3)_3)H$, $(CH_3)_5C_5Mo(CO)_2(P(C_6H_5)_2CH_3)H$, $(CH_3)_5C_5Mo(CO)(P(CH_3)_3)_2H$, $(CH_3)_5C_5Mo(CO)(P(C_6H_5)_2CH_3)_2H$, and $(CH_3)_5C_5Mo(CO)$-$(P(CH_3)_3)(P(C_6H_5)_2CH_3)H$ in a 1:1:3:5:10 ratio. The formation of the disubstituted products was shown to involve electron transfer with intermediate formation of the 19e and 17e radicals, $(CH_3)_5C_5Mo(CO)_{4-n}(^2D)_n^{\bullet}$ (n = 1, 2) and $(CH_3)_5C_5Mo(CO)_{3-n}(^2D)_n^{\bullet}$ (n = 0, 1, 2), respectively [102].

Reaction of the $[(CH_3)_5C_5Mo(CO)_3P(OCH_3)_3]^+$ cation with $[Ni((CH_3)_2PC_2H_4P(CH_3)_2)_2$-$H]PF_6$ in acetonitrile at room temperature resulted in reduction to cis-$(CH_3)_5C_5Mo(CO)_2$-$(P(OCH_3)_3)H$; use of $[Pt((CH_3)_2PC_2H_4P(CH_3)_2)_2H]PF_6$ as a reductant gave the formyl derivative cis-$(CH_3)_5C_5Mo(CO)_2(P(OCH_3)_3)CHO$; no net reaction was noted upon combination with $[Pt((C_2H_5)_2PC_2H_4P(C_2H_5)_2)_2H]PF_6$ [110].

1.5.1.4.2.5.3 Heterobimetallic Complexes

$(CH_3)_5C_5Mo(CO)_3CH_2CH_2CH_2Fe(CO)_2C_5H_5$ resulted from the treatment of $(CH_3)_5C_5Mo(CO)_3$-$(CH_2)_3Br$ with excess $Na[C_5H_5Fe(CO)_2]$ in THF at ambient conditions. The residue remaining after removal of solvent was chromatographed on an Al_2O_3 column. Elution with light petroleum provided a yellow fraction that furnished a 28% yield of the title complex as yellow crystals, m.p. 94 °C [59].

^1H NMR spectrum (C_6D_6): δ = 1.33 (dd, MCH_2; J_1 = 7, J_2 = 8 Hz), 1.60 (s, $(CH_3)_5C_5$), 1.81 (dd, MCH_2; J_1 = 7, J_2 = 8 Hz), 2.21 (m, central CH_2), 4.18 (s, C_5H_5) ppm. ^{13}C NMR spectrum (C_6D_6): δ = 10.7 ($(\mathit{C}H_3)_5C_5$), 10.8 (central CH_2), 18.4, 45.9 (both MCH_2), 85.8 (C_5H_5), 104.6 ($(CH_3)_5\mathit{C}_5$), 218.7 ($MoCO_{cis}$), 232.4 ($MoCO_{trans}$), 243.5 (FeCO) ppm. IR spectrum (light petroleum): 1911, 1919, 1953, 1999, 2009 (ν(CO)) cm^{-1} [59].

$(CH_3)_5C_5Mo(CO)_3HgX$ (X = Cl, Br, I). All three Mo−Hg-bonded complexes were made by reactions of $((CH_3)_5C_5Mo(CO)_3)_2Hg$ with a molar proportion of HgX_2 (X = Cl, Br, I). The resulting products were recrystallized from acetone or from CH_2Cl_2/hexane [55]. Selected physical data are given in the following table (NMR spectra in CH_2Cl_2/C_6D_6; ^{95}Mo spectra versus aqueous $Na_2[MoO_4]$ at pH 11; ^{199}Hg spectra versus neat $Hg(CH_3)_2$):

References on pp. 313/6

X	properties
Cl	^{1}H NMR: 2.09 ^{95}Mo NMR: −1567 ($\Delta\nu_{1/2}$ = 80 Hz) ^{199}Hg NMR: −645 ($\Delta\nu_{1/2}$ = 7 Hz) IR (Nujol): 1901, 1917, 1984 (CO)
Br	^{1}H NMR: 2.09 ^{95}Mo NMR: −1551 ($\Delta\nu_{1/2}$ = 160 Hz) ^{199}Hg NMR: −807 ($\Delta\nu_{1/2}$ = 12 Hz)
I	^{1}H NMR: 2.06 ^{95}Mo NMR: −1537 ($\Delta\nu_{1/2}$ = 120 Hz) ^{199}Hg NMR: −1060 ($\Delta\nu_{1/2}$ = 12 Hz)

For the three complexes (see table) and their lower methyl and phenyl homologues, $R_nC_5H_{5-n}Mo(CO)_3HgX$ ($R = CH_3$: n = 0, 1, 4; $R = C_6H_5$: n = 4; see there), the ^{95}Mo chemical shifts were observed to vary markedly with the number and nature of the substituents on the five-membered ring (the shielding decreasing in the order $C_5H_5 > CH_3C_5H_4 > (CH_3)_4C_5H > (CH_3)_5C_5 > (C_6H_5)_4C_5H$), but for a given cyclopentadienyl ligand they were little affected by changes in the X ligands (Cl, Br, I, and also $R_nC_5H_{5-n}Mo(CO)_3$). In contrast, the ^{199}Hg shifts were only slightly affected by the degree of ring substitution in the ring-methylated family $(CH_3)_nC_5H_{5-n}Mo(CO)_3HgX$ (n = 0, 1, 4, 5), but were strongly influenced by the ligands X, increasing in the order $X = (CH_3)_nC_5H_{5-n}Mo(CO)_3 < Cl < Br < I$ [55].

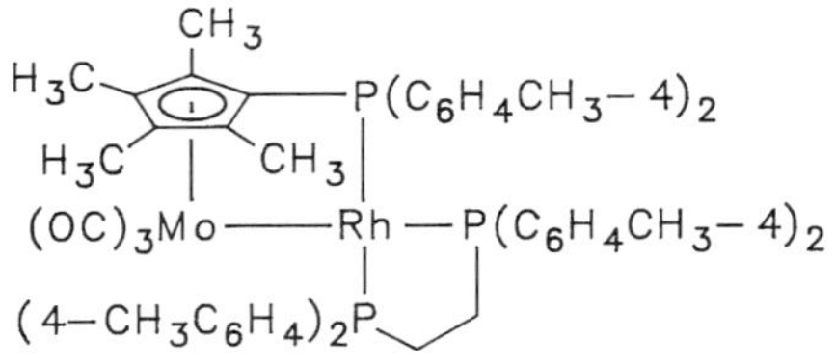

$$\begin{array}{c} (CH_3)_4C_5-P(C_6H_4CH_3\text{-}4)_2 \\ (OC)_3Mo—Rh—P(C_6H_4CH_3\text{-}4)_2 \\ (4\text{-}CH_3C_6H_4)_2P \end{array}$$

I

$Mo(CO)_3(\mu\text{-}(CH_3)_4C_5P(C_6H_4CH_3\text{-}4)_2)Rh((C_6H_4CH_3\text{-}4)_2PC_2H_4P(C_6H_4CH_3\text{-}4)_2)$ (Formula I) was prepared as follows: $Mo(CO)_6$ was added to a THF solution containing an equimolar quantity of in situ-generated $Li[(4\text{-}CH_3C_6H_4)_2P(CH_3)_4C_5]$. Refluxing of this mixture caused the formation of $Li[(4\text{-}CH_3C_6H_4)_2P(CH_3)_4C_5Mo(CO)_3]$ and $Li[Mo(CO)_5P(C_6H_4CH_3\text{-}4)_2C_5(CH_3)_4\text{-cyclo}]$ in an approximate 1:1 molar ratio. After cooling to −78 °C, half an equivalent of $(Rh(PR_2C_2H_4PR_2)(\mu\text{-}Cl))_2$ ($R = C_6H_4CH_3\text{-}4$) was added, the mixture was stirred at room temperature, and subsequently worked up by column chromatography (silica gel/toluene). A brown material was collected, which was further purified by preparative thin-layer chromatography (silica gel, hexane/ether (3:1)). Final recrystallization from toluene/pentane gave the title complex in 27% yield as an orange solid, m.p. >260 °C (dec.) [16].

^{1}H NMR spectrum (C_6D_6): δ = 1.44, 1.90, 1.96, 2.01, 2.13 (all s, all 2 CH_3), 6.83 ("t", 8 meta-H; "J" = 9 Hz), 7.17 (d, 4 meta-H; J = 9 Hz), 7.58 (dd, 4 ortho-H; J(H,H) = 8, J(P,H) = 9.6 Hz), 7.98 ("q", 8 ortho-H; "J" = 8 and 18 Hz) ppm; C_2H_4 resonances overlapped by the CH_3 signals. ^{13}C NMR spectrum (C_6D_6): δ = 11.9, 13.4 (both s, $(CH_3)_4C_5$), 21.0, 21.1, 21.4 (all s, all tolyl-CH_3), 104.3 (d, C-2,5 (?) of $(CH_3)_4C_5P$; J(P,C) = 11 Hz), 106.6 (d, C-3,4 (?) of $(CH_3)_4C_5P$; J(P,C) = 5 Hz), 232.1 (2 MoCO cis to Rh), 238.3 (MoCO trans to

Rh) ppm; C-1 of $(CH_3)_4C_5$ and C_2H_4 not detected; aromatic region complex and not assigned. ^{31}P NMR spectrum (C_6D_6; "P_A" = $(CH_3)_4C_5P$; "P_B" cis to "P_A", "P_C" trans to "P_A"): δ = 41.2 (ddd, P_A; $J(P_A,P_B)$ = 23, $J(P_A,P_C)$ = 351, $J(Rh,P_A)$ = 130 Hz), 51.4 (ddd, P_B; $J(P_B,P_C)$ = 32, $J(Rh,P_B)$ = 185 Hz), 67.3 (ddd, P_C; $J(Rh,P_C)$ = 148 Hz) ppm. IR spectrum (THF): 1803, 1834, 1915 (ν(CO)) cm^{-1} [16].

C_6D_6 solutions remained unchanged after heating at 72 °C for prolonged periods under 650 Torr H_2 pressure. Heating to 80 °C in toluene under 68 atm H_2 in the presence of added cyclohexene did not result in significant olefin hydrogenation [16].

References:

[1] King, R. B.; Bisnette, M. B. (J. Organomet. Chem. **8** [1967] 287/97).
[2] King, R. B.; Houk, L. W. (Can. J. Chem. **47** [1969] 2959/64).
[3] King, R. B.; Efraty, A. (J. Am. Chem. Soc. **93** [1971] 4950/2).
[4] King, R. B.; Efraty, A. (Proc. 5th Int. Conf. Organomet. Chem., Moscow, USSR, 1971, Vol. II, p. 14, Abstr. 235).
[5] King, R. B.; Efraty, A. (J. Am. Chem. Soc. **94** [1972] 3773/9).
[6] King, R. B.; Efraty, A.; Douglas, W. M. (J. Organomet. Chem. **60** [1973] 125/37).
[7] Ginley, D. S.; Wrighton, M. S. (J. Am. Chem. Soc. **97** [1975] 3533/5).
[8] Ginley, D. S.; Bock, C. R.; Wrighton, M. S. (Inorg. Chim. Acta **23** [1977] 85/94).
[9] King, R. B.; King, A. D., Jr.; Iqbal, M. Z.; Frazier, C. C. (J. Am. Chem. Soc. **100** [1978] 1687/94).
[10] Barnes, S. G.; Green, M. (J. Chem. Soc. Chem. Commun. **1980** 267/8).
[11] Malito, J. T.; Shakir, R.; Atwood, J. L. (J. Chem. Soc. Dalton Trans. **1980** 1253/8).
[12] Rakowski Du Bois, M.; Du Bois, D. L.; Van Derveer, M. C.; Haltiwanger, R. C. (Inorg. Chem. **20** [1981] 3064/71).
[13] Petrignani, J.-F.; Alper, H. (Organometallics **1** [1982] 1095/7).
[14] Hughes, R. P.; Lambert, J. M. J.; Reisch, J. W.; Smith, W. L. (Organometallics **1** [1982] 1403/5).
[15] Dubois, D. A.; Duesler, E. N.; Paine, R. T. (Organometallics **2** [1983] 1903/5).
[16] Casey, C. P.; Bullock, R. M.; Nief, F. (J. Am. Chem. Soc. **105** [1983] 7574/80).
[17] Malisch, W.; Maisch, R.; Meyer, A.; Greisinger, D.; Gross, E.; Colquhoun, I. J.; McFarlane, W. (Phosphorus Sulfur Relat. Elem. **18** [1983] 299/302).
[18] Dormond, A.; El Bouadili, A.; Moïse, C. (Tetrahedron Lett. **24** [1983] 3087/90).
[19] Slocum, D. W.; Johnson, S.; Matusz, M.; Duraj, S.; Cmarik, J. L.; Simpson, K. M.; Owen, D. A. (Polym. Mater. Sci. Eng. **49** [1983] 353/7).
[20] Maisch, R.; Barth, M.; Malisch, W. (J. Organomet. Chem. **260** [1984] C35/C39).
[21] Bruce, M. I.; Humphrey, M. G.; Matisons, J. G.; Roy, S. K.; Swincer, A. G. (Austral. J. Chem. **37** [1984] 1955/61).
[22] Matusz, M. (Diss. Southern Illinois Univ., Carbondale, IL, 1984; Diss. Abstr. Int. B **46** [1985] 835/6).
[23] Nolan, S. P.; Hoff, C. D.; Landrum, J. T. (J. Organomet. Chem. **282** [1985] 357/62).
[24] Alt, H. G.; Engelhardt, H. E.; Thewalt, U.; Riede, J. (J. Organomet. Chem. **288** [1985] 165/77).
[25] Cotton, J. D.; Kimlin, H. A. (J. Organomet. Chem. **294** [1985] 213/7).
[26] Gudat, D.; Niecke, E.; Malisch, W.; Hofmockel, U.; Quashie, S.; Cowley, A. H.; Arif, A. M.; Krebs, B.; Dartmann, M. (J. Chem. Soc. Chem. Commun. **1985** 1687/9).
[27] Dormond, A.; Moise, C. (Polyhedron **4** [1985] 595/8).
[28] Hughes, R. P.; Reisch, J. W.; Rheingold, A. L. (Organometallics **4** [1985] 241/4).

[29] Murray, R. C.; Blum, L.; Liu, A. H.; Schrock, R. R. (Organometallics **4** [1985] 953/4).
[30] Hughes, R. P.; Kläui, W.; Reisch, J. W.; Müller, A. (Organometallics **4** [1985] 1761/6).

[31] Kubas, G. J.; Wasserman, H. J.; Ryan, R. R. (Organometallics **4** [1985] 2012/21).
[32] Gudat, D.; Niecke, E.; Krebs, B.; Dartmann, M. (Chimia **39** [1985] 277/9).
[33] Slocum, D. W.; Duraj, S.; Matusz, M.; Cmarik, J. L.; Simpson, K. M.; Owen, D. A. (Met.-Containing Polym. Syst. **1985** 59/68).
[34] Brownlee, R. T. C.; O'Connor, M. J.; Shehan, B. P.; Wedd, A. G. (J. Magn. Reson. **61** [1985] 516/25).
[35] Moss, J. R.; Niven, M. L.; Stretch, P. M. (Inorg. Chim. Acta **119** [1986] 177/86).
[36] Kubas, G. J.; Ryan, R. R. (Polyhedron **5** [1986] 473/85).
[37] Kubas, G. J. (unpublished results quoted in [41]).
[38] Leoni, P.; Grilli, E.; Pasquali, M.; Tomassini, M. (J. Chem. Soc. Dalton Trans. **1986** 1041/3).
[39] Sweany, R. L. (Organometallics **5** [1986] 387/8).
[40] Gudat, D.; Niecke, E.; Krebs, B.; Dartmann, M. (Organometallics **5** [1986] 2376/7).

[41] Nolan, S. P.; Lopez de la Vega, R.; Hoff, C. D. (Organometallics **5** [1986] 2529/37).
[42] Moore, E. J.; Sullivan, J. M.; Norton, J. R. (J. Am. Chem. Soc. **108** [1986] 2257/63).
[43] Sweany, R. L. (J. Am. Chem. Soc. **108** [1986] 6986/91).
[44] Werner, H.; Roll, J.; Zolk, R.; Thometzek, P.; Linse, K.; Ziegler, M. L. (Chem. Ber. **120** [1987] 1553/64).
[45] Grobe, J.; Le Van, D.; Nientiedt, J. (Z. Naturforsch. **42b** [1987] 984/92).
[46] Breeze, P.; Ricalton, A.; Whiteley, M. W. (J. Organomet. Chem. **327** [1987] C29/C32).
[47] Asdar, A.; Lapinte, L. (J. Organomet. Chem. **327** [1987] C33/C36).
[48] Alt, H. G.; Engelhardt, H. E.; Kläui, W.; Müller, A. (J. Organomet. Chem. **331** [1987] 317/27).
[49] Alt, H. G.; Herrmann, G. S.; Engelhardt, H. E.; Rogers, R. D. (J. Organomet. Chem. **331** [1987] 329/39).
[50] Kreiter, C. G.; Wendt, G.; Sheldrick, W. S. (J. Organomet. Chem. **333** [1987] 47/59).

[51] Hitchcock, P. B.; Lappert, M. F.; Michalczyk, M. J. (J. Chem. Soc. Dalton Trans. **1987** 2635/42).
[52] Jutzi, P.; Meyer, U. (Chem. Ber. **121** [1988] 559/60).
[53] Dötz, K. H.; Rott, J. (J. Organomet. Chem. **338** [1988] C11/C13).
[54] Akita, M.; Kakinuma, N.; Moro-oka, Y. (J. Organomet. Chem. **338** [1988] 91/4).
[55] Kubicki, M. M.; Le Gall, J. Y.; Pichon, R.; Salaun, J. Y.; Cano, M.; Campo, J. A. (J. Organomet. Chem. **348** [1988] 349/56).
[56] Asdar, A.; Tudoret, M.-J.; Lapinte, C. (J. Organomet. Chem. **349** [1988] 353/66).
[57] Leoni, P.; Aquilini, E.; Pasquali, M.; Marchetti, F.; Sabat, M. (J. Chem. Soc. Dalton Trans. **1988** 329/33).
[58] Leoni, P.; Marchetti, F.; Pasquali, M.; Zanello, P. (J. Chem. Soc. Dalton Trans. **1988** 635/9).
[59] Bailey, N. A.; Dunn, D. A.; Foxcroft, C. N.; Harrison, G. R.; Winter, M. J.; Woodward, S. (J. Chem. Soc. Dalton Trans. **1988** 1449/56).
[60] Nolan, S. P.; Hoff, C. D. (Organomet. Synth. **4** [1988] 58/66).

[61] Ricalton, A.; Whiteley, M. W. (J. Organomet. Chem. **361** [1989] 101/8).
[62] Lyatifov, I. R.; Gasanov, T. Kh.; Petrovski, P. V.; Lutsenko, A. I. (J. Organomet. Chem. **361** [1989] 181/6).
[63] Roger, C.; Tudoret, M.-J.; Guerchais, V.; Lapinte, C. (J. Organomet. Chem. **365** [1989] 347/50).

[64] Alt, H. G.; Engelhardt, H. E.; Frister, T.; Rogers, R. D. (J. Organomet. Chem. **366** [1989] 297/304).
[65] Filippou, A. C.; Grünleitner, W.; Herdtweck, E. (J. Organomet. Chem. **373** [1989] 325/42).
[66] Bentley, G. W.; Hough, G.; Winter, M. J.; Woodward, S. (Polyhedron **8** [1989] 1861/2).
[67] Leoni, P.; Pasquali, M.; Salsini, L.; Di Bugno, C.; Braga, D.; Sabatino, P. (J. Chem. Soc. Dalton Trans. **1989** 155/9).
[68] Leoni, P.; Pasquali, M.; Braga, D.; Sabatino, P. (J. Chem. Soc. Dalton Trans. **1989** 959/63).
[69] Asdar, A.; Lapinte, C.; Toupet, L. (Organometallics **8** [1989] 2708/17).
[70] Franz, J. A.; Linehan, J. C.; Alnajjar, M. S. (unpublished work cited in [71]).

[71] Franz, J. A.; Bushaw, B. A.; Alnajjar, M. S. (J. Am. Chem. Soc. **111** [1989] 268/75).
[72] Tilset, M.; Parker, V. D. (J. Am. Chem. Soc. **111** [1989] 6711/7).
[73] Gibson, D. H.; Owens, K.; Mandal, S. K.; Sattich, W. E.; Franco, J. O. (Organometallics **8** [1989] 498/505).
[74] Sitzmann, H. (Chem. Ber. **123** [1990] 2311/5).
[75] Lang, H.; Leise, M.; Zsolnai, L. (J. Organomet. Chem. **386** [1990] 349/63).
[76] Lang, H.; Leise, M.; Zsolnai, L. (J. Organomet. Chem. **389** [1990] 325/32).
[77] Salsini, L.; Pasquali, M.; Zandomeneghi, M.; Festa, C.; Leoni, P.; Braga, D.; Sabatino, P. (J. Chem. Soc. Dalton Trans. **1990** 2007/12).
[78] Ryan, O. B.; Tilset, M.; Parker, V. D. (J. Am. Chem. Soc. **112** [1990] 2618/26).
[79] Mao, F.; Philbin, C. E.; Weakley, T. J. R.; Tyler, D. R. (Organometallics **9** [1990] 1510/6).
[80] Hwang, D.-K.; Chi, Y.; Peng, S.-M.; Lee, G.-H. (Organometallics **9** [1990] 2709/18).

[81] Tilset, M.; Parker, V. D. (J. Am. Chem. Soc. **112** [1990] 2843).
[82] Bullock, R. M.; Samsel, E. G. (J. Am. Chem. Soc. **112** [1990] 6886/98).
[83] Salsini, L.; Pasquali, M.; Leoni, P.; Braga, D.; Sabatino, P. (Gazz. Chim. Ital. **120** [1990] 465/70).
[84] Salsini, L.; Pasquali, M.; Leoni, P.; Braga, D.; Sabatino, P. (Gazz. Chim. Ital. **120** [1990] 823).
[85] Simoes, J. A. (Chemtracts: Anal., Phys., Inorg. Chem. **2** [1990] 360/3).
[86] Stein, D.; Sitzmann, H. (J. Organomet. Chem. **402** [1991] 249/57).
[87] Kisch, H.; Garn, D. (J. Organomet. Chem. **409** [1991] 347/54).
[88] Filippou, A. C.; Grünleitner, W.; Fischer, E. O.; Imhof, W.; Huttner, G. (J. Organomet. Chem. **413** [1991] 165/79).
[89] Gales, S. L.; Pennington, W. T.; Kolis, J. W. (J. Organomet. Chem. **419** [1991] C10/C13).
[90] Poli, R.; Gordon, J. C.; Desai, J. U.; Rheingold, A. L. (J. Chem. Soc. Chem. Commun. **1991** 1518/20).

[91] Cano, M.; Campo, J. A. (Polyhedron **10** [1991] 133/4).
[92] Adams, H.; Bailey, N. A.; Bentley, G. W.; Hough, G.; Winter, M. J.; Woodward, S. (J. Chem. Soc. Dalton Trans. **1991** 749/58).
[93] Benyunes, S. A.; Binelli, A.; Green, M.; Grimshire, M. J. (J. Chem. Soc. Dalton Trans. **1991** 895/904).
[94] Mao, F.; Sur, S. K.; Tyler, D. R. (Organometallics **10** [1991] 419/23).
[95] Koola, J. D.; Roddick, D. M. (Organometallics **10** [1991] 591/7).
[96] Trost, M. K.; Bergman, R. G. (Organometallics **10** [1991] 1172/8).
[97] Kubas, G. J.; Kiss, G.; Hoff, C. D. (Organometallics **10** [1991] 2870/6).

[98] Brombach, H.; Niecke, E.; Nieger, M. (Organometallics **10** [1991] 3949/51).
[99] Eisenberg, D. C.; Lawrie, C. J. C.; Moody, A. E.; Norton, J. R. (J. Am. Chem. Soc. **113** [1991] 4888/95).
[100] Parker, V. D.; Handoo, K. L.; Roness, F.; Tilset, M. (J. Am. Chem. Soc. **113** [1991] 7493/8).

[101] Benyunes, S. A.; Deeth, R. J.; Fries, A.; Green, M.; McPartlin, M.; Nation, C. B. M. (J. Chem. Soc. Dalton Trans. **1992** 3453/65).
[102] Tudoret, M.-J.; Robo, M.-L.; Lapinte, C. (Organometallics **11** [1992] 1419/22).
[103] Drago, R. S.; Wong, N. M.; Ferris, D. C. (J. Am. Chem. Soc. **114** [1992] 91/8).
[104] Pasquali, M.; Leoni, P.; Sabatino, P.; Braga, D. (Gazz. Chim. Ital. **122** [1992] 275/7).
[105] Tilset, M. (NATO ASI Ser. C **367** [1992] 109/29; C.A. **118** [1993] No. 111782).
[106] Gassman, P. G.; Sowa, J. R.; Mickelson, J. W. (U.S. 5245064 [1992/93] 10 pp.; C.A. **120** [1994] No. 54700).
[107] Niecke, E.; Metternich, H.-J.; Nieger, M.; Gudat, D.; Wenderoth, P.; Malisch, W.; Hahner, C.; Reich, W. (Chem. Ber. **126** [1993] 1299/309).
[108] Lai, R.; Bousquet, L.; Heumann, A. (J. Organomet. Chem. **444** [1993] 115/9).
[109] Saadeh, C.; Colbran, S. B.; Craig, D. C.; Rae, A. D. (Organometallics **12** [1993] 133/9).
[110] Miedaner, A.; DuBois, D. L.; Curtis, C. J.; Haltiwanger, R. C. (Organometallics **12** [1993] 299/303).

[111] Brammer, L.; Zhao, D.; Bullock, R. M.; McMullan, R. K. (Inorg. Chem. **32** [1993] 4819/24).
[112] Wang, J.-C.; Chi, Y.; Peng, S.-M.; Lee, G.-H.; Shyu, S.-G.; Tu, F.-H. (J. Organomet. Chem. **481** [1994] 143/52).
[113] Colbran, S. B.; Harrison, W. M.; Saadeh, C. (Organometallics **13** [1994] 1061/3).
[114] Chi, Y.; Su, C.-J.; Farrugia, L. J.; Peng, S.-M.; Lee, G.-H. (Organometallics **13** [1994] 4167/9).
[115] Abugideiri, F.; Brewer, G. A.; Desai, J. U.; Gordon, J. C.; Poli, R. (Inorg. Chem. **33** [1994] 3745/51).
[116] Desai, J. U.; Gordon, J. C.; Kraatz, H.-B.; Lee, V. T.; Owens-Waltermire, B. E.; Poli, R.; Rheingold, A. L.; White, C. B. (Inorg. Chem. **33** [1994] 3752/69).
[117] Lindsell, W. E.; McCullough, K. J.; Plancq, S. (J. Organomet. Chem. **491** [1995] 275/87).
[118] Schollhammer, P.; Pétillon, F. Y.; Pichon, R.; Poder-Guillou, S.; Talarmin, J.; Muir, K. W.; Manojlovic-Muir, Lj. (Organometallics **14** [1995] 2277/87).

1.5.1.4.2.6 Zwitterionic $^5LMo(CO)_3$ Complexes

1.5.1.4.2.6.1 Neutral Complexes

This section summarizes compounds of the $^5LMo(CO)_3$ type with ^5L dienyl rings substituted by formally cationic groups such as $-C(R')=N^{(+)}R_2$, $-P^{(+)}R_3$, $-S^{(+)}R_2$, and the like, which may be regarded as internally stabilized analogs of $[R_5C_5Mo(CO)_3]^-$ anions. Fulvene complexes containing $Mo(CO)_3$-bound $C_5R_4{=}CR'_2$ ligands with substituents R′ not suited to stabilize an exocyclic carbonium center (e.g., H, alkyl, or aryl) have been excluded from discussion under this heading as it has been shown that in these cases the fulvene system coordinates to $M(CO)_3$ fragments (M = Cr, Mo, W) in a ^6L fashion; see, e.g. [14, 22, 25].

General methods of synthesis for the complexes described hereafter and listed in Table 9 include the following procedures:

References on pp. 329/31

$N(C_3H_7\text{-}i)_2$ / $N(C_3H_7\text{-}i)_2$ / $N(CH_3)_2$

I II

Method I: From $Mo(CO)_6$ and the ^5L ligand by elimination of CO:

a. Refluxing equimolar quantities of $Mo(CO)_6$ and 6-dimethylaminofulvene or 6,6-bis(dimethylamino)fulvene in methylcyclohexane gave $(CH_3)_2NCHC_5H_4$-$Mo(CO)_3$ and $((CH_3)_2N)_2CC_5H_4Mo(CO)_3$ as insoluble precipitates which were recrystallized either from CH_2Cl_2/hexane or, preferably, from acetone/benzene mixtures [2]. The 5,6-bis(diisopropylamino)calicene complex cyclo-$((i\text{-}C_3H_7)_2NC)_2CC_5H_4Mo(CO)_3$ formed on refluxing a mixture of $Mo(CO)_6$ and calicene ligand (Formula I) in 1,4-dioxane [16].

b. Reasonable yields of $(C_6H_5)_3PC_5H_4Mo(CO)_3$, the first zwitterionic RC_5H_4-$Mo(CO)_3$ derivative described, were attained by treating the hexacarbonyl with triphenylphosphonium cyclopentadienylide in boiling dimethoxyethane or diethylene glycol dimethyl ether. The product was precipitated from the filtered solutions by adding water or hexane and was purified by recrystallization from $CHCl_3$/light petroleum [1, 3].

Method II: From $Mo(CO)_3(NCCH_3)_3$ and the ^5L ligand by elimination of CH_3CN:

a. A 1:1 mixture of $Mo(CO)_3(NCCH_3)_3$ and 6-dimethylaminofulvene in CH_3O-$C_2H_4OC_2H_4OCH_3$ was heated at 60 to 70 °C and then kept at room temperature. Precipitated crystals of $(CH_3)_2NCHC_5H_4Mo(CO)_3$ were filtered off and the remaining solution was diluted with hexane and allowed to stand at ambient conditions, thereby affording an additional crop of the product [17]. THF solutions containing equimolar mixtures of $C_6H_5(CH_3)N(CH{=}CH)_2CH{=}C_5H_4$ and $Mo(CO)_3(NCCH_3)_3$ deposited $C_6H_5(CH_3)N(C_2H_2)_2CHC_5H_4Mo(CO)_3$ when stirred at ambient temperature and subsequently diluted with heptane [28]. Stirring 6-dimethylamino-1-formylfulvene with a 10% excess of the tris(acetonitrile) complex in THF at room temperature gave $((CH_3)_2NC(H)C_5H_3$-$CHO\text{-}2)Mo(CO)_3$, which was separated from the evaporated mixture by repeated precipitation from THF/hexane [24].

b. Coordination of the 4-(dimethylamino)isodicyclopentafulvene ligand 4-$(CH_3)_2NCHC_{10}H_{10}$ (Formula II) to the $Mo(CO)_3$ fragment was achieved by heating $Mo(CO)_3(NCCH_3)_3$ with fulvene II in 1:1 stoichiometry in THF. After cooling and filtration, the filtrate was concentrated to give a red solid which was crystallized from acetone at −20 °C [29].

c. $C_5H_5NC_5H_4Mo(CO)_3$ formed when $Mo(CO)_3(NCCH_3)_3$ and pyridinium cyclopentadienylide were allowed to stand in 1:1 stoichiometry in $CH_3OC_2H_4O$-$C_2H_4OCH_3$ at 20 °C [15].

d. When a 1:1 molar mixture of mesoionic 1,3-diphenyltetrazolium-5-cyclopentadienide, 1,3-$(C_6H_5)_2N_4C^{(+)}-C_5H_4^{(-)}$, and $Mo(CO)_3(NCCH_3)_3$ was gently refluxed in benzene, 1,3-$(C_6H_5)_2N_4CC_5H_4Mo(CO)_3$ separated in high yield from the filtered and cooled solution [32].

e. Combination of $Mo(CO)_3(NCCH_3)$ with either $(C_6H_5)_3EC_5(C_6H_5)_4$ (E = P, As) or $(C_6H_5)_3PC_5H_3N_2Ar\text{-}2$ (Ar = p-$C_6H_4OCH_3$) in boiling THF gave good yields of $(C_6H_5)_3EC_5(C_6H_5)_4Mo(CO)_3$ and $((C_6H_5)_3PC_5H_3N_2Ar\text{-}2)Mo(CO)_3$, respec-

References on pp. 329/31

tively, of which the former could be purified by recrystallization from CH_2Cl_2/petroleum [6].

f. In the preparation of $(CH_3)_2SC_5H_4Mo(CO)_3$, the tris(acetonitrile) precursor was prepared in situ from $Mo(CO)_6$ and CH_3CN. The excess acetonitrile was distilled off and an equimolar quantity of dimethylsulfonium cyclopentadienylide in $CH_3OC_2H_4OC_2H_4OCH_3$ was added, causing the crystalline product to immediately precipitate from the reaction mixture [9].

Method III: To a toluene solution of the fulvene complex $H_2CC_5(C_6H_5)_4Mo(CO)_3$ were added two equivalents of $P(CH_3)_3$, $P(CH_3)_2C_6H_5$, or $P(C_6H_5)_2CH_3$. Dilution of the mixtures with pentane caused the adducts, $R_3PCH_2C_5(C_6H_5)_4Mo(CO)_3$, so formed to precipitate. In a similar preparation of $(n\text{-}C_4H_9)_3PCH_2C_5(C_6H_5)_4Mo(CO)_3$, CH_2Cl_2 was used as a solvent and the product was separated by crystallization at −78 °C after adding equal volumes of pentane and diethyl ether [25].

^1H and ^{13}C NMR spectroscopy were used in studying the rotational isomerism of 6-dimethylaminofulvene and its Group 6 complexes, $(CH_3)_2NCHC_5H_4M(CO)_3$ (M = Cr, Mo, W). It was demonstrated that coordination of the amino-substituted fulvene ligand raises the free energy for the barrier to rotation about the $(CH_3)_2N-C$ bond by increasing the degree of double-bond character between the carbon and the nitrogen atom. The free energy for the rotational barriers and, hence, the ligand polarity (as evidenced, inter alia, from the direct proportionality between $\Delta G^‡$ and the ^{13}C chemical shifts of the exocyclic carbon atoms) were shown to strongly depend on the nature of the central metal, increasing in the sequence M = W < Mo < Cr. It was also established that increasing the polarity of the solvent (acetone < nitromethane < dimethylformamide) raises the rotational barrier as a result of dipole-dipole interaction between the complexes and the solvent molecules, since the free energy for the barrier to rotation and the dipole moment of the solvent are directly proportional [18, 20, 31]. Kinetic parameters of rotation about the $(CH_3)_2N-C$ bond in 6-dimethylaminofulvene and its Group 6 metal carbonyl derivatives are given in the following table [18, 20, 23]; for practically identical values of $\Delta G^‡_{325}$, see [31]:

compound	solvent	E_a (kcal/mol)	$\Delta H^‡$ (kcal/mol)	$\Delta G^‡_{298}$ (kcal/mol)	$\Delta S^‡$ (cal·mol^{-1}·K^{-1})
$(CH_3)_2NCHC_5H_4$	acetone	13.9 ± 0.2	13.3 ± 0.2	14.0 ± 0.2	−2.5 ± 1.0
$(CH_3)_2NCHC_5H_4Cr(CO)_3$	acetone			17.9 ± 0.2	
$(CH_3)_2NCHC_5H_4Mo(CO)_3$	acetone	25.3 ± 0.7	24.7 ± 0.7	17.3 ± 0.4	25.0 ± 2.0
$(CH_3)_2NCHC_5H_4Mo(CO)_3$	CH_3NO_2	22.0 ± 0.6	21.3 ± 0.6	17.7 ± 0.4	12.0 ± 1.0
$(CH_3)_2NCHC_5H_4Mo(CO)_3$	$(CH_3)_2NCHO$	21.2 ± 0.6	20.5 ± 0.6	18.3 ± 0.4	7.4 ± 1.0
$(CH_3)_2NCHC_5H_4W(CO)_3$	acetone	24.9 ± 1.0	24.3 ± 1.0	16.2 ± 0.5	27.3 ± 2.0

Activation parameters for rotation about the exocyclic C−C bond in $(CH_3)_2NCHC_5H_4$-$M(CO)_3$ (M = Cr, Mo, W) and the free ^5L ligand were likewise obtained by dynamic NMR methods. Variable-temperature ^{13}C NMR spectra revealed that π-coordination diminished $\Delta G^‡_{421}$ of rotation about this bond by 7.7 kcal/mol from 22.1 kcal/mol in the uncomplexed fulvene molecule to 14.4 ± 0.1 kcal/mol in the metal derivatives without any dependence on the nature of the Group 16 metal involved. Kinetic parameters for rotation about the exocyclic C−C linkage in the three 6-dimethylaminofulvene complexes derived from dynamic ^1H NMR spectroscopy also did not vary significantly with the central metal. The values obtained for the Mo complex are E_a = 9.8 ± 0.4, $\Delta H^‡$ = 9.3 ± 0.4 kcal/mol, $\Delta G^‡$ = 13.0 ± 0.2, $\Delta S^‡$ = −12.3 ± 1.4 cal · mol^{-1} · K^{-1} [31].

^{13}C NMR data showed that π-coordination of $(C_6H_5)_3PC_5H_4$ and $(CH_3)_2SC_5H_4$ to $M(CO)_3$ fragments (M = Cr, Mo, W) not only enhanced the shielding of the ring carbon atoms but also caused considerable balancing of electron density in the cyclopentadienyl ring. The degree of shielding was observed to be determined by the nature of the central metal varying in the order Cr > W > Mo. For the triphenylphosphonium cyclopentadienylide derivatives, it was established that the $^{31}P,^{13}C$ coupling constants remained practically unaffected by complex formation demonstrating that the bond character in the π-ligand did not essentially change after coordination [13]. Dissolution of the 6-dimethylaminofulvene compounds $(CH_3)_2NCHC_5H_4M(CO)_3$ (M = Cr, Mo, W) in CH_3CO_2H/CF_3CO_2H and CF_3CO_2H/CH_2Cl_2 mixtures of differing acidity, H_0, of the medium resulted in reversible protonation at the metal atom, the rate of which was shown to be linearly dependent on the value of the acidity function. Complete protonation was established to take place at $H_0 = -3.03$ for M = Cr, -2.7 for M = Mo, and -2.03 for M = W. On the other hand, metal protonation did not occur at $H_0 = -2.51$ (M = Cr), ≥ 1 (M = Mo), and ≥ 1 to 3 (M = W). A quantitative comparison of the basicities of the three complexes (for details, see [17]) gave the basicity ratio Cr:Mo:W ≅ 1:150:1800 [17]. A similar conclusion was drawn for the series of triphenylphosphonium cyclopentadienylide derivatives, $(C_6H_5)_3PC_5H_4M(CO)_3$ (M = Cr, Mo, W), which likewise were found to dissolve readily in nonaqueous trifluoroacetic acid [3] or in CF_3CO_2H/CH_2Cl_2 mixtures [11] producing highly colored solutions, the ^1H NMR of which exhibited high-field resonances at $-5 > \delta > -8$, characteristic of the M–H group. The upfield shift of these resonances with increasing metal atomic weight was attributed to an increased charge density on the attached proton and, hence, to an increasing basicity of these compounds in the order M = Cr < Mo < W [3]. Tensimetric titrations of the three $(C_6H_5)_3PC_5H_4M(CO)_3$ complexes with BF_3 in CH_2Cl_2 also suggested this trend. Although the formation of 1:1 adducts was observed for all three compounds at −78 °C, the equilibrium constants for these reversible Lewis acid/Lewis base reactions at 0 °C increased in the order Cr ≪ Mo < W [3].

Kinetic data for the H/D exchange reactions between various onium-substituted chromium and molybdenum complexes $RC_5H_4M(CO)_3$ and C_2H_5OD in acetone at 50 °C showed that the exchange rates of the molybdenum derivatives were almost 140 times greater than those of the corresponding chromium compounds. For the $RC_5H_4Mo(CO)_3$ species studied, the exchange rate varied over one order of magnitude, increasing in the order $RC_5H_4Mo(CO)_3$ = $(CH_3)_2NCHC_5H_4Mo(CO)_3$ < $(C_6H_5)_3PC_5H_4Mo(CO)_3$ < $(CH_3)_2SC_5H_4Mo(CO)_3$ [12]; compare Nos. 1, 9, and 18.

Table 9
Zwitterionic $^5LMo(CO)_3$ Complexes.
An asterisk indicates further information at the end of the table.
For explanations, abbreviations, and units see p. X.

No.	^5L ligand	method of preparation (yield) properties and remarks
*1	$(CH_3)_2\overset{\oplus}{N}=CH-C_5H_4^{\ominus}$	Ia (90%) [2]; IIa (64%) [17] air-stable, red crystals decomposing at 230 to 240 °C without melting [2] ^1H NMR (acetone-d_6): 3.30, 3.41 (both s, CH_3), 5.51, 5.66 (both "t", C_5H_4; "J" = 2), 8.01 (s, =CH) [2] (similar in [17]); T = −65 °C: 3.29, 3.51 (both s, CH_3; coalescence temperature:

References on pp. 329/31

Table 9 (continued)

No.	^{5}L ligand	method of preparation (yield) properties and remarks
*1 (continued)		+48 °C), 5.40, 5.52, 5.61, 6.06 (all m, H-4,5,3,2 of C_5H_4; coalescence temperature: −24 °C), 8.21 (s, =CH) [18]; (CD_3NO_2, +24 °C): 3.23, 3.34 (both CH_3; coalescence temperature: +55 °C), 5.48, 5.62 (both C_5H_4), 7.71 (=CH) [19] ^{13}C NMR (acetone-d_6): 40.81, 47.77 (both CH_3), 86.36 (C-1), 90.08 (C-2), 91.41 (C-4), 94.15 (C-3), 96.56 (C-5), 155.09 (C-6), 229.84 (CO) [31] ^{13}C NMR (CP MAS, solid state): 37.84, 46.70 (both CH_3), 84.61 (C-1), 88.36 (C-2), 90.90 (C-4), 98.24 (C-3,5), 152.25 (C-6), 227.58, 229.35, 231.18 (all CO) [31] IR (KBr): 1624 (ν(C=N)), 1800, 1920 (CO); other bands at 462, 507, 597, 615, 630, 657, 778, 800, 940, 1060, 1133, 1170, 1250, 1287, 1345, 1365, 1410, 1450 [17]; see also [2]; (halocarbon mull): 1617 (ν(C=N)); 1785, 1815, 1908 (CO) [2]; (acetone): 1835, 1930 (CO) [12] UV (1,4-dioxane, ε) = 222 (12700), 253 (11500), 319 (8140), 442 (2910) [2]
2	$C_6H_5(CH_3)\overset{\oplus}{N}{=}CH(CH{=}CH)_2{-}C_5H_4^{\ominus}$	IIa (70%) [28] brownish purple powder decomposing above 150 °C [28] IR (KBr): 1813, 1840, 1925 (CO) [28] highly air-sensitive [28]
3	$[(CH_3)_2N]_2C^{\oplus}{-}C_5H_4^{\ominus}$	Ia (93%) [2] air-stable, yellow crystals, m.p. 281 to 284 °C (dec.) [2] ^{1}H NMR (acetone-d_6): 3.32 (CH_3); because of insufficient solubility, the C_5H_4 resonances remained unobserved [2] IR (halocarbon mull): 1525, 1571 (ν(C=N)); 1783, 1803, 1914 (CO); (KBr): 720, 796, 835, 843, 885, 887, 940, 1036, 1055, 1060, 1096, 1132, 1142, 1157, 1195, 1226, 1245, 1360, 1400, 1410, 1420, 1455, 1470 [2]
4	$[(i{-}C_3H_7)_2N]_2C_3^{\oplus}{-}C_5H_4^{\ominus}$	Ia [16] IR: 1527 (ν(C-N)) [21]

References on pp. 329/31

Table 9 (continued)

No.	^{5}L ligand	method of preparation (yield) properties and remarks
5	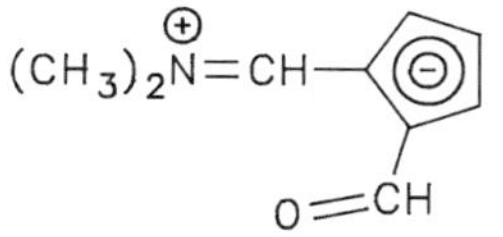 $(CH_3)_2\overset{\oplus}{N}=CH-C_5H_3^{\ominus}-CH=O$	IIa (33%) [24] dark red solid (not obtained in an analytically pure state) [24] IR (KBr): 1621 (ν(C=N)), 1662 (ν(C=O)), 1813, 1848, 1929 (CO) [24] hydrolysis in 5% aqueous NaOH at 50°C produced the $[1,2\text{-}(HC(O))_2C_5H_3Mo(CO)_3]^-$ anion which was isolated as its $[(C_6H_5)_3PNP(C_6H_5)_3]^+$ salt [24]
*6	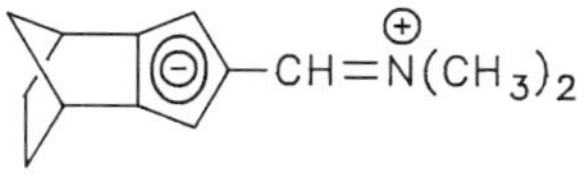 $-CH=\overset{\oplus}{N}(CH_3)_2$	IIb (77%) [29] dark red rods, m.p. > 230°C (dec.) [29] ^{1}H NMR (acetone-d_6): 1.25 (br s, 2 H), 1.46 (d, 1 H; J = 9), 1.96 (s (br), 2 H), 2.81 (d, 1 H; J = 10), 2.99 (d, 2 H; J = 8), 3.17, 3.30 (both s, $N(CH_3)_2$), 5.12, 5.71 (both br, ring-CH), 7.70 (s, CH=) [29] ^{13}C NMR (acetone-d_6): 39.36, 40.79, 47.37, 47.62, 81.06, 153.92, 230.86 [29] IR (KBr): 1625 (ν(C=N)), 1820, 1945 (CO); additional bands at 620, 1105, 1150, 1350, 1405, 1415, and 1445 [29]
7	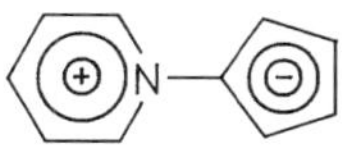 $C_5H_5N^{\oplus}-C_5H_4^{\ominus}$	IIc (20%) [15] violet crystals [15] ^{1}H NMR ($(CD_3)_2SO$): 5.30, 6.20 (both C_5H_4), 7.45, 8.17, 9.30 (all NC_5H_5) [15] IR (acetone): 1820, 1920 (CO) [15] mass spectrum: $[M]^+$, $[M-nCO]^+$ (n = 1, 2) [15] unstable in air, even in the solid state [15]
8	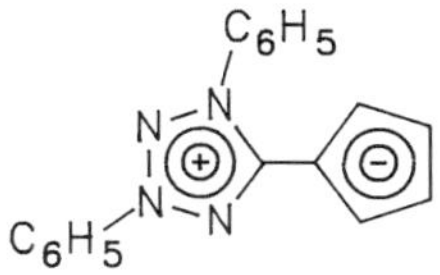 C_6H_5, N, N, N, N, C_6H_5	IId (95%) [32] dark brown solid, m.p. ca. 190°C (dec.) [32] ^{1}H NMR ($(CD_3)_2SO$): 5.39 (m, C_5H_4), 7.98, 8.35 (both m, C_6H_5) [32] ^{13}C NMR ($(CD_3)_2SO$): 76.9 (C of C_5H_4), 90.1, 90.2 (both CH of C_5H_4), 121.1, 127.1 (both ortho-C of C_6H_5), 130.5, 130.7 (both meta-C of C_6H_5), 132.3 (ipso-C of C_6H_5), 133.3, 133.4 (both para-C of C_6H_5), 134.8 (ipso-C of C_6H_5), 159.0 (N_4C), 231.3 (CO) [32] IR (KBr): 1794, 1912 (CO); other bands at 688, 764, 990, 1060, 1362, 1410, 1490, and 1546 [32] UV (CH_3CN, log ε): 213 (4.47), 260 (4.45), 413 (3.48) [32]

References on pp. 329/31

Table 9 (continued)

No.	^{5}L ligand	method of preparation (yield) properties and remarks
8 (continued)		mass spectrum (20 eV): $[M]^+$ (3%), $[M-3CO]^+$ (2%), $[M-Mo(CO)_3]^+$ (29%) [32] air-stable in crystalline form, but unstable in solution; attempted recrystallization caused partial decomposition [32]
*9	$(C_6H_5)_3\overset{\oplus}{P}$–$C_5H_4^{\ominus}$	Ib (ca. 60%) [1, 3]; for additional formations, see "Further information" fine golden yellow crystals [1, 3], m.p. 274 to 276 °C (dec.) [3] ^{1}H NMR ($CDCl_3$): 5.23, 5.49 (AA′BB′ system, C_5H_4; "J" ≅ 2.5) [3]; see also [12] ^{13}C NMR (CH_2Cl_2): C_5H_4 at 71.77 (d, C-1; J(P,C) = 113), 91.84 and 94.88 (both d, C-2,5 and C-3,4; J(P,C) = 15 and 12); C_6H_5 at 121.62 (d, C-1; J(P,C) = 93), 129.78, 134.03 (both d, C-2,6 and C-3,5; J(P,C) = 13 and 10), and 134.52 (s, C-4); CO at 230.43 (s) [13] IR (Nujol): 1791, 1808, 1904 (CO) [3]; (acetone): 1820, 1920 (CO) [12]; (CH_2Cl_2): 1811, 1911 (CO) [11]; ($CHCl_3$): 1812, 1920 (CO) [3] readily soluble in $CHCl_3$, less so in C_6H_6; insoluble in water [1]; for solutions in CF_3CO_2H, see "Further information"
10	$(C_6H_5)_3\overset{\oplus}{P}$–$C_5H_3^{\ominus}$–$N_2C_6H_4OCH_3$-4	IIe (60%) [6] purple solid [6] IR (KBr): 1819, 1830, 1924 (CO) [6] thermally unstable, decomposing on attempted recrystallization [6]
11	$(C_6H_5)_3\overset{\oplus}{P}$–$C_5(C_6H_5)_4^{\ominus}$	IIe (88%) [6] m.p. 234 °C [6] IR (CH_2Cl_2): 1813, 1920 (CO) [6] treatment with $[4\text{-}CH_3OC_6H_4N_2]PF_6$ (one equivalent in CH_2Cl_2) produced $(C_6H_5)_3PC_5(C_6H_5)_4Mo(CO)_2N_2C_6H_4OCH_3$-4 [6]
12	$(CH_3)_3\overset{\oplus}{P}CH_2$–$C_5(C_6H_5)_4^{\ominus}$	III (73%) [25] light yellow solid, m.p. 106 °C (dec.) [25] ^{1}H NMR (CD_2Cl_2): 1.05 (d, CH_3; J(P,H) = 13.0), 3.78 (d, CH_2; J(P,H) = 10.9), 6.72 to 7.49 (m, C_6H_5) [25] IR (KBr): 1780, 1869 (CO) [25] thermally unstable, decomposing with loss of $P(CH_3)_3$ [25]

References on pp. 329/31

Table 9 (continued)

No.	^5L ligand	method of preparation (yield) properties and remarks
13	$(n-C_4H_9)_3\overset{\oplus}{P}CH_2-C_5^{\ominus}(C_6H_5)_4$	III (82%) [25] yellow needles, m.p. 195°C (dec.) [25] ^1H NMR (CD_2Cl_2): 0.78 to 1.84 (m, C_4H_9), 3.59 (d, CH_2; J(P,H) = 11.9), 6.97 to 7.64 (m, C_6H_5) [25] ^{31}P NMR (CH_2Cl_2): 31.5 [25] IR (KBr): 1791, 1902 (CO) [25]
14	$C_6H_5(CH_3)_2\overset{\oplus}{P}CH_2-C_5^{\ominus}(C_6H_5)_4$	III (18%) [25] light yellow solid, m.p. 114°C (dec.) [25] ^1H NMR (acetone-d_6): 1.79 (d, CH_3; J(P,H) = 13.9), 3.98 (d, CH_2; J(P,H) = 12.5), 6.92 to 7.84 (m, C_6H_5) [25] ^{31}P NMR (CH_3CN): 31.1 [25] IR (KBr): 1780, 1893 (CO) [25]
15	$CH_3(C_6H_5)_2\overset{\oplus}{P}CH_2-C_5^{\ominus}(C_6H_5)_4$	III (29%) [25] light yellow solid, m.p. 138°C (dec.) [25] ^1H NMR (CD_3CN): 1.62 (d, CH_3; J(P,H) = 13.6), 4.25 (d, CH_2; J(P,H) = 11.6), 6.83 to 7.73 (m, C_6H_5) [25] ^{31}P NMR (CH_3CN): 17.6 [25] IR (KBr): 1782, 1898 (CO) [25]
16	$(C_6H_5)_3\overset{\oplus}{P}CH_2CH_2-C_5^{\ominus}(C_6H_5)_4$	2 equivalents of $(C_6H_5)_3P{=}CH_2$ were added to $H_2CC_5(C_6H_5)_4Mo(CO)_3$ in toluene/pentane (3:1); the adduct precipitated immediately (96%) [25] orange solid, m.p. 80°C (dec.) [25] IR (KBr): 1782, 1892 (CO) [25]
17	$(C_6H_5)_3\overset{\oplus}{As}-C_5^{\ominus}(C_6H_5)_4$	IIe (73%) [6] purple solid, m.p. 188 to 190°C [6] IR (KBr): 1819, 1830, 1924 (CO); (CH_2Cl_2): 1814, 1919 (CO) [6] thermally unstable, decomposing on attempted recrystallization [6]
*18	$(CH_3)_2\overset{\oplus}{S}-C_5^{\ominus}H_4$	IIf (38%) [9] light yellow crystals, m.p. 173 to 175°C (dec.) [9] ^1H NMR ($(CD_3)_2SO$): 3.16 (s, CH_3), 5.47, 6.09 (both "t", C_5H_4) [9, 10, 12] ^{13}C NMR ($(CD_3)_2SO$): 33.07 (CH_3), 83.36 (C of C_5H_4), 89.40, 91.42 (both CH of C_5H_4), 231.51 (CO) [13] IR (acetone): 1825, 1925 (CO) [9,12]; similar in [10]; additional bands in KBr at 465, 490, 510,

References on pp. 329/31

Table 9 (continued)

No.	^{5}L ligand	method of preparation (yield) properties and remarks
*18 (continued)		520, 610, 635, 790, 800, 835, 885, 900, 925, 970, 990, 1020, 1040, 1060, 1170, 1310, 1330, 1345, 1410, 1425, 2920, 3000, 3030, 3100, 3120 [9] readily soluble in $(CH_3)_2SO$, moderately soluble in acetone, and poorly soluble in C_6H_6, $(C_2H_5)_2O$, C_2H_5OH, and $CHCl_3$ [9, 10] mass spectrum (30 eV, 160°C): $[M]^+$, $[M-nCO]^+$ (n = 0 to 3), $[M-CH_3-nCO]^+$ (n = 0 to 3), $[MoC_5H_4SCH_2]^+$, $[MoS]^+$, $[(CH_3)_2SC_5H_4]^+$, $[CH_3SC_5H_4]^+$ [9]; for fragmentation scheme, see [10] air-stable in the solid state, less so in solution [9, 10]

*Further information:

$(CH_3)_2NCHC_5H_4Mo(CO)_3$ (Table **9**, No. **1**) proved to be sparingly soluble in common organic solvents, especially nonpolar ones [2]. The complex was also practically insoluble in acetic acid but was readily dissolved in nonaqueous mixtures of acetic and trifluoroacetic acids, in neat CF_3CO_2H, as well as in CH_2Cl_2 and nitromethane solutions of the latter. In these highly acidic media, No. 1 has been shown to undergo reversible protonation at its central metal, forming the $[(CH_3)_2NCHC_5H_4Mo(CO)_3H]^+$ cation at values of medium acidity, $H_0 \leq 1$ [17, 19].

The rate constant for H/D exchange between the C_5H_4 ring and C_2H_5OD in acetone solution at T = 50°C was determined as $k = 4.3 \times 10^{-6}\ s^{-1}$ [12].

Upon treatment with $Na[C_5H_5]$ (10% excess) in THF, the $[C_5H_4{=}CHC_5H_4Mo(CO)_3]^-$ anion was formed and isolated as its $[P(C_6H_5)_4]^+$ salt by precipitation with $[P(C_6H_5)_4]Br$ from ethanol/THF [30].

Reaction with $[O(AuP(C_6H_5)_3)_3]BF_4$ in THF in the presence of H_2O and K_2CO_3 between −5 and 0°C led to the formation of a Mo−Au-bonded formylcyclopentadienyl derivative containing two semibridging CO ligands, $HC(O)C_5H_4MoCO(\mu\text{-}CO)_2AuP(C_6H_5)_3$ [26, 27].

$(CH_3)_2NCHC_{10}H_{10}Mo(CO)_3$ (Table **9**, No. **6**). Single crystals grown from ethyl acetate/hexane were monoclinic, space group $P2_1/n$ (nonstandard setting of $P2_1/c$)−C_{2h}^5 (No. 14), with a = 5.959(1), b = 14.823(3), c = 17.107(3) Å, β = 94.47(1)°; Z = 4 molecules per unit cell, D_{calc} = 1.63 g/cm^3. A side view of the molecule (**Fig. 32**) shows the molybdenum atom to be located syn to the methano bridge of the norbornane fragment. The cyclopentadienyl-like ring is essentially planar. The $-CH{=}N^+(CH_3)_2$ group lies significantly out of this plane and in the direction of the metal atom. The bridgehead carbon atoms of the norbornane moiety also reside out of the plane defined by the five-membered ring but in the opposite direction [29].

$(C_6H_5)_3PC_5H_4Mo(CO)_3$ (Table **9**, No. **9**) was also observed as a product resulting (i) from dissociation of several of its weakly bonded Lewis acid adducts (see "Lewis Acid Adducts

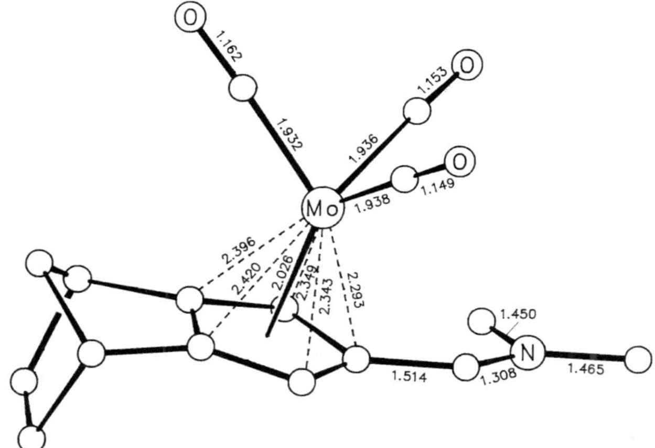

Fig. 32. The molecular structure of $(CH_3)_2NCHC_{10}H_{10}Mo(CO)_3$ [29].

of Zwitterionic $^5LMo(CO)_3$ Complexes"), (ii) from treating its BF_3 and $Al(CH_3)_3$ derivatives $(C_6H_5)_3PC_5H_4Mo(CO)_3^{\cdot}BF_3$ and $(C_6H_5)_3PC_5H_4Mo(CO)_2(\mu\text{-}CO)Al(CH_3)_3$ with trimethylamine, as well as (iii) from decomposing the latter with ethanol [3, 5]. Furthermore, high yields of the title compound were recovered when $[(C_6H_5)_3PC_5H_4Mo(CO)_3Br]PF_6$ or $[((C_6H_5)_3PC_5H_4\text{-}Mo(CO)_3)_2][ClO_4]_2$ (resulting from oxidation of No. 9 with bromine or tris(p-bromophenyl)-amminium perchlorate; see below) were treated with $NaBH_4$ in aqueous ethanol [6].

The complex dissolved readily in nonaqueous trifluoroacetic acid or in CF_3CO_2H/CH_2Cl_2 mixtures producing deep red solutions of the $[(C_6H_5)_3PC_5H_4Mo(CO)_3H]^+$ cation [3, 11].

Halogenation of $(C_6H_5)_3PC_5H_4Mo(CO)_3$ in CH_2Cl_2 solution at room temperature by treatment with a brief stream of chlorine or one equivalent of bromine or iodine yielded the highly crystalline salts $[(C_6H_5)_3PC_5H_4Mo(CO)_3X]X$ and/or $[(C_6H_5)_3PC_5H_4Mo(CO)_3X]X_3$ (X = Cl, Br, I), soluble in aqueous ethanol, from which they could be precipitated as their hexafluorophosphates [4, 6].

Tensimetric titration of CH_2Cl_2 solutions of the title compound with the hard Lewis acids BF_3 and BCl_3 at $-78\,^{\circ}C$ and $0\,^{\circ}C$ indicated the reversible formation of the metal-centered 1:1 adducts, $(C_6H_5)_3PMo(CO)_3^{\cdot}BX_3$ (X = F, Cl), of which the product of the BF_3 reaction could be isolated as a solid. On the other hand, there was little or no interaction with the soft acceptor B_2H_6, implying that $(C_6H_5)_3PC_5H_4Mo(CO)_3$ behaved as a hard Lewis base toward BX_3 Lewis acids [3]. Addition of the Lewis acids $AlCl_3$ [11], $GaBr_3$, $InBr_3$ [6], CdI_2, HgX_2 (X = Cl, Br, I) [4, 6], and $TiCl_4$ [11] to CH_2Cl_2 solutions similarly generated metal–metal-bonded adducts formulated as $(C_6H_5)_3PC_5H_4Mo(CO)_3^{\cdot}MX_n$ [11] or, alternatively, $[(C_6H_5)_3\text{-}PC_5H_4Mo(CO)_3^{\cdot}MX_{n-1}]X$ [4, 6].

$[((C_6H_5)_3PC_5H_4Mo(CO)_3)_2][PF_6]_2$, containing a metal–metal bond, was isolated from an unsuccessful attempt to nitrosylate the title complex with $NOPF_6$ in THF [6]. $SbCl_5$ (2 equivalents in CH_2Cl_2) behaved as a mild halogenating agent, smoothly producing $[(C_6H_5)_3\text{-}PC_5H_4Mo(CO)_3Cl]SbCl_6$ [6]. An ill-defined, weakly bonded SO_2 adduct was isolated from solutions of $(C_6H_5)_3PC_5H_4Mo(CO)_3$ in liquid sulfur dioxide or in methylene chloride saturated with SO_2 [6].

References on pp. 329/31

Differing from the reactions with inorganic electron-pair acceptors in CH_2Cl_2 (see above), combination of $(C_6H_5)_3PC_5H_4Mo(CO)_3$ with the organometallic acceptor $Al_2(CH_3)_6$ in toluene resulted in the formation of a CO-bridged adduct of the "isocarbonyl"-type, $(C_6H_5)_3PC_5H_4Mo(CO)_2(\mu\text{-}CO)Al(CH_3)_3$ [5].

When allowed to react with $C_5H_5Co(CO)I_2$ in refluxing CH_2Cl_2, $(C_6H_5)_3PC_5H_4Mo(CO)_3$ was transformed into the $[(C_6H_5)_3PC_5H_4Mo(CO)_3I]^+$ cation which was isolated as its PF_6^- salt [6]. Treatment with excess C_6H_5HgCl led to $(C_6H_5)_3PC_5H_4Mo(CO)_3 \cdot HgCl_2$ [4].

The rate constant for H/D exchange of the C_5H_4 moiety with C_2H_5OD in C_6H_6 at 75 °C, measured as $k = 2.7 \times 10^{-5}\ s^{-1}$, showed that coordination to the $Mo(CO)_3$ unit had little effect on the ability of the C_5H_4 ring of $(C_6H_5)_3PC_5H_4$ to enter into this exchange reaction in neutral medium ($k = 3.3 \times 10^{-5}\ s^{-1}$ for umcomplexed triphenylphosphonium cyclopentadienylide) [8]. H/D exchange with C_2H_5OD in acetone occurred with a rate constant of $8.2 \times 10^{-6}\ s^{-1}$ at 50 °C [12].

High yields of $(C_6H_5)_3PC_5H_4Mo(CO)_2N_2C_6H_4OCH_3$-4 were obtained from the reaction between equimolar amounts of $(C_6H_5)_3PC_5H_4Mo(CO)_3$ and $[4\text{-}CH_3C_6H_4N_2]PF_6$ in CH_2Cl_2 at room temperature [4, 6]. Oxidation with tris(p-bromo-phenyl)amminium hexachloroantimonate in CH_2Cl_2 gave the $[(C_6H_5)_3PC_5H_4Mo(CO)_3Cl]^+$ cation, which was readily isolated after conversion to its PF_6^- salt. Use of the corresponding amminium perchlorate, however, yielded a metal–metal-bonded homobinuclear salt, $[((C_6H_5)_3PC_5H_4Mo(CO)_3)_2][ClO_4]_2$ [6].

$(CH_3)_2SC_5H_4Mo(CO)_3$ (Table **9**, No. **18**). The rate of H/D exchange between the C_5H_4 protons and C_2H_5OD in acetone was found to be practically the same in the α- and β-positions of the ring at 50 °C, occurring with a rate constant, k, of $2.9 \times 10^{-5}\ s^{-1}$; for details of the deuterium distribution after exchange, see [12].

1.5.1.4.2.6.2 $[^5LMo(CO)_3H]^+$ Cations

$[(CH_3)_2NCHC_5H_4Mo(CO)_3H]O_2CF_3$ was observed to form in the strongly acidic CF_3CO_2H, CF_3CO_2H/CH_3CO_2H, CF_3CO_2H/CH_2Cl_2, and CF_3CO_2H/CH_3NO_2 solutions of $(CH_3)_2NCHC_5H_4Mo(CO)_3$, where complete protonation of the parent fulvene complex occurred at an acidity value of the medium of $H_0 = -2.7$ [17, 19].

^1H NMR spectrum (CF_3CO_2H): δ = −4.71 (MoH) ppm [17]; (CF_3CO_2H/CD_3NO_2 (1:500), 24 °C): δ = −5.04 (MoH), 3.65, 3.85 (both CH_3; coalescence temperature: 67 °C), 5.94, 6.12 (both C_5H_4), 8.57 (=CH) ppm [19]. IR spectrum (CF_3CO_2H/CH_2Cl_2 (1:50)): 1984, 2073 (CO) cm^{-1} [17].

Addition of triethylamine to solutions containing the protonated complex resulted in deprotonation with reformation of the starting fulvene compound [17].

$[(C_6H_5)_3PC_5H_4Mo(CO)_3H]O_2CF_3$ was identified as the deep red product formed in solution by protonation of $(C_6H_5)_3PC_5H_4Mo(CO)_3$ with neat CF_3CO_2H [3] or with trifluoroacetic acid dissolved in CH_2Cl_2 [11].

^1H NMR spectrum (CF_3CO_2H): δ = −5.34 (s, MoH) ppm [3]. IR spectrum (CF_3CO_2H): 1970, 2048 (CO) cm^{-1} [3]; (CF_3CO_2H/CH_2Cl_2 (1:100)): 1958, 1970, 2045 (ν(CO)) cm^{-1} [11].

Treatment in trifluoroacetic acid solution with excess CCl_4 or CBr_4 resulted in cationic halogeno complexes, $[(C_6H_5)_3PC_5H_5Mo(CO)_3X]^+$ (X = Cl, Br), which were isolated as their hexafluorophosphato salts by workup with aqueous ethanolic NH_4PF_6 [6].

References on pp. 329/31

1.5.1.4.2.6.3 Cationic Halogeno Complexes $[^5LMo(CO)_3X]^+$

$[(C_6H_5)_3PC_5H_4Mo(CO)_3Cl]PF_6$ was formed on treating a CH_2Cl_2 solution of $(C_6H_5)_3PC_5H_4$-$Mo(CO)_3$ with a brief stream of chlorine at room temperature. Removal of the solvent gave a red solid (presumably impure **$[(C_6H_5)_3PC_5H_4Mo(CO)_3Cl]Cl$**), which was dissolved in water containing ca. 20% ethanol. Addition of aqueous NH_4PF_6, followed by dilution with a large excess of water, precipitated the salt $[(C_6H_5)_3PC_5H_4Mo(CO)_3Cl]PF_6$, albeit in an analytically impure form [6]. Analytically pure $[(C_6H_5)_3PC_5H_4Mo(CO)_3Cl]PF_6$ was obtained from the action of excess CCl_4 (solution in THF) on the $[(C_6H_5)_3PC_5H_4Mo(CO)_3H]^+$ cation, generated in situ by dissolving $(C_6H_5)_3PC_5H_4Mo(CO)_3$ in nonaqueous CF_3CO_2H. The residue remaining after removal of the volatiles was extracted with water/ethanol (9:1), and the title complex was precipitated by adding aqueous NH_4PF_6. Recrystallization from methylene chloride/diethyl ether afforded the hexafluorophosphate salt in 85% yield [6]. Stirring a solution of $(C_6H_5)_3PC_5H_4Mo(CO)_3$ in CH_2Cl_2 with two equivalents of $SbCl_5$ in the same solvent resulted in the initial formation of **$[(C_6H_5)_3PC_5H_4Mo(CO)_3Cl]SbCl_6$** which was deposited as an oil after adding light petroleum and was further converted to the PF_6^- salt as described above [6]. Addition of solid tris(p-bromophenyl)amminium hexachloroantimonate to a CH_2Cl_2 solution of $(C_6H_5)_3PC_5H_4Mo(CO)_3$, followed by concentration of the solution in vacuum, afforded a light-orange solid, which was first recrystallized from CH_2Cl_2/ether and then converted to a hexafluorophosphate salt identical with the materials isolated from the procedures summarized above [6].

Light orange crystals, m.p. 214 to 215°C. IR spectrum (KBr): 1976, 2005, 2064 (CO) cm^{-1} [6].

$[(C_6H_5)_3PC_5H_4Mo(CO)_3Br]PF_6$. Similar to its chloro homologue, the complex was isolated in an analytically impure state from the reaction of $(C_6H_5)_3PC_5H_4Mo(CO)_3$ with one equivalent of bromine in CH_2Cl_2 solution at ambient conditions. The residue remaining after evaporation of the solvent (**$[(C_6H_5)_3PC_5H_4Mo(CO)_3Br]Br$** or **$[(C_6H_5)_3PC_5H_4Mo(CO)_3Br]Br_3$**) was dissolved in water containing added ethanol and was converted to the PF_6^- salt by precipitation with aqueous NH_4PF_6 as described for $[(C_6H_5)_3PC_5H_4Mo(CO)_3Cl]PF_6$ above [4, 6]. Pure title complex was prepared by adding a THF solution of carbon tetrabromide to a solution of $(C_6H_5)_3PC_5H_4Mo(CO)_3$ in neat CF_3CO_2H. Workup with aqueous ammonium hexafluorophosphate (see $[(C_6H_5)_3PC_5H_4Mo(CO)_3Cl]PF_6$) afforded the product in 90% yield [4, 6].

Salmon-pink crystals, m.p. 193 to 195°C. IR spectrum (KBr): 1982, 2008, 2063 (CO) cm^{-1} [6].

Treatment with excess $NaBH_4$ in water/ethanol regenerated $(C_6H_5)_3PC_5H_4Mo(CO)_3$ in high yield [6].

$[(C_6H_5)_3PC_5H_4Mo(CO)_3I]PF_6$. Distinct from the materials resulting from direct chlorination and bromination of $(C_6H_5)_3PC_5H_4Mo(CO)_3$, the product isolated by workup with aqueous NH_4PF_6 (see above) of the iodination reaction (one equivalent of I_2 in CH_2Cl_2 at room temperature) proved to be analytically pure $[(C_6H_5)_3PC_5H_4Mo(CO)_3I]PF_6$. After recrystallization from methylene chloride/ether, the complex was isolated in 93% yield. The salt was also formed when a solution of $(C_6H_5)_3PC_5H_4Mo(CO)_3$ in CH_2Cl_2 was added to a solution of $C_5H_5Co(CO)I_2$ in the same solvent and the mixture was heated briefly to the boiling point. The solvent was removed in vacuum and the red residue was converted to the title compound as described for $[(C_6H_5)_3PC_5H_4Mo(CO)_3Cl]PF_6$ [6].

Dark red crystals, m.p. 184 to 185°C. IR spectrum (KBr): 1976, 1997, 2052 (CO) cm^{-1} [6].

References on pp. 329/31

1.5.1.4.2.6.4 Adducts with Lewis Acids

The Lewis acid/Lewis base interactions dealt with in this section may yield two types of 1:1 adducts: (i) complexes in which the molybdenum atom serves as the donor toward the Lewis acidic acceptor MX_n, $^5LMo(CO)_3^{\cdot}MX_n$ or $[^5LMo(CO)_3MX_{n-1}]X$, and (ii) isocarbonyl-bridged derivatives, $^5LMo(CO)_2(\mu\text{-}CO)MX_n$, with the Lewis acid bonded to one of the carbonyl oxygen atoms. The reaction path depends on the nature of the reactants as well as the solvent employed [11]. Adduct formation at the carbonyl oxygen results in a pronounced decrease of the CO stretching frequency of the donating carbonyl group and a simultaneous increase by 20 to 50 cm^{-1} of those CO ligands which are not coordinated to the Lewis acid [5, 7, 11].

$(C_6H_5)_3PC_5H_4Mo(CO)_3 \cdot BF_3$ was formed on tensimetric titration of $(C_6H_5)_3PC_5H_4Mo(CO)_3$, dissolved in CH_2Cl_2, with BF_3 at −78 °C and 0 °C. The crude material was purified by rapid recrystallization from CH_2Cl_2/toluene or by washing with THF, which afforded the title complex as a deep red, air-stable solid [3].

^{11}B NMR spectrum (CH_2Cl_2): δ = −1.0 ppm. IR spectrum (Nujol): 1896, 1935, 1957, 1993, 2016 (CO) cm^{-1} [3].

The complex proved to be soluble in CH_2Cl_2 but insoluble in toluene and THF; treatment with the latter caused slow dissociation. Solutions in CH_2Cl_2 underwent gradual decomposition with the formation of an unidentified hydrido species (δ(MoH) = −4.85 ppm) [3].

The mass spectrum showed only the presence of $(C_6H_5)_3PC_5H_4Mo(CO)_3$ and its fragment ions [3].

No reaction occurred with trimethylamine in toluene. The same reaction in CH_2Cl_2 regenerated some of the starting $(C_6H_5)_3PC_5H_4Mo(CO)_3$ compound, but $(CH_3)_3N \cdot BF_3$ could not be isolated [3].

$(C_6H_5)_3PC_5H_4Mo(CO)_3^{\cdot}BCl_3$ was briefly mentioned as the 1:1 adduct resulting from tensimetric titration of the parent triphenylphosphonium cyclopentadienylide complex with BCl_3 in CH_2Cl_2 at −78 °C and 0 °C, but attempts to isolate and purify the product remained unsuccessful [3].

$(C_6H_5)_3PC_5H_4Mo(CO)_3^{\cdot}SO_2$ was assumed to form in the intense bloodred solutions of $(C_6H_5)_3PC_5H_4Mo(CO)_3$ in liquid sulfur dioxide or in methylene chloride saturated with SO_2. These solutions slowly decomposed on standing, yielding amorphous brown carbonyl-free materials. Rapid evaporation of freshly prepared solutions gave a deep red solid which quickly turned yellow. The yellow product proved to be unchanged $(C_6H_5)_3PC_5H_4Mo(CO)_3$. The red solid displayed two strong CO absorptions at 1968 and 1988 cm^{-1}, attributed to a reversibly formed, weakly bonded adduct, $(C_6H_5)_3PC_5H_4Mo(CO)_3 \cdot SO_2$ [6].

$(C_6H_5)_3PC_5H_4Mo(CO)_3 \cdot MX_n$ or **$[(C_6H_5)_3C_5H_4Mo(CO)_3MX_{n-1}]X$** ($MX_n$ = $AlCl_3$, $GaBr_3$, $InBr_3$, $TiCl_4$, CdI_2, $HgCl_2$, $HgBr_2$, HgI_2). The following table contains a number of Lewis acid/Lewis base adducts $^5LMo(CO)_3^{\cdot}MX_n$ with metal−metal bonds between donor and acceptor atoms, which formed when CH_2Cl_2 solutions of the Lewis basic ylide complex $(C_6H_5)_3PC_5H_4Mo(CO)_3$ were treated with the appropriate Lewis acid MX_n (generally 2 equivalents). The adducts formed with $AlCl_3$ and $TiCl_4$ were observed only in solution [11]; those obtained with $GaBr_3$, $InBr_3$, CdI_2, and HgX_2 (X = Cl, Br, I) were precipitated by adding ether and were recrystallized, with some difficulty, from CH_2Cl_2/ether. Yields of the isolated adducts were ca. 70% [6].

Since most of the metal−metal-bonded adducts listed in the table turned out to be too insoluble, even in polar solvents, for reliable conductivity measurements, no firm conclusions

References on pp. 329/31

as to the covalent ($^5LMo(CO)_3^{\cdot}MX_n$) or ionic ($[^5LMo(CO)_3MX_{n-1}]X$) nature of these complexes could be reached. $(C_6H_5)_3PC_5H_4Mo(CO)_3 \cdot HgI_2$, e.g., which proved to be reasonably soluble in acetone, had a molar conductance of 65 $cm^2 \cdot \Omega^{-1} \cdot mol^{-1}$ for a 10^{-3} M solution, rather high for a neutral complex but lower than expected for a 1:1 electrolyte [6].

The adducts $(C_6H_5)_3PC_5H_4Mo(CO)_3^{\cdot}MX_n$ (n = 2: M = Cd, Hg; X = Cl, Br, I. n = 3: M = Ga, In; X = Br) were described as moderately air-stable but light-sensitive [4, 6].

MX_n	properties and remarks
$AlCl_3$	IR (CH_2Cl_2): 1961, 2043 (CO) [11]
$GaBr_3$	white solid still containing some parent ylide complex after isolation [6] IR (KBr): 1903, 1929, 1994 (CO) [6]
$InBr_3$	white solid, m.p. 200 °C [6] IR (KBr): 1897, 1928, 1998 (CO) [6]
$TiCl_4$	IR (CH_2Cl_2): 1962, 2044 (CO) [11]
CdI_2	white solid, m.p. 205 °C [6] IR (KBr): 1870, 1900, 1975 (CO) [6]
$HgCl_2$	also obtained from the reaction of $(C_6H_5)_3PC_5H_4Mo(CO)_3$ with excess C_6H_5HgCl [4] yellow solid, m.p. 212 to 213 °C [6] IR (KBr): 1889, 1928, 1979 (CO) [6]
$HgBr_2$	yellow solid, m.p. 195 °C [6] IR (KBr): 1891, 1926, 1979 (CO) [6]
HgI_2	yellow solid [6] IR (KBr): 1903, 1927, 1994 (CO) [6]

$(C_6H_5)_3PC_5H_4Mo(CO)_2(\mu\text{-}CO)Al(CH_3)_3$, the first example of a complex containing a terminal carbonyl base [7], was prepared by adding $Al_2(CH_3)_6$ to a suspension of $(C_6H_5)_3PC_5H_4$-$Mo(CO)_3$ in toluene, which caused the starting material to dissolve readily. On dilution with hexane, the adduct precipitated as a brown solid that could be purified by rapid recrystallization from toluene/hexane. Successive crystallizations or continual washing with toluene ultimately resulted in complete dissociation [5].

IR spectrum (Nujol): 1665, 1845, 1932 (CO) cm^{-1}; absorptions due to coordinated $Al(CH_3)_3$ at 615, 690, and 1180 cm^{-1} [5].

Rapid exchange of bridging and terminal methyl groups was found by ^1H NMR spectroscopy to occur in toluene solutions of $(C_6H_5)_3PC_5H_4Mo(CO)_2(\mu\text{-}CO)Al(CH_3)_3/Al_2(CH_3)_6$ mixtures even at −80 °C, and was attributed to a rapid equilibrium "2 $(C_6H_5)_3PC_5H_4Mo$-$(CO)_3 + Al_2(CH_3)_6 \rightleftarrows 2\ (C_6H_5)_3PC_5H_4Mo(CO)_2(\mu\text{-}CO)Al(CH_3)_3$" [5].

Treatment with ethanol produced a quantitative yield of methane together with the parent ylide complex. Tensimetric titrations with trimethylamine indicated a 1:1 reaction from which pure $(C_6H_5)_3PC_5H_4Mo(CO)_3$ and $(CH_3)_3N \cdot Al(CH_3)_3$ could be isolated [5].

References:

[1] Abel, E. W.; Singh, A.; Wilkinson, G. (Chem. Ind. [London] **1959** 1067).
[2] King, R. B.; Bisnette, M. B. (Inorg. Chem. **3** [1964] 801/7).
[3] Kotz, J. C.; Pedrotty, D. G. (J. Organomet. Chem. **22** [1970] 425/38).

[4] Cashman, O.; Lalor, F. J. (J. Organomet. Chem. **24** [1970] C29/C30).
[5] Kotz, J. C.; Turnipseed, C. D. (J. Chem. Soc. D **1970** 41/2).
[6] Cashman, O.; Lalor, F. J. (J. Organomet. Chem. **32** [1971] 351/63).
[7] Shriver, D. F. (Chem. Br. **8** [1972] 419/21).
[8] Zdanovich, V. I.; Yurtanov, A. I.; Zhakaeva, A. Zh.; Setkina, V. N.; Kursanov, D. N. (Izv. Akad. Nauk SSSR Ser. Khim. **1973** 1375/6; Bull. Acad. Sci. USSR Div. Chem. Sci. [Engl. Transl.] **1973** 1334/5).
[9] Zdanovich, V. I.; Zhakaeva, A. Zh.; Setkina, V. N.; Kursanov, D. N. (J. Organomet. Chem. **64** [1974] C25/C26).
[10] Setkina, V. N.; Zdanovich, V. I.; Zhakaeva, A. Zh.; Nekrasov, Yu. S.; Vasyukova, N. I.; Kursanov, D. N. (Dokl. Akad. Nauk SSSR **219** [1974] 1137/9; Dokl. Chem. [Engl. Transl.] **214/219** [1974] 867/9).

[11] Lokshin, B. V.; Rusach, E. B.; Kolobova, N. E.; Makarov, Yu. V.; Ustynyuk, N. A.; Zdanovich, V. I.; Zhakaeva, A. Zh.; Setkina, V. N. (J. Organomet. Chem. **108** [1976] 353/61).
[12] Setkina, V. N.; Zhakaeva, A. Zh.; Zdanovich, V. I.; Kursanov, D. N. (Izv. Akad. Nauk SSSR Ser. Khim. **1976** 1766/8; Bull. Acad. Sci. USSR Div. Chem. Sci. [Engl. Transl.] **1976** 1664/6).
[13] Setkina, V. N.; Zhakaeva, A. Zh.; Panosyan, G. A.; Zdanovich, V. I.; Petrovskii, P. V.; Kursanov, D. N. (J. Organomet. Chem. **129** [1977] 361/9).
[14] Edelmann, F.; Behrens, U. (J. Organomet. Chem. **134** [1977] 31/6).
[15] Zhakaeva, A. Zh.; Orlova, T. Yu.; Zdanovich, V. I.; Setkina, V. N.; Kursanov, D. N. (Izv. Akad. Nauk SSSR Ser. Khim. **1977** 227/8; Bull. Acad. Sci. USSR Div. Chem. Sci. [Engl. Transl.] **1977** 202/3).
[16] Yoshida, Z.; Mitsubishi Chemical Industries Co., Ltd. (Jn. Kokai 78-37644 [1976/78] 4 pp.; C.A. **89** [1978] No. 109974).
[17] Setkina, V. N.; Strunin, B. N.; Kursanov, D. N. (J. Organomet. Chem. **186** [1980] 325/30).
[18] Strunin, B. N.; Bakhmutov, V. I.; Setkina, V. N.; Kursanov, D. N. (J. Organomet. Chem. **208** [1981] 81/8).
[19] Strunin, B. N.; Bakhmutov, V. I.; Setkina, V. N.; Kursanov, D. N. (J. Organomet. Chem. **219** [1981] 197/210).
[20] Bakhmutov, V. I.; Strunin, B. N.; Setkina, V. N. (Izv. Akad. Nauk SSSR Ser. Khim. **1981** 1652/3; Bull. Acad. Sci. USSR Div. Chem. Sci. [Engl. Transl.] **1981** 1341/2).

[21] Yoshida, Z. (Pure Appl. Chem. **54** [1982] 1059/74).
[22] Lubke, B.; Edelmann, F.; Behrens, U. (Chem. Ber. **116** [1983] 11/26).
[23] Strunin, B. N.; Bakhmutov, V. I.; Vasyukova, N. I.; Trembovler, V. N.; Setkina, V. N.; Kursanov, D. N. (J. Organomet. Chem. **246** [1983] 169/76).
[24] Edelmann, F.; Behrens, U. (Chem. Ber. **117** [1984] 3463/72).
[25] Drews, R.; Behrens, U. (Chem. Ber. **118** [1985] 888/94).
[26] Perevalova, E.; Grandberg, K.; Smyslova, E.; Vasyukova, N.; Strunin, B. (12th Int. Conf. Organomet. Chem., Vienna 1985, Abstr. p. 246).
[27] Strunin, B. N.; Grandberg, K. I.; Andrianov, V. G.; Setkina, V. N.; Perevalova, E. G.; Struchkov, Yu. T.; Kursanov, D. N. (Dokl. Akad. Nauk SSSR **281** [1985] 599/602; Dokl. Chem. [Engl. Transl.] **280/285** [1985] 106/9).
[28] Edelmann, F.; Behrens, P.; Behrens, S.; Behrens, U. (J. Organomet. Chem. **310** [1986] 333/55).
[29] Paquette, L. A.; Hathaway, S. J.; Schirch, P. F. T.; Gallucci, J. C. (Organometallics **5** [1986] 500/5).
[30] Härter, P.; Kiprof, P. (J. Organomet. Chem. **397** [1990] C1/C5).

[31] Bakhmutov, V. L.; Galakhov, M. V.; Strunin, B. N.; Nikanorov, V. A. (Metalloorg. Khim. **3** [1990] 1329/35; Organomet. Chem. USSR [Engl. Transl.] **3** [1990] 690/4).
[32] Araki, S.; Butsugan, Y. (Chem. Ber. **126** [1993] 1157/9).

1.5.1.4.2.7 Complexes with ^{5}L Ligands Different from Cyclopentadienyl Derivatives

1.5.1.4.2.7.1 $^5LMo(CO)_3$ Complexes with Open-Chain Pentadienyl Ligands

[K($CH_3OC_2H_4OC_2H_4OCH_3$)][2,4-$C_7H_{11}Mo(CO)_3$] (2,4-C_7H_{11} = Formula I) was obtained by addition of K[2,4-C_7H_{11}] in THF to a stirred solution of an equimolar quantity of $Mo(CO)_3$-($CH_3OC_2H_4OC_2H_4OCH_3$) in THF at −78 °C. After warming to room temperature and removal of solvent, the product was extracted with toluene/THF (4:1). Evaporation of the filtered extracts left the salt as yellow platelets, m.p.: 106 to 107 °C (dec.), in 80% yield. Single crystals were isolated by slow cooling of concentrated toluene/THF solutions to −20 °C [2].

^{1}H NMR spectrum (C_6D_6): δ = 1.57 (s, H_{endo}-1,5), 2.41 (s, CH_3-2,4), 3.08 (s, H_{exo}-1,5), 3.15 (s, CH_3O), 3.25 (m, C_2H_4), 5.28 (s, H-3) ppm. ^{13}C NMR spectrum (C_6D_6): δ = 30.7 (q, CH_3-2,4; J(C, H) = 125 Hz), 58.4 (t, CH_2O or CH_2-1,5; J(C, H) = 154 Hz), 59.1 (q, CH_3O; J(C, H) = 141 Hz), 69.6 (t, CH_2O or CH_2-1,5; J(C, H) = 142 Hz), 71.1 (t, CH_2O or CH_2-1,5; J(C, H) = 141 Hz), 85.7 (d, CH; J(C, H) = 159 Hz), 119.7 (s, C-2,4), 233.8 (s, CO) ppm. IR spectrum (Nujol): 1742, 1802, 1893 (ν(CO)); additional bands at 630, 795, 820, 833, 861, 935, 1005, 1031, 1070, 1108, 1133, 1200, 1245, 1260, 1350, and 1365 cm^{-1}. For m/e values and relative intensities of positive and negative fragment ions observed in the 17 eV FAB mass spectrum, see [2].

Structural characterization (crystal data and molecular parameters not reported) revealed a polymeric network involving a bridging coordination of the $CH_3OC_2H_4OC_2H_4OCH_3$ ligand and a close association of the potassium cation with one of the carbonyl ligands of the metalate anion: **Fig. 33** [3].

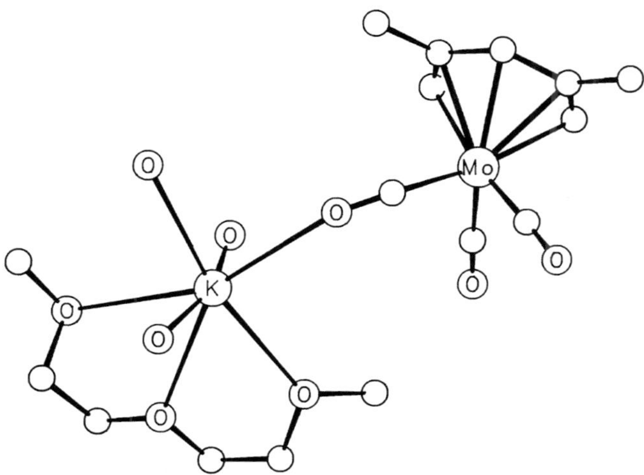

Fig. 33. Partial view of the three-dimensional structure of [K($CH_3OC_2H_4OC_2H_4OCH_3$)][$C_7H_{11}Mo(CO)_3$] [3].

References on p. 332

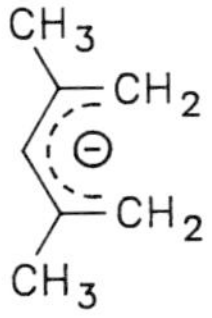

In THF or toluene at room temperature, the salt underwent clean reactions with I_2, AgI, and $Hg(CN)_2$ to give 2,4-$C_7H_{11}Mo(CO)_3I$, (2,4-$C_7H_{11}Mo(CO)_3)_2$, and (2,4-$C_7H_{11}Mo(CO)_3)_2Hg$, respectively [2]. (2,4-$C_7H_{11}Mo(CO)_3)_2$ was also obtained by oxidizing the [2,4-C_7H_{11}-$Mo(CO)_3]^-$ anion with ferrocenium salts [3] and with a number of alkyl and silyl halides; see below. Treatment with $C_5H_5Mo(CO)_3I$ gave the unsymmetric metal–metal-bonded complex 2,4-$C_7H_{11}Mo(CO)_3Mo(CO)_3C_5H_5$ [3].

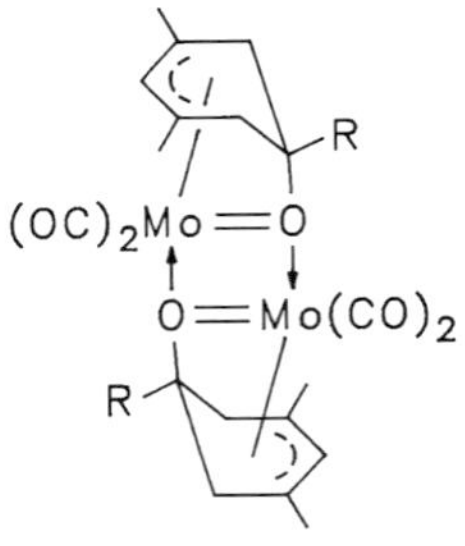

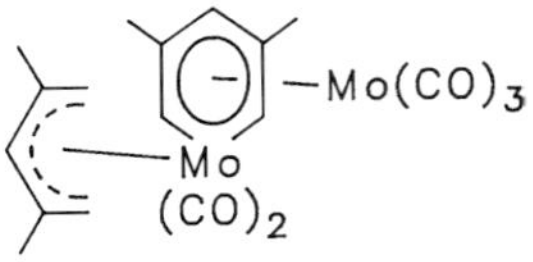

Addition of CH_3I (1 equivalent in THF at −78 °C) provided a binuclear product (Formula II; R = CH_3) via a multistep coupling process, during which one carbonyl ligand was converted to an alkoxide [1]. Ethyl iodide reacted similarly forming the coupled complex shown in Formula II with R = C_2H_5 [2]. Reaction with IC_2H_4I, however, gave a binuclear molybdabenzene derivative, Formula III. Spectral examination of the gaseous hydrocarbon products revealed the presence of ethene and the absence of ethane [2]. 1:1 reactions of the salt with the following alkyl and silyl halides in THF resulted in oxidation of the metalate anion to the (2,4-$C_7H_{11}Mo(CO)_3)_2$ homodimer: CCl_4, CH_2I_2, CHI_3, $(CH_3)_2CHI$, cyclo-$C_6H_{11}I$, $I(CH_2)_nI$ (n = 3, 4), and $Si(CH_3)_3I$ [2].

2,4-$C_7H_{11}Mo(CO)_3I$ (2,4-C_7H_{11} = Formula I) was isolated in 60% yield by reacting the potassium salt of the [2,4-$C_7H_{11}Mo(CO)_3]^-$ anion (see above) with an equimolar amount of I_2 in toluene or THF at ambient conditions and extracting the product with hexane after THF removal in vacuum. Subsequent concentration and cooling to −20 °C produced a 60% yield of orange air-stable crystals, m.p. 116 to 118 °C (dec.) [2].

^1H NMR spectrum (C_6D_6): δ = 1.28 (s, CH_3), 1.98 (d, H_{endo}-1,5; J = 2.3 Hz), 3.67 (d, H_{exo}-1,5; J = 1.2 Hz), 4.53 (s, H-3) ppm. ^{13}C NMR spectrum (C_6D_6): δ = 26.7 (q, CH_3; J(C, H) = 192 Hz), 77.8 (t, CH_2; J(C, H) = 162 Hz), 87.7 (d, CH; J(C, H) = 161 Hz), 119.2 (s, C-2,4); 211.0 (s, CO_{cis}), 229.5 (s, CO_{trans}) ppm. IR spectrum (Nujol): 1925, 1935, 1955, 1987, 2025 (ν(CO)); other characteristic bands at 771, 855, 890, 915, 937, 970, 1015, 1035, 1155, 1272, and 1565 cm^{-1}. For m/e values and relative intensities of the most abundant fragment ions observed in the 17 eV EI mass spectrum, see [2].

References:

[1] Kralik, M. S.; Hutchinson, J. P.; Ernst, R. D. (J. Am. Chem. Soc. **107** [1985] 8296/7).
[2] Kralik, M. S.; Rheingold, A. L.; Ernst, R. D. (Organometallics **6** [1987] 2612/4).
[3] Ernst, R. D. (Chem. Rev. **88** [1988] 1255/91).

1.5.1.4.2.7.2 $^5LMo(CO)_3$ Complexes with Cyclohexadienyl Ligands

[K($CH_3OC_2H_4OC_2H_4OCH_3)_n$][6,6-($CH_3)_2C_6H_5Mo(CO)_3$] (6,6-$(CH_3)_2C_5H_6$ = Formula I) was obtained by treatment of $Mo(CO)_3(CH_3OC_2H_4OC_2H_4OCH_3)$ with one equivalent of K[6,6-

$(CH_3)_2C_6H_5]$ in THF at $-78\,°C$. Warming the solution to room temperature and precipitation with hexane provided the title compound in 85% yield as a tan solid ($n \cong 0.7$), which could be freed from residual $CH_3OC_2H_4OC_2H_4OCH_3$ by careful washing with cold THF/pentane (1:5). The anion was also observed upon electrochemical reduction of the (6,6-$(CH_3)_2C_6H_5Mo(CO)_3)_2$ dimer occurring in THF/0.1 M $[N(C_4H_9\text{-}n)_4]BF_4$ at -0.60 V versus Ag|AgCl (in saturated $[N(CH_3)_4]Cl$/THF).

^1H NMR spectrum (CD_3CN): $\delta = 0.50$ (s, CH_3-exo), 1.18 (s, CH_3-endo), 2.73 (dd, H-1,5; $^3J(H, H) = 7.2$, $^4J(H, H) = 1.3$ Hz), 4.83 (dd, H-2,4), 5.21 (tt, H-3; $^3J(H, H) = 5.4$ Hz) ppm. IR spectrum (Nujol): 1745, 1790, 1890 (ν(CO)) cm^{-1}.

The salt dissolved readily in THF and CH_3CN giving brown solutions. Cyclic voltammetry in THF/0.1 M $[N(C_4H_9\text{-}n)_4]BF_4$ revealed no peaks down to -1.0 V; the return scan exhibited an electrochemically irreversible oxidation wave at 0.14 V (both potentials versus Ag|AgCl in saturated $[N(CH_3)_4]Cl$/THF) arising from oxidation to the rapidly dimerizing radical 6,6-$(CH_3)_2C_6H_5Mo(CO)_3^{\bullet}$. Oxidation of the air-sensitive complex in THF at $-78\,°C$ by I_2 produced 6,6-$(CH_3)_2C_6H_5Mo(CO)_3I$. Treatment with $Si(CH_3)_3I$ in THF led to decomposition. Attempts to substitute for CO in the anion using equimolar $P(CH_3)_3$ in CD_3CN resulted in no reaction over prolonged periods at $100\,°C$. The dimer (6,6-$(CH_3)_2C_6H_5Mo(CO)_3)_2$ was obtained by oxidizing the [6,6-$(CH_3)_2C_6H_5Mo(CO)_3]^-$ anion with $[(C_5H_5)_2Fe]X$ ($X^- = [B(C_6H_5)_4]^-$, PF_6^-) in THF, between -78 and $0\,°C$. The salt failed to undergo metalation with $ZnCl_2$ in THF but addition of HgI_2 to 2.0 equivalents of the title compound (THF, -78 to $0\,°C$) cleanly afforded (6,6-$(CH_3)_2C_6H_5Mo(CO)_3)_2Hg$.

The complex was recovered unchanged when subjected to protolysis by ROH (R = H, CH_3). Electrophilic attack by $[HO(C_2H_5)_2]BF_4$ (or CH_3I) at $-78\,°C$ in THF produced $Mo(CO)_6$ as the only tractable organometallic adduct. Combination with 4-$CH_3C_6H_4SO_2N(CH_3)NO$ in diethyl ether gave 6,6-$(CH_3)_2C_6H_5Mo(CO)_2NO$.

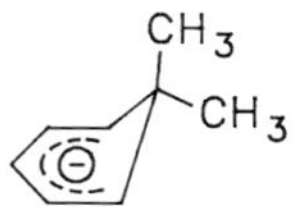

I

6,6-$(CH_3)_2C_6H_5Mo(CO)_3I$ (6,6-$(CH_3)_2C_5H_6$ = Formula I) was made by oxidizing the [6,6-$(CH_3)_2C_6H_5Mo(CO)_3]^-$ anion (K^+ salt; see above) with the stoichiometric amount of I_2 in THF at $-78\,°C$. After warming to $25\,°C$, removal of solvent, and extraction of the residue with hexane, the product was isolated by crystallization from the concentrated extracts in 10% yield. Yields of up to 65% were attained when (6,6-$(CH_3)_2C_6H_5Mo(CO)_3)_2Hg$ was halogenated using two equivalents of I_2 in CH_2Cl_2 at 0 to $25\,°C$.

Red-orange solid. ^1H NMR spectrum (C_6D_6): $\delta = 0.24$ (s, CH_3-exo), 1.09 (s, CH_3-endo), 3.52 (dd, H-1,5; $^3J(H, H) = 6.2$, $^4J(H, H) = 1.8$ Hz), 3.93 (dd, H-2,4), 5.04 (tt, H-3; $^3J(H, H) = 5.0$ Hz) ppm. ^{13}C NMR spectrum (CD_2Cl_2): $\delta = 29.29$, 39.05 (both CH_3), 36.43 (C-6), 83.96 (C-1,5), 90.73 (C-3), 100.48 (C-2,4), 215.83 (CO) ppm. IR spectrum (Nujol): 1948, 1959, 2015 (ν(CO)) cm^{-1}.

Reference:

DiMauro, P. D.; Wolczanski, P. T.; Parkanyi, L.; Hurrell Petach, H. (Organometallics **9** [1990] 1097/106).

1.5.1.4.2.7.3 $^5LMo(CO)_3$ Complexes with Cycloheptadienyl and Cycloheptatrienyl Ligands

When coordinated in a ^5L fashion, the cycloheptadienyl group, C_7H_9, adopts the conformation shown by Formula I.

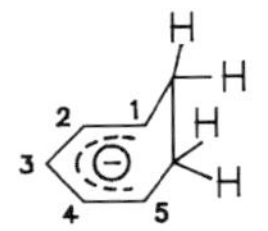

I

$K[C_7H_9Mo(CO)_3]$ was made by dropwise addition of $K[BH(C_2H_5)_3]$ (1.5 equivalents in THF) to a solution of the cycloheptatriene complex $C_7H_8Mo(CO)_3$ in diethyl ether at −78 °C. The resulting yellow suspension was stirred at dry-ice temperature until IR spectroscopy showed the $[C_7H_9Mo(CO)_3]^-$ anion as the only carbonyl-containing species present; yield: 95% [5].

IR spectrum (ether): 1743, 1799, 1895 (ν(CO)) cm^{-1} [5].

The salt proved to be highly air- and temperature-sensitive and converted readily to $C_7H_8Mo(CO)_3$ upon warming to ambient conditions. Attempts to isolate the $[C_7H_9Mo(CO)_3]^-$ anion by metathesis with $[(C_6H_5)_3PNP(C_6H_5)_3]Cl$ failed. Addition of equimolar amounts of trialkyltin chlorides, SnR_3Cl (R = CH_3, C_6H_5), to diethyl ether solutions of $K[C_7H_9Mo(CO)_3]$ at −78 °C afforded the corresponding trialkylstannyl derivatives, $C_7H_9Mo(CO)_3SnR_3$ (see below). Combination with N-methyl-N-nitroso-p-toluene sulfonamide in 1:1 stoichiometry in diethyl ether at −78 °C led to $C_7H_9Mo(CO)_2NO$ [5].

$C_7H_9Mo(CO)_3Sn(CH_3)_3$. $Sn(CH_3)_3Cl$ (1.5 equivalents) was added to a stirred solution of $K[C_7H_9Mo(CO)_3]$ in diethyl ether at −78 °C (see above). After filtration through Celite and removal of solvent, the stannyl derivative was extracted from the residue with hexane and isolated as a bright yellow, air- and temperature-sensitive powder after evaporation of the extracts to dryness (93%) [5].

^1H NMR spectrum (C_6D_6): δ = 0.42 ("t", CH_3; J(^{117}Sn, H) = J(^{119}Sn, H) = 46.1 Hz), 1.50 ("d", CH_2; "J" = 8.6 Hz), 1.90 (m, CH_2), 3.86 ("t", H-1,5; "J" = 3.2 Hz), 4.40 ("t", H-2,4; "J" = 4.3 Hz), 5.25 (t, H-3; J = 5.9 Hz) ppm. ^{13}C NMR spectrum (C_6D_6): δ = −6.1 (CH_3), 34.6 (CH_2), 87.4 (C-1,5), 95.8 (C-3), 100.0 (C-2,4), 221.3 (CO) ppm. IR spectrum (hexane): 1901, 1945, 2008 (ν(CO)) cm^{-1} [5].

$C_7H_9Mo(CO)_3Sn(C_6H_5)_3$ was prepared from $K[C_7H_9Mo(CO)_3]$ and $Sn(C_6H_5)_3Cl$ following the procedure outlined for its trimethyltin homologue above. Recrystallization of the evaporated hexane extracts from CH_2Cl_2/hexane afforded the product in 98% yield as a yellow powder pertinaciously retaining traces of unreacted $Sn(C_6H_5)_3Cl$ [5].

^1H NMR spectrum (C_6D_6): δ = 1.36 ("d", CH_2; "J" = 9.0 Hz), 1.78 (m, CH_2), 3.82 ("t", H-1,5), 4.43 ("t", H-2,4; "J" = 5.9 Hz), 5.75 (t, H-3), 6.89, 7.49, 7.76 (all m, C_6H_5) ppm. ^{13}C NMR spectrum (C_6D_6): δ = 34.5 (CH_2), 89.0 (C-1,5), 96.5 (C-3), 100.0 (C-2,4), 129.0, 129.3, 137.2 (all C_6H_5), 221.4 (CO) ppm. IR spectrum (hexane): 1911, 1956, 2013 (ν(CO)) cm^{-1} [5].

$C_7H_9Mo(CO)_3Cl$ was isolated from the protonation reaction occurring between tri-carbonyl(cycloheptatriene)molybdenum(0) and concentrated aqueous HCl in propionic anhydride. On stirring the mixture at room temperature and adding diethyl ether, the crude product separated from the solution as bright yellow crystals which were reprecipitated from toluene by adding a large excess of hexane and cooling to −30 °C; yield: 27% [1].

References on p. 336

IR spectrum (KBr): ν(CO) at 1840, 1940, and 2020/2040 (solid state splitting; ν(CO) pattern shown in the original literature); additional bands at 730, 870, 1035, 1050, 1100, 1160, 1230, 1250, 1270, 1370, 1380, 1400, 1440, 1450, 2860, and 2940 cm^{-1} [1].

The metathetical reaction with $AgBF_4$ in CH_3NO_2 resulted in the formation of "[C_7H_9Mo-$(CO)_3$]BF_4" together with much decomposed material [1].

[$C_7H_9Mo(CO)_3$]BF_4 (presumably better formulated as **$C_7H_9Mo(CO)_3FBF_3$**) was prepared by protonation of the cycloheptatriene complex $C_7H_8Mo(CO)_3$ using 50% aqueous HBF_4 in propionic anhydride. The product separated from the reaction mixture as orange crystals upon precipitation with diethyl ether; yield: 50%. The compound was also isolated from the reaction of $C_7H_9Mo(CO)_3Cl$ (see above) with $AgBF_4$ in nitromethane [1].

^{1}H NMR spectrum (CD_3NO_2): δ = 2.5 (m, CH_2), 5.7 (m, H-1,2,4,5), 7.2 (tt, H-3; ^{3}J(H, H) = 6, ^{4}J(H, H) = 2 Hz) ppm; spectrum shown in the original literature. IR spectrum (KBr): 1950, 1975, 2050 (ν(CO)); additional bands at 524, 536, 575, 600, 830, 1040, 1070, 1130, 1310, 1375, 1415, 1430, and 1465 cm^{-1} [1].

The solid was soluble in acetone and nitromethane and proved to be relatively air-stable. Treatment with $P(C_6H_5)_3$ in CH_2Cl_2 yielded [$P(C_6H_5)_3C_7H_9$]BF_4 rather than the expected [$C_7H_9Mo(CO)_3P(C_6H_5)_3$]$BF_4$ [1].

[$C_7H_9Mo(CO)_3CNC_4H_9$-t]BF_4. A CH_2Cl_2 solution of the cycloheptatriene compound C_7H_8Mo-$(CO)_2CNC_4H_9$-t was cooled to −78°C and purged with CO gas. Ethereal HBF_4 was added and the reaction mixture allowed to warm to room temperature while maintaining a constant flow of carbon monoxide through the solution. The product was separated from the concentrated mixture in 55% yield as a yellow solid by precipitation with ether followed by recrystallization from THF/ether [4].

^{1}H NMR spectrum ($CDCl_3$): δ = 1.69 (s, $C(CH_3)_3$), 2.23, 2.34 (both m, CH_2), 5.31 (m, H-1,5), 5.47 (m, H-2,4), 6.78 (m, H-3) ppm. ^{13}C NMR spectrum (CD_2Cl_2, −40°C): δ = 30.2 (C($C$$H_3)_3$), 32.8 ($CH_2$), 62.1 ($C$$(CH_3)_3$), 96.5 (C-3), 100.4, 100.7 (C-1,5, C-2,4), 141.8 (CN), 213.9, 219.3 (both CO) ppm. IR spectrum (CH_2Cl_2): 1991, 2014, 2070 (ν(CO)), 2196 (ν(CN)) cm^{-1}. Mass spectrum (FAB): $[M]^+$, $[M-CO]^+$, $[M-3CO-2H]^+$ [4].

Cis,cis-[$C_7H_9Mo(CO)_2(NCCH_3)_2CNC_4H_9$-t]$BF_4$ (C_7H_9 ligand bonded in a ^{3}L fashion) resulted from dissolution of the title complex in acetonitrile [4].

[$C_7H_9Mo(CO)_3P(C_6H_5)_3$]$BF_4$ was made by treatment of a CH_2Cl_2 solution of $C_7H_8Mo(CO)_2$-$P(C_6H_5)_3$ (C_7H_8 = cycloheptatriene) with [$HO(C_2H_5)_2$]BF_4 at −78°C in the presence of a vigorous stream of carbon monoxide. After warming to room temperature, the product was precipitated from the concentrated solution by adding diethyl ether and isolated as a salmon pink solid after recrystallization from acetone/diethyl ether (44%) [3, 4].

^{1}H NMR spectrum ($CDCl_3$): δ = 2.21, 2.37 (both m, CH_2), 5.22 (m, H-1,5), 5.26 (m, H-2,4), 6.14 (m, H-3), 7.31, 7.57 (both m, C_6H_5) ppm [3, 4]. ^{13}C NMR spectrum ($CDCl_3$): δ = 34.7 (CH_2), 99.1, 102.4, 108.2 (C-1 to 5), 130.0, 130.1, 132.3, 133.0, 133.2 (all C_6H_5), 219.2, 219.4 (both CO) ppm [3, 4]. IR spectrum (CH_2Cl_2): 1964, 2018, 2059 (ν(CO)) cm^{-1} [3, 4]. Mass spectrum (FAB): $[M]^+$, $[M-CO]^+$, $[M-3CO-2H]^+$ [4].

Acetonitrile reacted to form an incompletely characterized mixture of products of which the predominant component was cis-[$C_7H_9Mo(CO)_2(NCCH_3)_3$]BF_4 containing a ^{3}L cycloheptadienyl ligand [4].

$C_7H_7Mo(CO)_3H$ (C_7H_7 = Formula II) resulted from thermal rearrangement of $C_7H_8Mo(CO)_3$, when the parent cycloheptatriene complex was sublimed in vacuum at 28 to 45°C and then cocondensed with a matrix gas (Ar, N_2, CO, CH_4, or SF_6) at 12 K [2].

References on p. 336

II

IR spectrum (CO matrix, 12 K): 1909.2, 1956.8, 2042.3 (ν(CO)) cm^{-1}; spectrum drawn in [2].

References:

[1] Salzer, A.; Werner, H. (J. Organomet. Chem. **87** [1975] 101/8).

[2] Hooker, R. H.; Rest, A. J. (J. Organomet. Chem. **234** [1982] C23/C27).

[3] Hinchliffe, J. R.; Whiteley, M. W. (J. Organomet. Chem. **402** [1991] C50/C55).

[4] Beddoes, R. L.; Hinchliffe, J. R.; Whiteley, M. W. (J. Chem. Soc. Dalton Trans. **1993** 501/8).

[5] Wang, C.; Sheridan, J. B. (Organometallics **13** [1994] 3639/43).

1.5.1.4.2.7.4 $^5LMo(CO)_3$ Complexes with Azulene Ligands

Azulene ($C_{10}H_8$ throughout this chapter) has the structural Formula I in which the positions are numbered as shown.

$(OC)_3Cr$—$Mo(CO)_3$ $(OC)_3Mo$—$Fe(CO)_3$

I   II   III

$Mo(CO)_3(\mu$-$C_{10}H_8)Cr(CO)_3$ (Formula II; R = H) was prepared by reacting $C_{10}H_8Cr(CO)_3$ with $Mo(CO)_3(NCCH_3)_3$ in 1:1 stoichiometry in THF at 40 to 50°C. After evaporation to dryness, any unreacted $C_{10}H_8Cr(CO)_3$ was removed by washing the residue with petroleum ether and toluene. The heterobimetallic product was subsequently extracted into refluxing toluene or into cold CH_2Cl_2 and finally crystallized from toluene/hexane or CH_2Cl_2/hexane mixtures at −30°C; yield: 26% [2].

Black crystals decomposing above 200°C with loss of the azulene ligand. ^1H NMR spectrum ($CDCl_3$): δ = 3.58 (d, H-1,3; J(H-1,2) = J(H-2,3) = 2.6 Hz), 4.31 (d, H-4,8; J(H-4,5) = J(H-7,8) = 10.0 Hz), 4.63 (dd, H-5,7; J(H-5,6) = J(H-6,7) = 7.2 Hz), 5.33 (t, H-2), 6.25 (m, H-6) ppm. IR spectrum (KBr): 1842, 1880, 1907, 1927, 1942, 2015 (ν(CO)) cm^{-1}. Moderately soluble in toluene, acetone, and CH_2Cl_2; insoluble in hexane [2].

$Mo(CO)_3(\mu$-4,6,8-$(CH_3)_3C_{10}H_5)Cr(CO)_3$ (Formula II; R = CH_3) was isolated in 50% yield from the reaction between equimolar quantities of 4,6,8-$(CH_3)_3C_{10}H_5Cr(CO)_3$ and $Mo(CO)_3$-$(NCCH_3)_3$ in THF following the procedure outlined for $Mo(CO)_3(\mu$-$C_{10}H_8)Cr(CO)_3$ above [2].

Black-brown crystals with thermal properties and solubilities similar to those of Mo-$(CO)_3(\mu$-$C_{10}H_8)Cr(CO)_3$ (see above). ^1H NMR spectrum ($CDCl_3$): δ = 1.66 (s, CH_3-4,8), 2.74 (s, CH_3-6), 3.57 (d, H-1,3; J(H-1,2) = J(H-2,3) = 2.9 Hz), 4.59 (s, H-5,7), 5.61 (t, H-2) ppm.

References on p. 337

^{13}C NMR spectrum ($CDCl_3$): δ = 23.7 (CH_3-4,8), 26.4 (CH_3-6), 72.4 (C-1,3), 90.6 (C-2), 95.9 (C-4,8), 96.1 (C-9,10), 97.3 (C-5,7), 98.8 (C-6), 218.9 (MoCO), 235.6 (CrCO) ppm. IR spectrum (KBr): 1848, 1880, 1908, 1945, 1992 (ν(CO)) cm^{-1} [2].

$Mo(CO)_3(\mu$-1-$BrC_{10}H_7)Fe(CO)_3$, thought to possess the structure shown in Formula III, was briefly mentioned as the product resulting from heating a mixture consisting of equimolar portions of $Mo(CO)_6$, $Fe(CO)_5$, and azulene in $CH_3OC_2H_4OC_2H_4OCH_3$ at 100 °C for prolonged periods. Like various other azulene-containing homo- or heterobimetallics containing $M(CO)_n$ fragments (M = V, Mo, Mn, Fe, Ni), the complex could be used in depositing metallic mirrors on glass, glass cloth, resins, and other insulating supports by thermal decomposition in the vapor phase, and also exhibited excellent antiknock, antiwear, and lubricity-improving properties when used in fuels and lubricating oils, respectively [1].

References:

[1] Wilkinson, G.; Burton, R.; Ethyl Corp. (U.S. 3064023 [1959/62] 4 pp.; C.A. **58** [1963] 9135).
[2] Edelmann, F.; Töfke, S.; Behrens, U. (J. Organomet. Chem. **308** [1986] 27/34).

1.5.1.4.2.7.5 $^5LMo(CO)_3$ Complexes with Heterocyclic Pentadienyl Ligands

$[N(C_2H_5)_4][(3,5$-$(C_6H_5)_2C_5H_3SO)Mo(CO)_3]$ contains the anionic SO-bridged pentadienyl ligand 3,5-diphenylthiacyclohexadienyl 1-oxide (Formula I).The salt was originally made by treating the S-ylidic complex (3,5-$(C_6H_5)_2C_5H_3S(O)CH_3)Mo(CO)_3$ (see the section "$^5LMo(CO)_3$ Complexes with Heterocyclic Pentadienylide Ligands") with excess $LiP(C_6H_5)_2$ in THF at 0°C. Excess phosphide was destroyed by adding BrC_2H_4Br, the mixture filtered, and solvent evaporated. The crude lithium salt, **$Li[(3,5$-$(C_6H_5)_2C_5H_3SO)Mo(CO)_3]$**, remaining as an oily residue, was solidified by stirring it under petroleum ether. Subsequent treatment with a solution of $[N(C_2H_5)_4]Cl$ in CH_2Cl_2 followed by dilution of the filtered mixture with diethyl ether afforded the tetraethyl ammonium salt as an ocher precipitate, which was washed with water to remove any retained $[N(C_2H_5)_4]Cl$ and finally reprecipitated from CH_2Cl_2/ether; yield: 70% [1]. In a more convenient approach, (3,5-$(C_6H_5)_2C_5H_3S(O)CH_3)$-$Mo(CO)_3$ was demethylated using $Li[BH(C_2H_5)_3]$ in THF. The lithium salt of the [(3,5-$(C_6H_5)_2C_5H_3SO)Mo(CO)_3]^-$ anion so obtained was then converted to the title compound by adding a CH_2Cl_2 solution containing excess $[N(C_2H_5)_4]Cl$ (62%) [2]. The complex was also isolated from the reaction between (3,5-$(C_6H_5)_2C_5H_3S(O)CH_3)Mo(CO)_3$ and $Na_2[Fe(CO)_4]$ · 1.5 dioxane in THF. Workup of the initially formed sodium salt, **$[Na(OC_4H_8)_2][(3,5$-$(C_6H_5)_2C_5H_3SO)Mo(CO)_3]$** as outlined before furnished a 50% yield of the tetraethyl ammonium derivative [1].

Ocher solid, decomposing between 150 and 166°C. Conductivity (CH_2Cl_2, 7.6×10^{-5} M): Λ = 50.0 $cm^2 \cdot \Omega^{-1} \cdot mol^{-1}$. ^1H NMR spectrum (acetone-d_6): δ = 1.35 (br t, CH_3), 3.45 (q, CH_2; J = 7 Hz), 4.89 (d, H-2,6; J = 1 Hz), 6.14 (t, H-4), 7.27 to 7.83 (m, C_6H_5) ppm. ^{13}C NMR spectrum (acetone-d_6): δ = 7.60 (CH_3), 53.06 (CH_2), 67.23 (C-2,6), 87.85 (C-4), 111.12 (C-3,5), 128.67, 142.84 (both C_6H_5), 229.32 (CO) ppm. IR spectrum (Nujol): 1018, 1025 (ν(SO)), 1817, 1830, 1918 (ν(CO)); (CH_2Cl_2): 1818, 1853, 1923 (ν(CO)) cm^{-1} [1].

The salt proved to be soluble in CH_2Cl_2, less so in acetone and THF, but insoluble in ether and petroleum ether [1]. In CH_2Cl_2/0.1 M $[N(C_4H_9$-n$)_4]PF_6$, the anion underwent irreversible electrooxidation at $E_{ox} \cong 0.25$ V vs. $(C_5H_5)_2Co/[(C_5H_5)_2Co]Cl$ [3]. Halogenation of the air-stable complex by treatment with $C_6H_5ICl_2$, $Br_2 \cdot NC_5H_5$, or I_2 produced neutral halo derivatives, (3,5-$(C_6H_5)_2C_5H_3SO)Mo(CO)_3X$ (X = Cl, Br, I) [2]; see the following compound.

References on p. 338

$$\text{Formula I: } 3,5\text{-}(C_6H_5)_2C_5H_3S{=}O \text{ anion (thiabenzene oxide ring)}$$

I

(3,5-$(C_6H_5)_2C_5H_3SO)Mo(CO)_3X$ (^5L = Formula I; X = Cl, Br, I). The three halo complexes were prepared by reacting $[N(C_2H_5)_4][(3,5\text{-}(C_6H_5)_2C_5H_3SO)Mo(CO)_3]$ with each of the halogenating reagents $C_6H_5ICl_2$, $Br_2 \cdot NC_5H_5$, and I_2 in 1:1 stoichiometry in CH_2Cl_2 at −15 °C. Dilution of the mixtures with methyl cyclohexane caused oily by-products to precipitate, which were removed by filtration. Concentration of the filtrates with subsequent addition of petroleum ether induced the separation of the title compounds as ocher to brownish yellow solids; for yields, see the following table. $(3,5\text{-}(C_6H_5)_2C_5H_3SO)Mo(CO)_3Cl$ could not be obtained in an analytically pure form [2].

X	yield and properties
Cl	(5%) brownish yellow solid IR (Nujol): 1048, 1065 (ν(SO)), 1978, 2000, 2068 (CO); (CH_2Cl_2): 2000, 2020, 2074 (CO)
Br	(36%) ocher solid, m.p. 110 °C (dec.) ^{1}H NMR (acetone-d_6): 6.71 (d, H-2,6; J = 1), 7.44 to 8.09 (m, H-4 and C_6H_5) IR (Nujol): 1048, 1064 (ν(SO)),1960, 1978, 1995, 2064 (CO); (CH_2Cl_2): 1995, 2016, 2070 (CO)
I	(63%) ocher solid, m.p. 105 °C (dec.) ^{1}H NMR (acetone-d_6): 6.64 (d, H-2,6; J = 1), 7.44 to 8.07 (m, H-4 and C_6H_5) IR (Nujol): 1050, 1070 (ν(SO)), 1976, 1990, 2010, 2056 (CO); (CH_2Cl_2): 1990, 2010, 2062 (CO)

All three complexes are easily soluble in CH_2Cl_2 and $CHCl_3$, less so in toluene, ether, THF, and acetone, and insoluble in saturated hydrocarbons. They are thermally less stable (X = I > Br ≫ Cl) than their C_5H_5 analogs, $C_5H_5Mo(CO)_3X$, and are only moderately air-stable [2].

References:

[1] Weber, L. (Chem. Ber. **112** [1979] 99/106).
[2] Weber, L.; Wewers, D. (Z. Naturforsch. **37b** [1982] 68/72).
[3] Koelle, U.; Ding, T.-Zh.; Weber, L. (Organometallics **4** [1985] 302/5).

1.5.1.4.2.7.6 $^5LMo(CO)_3$ Complexes with Heteroaromatic ^5L Ligands

This subsection describes $^5LMo(CO)_3$ compounds where the ^5L ligand is represented by one of the heteroaromatic 6π-systems that follow. In addition, two salt-like complexes $[^5LMo(CO)_3]I$ with ^5L = N-methylquinoline or N-methylacridine were briefly mentioned as red

References on p. 343

crystalline products arising from reactions of $Mo(CO)_6$ with N-methylquinolinium iodide and n-methylacridinium iodide, respectively, but no characterizing data were reported [1].

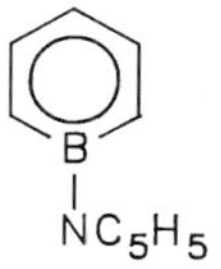

I

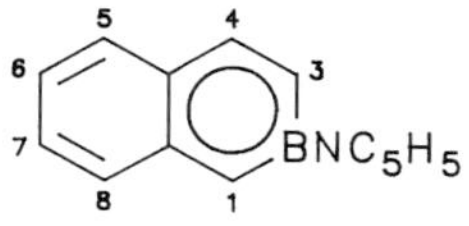

II

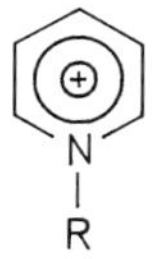

III

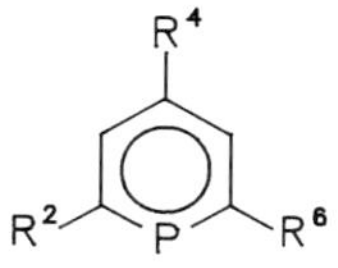

IV

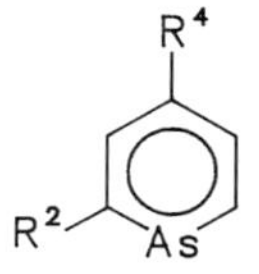

V

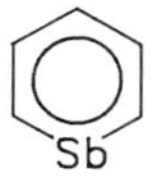

VI

The pronounced upfield shifts of the proton, carbon, and phosphorus signals of the respective ring atoms relative to those of free ligands (Formulas IV, V, and VI) displayed by the ^1H, ^{13}C, and ^{31}P NMR spectra of these $(R_nC_5H_{5-n}E)Mo(CO)_3$ compounds (E = P, As, Sb) are similar to those observed for (arene)metal complexes and thus clearly demonstrate π coordination of the heteroaromatic ligands [3, 5, 7].

Differing from the uncomplexed molecules (Formulas IV, V, and VI), their $Mo(CO)_3$ derivatives proved to be relatively air-stable as solids [3, 5, 7].

$(C_5H_5NBC_5H_5)Mo(CO)_3$ (^5L ligand = Formula I). Equimolar quantities of pyridine-borabenzene and $Mo(CO)_3(NCCH_3)_3$ were stirred in THF at ambient conditions. After filtration and concentration of the solution in vacuum, pentane was added to precipitate the crude complex which could be purified by recrystallization from CH_2Cl_2 at −30 °C; yield: 46% [8].

Orange-red crystals decomposing above 220 °C. ^1H NMR spectrum (CD_2Cl_2); BC_5H_5 ring: δ = 4.77 (d, H-2,6), 5.39 (t, H-4), 6.07 (m, H-3,5); NC_5H_5 ring: δ = 7.90 (t, H-3,5), 8.34 (t,

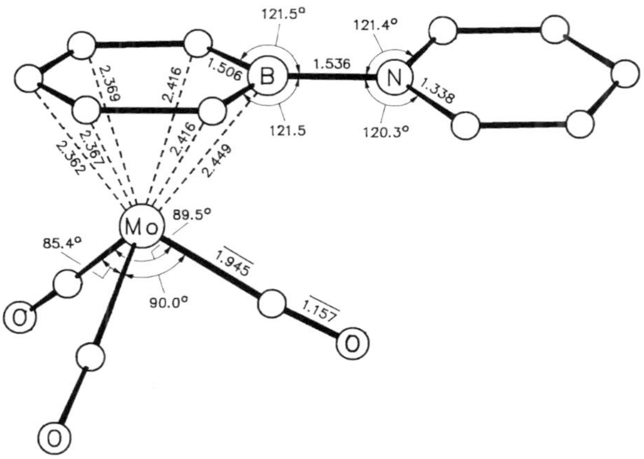

Fig. 34. The molecular structure of $(C_5H_5NBC_5H_5)Mo(CO)_3$ [8].

References on p. 343

H-4), 8.90 (d, H-2,6) ppm. ^{11}B NMR (CD_2Cl_2): δ = 23.5 ppm. IR (KBr): 1815, 1855, 1932 (ν(CO)) cm^{-1} [8].

Single crystals grown from THF were triclinic, space group $P\bar{1}-C_i^1$ (No. 2), with a = 7.329(1), b = 9.039(2), c = 10.492(2) Å, α = 95.60(1)°, β = 104.42(1)°, γ = 98.99(1)°; Z = 2 molecules per unit cell, D_{calc} = 1.69 g/cm^3. The structure determination proved the $C_5H_5NBC_5H_5$ to be bonded to the metal via the boron-containing six-membered ring which does not deviate significantly from planarity; see **Fig. 34** [8].

The air-stable complex remained unattacked when exposed to water for prolonged periods [8].

(2-$C_5H_5NBC_9H_7$)Mo(CO)$_3$ (^{5}L = Formula II) was isolated from the reaction between pyridine-2-boranaphthalene and $Mo(CO)_3(NCCH_3)_3$, conducted in 1:1 stoichiometry in 1,4-dioxane at room temperature. The crude product precipitated from the reaction solution and was recrystallized from CH_2Cl_2 (60%) [8].

Orange crystals, m.p. 215°C (dec.). ^{1}H NMR spectrum (CD_2Cl_2); 2-boranaphthalene system: δ = 5.19 (dd, H-3; J(H-1,3) = 2.6, J(H-3,4) = 9.6 Hz); 5.42 (d, H-1), 6.80 (d, H-4), 7.10, 7.29 (both dd, H-6,7; J(H-6,7) = 6.6, J(H-5,6) = J(H-7,8) = 8.5 Hz), 7.45, 7.64 (both d, H-5,8) ppm; NC_5H_5 ring: δ = 7.93 ("t", H-3,5; "J" = 7.2 Hz), 8.35 (t, H-4; J = 7.7 Hz), 9.02 (d, H-2,6; J = 5.2 Hz) ppm. ^{11}B NMR spectrum (CD_2Cl_2): δ = 26.8 ppm. IR spectrum (KBr): 1800, 1835, 1925 (ν(CO)) cm^{-1} [8].

Proof of coordination of the 2-$C_5H_5NBC_9H_7$ ligand via the BC_5 ring was derived from the upfield shift of the ^{11}B and ^{1}H (protons H-1,2, and 4) resonances observed on going from free to complexed pyridine-2-boranaphthalene. It was also confirmed by an X-ray structure analysis performed on the homologous chromium compound, (2-$C_5H_5NBC_9H_7$)Cr(CO)$_3$ [8].

[($CH_3NC_5H_5$)Mo(CO)$_3$]I (^{5}L ligand = Formula III; R = CH_3) was prepared by refluxing $Mo(CO)_6$ with 1-methylpyridinium iodide in THF. After removal of solvent, the complex was obtained from ethanol/petroleum mixtures as yellow crystals which decomposed at 120°C [1].

The compound, which other authors were unable to synthesize following the procedure outlined above [2], was described as sparingly soluble in nonpolar solvents but soluble in ethanol and water. Its aqueous solutions gave precipitates with reagents precipitating cyclopentadienyl and arene metal cations, such as silicotungstic acid and Reinecke's salt [1].

(2,4,6-$(C_6H_5)_3C_5H_2P$)Mo(CO)$_3$ (^{5}L ligand = Formula IV; R-2 = R-4 = R-6 = C_6H_5) was obtained from the reaction between equimolar quantities of (1,3,5-$(CH_3)_3C_6H_3$)Mo(CO)$_3$ and 2,4,6-triphenyl-λ^3-phosphorine in THF at room temperature. The residue remaining after removal of the solvent was dissolved in toluene, an equal volume of pentane was added, and the mixture was allowed to stand in a refrigerator, which caused the crystalline complex to separate in 45% yield [3]. The compound was also formed during thermal degradation of the σ complex $Mo(CO)_5PC_5H_2(C_6H_5)_3$-2,4,6 in refluxing dibutyl ether [3].

Dark red crystals, m.p. 152 to 153°C [3], 154°C [4]. ^{1}H NMR spectrum (THF): δ = 6.48 (d, C_6H_2; J(P, H) = 4.0 Hz), 7.42 (m, C_6H_5) ppm [3, 4] (spectrum depicted in [4]). ^{31}P NMR: δ = −3.4 ppm [3, 4]. IR spectrum (cyclohexane): 1933, 1942, 1996 (ν(CO)) cm^{-1} [3]; similar in [4]. Mass spectrum (70 eV, sketched in the original literature): $[M]^+$, $[M-n\ CO]^+$ (n = 1 to 3), $[(C_6H_5)_3C_5H_2P]^+$ [3].

(2,4-$R_2C_5H_3As$)Mo(CO)$_3$ (^{5}L ligand = Formula V; R-2 = H: R-4 = H, C_2H_5, t-C_4H_9, cyclo-C_6H_{11}, C_6H_5; R-2 = C_6H_5: R-4 = t-C_4H_9, cyclo-C_6H_{11}, C_6H_5). Known members of this family

References on p. 343

of compounds with λ^3-arsenine ligands (Formula V) bearing different substituents R-2 and R-4 are listed in the following table.

The parent arsabenzene derivative, $(C_5H_5As)Mo(CO)_3$, was isolated in small quantities from the pyrolysis reaction of the σ complex $Mo(CO)_5AsC_5H_5$ at 120 °C [5]. More convenient methods of synthesis were briefly reported to involve heating $Mo(CO)_6$ with arsabenzene in $CH_3OC_2H_4OC_4H_2OCH_3$ at temperatures above 120 °C [5, 6] or by the $(C_2H_5)_2O \cdot BF_3$-catalyzed ligand displacement from $Mo(CO)_3(NC_5H_5)_3$ [5].

In the preparation of the substituted derivatives, $(RC_5H_4As)Mo(CO)_3$ and $(R_2C_5H_3As)$-$Mo(CO)_3$, respectively, equimolar quantities of $Mo(CO)_3(NCCH_3)_3$ and the required λ^3-arsenine were heated in boiling dioxane. Evaporation of the solvent left oily to semisolid residues which were purified by column chromatography on silica gel using $C_6H_6/CHCl_3$ (1:1) as the eluent and, if necessary, by subsequent recrystallization from petroleum ether (monosubstituted complexes) or acetone (2,4-disubstituted compounds) [7].

R-2	R-4	yield, properties, and remarks
H	H	red crystals, m.p. 109 to 110 °C [5] 1H NMR ($CDCl_3$): 5.4 (H-3,5), 5.9 (H-2,6, H-4) [5] ^{13}C NMR ($CDCl_3$): 87.5 (C-4), 94.2 (C-3,5), 110.8 (C-2,6) [5] IR ($CHCl_3$): 1926, 1996 (CO) [5]
H	C_2H_5	(80%) [7] red oil [7] ^1H NMR ($CDCl_3$): 1.28 (q, CH_2; J = 7), 2.69 (t, CH_3); 5.4, 5.9 (AA′BB′ system, H-3,5, H-2,6; J(H-2,3) = 10, J(H-3,5) = 2; J(H-2,5) and J(H-2,6) not observed) [7] IR (film): 1892, 1972 (CO); ($CHCl_3$): 1910, 1982 (CO) [7]
H	t-C_4H_9	(83%) [7] orange-red crystals, m.p. 58 °C [7] ^1H NMR ($CDCl_3$): 1.41 (s, CH_3); 5.68, 5.85 (AA′BB′ system, H-3,5, H-2,6; J(H-2,3) = 9.00, J(H-2,5) = 0.70, J(H-2,6) = 1.25, J(H-3,5) = −2.50) [7] ^{13}C NMR ($CDCl_3$): 31.23 ($C(CH_3)_3$), 35.20 ($C(CH_3)_3$), 93.5 (C-3,5), 109.8 (C-2,6), 119.5 (C-4), 220.88 (CO); spectrum shown in [7] IR (KBr): 1827, 1957 (CO) [7]
H	cyclo-C_6H_{11}	(86%) [7] yellow crystals, m.p. 109 to 110 °C [7] ^1H NMR ($CDCl_3$): 1.14 to 2.44 (m, C_6H_{11}), 5.47, 5.86 (AA′BB′ system, H-3,5, H-2,6; J(H-2,3) = 9.80, J(H-2,5) = 0.70, J(H-2,6) = 1.60, J(H-3,5) = −2.60) (plus additional coupling (0.4 Hz) of H-3 and 5 with cyclohexyl-CH) [7] ^{13}C NMR ($CDCl_3$): 26.03, 26.77, 34.77 (all CH_2 of C_6H_{11}), 45.12 (CH of C_6H_{11}), 94.7 (C-3,5), 110.3 (C-2,6), 114.5 (C-4), 220.82 (CO) [7] IR (KBr): 1876, 1910, 1972 (CO); ($CHCl_3$): 1924, 1985 (CO) [7]
H	C_6H_5	(74%) [7] light orange crystals, m.p. 88 to 89 °C [7] ^1H NMR ($CDCl_3$): 5.88, 6.09 (AA′BB′ system, H-3,5, H-2,6; J(H-2,3) = 9.20, J(H-2,5) = 0.70, J(H-2,6) = 1.25, J(H-3,5) = −2.50); 7.36 to 7.66 (m, C_6H_5); spectrum shown in [7]

References on p. 343

R-2	R-4	yield, properties, and remarks
		^{13}C NMR ($CDCl_3$): 95.2 (C-3,5), 110.5 (C-2,6), 107.4 (C-4), 127.92, 128.78, 129.01 (C-2 to 6 of C_6H_5), 138.70 (C-1 of C_6H_5), 220.52 (CO) [7] IR (KBr): 1872, 1977 (CO) [7]
C_6H_5	t-C_4H_9	(57%) [7] orange-red crystals, m.p. 92 to 94°C [7] ^{1}H NMR ($CDCl_3$): 1.48 (s, CH_3), 5.72 (m, H-3; J(H-3,5) = 1.23, J(H-3,6) = 0.62), 5.90 (m, H-5; J(H-5,6) = 9.13), 6.02 (m, H-6), 7.33 ("s", C_6H_5); spectrum shown in [7] ^{13}C NMR ($CDCl_3$): 31.42 ($C(CH_3)_3$), 35.38 ($C(CH_3)_3$), 92.34 (C-3), 94.75 (C-5), 110.69 (C-6), 119.79 (C-4), 127.73, 128.66 (C-2 to 6 of C_6H_5), 133.43 (C-2), 141.98 (C-1 of C_6H_5), 221.12 (CO) [7] IR (KBr): 1885, 1890, 1905, 1960 (CO); (C_6H_6): 1915, 1980 (CO); spectra in THF, $CH_3OC_2H_4OCH_3$, CH_2Cl_2, and $CHCl_3$ also given in [7] UV (C_2H_5OH, log ε) = 254 (3.19), 290 (2.85), 336 (3.22); spectrum plotted in [7] mass spectrum (70 eV): $[M]^+$ (43%), $[M-n\ CO]^+$ (n = 1 (24%), 2 (45%), 3 (100%)); for further fragment ions with minor quantities, see [7]
C_6H_5	cyclo-C_6H_{11}	(63%) [7] bright orange crystals, m.p. 134 to 136°C [7] ^{1}H NMR ($CDCl_3$): 1.10 to 2.79 (m, C_6H_{11}), 5.51 (m, H-3; J(H-3,5) = 1.10, J(H-3,6) = 0.38, J(H-3, CH of C_6H_{11}) ≅ 0.4), 5.70 (m, H-5; J(H-5,6) = 9.14, J(H-5, CH of C_6H_{11}) ≅ 0.4), 6.05 (m, H-6), 7.39 ("s", C_6H_5) [7] ^{13}C NMR ($CDCl_3$): 26.09, 26.83, 34.70, 35.20 (all CH_2 of C_6H_{11}), 45.30 (CH of C_6H_{11}), 93.39 (C-3), 96.00 (C-5), 111.31 (C-6), 114.78 (C-4), 127.73, 128.66 (C-2 to 6 of C_6H_5), 133.86 (C-2), 221.12 (CO) [7] IR (KBr): 1834, 1910, 1972 (CO); (CH_2Cl_2): 1912, 1982 (CO); ($CHCl_3$): 1920, 1985 (CO); (CCl_4): 1914, 1929, 1989 (CO) [7] mass spectrum (70 eV): $[M]^+$ (59%), 452 $[M-CO]^+$ (26%), $[M-2CO-H_2]^+$ (77%), $[M-3CO]^+$ (100%); for additional, but less frequent fragments, see [7]
C_6H_5	C_6H_5	(52%) [7] orange crystals, m.p. 99 to 102°C [7] IR (KBr): 1896, 1924, 1983 (CO); (CCl_4): 1938, 1992 (CO) [7]

$(C_5H_5Sb)Mo(CO)_3$ (^{5}L ligand = Formula VI). Similar to its arsabenzene homologue, the stibabenzene complex was easily prepared by the boron trifluoride etherate catalyzed reaction of stibabenzene (Formula VI) with $Mo(CO)_3(NC_5H_5)_3$ [5, 6].

Red-brown crystals, m.p. 106 to 108°C. ^{1}H NMR spectrum ($CDCl_3$): δ = 5.7 (H-3,5), 6.1 (H-2,6, H-4) ppm. ^{13}C NMR spectrum ($CDCl_3$): δ = 88.0 (C-4), 95.5 (C-3,5), 114.6 (C-2,6) ppm. IR spectrum ($CHCl_3$): 1919, 1990 (ν(CO)) cm^{-1}. Mass spectral parent ion observed [5].

The complex proved to be much more stable than free stibabenzene [5].

References on p. 343

References:

[1] Moore, B.; Wilkinson. G. (Proc. Chem. Soc. London **1959** 61).
[2] Fischer, E. O.; Öfele, K. (Z. Naturforsch. **14b** [1959] 736/7).
[3] Deberitz, J.; Nöth, H. (Chem. Ber. **106** [1973] 2222/6).
[4] Nöth, H. Deberitz, J. (Kem. Kozlem. **40** [1973] 9/21).
[5] Ashe, A. J., III; Colburn, J. C. (J. Am. Chem. Soc. **99** [1977] 8099/100).
[6] Colburn, J. C. (Diss. Univ. Michigan, Ann Arbor, Mich., 1978; Diss. Abstr. Int. B **39** [1978] 745).
[7] Märkl, G.; Baier, H.; Lierl, R.; Mayer, K. K. (J. Organomet. Chem. **217** [1981] 333/56).
[8] Boese, R.; Finke, N.; Keil, T. H.; Paetzold, P.; Schmid, G. (Z. Naturforsch. **40b** [1985] 1327/32).

1.5.1.4.2.7.7 $^5LMo(CO)_3$ Complexes with Heterocyclic Pentadienylide Ligands

The ligand systems bound to the $Mo(CO)_3$ fragment dealt with under this heading comprise P-ylidic λ^5-phosphorines (Formula I), as well as S-ylidic heterocycles derived from λ^4-thiines (Formula II), and the corresponding S-oxides (Formula III).

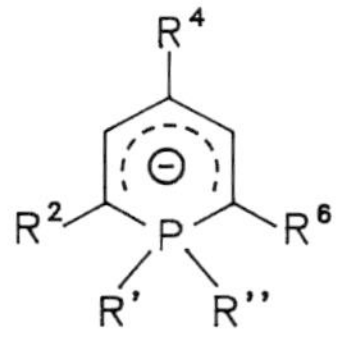

I 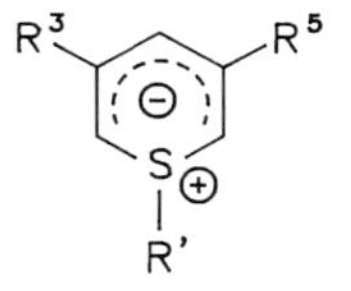 II 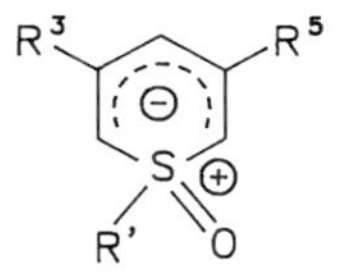 III

For the S-heterocycles, the ylidic character is much more pronounced than for the λ^5-phosphorines, the electronic system of which is best regarded as a superposition of ylidic and aromatic limiting structures with the former prevailing [4, 12]. Thus, 2,4,6-$(C_6H_5)_3$-$C_5H_2P(OCH_3)_2$ is planar [1], whereas 3,5-$(C_6H_5)_2C_5H_3S(O)CH_3$ has been shown to deviate significantly from planarity [5]. These ring conformations are not changed upon complexation to $M(CO)_3$ fragments (M = Cr, Mo, W) as established by X-ray structure determinations carried out for the chromium derivatives (2,4,6-$(C_6H_5)_3C_5H_2P(OCH_3)_2$)$Cr(CO)_3$ (planar heterocycle [3]), (3,5-$(C_6H_5)_2C_5H_3S(O)CH_3$)$Cr(CO)_3$ (two isomers with nonplanar rings differing in the orientation of S=O and $S-CH_3$ with respect to the $Cr(CO)_3$ moiety [5]), and also (3,5-$(C_6H_5)_2C_5H_3SCH_3$)$Cr(CO)_3$ (nonplanar ring; $S-CH_3$ anti to the $Cr(CO)_3$ group [8, 14]). Yet, a considerable body of structural and NMR spectroscopic evidence compiled for a series of differently substituted chromium complexes (2,4,6-$R_3C_5H_2P(R')R''$)$Cr(CO)_3$ has provided convincing arguments that the Group 6 $M(CO)_3$ derivatives of λ^5-phosphorines, (2,4,6-$R_3C_5H_2P(R')R''$)$M(CO)_3$ (M = Cr, Mo, W), should also be looked at as inner salts of the phosphonium ylide type rather than heteroarene complexes; for details, see [11].

The ^5L ligands of (3,5-$R_2C_5H_3SR'$)$Mo(CO)_3$ (Nos. 4 to 6) were assigned a conformation with the $Mo(CO)_3$ and $S-R'$ moieties in mutual anti orientation [8, 14, 17]. In contrast to (3,5-$(C_3H_5)_2C_5H_3S(O)CH_3$)$Cr(CO)_3$, shown to exist as two isomers containing the $S-CH_3$ in an anti or syn orientation relative to the $Cr(CO)_3$ fragment [5], such isomers were not observed for the molybdenum and tungsten homologues ($R_2C_5H_3S(O)R$)$M(CO)_3$, with the derivatives (3,5-(t-$C_4H_9)_2C_5H_3S(O)CH_3$)$M(CO)_3$ being noteworthy exceptions [16]. For all the other compounds of this type listed in Table 10, a conformation of the ^5L ligand with $Mo(CO)_3$ and $S-R'$ in mutual syn orientation was derived from the spectroscopic data available [5, 7, 9, 16].

References on p. 351

The compounds listed in Table 10 were prepared by one of the following Methods:

Method I: The required λ^5-phosphorine was reacted with an excess of $Mo(CO)_6$ in refluxing toluene or with $Mo(CO)_3(NCCH_3)_3$ in boiling 1,4-dioxane. After cooling, the solutions were concentrated and chromatographed on silica gel columns eluting with petroleum ether/acetone (2:1). The eluates were evaporated and the products (Nos. 1 to 3) crystallized from ethanol [2, 10].

Method II: The requisite thiinium salts (Formula IV; $R' = CH_3$, C_2H_5) were deprotonated with KOC_4H_9-t in dimethyl sulfoxide in the presence of an equimolar quantity of tricarbonyl(cycloheptatriene)molybdenum(0). The evaporated mixtures were extracted into benzene and the extracts diluted with petroleum ether to precipitate the products as oils which solidified on stirring under petroleum ether. The crude materials were purified by repeated reprecipitation from benzene/petroleum ether to yield Nos. 5 and 6 [8, 14].

Method III: $(3,5\text{-}(t\text{-}C_4H_9)_2C_5H_3SCH_3)Mo(CO)_3$ was prepared by deoxygenation of $(3,5\text{-}(t\text{-}C_4H_9)_2C_5H_3S(O)CH_3)Mo(CO)_3$, which occurred smoothly when the anti isomer of the latter λ^6-thiabenzene S-oxide complex was combined with $Na[AlH_2(OC_2H_4OCH_3)_2]$ (solution in toluene) in 1:2 stoichiometry in benzene at ambient conditions. The excess reducing agent was destroyed by adding methanol and the complex was isolated from the concentrated mixture by chromatography on a silica gel column using benzene [17].

Method IV: The S-methyl compounds $(3,5\text{-}R_2C_5H_3S(O)CH_3)Mo(CO)_3$ ($R = CH_3$, $t\text{-}C_4H_9$, C_6H_5) were synthesized by heating the required 3,5-disubstituted 1-methyl-thiabenzene 1-oxide with excess $Mo(CO)_3(NCCH_3)_3$ in 1,4-dioxane at 60 °C. The tarry residues remaining after evaporation to dryness were subjected to chromatography on Al_2O_3 columns using $CHCl_3$ [16] or a 1:1 mixture of $CHCl_3$/petroleum ether (only employed for isolation of the diphenyl-substituted complex, $(3,5\text{-}(C_6H_5)_5C_5H_3S(O)CH_3)Mo(CO)_3$, which was further purified by recrystallization from CH_2Cl_2/methylcyclohexane [5]).

Method V: $(3,5\text{-}(C_6H_5)_2C_5H_3S(O)CH_3)Mo(CO)_3$ was metalated at its methyl group by treatment with n-butyl lithium (1 equivalent in hexane) in THF solution at −70 °C. Addition of excess allyl or benzyl bromide, followed by warming to room temperature, quenching the mixtures with water, and removing the volatiles in vacuum, gave the crude products as oily materials which were worked up by column chromatography on Al_2O_3. Elution with $CHCl_3$/petroluem ether (1:1) resulted in separation of the mono- and disubstituted derivatives formed during alkylation, $(3,5\text{-}(C_6H_5)_2C_5H_3S(O)CH_2R'')Mo(CO)_3$ and $(3,5\text{-}(C_6H_5)_2C_5H_3S(O)CHR''_2)Mo(CO)_3$ ($R'' = CH_2CH{=}CH_2$, $CH_2C_6H_5$), the latter eluting first. The complexes were purified further by recrystallization from CH_2Cl_2/petroleum ether [7]. The crude products resulting from treatment of the lithiated intermediate with $Si(CH_3)_3Cl$ (excess; −196 °C → −70 °C → +20 °C) were chromatographed on silanized silica gel eluting with $CHCl_3$/petroleum ether (1:4) to yield $(3,5\text{-}(C_6H_5)_2C_5H_3S(O)CH(Si(CH_3)_3)_2)Mo(CO)_3$ as the first fraction, followed by a second zone containing $(3,5\text{-}(C_6H_5)_2C_5H_3S(O)CH_2Si(CH_3)_3)Mo(CO)_3$. Final purification was achieved by crystallization from CH_2Cl_2/methylcyclohexane [9].

Method VI: $(3,5\text{-}(C_6H_5)_2C_5H_3S(O)CH_3)Mo(CO)_3$ was alkylated with $(CH_3O)_2SO_2$, $CH_2CH{=}CH_2Br$, or $C_6H_5CH_2Br$ under phase-transfer conditions employing the two-phase system CH_2Cl_2/50% aqueous NaOH as the reaction medium and $[N(C_4H_9\text{-}n)_4]I$ as the phase-transfer catalyst. After stirring at ambient conditions,

References on p. 351

the aqueous phase was frozen out at −30 °C, the organic phase decanted, and the volatile materials evaporated in vacuum. Further workup was accomplished as delineated under "Method V" above. Phase-transfer alkylation with dimethyl sulfate and allyl bromide afforded the monosubstituted derivatives (3,5-$(C_6H_5)_2$-$C_5H_3S(O)CH_2CH_3)Mo(CO)_3$ and (3,5-$(C_6H_5)_2C_5H_3S(O)CH_2CH_2CH{=}CH_2)Mo(CO)_3$ as the exclusively formed organometallic species, whereas both (3,5-$(C_6H_5)_2$-$C_5H_3S(O)CH_2CH_2C_6H_5)Mo(CO)_3$ and (3,5-$(C_6H_5)_2C_5H_3S(O)CH(CH_2C_6H_5)_2)$Mo-$(CO)_3$ resulted from treatment with benzyl bromide under the conditions of phase-transfer catalysis [7].

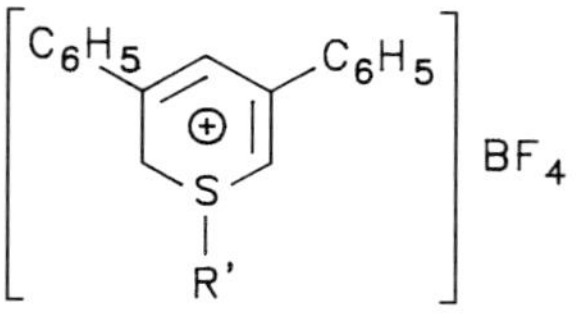

$$\left[\text{3,5-}(C_6H_5)_2C_5H_3S^{\oplus}R'\right]BF_4$$

IV

The complexes (2,4,6-$(C_6H_5)_3C_5H_2P(R')R'')Mo(CO)_3$ (Nos. 1 to 3) are highly crystalline, yellow to orange, air-stable compounds [2, 10]. (3,5-$R_2C_5H_3SR')Mo(CO)_3$ (Nos. 4 to 6) are freely soluble in THF, acetone, dimethyl sulfoxide, and CH_2Cl_2, sparingly soluble in benzene and ether, but insoluble in saturated hydrocarbons. They are stable toward air for short periods, both in the solid state and in solution [8, 14, 17]. The complexes (3,5-$R_2C_5H_3$-$S(O)R')Mo(CO)_3$ (Nos. 7 to 19) form air-stable dark red crystals, freely soluble in the more polar solvents, such as CH_2Cl_2, $CHCl_3$, acetone, 1,4-dioxane, and THF, sparingly soluble in diethyl ether, and insoluble in saturated hydrocarbons [5, 7, 16]. In CF_3CO_2H solution, the S-methylated derivatives, (3,5-$R_2C_5H_3S(O)CH_3)Mo(CO)_3$, underwent protonation at C-2 forming $[(3,5\text{-}R_2C_5H_4S(O)CH_3)Mo(CO)_3]O_2CCF_3$ [15].

Table 10
$^5LMo(CO)_3$ Complexes Containing Heterocyclic Pentadienylide Ligands.
An asterisk indicates further information at the end of the table.
For explanations, abbreviations, and units see p. X.

No.	substituents	method of preparation (yield) properties and remarks
compounds of the type (2,4,6-$(C_6H_5)_3C_5H_2P(R')R'')Mo(CO)_3$ (Formula I)		
R'/R'' =		
1	CH_3/CH_3	I (65%) [10] m.p. 217 to 219 °C (dec.) [10] ^{1}H NMR (acetone-d_6): 1.43 (d, CH_3-exo; J(P, H) = 12), 2.02 (d, CH_3-endo; J(P, H) = 12), 6.53 (d, H-3,5; J(P, H) = 20) [10] ^{31}P NMR (acetone-d_6): 12.7 [10] IR (KBr): 1832 (A'), 1854 (A'), 1940 (A'') (CO) [10] mass spectrum: $[M]^+$ (2.4%), $[M-nCO]^+$ (n = 1 (23%), 2 (5%), 3 (56%)), $[M-3CO-2CH_3]^+$ (86%); for additional peaks, see [10]

References on p. 351

Table 10 (continued)

No.	substituents	method of preparation (yield) properties and remarks
1 (continued)		treatment with LiC_4H_{9-t} or $LiC(C_6H_5)_3$ in THF followed by alkylation with CH_3I yielded $(2,4,6-(C_6H_5)_3C_5H_2P(CH_3)C_2H_5)Mo(CO)_3$ [12]
2	OCH_3/OCH_3	I (6%) [2, 10] m.p. 197 to 203°C (dec.) [2, 10] ^1H NMR (acetone-d_6): 3.75 (d, OCH_3-exo; J(P, H) = 13.5), 3.78 (d, OCH_3-endo; J(P, H) = 9.5), 6.72 (d, H-3,5; J(P, H) = 26.3) [2, 10] IR (KBr): 1830 (A′), 1870 (A′), 1940 (A″) (CO) [10]; (cyclohexane): 1845, 1885, 1950 (CO) [2] mass spectrum: $[M]^+$ (14%), $[M-3CO]^+$ (33%), $[M-3CO-2OCH_3]^+$ (7%); for additional peaks, see [10]
3	CH_3/C_2H_5	I (2%) [10, 12]; also in 30% yield from $(2,4,6-(C_6H_5)_3C_5H_2P(CH_3)_2)Mo(CO)_3$ by lithiation (LiC_4H_9-t or $LiC(C_6H_5)_3$ in THF) and subsequent alkylation with CH_3I [12] m.p. 220°C (dec.) [10] ^1H NMR (acetone-d_6): 1.0 to 1.45 (m, C_2H_5), 1.92 (d, CH_3; J(P, H) = 12.5), 6.31 (d, H-3,5; J(P, H) = 19.5) [10] IR (KBr): 1830 (A′), 1855 (A′), 1940 (A″) (CO) [10] mass spectrum: $[M]^+$ (41%), $[M-nCO]^+$ (n = 2 (27%), 3 (42%)), $[M-3CO-CH_3-C_2H_5]^+$ (91.5%); for additional peaks, see [10]

compounds of the type $(3,5\text{-}R_2C_5H_3SR')Mo(CO)_3$ (Formula II)

R/R′ =

No.	substituents	method of preparation (yield) properties and remarks
4	$t\text{-}C_4H_9/CH_3$	III (31%) [17] orange-yellow crystals, m.p. 161 to 163°C (dec.) [17] ^1H NMR ($CDCl_3$): 1.25 (s, $C(CH_3)_3$), 1.80 (s, SCH_3), 2.97 (d, H-2,6; J = 1.5), 6.01 (t, H-4) [17] IR (Nujol): 1828, 1855, 1933 (CO); (CH_2Cl_2): 1837, 1872, 1948 (CO) [17] mass spectrum (70 eV, 140°C): $[M]^+$, $[M-CH_3]^+$, $[M-CO]^+$, $[M-CH_3-n\ CO]^+$ (n = 1 to 3) [17]
5	C_6H_5/CH_3	II (42%) [8, 14] dark red to red-violet crystals, m.p. 169°C (dec.) [8, 14] ^1H NMR (acetone-d_6): 2.33 (s, CH_3), 3.76 (d, H-2,6; J = 1.5), 6.74 (t, H-4), 7.36 to 7.44, 7.70 to 7.77 (both m, C_6H_5) [14]

References on p. 351

Table 10 (continued)

No.	substituents	method of preparation (yield) properties and remarks
		^{13}C NMR (acetone-d_6): 28.76 (C-2,6), 41.58 (CH_3), 90.56 (C-4), 109.40 (C-3,5), 128.74, 129.15, 129.77, 141.00 (all C_6H_5), 226.99 (CO) [14] IR (Nujol): 1837, 1874, 1941 (CO); (CH_2Cl_2): 1855, 1888, 1958 (CO) [14] mass spectrum (70 eV, 150 °C): $[M]^+$, $[M-CH_3]^+$, $[M-n\ CO]^+$ (n = 1 to 3), $[M-n\ CO-CH_3]^+$ (n = 1 to 3), $[(C_6H_5)_2C_5H_3Mo]^+$ [14] lithiation (LiC_4H_9-t/THF) of the CH_3S group, followed by alkylation with CH_3I afforded No. 6 [14]
6	C_6H_5/C_2H_5	II (15%) [14]; also by metalation of No. 5 with LiC_4H_9-t in THF at −70 °C and subsequent addition of excess CH_3I (44%) [14] dark red crystals, m.p. 164 to 168 °C [14] ^1H NMR (acetone-d_6): 1.13 (t, CH_3; J = 7.5), 2.65 (q, CH_2), 3.78 (d, H-2,6; J = 1.5), 6.71 (t, H-4), 7.38 to 7.42, 7.74 to 7.80 (both m, C_6H_5) [14] IR (Nujol): 1836, 1854, 1935 (CO); (CH_2Cl_2): 1853, 1885, 1956 (CO) [14] mass spectrum (70 eV, 160 °C): $[M]^+$, $[M-n\ CO]^+$ (n = 1 to 3), $[M-3CO-C_2H_5]^+$, $[(C_6H_5)_2C_5H_3Mo]^+$ [14] in CH_2Cl_2/0.1 M $[N(C_4H_9\text{-}n)_4]PF_6$, irreversible electrooxidation was observed to occur at +0.58 V vs. $(C_5H_5)_2Co/[(C_5H_5)_2Co]Cl$ [18]

compounds of the type $(3,5\text{-}R_2C_5H_3S(O)R')Mo(CO)_3$ (Formula III)

R-3/R-5/R′ =

No.	substituents	method of preparation (yield) properties and remarks
7	$CH_3/CH_3/CH_3$	IV (72%) [16] canary crystals, m.p. 230 °C (dec.) [16] ^1H NMR ($CDCl_3$): 2.31 (s, CCH_3), 3.66 (s, SCH_3), 4.50 (d, H-2,6; J = 1.1), 5.22 (t, H-4) [16] IR (Nujol): 1177 (ν(SO)); (CH_2Cl_2): 1873, 1961 (CO) [16] mass spectrum (70 eV, 155 °C): $[M]^+$, $[M-nCO]^+$ (n = 1, 3), $[M-3CO-O]^+$, $[(CH_3)_2C_3H_5SMo]^+$, $[M-3CO-O-2CH_3]^+$, $[(CH_3)_2C_3H_5Mo]^+$ [16] in CH_2Cl_2/0.1 M $[N(C_4H_9\text{-}n)_4]PF_6$, the complex was irreversibly oxidized at +0.715 V vs. $(C_5H_5)_2Co/[(C_5H_5)_2Co]Cl$ [18]
8	t-C_4H_9/t-C_4H_9/CH_3 (syn)	IV (64%) [16] intense yellow crystals, m.p. 200 °C (dec.) [16] ^1H NMR ($CDCl_3$): 1.29 (s, $C(CH_3)_3$), 3.63 (s, SCH_3), 4.64 (s, H-2,6), 5.46 (s, H-4) [16]

References on p. 351

Table 10 (continued)

No.	substituents	method of preparation (yield) properties and remarks
8 (continued)		IR (Nujol): 1174 (ν(SO)); (CH_2Cl_2): 1864, 1952 (CO) [16] mass spectrum (70 eV, 330°C): $[M]^+$, $[(C_4H_9)_2C_5H_3S]^+$, $[C_4H_9]^+$ [16] irreversible electrooxidation occurred at +0.740 V vs. $(C_5H_5)_2Co/[(C_5H_5)_2Co]Cl$ (CH_2Cl_2/0.1 M $[N(C_4H_9\text{-}n)_4]PF_6$); E_{red} = +0.660 V (same conditions) [18]
9	t-C_4H_9/t-C_4H_9/CH_3 (anti)	IV (9%) [16] orange crystals, m.p. 152°C (dec.) [16] ^{1}H NMR ($CDCl_3$): 1.27 (s, $C(CH_3)_3$), 2.71 (s, SCH_3), 3.37 (s, H-2,6), 5.65 (s, H-4) [16] IR (Nujol): 1193 (ν(SO)); (CH_2Cl_2): 1854, 1894, 1968 (CO) [16] mass spectrum (70 eV, 330°C): $[M]^+$, $[(C_4H_9)_2C_5H_3S(CH_3)Mo]^+$, $[(C_4H_9)_2C_5H_3S(O)CH_3]^+$, $[(C_4H_9)_2C_5H_3SO]^+$, $[(C_4H_9)_2C_5H_3S]^+$, $[C_4H_9]^+$ [16] deoxygenation with the formation of (3,5-(t-C_4H_9)$_2C_5H_3SCH_3$)$Mo(CO)_3$ occurred on combination with $Na[AlH_2(OC_2H_4OCH_3)_2]$ in benzene at room temperature [17]
10	$CH_3/C_6H_5/CH_3$	IV (48%) [16] red-brown crystals, m.p. 193°C (dec.) [16] ^{1}H NMR ($CDCl_3$): 2.42 (s, CCH_3), 3.75 (SCH_3), 4.61, 4.94 (both dd, H-2,6; J(H-2,6) = 4.1, J(H-2,4) = J(H-4,6) = 1), 5.64 (t, H-4), 7.34 to 7.52 (m, C_6H_5) [16] IR (Nujol): 1187 (ν(SO)); (CH_2Cl_2): 1878, 1960 (CO) [16] mass spectrum (70 eV, 195°C): $[M]^+$, $[M-nCO]^+$ (n = 1 to 3), $[M-3CO-CH_3]^+$, $[M-3CO-O]^+$, $[M-3CO-O-CH_4]^+$, $[CH_3(C_6H_5)C_5H_3Mo]^+$ [16]
11	t-$C_4H_9/C_6H_5/CH_3$	IV (67%) [16] orange-red crystals, m.p. 165°C (dec.) [16] ^{1}H NMR ($CDCl_3$): 1.36 (s, $C(CH_3)_3$), 3.73 (SCH_3), 4.66, 5.03 (both dd, H-2,6; J(H-2,6) = 4.1, J(H-2,4) = J(H-4,6) = 1), 5.74 (t, H-4), 7.35 to 7.52 (m, C_6H_5) [16] IR (Nujol): 1183 (ν(SO)); (CH_2Cl_2): 1877, 1960 (CO) [16] mass spectrum (70 eV, 210°C): $[M]^+$, $[M-nCO]^+$ (n = 1, 3), $[M-3CO-O]^+$, $[C_4H_9(C_6H_5)C_5H_3SMo]^+$, $[C_4H_9(C_6H_5)C_5H_3Mo]^+$ [16]

References on p. 351

Table 10 (continued)

No.	substituents	method of preparation (yield) properties and remarks
*12	$C_6H_5/C_6H_5/CH_3$	IV (79%) [5] m.p. 170 to 190°C (dec.) [5] ^{1}H NMR (acetone-d_6): 4.07 (s, CH_3), 5.79 (d, H-2,6; J = 1), 6.45 (t, H-4), 7.40 to 7.60, 7.70 to 7.90 (both m, C_6H_5) [6]; ($CDCl_3$): 3.84 (s, CH_3), 5.06 (d, H-2,6; J = 1.2), 6.06 (t, H-4) 7.44 (m, C_6H_5) [5] ^{13}C NMR (acetone-d_6): 43.06 (CH_3), 61.77 (C-2,6), 88.33 (C-4), 108.96 (C-3,5), 128.63, 129.59, 130.52, 139.06 (all C_6H_5), 224.92 (CO) [5, 6] IR (Nujol): 1185, 1195 (ν(SO)), 1854, 1902, 1954 (sh at 1825, 1948, and 1986) (CO); additional bands at 437, 460, 471, 495, 512, 530, 561, 591, 615, 633, 674, 707, 730, 750, 779, 838, 858, 869, 935, 1030, 1083, 1162, 1243, 1316, 1408, 1456, 1581, 1600, and 3052; (CH_2Cl_2): 1881, 1965 (CO) [5] UV (CH_3OH, log ε) = 211 (4.6), 275 (4.3), 327 (3.9), 404 (3.8) [5]
13	$C_6H_5/C_6H_5/CH_2CH_3$	VI (74%) [7] m.p. 220 to 223°C (dec.) [7] ^{1}H NMR (acetone-d_6): 1.71 (t, CH_3; J = 7.5), 4.29 (q, CH_2), 5.66 (d, H-2,6; J = 1.2), 6.39 (t, H-4), 7.36 to 7.52, 7.66 to 7.85 (both m, C_6H_5) [7] IR (Nujol): 1182 (ν(SO)); (CH_2Cl_2): 1879, 1962 (CO) [7]
14	$C_6H_5/C_6H_5/CH_2CH_2CH{=}CH_2$	V (42%) [7]; VI (6%) [7] m.p. 183 to 186°C (dec.) [7] ^{1}H NMR (acetone-d_6): 2.83 to 3.07 (m, $\mathit{CH_2}CH{=}CH_2$), 4.29 to 4.45 (m, S(O)CH_2), 5.18 to 5.50 (m, $CH_2CH{=}\mathit{CH_2}$), 5.69 (d, H-2,6; J = 1.2), 5.88 to 6.30 (m, $CH_2\mathit{CH}{=}CH_2$), 6.42 (t, H-4), 7.34 to 7.60, 7.66 to 7.84 (both m, C_6H_5) [7] IR (Nujol): 1178 (ν(SO)); 1642 (ν(C=C)); (CH_2Cl_2): 1882, 1962 (CO) [7]
15	$C_6H_5/C_6H_5/CH_2CH_2C_6H_5$	V (44%) [7]; VI (17%) [7] m.p. 179 to 183°C (dec.) [7] ^{1}H NMR (acetone-d_6): 3.49, 4.57 (AA′BB′ system, CH_2CH_2; J(AA′) = J(BB′) = −15.3, J(A, B) = 6.4, J(A, B′) = 9.3), 5.39 (d, H-2,6; J = 1.2), 6.38 (t, H-4), 7.30 to 7.78 (m, C_6H_5) [7] IR (Nujol): 1183, 1190 (ν(SO)); (CH_2Cl_2): 1882, 1963 (CO) [7]
16	$C_6H_5/C_6H_5/CH_2Si(CH_3)_3$	V (8%) [9] m.p. 159 to 161°C [9]

References on p. 351

Table 10 (continued)

No.	substituents	method of preparation (yield) properties and remarks
16	(continued)	^{1}H NMR ($CDCl_3$): 0.50 (s, CH_3), 3.69 (s, CH_2), 5.08 (d, H-2,6; J = 1.1), 6.11 (t, H-4), 7.39 to 7.77 (m, C_6H_5) [9] IR (Nujol): 1178 (ν(SO)); (CH_2Cl_2): 1880, 1960 (CO) [9]
17	$C_6H_5/C_6H_5/CH(CH_2CH{=}CH_2)_2$	V (19%) [7] m.p. 173 to 175 °C (dec.) [7] ^{1}H NMR (acetone-d_6): 2.73 to 3.28 (m, $CH_2CH{=}CH_2$), 4.60 (qui, S(O)CH; J = 6.3), 5.25 to 5.53 (m, $CH_2CH{=}CH_2$), 5.56 (d, H-2,6; J = 1.2), 5.89 to 6.35 (m, $CH_2CH{=}CH_2$), 6.36 (t, H-4), 7.34 to 7.57, 7.57 to 7.87 (both m, C_6H_5) [7] IR (Nujol): 1175 (ν(SO)); 1636, 1641 (ν(C=C)); (CH_2Cl_2): 1879, 1961 (CO) [7]
18	$C_6H_5/C_6H_5/CH(CH_2C_6H_5)_2$	V (20%) [7]; VI (9%) [7] m.p. 209 to 211 °C (dec.) [7] ^{1}H NMR (acetone-d_6): 3.21, 3.70, 5.10 ($A(MX)_2$ system, $CH(CH_2)_2$; J(A, M) = J(A, X) = 7.5, J(M, X) = −14.3), 4.79 (d, H-2,6; J = 1.2), 6.35 (t, H-4), 7.34 to 7.60 (m, C_6H_5) [7] IR (Nujol): 1179, 1188 (ν(SO)); (CH_2Cl_2): 1879, 1960 (CO) [7] $[(3,5\text{-}(C_6H_5)_2C_5H_3S(O)CH(CH_2C_6H_5)_2)\text{-}Mo(CO)_2NO]PF_6$ resulted from treatment with $NOPF_6$ in CH_2Cl_2 at −70 °C [9]
19	$C_6H_5/C_6H_5/CH(Si(CH_3)_3)_2$	V (27%) [9] m.p. 176 to 178 °C [9] ^{1}H NMR (CD_2Cl_2): 0.53 (s, CH_3), 3.37 (s, CH), 5.11 (d, H-2,6; J = 1.2), 6.08 (t, H-4), 7.35 to 7.65 (m, C_6H_5) [9] IR (Nujol): 1171 (ν(SO)); (CH_2Cl_2): 1877, 1958 (CO) [9] $[(3,5\text{-}(C_6H_5)_2C_5H_3S(O)CH(Si(CH_3)_3)_2)\text{-}Mo(CO)_2NO]PF_6$ was obtained from the reaction with $NOPF_6$ in CH_2Cl_2 at −70 °C [9]

* Further information:

$(3,5\text{-}(C_6H_5)_2C_5H_3S(O)CH_3)Mo(CO)_3$ (Table **10**, No. **12**). Demethylation reactions with $Li[BH(C_2H_5)_3]$ [13], $LiP(C_6H_5)_2$, or $Na_2[Fe(CO)_4] \cdot 1.5\ O(C_2H_4)_2O$ [6] in THF, followed by cation metathesis using $[N(C_2H_5)_4]Cl$ yielded $[N(C_2H_5)_4][(3,5\text{-}(C_6H_5)_2C_5H_3SO)Mo(CO)_3]$; from the latter two reactions, $CH_3P(C_6H_5)_2$ and $Na[CH_3Fe(CO)_4]$ were isolated as by-products [6]. Treatment with tertiary phosphanes, e.g., $P(C_6H_5)_2CH_3$ or $P(C_6H_5)_3$ in refluxing THF, resulted in fission of the metal−ring linkage [7, 9]. For metalation and alkylation reactions at the CH_3 group yielding the $(3,5\text{-}(C_6H_5)_2C_5H_3S(O)R')Mo(CO)_3$ [7, 9], see Methods V and VI.

References on p. 351

References:

[1] Thewalt, U. (Angew. Chem. **81** [1969] 783/4; Angew. Chem. Int. Ed. Engl. **8** [1969] 769).
[2] Lückoff, M.; Dimroth, K. (Angew. Chem. **88** [1976] 543/4; Angew. Chem. Int. Ed. Engl. **15** [1976] 503).
[3] Debaerdemaeker, T. (Angew. Chem. **88** [1976] 544/5; Angew. Chem. Int. Ed. Engl. **15** [1976] 504).
[4] Schäfer, W.; Schweig, A.; Dimroth, K.; Kanter, H. (J. Am. Chem. Soc. **98** [1976] 4410/8).
[5] Weber, L.; Krüger, C.; Tsay, Y.-H. (Chem. Ber. **111** [1978] 1709/20).
[6] Weber, L. (Chem. Ber. **112** [1979] 99/106).
[7] Weber, L. (Chem. Ber. **112** [1979] 3828/41).
[8] Weber, L. (Angew. Chem. **93** [1981] 305/7; Angew. Chem. Int. Ed. Engl. **20** [1981] 279).
[9] Weber, L. (Chem. Ber. **114** [1981] 1/13).
[10] Dimroth, K.; Lückoff, M.; Kaletsch, H. (Phosphorus Sulfur Relat. Elem. **10** [1981] 285/94).

[11] Dimroth, K.; Berger, S.; Kaletsch, H. (Phosphorus Sulfur Relat. Elem. **10** [1981] 295/304).
[12] Dimroth, K.; Berger, S.; Kaletsch, H. (Phosphorus Sulfur Relat. Elem. **10** [1981] 305/16).
[13] Weber, L.; Wewers, D. (Z. Naturforsch. **37b** [1982] 68/72).
[14] Weber, L.; Boese, R. (Chem. Ber. **115** [1982] 1175/86).
[15] Weber, L. (Angew. Chem. **95** [1983] 539/51; Angew. Chem. Int. Ed. Engl. **22** [1983] 516).
[16] Weber, L.; Wewers, D. (Chem. Ber. **116** [1983] 1327/35).
[17] Weber, L. (Chem. Ber. **116** [1983] 2022/7).
[18] Koelle, U.; Ding, T.-Zh.; Weber, L. (Organometallics **4** [1985] 302/5).

1.5.1.4.2.7.8 $^5LMo(CO)_3$ Complexes with Metalaheterocyclic ^5L Ligands

$(3,5\text{-}(CH_3)_2C_5H_3Ir(P(C_2H_5)_3)_2Y)Mo(CO)_3$ (Formula I; Y = CO, $P(CH_3)_3$, $P(C_2H_5)_3$). Two preparative approaches have been described: (I) The preformed ligand-substituted iridabenzene derivative, $3,5\text{-}(CH_3)_2C_5H_3Ir(P(C_2H_5)_3)_2Y$, was reacted with an equimolar amount of $1,4\text{-}(CH_3)_2C_6H_4Mo(CO)_3$ in THF at room temperature. The residues remaining after removal of solvent in vacuum were dissolved in the minimal quantity of acetone (Y = $P(CH_3)_3$, $P(C_2H_5)_3$) or diethyl ether (Y = CO) and crystallized at −30 °C [1, 2]. (II) The tris(triethylphosphine) complex $(3,5\text{-}(CH_3)_2C_5H_3Ir(P(C_2H_5)_3)_3)Mo(CO)_3$ was refluxed in acetone in the presence of excess $P(CH_3)_3$. $(3,5\text{-}(CH_3)_2C_5H_3Ir(P(C_2H_5)_3)_2P(CH_3)_3)Mo(CO)_3$ was isolated from the evaporated mixture by low-temperature (−30 °C) crystallization from acetone [2]. $(3,5\text{-}(CH_3)_2\text{-}$

I

References on p. 356

$C_5H_3Ir(P(C_2H_5)_3)_3)Mo(CO)_3$ was refluxed in THF while carbon monoxide was bubbled vigorously through the solution. After removal of solvent, $(3,5-(CH_3)_2C_5H_3Ir(P(C_2H_5)_3)_2CO)-Mo(CO)_3$ was separated by redissolution of the residue in diethyl ether, followed by cooling to −30°C [2].

Yields and selected properties are listed in the following table. In solution, the iridabenzene ligand was found to rotate with respect to the $Mo(CO)_3$ fragment. The barrier for this process ($\Delta G^{\ddagger}$) was estimated to be less than 8 kcal/mol [2].

Y	method of synthesis (yield), properties, and remarks
CO	I; IIb (80%) [2] reddish orange needles [2] ^{1}H NMR (THF-d_8, 17°C): 0.87 to 1.22 (m, $P(CH_2CH_3)_3$), 1.67 to 2.13 (m, $P(CH_2CH_3)_3$), 2.05, 2.12 (both s, ring-CH_3), 6.28 (s, H-4), 7.75 (s, H-2), 8.52 (br d, H-6; cis-$^3J(P_{basal}, H)$ = 22.8) [2] ^{13}C NMR (THF-d_8, 17°C): 8.0, 8.8 (both s, $P(CH_2CH_3)_3$), 20.5, 23.5 (both d, $P(CH_2CH_3)_3$; J(P, C) = 27.3 and 31.9), 28.8 (m, ring-CH_3), 99.3 (s, C-4), 107.4, 110.4 (both s, C-3,5), 125.1 (d, C-2; trans-^2J(P, C) = 66.7), 143.9 (d, C-6; cis-$^2J(P_{basal}, C)$ = 8.3), 185.6 (d, IrCO; J(P, C) = 12.8), 228.0 (s, MoCO) [2] ^{31}P NMR (THF-d_8, 17°C): −13.0 (s, P_{basal}), 16.7 (s, P_{axial}) [2] IR (THF): 1846, 1871, 1929, 1977 (CO) [2]
$P(CH_3)_3$	I; IIa (83%) [2] reddish orange needles [2] ^{1}H NMR (THF-d_8, 17°C): 0.84 to 1.29 (m, $P(CH_2CH_3)_3$), 1.74 (d, $P(CH_3)_3$; J(P, H) = 7.2), 1.61 to 2.12 (m, $P(CH_2CH_3)_3$), 2.09 (s, ring-CH_3), 6.32 (s, H-4), 7.83 (d, H-2; cis-$^3J(P_{basal}, H)$ = 20.8), 8.35 (d, H-6; cis-$^3J(P_{basal}, H)$ = 21.7) [2] ^{13}C NMR (THF-d_8, 17°C): 8.7, 10.0 (both s, $P(CH_2CH_3)_3$), 20.8 (d, $P(CH_3)_3$; J(P, C) = 27.3), 21.3, 24.0 (both d, $P(CH_2CH_3)_3$; J(P, C) = 24.2 and 28.3), 29.1, 29.5 (both d, ring-CH_3; J(P, C) = 6.5 each), 99.9 (s, C-4), 106.9, 108.4 (both s, C-3,5), 132.5 (dd, C-2, cis-$^2J(P_{basal}, C)$ = 7.2, trans-^2J(P, C) = 71.1), 140.1 (br d, C-6; trans-^2J(P, C) = 73.2), 230.2 (s, CO) [2] ^{31}P NMR (THF-d_8): −49.9 (dd, $P(CH_3)_3$; $^2J(P_{basal},P_{axial})$ = 8.9, $^2J(P_{basal},P_{basal})$ = 15.0), −15.9 (d, basal $P(C_2H_5)_3$), 18.4 (d, axial $P(C_2H_5)_3$) [2] IR (THF): 1835, 1918 (CO) [2]
$P(C_2H_5)_3$	I (87%) [1, 2] reddish orange needles [2] ^{1}H NMR (THF-d_8, 17°C): 0.82 to 0.95 (m, 1 $P(CH_2CH_3)_3$), 1.18 to 1.35 (m, 2 $P(CH_2CH_3)_2$), 1.62 to 1.80 (m, 1 $P(CH_2CH_3)_3$), 1.98 to 2.22 (m, 2 $P(CH_2CH_3)_3$), 2.07 (s, ring-CH_3), 6.25 (s, H-4), 8.08 (d, H-2,6; cis-$^3J(P_{basal},H)$ = 20.7) [2]; similar in acetone-d_6 [1] ^{13}C NMR (THF-d_8, 17°C): 9.0, 10.4 (both s, $P(CH_2CH_3)_3$), 21.6 to 22.0, 24.3 to 24.8 (both m, $P(CH_2CH_3)_3$), 29.3 (s, ring-CH_3), 99.5 (s, C-4), 107.9 (s, C-3,5), 135.7 (dd, C-2,6, cis-$^2J(P_{basal},C)$ = 9.4, trans-^2J(P,C) = 74.5), 229.8 (s, CO) [1, 2] ^{31}P NMR (THF-d_8): −19.1 (d, 2 P_{basal}; ^2J(P,P) = 3.5), 18.8 (t, P_{apical}) [2]; similar in acetone-d_6 [1]

References on p. 356

Y	method of synthesis (yield), properties, and remarks
	IR (THF): 1836, 1918 (CO) [1, 2] for clean replacement of one basal $P(C_2H_5)_3$ ligand by CO and $P(CH_3)_3$ [2], see preparative Method II

$(3,5\text{-}(CH_3)_2C_5H_3Ir(P(C_2H_5)_3)_3)Mo(CO)_3$ crystallizes in the monoclinic space group $P2_1/n$ (nonconventional setting of $P2_1/c)-C_{2h}^5$ (No. 14) with a = 9.897(1), b = 16.213(3), c = 20.937(3) Å, β = 96.68(1)°; Z = 4 molecules per unit cell, D_{calc} = 1.632 g/cm³ [1, 2]. $(3,5\text{-}(CH_3)_2C_5H_3Ir(P(C_2H_5)_3)_2P(CH_3)_3)Mo(CO)_3$ crystallizes in the monoclinic space group $P2_1/n$ (nonconventional setting of $P2_1/c)-C_{2h}^5$ (No. 14) with a = 10.005(2), b = 18.012(3), c = 17.055(4) Å, β = 93.33(21)°; Z = 4 molecules per unit cell, D_{calc} = 1.638 g/cm³ [2]. The molecular structures revealed an essentially planar iridabenzene system, with the molybdenum atom being strongly π-complexed to all six atoms of the ring; see **Figs. 35** and **36**.

Treatment of all three compounds with $[HO(C_2H_5)_2]BF_4$ led to protonation of the metal centers and production of H-bridged heterobimetallics, $[(3,5\text{-}(CH_3)_2C_5H_3Ir(P(C_2H_5)_3)_2Y)\text{-}(\mu\text{-}H)Mo(CO)_3]BF_4$ (Formula II) [2].

$[(3,5\text{-}(CH_3)_2C_5H_3Ir(P(C_2H_5)_3)_3Y)(\mu\text{-}H)Mo(CO)_3]BF_4$ (Formula II; Y = CO, $P(CH_3)_3$, $P(C_2H_5)_3$). Excess $[(C_2H_5)_2OH]BF_4$ was added to THF solutions of the parent $(3,5\text{-}(CH_3)_2C_5H_3\text{-}Ir(P(C_2H_5)_3)_2Y)Mo(CO)_3$ complexes cooled to −30 °C. After stirring briefly, the solutions were

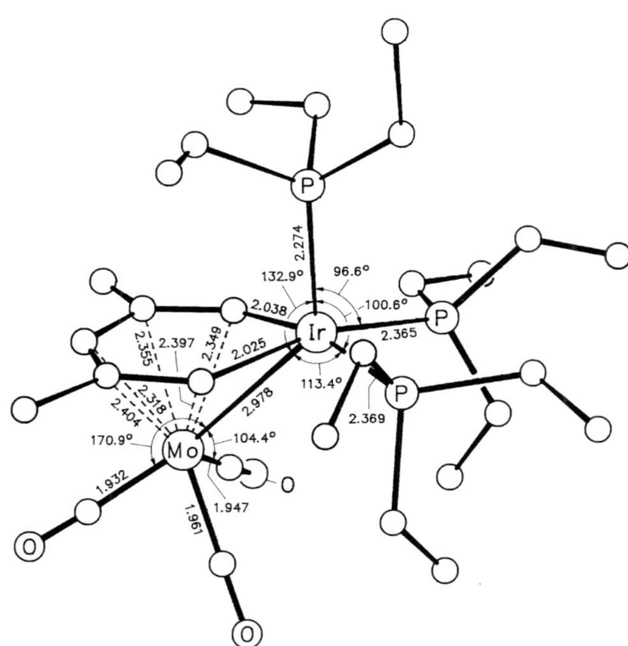

Fig. 35. The molecular structure of $(3,5\text{-}(CH_3)_2C_5H_3Ir(P(C_2H_5)_3)_3)Mo(CO)_3$ [1, 2].

References on p. 356

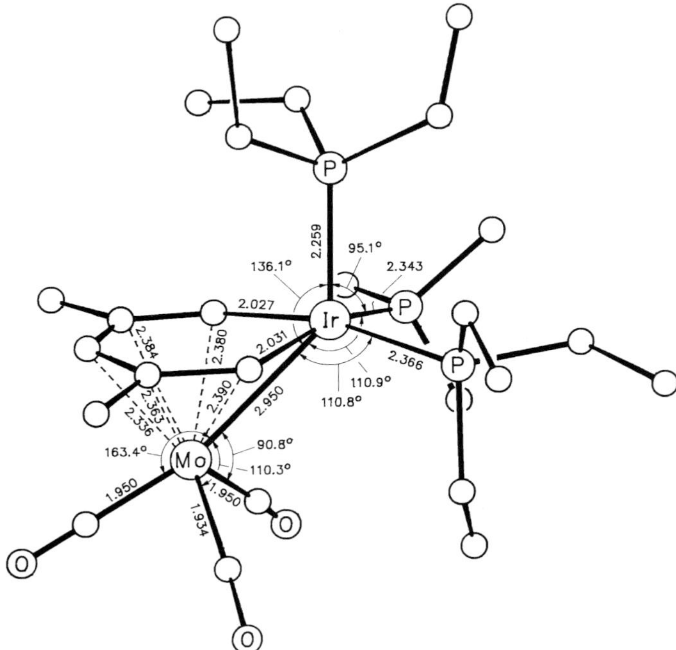

Fig. 36. The molecular structure of $(3,5\text{-}(CH_3)_2C_5H_3Ir(P(C_2H_5)_3)_2P(CH_3)_3)Mo(CO)_3$ [2].

$$\left[(3,5\text{-}(CH_3)_2C_6H_3)Ir(P(C_2H_5)_3)_2(Y)(\mu\text{-}H)Mo(CO)_3 \right] BF_4$$

II

stored at $-30\,°C$. $[(3,5\text{-}(CH_3)_2C_5H_3Ir(P(C_2H_5)_3)_2CO)(\mu\text{-}H)Mo(CO)_3]BF_4$ separated as an oil which solidified upon repeated treatment with diethyl ether (60%). $[(3,5\text{-}(CH_3)_2C_5H_3\text{-}Ir(P(C_2H_5)_3)_2P(CH_3)_3)(\mu\text{-}H)Mo(CO)_3]BF_4$ crystallized as a 1:1 solvate with THF which could be removed by evacuation (69%). $[(3,5\text{-}(CH_3)_2C_5H_3Ir(P(C_2H_5)_3)_3)(\mu\text{-}H)Mo(CO)_3]BF_4$ precipitated from the mixture and was recrystallized from acetone/diethyl ether (63%) [2]. For physical properties, see the following table.

Y	properties
CO	purple-red microcrystalline powder ^1H NMR (CD_3CN, 19 °C): -12.91 (dd, μ-H; cis-$^2J(P,H) = 10.0$, trans-$^2J(P,H) = 45.8$), 0.90 to 1.04, 1.10 to 1.24 (both m, $P(CH_2CH_3)_3$), 1.82 to 2.20 (m, $P(CH_2CH_3)_3$), 2.13, 2.14 (both s, ring-CH_3), 6.97 (s, H-4), 7.49 (d, H-6; cis-$^3J(P_{basal},H) = 18.1$), 8.19 (s, H-2)

References on p. 356

Y	properties
	^{13}C NMR (CD_3CN, 19 °C): 8.2, 8.8 (both d, $P(CH_2CH_3)_3$; J(P,C) = 4.9 each), 18.9, 22.0 (both d, $P(CH_2CH_3)_3$; J(P,C) = 30.6 and 36.7), 27.4 (d, ring-CH_3; J(P,C) = 6.2), 27.9 (s, ring-CH_3), 99.0 (s, C-4), 114.0, 117.1 (both s, C-3,5), 124.1 (d, C-2, trans-$^2J(P,C)$ = 62.5), 125.5 (t, C-6; cis-$^2J(P_{axial},C)$ = cis-$^2J(P_{basal},C)$ = 6.6), 173.1 (t, IrCO; cis-$^2J(P_{axial},C)$ = cis-$^2J(P_{basal},C)$ = 6.8), 219.4 (br, MoCO); (CD_3CN, −30 °C): 217.9, 219.3, 219.9 (all s, MoCO) ^{31}P NMR (CD_3CN, 19 °C): −16.9 (d, P_{basal}; J(P,P) = 12.7), 6.5 (d, P_{axial}) IR (CH_3CN): 1931, 1956, 2021, 2046 (CO)
$P(CH_3)_3$	red needles (THF solvate) ^1H NMR (CD_3CN, 19 °C): −13.15 (dt, μ-H; cis-$^2J(P_2,H)$ = 11.4, trans-$^2J(P,H)$ = 42.0), 0.99 to 1.13, 1.25 to 1.38 (both m, $P(CH_2CH_3)_3$), 1.80 (d, $P(CH_3)_3$; J(P,H) = 8.4), 1.77 to 1.97, 2.00 to 2.28 (both m, $P(CH_2CH_3)_3$), 2.21, 2.23 (both s, ring-CH_3), 7.06 (s, H-4), 7.92 (d, H-6; cis-$^3J(P_{basal},H)$ = 15.0), 7.98 (d, H-2; cis-$^3J(P_{basal},H)$ = 18.0) ^{13}C NMR (CD_3CN, 19 °C): 8.7, 9.2 (both s, $P(CH_2CH_3)_3$), 18.9, 19.1, 21.9 (all d, $P(CH_3)_3$ and $P(CH_2CH_3)_3$; J(P,C) = 28.6, 32.0, and 34.0), 28.2, 28.3 (both s, ring-CH_3), 98.8 (s, C-4), 114.7 (s, C-3,5), 132.5 (br d, C-2; trans-$^2J(P,C)$ = 71.0), 134.0 (br d, C-6; trans-$^2J(P,C)$ = 72.3), 220.3 (br, CO); (CD_3CN, −30 °C): 219.9, 220.2, 220.8 (all s, MoCO) ^{31}P NMR (CD_3CN, 19 °C): −49.0 (dd, $P(CH_3)_3$; $^2J(P_{basal},P_{axial})$ = 18.4, $^2J(P_{basal},P_{basal})$ = 12.2), −22.3 (dd, basal $P(C_2H_5)_3$; $^2J(P_{basal},P_{axial})$ = 16.5), 1.2 (dd, axial $P(C_2H_5)_3$) IR (CH_3CN): 1913, 1937, 2010 (CO)
$P(C_2H_5)_3$	reddish needles ^1H NMR (CD_3CN, 19 °C): −13.30 (dt, μ-H; cis-$^2J(P_2,H)$ = 11.3, trans-$^2J(P,H)$ = 42.1), 1.02 to 1.13 (m, 1 $P(CH_2CH_3)_3$), 1.26 to 1.35 (m, 2 $P(CH_2CH_3)_3$), 1.95 to 2.10 (m, 1 $P(CH_2CH_3)_3$), 2.12 to 2.28 (m, 2 $P(CH_2CH_3)_3$), 2.20 (s, ring-CH_3), 7.01 (s, H-4), 7.89 (br d, H-2,6; cis-$^3J(P_{basal},H)$ = 14.6) ^{13}C NMR (CD_3CN, 19 °C): 9.3, 9.5 (both s, $P(CH_2CH_3)_3$), 19.2, 22.5 (both d, $P(CH_2CH_3)_3$; J(P,C) = 30.3 and 34.3), 28.3 (s, ring-CH_3), 98.4 (s, C-4), 114.5 (s, C-3,5), 132.9 (br d, C-2,6, trans-$^2J(P,C)$ = 70.7), 220.0 (br, CO); (CD_3CN, −30 °C): 219.6 (s, 2 MoCO), 220.7 (s, 1 MoCO) ^{31}P NMR (CD_3CN, 19 °C): −24.2 (d, P_{basal}; $^2J(P,P)$ = 16.4), −2.4 (t, P_{apical}) IR (CH_3CN): 1913, 1937, 2009 (CO)

Line shape simulations of variable-temperature ^{13}C NMR spectra recorded in the MoCO region over the temperature range of −30 to +50 °C established free energies of activation for iridabenzene ligand rotation in the three $[(3,5\text{-}(CH_3)_2C_5H_3Ir(P(C_2H_5)_3)_2Y)(\mu\text{-}H)Mo(CO)_3]^+$ cations as $\Delta G^{\ddagger}$ = 13.7(4) kcal/mol for Y = $P(C_2H_5)_3$, and 13.5 to 15 kcal/mol for Y = $P(CH_3)_3$ and CO [2].

Crystals of $[(3,5\text{-}(CH_3)_2C_5H_3Ir(P(C_2H_5)_3)_2P(CH_3)_3)(\mu\text{-}H)Mo(CO)_3]BF_4 \cdot OC_4H_8$ grown from THF belonged to the orthorhombic system, space group $P2_12_12_1-D_2^4$ (No. 19) with the lattice parameters at 123 K: a = 10.200(4), b = 16.467(9), c = 22.023(7) Å; Z = 4 molecules per unit cell, D_{calc} = 1.684 g/cm³. **Fig. 37** shows the molecular structure of the cation [2].

References on p. 356

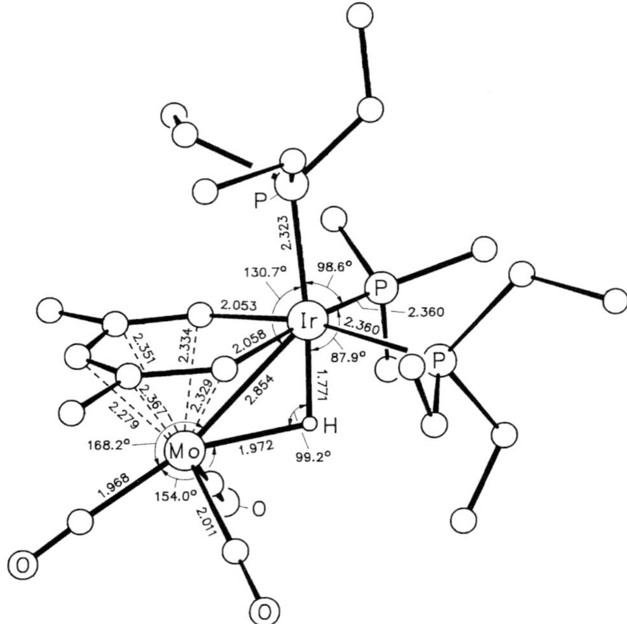

Fig. 37. The molecular structure of $[(3,5\text{-}(CH_3)_2C_5H_3Ir(P(C_2H_5)_3)_2(P(CH_3)_3)(\mu\text{-}H)Mo(CO)_3]^+$ [2].

References:

[1] Bleeke, J. R.; Xie, Y.-F.; Bass, L.; Chiang, M. Y. (J. Am. Chem. Soc. **113** [1991] 4703/4).

[2] Bleeke, J. R.; Bass, L. A.; Xie, Y.-F.; Chiang, M. Y. (J. Am. Chem. Soc. **114** [1992] 4213/9).

Empirical Formula Index

In the following index the compounds are listed by their empirical formulas in the order of increasing carbon content (first column). Salts are identified by the empirical formula of the molybdenum-containing cation or anion. The polymers nos. 217 to 219 in table 6 are not included.

The second column contains the linearized structure, wherein cyclic ligands are partly denoted also by their empirical formula or a linear sequence of ring atoms. In many cases the formulas given in the handbook had to be modified. This became necessary in order to make the index data compatible with the Gmelin Formula Index.

In the third column, page references are printed in ordinary types, table numbers in bold face, and compound numbers within the tables in italics.

	$(C_5H_5)Mo(CO)_3\text{-}C_3H_7\text{-i}$	24, **1**, *4*
	$(C_5H_5)Mo(CO)_3\text{-}C_3H_7\text{-n}$	24, **1**, *3*
$C_{11}H_{12}MoO_4$	$(C_5H_5)Mo(CO)_3\text{-}CH_2CH_2\text{-}OCH_3$	64, **1**, *140*
	$(HO\text{-}CH_2CH_2CH_2\text{-}C_5H_4)Mo(CO)_3H$	182, **6**, *86*
	$[CH_3\text{-}CH(OH)\text{-}C_5H_4]Mo(CO)_3\text{-}CH_3$	200, **6**, *164*
$C_{11}H_{12}MoO_4S$	$[1,3,5\text{-}(CH_3)_3\text{-}SC_5H_3(=O)\text{-}1]Mo(CO)_3$	347, **10**, *7*
$C_{11}H_{12}MoO_6S$	$(C_5H_5)Mo(CO)_3\text{-}CH_2CH_2\text{-}S(O)_2\text{-}OCH_3$	65/6, **1**, *144*
$C_{11}H_{13}IMoO_3Si$	$[(CH_3)_3Si\text{-}C_5H_4]Mo(CO)_3I$	191, **6**, *135*
$C_{11}H_{13}MoNO_3$	$(C_5H_5)Mo(CO)_3\text{-}CH_2\text{-}N(CH_3)_2$	57, **1**, *114*
$C_{11}H_{13}MoO_3S^+$	$[(C_5H_5)Mo(CO)_3\text{-}CH_2\text{-}S(CH_3)_2][BF_4]$	60/1, **1**, *125*
$C_{11}H_{13}MoO_3Si^-$	$Na[((CH_3)_3Si\text{-}C_5H_4)Mo(CO)_3]$	178, **6**, *71*
$C_{11}H_{13}MoO_3Sn^-$	$Li[((CH_3)_3Sn\text{-}C_5H_4)Mo(CO)_3]$	178, **6**, *74*
$C_{11}H_{13}MoO_4S^+$	$[(CH_3\text{-}C_5H_4)Mo(CO)_3\text{-}OS(CH_3)_2][(CH_3\text{-}C_5H_4)Mo(CO)_3]$	168, **6**, *14*
$\mathbf{C}_{12}H_5ClMoN_2O_3$	$(C_5H_5)Mo(CO)_3\text{-}CCl=C(CN)_2$	120, **2**, *5*
$C_{12}H_5ClMoO_5$	$(C_5H_5)Mo(CO)_3\text{-}C[\text{-}C(O)\text{-}C(O)\text{-}C(Cl)=]$	121, **2**, *9*
$C_{12}H_5F_3MoN_2O_3$	$4\text{-}(C_5H_5)Mo(CO)_3\text{-}1,2\text{-}N_2C_4F_3$	144, **4**, *2*
$C_{12}H_5F_7MoO_4$	$(C_5H_5)Mo(CO)_3\text{-}C(O)\text{-}C_3F_7\text{-n}$	135, **3**, *7*
$C_{12}H_6F_6MoO_3$	$(C_5H_5)Mo(CO)_3\text{-}C(CF_3)=CH\text{-}CF_3$	120, **2**, *6*
$C_{12}H_6MoN_2O_3$	$(C_5H_5)Mo(CO)_3\text{-}C(CN)=CH\text{-}CN$	119/20, **2**, *2*
	$(C_5H_5)Mo(CO)_3\text{-}CH=C(CN)_2$	120, **2**, *3*
$C_{12}H_7BrMoO_3$	$(C_9H_7)Mo(CO)_3Br$	261, **7**, *9*
$C_{12}H_7ClMoO_3$	$(C_9H_7)Mo(CO)_3Cl$	261, **7**, *8*
$C_{12}H_7IMoO_3$	$(C_9H_7)Mo(CO)_3I$	261, **7**, *10*
$C_{12}H_7MoO_3^-$	$K[(C_9H_7)Mo(CO)_3]$	260, **7**, *3*
	$Li[(C_9H_7)Mo(CO)_3]$	260, **7**, *1*
	$Na[(C_9H_7)Mo(CO)_3]$	260, **7**, *2*
	$[C_2(N(CH_3)_2)_4][(C_9H_7)Mo(CO)_3]_2$	258
$C_{12}H_8MoO_5$	$(C_5H_5)Mo(CO)_3\text{-}C[=CH\text{-}CH_2\text{-}C(O)\text{-}O\text{-}]$	123, **2**, *18*

Transition Metal Reference

All organomolybdenum compounds containing transition metals are listed below in alphabetical order of these metals. Pages are printed in ordinary type, table numbers in bold face, and compound numbers within the tables in italics.

Au-containing compound

$(OCH-C_5H_4)Mo(CO)_3-Au-P(C_6H_5)_3$ 231/2

Cd-containing compounds

$(CH_3-C_5H_4)Mo(CO)_3-CdBr$ 232
$(C_5H_4-CH_3)Mo(CO)_3-Cd[-P(C_6H_5)_2-CH_2-P(C_6H_5)_2-Fe(CO)_3(Si(OCH_3)_3)-]$ 251/2
$[(C_6H_5)_3P-C_5H_4]Mo(CO)_3 \cdot CdI_2$ 328/9
$[((C_6H_5)_3P-C_5H_4)Mo(CO)_3CdI][I]$ 328/9

Co-containing compounds

$(C_5H_5)Mo(CO)_3-CH_2-C_2Co_2(CO)_6(C_6H_5)$ 37, **1**, *48*
$[(C_5H_5)_2Co][(9,10-(O)_2-C_{14}H_7-2-C(O)NHCH_2CH_2-C_5H_4)Mo(CO)_3]$ 175, **6**, *55*
$(CH_3-C_5H_4)Mo(CO)_3-Co(CO)_4$ 228
$[CH_3C(O)-C_5H_4]Mo(CO)_3-Co(CO)_4$ 228
$[C_2H_5-OC(O)-C_5H_4]Mo(CO)_3-Co(CO)_4$ 228

Cr-containing compounds

$(C_5H_5)Mo(CO)_3-CH_2-C_6H_4-2-NC[Cr(CO)_5]$ 43, **1**, *69*
$Mo(CO)_3(C_5H_4-C_5H_4)Cr(CO)_3$ 235
$(CO)_3Mo(C_{10}H_8)Cr(CO)_3$ 336
$(CO)_3Mo[4,6,8-(CH_3)_3-C_{10}H_5]Cr(CO)_3$ 336/7

Fe-containing compounds

$[(C_5H_5)Mo(CO)_3(CH_2CHCH_2)Fe(CO)_2(C_5H_5)][PF_6]$ 28/9, **1**, *18*
$[(C_5H_5)Mo(CO)_3-CH_2-C(=C(CN)_2)-C_5H_4]Fe(C_5H_5)$ 32/3, **1**, *30*

Hg-containing compounds

Ir-containing compounds

Mn-containing compounds

Ni-containing compounds

Pd-containing compounds

Pt-containing compounds

Re-containing compounds

Physical Constants and Conversion Factors

Avogadro constant N_A (or L) $= 6.02214 \times 10^{23}$ mol^{-1}
Faraday constant $F = 9.64853 \times 10^{4}$ C/mol
molar gas constant $R = 8.31451$ $J \cdot mol^{-1} \cdot K^{-1}$
molar volume (ideal gas) $V_m = 2.24141 \times 10^{1}$ L/mol
(273.15 K, 101325 Pa)

Planck constant $h = 6.62608 \times 10^{-34}$ $J \cdot s$
elementary charge $e = 1.60218 \times 10^{-19}$ C
electron mass $m_e = 9.10939 \times 10^{-31}$ kg
proton mass $m_p = 1.67262 \times 10^{-27}$ kg

1 kg $= 2.205$ pounds
1 m $= 3.937 \times 10^{1}$ inches $= 3.281$ feet
1 m^3 $= 2.642 \times 10^{2}$ gallons (U.S.)
1 m^3 $= 2.200 \times 10^{2}$ gallons (Imperial)

Force	N	dyn	kp
1 N	1	10^{5}	1.019716×10^{-1}
1 dyn	10^{-5}	1	1.019716×10^{-6}
1 kp	9.80665	9.80665×10^{5}	1

Pressure	Pa	bar	kp/m^2	at	atm	Torr	lb/in^2
1 Pa = 1 N/m^2	1	10^{-5}	1.019716×10^{-1}	1.019716×10^{-5}	9.86923×10^{-6}	7.50062×10^{-3}	1.450378×10^{-4}
1 bar = 10^6 dyn/cm^2	10^{5}	1	1.019716×10^{4}	1.019716	9.86923×10^{-1}	7.50062×10^{2}	1.450378×10^{1}
1 kp/m^2 = 1 mm H_2O	9.80665	9.80665×10^{-5}	1	10^{-4}	9.67841×10^{-5}	7.35559×10^{-2}	1.422335×10^{-3}
1 at (technical)	9.80665×10^{4}	9.80665×10^{-1}	10^{4}	1	9.67841×10^{-1}	7.35559×10^{2}	1.422335×10^{1}
1 atm = 760 Torr	1.01325×10^{5}	1.01325	1.033227×10^{4}	1.033227	1	7.60×10^{2}	1.469595×10^{1}
1 Torr = 1 mm Hg	1.333224×10^{2}	1.333224×10^{-3}	1.359510×10^{1}	1.359510×10^{-3}	1.315789×10^{-3}	1	1.933678×10^{-2}
1 lb/in^2 = 1 psi	6.89476×10^{3}	6.89476×10^{-2}	7.03069×10^{2}	7.03069×10^{-2}	6.80460×10^{-2}	5.17149×10^{1}	1

Work, Energy, Heat	J	kW·h	kcal	Btu	eV
1 J = 1 W·s = 1 N·m = 10^7 erg	1	2.778×10^{-7}	2.39006×10^{-4}	9.4781×10^{-4}	6.242×10^{18}
1 kW·h	3.6×10^{6}	1	8.604×10^{2}	3.41214×10^{3}	2.247×10^{25}
1 kcal	4.1840×10^{3}	1.1622×10^{-3}	1	3.96566	2.6117×10^{22}
1 Btu (British thermal unit)	1.05506×10^{3}	2.93071×10^{-4}	2.5164×10^{-1}	1	6.5858×10^{21}
1 eV	1.602×10^{-19}	4.450×10^{-26}	3.8289×10^{-23}	1.51840×10^{-22}	1

$1\ cm^{-1} \triangleq 1.239842 \times 10^{-4}$ eV $1\ Hz \triangleq 4.135669 \times 10^{-15}$ eV

2 Rydberg (Ry) = 1 hartree = 27.2114 eV $1\ eV \triangleq 23.0578$ kcal/mol

Power	kW	hp	$kp \cdot m \cdot s^{-1}$	kcal/s
1 kW = 10^3 J/s	1	1.35962	1.01972×10^{2}	2.39006×10^{-1}
1 hp (horsepower, metric)	7.3550×10^{-1}	1	7.5×10^{1}	1.7579×10^{-1}
1 $kp \cdot m \cdot s^{-1}$	9.80665×10^{-3}	1.333×10^{-2}	1	2.34384×10^{-3}
1 kcal/s	4.1840	5.6886	4.26650×10^{2}	1

References:

Mills, I. (Ed.), International Union of Pure and Applied Chemistry, Quantities, Units and Symbols in Physical Chemistry, Blackwell Scientific Publications, Oxford 1988.

The International System of Units (SI), National Bureau of Standards Spec. Publ. 330 [1972].

Landolt-Börnstein, 6th Ed., Vol. II, Pt. 1, 1971, pp. 1/14.

ISO Standards Handbook 2, Units of Measurement, 2nd Ed., Geneva 1982.

Cohen, E. R., Taylor, B. N., Codata Bulletin No. 63, Pergamon, Oxford 1986.